THE HOME OF THE HONEY-BEES IN 1878.

THE HOME OF THE HONEY-BEES IN 1884.

THE HOME OF THE HONEY-BEES IN 1887.

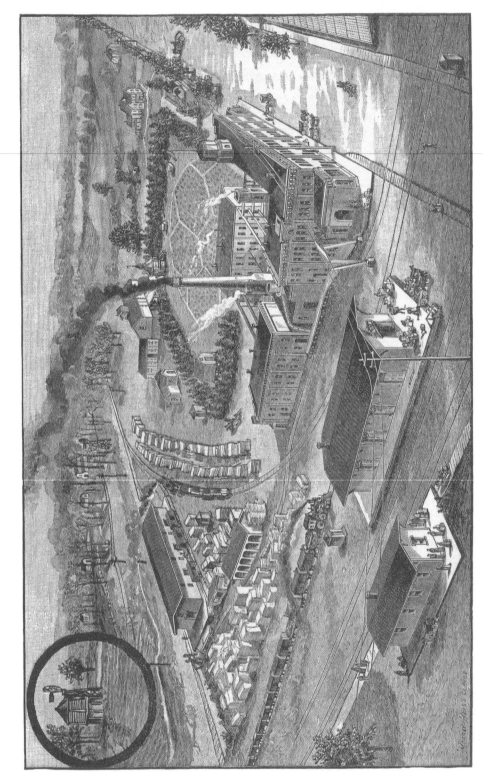

THE HOME OF THE HONEY-BEES IN 1891.

THE HOME OF THE HONEY-BEES IN 1891.

Upper View Shows The A.I. Root Company's Office and Publishing House. Lower View Shows the Interior of the Printing Department. 1908

The ABC & XYZ of Bee Culture

"A Cyclopedia of Everything Pertaining to the Care of the Honey-bee; Bees, Hives, Honey, Implements, Honey Plants, etc. Facts Gleaned from the Experience of Thousands of Beekeepers and Afterward Verified in Our Apiary."

A.I. Root 1879

Library of Congress Cataloging-in-Publication Data

Root, A.I. (Amos Ives), 1839-1923.
 The ABC & XYZ of bee culture: a cyclopedia of everything pertaining to the care of the honey-bee: bees, hives, honey, implements, honey plants, etc.: facts gleaned from the experience of thousands of beekeepers and afterward verified in our apiary – 41st ed. / written and edited by Hachiro Shimanuki, Kim Flottum, Ann Harman: with associate editor: Sharon Garceau.
      p. cm.
 "Original work by Amos Ives Root."
 "Revised editions edited by E.R. Root . . ."
 Includes index.
 ISBN-13: 978-0-936028-22-4
 ISBN-10: 0-936028-22-X
  1. Bee culture – Encyclopedia. I. Shimanuki, H. II. Flottum, Kim. III. Harman, Ann. IV. Garceau, Sharon. V. Root, E.R. (Ernest Rob), 1862-1953. VI. Title. VII. ABC and XYZ of bee culture. VIII. Title: Cyclopedia of everything pertaining to the care of the honey bee.
SF523.R7 2007
638'.1--dc22

2006038381

The ABC & XYZ of Bee Culture
41st Edition – 1st Printing

An encyclopedia pertaining to the scientific and practical culture of honey bees

Original work by Amos Ives Root
Founder of the A.I. Root Company and Gleanings In Bee Culture

Revised Editions Edited by
E.R. Root
H.H. Root
M.J. Deyell
J.A. Root
Dr. R.A. Morse
K. Flottum

The 41st Edition Written and Edited by
Dr. Hachiro Shimanuki
Kim Flottum
Ann Harman

Published by The A.I. Root Company
ISBN-13: 978-0-936028-22-4
ISBN-10: 0-936028-22-X

Copyright © 2006 by A.I. Root Companny
All Rights Reserved

Published in 2007 by
The A.I. Root Company
623 West Liberty Street
Medina, Ohio 44256
www.beeculture.com

Printed in the U.S.

Preface

In preparing this work, I have been much indebted to the books of Langstroth, Quinby, Prof. Cook, King, and some others, as well as to all the Bee-Journals; but, more than to all these, have I been indebted to the thousands of friends scattered far and wide, who have so kindly furnished the fullest particulars in regard to all the new improvements, as they have come up, in our beloved branch of rural industry. Those who questioned me so much, a few years ago, are now repaying by giving me such long kind letters in answer to any inquiry I may happen to make, that I often feel ashamed to think what meager answers I have been obliged to give them under similar circumstances. A great part of this ABC book is really the work of the people, and the task that devolves on me is to collect, condense, verify, and utilize, what has been scattered through thousands of letters, for years past. My own apiary has been greatly devoted to carefully testing each new device, invention, or process, as it came up; the task has been a very pleasant one, and, if the perusal of the following pages affords you as much pleasure, I shall feel amply repaid.

Medina, Ohio, November 1877
Amos Ives Root

Introduction

About the year 1865, during the month of August, a swarm of bees passed overhead where we were at work; and my fellow-workman, in answer to some of my inquiries respecting their habits, asked what I would give for them. I, not dreaming he could by any means call them down, offered him a dollar, and he started after them. To my astonishment, he, in a short time, returned with them hived in a rough box he had hastily picked up, and, at that moment, I commenced learning my A B C in bee culture. Before night I had questioned not only the bees, but every one I knew, who could tell me anything about these strange new acquaintances of mine. Our books and papers were overhauled that evening; but the little that I found only puzzled me the more, and kindled anew the desire to explore and follow out this new hobby of mine; for, dear reader, I have been all my life *much* given to hobbies and new projects.

Farmers who had kept bees assured me that they once paid, when the country was new, but of late years they were of no profit, and everybody was abandoning the business. I had some headstrong views in the matter, and in a few days I visited Cleveland, ostensibly on other business, but I had really little interest in any thing until I could visit the book-stores and look over the books on bees. I found but two, and I very quickly chose Langstroth. May God reward and for ever bless Mr. Langstroth for the kind and pleasant way in which he unfolds to his readers the truths and wonders of creation, to be found inside of a beehive.

What a gold-mine that book seemed to me, as I looked it over on my journey home! Never was romance so enticing; no, not even Robinson Crusoe; and, best of all, right at my own home I could live out and verify all the wonderful things told therein. Late as it was, I yet made an observatory-hive, and raised queens from worker-eggs before winter, and wound up by purchasing a queen of Mr. L. for $20.00. I should, in fact, have wound up the whole business, queen and all, most effectually, had it not been for some timely advice toward Christmas, from a plain practical farmer near by. With his assistance, and by the purchase of some more bees, I brought all safely through the winter. Through Mr. L., I learned of Mr. Wagner; shortly afterward he was induced to re-commence the publication of the *American Bee Journal*; and through this I gave accounts monthly of my blunders and occasional successes.

In 1867, news came across the ocean from Germany, of the honey-extractor; and with the aid of a simple home-made machine I took 100 lbs. of honey from 20 stocks, and increased them to 35. This made quite a sensation, and numbers embarked in the new business; but when I lost all but 11 of the 35 the next winter, many said, "There! I told you how it would turn out."

I said nothing, but went to work quietly, and increased the 11 to 48, during the one season, not using the extractor at all. The 48 were wintered entirely without loss, and I think it was, mainly, because I took care and pains with each individual colony. From the 48, I secured 6,162 lbs. of extracted honey, and sold almost the entire crop for 25 cents per lb. This capped the climax, and inquiries in regard to the new industry began to come in from all sides; beginners were eager to know what hives to adopt, and where to get honey-extractors. As the hives in use seemed very poorly adapted to the use of the extractor, and as the machines offered for sale were heavy and poorly adapted to the purpose, besides being "patented," there really seemed to be no other way before me than to manufacture these implements. Unless I did this, I should be compelled to undertake a correspondence that would occupy a great part of my time, without affording any compensation of any account. The fullest directions I knew how to give for making plain simple hives, etc., were from time to time published in the *American Bee Journal*; but the demand for further particulars was such that a circular was printed, and, shortly after, a second edition; then another, and another. These were intended to answer the greater part of the queries; and from the cheering words received in regard to them, it seemed the idea was a happy one.

Until 1873, all these circulars were sent out gratuitously; but at that time it was deemed best to issue a quarterly at 25¢ per year, for the purpose of answering these inquiries. The very first number was received with such favor that it was immediately changed to a monthly, at 75¢. The name given it was "*Gleanings in Bee Culture*," and it was gradually enlarged until, in 1876, the price was changed to $1.00. During all this time, it has served the purpose excellently of answering questions as they come up, both old and new; and even if some new subscriber should ask in regard to something that had been discussed at length but a short time before, it was an easy matter to refer him to it, or send him the number containing the subject in question.

After *Gleanings* was about commencing its fifth year, inquirers began to dislike being referred to something that was published a half-dozen years ago. Besides, the decisions that were then arrived at perhaps needed to be considerably modified to meet present wants. Now, if we go over the whole matter again every year or two, for the benefit of those who have recently subscribed, we shall do our regular subscribers injustice, for they will justly complain that *Gleanings* is the same thing over and over again, year after year.

Now you can see whence the necessity for this A B C book, its office, and the place we purpose to have it fill. In writing it I have taken pains to post myself thoroughly in regard to each subject treated, not only by consulting all the books and journals treating of bee culture, which I have always ready at hand, but by going out into the fields, writing to those who can furnish information in that special direction, or by sacrificing a colony of bees, if need be, until I am perfectly satisfied. Still further: this book is all printed from type kept constantly standing; and as the sheets are printed only so fast as wanted, any thing that is discovered, at any future time, to be an error, can be promptly righted. For the same reason, all new inventions and discoveries that may come up – they are coming up constantly – can be embodied in the work just as soon as they have been tested sufficiently to entitle them to a place in such a work. In other words, I purpose it to be never out of date or behind the times. – December 1878,

Amos Ives Root.

The Home of the Honey-bees, and the Growth of the Beekeeping Industry.

A glance at the frontispiece engravings, showing the buildings and the lumber-yards that go to make up the Home of the Honey bees, covering over six acres of ground, gives some idea of the demands of bee-keepers; and when it is remembered that The A.I. Root Co.'s manufacturing plant is only one, notwithstanding it is largest of several, one can get some idea of the magnitude of the beekeeping industry as a whole.

In one year's time there are made and sold anywhere from 40 to 60 millions of section honey-boxes in the United States alone. Estimating that there are about fourteen ounces of honey to each section sold, on the average, there is marketed annually in the United States something like 50 million pounds of comb honey; and as there is twice as much again of extracted produced as comb honey, the total aggregate would reach 100 to125 million pounds of honey all told, or represent a money value of from 8 to 10 million dollars.

Perhaps it would be interesting to trace the development of just one manufacturing plant – a plant where every thing is made a bee-keeper can possibly require, all the way from a queen-cage to two, four, six, eight, and even twenty-five frame steam-power honey-extractors; bee-hives by the twenty-five thousands; smokers by the tens of thousands; perforated zinc by the thousands of square feet; sections by the tens of millions – making an aggregate of thousands of tons of freight every year.

As already explained in the Introduction, the nucleus of this enterprise was a swarm of bees that went over the jewelry shop of A.I. Root in 1865. From this one swarm there developed a little apiary of some fifty or sixty colonies, and a bee-man who was destined to influence the whole bee-keeping world. This man began writing for the *American Bee Journal* under the *nom de plume* of "Novice;" and the result was, there came in inquiries from all over the United States, asking how to make hives, extractors, and where to get them. At that time there was no factory devoted exclusively to making bee-supplies in the world. But A.I. Root at his jewelry shop had a windmill, and pretty soon put in operation a buzz-saw which he hitched on to the mill; and well do I remember the time how we waited and waited for one of the most uncertain of all things – wind, just a little wind – to fill pressing orders for hives and other beekeepers' appliances; also do I remember how we used to sleep in the shop, father and I, in order that we might be awakened by the rumbling of the shafting and creaking of the belting when the wind did come, so we could make hives by lamplight while the power lasted, for in those days it was not wise to wait till daylight, for the breeze might go down. Later a foot-power buzz-saw was purchased – yes, two of them – to "help us out." The orders began to come in until a 4½-horse-power engine was ordered; and if ever a youth reached the very height of his ambition it was when the writer of this, then a lad of about fifteen, was installed as engineer of the little engine. My! But didn't the buzz-saws whir: and didn't we get the goods off?

By and by even the little engine began to groan under its load, for it had two buzz-saws and a planer to run, and it became necessary to run the little jewelry shop "up town" night and day; but this shop had been converted into a bee-hive establishment. It was easy to be seen that a new building would soon have to be erected near the depot, and so plans were laid for one 40x100, two stories and basement, metal roof. The old jewelry stock was sold out at auction, and the "up-town" store sold. The undertaking, involving the purchase of 18 acres of valuable land and the erection of so large a building, was tremendous for those days, and it nearly exhausted A.I. Root's good credit to pay his debts, and many were the speculations that he would "go under." But he did not. The 40-horse-power engine that had been installed, and the dozen or so buzz-saws, planers, etc., had all they could do to take care of the trade that had more than quadrupled. This was in 1880. The business continued to grow until it became necessary to add on a wing, 40x85 on the west end.

About this time the industry had begun to assume, as we then thought, massive proportions. Two shorthand writers were constantly employed, each one supplied with the latest improved typewriter. The business continued to grow at such a rate that the proprietor himself was almost demoralized by the mass of business that was poured down upon him.

Still the little bee seemed to be able to make a bigger stir than ever throughout the world, and in 1886 another building, 44x96, was added to the works. The old 40-horse-power engine was supplanted by a new and modern 90-horse-power automatic. Besides that, there was 250 feet of line shafting, with its attendant lot of machinery. Again, in 1888 the works had to be again enlarged; a smaller structure was put on. In 1889 another 60-horse-power steam-boiler was added, plus a 90-foot smoke-stack. Besides this, a good deal of additional machinery was put in. Still again, in 1890 the trade had nearly doubled over former years, and we were compelled to extend our works by the addition of another brick building, two stories and basement, 37x98. In that same year other improvements were also introduced, such as electric lights, Grinnell automatic sprinklers, a huge fire-pump, and another large boiler. During that time an east and west railroad was also put through, close to our works – in fact, right through our grounds – thus bringing more and better shipping facilities. Again, in 1891 a three-story warehouse, in which to store goods, was erected. Four years later a third story was added to the wood-working building.

Again, in 1896 a lumber-shed, covered with iron, 60x120, was put up. This building, the largest of the entire group, is of sufficient capacity to hold nearly a million feet of basswood lumber for making section honey-boxes; and yet, as large as it is, we have used all the lumber out of it inside of three months, just for section honey-boxes.

In 1897 we were obliged to run night and day, and yet we were not able to take all the trade by considerable. We had to refuse money-orders, and turn away a good deal of other desirable trade. We hardly thought that, after such a heavy run of business, it would be necessary to run again nights; but in 1898 we were compelled to make double turns again, and for a much longer period of time, continuing clear up to the middle of July.

It became evident, by this time, that there would have to be a substantial enlargement, and more machinery, if we would keep up with our rapidly growing trade. Accordingly, during the latter part of 1898, we installed about $20,000 worth of improvements and enlargements – a 400-horse-power engine and a 400-horse-power boiler, the latter of the new water-tube type, a 135-horse-power electric-transmission equipment, the latter to carry power to distant points in our manufacturing plant. This, together with the electric apparatus that we already had in, made an investment in electric equipment of something like $4,000. The entire outfit comprises two dynamos, one of 100 horse-power, and the other 35. There are scattered over or plant 21 different electric motors, all operated by the two dynamos referred to, or what is technically called "generators." The machinery immediately adjacent to the big engine is operated by belting and shafting, so that, all in all, we now have one of the latest and best equipped power plants of its size that can be found in the world.

All of this necessitated the rebuilding and enlarging and remodeling of the engine and boiler rooms. An annex, operated entirely by electricity, was also put on to the end of one of the big buildings, the sole purpose of which was to take in the big planer, costing $1,000, and some other special machinery.

Something like a dozen clerks are employed almost constantly in our main home office in taking care of the general business, answering letters, keeping the books, and doing general office work. From three to four stenographers are required to take dictation from the members of the firm; and six typewriters are kept in use the greater part of the time.

There are scattered over the various portions of the United States ten branch offices under the name of "The A.I. Root Co." Besides this there are something like fourteen or fifteen large agencies that handle goods by the carload, to say nothing of smaller agencies that handle supplies in smaller quantities. All these branch offices and agencies keep in close touch with the home office.

E.R. Root, January 1905

Preface to Later Editions

It was not thought necessary to reproduce the prefaces of each succeeding edition of this work. All told, there have been 40 editions, with approximately a million copies of all editions printed to date. If we include the editions printed in foreign languages, the number would be over a million. It therefore transpires that the book written by A.I. Root for beginners, has finally developed into the *ABC and XYZ of Bee Culture.*

A.I. Root's first book was designed primarily for beginners, and hence the name, *ABC of Bee Culture.* Later on there came a call for a book which would cover the need of the more advanced as well as that of the beginner. In the meantime, Mr. Root's health had begun to fail and it became necessary for him to call in an assistant to do the work which he was physically unable to carry through. Mr. Root's son, Ernest Root, assumed the authorship of subsequent editions. The junior Mr. Root was present at the early beginnings of his father, who Dr. E. F. Phillips says, more than any other man blazed the way for practical bee culture. He was, in fact, says Phillips, the evangelist who pointed the way to methods of keeping bees that revolutionized the industry. Not that he invented or discovered some of the new devices or practices which he adopted, but rather that he made practical and workable those ideas advanced by others, so that beekeeping might become a commercial possibility as well as a pastime for back-lotters.

A.I. Root did not invent the modern hive and, much less, movable frames of Langstroth, but he was among the first in the field to point out that the Langstroth hive and frame were superior to all those of others preceding him. He was the first to proclaim that a hive with relatively shallow frames was better than one with deep frames. He was the first to announce that shallow hives could be tiered up, one on top of the other, until they would reach a size equal to that of our large colonies of bees at the present time. On the basis of these ideas he built a factory to make hives and frames and other equipment for general bee culture; the first of its kind in the world.

He was the first to see the importance of having the outside of a beehive smooth clear around with no projections of any sort. He was first to abandon the portico of the original Langstroth hive as a place where bees assemble on hot days and where, too, spiders would sometimes build cobwebs. So his first hive had hand-holes cut in the sides and ends – no cleats anywhere and the hives could be loaded on to a wagon or truck without any projections.

He was the first to recognize the importance of a standard frame throughout the United States. In a series of articles he wrote he proclaimed the advantages of standard equipment. He had had experience with different kinds of hives and frames in his own beeyard and saw the difficulty in interchanging the frames or the supers. As a result of this teaching in the early days, we have in this country one size (Langstroth) hive and frame.

He and his son, H.H. Root were the first to point out the differences in beeswax. First, there is capping wax which is pure wax, and then there is the wax that occurs in the combs, on the sides of the hives, and on the frames. This contains a great deal of propolis. If old hives and frames are scraped too much and the propolis goes in with the rendering of old combs, this wax will stretch more in comb foundation than wax from cappings. (For his early work on this subject, see *Gleanings in Bee Culture*, page 280, June 1944.)

Mr. Root did not invent comb foundation, but he was the first in the world, according to C.P. Dadant, to develop an article that would be alike useful and commercially possible. He did finally, with the help of his assistant, A. Washburn, develop the first commercially successful comb foundation.

Following this, A.I. Root, finding that comb foundation would stretch more or less after it was built into combs, conceived the idea of wiring his frames. Foundation was placed between the vertical and the V-shaped wires which were then imbedded in the wax. This was in 1878, years before others saw the importance of supporting comb foundation in the frames.

Mr. Root did not invent honey extractors, but he made the first all-metal commercial machine, which he called the Novice, that was ever put before the public, and of these machines he sold many thousands. The extractor, plus comb foundation, revolutionized bee culture commercially for the first time in dollars and cents.

He did not even invent the section honey box, but he was the first to manufacture one-pound sections by the thousand. He pointed out the way to produce comb as well as extracted honey.

He did not invent the bee smoker. But he, with T.F. Bingham, improved the implement put out by Moses Quinby, with the result that the early smokers of these two men were practically the same in principle as those in use today.

He was among the very first to send queen bees through the mail in a cage filled with a mixture of honey and powdered sugar. Not only that but he sold queens by the thousand for an even dollar each. Wiseacres of the day said this was impossible.

Next to his commercial comb foundation, perhaps his greatest contribution to bee culture was the sending of bees without combs, by mail and express in wire cages, in quarter, half and one pound lots. By the late 40s something like a half million pounds of bees without combs are sent from the South to the North annually. A large industry has developed so that literally hundreds of people are shipping bees North.

During 1894 and on to the present century there came an important epoch or era in beekeeping pioneered first by Charles Dadant and his son, C.P., advocating a brood nest for a queen, larger than that provided for either an eight or 10 frame Langstroth hive. In the early part of the 1870s, Charles Dadant began using an 11-frame Quinby hive which he called the Modified Dadant. He claimed that it would produce more honey with less swarming than the smaller brood nest in a 10-frame Langstroth. Both he and his son in the early 1890s actually proved that there would be less swarming. In fact, swarming was reduced down to less than two percent. Their crops of honey per colony proved that there was merit in a large brood nest.

In 1894 E.R. Root, A.I. Root's son, and an early author of this work, noticing these results, began to raise the question whether or not a two-story eight or 10 frame L. hive with a combined breeding capacity still larger would not accomplish the same results. (See page 343, 1894 *Gleanings*.) It would be impossible, he felt, for the beekeepers over the country to change from the regular Langstroth size of hive, either eight or 10 frame, to a larger Quinby. It would cost too much and neither would it fit the old Langstroth size equipment. By using two of these units Mr. Root believed he could get the same results as secured by the Dadants with their single large brood chamber, 11-frame Quinby hive. The seasons following for the next 10 years proved Mr. Root's theory correct in practice. This was brought out in a series of articles in *Gleanings* by the Dadants and Mr. Root in 1894 and on to the present century.

Mr. George Demuth, former editor of *Gleanings in Bee Culture*, a thorough believer in the double brood chamber, in the early part of last century pioneered another era, viz, the food chamber, consisting of two Langstroth hive bodies, one of which would be filled with honey from the tall end of the crop. Heretofore, we had been trying to winter in a single brood chamber with only about 20 pounds of sugar syrup. Good honey sealed in the combs, as he said, was the best insurance against loss in wintering and the best preparation for a good season of honey flow. These two ideas, that is a larger brood nest and the food chamber, revolutionized beekeeping practices.

A.I. Root's First Barrel of Honey from One Colony

In the meantime, Mr. Root was learning how to build up colonies in the spring and one year, 1878, he built up his colonies to enormous strength. One colony in particular was headed by a queen he called "Giantess." From this colony he extracted a barrel of honey, much of which was basswood. The amazing story of how he did it, set the bee world afire. Old bee masters said that such a thing was impossible.

When he produced his barrel of honey he did not, at the time, see that a two and three-story colony where the queen had unlimited brood-rearing space could produce a big crop of honey where a medium colony would barely get a living. It was only in later years that his son E.R. Root saw the importance of enormous colonies and that any good queen could fill two stories high with brood and bees in sufficient number to produce a crop of honey if nectar was to be had in the field.

When A.I. Root secured this barrel of honey, he had gone through another serious experience that cost him heavily the year before. His crops of honey were becoming so large that he borrowed wash boilers and pans of all sorts, and then began to talk about a cistern to hold his honey. Honey was coming in very rapidly and, not having time to uncap the combs, he extracted the honey just the same, believing that the honey standing in tanks would evaporate and "get good," but it did not. It began to foam and ferment even in the bottles, and much of it came back on his hands. The next year he made up his mind that the bees knew better than man how to ripen honey, and it was then that he secured the barrel of fully ripened honey.

In the years that have gone by, the reader who was charmed by the first edition of the *ABC of Beekeeping*, will perhaps see that there is very little left of A.I. Root's old writings in the later editions, especially in the ones bearing the title, *ABC and XYZ of Bee Culture*. His style of writing, however, has been updated and modernized, but his early contributions on many subjects were fundamental, and little revision has been needed.

The Later Editions of the *ABC and XYZ of Bee Culture*

As the industry developed, it required a larger work, so in 1910, the *ABC of Bee Culture* became the *ABC and XYZ of Bee Culture*. In 1923 A.I. Root's modest book of 200 pages, had developed into one of nearly 1,000 pages, comprising many new subjects, as well as new treatment in the handling of bees. While it is true that only a small percentage of A.I. Root's writings remained in this book, and that most of it was the work of his son, E.R. Root, A.I. Root's name and face will continue to appear, because he laid the foundation of modern bee culture.

E.R. Root
1950

Later editions of *ABC* were updated by a younger generation in the Root Family. E.R. Root's son Alan took over the management of the company after E.R. and H.H. moved on, and Alan's son, John, took over as Editor of several Editions of *ABC*, and also *Gleanings in Bee Culture*. John worked on several editions, then when he became President of the Root Company when Alan retired, Larry Goltz edited both the magazine and a few editions of *ABC*.

With the 40th Edition, published in 1990, The Root Company broke tradition and asked Dr. Roger Morse, Professor of Entomology at Cornell University, in Ithaca, New York, to take the lead, with assistance from Kim Flottum, the magazine's Editor. Dr. Morse rewrote much of what had been produced and brought in 20 or so writers for contributions in their various fields. Editor Flottum provided final edit, and selected or supplied many of the photos.

This edition, the 41st again brought in outside expertise. Dr. Hachiro Shimanuki, Research Leader of The USDA Honey Bee Lab in Beltsville, MD, just retired, updated much of the existing text, Ms. Ann Harman provided additional information and editing and several others provided specialized input and are acknowledged at the end of their chapters with initials, identified elsewhere in this book.

Finally, Kim Flottum, *Bee Culture* Editor brought in additional information, many new photos and resources and expanded many sections of this edition.

A.I. Root started his company in 1865, published the first edition of his magazine in 1872, and the first edition of *ABC* in 1877. For over 140 years the A.I. Root Company has been providing information and assistance for the beekeepers of the world. We are pleased to continue that tradition with this work.

Kim Flottum
Medina, Ohio 2007

AUTHORS

Many authors have made contributions to this work so we are able to offer the best in the many fields of science the honey bee is a part of, and the latest in beekeeping information from a practical aspect. Many beekeeping businesses assisted also, supplying their latest product information. We are indebted to all of these people for their contributions, suggestions and assistance.

Some authors made direct contributions, and their initials appear at the end of the sections they provided. But many assisted with edits, corrections, updates and opinions. Their names are listed here also. Too, we have borrowed sections from *Gleanings In Bee Culture*, *The American Bee Journal* and the book *Beekeeping Associates* for biographical information on many of the personalities include in this edition.

We thank all of these friends of beekeeping for their contributions.

RB	Robert Berthold, Ph.D, Professor, Delaware Valley College, Doylestown, PA
BB	Bruce Boynton, CAE, Chief Executive Officer, National Honey Board, Longmont, CO
JB	Jordi Bosch, Ph.D, Research Assistant Professor, Dept. of Biology, UT State Univ., Logan, UT
JJB	Jerry J. Bromenshenk, Ph.D, Research Professor, Univ. of Montana-Missoula, Missoula, MT
DMB	D. Michael Burgett, Ph.D, Professor Emeritus of Entomology, Oregon State Univ., Corvalis, OR
NWC	Nicholas W. Calderone,. Ph.D, Professor of Entomology, Cornell Univ., Ithaca, NY
SC	Scott Camazine, M.D., Ph.D.
JHC	Jim H. Cane, Ph.D, Research Entomologist, U.S.D.A., Logan, UT
HC	Heather Clay, B.S., National Coordinator, Canadian Honey Council, Calgary, Alberta, Canada
SWC	Susan W. Cobey, Staff Apiarist, Rothenbuhler Honey Bee Lab, OH State Univ., Columbus, OH
RWC	Robert Currie, Ph.D, Associate Professor, Dept. of Ent., Univ. of Manitoba, Winnipeg, Manitoba, Canada
LDG	Lilia DeGuzman, Ph.D, Research Entomologist, U.S.D.A., Baton Rouge, LA
PJE	Patti Elzen, Ph.D, Research Entomologist, U.S.D.A., Weslaco, TX, Deceased
EHE	Eric H. Erickson, Ph.D, U.S.D.A., Retired, Tucson, AZ
MSF	Michael S. Ferracane, M.S. Graduate Student, Cornell Univ., Ithaca, NY
KF	Kim Flottum, Editor, *Gleanings in Bee Culture*, The A.I. Root Co., Medina, OH
RDG	Russell D. Goodman, Dept. Hort., Apicultural Scientist, Dept. of Primary Industries, Victoria, Australia
EG	Ernesto Guzmán-Novoa., Ph.D, Univ. of Guelph, Guelph, Ontario, Canada
JRH	John R. Harbo, Ph.D, Research Entomologist, U.S.D.A., Baton Rouge, LA, Retired
AWH	Ann W. Harman, B.S., Contributing Editor, *Bee Culture*, Flint Hills, VA
CBH	Colin B. Henderson, Ph.D, Professor, Research Assistant Professor, Univ. of MT-Missoula, Missoula, MT
WMH	Wm. Michael Hood, Ph.D, Professor of Entomology, Clemson Univ., Clemson, SC
ZYH	Zachary Huang, Ph.D, Assistant Professor, Dept. of Ent., MI State Univ., East Lansing, MI
RRJ	Rosalind R. James, Ph.D, Research Entomologist, U.S.D.A., Logan UT
RJ	Richard Jones, M.ED., Director, International Bee Research Association, Cardiff, Wales, UK
WPK	William P. Kemp, Ph.D, Research Leader, U.S.D.A., Logan UT
BM	Bill Mares, A.B., Author, Teacher, Burlington, VT
RAM	Roger A. Morse, Ph.D, Professor of Entomology, Cornell Univ., Ithaca, NY, Deceased
ECM	Eric C. Mussen, Ph.D, Extension Apiculturist, Univ. of California, Davis, CA
DLN	Don L. Nelson, Ph.D, Research Scientist, Agr. & Agri-Food Canada, Ret., Beaverlodge, Alberta, Canada
RN	Richard Nowogrodzki, Ph.D, Ithaca, NY
REP	Robert E. Page, Jr., Ph.D, Professor of Entomology, Univ. of California, Davis, CA
TLP	Theresa L. Pitts-Singer, Ph.D, Research Entomologist, U.S.D.A., Logan, UT
JQ	Javier G. Quezada-Euán, Departamento De Apicultura, FMVZ, Universidad De Yucatan, Merida, Mexico
MR	Murray Reid, National Manager Apiculture, AgriQuality NZ Ltd, Hamilton, New Zealand
GER	Gene E. Robinson. Ph.D, Professor of Entomology, University of Illinois, Urbana, IL
JAR	John A. Root, Chairman, The A.I. Root Co., Medina, OH
MTS	Malcolm T. Sanford, Ph.D, Professor Emeritus, University of Florida, Gainesville, FL
HS	Hachiro Shimanuki, Ph.D, U.S. Department of Agriculture, Retired, Lake Placid, FL
DRRS	Deborah Ruth Roan Smith, Ph.D, Associate Professor of Ent., Univ. of KS, Lawrence, KS
HAS	H. Allen Sylvester, Ph.D, Research Entomologist, U.S.Dept. of Agriculture, Baton Rouge, LA
RT	Richard Taylor, Ph.D, Contributing Editor, *Gleanings in Bee Culture*, Trumansburg, NY, Deceased
AWV	Allen W. Vaughan, Entomologist, U.S. Environmental Protection Agency, Washington, DC
PKV	P. Kirk Visscher, Ph.D., Assistant Professor of Entomology, Univ. of California, Riverside, CA
DW	Danny Weaver, J.D., B Weaver Apiaries, Navasota, TX
LSW	Lois Schertz-Willett, Ph.D., Professor of Food & Resource Economics, Univ. of FL, Fort Pierce, FL
TY	Tadaharu Yoshida, Professor, Honeybee Science Research Center, Tamagawa University, Tokyo, Japan
SZ	Steve Zimmerman, B.A. Owner, Dawes Hill Honey, Co., Newfield, NY
SS	Stanley Schneider, Ph.D, Univ. of NC, Charlotte, NC
	Wyatt Mangum, Ph.D, Univ. of Mary Washington, Fredericksburg, VA
	Diana Sammataro, Ph.D, Research Entomologist, USDA, Tucson, AZ
	Adrian Wenner, Ph.D, Univ. of CA, Santa Cruz, CA, Retired
	Eric Wenger, Quality Control, Golden Heritage Foods, Hillsboro, KS
	Walter Rothenbuhler, Ph.D, The OH State Univ., Columbus, OH, Deceased

Editors

Hachiro Shimanuki, microbiologist, was born in Hawaii and received his BA and PhD degrees from the University of Hawaii and Iowa State University, respectively. Dr. Shimanuki spent his entire career with the USDA, two years in Laramie, Wyoming and 34 years in Beltsville, Maryland where he served as Investigations Leader for Bee Diseases and also as the Head of the bee laboratory. Dr. Shimanuki is noted primarily for his research on the diagnoses and control of bee diseases and pests. He also provided technical assistance to various Federal agencies including the USDA Animal and Plant Health Inspection Service, the Environmental Protection Agency and US Food and Drug Administration. Internationally, Dr. Shimanuki has been an invited lecturer to conferences and workshops, and served as consultant and advisor for beekeeping in North America, Europe, South and Central America, Africa, the Middle and Far East, and New Zealand. He was the English editor for Apidologie and also served as president of the International Bee Research Association. Dr. Shimanuki and his wife, Susan retired to Lake Placid, Florida.

Kim Flottum received his B.S. Degree from the University of Wisconsin, Madison. He worked for the State Extension Fruit Entomologist, studying pest control for a wide variety of pests on horticultural crops. After graduation he worked as a Research Technician at the USDA Honey Bee Research Lab in Madison. There he studied soybean culture and pollination, plus the honey bee, sweet corn, pesticide relationship.

He became Editor of *Bee Culture* in 1986 and while in that position assisted Roger Morse with the 40th Edition of *ABC*, and *Honey Bee Pests, Predators and Diseases*, which was a Gold Medal winner at Apimondia, 1999. He has produced over a dozen beekeeping books and instructional videos, and written his own book, *The Backyard Beekeeper*.

Although Ann Harman spent part of her life as a research chemist, honey bees have created a second career as an International Consultant, teaching beekeeping skills and modern management techniques in Third-World countries. To date she has worked in 26 countries on five continents for a total of 49 assignments. Here in the United States she is an Eastern Apicultural Society Certified Master Beekeeper. She put that to use teaching in short courses and lecturing at beekeeping association meetings. Ann is a judge of honey and hive products: local, state and national, also a Certified Honey Judge, Wales, UK, Beekeeping Institute. Participation in industry activities includes being an Alternate of the National Honey Board. In addition she writes a monthly article for *Bee Culture,* regular articles for *Beekeepers Quarterly*, and frequently for *Bee Craft*. Her hives now serve as teaching hives, not only for beekeepers, but also for teaching youth so that our pollination needs continue to be met in the years to come.

The Editors and Authors extend their appreciation to The A.I. Root Company, and John Root, Chairman and President, for the support and encouragement extended during the production of this Edition of *The ABC and XYZ of Bee Culture*. This work could not have been completed without The A.I. Root Company's continued dedication to the beekeeping industry.

The Publisher and Editors are indebted to the publishers of the books listed below for allowing the use of their information in this Edition. Additional biographical information was used that was originally published in *Gleanings In Bee Culture* and very early editions of *The American Bee Journal*.

Anatomy and Dissection of the Honeybee
H.A. Dade
published by IBRA

A Scanning Electron Microscope Atlas of the Honey Bee
Eric H. Erickson, Jr.
Stanley D. Carlson
Martin B. Garment
published by The Iowa State University Press

Form and Function in the Honey Bee
Lesley Goodman
published by IBRA

Some Beekeepers and Associates
Joseph O. Moffett
published by Joseph O. Moffett

Anatomy of The Honey Bee
R.E. Snodgrass
published by Northern Bee Books w/permission from IBRA

The Publisher, The Editors and The Authors wish to acknowledge and thank Kathy Summers, who coordinated content organization, layout and design, photography and indexing.

And Sharon Garceau, Chief Copy Editor, fact checker and Glossary organizer.

And for proofreading, encouragement and support, thank you to Susan Shimanuki.

ABNORMAL BEES – Abnormal, or different, bees can be found in a hive. Most abnormal bees are different for one of two reasons. The differences may be genetic, as in bees with white or pink eyes, different body or hair color, or other mutations. Bees with light-colored eyes are noticed most often since they are strikingly different. More than 30 mutations have been found in European honey bees in North America. An oddity that turns up every so often is the cycloptic bee whose head width is less than normal and whose two compound eyes join at the top to make one; cycloptism is usually believed to be genetic in origin (See also GYNANDROMORPHIC BEES).

A second reason for the existence of abnormal bees is environmental. Chilling or overheating of the brood, malnutrition, drugs, pesticides, mites or disease can bring about changes during larval or pupal life that are seen when the bees mature and become adults. One of the common problems in colonies where *Varroa* mites are present is that bees may have deformed wings or legs. This occurs when mites feed on honey bee pupae and transmit deforming, and often lethal viruses.

Abnormal bees occur, or at least are seen, rarely. Bees usually remove abnormal individuals from their midst and carry them outside of the hive where they die.

ABRAMS, GEORGE (1902-1965) – Served as Assistant Professor, Extension Apiculturist and Supervisor of bee disease control in Maryland from 1929 to 1965. He graduated from University of Maryland with a B.S. in 1927 and M.S. in 1929. He organized the second Tri-State Beekeepers meeting in 1955, where the organization officially became the Eastern Apicultural Society. He was president of EAS again in 1965, shortly before his death.

George Abrams

ABSCONDING SWARM – Absconding means the abandonment of the nest by the entire adult population because of disease, lack of food or other unfavorable conditions. In contrast to swarms that occur as a division, absconding does not result in an increase in colony numbers. Typically absconding swarms leave behind no adult bees and may abandon large food stores and brood. (see SWARMING IN HONEY BEES)

ACARINE DISEASE (*ACARAPIS WOODI*) – (see MITES AFFECTING HONEY BEES)

ADDLED BROOD – This term is not used today but may be found in some of the older literature. It describes sick brood, especially larvae, whose condition cannot be ascribed to any of the known causes of disease. In the U.S. addled brood was once used to describe brood afflicted with toxic pollen or nectar. California buckeye, *Aesculus californica*, which is abundant in much of California and produces a toxic nectar, was said to cause addled brood. In Europe, the term was used to describe larvae with a genetic disorder. Addled brood is one of those unclear and poorly-defined terms that is best discarded.

ADEE HONEY FARMS – Based in Bruce, South Dakota, but with operations in North Dakota, Minnesota, Kansas, Nebraska, California and Mississippi, this is the largest beekeeping business in the United States, approaching 100,000 colonies. Honey production and pollination are the principal businesses, but beeswax, production of some beekeeping equipment, nuc and queen production add to the mix.

Started by Vernon Adee and four of his brothers in 1957, Adee Honey Farms

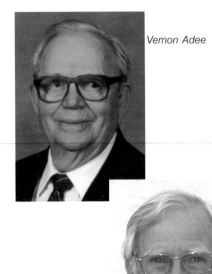

Vernon Adee

Richard Adee

now is operated by Vernon's son, Richard Adee, his two sons, Brett and Kelvin and a score of dedicated, long-term employees.

AEBI, HARRY J. (1891-1986) and **AEBI, ORMAND** (1916-2004) – Harry and his son, Ormand, were hobby beekeepers living in Santa Cruz, California. Together, they held the world's record in the Guinness Book of Records (from 1976-1984) for the most honey produced from a single hive in a single location with a single queen – 404 pounds. The record was broken by a multiple-queen colony. Together, they coauthored *The Art And Adventure of Beekeeping*, and *Mastering The Art Of Beekeeping*.

AFRICA – (see BEEKEEPING IN VARIOUS PARTS OF THE WORLD, Africa)

African Honey Guide Bird

AFRICAN HONEY GUIDES – An African bird, *Indicator indicator*, the Greater Honey Guide, makes contact with people by flying close to them, making a double-noted chirping sound and then flying in the direction of a honey bee nest they have found. In this way natives in East Africa have learned that they may find a nest, take the honey for themselves and, by leaving some wax and brood, pay the bird for its efforts.

These interrelations between birds and men were reported over two hundred years ago but seemed so fantastic that they were usually doubted until a thorough study was made of the matter and reported in a Bulletin of the U.S. National Museum in 1955. More recent research is even more detailed and reads almost like science fiction.

The Boran people, a nomadic tribe in the dry bush country of Northern Kenya, routinely hunt for bees and honey. In unfamiliar areas, finding a honey bee nest without help may take nearly nine hours. By using the greater honey guides that time is cut to a little over three hours. In addition to making appropriate sounds, and flying in the direction of the nest, the birds that are guiding honey hunters display their outer white tail feathers. A Boran bee hunter responds to the bird by whistling, banging on wood and talking loudly to the bird so that the bird is aware that it is being followed. It is not always a question of birds finding bee hunters; the hunters may seek out birds by attracting them from a distance of as much as half a mile (one kilometer) with a shrill whistle.

Researchers following the birds plotted the directional routes over which they were guided. Once they found the bee's nest they did not destroy it but watched to confirm the activities of the bird. When the bird arrives at the nest site two changes are apparent in its behavior. It first produces a different sound for a few calls and then remains silent. It then flies to a nearby perch and sometimes flies around the nest to indicate its location.

Some who have been skeptical of this intricate behavioral exchange between men and birds have suggested that the birds guide people until they find nests by accident. However, camouflaged observers found birds would monitor nests they had found before they guided people to them. On cloudy and cool days when the bees were not active, the birds would fly to the nest entrance and look inside.

Once the Boran people find a nest they use fire and smoke while they open the nest and take the honey. It was observed that 96 percent of the nests were inaccessible to the birds until opened by the honey hunters. The smoke also served to protect the birds while they fed on wax and brood. The article concludes by stating that the Boran people are indeed great observers of nature. (Isack, H.A. and H.U. Reyer. "Honeyguides and honey gatherers: interspecific communication in a symbiotic relationship." *Science* 243: 1343-1346.1989.)

AHB stinging leather target during defense test.

AHB stinging camera straps, a favorite target.

AFRICANIZED HONEY BEES (AHB) –

Professor Warwick E. Kerr of Brazil imported 170 queens from Africa in 1956. Approximately 50 queens of *Apis mellifera scutellata* survived the journey. Most of the queens came from an apiary near the South African city of Pretoria. Kerr was doing nothing more than agricultural researchers have been doing for years: moving plants and animals around the earth to find those that will do best and serve to improve agricultural production in new locations. The defensiveness of the African honey bees was well known but it was expected that this characteristic could be eliminated from the imported stock without difficulty. Furthermore, it was well known that queen bees from Africa had been carried to other countries without incident. Apparently, however, an importation on the scale tried by Kerr had not been attempted earlier.

The result of this importation is that today there are AHB in tropical and subtropical South and Central America and in North America. In these countries they are called the Africanized honey bees (AHB). They were given this name to differentiate them from the African bees in Africa and the stingless bees in Brazil.

The first swarm of AHB in the U.S. was captured in Hidalgo, Texas in October 1990. The swarm was believed to be the result of natural range expansion of the AHB and not a man-assisted move. A number of states and the federal government instituted programs to capture swarms, monitor airports and seaports, and to monitor movements of AHB intra-state. In addition some states regulate the movement of AHB colonies.

Apis mellifera scutellata in Africa, and the New World hybrids, show variations in their behavior and physiology that relate to the African environment in which they live. Because of the presence of predators that can destroy colonies, like ants, honey badgers, and man guided by honey guide birds, AHB have

AHB stinging bee gloves.

developed more intense levels of colony defense. The tropical and subtropical patterns of plant flowering, which often follow rain patterns and are spread throughout the year, selected for increased absconding and swarming. Colonies left areas where blooming was finished and moved to locations where flowers could still be found. Swarms produced during much of the year could survive, unlike in temperate areas where late summer swarms may perish from lack of any honey sources. No swarming occurs during cold months in temperate-climate bees.

Many of these differences made the African type of honey bee more suited to the tropical and subtropical areas of the Americas and allowed for the successful expansion of the AHB into areas with no *Apis mellifera* or European varieties. Beekeepers whose operations changed from EHB to AHB over several years often gave up beekeeping because they and their neighbors did not care for such defensive colonies. And the increased production of swarms from strong colonies drastically reduced honey production. However, with proper management these differences have been overcome. Many beekeepers were able to adapt their practices to the different biological rhythms of the AHB and prospered. In addition, a new type of honey harvesting developed in remote areas. Empty colonies were moved in, allowed to be colonized by swarms and the honey harvested with no care for preserving the bees. A sort of non-management beekeeping.

Some of the differences are more positive. In many areas of South America, beekeepers do not treat their AHB colonies for *Varroa* mites. In addition, the tracheal mites, American foulbrood and sacbrood are seldom seen.

Absconding – The fact that African bees and AHB swarm more frequently and are more likely to produce afterswarms has posed problems for beekeepers where these bees are found. For several years commercial beekeepers in South Africa reported that they made up much of their increase by using bait hives and that the captured bees were later re-hived and the colonies moved for pollination or honey production. After using this method for several years the South African beekeepers reported that

Working AHB in Venezuela.

the absconding rate among these re-hived bees was too high to make the venture profitable. However, it was soon learned that if these colonies were requeened with selected queens the absconding rate would be reduced.

This experience was repeated in the New World as AHB spread into new areas and the beekeepers found that captured swarms had a greater inclination to abscond than the colonies to which they were accustomed. Requeening with European or hybrid stock or even selected Africanized queens made a difference and has now become a standard practice. Absconding is a behavioral trait that can be selected for or against.

Colony invasion by migrating swarms – Several published accounts have told of swarms of Africanized honey bees invading and taking over colonies of European honey bees. Camazine interviewed an experienced Brazilian queen breeder. (Camazine, S. "Queen rearing in São Paulo state, Brazil: A beekeeping experience of over 20 years." *American Bee Journal* 126:414-416. 1986.) During 20 years this beekeeper had rarely, two or three times a year, seen small, queenright swarms of Africanized honey bees land at the entrance of nucleus colonies of European bees. In all cases the Africanized swarms had fewer than 3000 bees and the invaded colony of European bees was also low in population. The Africanized bees balled the European queens, and if left to their own devices the Africanized queens would replace the European queens.

Defensiveness – AHB bees do an excellent job of defending their homes. Once a colony is aroused it may remain defensive all day. A defensive colony will arouse others in the same bee yard. Under such conditions it is difficult to make controlled, comparative tests of defensiveness.

Note the directional shield.

Left: 4"x10" smoker. Right: smoker used for Africanized bees.

Researchers A. Stort in Brazil, A. Collins in Venezuela and others devised a variety of methods of measuring defensiveness. Some of the techniques include counting the number of stings in black leather or felt objects of a given size held or waved in front of a colony entrance for a given period of time. One popular test involved arousing a colony and measuring the distance defensive bees would follow the beekeeper that walked or ran away in a straight line. Some AHB were found to pursue a person for more than a mile and remain aroused for days.

Defensiveness in honey bees is a genetically-based behavioral trait that can be emphasized or de-emphasized in a bee breeding program. Some beekeepers in the Americas have developed hybrid bees that are less defensive but still good honey producers.

Dressing for defensive bees – It is a well-established fact that all stinging and biting insects are much less inclined to sting or bite through light-colored, smooth-finished clothing. Honey bees respond this way as well. Perhaps textiles with rough, dark surfaces resemble the exteriors of animals that might attack beehives.

Beekeepers working with AHB traditionally wear khaki or white cotton clothing, usually coveralls, and sometimes in several layers. Boots are usually six to eight inches high (15 to 20 cm) and are made of light-colored leather since it is the most comfortable material. Some beekeepers use white rubber boots. The trouser legs and arm length sleeves are fitted with elastic or are tied down to prevent bees from crawling underneath.

A properly dressed beekeeper will be stung less. In dealing with defensive bees this is especially important. An innovative suit that incorporates outer and inner layers of mesh separated by spongy padding allows a beekeeper to work in comfort in hot, humid areas yet prevents stings from penetrating to the skin.

Veil construction has been found to be an important aspect as well. Hats or hoods that prevent the veils from being forced against a beekeeper's face or neck by large numbers of defending or hanging bees are necessary. Traditional screens made of black material, easier to see through, attract more bees to sting. If the outside of the veil is painted white or aluminum, this tendency is reduced.

Many beekeepers have stopped using leather gloves, as they are hot, clumsy, and retain odors (alarm pheromone) from previous stingings. With the high level of stinging by AHB, frequent penetration by stings weakened the protective factor of the leather. The leather gloves have been replaced by lightweight, light-colored gloves such as rubber dish gloves or layers of cotton mesh gloves under rubber ones. The rubber ones are easy to keep clean and are stung less often due to the slick surface. The cotton mesh provides for sweat absorption.

Effect of feral colonies – Swarms of AHB bees will accept smaller nest cavities than European bees including ground nest-sites like caves and a wide variety of man-made holes, such as old rubber tires and decaying walls. In tropical areas even very large open-air colonies may be found hanging from limbs, bridges and water towers. Feral colonies compete for food; they furnish drones that may or may not be of undesirable stock.

Requeening and finding queens – African and Africanized bees have the bad habit of running on the combs. In this manner they behave much like the German black bees that were once commonly used in North America. As one examines a comb the bees will move to one end where they form a large ball which frequently drops back into the hive.. Under such circumstances finding a queen is nearly impossible. The easiest way to cope with the running, if one is merely examining brood, is to gently shake the bees onto the grass or dirt in front of the hive.

Africanized queens are slightly smaller and darker than their European counterparts. It is much more difficult to find a dark queen. These queens are also very runny like workers and have even been seen to dump eggs and fly off combs when disturbed. One of the great virtues of the European stock of bees is that they usually remain calm and evenly distributed on a comb when one is examining it.

When finding queens of any stock it is best to use a minimum of smoke since it disturbs the bees and increases their tendency to run. While Africanized bees are more defensive, and need more smoke when examining colonies, moving slowly and carefully when removing combs will reduce the need for smoke. This, in turn, makes queen finding easier. One trick that often works with AHB colonies is to smoke the entrance several times and then wait for a few minutes before opening the hive. If the inner cover is removed and quickly inverted, the queen may be found there after running up through the hive to escape the smoke.

Where beekeepers are using hybrid queens, those of European origin, but mated with Africanized drones, annual requeening is advised. Finding queens is always a slow process. Many beekeepers have boxes with queen excluders on the bottoms. They shake the bees into the boxes and force them through the excluder with smoke. The queens are left on top where they are easily found.

Requeening Africanized colonies is no different from requeening colonies of any race. Small colonies accept queens more readily than do strong colonies. If one introduces a queen into a large colony of Africanized bees using the queen cage technique, the failure rate will be quite high. The best way to requeen a strong colony of Africanized bees is to find and kill the old queen and place a small, nucleus colony with the new queen on top. A sheet of newspaper with a couple of slits is placed between. Push-in cages and the queen-cage technique work very well in the case of smaller, less-populous colonies (see REQUEENING).

Honey production – Reports from Africa of large honey crops produced over a year led to a reputation for these bees to be better honey producers than European bees. In fact *Apis mellifera scutellata* has behaviors that are better adapted to the tropical pattern of flowering, and make efficient use of the almost year-round availability of some kind of nectar. Dr. T. Rinderer and colleagues have shown that AHB workers tend to forage more independently and dance less, and to accept lower quality nectar. In this way, the foragers spread out across a wider range of floral types during times when many plants are blooming. A high proportion of the honey is turned into more bees to produce swarms. During periods of poor nectar flows, the AHB continue to bring in enough of the poor quality nectar for the

Africanized Honey Bees

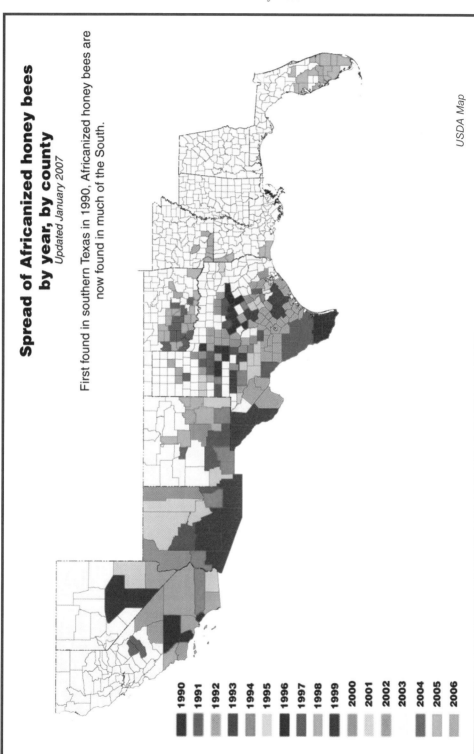

colony to just survive. European bees in tropical areas will starve during rainy periods because they will not collect the low sugar concentration nectar that is available. Successful keepers of AHB shift their honey harvesting to take advantage of the bees ability to survive and produce honey under tropical conditions. In contrast, AHB behavior is at a disadvantage in temperate areas, where honey flows are often brief. Temperate-climate foragers do best by collecting the most sugar-rich nectars, and store large amounts of honey to survive over winter.

Identification of Africanized honey bees – Africanized honey bees cannot be distinguished from European honey bees by any simple method. Because of the importance of correctly identifying possible Africanized colonies, much effort has gone into developing reliable identification methods. USDA-ID, the official method recognized by the USDA Animal and Plant Health Inspection Service, uses morphometrics, the measurement of the lengths and angles of several body parts from 10 bees and the analysis of these measurements by a computer program. This approach is successful because the AHB are about 10 percent smaller than their European counterparts. This method was first developed by Howell V. Daly at the University of California at Berkeley. It was further refined by Thomas E. Rinderer and H. Allen Sylvester of the USDA, ARS, Honey Bee-Breeding, Genetics and Physiology Laboratory in Baton Rouge, LA. Because of the complexity of this method and the equipment required, they then developed a much simpler field-screening method, Fast Africanized Bee Identification System or FABIS, to analyze large numbers of samples and eliminate those samples that are clearly not Africanized. FABIS measures only the forewing lengths and the weight of a sample of bees. Those samples that might be Africanized are then analyzed by the full morphometric procedure in a laboratory. These remain the official methods because their accuracy and reliability have been verified by the analysis of a very large number of samples.

Cuticular hydrocarbon analysis compares the ratios of waxes and other compounds from the surface of honey bees. However, these ratios also change as bees age and reliable differences between Africanized and European honey bees of unknown ages were never reported. Some enzymes that show genetic variation, known as allozymes, differ in the ratios of the variants between Africanized and European honey bees. However, no known variants are found only in all Africanized or only in all European honey bees. Several nuclear and mitochondrial DNA markers have been reported to differ between Africanized and European honey bees. Mitochondrial markers are only inherited from the mother and so cannot distinguish offspring of Africanized drones. Mitochondrial DNA is used to distinguish the Africanized honey bees from European honey bees. -HAS

Range expansion of Africanized honey bees – The range expansion of the AHB has been principally in a northwesterly direction since the first capture in Hidalgo, Texas. As of this writing, the AHB are established in all of Texas, New Mexico, and Arizona, the southern half of California, and parts of Nevada, Arkansas, Utah, Oklahoma, Mississippi, Louisiana, and Florida. It is expected that feral colonies of the AHB will establish themselves in sections of all of the southernmost states and along some of the coastal states that border the Pacific and Atlantic Oceans. The ultimate range of the AHB is unpredictable and to some degree dependent on temperature and annual rainfall. See Villa, J.D., T.E. Rinderer, and J.A. Stelzer. "Answers to the puzzling distribution of Africanized bees in the United States." *American Bee Journal* 142: 480-483. 2002 for their hypothesis. The authors state that their "coarse-grained data" is not based on degrees of hybridization as determined by morphometric and DNA analysis.

Further reading – Much has been written on the Africanized honey bees from the serious to the ridiculous. For the serious reader, one of the best sources of information is *The "African" Honey Bee*. Edited by M. Spivak, D.J.C. Fletcher, and M.D. Breed. (Westview Press, Boulder, CO, 435 pages. 1991.) For a history of AHB in the Americas see Caron, D.M. *Africanized Honey Bees in The Americas*. (A.I. Root, Medina, OH. 228 pages. 2001.)

AFTERSWARMS – (see SWARMING IN HONEY BEES)

AGING IN HONEY BEES – (see LENGTH OF LIFE OF THE HONEY BEES and CASTES)

ALARM ODOR – (see PHEROMONES)

ALFALFA LEAFCUTTING BEE – (see LEAFCUTTING BEE)

ALIMENTARY SYSTEM – (see ANATOMY AND MORPHOLOGY OF THE HONEY BEE)

ALKALI BEE – The alkali bee, *Nomia melanderi*, is native to deserts and semi-arid desert basins of the western United States. This medium-sized non-social, gregarious bee bears striking pearly green or orange abdominal bands. It is a very effective and manageable pollinator for the production of seed in alfalfa (lucerne) and some other crops, such as onion. It is the only intensively-managed ground-nesting bee in the world. Originally, growers placed alfalfa seed fields adjacent to natural populous nesting sites. In and around the Great Basin, alfalfa seed growers now construct large sub-irrigated silty nest sites for this bee, supplying them with salt-crusted surfaces. Minimal subsequent maintenance makes this the most cost-effective pollinator of alfalfa. Densities of 400 nests/m^2 over a hectare (2.4 acres) or more can be obtained with this gregarious bee. The largest managed nesting aggregations contain 1.5 million nests. Nest sites can remain populous for more than 50 years with vigilant moisture and pest control.

Adults emerge near midsummer in the state of Washington. Male emergence generally precedes that of females. Males initially patrol the nesting aggregation for emerging virgin females. Once mated, each female excavates a shallow, individual subterranean nest. On average, one nest cell is prepared daily and provisioned with a flattened round pellet composed of several million pollen grains moistened with nectar. Females must visit more than 5000 alfalfa flowers to obtain sufficient pollen for each nest cell, although they will use diverse floral hosts if available. Nesting persists for about 30 days. In her lifetime, each female's foraging activities in alfalfa results in an estimated 1/3 pound (0.15 kg) of seed being produced.

New nesting sites can be created using soils rich in silt or fine sands. Sub-irrigation is provided using a grid of perforated corrugated drainpipe laid 1.5-2 feet (45 to 60 cm) underground. Salt is applied to the surface. Before May, several thousand cubic-foot soil cores are punched from a populous aggregation, boxed, and set in trenches in the new nesting bed. The cardboard is burned away, the cores wedged into place with soil, and sub-irrigation water supplied. One of the oldest extant nesting sites was created in 1955, testimony to their longevity.

Populations annually increase several-fold unless checked by sub-optimal conditions. These include: overstocking of bees, inappropriate insecticide application during bloom, inadequate sub-irrigation, daytime cloudbursts during nesting, or build-ups of pests such as blister beetles (Meloidae) or bomber flies (Bombyliidae), both of which can be managed using simple physical controls. For further information see Johansen, C., and D. Mayer, Alkali bees: their biology and management for alfalfa seed production in the Pacific Northwest. *PNW 155*. 19 pages. 1976. -JHC

ALLERGIES TO STINGS – (see REACTIONS TO BEE AND WASP STINGS)

AMITRAZ – In the early 1990s, U.S. beekeepers also had available another EPA registered acaricide for *Varroa* control. This compound, amitraz, sold as the product Miticur®, was formulated as a strip and inserted between brood frames, similar to Apistan® and CheckMite+®. Amitraz is in a pesticide class entirely different than fluvalinate, coumaphos or formic acid; it is classified as a formamidine pesticide. Amitraz was quickly removed from sale to beekeepers, however, when a small number of beekeepers felt they were experiencing elevated bee kills during amitraz use in their hives. Unfortunately *Varroa* mites that are resistant to fluvalinate are also resistant to amitraz. -PJE

ANATOMY AND MORPHOLOGY OF THE HONEY BEE – A honey bee has three body parts and three pair of legs, typical for an insect. However, the honey bee body is modified both internally and externally, reflecting its special relationship to flowers and the fact that bees live exclusively on pollen, honey and water.

The bodies of all bees, not just the honey bees but the solitary and subsocial bees as well, are covered with plumose (branched) hairs in which pollen is easily trapped and thus carried from one plant to another. The crop (honey stomach) of the worker honey bee is modified for carrying nectar or water and the hind legs are made to carry pollen or propolis. The glandular systems of workers, and to a lesser extent those of queens, are geared to the special needs of the colony. The reproductive systems of drones, queens and workers are very different from each other. Also, drones have far more elaborate and well-developed eyes and antennae than the two female castes. The workers have rudimentary ovaries and ovarioles and are not large enough to mate; however, they can lay eggs under special circumstances. The queen may properly be called an egg-laying machine and her abdomen is crammed with the associated organs that enable her to lay her own weight in eggs in a day.

On the head of the honey bee are the

proboscis, with the other mouth parts, the antennae and the eyes. Within the head are special glands that produce food for the larvae and the queen, produce the secretions needed to make honey from nectar, and secrete pheromones used in communication. As in most insects the thorax is filled with muscles that drive the legs and the wings. Externally, the wax glands are found on the underside of the abdomen and the sting and its associated glands are contained within the tip of the abdomen in the females.

In thinking about honey bee anatomy and morphology it is well to remember how insects are different from mammals and other animals with which we are more familiar. Insects do not have an internal skeleton as we do but rather an exoskeleton (see Exoskeleton). The muscles in an insect's body attach to the inside of the exoskeleton rather than to bones as in our own body. The oxygen delivery system is very different. Bees have no lungs and oxygen is not carried to the cells in the blood. Rather the oxygen is delivered by tubes, called tracheae, which open to the outside of the body and that branch throughout the body and thus carry oxygen to every cell. It is not a very efficient system; it is not a system that would lend it itself to use by a larger animal.

In this portion of this book are found a number of the classical drawings of honey bees by Robert E. Snodgrass who wrote extensively on honey bee anatomy and morphology. His textbook (*Anatomy of the Honey Bee*), which has been reprinted many times, is still in print. Snodgrass also wrote the classic textbook on general insect anatomy and morphology. The scanning electron microscope (SEM) allows us to see the fine detail of honey bee anatomy and many are used here, also. (See Erickson, E.H. Jr., S.D. Carlson and M.B. Garment. *A Scanning Electron Microscope Atlas of the Honey Bee. Iowa State Univ. Press, Ames.* 292 pages. 1986.) This text contains several hundred SEM photographs, many of which have been reproduced here in conjunction with the Snodgrass drawings. Additional resources used here with permission include the excellent IBRA text *Form and Function in the Honey Bee*, published in 2003, written by Lesley Goodman. This text incorporates line art, paintings and additional SEM photos to illustrate the anatomical connections to honey bee behavior. The 1977 edition of *Anatomy and Dissection of the Honey Bee* by H.A. Dade, also originally published by IBRA, is an excellent text. It, too, has excellent drawings and the best information we know on making dissections . Some of this work is included here, also.

Average Development Times For Queens, Workers and Drones

	Egg	Larva	Pupa	Total Development Time	
Queen	3 days	5½ days	7½ days	16 days	
Worker	3 days	6 days	11 days	20 days	
Drone	3 days	6½ days	14½ days		24 days

Anatomy and Morphology of the Honey Bee

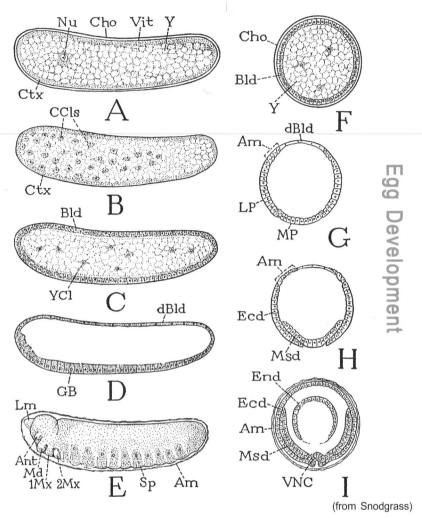

(from Snodgrass)

Early development of the honey bee egg and the young embryo (diagrams based on Nelson, 1915). A, lengthwise section of egg before development, surrounded by chorion. B, cleavage cells formed by division of egg nucleus, scattering through the yolk and pressing into the cortex. C, cleavage cells mostly in the cortex, forming the blastoderm. D, blastoderm thickened on ventral side to form the germ band. E, young embryo 52 to 54 hours from beginning of development. F, cross section of egg in blastula stage. G, cross section of germ band differentiated into median plate and lateral plates. H, same, median plate becomes mesoderm, lateral plates ectoderm. I, cross section of young embryo in amnion.

Explanation of Abbreviations

Am, amnion
Ant, antenna
Bld, blastoderm
CCls, cleavage cells
Cho, chorion
Ctx, cortical cytoplasm, periplasm
dBld, dorsal blastoderm
Ecd, ectoderm
End, endoderm
GB, germ band
Lm, labrum

LP, lateral plate of germ band
Md, mandible
Msd, mesoderm
1Mx, first maxilla
2Mx, second maxilla
Nu, nucleus
Sp, spiracle
Vit, vitelline membrane
VNC, ventral nerve cord
Y, yolk
YCl, yolk cell

Anatomy and Morphology of the Honey Bee

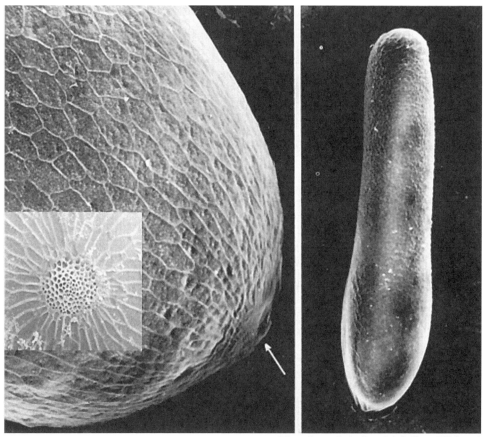

Micropile and chorion detail. (from Erickson)

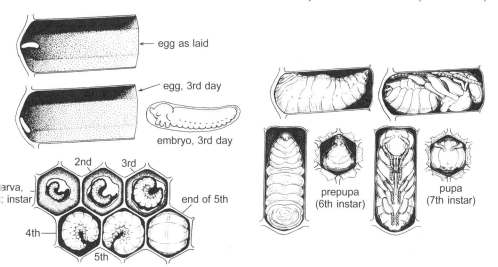

Stages in development from egg to pupa (after J.A. Nelson, A.P. Sturtevant, & B. Lineburg, 1924).
(from Dade)

Anatomy and Morphology of the Honey Bee

Eggs.

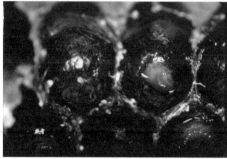
Young larva and egg.

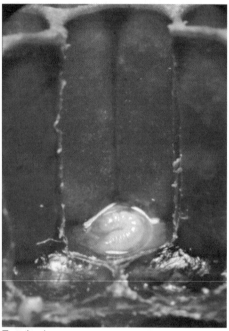
Two-day larva.

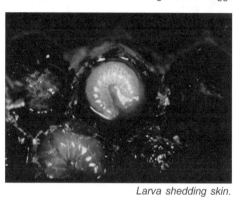
Larva shedding skin.

Worker prepupa.

(Jaycox photos)

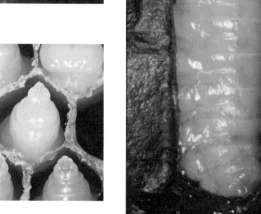
Drone prepupa.

Anatomy and Morphology of the Honey Bee

Final development stages in the honey bee.

Larvae to adult, but still in cells, caps removed.
(Jaycox photos)

Purple eye, brown leg line pupa – about 16 days.

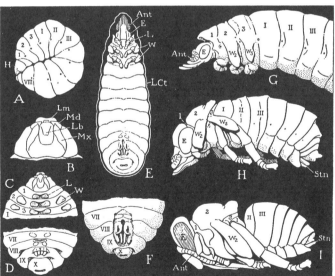

Explanation of Abbreviations
Ant, antenna
E, compound eye
L, leg
Lb, labium
LCt, larval cuticle
Lm, labrum
Md, mandible
Mx, maxilla
Stn, sting
W, wing (W_2, W_3, mesothoracic and metathoracic wings)

External changes during metamorphosis from larva to pupa.
A, mature larva coiled in cell. B, ventral surface of larval head. C, ventral surface of head and thorax of mature larva, leg and wing buds everting beneath larval cuticle. D, terminal segments of worker larva, ventral, showing rudiments of sting. E, early stage of propupa in larval cuticle. f, terminal segments of propupa with rudiments of sting more advanced. G, lateral view of propupa. H, late stage of propupa just before shedding larval cuticle. I, pupa after ecdysis. (From Snodgrass)

Final Development Stages

Anatomy and Morphology of the Honey Bee

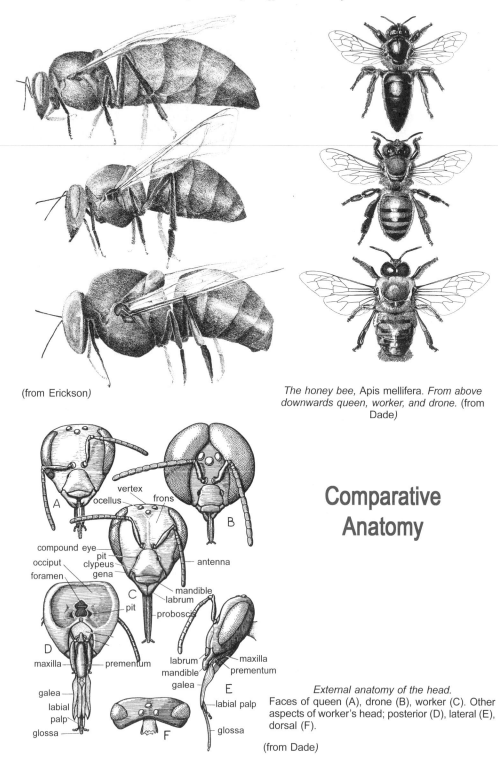

(from Erickson)

The honey bee, Apis mellifera. From above downwards queen, worker, and drone. (from Dade)

Comparative Anatomy

External anatomy of the head. Faces of queen (A), drone (B), worker (C). Other aspects of worker's head; posterior (D), lateral (E), dorsal (F).

(from Dade)

Anatomy and Morphology of the Honey Bee

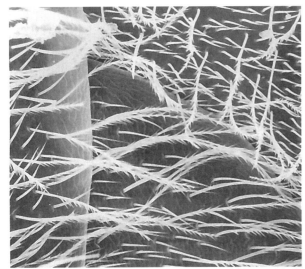

Portions of two terminal abdominal segments. Of interest is the cuticular relief that extends over these segments and the fringed and glabrous cuticular hairs that are abundant over the entire body. (From Erickson)

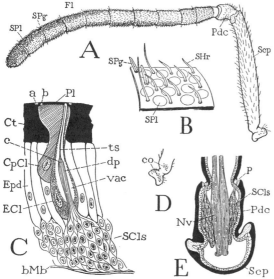

Last five segments of the queen antenna. The antenna surface is profusely covered with a variety of uniformly distributed sense organs. On each segment, usually at the proximal and distal border, are small clusters of pit organs, which appear as bright spots. (From Erickson)

Antennal sense organs

A, left antenna of worker, showing plate organs, peg organs, and pits of organ of Johnston between bases of flagellum and pedicel. B, part of antennal surface with sensory hairs, pegs, and plate organs. C, diagrammatic vertical section of a plate organ. D, campaniform organs on base of scape of antenna. E, lengthwise section of antennal pedicel containing organ of Johnston. a, outer ring of plate organ; b, inner groove of same; c, ends of fibers in terminal strand; p, pit. (from Snodgrass)

Explanation of Abbreviations

bMb, basement membrane
CpCl, cap cell of sense organ
Ct, cuticle
dp, distal process of sense cell
ECl, enclosing cell of sense organ
Epd, epidermis
Fl, antennal flagellum
Nv, nerve
Pdc, antennal pedicel

Pl, plate of plate sense organ
SCls, sensory nerve cells
Scp, scape of antenna
SHr, sensory hair
SPg, sensory peg
SPl, sensory plate
ts, terminal strand of attachment fibers of sense cells
Vac, vacuole of sense organ

Muscles, Wings, and Landing

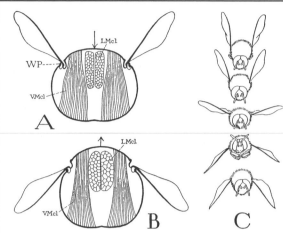

The wing movements.
A, the wings turned upward on the pleural wing processes (*WP*) by depression of the notum caused by contraction of the vertical indirect wing muscles (*LMcl*). B, wings turned down by upward curvature of the notum produced by contraction of the longitudinal indirect wing muscles (*LMcl*). C, successive positions of the wings of a drone in flight (from photographs by Stellwaag, 1910).
(from Snodgrass)

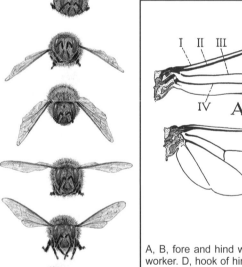

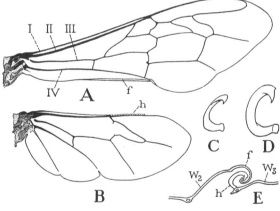

The wings.
A, B, fore and hind wing of a drone. C, hook of hind wing of a worker. D, hook of hind wing of a drone. E, the interlocked wing margins. f, fold on posterior margin of forewing; h, hooks on anterior margin of hind wing; I-IV, main veins of wing.
(from Snodgrass)

(from Goodman)

FLIGHT

Anatomy and Morphology of the Honey Bee

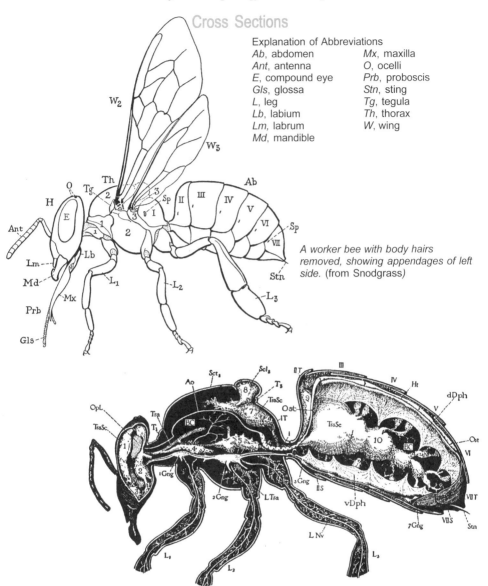

Explanation of Abbreviations
Ab, abdomen
Ant, antenna
E, compound eye
Gls, glossa
L, leg
Lb, labium
Lm, labrum
Md, mandible
Mx, maxilla
O, ocelli
Prb, proboscis
Stn, sting
Tg, tegula
Th, thorax
W, wing

A worker bee with body hairs removed, showing appendages of left side. (from Snodgrass)

Body of worker bee cut longitudinally, muscles and alimentary canal removed, exposing dorsal blood vessel, diaphragms, tracheae, air sacs, and ventral nerve cord. (From Snodgrass)

Explanation of Abbreviations
Gng, ganglion
Ht, heart
LNv, leg nerve
LTra, leg trachea
OpL, optic lobe
Ost, ostium of heart (*1Ost*, first ostium)
Scl, scutellum
Sct, scutum
Stn, sting
T, tergum (*IT*, tergum of propodeum)
Tra, trachea
TraSc, tracheal air sac
vDph, ventral diaphragm

21

Anatomy and Morphology of the Honey Bee

Explanation of Abbreviations
AntNv, antennal nerve
Br, brain (*1Br*, protocerebrum; *2Br*, deutocerebrum; *3Br*, tritocerebrum)
E, compound eye
Endst, endosternum; *Endst$_{2+3}$*, composite endosternum of mesothorax and metathorax
Gng, ganglion
O, ocellus, or ocellar rudiment in epidermis
OpL, optic lobe of brain
WNv, wing nerve

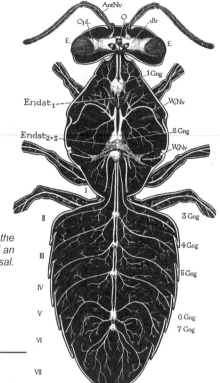

General view of the nervous system of an adult worker bee, dorsal.

Longitudinal Looks

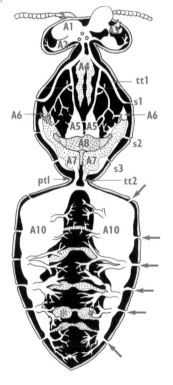

In the adult bee, the longitudinal tracheal trunks are expanded into large air sacs. The large tracheal trunks (tt1) arising from the first spiracle (s1) pass forward into the head and expand into three air sacs covering the top of the brain (A1), supplying the compound eyes and optic lobes (A2) and the underside of the brain and mouthparts (A3) (part of A1 has been moved forward on the right to reveal the underlying A3). From these air sacs numerous tracheae penetrate the tissues and subdivide to form fine tracheoles. Posteriorly running branches also arise from the first spiracular tracheal trunks, giving rise to a complex of interconnected air sacs (A4, A5, A6, A7 and A8) supplying the thorax. Spiracles 2 and 3 (s2, s3) also open into these air sacs. Two large tracheal trunks (tt2) run from the thoracic complex through the petiole (ptl) into the abdomen where they expand into two enormous lateral air sacs (A10). Two small air sacs (A9) lie just inside the abdomen (not visible here). Abdominal spiracles (arrows) open into these sacs which are connected to each other via transverse commissures (star). The first abdominal spiracles (s3) are situated on the propodeum; the eighth abdominal spiracles are hidden within the sting chamber. Tracheae from these sacs ramify among the abdominal tissues. Stippled areas lie ventrally. (From Goodman)

Anatomy and Morphology of the Honey Bee

Longitudinal Looks

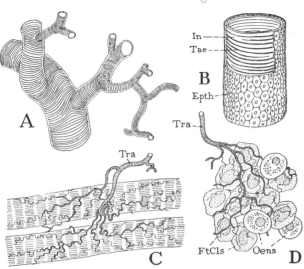

Explanations of Abbreviations
Epth, tracheal epithelium
FtCls, fat cells
In, tracheal intima
Oens, oenocytes
Tae, taenidium
Tra, trachea

Details of tracheal structure. A, piece of branching trachea. B, structure of a tracheal tube. C, trachea and branches ending in tracheoles on muscle fibers. D, tracheae branching to fat cells, but not on oenocytes. (from Snodgrass)

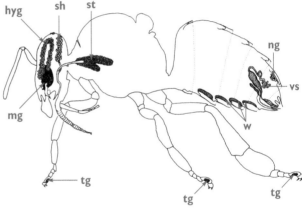

Worker honey bee showing the position of some of the major glands. Hypopharyngeal gland (hyg), wax glands (w), Nasonov gland (ng), mandibular gland (mg), venom sac (vs), head (sh) and thoracic (st) salivary glands, and tarsel glands (tg). (from Goodman)

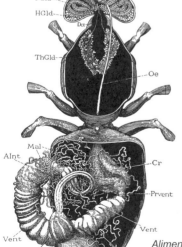

Explanation of Abbreviations
AInt, Anterior intestine
Cr, crop ("honey stomach")
FGld, food gland
HGld, head salivary gland
Mal, Malpighian tubules
Oe, oesophagus
Prvent, proventriculus
Rect, rectum
rp, rectal pad
ThGld, thoracic salivary gland
Vent, ventriculus (mesenteron of embryo)

Alimentary canal and glands of the head and thorax of a worker bee, dorsal. (from Snodgrass)

DRONES – Head & Thorax

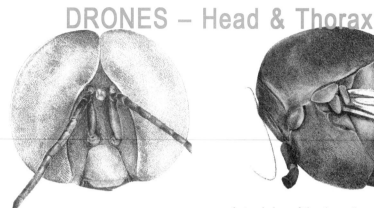

Frontal view of the drone head. (from Erickson)

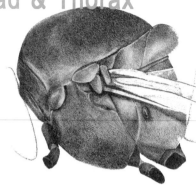

Lateral view of the drone thorax. (from Erickson)

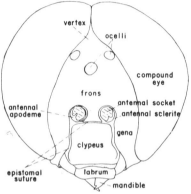

Drone, Frontal View (from Erickson)

Close-up of the wing hooks. These are somewhat larger than those of the worker. Socketed small peg organs are abundant on the surface engaged by the trailing edge of the forewing. (from Erickson)

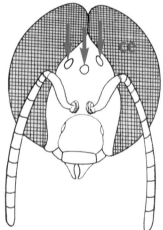

Frontal view of the head of a drone showing the enormous development of the compound eyes (ce) which meet over the top of the head. The ocelli (arrows) are displaced forward. (from Goodman)

Lateral view of the pretarsus and last (fifth) tarsal segment showing the contracted nature of the arolium and the upright orientation of the medial sclerite. The five trichoid sensilla arise from the medial sclerite at different levels. (from Erickson)

DRONES – Abdomen

Lateral view of the drone gaster (abdomen). (from Erickson)

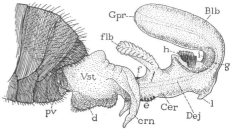

Male genital parts concerned with mating. The fully everted penis with gonopore (Gpr) at apex of bulb. (from Snodgrass)

Explanation of Abbreviations
Blb, bulb of penis
Cer, cervix of penis
crn, cornua of penis
Dej, ductus ejaculatorius
Gpr, gonopore
pv, penis valve (mesomere)
Vst, vestibulum of inverted penis

In a newly emerged drone, the pair of testes are large and full of sperm. The white dots are fat bodies. (USDA/Virginia Williams)

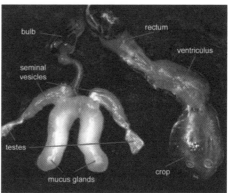

In this view, the digestive system (right: crop, ventriculus and rectum) has been spread out separately from the reproductive system. The white mucus glands are very distinct from the tan seminal vesicles filled with semen. (USDA/Virginia Williams)

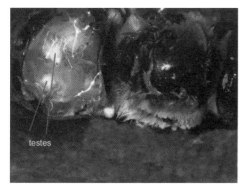

In a mature drone, all of the sperm have moved into the seminal vesicles, and the testes shrink. (USDA/Virginia Williams)

Anatomy and Morphology of the Honey Bee

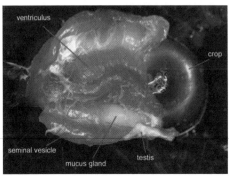

When the outer cuticle is removed, you can see the digestive and reproductive systems folded up together to fit into the abdomen. (USDA/Virginia Williams)

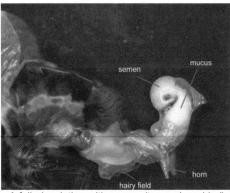

A full ejaculation with semen (tan and marbled) and mucus (white) on the tip of the penis. (USDA/Virginia Williams)

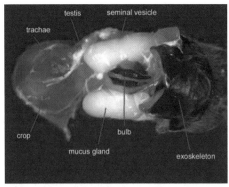

This is a view of the abdominal contents from the other side. Some of the cuticle (exoskeleton) has been left in place at the tip of the abdomen. The bulb of the penis can be seen nestled between the full mucus glands. This is a mature drone, so the testes you see are very small. (USDA/Virginia Williams)

DRONES –
Reproductive

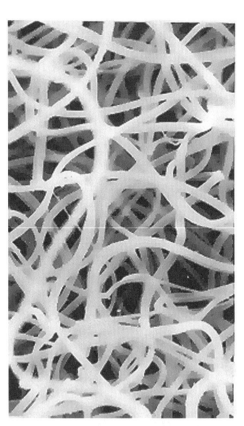

Massed, intertwined spermatozoa. (from Erickson)

Anatomy and Morphology of the Honey Bee

WORKER

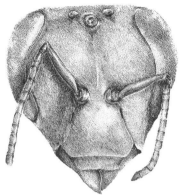

Frontal View. (from Erickson)

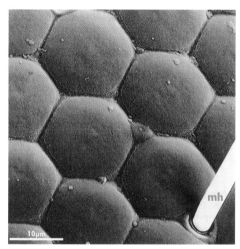

The compound eye is composed of many small lenses overlying groups of photoreceptor cells. They are packed closely together in the bee, and the lenses are hexagonal. Part of one of the mechanosensory hairs (mh) is shown. These hairs are set in sockets and innervated at the base of the shaft. The neuron is stimulated by movement so that the bee is aware of anything touching the surface of the eye. (from Goodson)

Head

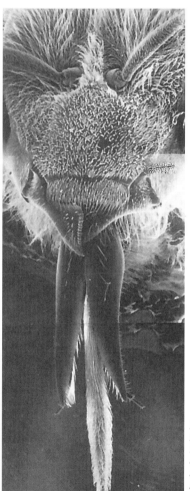

Photomontage of the glossa. The glossa is extended when feeding: here it is extending between the pendulous galeae, the terminal segments of the underlying labial palps are nearly covered by the galeae, and only the tips are exposed. The arrow indicates the area further magnified in the top right micrograph. (from Erickson)

Anatomy and Morphology of the Honey Bee

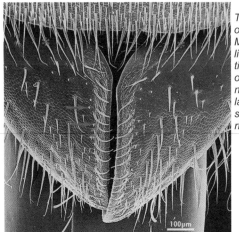

The cutting surface of the mandible is ridged and the opposing surfaces can just slide past each other. Mechanoreceptive hairs (mh) lining the ridges are believed to register movement of the cutting edges relative to each other. Other long hairs with socketed bases on the surface of the mandible probably have a mechanoreceptive function, as do those on the edge of the labrum (lb) overhanging the mandibles (stars). Small sensory pegs, or basiconic sensillae, lie alongside each ridge (arrows). (from Goodman)

Worker, Frontal View (from Erickson)

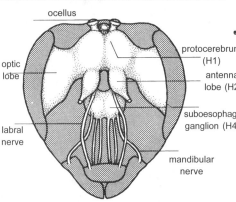

Worker, Posterior View (From Erickson)

The brain and principal nerves of the head, anterior aspect. The labral nerves come from the very small tritocerebrum (H3), concealed behind the antennal lobes. The mandibular nerve comes from H4. Two other pairs of nerves (shown but not flagged) come from H5 and H6, and go to the maxillae and labium respectively. (from Dade)

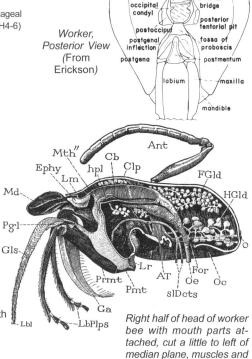

Right half of head of worker bee with mouth parts attached, cut a little to left of median plane, muscles and nerve tissue omitted. (from Snodgrass)

Explanation of Abbreviations
Ant, antenna
AT, anterior tentorial arm
Cb, cibarium
Clp, clypeus
Ephy, epipharynx
FGld, hypopharyngeal food bland
For, occipital foramen
Ga, galea
Gls, glossa (tongue)
HGld, head salivary gland
hpl, hypopharyngeal plate
Lbl, labellum
LbPlp, labial palpus
Lm, labrum
Lr, lorum
Md, mandible
Mth", functional mouth
O, ocellus
Oc, occiput
Oe, oesophagus
Pgl, paraglossa
Pmt, postmentum
Prmt, prementum
slDct, common salivary duct

Anatomy and Morphology of the Honey Bee

WORKER – Thorax

Lateral view of the worker thorax. (from Erickson)

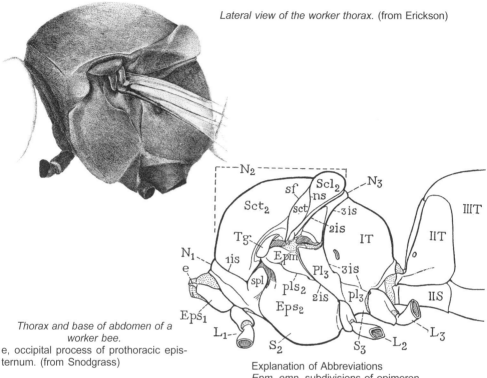

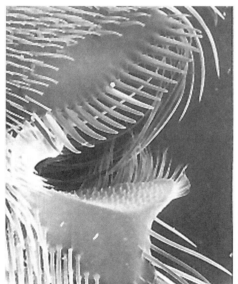

Thorax and base of abdomen of a worker bee.
e, occipital process of prothoracic episternum. (from Snodgrass)

Explanation of Abbreviations
Epm, emp, subdivisions of epimeron
Eps, episternum
1is, 2is, 3is, first, second, and third intersegmental grooves of thorax
L, leg
N, notum
ns, external notal sulcus
Pl_3, pl_3, pleural plates of metathorax
pls, pleural sulcus
S, sternum (*IS*, propodeal sternum; *IIS-VIIS*, abdominal sterna)
Scl, scutellum
Sct, scutum, sct_2, subdivisions of mesoscutum
sf, scutal fissure
spl, spiracular lobe of pronotum
T, tergum (*IT*, propodeal tergum; *IIT, VIIT*, abdominal gerga)
Tg, tegula

Mediolateral view of the pollen press. The floor of the press is edged with fine hairs, and its surface is covered with denticlelike cuticular spines or scales. Long, curved hairs from the tibia bend down and lie over the press and a picket of shorter, stiff spines (rastellum) lines the dorsomedial margin of the press. Small mechanoreceptor hairs are visible at the leading edge of the spatulate hairs (upper left). (from Erickson)

Anatomy and Morphology of the Honey Bee

Apis mellifera *worker legs.* (from Erickson)

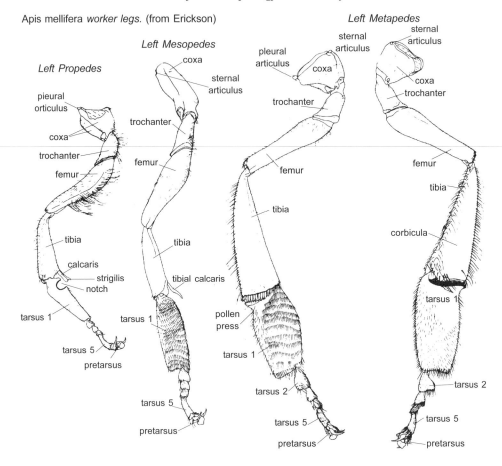

Hind leg of worker and drone and pollen press of worker. (from Snodgrass)

A, right hind leg of worker, posterior (inner) surface. B, ends of tibia and basitarsus of hind leg, separated, posterior. C, pollen basket (corbicula) on outer surface of hind tibia of a worker. D, pollen press between tibia and tarsus of hind leg of worker, dorsal. E, same with tibial hairs removed. F, right hind leg of drone, posterior.

Explanation of Abbreviations
au, auricle
Btar, basitarsus
Cbl, pollen basket, corbicula
Cx, coxa
Fm, femur
Pr, pollen press
ras, rastellum, rake
Tb, trochanter

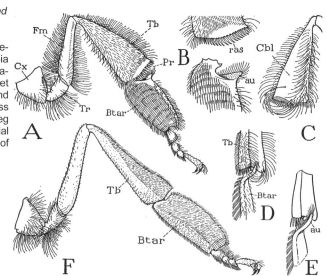

Anatomy and Morphology of the Honey Bee

WORKER – Thorax

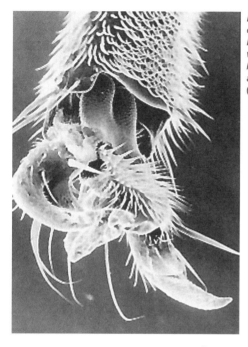

Last tarsomere and pretarsus. Claws are on either side of the wrinkled arolium. Above the arolium is the heavily bristled planta; above the planta is the unguitractor, which is without sets but has rows of low-lying cuticular scales. This structure may function like the heel of a hand, providing an opposing surface for the claw. (from Erickson)

Last tarsomere and pretarsus of a worker. A, dorsal; B, ventral; C, lateral. f, marginal flange of fifth tarsomere. (from Snodgrass)

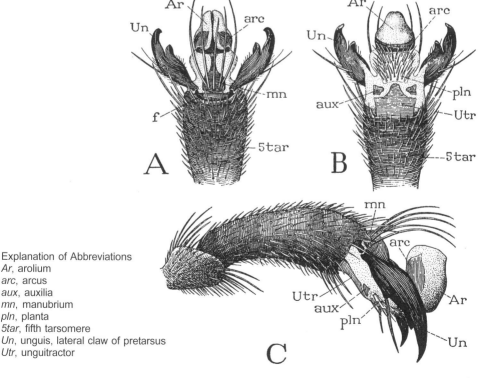

Explanation of Abbreviations
Ar, arolium
arc, arcus
aux, auxilia
mn, manubrium
pln, planta
5tar, fifth tarsomere
Un, unguis, lateral claw of pretarsus
Utr, unguitractor

31

Anatomy and Morphology of the Honey Bee

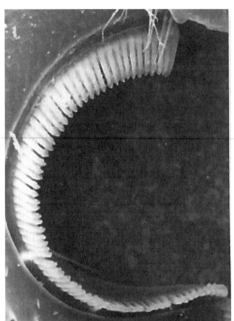

"Fan" of the seventy spinelike hairs of the tarsal comb. These hairs are responsible for cleaning the outer surface of the antenna. (from Erickson)

WORKER – Thorax

Explanation of Abbreviations
Btar, basitarsus
Tb, tibia

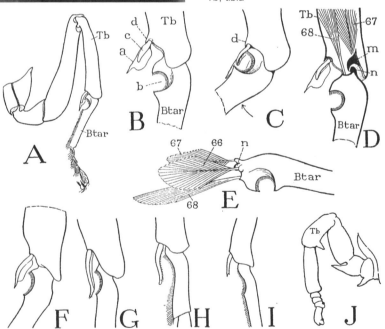

The antenna cleaner.
A, left foreleg of worker, showing antenna cleaner at base of tarsus. B, antenna cleaner open. C, same, closed. D, tibiotarsal joint of front leg and tarsal muscles in tibia. E, base of first tarsomere with attached muscles, anterior. F, antenna cleaner of *Halictoides calochorti* Ckll. G, same of *Vespula maculata* (L.). H, same of *Trogus vulpinus* (Grav.). I, same of *Orussus sayi* Westw. J, foreleg of honey-bee pupa, showing tibial spur. *a,* closing lobe (fibula) of antenna cleaner; *b,* notch of antenna cleaner; *c,* anterior lobe of fibula; *d,* tibial process at base of fibula; *m,* articular process of tibia; *n,* articular process of basitarsus; *66, 67,* extensor muscles of tarsus; *68,* flexor muscle of tarsus. (from Snodgrass)

Anatomy and Morphology of the Honey Bee

WORKER – Abdomen

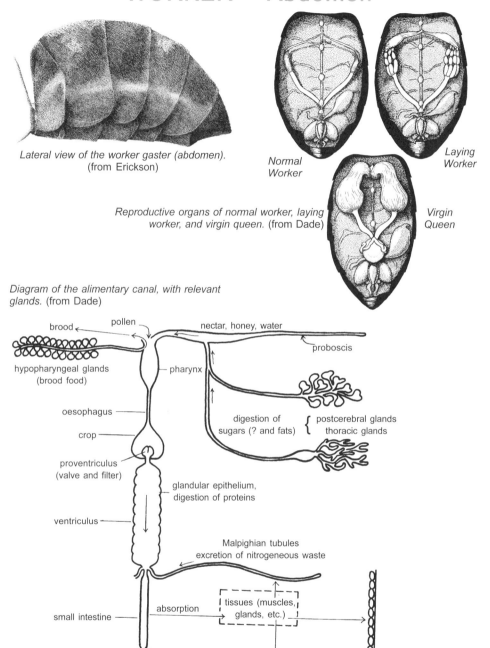

Lateral view of the worker gaster (abdomen). (from Erickson)

Reproductive organs of normal worker, laying worker, and virgin queen. (from Dade)

Normal Worker

Laying Worker

Virgin Queen

Diagram of the alimentary canal, with relevant glands. (from Dade)

brood ← pollen

nectar, honey, water

proboscis

hypopharyngeal glands (brood food)

pharynx

oesophagus

digestion of sugars (? and fats)

postcerebral glands
thoracic glands

crop

proventriculus (valve and filter)

glandular epithelium, digestion of proteins

ventriculus

Malpighian tubules
excretion of nitrogenous waste

small intestine — absorption

tissues (muscles, glands, etc.)

rectum

tracheae

fat body, food storage

excretion of CO_2

Anatomy and Morphology of the Honey Bee

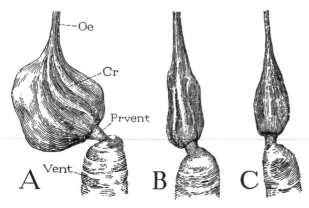

The crop (honey stomach) and proventriculus.
A, crop, proventriculus, and upper end of ventriculus of a worker. B, same of queen. C, same of a drone. D, wall of crop mostly cut away, exposing mouth of proventriculus projecting into the crop. E, cross section of proventriculus (from Trappmann, 1923). m, mouth of proventriculus. (from Snodgrass)

Explanation of Abbreviations
cmcl, circular muscles
Cr, crop ("honey stomach")
elmcl, external longitudinal muscles
ilmcl, inner longitudinal muscles
Oe, oesophagus
Prvent, proventriculus
Vent, ventriculus (mesenteron of embryo)

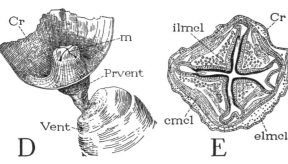

The proventriculus
A, anterior aspect, lips closed; B, ditto, lips open to show the short spines and long hairs of the lips, and the lumen partly closed by the muscles below the lips, also the pouches which open into the lumen. (from Dade)

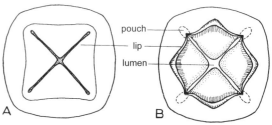

The scent gland
A, abodominal tergum VII of a worker. B, same of a queen. C, lengthwise section through base of tergum VII of worker, covered by tergum VI. D, group of scent-gland cells and ducts (from McIndoo, 1914a). E, a single gland cell and its duct (from McIndoo, 1914a). a, line of strong submarginal inner ridge on base of tergum VII of worker; b,c, margins of elevation on tergum VII of worker. (from Snodgrass)

Explanation of Abbreviations
Ac, antecosta
Amp, ampulla
Can, canal on tergum VII of worker
Epd, epidermis
FtCls, fat cells
Mb, intersegmental membrane
SntGld, scent gland

WORKER – Abdomen

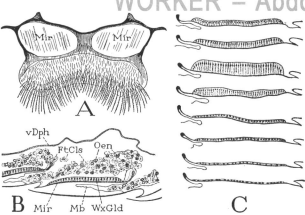

The wax glands.
A, sternum of segment VI of worker, ventral, showing polished "mirrors" beneath wax glands. B, lengthwise section through two wax glands with overlying masses of fat cells and oenocytes (from Rösch, 1930). C, stages in the development and regression of a wax gland (from Rösch, 1927a). *(From Snodgrass)*

Extruding wax flakes from the wax glands. (File photo)

Explanation of Abbreviations
FtCls, fat cells
Mb, intersegmental membrane
Mir, "mirror" under wax gland
Oen, oenocytes
vDph, ventral diaphragm
WxGld, wax gland

Left: Abdominal sternites, ventral view. The glabrous surfaces are the wax "mirrors," or plates. Beneath these plates are the wax glands. Wax permeates the mirrors and hardens into the visible wax scales from which the honey bees construct comb. The rastellum, not the wax spur, on the hind leg is used to remove these scales when fully formed (rastellar marks are visible on the medial edges of the scales). Right: Fully formed wax scale. (from Erickson)

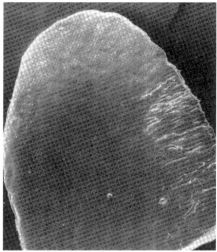

WORKER – The Sting

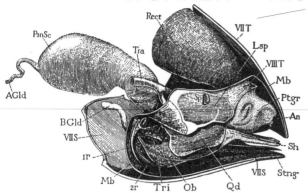

Explanation of Abbreviations
AGld, poison gland of sting
ap, apodeme of tergum IX of drone
Bgld, accessory gland of sting
blb, bulb of stylet of sting
Frc, furcula
Let, lancet of sting
Lsp, lamina spiracularis of segment VIII
Mb, intersegmental membrane
Ob, oblong plate of sting
PsnSc, poison sac of sting
Ptgr, proctiger
Qd, quadrate plate of sting
Sh, sheath lobe of sting
Stl, stylet of sting
Tri, triangular plate of sting
Vlv, valve on lancet of sting

The sting of a worker and associated structures in the sting chamber, exposed by removal of left wall of segment VII. (from Snodgrass)

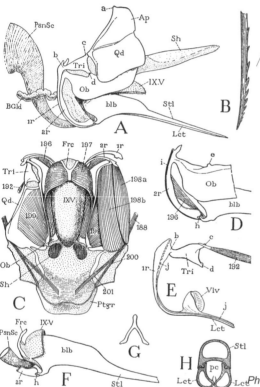

Structural details of the sting of a worker. A, outline of left side of sting. B, barbed end of a lancet. C, muscles of sting, dorsal. D, base of oblong plate and bulb connected by second ramus. E, triangular plate and base of lancet connected by first ramus. F, bulb of stylet articulated on second ramus with the associated furcula. G, furcula. H, cross section of distal part of sting shaft. (from *Snodgrass*)

Photomontage of the sting, dorsal view. Visible are the first and second rami on the medial and lateral sides respectively of the double-arched extension of the triangular plate. This proximal part of the sting connects the lancets and stylet to the sting protractor muscles. A hairy lobe embraces the base of the lancets; this structure is believed to be the ventral wall of abdominal segment nine. (from Erickson)

QUEEN – Head

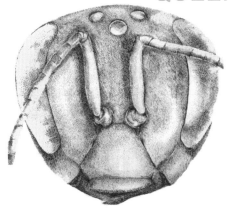

Frontal view of the queen head. (from Erickson)

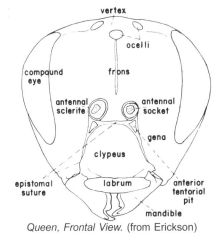

Queen, Frontal View. (from Erickson)

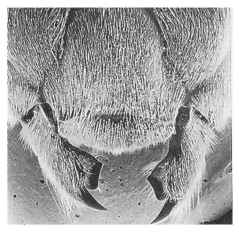

Queen mouthparts, dorsoanterior view. Here the mandibles are better outlined and part of their articulation with the genae is visible. At this viewing angle the labrum completely covers the galea and labial lobes.

and Thorax

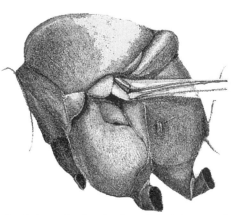

Posterior region bearing the relatively large oval spiracle (arrow) is actually the first segment of the abdomen although it is broadly fused to the thorax. (from Erickson)

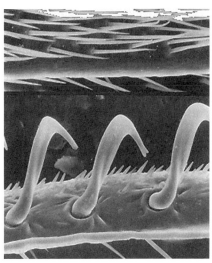

Hooks on the leading edge of the hind wing, each with a slightly forked terminus. The bent and twisted nature of the hook is apparent from this angle; in three dimensions the hook extends in two directions. (from Erickson)

QUEEN Abdomen

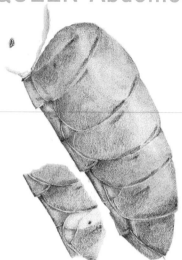

Photomontage of the middle, or mesothoracic (left), and hind, or metathoracic (right) legs. At the leg base the first segment, the coxa, is almost completely obscured by the hairs of the mesothoracic pleurites. The next segment, the trochanter, is visible, extending (horizontally) to articulate with the larger femur. The tibia joins the femur (at the "knee") and extends downward. The mesothoracic tibia is about as wide as the femur, but the hind tibia is flattened and much broader. The second downward-projecting segment is the basal tarsomere, which is clearly much larger than the other, more distal tarsal segments. In the queen the metathoracic basal tarsomere lacks the pollen collection apparatus of its counterpart in the worker. Four remaining tarsal segments are present; the last one (pretarsus) is elongate and bears claws. These first four tarsal segments have no muscles, but a common tendon traverses all of them and inserts into the flexor muscle of the pretarsus. (from Erickson)

Gaster (abdomen), which consists of a first, rather indistinct segment (attached to the thoracic region) with a prominent spiracle, the petiolé (waist), and the remaining posterior gaster segments. The dark spots indicate the relative position of spiracles on the first abdominal segment and on that portion of the abdomen behind the petiolé. There is one spiracle on each side of the first seven segments. (from Erickson)

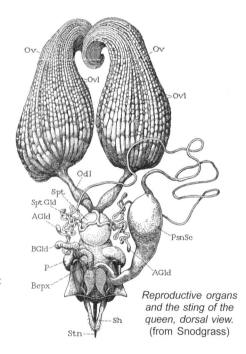

Explanation of Abbreviations
AGld, poison gland of sting
Bcpx, bursa copulatrix
Ov, ovary
P, lateral pouch of bursa copulatrix
Sh, sheath lobes of sting
Spt, spermatheca
SptGld, spermathecal gland
Stn, shaft of sting

Reproductive organs and the sting of the queen, dorsal view. (from Snodgrass)

QUEEN – Abdomen, Reproductive Organs

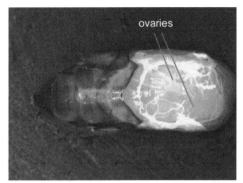

A good laying queen will have an abdomen filled by her ovaries. (USDA/Virginia Williams)

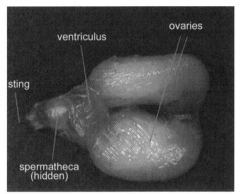

The rest of the abdomen has the digestive tract and the sperm storage organ, the spermatheca. (USDA/Virginia Williams)

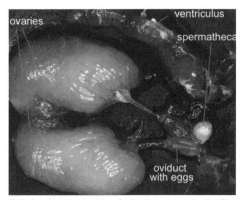

This is what the inside of a laying queen looks like when the organs are spread out. Notice that you can see the individual segments, or ovariole, of the ovary, each with a row of eggs moving down to the oviducts. Several eggs are near the spermatheca, about to be laid. (USDA/Virginia Williams)

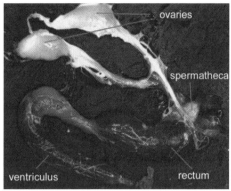

When a queen has been caged and held in a queen bank for some time, the eggs are reabsorbed and the ovaries shrink. (USDA/Virginia Williams)

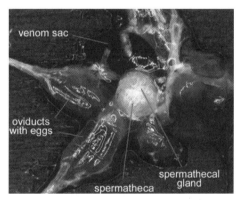

This is a close-up view of the end of the reproductive tract, where the eggs that are bring laid move past the duct from the spermatheca to be fertilized. (USDA/Virginia Williams)

The spermatheca of a virgin queen. With no sperm present, it appears clear. (USDA/Virginia Williams)

Anatomy and Morphology of the Honey Bee

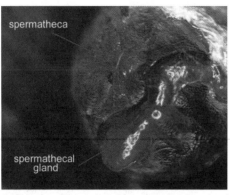

A close-up of the spermatheca showing the gland that is attached. Without the gland producing seminal fluid, the stored sperm will die. (USDA/Virginia Williams)

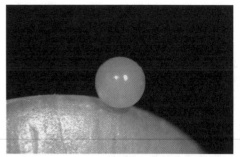

The cloudy appearance of this spermatheca indicates the queen was poorly mated or has used most of her stored sperm. She is or soon will become a drone layer. (Sue Cobey photo)

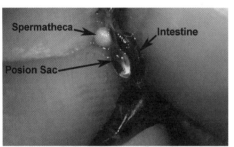

In the pulled abdomen, the clear poison sac, spermatheca and intestine are exposed. (Sue Cobey photo)

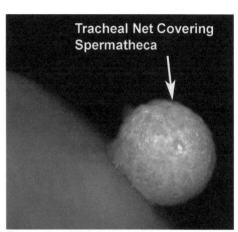

The rough, whitish texture of the permatheca is due to a covering of tracheal net. This is removed by gently rolling the spermatheca between your fingers. (Sue Cobey photo)

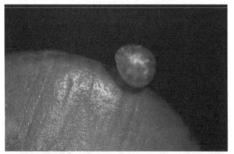

The spermatheca of a mated queen is the same color as fresh drone semen. Notice the light and darkish swirl pattern of densely packed sperm. (Sue Cobey photo)

Comparison of the spermatheca of a virgin and mated queen. Tracheal net coverings have been removed. (Sue Cobey photo)

QUEEN Reproductive

Photos by Elbert Jaycox, Extension Specialist in Apiculture, University of IL, are also used and so labeled. These are the best we've seen on larval development. Dr. Jaycox used some of these in his books, but many remained unpublished until after his death. We are fortunate to have them.

We have provided several glimpses of honey bee anatomy and morphology in the following sections. The photos, drawings and line art used here with permission, come from several excellent sources. The generalized topics include development, gross anatomy, sensory organs, flight, some internal anatomy, and then more specific details on the drone, workers and queens.

This section is not intended to be a complete text on this subject. Rather, it is only an introduction to the science, with emphasis on those aspects generally considered most important.

Knowing the fundamentals of biology better prepares a beekeeper to understand the many variables in honey bee behavior, honey bee pest and disease problems and practical seasonal management.

Once you become familiar with the fundamentals you may want to explore more sophisticated resources. We recommend several at the end of this section.

ANDERSON, EDWIN J. (1900-1995) – Anderson was a Professor of Apiculture at Penn State University. He received a B.A. degree from Penn State in 1924. He received an M.S. degree from Cornell University in 1925 and taught entomology at Clemson College before returning to Penn State in 1926 as an extension specialist worker in beekeeping and entomology.

In 1927 he was named editor of the Pennsylvania Beekeeper, which he continued after his retirement. He published more than 100 papers in national journals and in 1963 was cited by the Pennsylvania Beekeepers Association for his service to the industry.

He was a member of the American Association for Advancement of Science and Bee Research and was a member of Delta Theta Sigma, Epsilon Sigma Phi and The Society of the Sigma Xi. He was a member of the American Association of Economic Entomologists, the Honey producers League, the Pennsylvania State Beekeepers Association and the Eastern Apicultural Society which he served as president.

His research was concerned with bees and included studies of diseases affecting bees, effect of high humidity on winter losses among bees, and the commercial uses of honey.

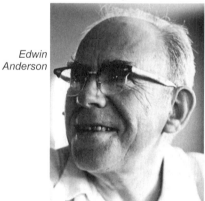
Edwin Anderson

ANIMAL AWARENESS – Do animals think? Are animals aware of what is taking place around them, and can they make decisions? These questions are raised here because honey bees are often mentioned in the debate that has surrounded these questions for many decades.

One of the problems in answering

these questions concerns our inability to determine if an animal is responding to a situation consciously or unconsciously. For example, if we unknowingly touch a hot pan we withdraw our hand automatically without thinking about it. However, if we know a pan is hot we think about it and handle it so we do not burn ourselves.

An interesting aspect of this question revolves around brain size. Some say that the honey bee brain is so small that it has no room for cells for thinking purposes. The opposite opinion is that a honey bee brain is so small that bees must have a thinking system; there are not enough cells in the bee's small brain to store all of the information that would otherwise be needed.

Some examples to support the belief that honey bees think and make decisions are as follows: when a foraging worker bee lands on a flower searching for nectar she must probe the area of the nectary for three to four seconds to determine if food is present. Does this process require thought? At what point does a bee give up searching an individual flower? If nectar is found is it acceptable? Honey bees appear to be measuring the sugar concentration and quantity of a food supply and will be found foraging on the richest source.

Another example is a scout bee's ability to measure various aspects of a new home site. As discussed under bait hives, when we offer bees a choice of two hives that vary in only one item, the bees make what appears to be a reasonable choice. For example, they always select a nest with a relatively small entrance that is easy to defend.

Perhaps the most important aspect of this question concerns the honey bee dance language. It has been said that humans differ from other animals because we can communicate with one another. However, we observe that the honey bee dance can convey information about direction, distance, quality, and the type of food to be found even a mile (1.6 km) or more from the hive. Without the use of a compass and a machine to measure distance, we might find it difficult to convey such information to another human ourselves. Obviously the honey bee dance language has certain deficiencies, such as an inability to convey information about color, shape, and upward or downward direction. Still, it is probably correct to state that only human language is superior to that of a honey bee. Those who wish to pursue this question will find several books on the subject, including: Griffin, D.R. *Animal Thinking* (Harvard University Press, Cambridge. 237 pages, 1984) and Wenner and Wells, *Anatomy Of A Controversy* (Columbia University Press, New York. 400 pages. 1984). (see also LEARNING IN THE HONEY BEE).

ANNUAL CYCLE OF HONEY BEE COLONIES – Honey bees originated in the tropics, but *Apis mellifera* now inhabits a wide range of environments up to 60° north latitude. One of the most striking adaptations to survival in colder climates is the intensive thermoregulation of whole colonies. Honey bees alone among temperate-climate insects confront the cold by producing heat to maintain warmth throughout the winter, rather than hibernating in a dormant state, like the vast majority of insects, or migrating to warmer regions as a few insects do. This winter thermoregulation, in turn, is supported by the timing of events throughout the bee's entire annual cycle, since winter survival depends heavily on the honey stores and bee populations established during the preceding warm season.

A honey bee colony is founded as a swarm in the springtime, gathers a large mass of food in the relatively short summer, consumes this food steadily throughout the cold season, finally reproducing itself early in the foraging season. Bees accomplish winter heat production by consuming honey and vibrating their wing muscles (much like our shivering). This consumes a large quantity of honey, about 50 pounds (23 kg) each winter for a small natural colony. Large colonies kept by beekeepers use more than this, about 60 to 70 pounds (27 to 32 kg).

The collection of nectar is extremely variable. In a study in Connecticut, in only 16 weeks of the year did a colony show a net increase in honey stores, and 50 percent of the total honey collected was harvested in just the three best weeks of the year. Because of this pattern it is essential, both to the survival of bees living on their own, and to the honey production of a beekeeper's colonies, that colonies achieve a high population by the time honey stores can be gathered. For a newly-established swarm of bees, the earlier the swarm leaves the nest, the more time it will have to build up its population and gather nectar, and the more opportunity it will have to gather the food from the few rich honey flows in its environment.

The annual cycle of brood rearing reflects the value of early swarming. Brood rearing begins in mid-winter (end of December to January in temperate climates), accelerates in late winter and early spring, reaches a peak soon after the first forage becomes available, reduces later in the summer, and ceases entirely in autumn. The striking thing about this pattern is that it is so different from the pattern of when forage is available. Winter brood rearing requires a colony to maintain a warmer cluster, and thus to use more honey than it would with no brood rearing. The importance of winter brood rearing is that it allows the colony to reach a population large enough to cast a swarm early in the season, increasing the chance of the survival of the swarms produced (see SWARMING IN HONEY BEES). -PKV

ANTIBIOTICS AND OTHER CHEMICALS FOR BEE DISEASE CONTROL – Strictly speaking, antibiotics are substances produced by living organisms that will kill or prevent the growth of bacteria. Today, however, with our loose way of using words, any substances that might adversely affect any microorganism such as bacteria are considered antibiotics.

It was known as early as the late 1800s that certain microorganisms would inhibit the growth of others. However, it was not until 1941 that researchers discovered and successfully used penicillin, the first true antibiotic, in humans. Only a few years earlier, in 1935, German scientists announced that a dye, later named sulfanilamide, was active against certain bacteria. For the beekeeping industry, it was not until 1944 that sulfa drugs were demonstrated to control one bee disease, American foulbrood (AFB). Antibiotics approved for use in the beekeeping industry are oxytetracycline (Terramycin®) and fumagillin (see Fumidil-B®). Tylosin, sold as Tylan® is the latest approved antibiotic for the control – not prevention or cure – for American foulbrood. Tylan is now the drug of choice since AFB bacteria have become resistant to Terramycin. We have included instructions on its application elsewhere in this book.) The first is effective against both European foulbrood (EFB) and American foulbrood (AFB), and the second controls nosema.

All drugs, pesticides and other chemicals usually have two names, one the generic or technical name, the other the trade name. Since a drug may be packaged and sold by more than one company, it may have more than one trade name. This is why manufacturers are required to place the generic name and often the chemical make-up of the material on the label. An example is the drug (antibiotic) used to treat nosema disease of honey bees. Its generic name is fumagillin, but it is sold under the trade names Fumidil-B and Fumigillin-B. The trade name is capitalized, while the generic name is not.

Fumagillin – Drs. H. Katznelson and C.A. Jamieson of the Canada Department of Agriculture discovered in 1952 that fumagillin would effectively suppress *Nosema apis*, the organism responsible for nosema disease. In North America the drug is sold under the trade names Fumidil B and Fumigillin-B. The only FDA-approved and satisfactory way to feed this drug is to mix it with sugar syrup that is fed preferably in the fall and spring for over-wintering colonies. Fumagillin is not effective when fed in a dust formulation or in an extender patty. However, when used in queen cage candy, it has been shown to protect queens in shipment. Shipping is stressful and nosema is a stress-related disease. Since some package bee and queen producers routinely feed fumagillin to suppress the disease in the bees they ship it is desirable to continue the treatment when package bees are newly installed. Label directions should be followed closely.

Oxytetracycline – In 1951, Dr. T.A. Gochnauer, then at the University of Minnesota and later with the Canada Department of Agriculture, found that oxytetracycline was effective in controlling both American and European foulbrood. Because of the shorter life of the new drug, sulfathiazole remained the most popular material to use to treat AFB. It was not until 1979 that oxytetracycline finally became the drug of choice to treat both EFB and AFB in the U.S. Oxytetracycline is available under the trade name Terramycin®. Several formulations of

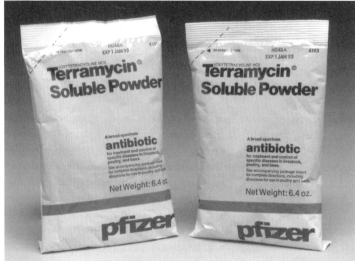

Terramycin was long used in the control of both American and European foulbrood. It is being replaced by other drugs for the control of AFB, however. (File photo)

Terramycin and Tylan are packaged for use in animals as well as the honey bee. Terramycin for honey bees can be obtained from bee supply companies. Beekeepers are cautioned to read the labels carefully since the formulations for different animals vary in the concentration of the active ingredient and the carrier. The carrier (inert ingredient) affects the suitability of formulations for dust.

After over 40 years of use, resistance of *P. larvae* subsp. *larvae* to oxytetracycline was reported in the U.S. Subsequently, *P. larvae* subsp. *larvae* resistant to oxytetracycline was also found in Canada.

Tylosin – In the late 1990s strains of American foulbrood resistant to control with oxytetracycline were being discovered. Since this was the only drug registered for use in beehives to control this disease, researchers stepped up activities to identify other compounds that were both effective to use and safe for food use. Earlier studies found that the antibiotic tylosin, as tartrate, met these criteria.

Continued research found this product to be somewhat stable in honey, and recommendations were for applications to be made only as a powder mixed with powdered sugar, prior to honey flow, and only as a control, as opposed to prophylactic treatments.

Sulfathiazole – It was demonstrated in 1944 that sulfathiazole effectively prevented and controlled AFB. Interestingly, it is not effective against European foulbrood or any other bee disease. Since good methods of controlling AFB already existed (searching for and burning infected colonies) the industry was slow to use drugs for disease control. However, as migratory beekeeping increased in the U.S. and labor became more expensive, beekeepers depended increasingly on drugs. Although there was some concern over contamination of food products, methods of detecting residues were crude. All this changed when residue analysis became a more sophisticated science and the formal registration of drugs for bee disease control was demanded. At that time, the use of sulfathiazole for the treatment of AFB ceased in the U.S.

Because sulfa drugs are not registered for use to control bee diseases in the U.S., a "zero tolerance" exists, that is, it is not permissible in this country to sell honey containing any level of sulfa. The last year sulfathiazole was advertised and sold through the U.S. bee supply companies was 1978.

Canadian beekeepers continued to use sulfa drugs long after its sale and use against American foulbrood was halted in the U.S. The Canadian federal government set a tolerance of one part per million for sulfa in honey, but the Province of Quebec set a lower level. As a result, in 1987, contaminated Canadian honey was seized in both Quebec, because of the lower tolerance there, and the U.S., because of zero tolerance. Honey-importing countries have since been demanding sulfa-free honey (see Stecyk, T. "Drug Free Honey Hard on Canadian Beekeepers." *American Bee Journal* 127: 760, 802. l987). However, even as late as 2005, residues were found in Canadian honey when imported into the E.U., according to reports from German importers.

In view of the industry's concern for the purity of its product, the public's desire for uncontaminated food, and the fact that good methods of AFB control are known, the use of sulfa drugs to control bee diseases should be discontinued everywhere.

Government regulation and monitoring of pesticides and drugs – There are two major laws that govern the regulation and use of pesticides and drugs in the U.S.: the Federal Insecticide, Fungicide, and Rodenticide Act (FIFRA), and the Federal Food, Drug, and Cosmetic Act (FFDCA). FIFRA was first enacted in 1947 by the Congress and has been amended periodically since that time. FFDCA, passed in 1938, has also been amended several times. In 1962, Congress amended FFDCA to require that the Food and Drug Administration (FDA) evaluate new drugs for effectiveness as well as safety. The Food Quality Protection Act of 1996 amended both these laws to provide additional assessment of pesticides in food and feed items.

In 1970 Congress created the Environmental Protection Agency (EPA). The EPA is responsible, under FIFRA, for the registration of all pesticides sold or distributed in the U. S. Under FFDCA, the EPA is required to set pesticide tolerances for all pesticides used in or on food. A tolerance is the maximum permissible level for pesticide residues allowed in or on human food or animal feed items. It is EPA's responsibility, in cooperation with the states, to monitor pesticide use and ensure that tolerances are not exceeded.

The FDA regulates the use of antibiotics for the prevention and control of bee diseases. It is FDA's responsibility to establish tolerances for antibiotics in food/feed items, and to ensure that those tolerances are not exceeded. This is done through monitoring and surveillance programs.

In addition to the activities at the federal level, there may also be regulation and monitoring by the states. States may place further restrictions on pesticides sold or used within their own jurisdictions. States may also apply for temporary special regulations if emergency situations exist or a "special local need" can be demonstrated.

It is important to understand some basic rules concerning the use of pesticides. For example, one may use a pesticide only as indicated on the label. A material that is registered for use on one plant or pest cannot be used for the treatment of another unless the other plant or pest is explicitly named on the label. For example, if an insecticide is labeled for use on corn to control a particular insect pest, it cannot be used to control that pest on another crop unless that crop is also included on the label. This rule is very clear and unfortunately is sometimes violated. Another rule is that pesticides must be used at the rate(s) indicated on the label. There are also legal requirements for storing pesticides, and for disposal of unused portions and empty containers.

Several pesticides and drugs that were used in the past for various purposes in beekeeping operations are no longer registered for use. These include cyanide

Apply antibiotic dusts to top bars, avoiding the brood area below. Follow label instructions carefully to avoid contaminating your honey crop.

(to protect stored combs), sulfathiazole (to control American foulbrood), carbolic acid (a bee repellent for harvesting honey and other purposes), and propionic anhydride (a bee repellent). Several products are under examination and/or are being registered to aid in the control of diseases and pests of honey bees. Changes in the areas of disease and pest control are taking place rapidly, too much so for us to make many recommendations for specific materials. Every beekeeper must read the labels carefully and keep in close touch with state and regulatory agencies.

There are numerous sources of information for persons who wish to learn more about pesticides and their use. The following would be good places to begin: National Pesticide Information Center (NPIC), Oregon State University, 33 Weniger, Corvalis, OR 97331-6502. http://npic.orst.edu. Or, USEPA, Office of Pesticide Programs (OPP), Communication Services Branch, FEAD/OPP (7506C), 1200 Pennsylvania Ave., NW, Washington, DC 20460-0001. www.epa.gov/pesticides. -AWV

Feeding antibiotics – The FDA has approved different methods of feeding oxytetracycline (Terramycin) and fumagillin and one method for tylosin. Each of the methods has its advantages and disadvantages. Since there are many formulations of Terramycin, it is important that beekeepers read and follow the labels carefully. No antibiotic should be fed to colonies when there is a danger of contaminating the surplus honey crop. Feeding Terramycin and Fumidil should be terminated 45 days before the beginning of the surplus honey flow, while Tylan, according to label information, should be terminated 28 days before the honey flow.

Bulk feeding – Here the antibiotic is mixed with sugar syrup and fed to the colonies in a variety of methods (see feeding bees). Bulk feeding may be the method of choice if you feed sugar syrup as a routine management practice. The advantage of this method is that it allows for a good dosage control and it yields rapid results. The disadvantages are that this method has the highest risk of contaminating the surplus honey. The rate of uptake of the medicated sugar syrup is dependent on the availability of nectar. Oxytetracycline is least stable in sugar syrup. However this instability could also be a benefit as it would reduce the risk of contaminating the surplus honey. The largest disadvantage is that three treatments are required at four to five day intervals.

Dusting – Both oxytetracycline and Tylan are best fed in a dust formulated with powdered (confectioner's) sugar. Applying dust treatments does not require any special feeders; the results are rapid and the risk of contaminating the surplus honey is lower than bulk feeding. The disadvantage of dusting is that dose application is uneven. It is important not to get the dust formulation in the open cells of brood as it could kill the developing larvae. Like bulk feeding, three treatments are required at four to five day intervals.

Extender patties – Only oxytetracycline can be given in a carrier of solid vegetable oil and sugar. In this case, the resultant patty is placed on wax paper or plastic film on the top bars of frames where it would be consumed slowly by the bees. The sugar in the patty encourages the bees to feed while they are discouraged from feeding by the vegetable oil. The patty is consumed slowly over a period of many weeks thus avoiding the repeated trips required to apply bulk feed or dust. The risk of contaminating the surplus honey crop is lowest of all the methods of drug feeding. Antibiotic extender patties are less effective in colonies with fewer bees. Where bulk feeding and dusting are used for the prevention and control of American and European foulbrood, the antibiotic extender patty is effective only for American foulbrood disease. The uneven distribution and dose response that results from using extender patties is thought to be the reason the AFB bacteria are now resistant to Terramycin®. As a result, they are no longer recommended as a reliable treatment. Tylan cannot be applied by this method.

Pre-mixes – There are several pre-mixes available from bee supply companies for applying antibiotics to honey bee colonies. Some formulations contain a pollen substitute, sugar and the drugs, while others are simply a pre-mix of oxytetracycline and confectioner's sugar. These are safe and can be effective but may lead to resistant bacteria.

Honey flows and bee disease control – Beekeepers have known for many years that a honey flow does much to eliminate many disease symptoms and to improve the health of a colony. However, it has not always been so easy to explain why this is true. Some parts of the question are obvious and others have been made clear, or at least clearer, by recent research.

A honey flow stimulates bees to clean cells for food storage. This involves the removal of debris and, in most races of honey bees, the use of more propolis. Propolis contains substances that have a long-lasting antibacterial effect. As fresh propolis is distributed it tends to aid in the control of disease-causing microbes.

Another factor is that a honey flow, which is often accompanied by a pollen flow, also stimulates brood rearing. This results in an increase of young un-infested and uninfected bees that first turn their attention to cell cleaning. This young population dilutes the population of older bees that are the carriers of some disease agents.

Perhaps the most important reason that a honey flow aids in disease control is that it causes the bees to fly and work hard causing their death at a relatively early age, a good thing for the health of the colony. It is popularly thought that honey bees always work very hard but this is not true. They work hard when there is work to do, such as during a honey flow. When there is little or no work to do the bulk of the bees are generally idle and thus conserve their energy and bodies for the time when they will be needed. Some worker bees live to an older age, on average, and certain disease-causing organisms, such as tracheal mites, may accumulate in large numbers in older bees.

Another factor is that a honey flow means that there is adequate to abundant food available in the colony. Pollen, especially from a variety of sources, eases the stress of feeding and boosts the nutritional value of what the bees are eating – both adults and larvae.

This discussion has a direct bearing on how beekeepers should view bee diseases and when they should treat colonies to control the disease-causing organisms most effectively. Since a honey flow will aid in disease control it would not be appropriate to use chemicals for disease control prior to and during a honey flow. The medicines could also create potential honey contamination problems. It is after the honey flow, especially in the autumn when colony populations are dwindling and worker bees are headed into a time when they will live longer because of doing less work, that one can apply chemicals for disease control. In much of good beekeeping, timing is critical.

ANTIQUITY, BEEKEEPING IN - (see BEEKEEPING IN ANTIQUITY and BEEKEEPING IN VARIOUS PARTS OF THE WORLD)

ANTS – A PROBLEM FOR HONEY BEES – In many parts of the world, especially tropical and subtropical areas, ants can be a nuisance and can even destroy honey bee colonies. All ant species are social, although colony population size may vary greatly. Because of their social nature, ants may raid in groups and honey bees are often unable to defend themselves and their colonies. For example, in the southern

Ants that live near a colony can be just as irritating to a beekeeper. These are Red Imported Fire Ants, found in the Southern U.S. Standing on one of their nests while working a colony can be a short, but exciting experience. Further, having fire ants on a pallet or hive when moving bees from one state to another can cause regulatory problems. Treating beeyards for this pest is necessary.

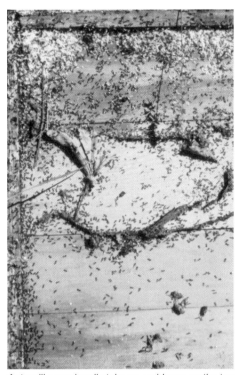

Ants will occasionally take up residence on the top of an inner cover. Bees, seldom visiting this location in a small colony, leave well enough alone. Strong colonies may battle an ant colony such as this. If you discover an ant nest such as this, simply remove the inner cover and shake off the ants, eggs, larvae and nest material several yards away.

U.S. entire colonies have been killed and their brood, stored pollen and honey removed within one day.

Killing ant colonies is possible but often difficult and too time-consuming, especially in rural areas where ant colonies migrate and/or new colonies are started routinely. At one time some beekeepers used soil insecticides to control ants, but these compounds often have adverse environmental side effects that make their use questionable.

A favorite way to protect against ants in many warm climates is to use an oil barrier. A hive stand is used with the legs immersed in oil. Most heavyweight oils evaporate slowly and a hive stand needs little attention. Hive stands, in addition to protecting against ants, have other virtues: they keep colonies off cold, damp ground. The raised entrances are above weed and ground vegetation that might otherwise deter honey bee flight.

APIARY LOCATIONS – Choosing a good site for an apiary is one of the most important considerations in beekeeping. First, it is important to determine how many colonies the honey plants in an area will support. Second, one must find a location that has the proper physical and ecological attributes.

Beekeepers generally do not grow plants for honey production. Rather, they move their colonies to areas where honey plants abound and the soil and weather conditions are appropriate for the plants to produce nectar and pollen. One may grow a few honey plants in order to follow their flowering, or to watch bees work, but it is usually not practical to grow a crop only for the nectar it may produce. This, however, is changing, as some beekeepers are practicing landbased honey production (see Landbased Honey Production).

A single colony of honey bees may usually be kept almost anywhere people can live in comfort. Even forested areas will often have sufficient forage for a few

A backyard location. Wind break, storage and lots of forage.

hives of bees, although the amount of surplus honey produced may be limited. Commercial beekeepers need to keep 30, 40 or more colonies in a location for it to be profitable. Small apiaries in different locations force the beekeeper to spend too much time on the road moving from one location to another. Measuring the number of colonies that may be kept in an area may take several years because weather and climatic conditions vary from one season to another and will have a profound effect on honey flows. Apiaries should be measured one against the other and colonies moved to find the most profitable arrangement.

In the tropics, shade during the hottest part of the day is a good idea.

Being able to drive between two rows of colonies reduces your work by half.

Articles in the bee journals have described beekeeping conditions in many states. Libraries at some of the state universities have bound volumes of these periodicals which may be searched for such literature. These records can be of great value when one is seeking new locations, but they must be read carefully because great changes may have occurred in agriculture since they were written. For example, great fields of buckwheat once grew in the northeastern states, resulting in tons of honey produced by bees in that area. Buckwheat is now out of favor as a crop, and thus beekeeping in the buckwheat area has changed.

Another example is in Florida. Beekeepers in Florida know that severe freezes in that state in recent years killed thousands of orange trees in what was once the northern end of the citrus belt. Beekeeping has changed drastically in Florida as a result. In California, almond acreage has expanded to the point that more than a million colonies of honey bees are needed to pollinate them each spring.

Apiary sites need to take into consideration the climate of the area as well as the type of beekeeping. Hobbyist beekeepers with hives at their homes will have a different situation from that of the sideliner or commercial beekeeper.

Primary considerations for an apiary site include year-round vehicle access (especially very wet springtime); protection from winter winds; either distance from vandal attacks or restricted access to the location;

For outyards, easy access with your truck, dry, neighbors close to keep an eye out for trouble, and colonies close enough to work easily keep your efforts to a minimum. Make sure you have access to the rear of your colonies so working them is easy.

Apiary Locations

Make sure you have year-round access, and that the land owner doesn't decide to plow you in.

protection from other large predators (bear) by proximity to dogs, people or fencing; nearness of year-round fresh water, good light exposure and ability to service the yard directly with a truck and/or pallet-loading equipment such as a forklift or bobcat.

Special consideration must be taken when placing an apiary in an urban or suburban location. All of the above topics are important, certainly, but probably more important are good neighbor practices (see GOOD NEIGHBOR PRACTICES).

With African honey bees becoming routine in the southern U.S., and migratory and package producers handling this hybrid in their operations, the opportunity for some level of defensive honey bee behavior to show up in any colony, anywhere, is possible. This makes the choice of an apiary even more important, and management, queen replacement, seclusion and even colony removal must be possible at all times. All of these make the choice of a location important.

Protection, but good exposure, easy access with your truck or wagon, and easy to work around.

APIMONDIA – (see ASSOCIATIONS FOR BEEKEEPERS)

APIMYIASIS – Several species of flies have larvae that will feed internally on live honey bees ("myiasis" is the infestation of an animal by fly larvae). The female flies that deposit the eggs apparently attach to or attack foraging bees, though sometimes they pursue those near a hive entrance. It appears that the honey bees are not capable of defending themselves against these attacks. Many of these flies larvaeposit, that is the eggs hatch within the female fly and larvae that are deposited are capable of immediately burrowing into the bee's body through the intersegmental membranes. The problem is rare, or at least rarely seen. The flies that may cause this problem belong to three fly families: Conopidae, Tachinidae and Sarcophagidae. Only adult bees are attacked by these flesh-eating flies; the larvae and pupae in the hive are protected from these predators by guard and house bees.

***APIS*, DISTRIBUTION AND SPECIES** – Honey bees (the genus *Apis*) are a small, distinct group of insects that have no close relatives. This group is primarily Asian, but one species, the western honey bee, *A. mellifera*, is native to Europe, Africa and the Near East, and has been transported around the world by humans.

All the honey bees share several characteristics in common. For example, all honey bees construct wax combs with hexagonal cells, all control their brood nest temperature, separate their brood and food, store their food above their brood, and live exclusively on pollen and honey. All are characterized by a complex social system, consisting of a single egg-laying queen who mates with numerous males, or drones, and large numbers of worker daughters who engage in nest construction, brood care, foraging and nest defense. All apparently use a dance language, and probably odor cues as well, to convey information about the direction and distance of food resources.

At one time the genus *Apis* included just four species of honey bees. Today, around 11 species are known – not because new species have recently evolved but because the diverse honey bees of Asia are receiving more investigation. Honey bees can be divided into three groups: the dwarf or small honey bees, the giant honey bees, and the medium sized cavity-nesting honey bees. The dwarf bees are believed to be the most basal, or primitive of the living honey bees, and the cavity-nesting honey bees the most derived, or advanced group.

Dwarf Honey Bees – These are the smallest of the honey bees, with workers about the size of a common house fly.

Left to right – *Apis florea, A. dorsata, A. cerana, A. mellifera.* (Koeninger photo!)

53

A. florea on honey comb. (Burgett photo)

Two species are known, *A. florea* and *A. andreniformis*, also known as the red and black dwarf honey bees, because the first two abdominal tergites are reddish brown in *A. florea*, black in *A. andreniformis*. The ranges of the two species overlap extensively, though *A. florea* has by far the broader distribution.

Apis andreniformis is known from southern Yunnan province in China, southwards through Laos, Thailand, Vietnam, peninsular Malaysia, the Indonesian islands of Sumatra and Java, and from Borneo and the nearby Philippine island of Palawan. Although it has not been collected from Cambodia or Myanmar, it almost certainly occurs in those places.

Florea nests for sale. (Burgett photo)

Florea nests sans workers. (Burgett photo)

The range of *Apis florea* extends from the eastern Arabian peninsula and Iraq in the west, through the Indian subcontinent to Indochina in the east, and south through Thailand. It is absent from peninsular Malaysia and most islands of Indonesia, and absent from Borneo and Palawan. Outlying or "disjunct" populations are known from the Indonesian island of Java, and from the Sudan in Africa. It was almost certainly transported to Java and Africa by humans, probably accidentally on ships.

Unlike the nests of the familiar *Apis mellifera*, the nests of dwarf bees consist of a single, small vertical comb, about the size of a person's outspread hand. The comb is attached to the branch of a tree or shrub. Brood cells and pollen stores are in the comb hanging below the branch, while the cells holding honey stores surround the sides and upper surface of the branch. The nests of the two species are similar, though that of *A. florea* is larger than that of *A. andreniformis*.

In *A. florea* nests the honey storage cells appear to radiate outwards from the supporting tree branch and form a rounded platform on the top of the

A. florea. (Burgett photo)

branch; many cells are irregular in shape. In *A. andreniformis* nests the honey storage cells are regular in shape, the bases of cells on opposite sides of the comb meet to form a midrib, and the whole honey storage area forms a crest above the branch.

Although the nests are not hidden in a cavity or hollow tree, they are protected by a curtain of worker bees that covers the comb. Nests of *A. florea* are often collected and sold in markets, especially in Thailand, where the honey, larvae and pupae are relished.

*A. dorsata in tree (*Burgett photo*)*

Natural nest. (Burgett photo)

Incipient dorsata (Burgett photo)

The dance language has been more thoroughly investigated in *A. florea* than in *A. andreniformis*. Unlike the giant bees or cavity-nesting bees, these bees dance on the horizontal surface at the top of the comb, and thus the straight run of the dance can point directly to the food source. These bees use celestial cues to orient their dances, and on overcast days they can use local landmarks around the nest. Unlike the other *Apis* species, gravity does not influence their dances.

Giant Honey Bees – Workers of the giant honey bees are about twice the size of *A. mellifera* workers. Giant bees are found from Iran in the west to the Republic of the Philippines in the east, from Pakistan and the foothills of the Himalayas in the north, to the islands of Indonesia in the south.

Apis laboriosa is found in the

Dorsata curtain. (Burgett photo)

Himalayan region, where it lives at 4000 to 11,000 feet (1200 to 3300 meters) elevation, and in parts of northern Vietnam. It is slightly larger than *Apis dorsata* and its body is much more densely covered with hair than any other *Apis* species. Close observations of this bee have been made in Nepal and there is still much to learn about it.

Apis breviligula is found in the Philippines (excluding Palawan), and *A. binghami* is found only on the Indonesian island of Sulawesi. Because *A. dorsata*, *A. breviligula* and *A. binghami* do not overlap in their distributions, it is difficult to determine if they are capable of successful interbreeding or if they would be reproductively isolated when brought together.

Like that of the dwarf honey bees, the nest of the giant honey bees consists of a single, exposed comb – but much, much larger. While an average comb might be two feet (60 cm) across and two feet (60 cm) deep, enormous combs over five feet (1.5 meter) from top to bottom have been found. The combs are attached to the underside of tree limbs, under rock ledges, and quite often in window frames and on other human constructions. Unlike the nests of the dwarf bees, the combs of giant bees do not wrap around the branches or other substrate.

Giant bees have been called the most ferocious stinging insect on earth. It is not uncommon for as many as 5000 of these bees to attack an enemy at once. Colonies of *Apis dorsata* may contain up to 70,000 bees, but most are occupied by forming a blanket or curtain around the nest to control nest temperature and humidity; thus, many of these bees are not free to forage.

Giant bees show several intriguing behaviors. Nests of *Apis dorsata* and *Apis laboriosa* are often aggregated on particular trees, cliffs or buildings. As many as 186 *A. dorsata* nests have been reported from a single, large Banyan tree in southern India. These bees are also known for their seasonal migratory behavior. *Apis dorsata* colonies may travel hundreds of kilometers annually, apparently in response to seasonal patterns of nectar and pollen abundance. *A. laboriosa* is known for altitudinal migrations apparently associated with surviving colder temperatures. Genetic evidence indicates that migrating populations return to the same aggregation sites they abandoned earlier in the year. On the other hand, neither *A. breviligula* nor *A. binghami* are known to aggregate their nests, or to show migratory behavior.

The foragers of giant bees carry out their dances on the vertical surface of the nest, like the workers of the familiar *A. mellifera*. Although the dancer can usually see the sky when it dances, it can also orient on overcast days with the aid of landmarks. *Apis dorsata* and *A. breviligula* workers have even been observed to forage on moonlit nights.

Cavity-Nesting Honey Bees – As their name implies, cavity-nesting honey bees typically build their nests in hollow trees, small caves, or in man-made structures. Mature nests contain several vertical wax combs arranged parallel to one another, an arrangement that is mimicked by the beekeeper's Langstroth hive.

The cavity-nesting habit has several important consequences for honey bees. First, it enables these bees to control the nest temperature easily, and perhaps to help defend their stored food resources. The ability to control nest temperature and defend large pollen and honey supplies may have enabled *A. mellifera* and *A. cerana* to extend their ranges far outside the tropical and semitropical regions inhabited by most species of honey bees. Finally, foraging workers performing a waggle dance cannot see the sky from inside the enclosed nest. While foragers of the dwarf and giant honeybees usually can see the sky when they dance, cavity-nesting bees dance in the dark on the vertical surfaces of their combs. They use gravity ("downwards") to represent the position of the sun.

A. cerana. (Burgett photo)

The cavity-nesting species include *A. cerana*, *A. koschevnikovi*, *A. nigrocincta* and *A. nuluensis* in Asia, and the familiar *A. mellifera* in the west. Of these, *A. mellifera* and *A. cerana* have the broadest distributions, while the others have quite limited ranges. *Apis mellifera* is a native of Europe, Africa and the near East including western Iran. It has been spread deliberately to North and South America, Australia, New Zealand, and Asia, and is the honey bee of commerce in most parts of the world.

Apis cerana occurs over most of temperate and tropical Asia, from eastern Iran in the west to Japan in the east, and south to the islands of Indonesia, Malaysia and the Philippines. It is the second best-known honey bee species in the world because it is used commercially in several Asian countries, including India and Thailand. It adapts readily to a small version of the Langstroth hive and may be managed as far as swarm control, queen rearing, and honey production are concerned.

A. cerana. (Burgett photo)

Like the western honey bee, *A. cerana* varies in size, color and behavior across its enormous range. Many subspecies have been proposed; current genetic evidence suggests that there are at least four or five main groups of *A. cerana*

A. cerana. (Burgett photo)

populations: mainland Asian bees, "Sundaland" bees (including bees in peninsular Malaysia, the islands of Sumatra, Java, Bali and Borneo); Palawan island bees, and bees of the other Philippine islands.

The other three cavity-nesting species are restricted to southern Asia.

A. koschevnikovi – also called the Sundaland bee – has been found in Borneo, peninsular Malaysia, Java and Sumatra. *A. nigrocincta* is known from the Indonesian islands of Sulawesi and Sangihe; *Apis nuluensis* is so far known only from Borneo.

The Asian honey bees are of interest to both the practical beekeeper and the honey bee scientist. They provide a glimpse of the history and diversity of honey bees, and provide clues to the evolution of traits such as the honey bee's complex dance communication.

They are also the source of many actual and potential threats to *A. mellifera* beekeeping, such as brood mites. The dwarf honey bees are the natural hosts of mites in the genus *Euvarroa*. *Apis florea* is host to the brood-infesting mite *Euvarroa sinhai*, *Apis andreniformis* is host to the mite *E. wongsirii*. Giant bees are the natural hosts of *Tropilaelaps* brood mites, *T. clarae*, and *T. koenigerum*. The Asian cavity-nesting bees are the natural hosts of the *Varroa* mites, including *V. destructor, V. jacobsoni, V. rindereri* and *V. underwoodi*.

These mites generally cause little damage to their natural hosts. However, one species, *Varroa destructor* (formerly thought to be *V. jacobsoni*), was able to colonize the western honey bee, *Apis mellifera*, introduced to Asia for commercial apiculture. For this reason, Asian cavity-nesting bees are of great interest to those who study bee parasites and diseases. *Euvarroa sinhai* and *T. clarae* have been found on *Apis mellifera* in Asia, and no one can predict which of these or other Asian mites may become the next "new" pest of *Apis mellifera*. It is important to be prepared for new threats to beekeeping and this means understanding the lives of the Asian honey bees and their natural defenses against parasites and diseases. (For more on European honey bees see RACES OF HONEY BEES.) -DRRS

APISTAN® – (see FLUVALINATE)

APITHERAPY – The use of various honey bee and hive products for medicinal or therapeutic purposes is called apitherapy. Proponents of apitherapy usually seek to use natural products that are more in keeping with a natural life style. Some of the claims made for bee venom, propolis, pollen and royal jelly may be overstated but arc often supported by a number of testimonials (see NEUTRACEUTICALS).

ARTIFICIAL INSEMINATION – (see INSTRUMENTAL INSEMINATION)

ARTIFICIAL SWARMING – This is the term that was used when one divided a colony by removing a portion of the bees either with or without the old queen. Today the terms more frequently heard used are "divides" and "splits." In a natural swarm the old queen usually accompanies the swarm (see SWARMING IN HONEY BEES and SWARMING, and SUPERSEDURE).

ASSOCIATIONS FOR BEEKEEPERS – Everywhere bees are kept one will find beekeepers' associations. Most of these groups will have regular meetings; some offer their members instruction with formal courses while others have outside lecturers, videos, honey shows and demonstrations during the active season. A group in a London (England) suburb owns its own land with a club apiary, a modest lecture hall and a shop. Bee clubs often cooperate in buying journals and books for their members at reduced rates. Many clubs have libraries and circulate books and videos among their members; many have regular hard copy and electronic newsletters, and many have their own web pages. In the U.S. some bee clubs work closely with the agricultural extension service. The primary function of most beekeepers' associations is education though occasionally they become concerned with matters such as pesticide misuse, disease control, legislation, and honey sales.

The American Association of Professional Apiculturists (AAPA) – Apiculturists at the March 1978 Mid-West Bee Diseases Clinic in Springfield, Illinois, met the day after the conference to discuss various topics of interest. Drs. Elbert Jaycox (IL) and Basil Furgala (MN) stated that there appeared to be a need for a U.S. professional apicultural organization, similar to the Canadian Association of Professional Apiculturists (CAPA). Dr. Martin (MI) was chosen to devise and distribute a questionnaire to determine the interest of the apicultural researchers, extension specialists, and apiary inspectors. It was mailed in May of 1978.

The "ballot" included reasons for organizing: 1.a chance to discuss important industry concerns, 2.a mechanism to promote apiculture, 3.no currently existing organization was appropriate for this purpose, and 4.lack of travel funding limited the number of meetings apiculturists could attend.

Three reasons were given for not organizing: 1.there is too much heterogeneity for one meeting to be attractive to all potential members, 2.the Apiary Inspectors of America (AIA) had their own identity and program, and 3.administrative restrictions might be prohibitive.

The results of the survey demonstrated a desire for such an organization. On March 11, 1982, Dr. Eric Mussen (CA), serving as Corresponding Secretary, sent a letter written by Dr. Malcolm T. Sanford (FL) to identified potential members and requested a dues payment of $10.00. Dr. Sanford began publishing the AAPA newsletter. The first formal meeting of the American Association of Professional Apiculturists (AAPA) was held in January, 1983, in Orlando, Florida.

The American Beekeeping Federation (ABF) – The American Beekeeping Federation was founded in 1943, bringing together all persons interested in bees and beekeeping. The purpose of the Federation, as stated in the Bylaws, is to engage in any lawful activity that will promote the common interests and the general welfare of the diverse segments of the United States beekeeping industry. Membership is open to all. Owning bees is not a prerequisite. Membership in the Federation includes hobbyists, commercial beekeepers, honey packers, bee supply manufacturers, teachers, scientists, publishers and regulators.

The Federation has its roots in a series of earlier organizations with much the same purpose and goals. According to Pellett (Pellett, F.C. *History of American Beekeeping*. 1938. Collegiate Press, Inc., Ames Iowa. 213 pages) attempts to found a national beekeepers' organization were made as early as 1859, but the Civil War delayed this effort. Professor A.J. Cook, who was well known for his writings and books on beekeeping, was a strong stimulus in the formation of a national organization and an annual meeting of beekeepers. As a result of his efforts, beekeepers from several states met on December 21, 1870, in Indianapolis. A second national organization was founded soon thereafter but the two amalgamated the following year. Since that time a national meeting of beekeepers has occurred every year, though the name of the organization has changed.

There is a national headquarters staffed by the Executive Secretary. Officers and Directors are elected at the annual conference. A newsletter is issued six times a year. The Federation has championed a number of causes over the years and actively lobbied at the national level. Information on the activities and the annual conference, usually held in January, is found in the newsletter, the beekeeping journals and on its website, www.ABFnet.org.

The American Bee Research Conference (ABRC) – The American Bee Research Conference held its first annual meeting in Baton Rouge, LA, on October 7–8, 1986. The purpose of the organization is to provide an opportunity for scientists to present their current work to those attending the meeting, as well as to a worldwide audience through the publication of abstracts of the proceedings. Anyone is welcome to attend the meetings and researchers from all countries are invited to present papers. A second objective is to foster cooperative work and communication among scientists by organizing the meeting to consist of informal social settings as well as a more formal structure of scientific presentation and discussion.

The American Honey Producers' Association (AHPA) – The organization was founded in 1969 and held its first national meeting in January, 1970. Membership is open to all beekeepers. The organization seeks to promote the common interest and general welfare of the honey producers and beekeepers of the U.S. An annual convention is held that features outstanding speakers, such as bee scientists, commercial beekeepers, politicians, lawyers and others. The organization has been concerned with pesticides and their effects on honey bees, the price support program when it existed, contaminated honey problems, research, the world honey market and all other matters of concern to beekeepers. A newsletter, *Honey Producer*, is published quarterly. Their web page is www.AmericanHoneyProducers.org.

Apiary Inspectors of America (AIA) – Many state Departments of Agriculture hire one or more specialists in beekeeping. The initial concern of these individuals, and those they in turn hire to inspect honey bee colonies, is the control of American foulbrood. In some states these individuals conduct educational programs to train beekeepers in disease recognition and to help cope with bee disease problems. The responsibilities of the state apiary inspection have expanded as new pests were detected in the U. S. i.e. tracheal mites, *Acarapis woodi*, in 1984, *Varroa destructor* in 1987, the Africanized honey bees in 1990 and *Aethina tumida*, the small hive beetle, in 1998.

In 1929 state and federal apiculturists gathered in Sioux City, Iowa, and formed the Association of Apiary Inspectors of America; the name was later changed to Apiary Inspectors of America. The group has met annually since that date, often in conjunction with one of the national beekeepers' organizations, and at state and federal bee research laboratories. The members exchange information on the incidence of disease in their respective states and discuss methods of control. Membership is limited to those who play an official role in bee inspection but parts of the meetings and lectures are often open to other interested persons. An annual proceedings is published as an update on each state's conditions.

Apimondia – This is the international beekeepers' organization whose members are national organizations, regional organizations and even individual beekeepers. Apimondia sponsors a regular conference every other year and periodically sponsors special symposia not necessarily in conjunction with their regular meetings. A proceeding is published that abstracts the lectures, awards, special symposia and field trips that are held. Bee supply dealers from around the world attend these meetings and thus there is always a good display of beekeeping and honey processing equipment. Apimondia meetings, especially in recent years, have been attended by several thousand people. The last Apimondia meeting in North America was in Vancouver, Canada, in 1999.

Beekeeping Association Directory – With the advent of the world wide web locating and contacting federal, state and local beekeeping contacts has become much easier. *Bee Culture* magazine (www.BeeCulture.com) maintains a fairly current listing of these associations under the titles of The Science of Beekeeping, and Who's Who In Apiculture.

Canadian Association of Professional Apiculturists (CAPA) – The Canadian Association of Professional Apiculturists is a group of professionals who are employed as apiculturists in Canada. Membership in CAPA is restricted to individuals employed as apiculturists in extension, apiary inspection, teaching or research. Members of the organization include provincial apiarists, university researchers, federal researchers, apiary inspectors and apicultural technicians. Associate members include representatives from Agriculture Canada, the Canadian Honey Council, the American Association of Professional Apiculturists and the Apiary Inspectors of America.

The objectives of the Association are to promote, and develop cooperation among the different groups of professional apiculturists within and outside Canada. CAPA also serves as an important vehicle to facilitate discussion between groups of professionals on problems of common interest and to help coordinate extension, regulatory and research activities. Representatives from the Canadian beekeeping industries national organization, the Canadian Honey Council, also attend these meetings to facilitate consultation between the beekeeping industry and the professionals that serve it.

The Association also plays a major role in developing extension materials for the industry. Publications that are written, produced and published by CAPA include *Honey Bee Diseases and Pests* and *A Guide for Managing Bees for Crop Pollination*. -RWC

Canadian Honey Council (CHC) – The Canadian Honey Council is the national association of Canadian honey producers, beekeepers and co-op packers. Formed in 1940 it provides liaison between the federal government, provincial governments, beekeeper associations and the honey industry. It is financially supported entirely through memberships of individuals and provincial associations. The national coordinator is given guidance by the board of directors which is composed of eight voting delegates. The decision makers represent the six larger provinces, plus one delegate from the three maritime provinces and one delegate from the co-op packer. The CHC works closely with the Canadian Association of Professional Apiculturists (CAPA) and the president of CAPA is a non-voting director of the CHC. The annual meetings of CHC and CAPA are held at the same time and at this event the CHC hosts an annual joint research symposium. The CHC also administers the Canadian Bee Research Fund jointly with CAPA. This unique arrangement allows researchers to access grant money for projects according to priorities identified by industry. The CHC also maintains a website with apiculture information including a list of apiculture contacts in Canada. Members of the CHC receive *Hivelights* magazine, a quarterly publication with current apiculture news and events in Canada. Beekeepers outside of Canada can subscribe to the magazine and view selected past articles through the website www.honeycouncil.ca. -HC

Eastern Apicultural Society (EAS) – The Eastern Apicultural Society of North America is the largest regional beekeeper's organization in the United States. Founded in 1955 at the University of Maryland, it encompasses 27 eastern states and the five eastern Canadian provinces. It has approximately one thousand members. The primary aim of the EAS is to "promote the art and science of beekeeping." Its membership includes primarily hobbyists, but professional apiculturists, commercial beekeepers, and other bee-related professionals also belong. Membership is not restricted.

Each summer the society holds a week-long conference at a regional college campus or conference center.

The first three days of the meeting are dedicated to an intense beekeeping short course. Attendance is limited and the classes are usually high level with a wealth of academic instructors. The midweek day, Wednesday, is an overlap day where the short course and main conference overlap. This is usually the most intense day of the week. The conference that follows consists of lectures and workshops, social events and entertainment.

EAS also sponsors the only multi-state Master Beekeeper organization. Lab, written, oral and field exams are administered each year at the conference. Passing all of these is required to become an EAS Master Beekeeper.

The EAS Journal is published four times each year, and the EAS web page (www.EasternApiculture.org) is interactive for conference registration, Master Beekeeper information, Honey Show guidelines, director contacts, the annual conference program and other information.

EAS is governed by a Chairman of the Board, an Executive committee consisting of a vice chairman, secretary, treasurer, past chairman and past president. Directors are elected by each member state and serve on the board up to two, four-year terms. The Master Beekeepers also have a director on the Board representing their group. EAS is incorporated, and is a 501(3C) non-profit organization.

Government Associations:

Cooperative Extension Service – Many counties have an extension service that is financed cooperatively by the federal, state and county governments. This group is listed as the Cooperative Extension Service. In the past, many county beekeepers' organizations worked closely with the extension service, though these connections are becoming scarce. The Cooperative Extension Agents often assisted bee clubs in arranging programs and special events. However, there are fewer and fewer extension personnel with any background in beekeeping. Further, restrictive budgets have reduced county resources severely.

The legislation that is the basis for the extension service was passed as an act of Congress in 1914; the primary function of the service was the extension of information, especially research on plants and animals. At one time, many years ago, extension dealt almost exclusively with farmers and growers. Today, Cooperative Extension, in addition to concerns of agriculture, also devotes time and energy to home affairs, nutrition and other programs of interest and value to a wide audience that is both rural and urban.

State Colleges and Universities – Almost every state has an agricultural college whose mission is teaching, research and extension. Those at the state extension level usually work closely with the county extension services, but this is not always possible. The state colleges of agriculture have their basis in the National Land Grant Act of 1862. This act provided land in the central and western states which could be sold to raise money to help finance new or existing colleges of agriculture, one in each state. Some state colleges have at least one entomologist and a few have a part time apiculturist with obligations of teaching, research and extension in apicultural science and related subjects.

Currently there are fewer than a dozen even part-time apicultural positions in the U.S. and Canada. Moreover, there are even fewer state-level Extension Specialists in apiculture. There are several who have responsibility in this area but their training and expertise in the field is limited. This reflects, certainly, the decline in the number of hobby, sideline and commercial beekeepers since the heyday of the 1940s and 1950s, the willingness of the public to support these positions, and the ability of the industry to make meaningful financial contributions.

Heartland Apiculture Society (HAS) – HAS, the youngest of the U.S. regional beekeeping associations first met in 2002. Like WAS, it is loosely based on EAS in structure and goverance and was initially funded by EAS. It's member states encompass the very western areas of EAS and the rest of the central states.

They meet once per year for a two or three day conference on a University campus in a member state. The Board, like EAS and WAS, consists of elected members from member states.

International Bee Research Association (IBRA) – For the beekeeper, teacher, researcher and anyone interested in extending knowledge about bees, beekeeping, bee science and any related bee subject to others, an important source of information for many years was the International Bee Research Association (www.ibra.org.uk). This organization has a long history dating back to the formation of the Apis Club in England in 1919. The Apis Club began publication of the journal *Bee World* that was taken over by the Bee Research Association (BRA) when it was established by Dr. Eva Crane in 1949. Although international in outlook from the very beginning that word did not officially figure in its title until 1976 when BRA became IBRA.

IBRA's extensive library of scientific papers, books, journals, and reprints is maintained and accessible through the National Library Library of Wales (links on IBRA websites) while the collection of charts, slides, photographs, pamphlets and historical artifactsare being digitized so as to be available on computer by way of CD-ROM or the internet.

An important aspect of IBRA's mission

Home of Dr. Eva Crane, Gerrards Cross, England.

is to promote appropriate beekeeping development as a practical, sustainable economic activity in the Developing World.

In short, IBRA aims to make available all that is known on bees, and has done so for many years. To this end it has a book shop from which it sells not only its own publications but many that are prepared elsewhere. Twelve dictionaries of beekeeping terms in different languages have been prepared by IBRA, and are indispensable resources.

The Journals IBRA published included:

Bee World, started in 1919, at four times a year. It is devoted to beekeeping news, upcoming events, and both short and in-depth articles, especially review papers, on a variety of subjects. It is the bridge between bee science and practical beekeeping.

The Journal of Apicultural Research was started in 1962; it contains reviewed articles of scientific interest on all aspects of apiculture, including studies on solitary and subsocial bees. Four issues are published annually. Articles on selected themes, i.e. pollination are available on CD-ROM and individual papers from both these journals can be accessed on the Internet. All papers in both these learned journals are subject to peer review.

Apiculture Abstracts, originally a part of *Bee World*, was started in 1950 but became a separate publication in 1962. Four issues plus an index are published each year; thousands of papers, bulletins, circulars, and books relating to the field have been abstracted in this journal. Five-year blocks of this definitive bee literature abstracting service are available on CD-ROM and quarterly updates were also produced on CD-ROM. Similarly bibliographies of abstracts on specific themes such as Beekeeping Management have been produced by selecting relevant material over a 10-year period.

The IBRA was founded in 1949 and was directed by Dr. Eva Crane until her retirement in 1984. The current IBRA Director is Richard Jones who has held the office since 1996.

In late 2005, IBRA was transitioning from its former labor-intensive search process to a more streamlined organization. The many thousand books, papers, journals, reprints, photos, videos, negative and magic lantern slides from the Eva Crane library will be moved to the National Library of Wales, at Aberystwyth.

The extensive collection of artifacts has been photographed and put in virtual form on CD for purchase or review. The collection itself was moved to the International Beekeeping Visitor Centre at Eeblo, Belgium, where it will be displayed.

The, activities IBRA will pursue in the future, though not perfectly clear, appear to be as a grant-giving trust, and after additional review, publisher of research papers and information. (For additional information see CRANE, EVA and IBRA, HISTORY.)

Western Apicultural Society of North America (WAS) – This is an association of western North American beekeepers, patterned after the Eastern Apicultural Society, held its first annual meeting in August, 1978, on the Davis campus of the University of

California. The motivation for the formation of WAS occurred during the summer of 1977 when a group of beekeepers met to discuss the formation of an association of western beekeepers, based on the structure of EAS, and initially funded by EAS.

The purpose of WAS is to provide a beekeeping organization located in the western U.S. for educating beekeepers and members of the public on all matters relating to the field of apiculture. Membership is open to any individual within 12 western states of the U.S. (including Hawaii and Alaska) and the five western provinces of Canada (including the Yukon). WAS is an apolitical organization which purposely avoids involvement in topics which have the potential to lead to the politicalization or polarization of beekeepers. Annual meetings usually are held in July or August of each year on a campus of a western college or university or at a non-academic venue. In addition to the annual meeting, the Society produces a journal and proceedings for past meetings. -DMB & ECM

ATCHLEY, JENNIE (1857-1927) – She lived most of her life in Beeville, Texas, where she ran the Jennie Atchley Company, a queen production operation. She also published *The Southland Queen*, a pioneering Texas oriented beekeeping publication which ran from 1896-1904.

ATKINS, E. LAURENCE – He was a graduate of the University of Illinois, with a B.S. in medical/veterinary entomology studies and a M.S. in morphology entomology study. He joined the U.C. Riverside "Citrus Research Center and Agricultural Experiment Station" in 1952, to work on the biology and control of orange worms. During these studies he experimented in all phases of control, centering on parasites,

Laurence Atkins

predators, cultural-trimming "skirts," and chemicals. He started bee research in 1952 when Loren Anderson's assistant was hospitalized, for the fourth time, by bee-sting-induced anaphylaxis. Atkins took over the laboratory component and helped Anderson with field studies in bee toxicity in 1954, becoming responsible for the whole bee toxicity program in 1958. He has 60 publications on insect control in citrus, 90 publications on toxicity of agricultural chemicals to bees, and has been involved in about 60 court cases dealing with bee kills. He revised the text for a U.C. Extension publication entitled "Reducing Pesticide Hazard to Honey Bees."

AUSTRALIA – (see BEEKEEPING IN VARIOUS PARTS OF THE WORLD)

References:
Dade, H.A., *Anatomy and Dissection of the Honeybee*, 1962.
Erickson, E.H., S.D. Carlson, M.B. Garment, *Atlas of the Honey Bee*, 1986.
Goodman, L., *Form and Function in the Honey Bee*, 2003.
Snodgrass, R.E., *Anatomy of the Honey Bee*, 1956.

BACON, MILO R. – Appointed Chief Apiary Inspector for the Commonwealth of Massachusetts, he served in that position from 1952 to 1967.

Milo and several other beekeepers organized the Norfolk County Beekeepers' Association in 1959, and he served as President of the Association from 1975 to 1979.

He was elected Secretary-Treasurer of the Massachusetts Federation of Beekeepers in 1971 and he remained an officer of the federation until the time of his death.

He was a charter member of the Eastern Apicultural Society. He also served on the Massachusetts Federation of Beekeepers Committee which hosted the Eastern Apicultural Society when it held its annual meeting at the Massachusetts Maritime Academy in Bourne, Massachusetts in 1975. He died in 1981.

Seymour Bailey

BAILEY, SEYMOUR E. (1908-1982) – Mr. Bailey was the State Apiarist of Ohio for 28 years. He was a graduate of The Ohio State University and was named Ohio Beekeeper of the Year in 1975.

BAIT HIVES – In many areas it is possible to use bait hives to capture stray swarms of honey bees. A bait hive is a simple box or other receptacle in which bees might build a nest. While there is not yet a perfect bait hive that a swarm cannot refuse, we know many of the parameters that bees in swarms prefer when given a choice. This list pertains to European honey bees; Africanized honey bees have a much less restrictive list and are known to use cavities that are smaller and closer to the ground. See Africanized Honey Bees for more detail.

For European Honey Bees:
1. Height: about 10 to 15 feet (three to five meters) above the ground.
2. Shade-Visibility: well shaded, but highly visible; if the sun hits the bait hive the bees will probably leave if no brood is present.
3. Distance from parent nest: not important.
4. Entrance size: about 1-1/4 inch (three cm) in diameter.
5. Entrance shape: not important.

Bait hives come in a variety of styles. Available in some locations, pressed wood pulp treated with weather resistant materials are often used, but the manufacturing process tends to be environmentally unfriendly. A deep hive body, baited with old comb, 10' - 30' above ground, secure, dry and with a single, small entrance is ideal.

6. Entrance position: near the floor of the hive.
7. Entrance direction: facing south, but not absolutely required.
8. Cavity volume: 40 liters or about 1.4 cubic feet. This is almost equivalent to the volume of a 10-frame deep hive body. Hive bodies fitted with tight covers, so that light does not enter, make good bait hives.
9. Cavity shape: not important.
10. Dryness and draftiness: dry and snug, especially the top.
11. Color: no data, but dark colors may reduce vandalism.
12. Type of wood: many types of trees have been occupied, but bees may avoid new wood.

Although beekeepers successfully have used cardboard boxes and even metal containers for bait hives successfully a commercially available pulp bait hive is available from equipment suppliers.

Success in using bait hives depends

upon proper site selection. Swarms have moved into bait hives only to reject them an hour or a day later. Scouts, it seems, can make errors. Once bees have built combs and have brood, they will rarely abandon a new home.

There are a couple of techniques to use to increase the attractiveness of the bait hive to curious scout bees looking for new quarters.

The pheromone lure containing a Nasonov-like chemical is available from bee supply companies. Fastened inside the bait hive and allowed to slowly evaporate, it will attract scouts over a period of time.

An old frame with comb also acts as an attractant complete with a bit of honey. Of course early removal, before wax moth moves in, is recommended.

The U.S. and Canada require that bees be kept in movable frame hives so that they may be inspected for disease, so from a practical point of view it is best to transfer the bees to a movable frame hive as soon as possible. This may be done in a variety of ways if the swarm has been captured in a hive body fitted with frames. Drumming (see DRUMMING) allows the bees to move up naturally into a new hive body placed above the bait hive; removing the bottom and inverting the bait hive allows the bees to move upward. Another method is to remove naturally drawn combs one by one, saving them by tying them (string or rubber bands) into a frame. This last method can be difficult when the bees have only recently occupied the nest since the comb is new and fragile. Obviously, the best method is to check bait hives frequently during swarming season and move them into standard equipment, thus freeing up the bait hive for additional swarms.

Capturing swarms in bait hives late in the summer can be a problem because the bees may not have time to store sufficient food for winter. You will have to feed these bees most of their overwintering food, or just join them with another colony.

BALD BROOD – Larvae or pupae whose heads are partially or completely missing in brood cells that have recently been uncapped. Bald brood is usually the result of lesser wax moth damage especially if fecal matter is also found on the bald brood. In rare instances this could also be the result of cannibalism.

BALLING – A large mass of worker bees, 25 to 50, may sometimes form a firm or hard cluster around a queen, especially a young queen; this is called balling. On rare occasions an old queen may be balled, too. Balling sometimes occurs when one is examining a colony, and there may be no apparent explanation for what is taking place. When balling occurs the best thing to do is to close the hive; usually the colony will return to normal and the queen will be alive when the colony is examined several hours or a day later. A ball of bees may sometimes roll down the side of comb when it is being examined and even fall onto the ground. If this happens, a comb should be placed nearby so that the bees, and especially the queen, may crawl onto it and be put back into the hive.

Balling does not appear to be a killing process, though queens are often killed when it takes place. Huber, the Swiss observer recorded that he and his assistant once watched a queen being balled for 17 hours, but that after that time she emerged unscathed. The worker bees in a ball will clamp onto the queen's legs, wings, mouthparts or wherever they can gain a hold. The queen is thus immobilized. Perhaps balling bees are only trying to prevent queens from attacking and killing each other. At times balling appears to be a holding process, when two or more queens are present in a hive or a swarm and the bees may only seek to keep both alive until some decision is made.

Worker bees in a ball will protrude their stings slightly. These may, apparently accidentally, catch into a queen and result in her death, or the same thing may happen to other workers in the ball. Beekeepers have stated that queens that are balled appear to move rapidly, or in a nervous manner, over a comb. That thought is anecdotal only, and no data support it. Bees, when balling, can raise the temperature of the ball to lethal temperatures. This tactic is used when intruders, such as hornets, invade the hive.

All the causes of balling are not clear. Several circumstances can give us some clues. For example, if one cages and removes a queen from a hanging swarm and places her nearby, the bees will find her and the swarm will move to her. If a second queen that is foreign to the swarm is also caged and placed nearby, some bees will find her and form a ball around her. These bees will usually not kill the foreign queen. While the bees are holding or balling her, the swarm itself will move to their own queen. If their queen is removed from the scene, the swarm will eventually move to the foreign queen and after a day or so accept her as theirs. This appears to be one of those cases in which the bees are holding the foreign queen for some unknown purpose.

Bees will also form balls around queens in a queen bank (see QUEEN REARING: GRAFTING AND COMMERCIAL QUEEN PRODUCTION – Bank colonies). If young bees are added to the bank, accomplished by adding a frame of mature brood about every five days, the queens will be carefully attended and not balled. It is, then, the older bees that form balls around queens. Queens held in queen banks with only older bees will soon lose body parts, such as legs, wings and mouthparts, if the workers have access. This type of balling has lead people to believe that balling is, at least in part, a hostile act. Queens held in banks in cages under these conditions may lose foot pads, antennae segments or hairs.

If a queen is washed in a pure alcohol solution and the solution is placed on a small piece of cork or styrofoam and allowed to dry, worker bees will surround the object and ball it. Some people have mistakenly believed this happens because the piece is attractive to the bees, but it is not the case. If this action is observed closely one will note that the bees have their mandibles spread and may even protrude their stingers slightly. These actions appear to be aggressive.

BEARS – These animals often pose a problem for beekeepers. Bears like to eat both brood and honey, but brood seems to be preferable. Bears are a threat to honey bee colonies not only in North America but anywhere they are found.

Usually a bear that is feeding in an apiary will take or attack only one or two hives each night. If the bear is not detected and deterred it may destroy an entire apiary. Bears will often pick up a hive body or super and carry it as much as several hundred feet (20 to 100 meters) from the apiary. When this is done the guard bees will fly back to the location from which the hive body or super was taken because that is the area

(Cella photo)

A stout, strong electric fence is the only sure deterrent to stop bear from feasting on your hives. Fence designs are many, and supplies are readily available at farm stores. Basics for a good fence include a sure and reliable source of power (solar rechargers work), a sturdy fence (wires, panels or grid) and a good ground. Check with your state Natural Resources Department (or similar department) for fence plans and guidelines. Here are four options ranging from the temporary plastic, to panels, to metal posts to wooden posts.

ELECTRIC BEAR FENCE

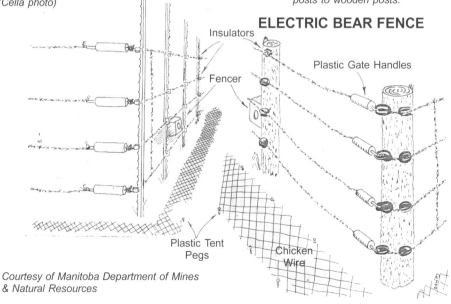

Courtesy of Manitoba Department of Mines & Natural Resources

Bear cages are often moved to a location where troublesome bear are known to wander. Baited with a variety of tempting treats – donuts, bacon and the like, the theory is that a bear will wander in and trip the door. Captured bears are moved, or dispatched.

they defend. The bear thus escapes many stings. However, it is obvious that in the process of feeding on colonies bears receive many stings. Yet bears like brood and honey so much that even stings in the mouth and throat do not deter their feeding.

To feed on the brood and honey, bears will often turn the boxes upside down and rake out the bottom bars with their claws. Bears may also pick up hive parts and throw them onto the ground to break them apart.

Some states and Canadian provinces compensate beekeepers for losses from bears. Certain states or provinces stipulate that they will not pay compensation for bear damage done in the same apiary a second year. Others will pay only when the bees are close to a house or other buildings. Bear populations are growing in many states, posing more problems to beekeepers. This is because of reduced hunting, and increased development, squeezing more bear into smaller areas.

Electric fences around apiaries offer some protection against bears, though it is difficult to discourage bears that have fed on honey and brood. They may dig under, jump over, or somehow attempt to destroy an electric fence. Beekeepers have hung various materials such as hair, dung, and odoriferous soap in apiaries to discourage bears, but without success. One experiment involved mixing an emetic, lithium chloride, with honey and brood to make the bears sick. However, it was found that when the bears recovered, the experience did not discourage them from feeding in bee hives again. - WMH

Bear damage can range from a single honey super on one colony to the destruction of every colony.

BECK, DR. BODOG F. (1871-1942) – Born in Budapest, Hungary, Dr. Beck graduated from The Royal Hungarian University in 1894 and studied in several prominent surgical clinics in Europe afterwards. In 1901 he moved to New York, New York. He advocated increased use of honey in the diet, and studied extensively the use of bee venom for the treatment of arthritis and rheumatism. His book *Bee Venom Therapy*, published in 1935 continues to be used, and his book *Honey and Health*, published in 1938 is also still used.

BEE BEARDS – Making a beard, using two or three or even 10 pounds (one to four kg) of bees has been a favorite trick of beekeepers for decades. Properly applied, a bee beard is easy to set up and the procedure is generally regarded as safe. There is little danger of being stung severely, however, being stung on the eyeball and other sensitive places around the face or ears is dangerous, and some people refuse to make bee beards. Everyone should be aware of this danger. It is also dangerous to walk or work in an apiary without a bee veil even when the bees are calm.

Before you begin, place cotton balls in the nostrils and ears of the beard wearer to keep bees from wandering into these cavities. Remove a shirt if it has a collar, bees can become trapped under it. Glasses, too, should be removed.

When building a bee beard it is important to remember three aspects of bee biology: First, well-fed (engorged) bees are not inclined to sting. Second, bees away from their brood and food are intent on remaining with their queen, and do not have a nest to defend. Third, temperature has a strong effect on bee behavior. One should never try to make a bee beard unless the temperature is favorable for bee flight and the sun is shining.

A simple way to make a bee beard is to find, remove and cage the queen from a colony. Next, move the colony away from its stand for at least a couple of hours, so the older field bees return to

Bee Beards

Probably the most notable bee beard presenter is Dr. Norm Gary, UC Davis, retired. His company, The B-Team, works with movies and TV as Bee Wranglers. Dr. Gary, here, is obviously playing his clarinet in the key of 'B'.

Stuff ears and nostrils with cotton. Put queen in cage under chin. Hold support board level and steady. Gradually add bees. Continue adding bees until all are on board. Assist bees in surrounding queen cage, and keep from straying. Gradually remove board so bees cascade down shirt.

When complete, stand quietly so as not to jostle the bees too much, turn and face the crowd. Smile. Lots of people are taking your picture. Find the requisite lady to kiss, and you're done. To remove bees, use a vacuum with bee cage attached. If you are to use these bees again, feed, feed, feed.

the original location, not the colony. Next it is important to force the bees to engorge by feeding with 2:1 sugar syrup. The bees, if allowed to feed uninhibited, will be engorged within an hour or two if the weather conditions are good.

To begin, the queen cage is tied under the chin. The bees are then gently shaken off the frame onto a platform held next to the chin. The bees will move from the platform to the queen in her cage since these are all nurse bees. The bees can be gently pushed off the platform toward the queen cage using a stiff card. The same can be used to move bees away from eyes.

To remove the beard, first remove the queen cage. It can be placed on the ground and the wearer jumps high and lands hard, jarring the bees loose so they fall on the ground, on the queen cage. The bees will again join the queen. Often bees are removed from the wearer using a bee-vacuum device, designed so bees are not injured in the process. This spares the exciting event of jumping and causing bees to fly.

Caution: Building a bee beard can be fun and has been done by many beekeepers. However, it is not a trick to be practiced by anyone who is not thoroughly familiar with bee behavior. The bees must be well fed by an expert who understands how to feed the bees and the effects of feeding.

Bee beards are often used as an attention-getting device at fairs and other public gatherings. This procedure, or any honey bee/human interaction for public display, is not recommended. The risk is high, and the rewards are few.

Basically, there are three schools of thought on this controversial subject.

One argument is that applying and wearing a bee beard at a public event demonstrates that bees are kind and gentle creatures, and should not be feared.

A second argument suggests that bee beards attract attention for the same reasons that people watch car races or boxing matches – will somebody have that spectacular accident? Will blood be spilled?

Finally, bee beards done for other beekeepers as the crowd are fine, since there's no death-wish in the crowd, and no promotional agenda by the wearer.

BEE BEHAVIOR – (see COMMUNICATION AMONG HONEY BEES – The dance language and LEARNING IN THE HONEY BEE)

BEE BOLES – In Europe, especially Great Britain and Ireland, it was popular in centuries past to place colonies of honey bees (usually skep-style hives) in indentations or alcoves in walls or the sides of houses and castles. These spaces presumably gave the bees and the fragile skeps protection against the weather and still kept them near their owner. Bees were sometimes used as offensive weapons and were thrown down over the sides of castle walls to drive off attackers. A colony kept in a protected spot was always ready should it be needed. The International Bee Research Association, with headquarters in England, has a registry of bee boles. Many studies of these unique structures have been made, and several books on location, design and styles have been published in the U.K.

BEE BREAD – This term is sometimes given to pollen that is stored in a comb. The origin of the term is not clear. (see POLLEN)

BEE, DEFINITION OF – Honey bees constitute a small genus of highly derived bees. They are no more representative of all bees than are humans representative of all mammals. Bees as a group are ubiquitous, often predominant wild-land pollinators. Their ancestors arose with the earliest flowering plants, in the late Cretaceous period, when dinosaurs still roamed. Hapless individuals entombed in amber from 40 million years ago are readily recognizable as bees. Today, 16,000 species of bees are known; many more remain undescribed. They have proliferated wherever flowers grow, from Arctic and alpine tundra to parched deserts to steamy rainforest.

Several hundred species of bees can

usually be found at most locales. Bees are distinct from nearly all other flower-visiting insects, however, for bees gather pollen, which they bring back to a nest to feed their grub-like progeny. Any flower visitor carrying a discrete external pollen load is a bee. Other diagnostic features, such as plumose (branched) body hairs, are adaptations for pollen collection. Bees, then, are vegetarian, powered by nectar sugars and built from the proteins, fats and minerals found in pollen. Data gathered in 2002 showed that $14.6 billion worth of U.S. agricultural crops were pollinated by honey bees. These crops included many fruits, seed crops like sunflower and alfalfa, numerous vegetables and almonds.

Sociality is unusual among bees outside of equatorial regions. In North America, perhaps 5% of our estimated 3-4,000 bee species are social, where sterile daughters largely forego reproduction in aiding their mother to build a colony. In North America, only the introduced Old World honey bee has perennial colonies. Nests of bumble bees and social sweat bees are founded in the spring by lone adult females that mated the preceding autumn. Among bees, only honey bees, bumble bees and our largest non-social species pack an authoritative sting. Except for those, bees are little or no threat.

Adult females of most bee species that grace our gardens and wild lands are fertile, each independently building and maintaining her own nest during her final few weeks of life. Thus, these bees are solitary, not social. Most fashion linear or simple branching networks of underground tunnels. At the surface, the only visible clue to the nest's existence will be a soil heap (tumulus) shaped like a delta, volcano or tiny mound, with a perfectly cylindrical hole. If bees occupy a soil bank, then only the hole is evident. Other species nest aerially in a pithy stem or a tunnel left in a log or branch by a wood-boring beetle. Some of these cavity-nesting solitary species are being developed and used as manageable pollinators for select fruit and seed crops.

Over the course of a day, a mother bee comes and goes on five to 40 foraging trips, returning home each time with pollen and nectar. From this larder she gradually molds a typically doughy pellet or pasty mass, although some species accumulate a much more soupy provision. When completed, she lays a single egg on this provision and seals off the cell. The emerging grub-like larva will dine for a few weeks in unlit solitude, having been given enough food to develop into an adult bee without further intervention by its mother.

One bee's work may be another bee's booty, however. Females of some bee species are nest parasites. The parasitic bees never gather pollen, and only seek nectar for their own sustenance. These often waspy-looking "cuckoo" bees skulk into the nests of host bee species, laying eggs in provisioned nest cells. Each egg hatches into an assassin larva that kills the host grub and proceeds to devour the cached provision of pollen and nectar.

Although most bee species are not social, individual ground-nesting females may nonetheless nest gregariously in populous "villages." In suburban lawns and gardens, these bee "villages" can become a noticeable array of soil volcanoes, eliciting dismay or wonder from homeowners. Being non-social, the bees in these aggregations lack the potent stings and venom of social bees. Because of their

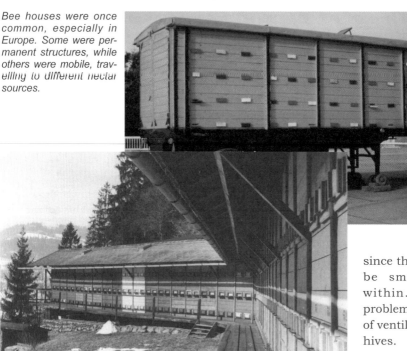

Bee houses were once common, especially in Europe. Some were permanent structures, while others were mobile, travelling to different nectar sources.

concentrated numbers, bees that nest gregariously can be important agricultural pollinators. - JHC

BEE ESCAPES – (see REMOVING BEES FROM SUPERS)

BEE HOUSES – In parts of Europe it is popular to keep several colonies of honey bees together in a small house. The colonies are kept in a structure with the entrances facing outward so that the bees always have free flight. The colonies are opened from within the building. Beekeepers report that they receive few stings in the building, since any flying bees go to openings in the house where they see light and escape.

Bee houses have several advantages. The hives are protected against the weather, and extra equipment may be kept in the building. One problem is that the inside of the building becomes smoky since the bees must be smoked from within. Another problem is the lack of ventilation for the hives.

Colonies are often stacked vertically, and supering is difficult or impossible. Often colonies are worked from the rear, and several combs must be removed and placed in a special rack designed to hold them to get to the brood nest. Since bees do not prosper in a horizontal nest, the colony populations are never too large.

Because colonies are close together, drifting may be a problem. For this reason colony entrances may be painted different colors so that the bees may distinguish between them. Bee houses and often wagons used for similar purposes are usually picturesque and are very much a part of rural scenery in many parts of Europe.

BEEKEEPER INDEMNITY PAYMENT PROGRAM – This program was passed by the U.S. Congress in 1970 as part of the Agricultural Act. Under this program, which was retroactive to 1967, beekeepers were paid for any losses of their bees due to pesticides. The program was administered by the USDA Agricultural Conservation and Stabilization Service. When it was signed into law the program was hailed by beekeepers as a major breakthrough in their efforts to force government and pesticide manufacturers and users to recognize that pesticides were being misused and causing widespread damage to bees. Upon its termination in 1980, nearly 40 million dollars had been paid to beekeepers, a sum exceeding all the money spent on bee research by the federal government during the same time period.

Beekeepers did not voice much dissatisfaction over the loss of the program because they recognized that a small number of beekeepers had misused it. They were also aware that bureaucratic problems arose occasionally with the administration of the program. Most important, they came to understand that paying a beekeeper for a loss caused by someone else did nothing to correct the basic problem, which was pesticide misuse. Under this program beekeepers, pesticide manufacturers, pesticide applicators, and government regulators were all excused from taking steps to correct any aspect of the problem. The colonies of bees that suffered losses usually recovered, or were easily replaced, and while they did not produce honey in the year or season they were abused, they were able to do so the following year. Because affected colonies were weakened and did not store surplus honey, pesticide contamination of honey did not occur. This saved everyone involved much embarrassment but only masked the severity of the problem.

During the 1970s, the Environmental Protection Agency, together with several other state and federal groups, worked toward better control of the manufacture, distribution and application of pesticides. Pesticide applicators were required to be certified when using the more toxic materials. Warning labels were required on all pesticide containers so that the public could be aware of any potential problems. The public in general became more concerned about proper use of chemicals. Thus, by the time the indemnity program was discontinued, other measures were in place to give beekeepers protection against losses. Especially important is the understanding that the applicator has a strong responsibility to use pesticides in such a way that they will not adversely affect others. At the same time, beekeepers learned that they should take steps to move their bees away from areas where losses might occur. While pesticide-related losses of honey bees still occur in the U. S., they are much less common today than they once were. RAM

BEEKEEPING IN ANTIQUITY – Humans have had a long association with the honey bee. The importance of honey and the honey bee to ancient peoples becomes clearer when we realize that sugar cane and sugar beets, our chief

Beehives made from sections of trees. From a museum in Europe.

sources of natural sweets today, are quite recent discoveries. Sugar cane originally grew only in the South Pacific. The Chinese built the first sugar mill about 2,200 years ago, but sugar cane was not introduced to the Mediterranean area from China and India until about 1,200 years ago. Sugar beets were first selected and cultivated in the 1800s. Before the discovery of sugar cane honey was the chief sweet.

Precise data on sugar production and consumption, which may be different due to wastage and use of sugar in animal feeding, are difficult to find. Annual sugar consumption for England, the country for which we have the best data, was only about four pounds (2 kg) per person in 1700. By 1800 this had risen to 18 pounds (40 kg) and was 85 pounds (190 kg) in 1900. In the United States today, about 140 pounds (300 kg) are produced per person and consumption is somewhat less, about 120 pounds (260 kg) per person. If people in the past had as much of a sweet tooth as today, the importance of honey to our ancestors becomes much clearer.

The earliest records of our association with the honey bee are found in rock paintings in Spain, which are 6,000 to 8,000 years old. One in particular depicts a person collecting honey combs from a cave on the face of a cliff in eastern Spain. Although this painting is commonly reproduced in books, ancient rock paintings exist in other parts of the world, especially southern Africa. These are cited and discussed more fully in Crane's book, listed below.

The Egyptians were the world's first practical, extensive beekeepers though we have little direct evidence of what took place in their civilization. Only four illustrations, one from a temple and three from tombs dating from 2400 to 600 B. C., tell us about beekeeping in that country. It is known that the honey bee was the symbol of lower Egypt and was later used to document, together with the symbol of upper Egypt, that upper and lower Egypt were one country.

Only a few Egyptian papyri mention bees and beekeeping, but one, produced in 256 B. C., tells of a beekeeper with 5000 hives. This suggests that beekeeping was practiced extensively. We also know that honey was an ingredient in over 500 Egyptian medicines. Beeswax and propolis were also important to these peoples. The best account of medicine in ancient Egypt is by Majno (Majno, G. *The Healing Hand, Man and Wound in the Ancient World.* Harvard University Press, Cambridge. 571 pages. 1975.)

In Egypt today modern beekeeping is practiced with Langstroth hives. The cylindrical mud hives that strongly resemble those depicted in one of the ancient tombs are rarely found today. An Englishman working in Egypt wrote a great deal in 1929 about how colonies were divided and new ones made. He also recorded the quantity of honey produced per colony. While yields were low, beekeepers would often keep several hundred colonies in an apiary, and collectively they would produce a large

Early skeps, and skep-like hives from a museum in Europe.

quantity of honey. What is known about mud hive beekeeping in Egypt is discussed in an article (Morse, R.A. "Beekeeping in Egypt." *Gleanings in Bee Culture* 112:497-499. 1984.) (see BEEKEEPING IN VARIOUS PARTS OF THE WORLD - Egypt)

The Greeks and the Romans, whose civilizations followed the Egyptians, practiced beekeeping also. The lands around the present capital cities of both countries are good honey-producing areas today, and presumably for at least the past 3,000 years. People in both of these civilizations were more interested in warfare than in science and agriculture, and thus neither of these latter two areas of study was given much serious consideration. Unfortunately what records and books did exist in these two countries were mostly destroyed by the Christians, who did away with all of the libraries and books in the areas they conquered or where they were present. The manuscripts we do have from these early people survived in Persia (Iran) and further east and were brought back to Europe much later in the Middle Ages.

It is very curious that the beekeepers in these ancient civilizations used horizontal hives almost exclusively. We now know, from our studies of natural nests, that honey bees prefer vertical nests and that more honey is produced in such nests.

Records of forest and skep beekeeping in northern Europe from about 2,000 to 3,000 years ago to the present are more extensive. Again, it must be remembered that honey and a very few fruits were the chief sweets available to humans in these areas until one or two hundred years ago. Many books and articles have been written on beekeeping in ancient times, but two are most complete and worthy of further study. (Ransome, H. *The Sacred Bee in Ancient Times and Folklore.* George Allen and Unwin, London. 320 pages. l937 and the reprint by Dover Publications, 2004; Crane, E. *The Archaeology of Beekeeping.* Duckworth, London. 360 pages. 1983. Crane, E. *The World History of Beekeeping and Honey Hunting.* Duckworth, London. 682 pages. 1999).

BEEKEEPING IN VARIOUS PARTS OF THE WORLD – In this section two questions arise: what are the major honey-producing areas in the world? And, what countries have the greatest influence on the world price of honey? Changes are slowly taking place in answer to both of these questions. China entered the international honey market in the early 1960s and today produces the most honey of any country. Countries that are currently considered the major honey producers are: China, the United States of America, Argentina, Mexico, Australia and Canada. New countries are constantly emerging in the international honey markets. To name a few these include Thailand, India, Uruguay, Chile, Brazil and Vietnam.

A second major change that is taking place slowly is that some areas are abandoning agricultural land and allowing it to grow first into brush and then to trees or for growth of cities and suburbs. At the same time forests are being destroyed in many places, such as South America. All these changes can have a strong effect on the honey plants that grow in an area.

Africa – Africa has several subspecies of honey bees and diverse climates. Many African countries have a long history of successful beekeeping on an extensive scale. Egypt is one of these and the one about which we know the most in antiquity.

The area called East Africa, including the countries of Ethiopia, Somalia, Kenya, Tanzania and Uganda, is of special interest. In this area, whole villages may devote themselves to beekeeping. It is not unreasonable to believe that the Egyptians, who were trading along the coast of East Africa thousands of years ago were also bartering for beeswax and perhaps honey with the ancestors of those living there today. The beeswax that is harvested in East Africa is usually of good quality since much of it comes from virgin (new) comb.

Often, the primary use for the honey in East Africa is to make honey beer, which is a local drink with considerable tradition. The methods for making the beer are simple: honey, including comb

Log hives in Africa are suspended high in trees to protect them from predators.

and sometimes brood, is mixed with grain and water. The mixture is placed in open-headed barrels that have been used before and therefore contain a good yeast culture. In the warm climate of East Africa the fermentation is rapid and the brew is ready and sold after three or four days. This is too short a period of time for bacteria to convert the alcohol to vinegar and the beer contains about four to six percent alcohol that is not too different in alcoholic content of many beers.

The traditional beehive in East Africa is a hollow log though some are made of bark. Like their neighbors the Egyptians, East African beekeepers hang their hives horizontally. However, hives now in the area are of greater diameter than those formerly used in Egypt and thus honey production per colony is greater. At present there is an effort in East Africa to use top bar hives in place of log hives but these too are not a good design for honey production. Many countries are changing to the Langstroth hive design and using supers for honey. (See HIVES, TYPES OF).

The traditional log hives are hung in trees where they are shaded, but visible. The hives are sometimes fairly high above the ground, three to five yards (three to five meters), heights that bees prefer. Under these circumstances the hives attract swarms. The beekeepers know when the honey flows occur. They harvest the honey at night using smoke from burning brands. Harvesting the horizontal hive has an advantage. Bees will build their brood nest near the end of the hive where there is an entrance. When smoke is applied to the opposite end the bees are easily driven off of the combs of honey that are stored separately behind the brood. Beekeepers will frequently over-harvest since they are not worried about the bees starving, so some brood may be found in the combs. Colonies will often abscond, that is abandon their nest, during a dearth of nectar; however, since swarms will return the following year this is not of concern to the beekeepers. Beekeeping is also carried on in West Africa but much less has been written about it both historically as well as recently. In South Africa both traditional beekeeping and a well-developed commercial industry are found. Many of the beekeepers in that country are migratory. However, the industry is struggling with dealing with *Apis mellifera capensis*, the Cape bee. This race of honey bee invades a *A.m. scutellata* colony and destroys the queen. The colony eventually dies. (See RACES OF HONEY BEE, CAPE BEES)

Diseases of honey bees have been little studied in Africa. It is noteworthy that those who have visited various African countries and who have examined colonies do not report any serious disease problems. African honey bees appear to be resistant to many of the common diseases found in Europe; however colonies are not moved long distances around the continent as they are in North America, South America or parts of Asia. (See also BEEKEEPING IN VARIOUS PARTS OF THE WORLD-South Africa)

Argentina – Some of the world's finest light, mild table honey is produced in Argentina principally from thistle, clover, alfalfa and soybean. This high quality honey is much sought after by honey packers around the world. Argentina is one of the world's major exporters of honey.

Asia, Southeast – Southeast Asia is defined as including the countries of

Apis dorsata. (Burgett photo)

Bangladesh, Myanmar (Burma), Thailand, Laos, Cambodia, Vietnam, Indonesia and Malaysia. This area possesses the greatest diversity of honey bee species anywhere in the world. Several species of true honey bees in the genus *Apis,* are found living there today and this includes the recently introduced European or western honey bee *Apis mellifera*. With such a diversity of honey bee species, wide variations in beekeeping exist in Southeast Asia. (See *APIS,* DISTRIBUTION AND SPECIES)

Beekeeping as practiced today in Southeast Asia, includes some thoroughly modern and highly productive operations that solely depend on the introduced western honey bee, *A. mellifera*. On the other end of this spectrum are the primitive honey-hunting practices found throughout Southeast Asia that target giant honey bees (*Apis dorsata* and *A. laboriosa*) and the dwarf honey bees (*Apis florea* and *A. andreniformis*). Also seen is what could be termed "backyard" beekeeping with the traditional hive bee of Asia, *Apis cerana*. Each of these approaches to beekeeping will be discussed separately.

Apis mellifera, the western honey bee, is a relatively recent introduction to Southeast Asia. Over the past century numerous small-scale introductions by European colonialists have attempted to bring the familiar *A. mellifera* to tropical Southeast Asia. Several larger scale programs have been tried and have met with few successes and many failures. It is important to remember that European *A. mellifera* is ecologically adapted to the temperate climates of Europe so bringing it to Southeast Asia,

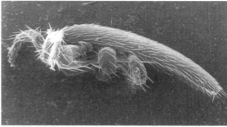
Young female T. clareae moving freely on comb.

Mated female T. clareae on brood.

with its sub-tropical to classically tropical climates, is fraught with difficulties.

The most successful use of *A. mellifera* in the region has been in Thailand, where approximately 150,000 colonies of the western honey bee are kept by some 300 commercial beekeepers, primarily located in northern Thailand. A critical contribution to this success is a reliable nectar source. The plant most responsible is the tropical fruit longan (*Dimocarpus longan*), which is grown in large monocultures throughout northern

Thailand. Other attempts at placing *A. mellifera* in the region have not met with the same success seen in Thailand. In the early 1980s the U. N. Food and Agriculture Organization (FAO) sponsored a major program in Myanmar (Burma) that utilized *A. mellifera*. Today there is little carryover of this program. Another FAO-sponsored effort was in Indonesia, which utilized not only *A. mellifera*, but included a component that concentrated on the native hive bee, *A. cerana*. This program is characterized as having limited success for utilizing *A. mellifera*. Vietnam is another country that claims success with *A. mellifera*. The Vietnamese program has been carried out with little international aid assistance. In recent years both Thailand and Vietnam have become exporters of *A. mellifera* honey to world markets.

Apis mellifera beekeeping in the region experiences major problems with parasites, especially *Tropilaelaps clareae*, the brood mite of the giant honey bee, *A. dorsata*. *T. clareae* readily accepts *A. mellifera* as an alternative host bee. *T. clareae* is more pathogenic to the western honey bee than *Varroa destructor*. Frequent use of miticides is necessary to maintain productive colonies of European honey bees throughout S. E. Asia.

Honey hunting – Long before the limited historical beekeeping records for Southeast Asia were kept, humans have "hunted" the nests of the native honey bees. To this day bee hunters roam everywhere in the region seeking honey, wax and brood from the indigenous honey bee species. The primary target species are giant honey bees and the dwarf or small honey bees. Individuals normally specialize in one of the two bee "types."

Most spectacular and hazardous is the hunting of *A. dorsata* nests. The giant honey bees occur throughout Southeast Asia and can often account for a major portion of honey produced in a given region. In addition to honey, the bee hunters are also interested in the wax and very often the brood as a human food. In Thailand extensive networks of bee hunters, transitory middlemen and retail vendors exist that bring giant honey bee products to local and regional markets. Such is the case in other areas of Southeast Asia where there are concentrated cadres of giant honey bee hunters.

Giant honey bee hunters are specialists and the tradition is normally generational within bee-hunter groups. It requires skill and knowledge of the migratory habits of these magnificent animals. Limited management is practiced in a few areas in the form of "rafter" beekeeping where artificial nest substrates, such as tree limbs, are

Traditional A. cerana *hive.* (Burgett photo)

placed throughout areas where giant honey bees are known to be seasonally found. Rafter beekeeping is practiced in Vietnam, Malaysia and Indonesia. This is normally practiced in river delta regions where naturally occurring tall trees are at a premium and the bees will normally nest relatively close to the ground.

Collection of the nest of the small honey bees (*A. florea* & *A. andreniformis*) does not require the skills of giant honey bee predation. The small honey bee normally constructs a single-comb, exposed nest close to the ground. The defensive capabilities of these, the smallest of all honey bee species, are minimal, especially when compared to other indigenous bee species. One needs only to grasp the small branch on which the colony is attached and give a vigorous shake to dislodge the bees. The entire nest with honey and brood but without the adult bees is usually the marketed unit.

Apis cerana is the closest relative to our familiar *A. mellifera*. It is a multiple comb, cavity-nesting species, intensively cultivated in Southeast Asia for probably as long as *A. mellifera* in European and Mediterranean cultures. *A. cerana* is also known as the eastern, or Indian honey bee.

There are many types of beekeeping with the eastern honey bee, from the traditional "hunting" of feral colonies, to management in constructed hives. Most often *A. cerana* beekeepers maintain what would be called "backyard" operations, *e.g.*, a few colonies kept in either log hives, or any of a wide variety of constructed hives. These range from fixed comb units to moveable frame equipment.

Honey production from the eastern honey normally is no more than five to ten kilograms (12 to 22 pounds) of honey each year. The reasons for the low production compared to the western honey bee are many but a major feature of *A. cerana* behavior is its natural propensity to migrate (seasonally abscond). Historically, honey from the eastern honey bee has sold for a relatively high price in local economies and its consumption has for medicinal purposes rather than as a sweetener. Prior to the arrival of the western honey bee in Southeast Asia, honey from the eastern honey bee accounted for the majority of honey production throughout the region.

For some of Southeast Asia *A. cerana* represents the "standard" form of beekeeping as defined by the management of a bee species for honey production. It is being replaced by the western honey bee in many areas.

Australia – Australia is one of the top ten honey producers in the world and a major exporter of honey. As of 1998 there were approximately 672,557 colonies, 466,684 of which were managed by beekeepers classed as commercial producers owning a minimum of 200 colonies each. There are also several thousand hobbyists keeping bees in Australia as well. The national average for yearly honey production across Australia has varied only slightly, remaining fairly constant over the last two decades. For 1998/99, the average production per productive colony was 144.1 pounds (65.5 kilograms). Many commercial operators achieve an average production of 220 pounds (100 kilograms) and some, an average of 286 pounds (130 kilograms). In Western Australia, annual honey production levels of 440 pounds (200 kilograms) per hive are not uncommon. An estimate of total production of 32,675 tons for 1998/99 was based on an annual average production of 154 pounds (70 kilograms) per hive (see Gibbs, D. M. H. and Muirhead I. F. 1998. The economic value and environmental impact of the Australian beekeeping industry. A report prepared for the Australian Beekeeping Industry). The estimated production of beeswax for that year was 545 tons based on the assumption of 2.2 pounds (one kilogram) of wax for every 132 pounds (60 kilograms) of honey produced.

Australia presently exports about one-half of its honey production. The major importers of Australian honey are the United Kingdom and Germany. Other significant importers include Japan, Spain, Portugal, Singapore, Malaysia, and Saudi Arabia. About half of the beeswax produced in Australia is exported.

Many species of Eucalyptus produce honey crops in Australia. This is E. globlus.

Beekeepers in Australia are regulated by their state government departments of agriculture, which provide information and advice, register beekeepers, oversee apiary inspection programs, and provide health certificates for movement of colonies across state boundaries. The Australian Honey Bee Industry Council represents the interests of industry at the Federal level. Funding for the Council is generated by a voluntary levy on honey sold by commercial and sideline beekeepers.

The primary sources of nectar in most of Australia are various species in the genus *Eucalyptus*. Many of the trees that make up this genus are prolific producers of nectar and yields of up to 440 pounds (200 kilograms) per colony have been recorded. Many of the species produce significant amounts of nectar only after reaching maturity at 20 to 25 years of age and then often yield copious nectar at intervals of two, three, five or more years. This seasonal variability has contributed to the migratory nature of Australia's commercial beekeepers. The color of the honey from the various species of *Eucalyptus* ranges from dark amber for some "stringy bark" species to almost water white for red gum or yellow box. Each type of *Eucalyptus* honey also has its own distinctive flavor with some honeys pleasant and mild-flavored and others more strongly flavored. Much of the stronger-flavored honey produced may be blended with milder tasting honey, although a number of honeys are marketed as true to floral type. Some have strong medicinal properties, and are being produced, and sold with that market in mind. Bees do not collect pollen from several species of *Eucalyptus*, such as yellow box and red iron bark. If other pollens are unavailable colonies may be fed supplementary protein to ensure ongoing vigor.

Other sources of honey include a number of native plant species as well as several that are introduced. The various clover species sometimes give good yields but only in years with sufficient spring rainfall. Alfalfa is also an important nectar source in certain areas. Other plants that are important either for build-up of colonies or for harvest of honey include canola, salvation Jane, cape weed, blackberry, banksia, and tea tree.

The major areas of honey production in Australia are located away from the hot arid center. These areas include the mountains and eastern coastal plain of Queensland and New South Wales, the southern coastal plain of New South Wales, Victoria, and South Australia, the southwest coast of Western Australia, and the western half of Tasmania. The leading state for honey production is New South Wales accounting for about 45 percent of the current total honey production.

The first honey bees were successfully introduced to Australia around 1822; in Tasmania, probably in 1832 from England. These were German black bees that yielded only meager returns of honey and were prone to severe infestation by wax moths. Italian bees first made their appearance in Australia in 1862 but did not survive long. In 1880 and again in 1883, new introductions of Italian bees were made and this time the bees flourished. The hardiness of the new Italian bee and the adoption of modern apicultural practices and equipment, such as the Langstroth hive, soon made beekeeping in Australia a profitable enterprise and allowed for commercialization. Further introductions of other bee strains such as Carniolans and Caucasians have been made from time to time, but overall the Italian bees appear to be the most popular. Today, introductions of breeding stock occur from time to time from approved countries, subject to strict quarantine regulations that require all imports to enter through a special honey bee quarantine facility in New South Wales.

In the early 1900s, commercial apiaries in Australia rarely contained more than several hundred colonies. Today some apiarists manage 2000 or more colonies. Commercial beekeepers in Australia practice migratory beekeeping and while a few beekeepers may move their colonies as much as 1000 miles (1600 kilometers), or more, most move their colonies no more than 200–300 (320-480 km) miles. The bees are moved in order to place them in the best nectar- and pollen-producing areas and to take advantage of the differences in peak flowering times in different areas. This practice helped an Australian beekeeper to set the still-standing world record for honey production during the 1953–54 season with an average of 784 pounds (356 kg) per colony for 450 colonies.

Although blessed with bountiful nectar plants, Australia is burdened with many of the same bee diseases as the rest of the world. American foulbrood (AFB), European foulbrood (EFB), chalkbrood, sacbrood and nosema are all present and can cause problems. As elsewhere in the world, American foulbrood is the most serious bee disease in Australia. State apiary inspection programs in partnership with apiarists are responsible for inspecting colonies for AFB. The use of antibiotics as a preventive or treatment of AFB is not permitted in mainland states but under some circumstances may be permitted in Tasmania. When infected colonies are found the bees plus infected combs and other hive components are destroyed by fire. Sterilization by gamma-irradiation of sound hive components including empty combs, but not combs containing brood or honey, is in some cases a more economical alternative to destruction by fire.

In 2005 the U.S. regulations on importing bees were relaxed, and for the first time in over 80 years bees were allowed into this country. The first major shipments of packages and queens were brought into the U.S. from Australia to supplement almond pollination in the spring of that year.

It is expected that this practice will continue since packages can arrive in the U.S. in the late fall, build up during the winter months in the southern U.S., be divided, and be ready for pollination duty in the spring. This "double duty" goes a long way toward paying for the high cost of these packages (double what a U.S. package cost at the same time).– RDG & RAM

Brazil – The largest country in South America, Brazil has gained notoriety in the beekeeping world because Professor Warwick E. Kerr brought bees from Africa into that country in 1956. Kerr knew that bees from Africa were good honey producers and pollinators. European honey bees, which evolved in a more temperate climate, were only moderately successful in Central and Southern Brazil. There was almost no beekeeping and no beekeeping industry in northern, tropical Brazil. The African bees have done what was expected, the number of colonies and honey production have increased throughout the country after a temporary decrease. Today there are about 300,000 beekeepers and 2,500,000 managed colonies and exports are increasing annually.

Canada – For a great number of years Canada has been one of the major honey exporting countries. Apimondia held its convention in Vancouver, British Columbia, in September 1999. Canada traditionally ranks in the top ten honey producers and today its 13,000 beekeepers produce about 33,000 tons of honey. While there are many important honey-producing areas across the country, the best known is the Peace River district that stretches across the northern portion of the prairie provinces of Saskatchewan and Alberta. At one time beekeepers in the area used package bees from California to grow new colonies each year because the bees in the producing colonies were killed at the end of each season. In recent years it has been found that with careful feeding of sugar syrup and appropriate winter packing it is possible to winter bees in the area, and beekeepers are doing so.

When tracheal mites were found in the U.S., the Canadian border closed and U.S. bees were not allowed into the country. That ban was lifted in 2004 and honey bees can again be imported into Canada from the U.S. Many plane loads of bees were flown into Canada from New Zealand and Australia with varying degrees of success during the border closure with the U.S. Though initially strong, these imports diminished over the years as overwintering and queen production increased.

China – China has about seven million colonies, five million are introduced European honey bees, *A. mellifera,* and the remaining two million colonies are the native Asian honey bees, *A. cerana.* The average honey production per year is about 60 lb (27 kg) per colony for *A. mellifera* colonies and 15–20 lb (seven to 10 kg) for *A. cerana* colonies. Together, the European and Asian honey bees in China produce an average of 420 million lb (190 million kg) of honey per year, which is approximately one-fifth of the total world production.

A. cerana colonies are mostly managed in movable frame hives, that are about 20% smaller than the standard Langstroth hive, but in South China some beekeepers still use some traditional hives with non-movable frames. The honey yield of *A. cerana* is considerably less than that of *A. millifera* due to smaller colony size and a higher tendency for swarming and absconding.

China is also the world leader in the production of royal jelly (two million lb [1 million kg] yearly), bee collected pollen (1.6 million lb [0.7 million kg]), and beeswax (6 million lb [2.7 million kg]). Royal jelly production is especially attractive in China because of the relatively low labor costs, larger local demand for the product (currently royal jelly is priced up to 100 times more than honey). In addition, bees have been bred specifically for increased royal jelly production. France now imports these bees for royal jelly production to satisfy their local market.

China probably leads the world in breeding programs with *A. cerana.* Bee breeders in China are using artificial insemination and selection methods developed for *A. mellifera* and applying them to *A. cerana.*

In addition to *A. cerana,* China has four other native species of bees: the giant honey bees *A. dorsata* and *A. laboriosa*; the dwarf honey bees *A. florea* and *A. andreniformis.* The latter four species are not managed in man-made hives but mostly hunted by man for their honey.

Migratory beekeeping has become more popular in the last decade with development of an improved highway system and the beginning of privately-owned transportation. Traditionally beekeepers used trains to transport their beehives. Unfortunately trains lacked flexibility in scheduling and routes and created some public relations problems. Since the bees and passengers had to share the same railroad cars there were problems with escaping bees. Train travel was sometimes catastrophic for bees from overheating of the hives.

In China, *Varroa destructor* is referred to as the "greater mite", and *Tropilaelaps clareae,* the "lesser mite," even though the latter mite is more damaging to the bees in China. Just as the host range of *V. destructor* extended from only *A. cerana* to *A. mellifera, T. clareae* also expanded its host range from *A. dorsata* to *A. mellifera.* In doing so, *T. clareae* causes considerable damage to the five million *A. mellifera* colonies.

Even though T. *clareae* has been associated with *A. mellifera* since 1960, the mite is more damaging to bees than *Varroa* in temperate and subtropical areas. Unlike *Varroa*, *T. clareae* cannot survive for more than two days on adult workers. Therefore its distribution range is limited to areas where the average monthly temperature is higher than 34°F (1°C). Because the winter survival of *T. clareae* depends on a supply of continuous brood-rearing, a relatively easy way to curtail its population is to break the brood cycle by temporarily removing the queen or restricting her egg laying to a particular comb and then removing the comb with the eggs.

The earliest record of *Varroa* in *A. mellifera* colonies in China was 1956. In spite of the relatively long period of coexistence between bees and mites, China yet has to discover mite-resistant bee strains, either by natural selection or man-assisted selection.

In China, the sacbrood virus of *A. cerana* (later termed Thai sacbrood virus) was the cause of a serious disease which killed nearly one million cerana colonies in China in 1972. The cerana population has rebounded and is now seemingly coping well with this disease unlike the experience with the mites.

The poorest developed area of beekeeping knowledge in China is probably pollination. Many people mistakenly believe that honey is made from fermentation of bee-collected pollen and therefore farmers assume that bees are taking things "away" from the plants. As recent as the last two to three years, some growers still use "human-bees" to hand-pollinate apple blossoms. Growers charge beekeepers a fee when good honey is produced from the plants (such as litchi) even though bees provide vital pollination service for the crop. - ZYH

Egypt, traditional beekeeping in – Civilization began in Egypt. It was in this country that the first cities and city-states were organized, that mathematics and astronomy had their beginnings. In agriculture the first animal selection and breeding programs, at least on a large scale, started here. The first practical beekeeping methods began here too, probably about 8000 to 10,000 years ago. In recent years there has been such a strong effort to introduce the Langstroth hive, wax foundation and other modern beekeeping techniques that traditional beekeeping has been overlooked. Historically, keeping bees in long cylindrical mud hives was a major advance over harvesting honey and wax from colonies in hollow trees and caves. A very few traditional beekeepers still exist in Egypt and it is from these men that we learn their techniques and the management cycle of the year.

The traditional Egyptian bee hives served this early civilization very well. As history shows, early Egypt had an abundance of honey. However, the low production from these hives is also a lesson for those who would design new hives. Horizontal top bar hives recommended in many parts of Africa have the same problems as the mud hive and are not practical for producing large quantities of honey.

Europe – Most countries in Europe import honey. A few European countries, especially Germany, import honey that is repacked and exported again. There are few commercial beekeepers and honey producers in most European countries, though in certain parts of that continent there are some unusual honey plants that are good nectar producers. On average the honey produced in Europe is darker and stronger than the table honey used in the U.S.

In general terms there is less honey produced in Europe because of a higher human population per unit of land than in the U.S. Therefore, the land is more intensively cultivated and managed so there are almost no weed plants on which bees might feed. In the U.S. quite the reverse is true and it is estimated that at least half of our honey is produced from plants generally considered weeds or wild. Hobby beekeeping is widely practiced in Europe. Possibly parts of Europe have too many colonies of bees which may adversely affect honey production.

Some of the centers for honey production follow: heather is an important honey plant in many parts of Britain, northern Germany and Poland. Forest honey (honeydew) is often produced in quantity and is much sought after in parts of Germany and Switzerland. Climate has a strong effect on the populations of aphids that produce the honeydew and as a result, production varies greatly from one year to the next. In France, and certain other countries, a number of mints and plants grown for perfume are excellent nectar producers. Northern Italy has been known for many decades as an excellent queen-rearing area and some honey is produced there. Certain parts of Greece and the Greek Islands have been known, even thousands of years ago, as excellent honey producing areas; thyme is especially abundant and an excellent producer of nectar. Around the Mediterranean Sea, several countries, most notably Spain and Israel, produce large quantities of citrus honey that is often exported. Several of the Balkan countries are noted for certain specialty honeys; acacia honey from Hungary is especially well known. To the east, Russia is a major honey-producing country.

Japan – The first description of honey bees in Japan appeared about 1400 years ago, telling of a general who tried to keep bees at a certain holy place but failed. This colony seemed to be the native Japanese honey bee, *Apis cerana japonica,* which is distributed in most of Japan except Hokkaido. They were kept in boxes or hollowed logs. Modern beekeeping in Japan started in 1877 when European honey bees in movable-frame hives were introduced.

The number of bee colonies registered in Japan as of January, 2002, is 175,281 kept by 4,796 beekeepers. In addition 92,853 colonies are used for greenhouse pollination (mainly for strawberries) and 33,154 colonies were used for non-greenhouse pollination.

Giant Hornets are the main parasite and predator of Japan's honey bees.

These are unregistered *Apis mellifera* and *Apis cerana japonica* colonies kept by hobby beekeepers.

Domestic honey production had peaked (ca. 8500 tons) in the 1960s, and then decreased gradually to 2687 tons in 2001. On the other hand, honey importation increased to 40,188 tons in 2001, 94% of the total national consumption. Honey comes mainly from China (92%) and Argentina (4%), and the rest from 28 countries. Since Japanese people prefer honey of light color and mild taste, Chinese milk vetch (*Astragalus sinicus*) and black locust (*Robinia pseudoacacia*) honey are regarded as the best. The recent invasion of the alfalfa weevil, *Hypera postica* has severely reduced honey production from Chinese milk vetch in western Japan. Orange, clover, and horse chestnut honey are also popular in Japan.

Domestic royal jelly production in 2001 was four tons, whereas almost 550 tons were imported from China (94%). The popularity of royal jelly increased among Japanese consumers starting in the 1950s resulting in an increase in royal jelly importation and ultimately making Japan the world's highest consumer. In Japan the production of beeswax is 37 tons and pollen production is three tons. Recently propolis became popular as a health food (market price is close to honey and royal jelly) and is imported mainly from Brazil and China.

American and European foulbrood are designated by law as livestock contagious diseases and approximately 0.5% of the colonies are incinerated annually because of AFB. In October 1999 the Japanese government approved a new antibiotic, Apiten, active ingredient mirosamicin for the prevention of American foulbrood. In 1998, the law was amended according to the OIE Code to include four additional conditions *Varroa*, tracheal mite, chalkbrood and nosema. The tracheal mite has not been reported in Japan. *Varroa* and giant hornets are the main parasite and predator of honey bees in Japan. Japanese beekeepers treat hives with Danikoropar and Apistan once or twice a year to control *Varroa*.

The National Honeybee Research Laboratory is part of the National Institute of Livestock and Grassland Science, Department of Animal Breeding and Reproduction, Apiculture Laboratory. The Tamagawa University, Honeybee Science Research Center, acts as the information center. The quarterly journal "Honeybee Science" has been published by the Honeybee Research Center since 1980. The Japan Beekeeping Association is a nationwide organization that has 3400 member beekeepers, from beekeepers' associations in each prefecture.

Japan has a number of laws and regulations regarding beekeeping and the distribution of honey products. The most important is the Beekeeping Promotion Law established in 1955. This law provides for the registration of the number of colonies and their location, controls on moving colonies, protection of honey plants, labels for honey containers, etc. The Fair Trade Conference for Honey and Royal Jelly works to maintain the quality level of each product. The new standards of honey started in 2002, in accordance to CODEX ALIMENTARIUS. - TY

Mexico – Beekeeping in Mexico is an ancient and important industry from both the economic and the social standpoint. The Indian cultures that existed before the arrival of European settlers kept stingless bees to produce honey (see STINGLESS BEES). Honey was then considered a valuable merchandise used for trading and religious ceremonies. Mexico's annual production of honey and wax is worth more than 100 million U.S. dollars. Over 50,000 beekeepers are organized in about 120 beekeeping associations and cooperatives throughout the country. Most of them produce honey as their primary product. There are a few large operations in Mexico, but the majority of the beekeepers are peasants with low income that have found beekeeping an alternative to improve their living standard. Besides honey production, more than 150,000 colonies are rented every year to pollinate crops such as melons, pumpkins, cucumbers, apples, pears, and almonds. The annual value of the crops pollinated with commercial honey bee colonies is worth over two billion U.S. dollars. In addition to the above, about 400,000 queen honey bees are reared and sold by about 50 queen breeders.

Mexico's honey comes from different floral sources and regions, and therefore its quality and appearance varies. Light and dense honeys are produced in the high plateau and north of the country, whereas dark and more moist honeys are harvested in the coastal regions and the Yucatan peninsula. The Yucatan peninsula has several major honey plants and is the most productive region, yielding more than 30% of the country's honey. Over 95% of the Yucatan honey is exported to Germany, where it is blended with lighter-colored honeys.

Mexico became the largest exporter of honey in the world between the seventies and the early eighties. During those years Mexico was the fourth largest honey producing country in the world with over 67,000 tons of honey a year, of which more than 55,000 tons were exported. However, during the nineties, honey production and exports have decreased. Production has decreased by about 30%, whereas exports have decreased by more than 45%. Several factors including *Varroa* mites and low honey prices have contributed to this decline, but the major factor affecting the Mexican beekeeping industry is the presence of Africanized honey bees.

Africanized bees (see Africanized bees) arrived in Mexico in 1986 and were established in all states of the country by 1996. These bees have been less productive than European bees. A study conducted with more than 400 colonies (Uribe et al. 2002, *Veterinaria Mexico*) showed that colonies of African ancestry produced 35% less honey than colonies of European ancestry. Beekeepers complain that the honey yield per colony has decreased and they attribute this decrease in yield to swarming, absconding, and competition from feral colonies (feral colonies have increased in number since the arrival of Africanized bees). However, the high defensiveness of Africanized bees has been the most undesirable characteristic. This trait has resulted in the death of thousands of animals and more than 400 people. Beekeepers usually pay for the medical expenses and cover the costs of lost animals when it can be proven that their bees were responsible for an attack. Normally, the injured party contacts the beekeeper directly and they settle "out of court" without incurring additional legal costs.

Because of constant stinging incidents, many old and commercial beekeepers have quit keeping bees or have reduced their number of colonies as a result of difficulties associated with finding suitable sites to relocate their apiaries, higher production costs, and unpleasant changes in management compared to European bees. For example, Miel Carlota, a company that once managed more than 40,000 colonies went out of busines. Today they only pack honey purchased from other beekeepers but do not have a single hive to produce it. In contrast, many new, small-sized and hobby beekeepers are arising as a result of the higher availability of feral colonies and swarms. These new beekeepers never worked with European bees and have learned the new management techniques that reduce the risk of stinging incidents, have lower costs than commercial beekeepers (no extra labor and transportation expenses), sell their honey at retail prices, and are better prepared to deal with the problems associated with Africanized bees.

The presence of Africanized bees in Mexico has forced many changes in management and breeding practices that have increased the production costs of commercially-managed colonies by about 30%. These increased costs are due primarily to the following factors: 1) relocation of apiaries to more isolated areas (resulting in higher fuel costs and wear and tear on vehicles), 2) labor costs (each laborer works fewer hives per day because apiaries are spread over greater distances and because each colony requires more visits a year), 3) queen replacement costs and breeding programs. Before Africanization few beekeepers requeened regularly. Today, most beekeepers replace the queens of the most defensive colonies, and some larger beekeepers are investing in breeding programs with the aim of selecting more productive and less defensive bees, 4) other costs include better protective equipment and more sugar to feed the colonies in order to decrease colony losses due to absconding. Increased production costs are affecting larger beekeeping operations more adversely than smaller ones. As mentioned, small beekeeping operations are on the increase and this trend toward smaller businesses is expected to continue.

In addition to the above, the Mexican government is subsidizing small beekeepers with up to 50% of the amount of money required to purchase queens, nuclei, medicines, and beekeeping equipment. Moreover, several universities and research institutes are dedicating the effort of several scientists to study the defensive behavior of Africanized bees, and to try to find solutions and technology that help the Mexican beekeeping industry to recover. In summary, beekeeping in Mexico today is more expensive than it was before Africanization due to management and breeding changes that require a greater financial investment. However, beekeepers believe that these problems can be overcome with better technology, management, breeding, and, above all, better honey prices. - EG

New Zealand – New Zealand is a small nation, the size of the state of Colorado, in the southwest Pacific Ocean. New Zealand produces an average honey crop of 18,000 pounds (8200 kg) per year. Twenty to thirty percent of the crop is exported, much of it as comb honey or as retail packs of individual floral types. Average colony production is 64 pounds (29 kg) per year.

Most of the honey produced in New Zealand is from white clover and other pasture plants, although unique honeys are also harvested from native trees such as rewarewa, rata, and kamahi. Significant amounts of honeydew are produced from native beech forests, most of which is exported to Europe. The most famous honey produced in New Zealand is manuka honey, which is in demand for treating wounds and burns because of its unique antibacterial properties

New Zealand has a small queen and package bee industry and exports about 14,000 queen bees and 18,000 two-pound (1 kg) packages each year, mainly to Korea, Japan, Canada and Europe. Many beekeepers actively harvest propolis and pollen for use in the therapeutic industry. It is estimated that about 100,000 colonies are used for paid pollination services each year in the pip and stone fruit, berry fruit, kiwifruit, cucurbit and small seeds industries.

Currently there are 3,840 registered beekeepers in New Zealand keeping over 290,000 colonies of bees. The number of beekeepers has declined from 5,000 in April 2000, when *Varroa* was discovered, and is expected to decline further as the mite spreads throughout the country, though it was still isolated on the north island as late as 2005.

Honey bees were introduced into New Zealand from England in 1839 by Miss Bumby, a missionary's sister, who brought in two skeps of British black bees. Many other stocks of bees were imported over the years and beekeeping became a popular hobby. Italian queens were first imported in 1880 and have since become the predominant strain used in New Zealand. Isaac Hopkins, a pioneer in New Zealand beekeeping, first used the Langstroth beehive in 1874 and started commercial beekeeping in New Zealand. He helped to create the New Zealand Beekeepers' Association in 1884 and the first beekeeping journal in the southern hemisphere, "The New Zealand and Australian Bee Journal," in 1883.

Hopkins was a vocal proponent for legislation to help control American foulbrood, which had become a major problem. In 1905 he was appointed Government Apiarist and the following year the Apiaries Act was passed making the keeping of bees in box hives illegal. An active apiary inspection program was started in 1908 with the appointment of two apiary inspectors, one for each island. Currently, beekeeping is regulated by the industry itself under its Biosecurity (National American Foulbrood Pest Management Strategy) Order 1998 and the Biosecurity Act 1993. Under the Pest Management Strategy beekeepers are required to register all apiaries, demonstrate competency in recognizing and controlling American foulbrood (AFB), and destroy all colonies found with foulbrood. Drugs are not fed for controlling AFB in New Zealand.

Reporting American foulbrood is mandatory and the annual declared incidence is just under 4% of apiaries and 0.5% of hives. Infected colonies must be destroyed but infected hive parts can be sterilized by boiling in paraffin wax heated to 160°C (320°F) for 10 minutes.

Varroa was found in Auckland in April 2000, and has since spread through most of the North Island. Movement controls will remain in place to try and keep *Varroa* out of the South Island.

Other bee diseases present include chalkbrood, nosema, sacbrood and other common bee viruses. European foulbrood, tracheal mite (*Acarapis woodi*), Asian mite *(Tropilaelaps clareae)*, the African or Africanized honey bee, the Cape Bee, the Small Hive Beetle, and the wingless fly *(Braula)*, have not been reported in New Zealand.

The National Beekeepers' Association of New Zealand has been funded by compulsory levies since 1978, but is currently moving towards a voluntary association. Funding for the industry's AFB control program will likely continue using powers available under the Biosecurity Act of 1993. - MR

South Africa – The three races of *Apis mellifera* in South Africa are *A. m. scutellata*, *A. m. capensis* and *A .m. litorea*. *Apis mellifera scutellata* is the bee of commerce in most of South Africa. The African bees that were imported to Brazil were from *A. m. scutellata* from South Africa.

The Cape bee is most noted for its thelytokous parthenogenisis (see PARTHENOGENESIS). This trait has caused a serious problem in *A. m. scutellata* when worker bees from *A.m. capensis* invade colonies of *A. m. scutellata* in the North. The natural range of the Cape bee, *A. m. capensis,* is approximately the same as the fynbos vegetation of the Cape.

The small hive beetle, *Aethina tumida,* was first described in South Africa. The beetle is widespread in South Africa and very similar to the small hive beetle found in the U.S.

Approximately half of the honey crop in South Africa is from *Eucalyptus spp.* Another major source of nectar is Aloe which is also an important source of pollen. Other nectar sources are citrus, canola, alfalfa and sunflower.

South Africa has a rich tradition in beekeeping. The South African Beekeepers' Association was formed in 1907. In October 2001, The South African Federation of Beekeepers' Association hosted the Apimondia convention in Durban, South Africa. For further information on beekeeping in South Africa see M.F. Johannsmeier. *Beekeeping in South Africa.* Third Edition. (ARC-Plant Protection Research Institute. Pretoria, South Africa. 288 pages. 2001.)

BEEKEEPING JOURNALS – Soon after the discovery of bee space in 1851 an impressive number of books and journals devoted exclusively to bees and beekeeping began to appear. A few books on the subject had been published earlier in the U.S. and many had appeared in Europe. However, monthly and/or weekly bee journals were not as common. At least 100 journals appeared in the U.S. alone, some of which survived for only a few issues and others that ran for decades in the late 1800s. Many were also founded in Europe. Collecting these old journals has become a passion for a few hobby beekeepers.

Two bee journals, *American Bee Journal*, which was started in 1861 but discontinued during the Civil War, and *Gleanings in Bee Culture*, the first issue of which appeared in 1872, have survived to the present. The name *Gleanings in Bee Culture* was changed to *Bee Culture* in 1993. In 1972 *The Speedy Bee*, a bee trade journal, made its appearance. All three of the publications record the history and rise of commercial beekeeping in North America. Over the years, the journals have included several articles about beekeeping in various countries; these alone make them worthwhile reference items. These journals are often useful in studying the history of a management scheme or product. One may follow, for example, the appearance and impact of the parasitic mites, Africanized honey bees, and the small hive beetle on U.S.

beekeeping. In addition, one can follow the beginning and demise of the beekeepers' indemnity payment program that compensated beekeepers who lost bees because of pesticides; sugar rationing and the resulting increased interest in beekeeping during both of the two world wars, etc. Yearly indices are published in the December issue of the *American Bee Journal* and *Bee Culture*.

While these journals are devoted almost exclusively to beekeeping, their articles occasionally touch on other subjects. A.I. Root, for example, was the first to record, in *Gleanings in Bee Culture*, the first air flight by the Wright Bros. in Ohio in 1905. The issue is a collector's item for those interested in aviation.

Worldwide, the most important journal has been *Bee World*, founded in 1919. An index for the years 1919–1949 was published. Since 1950, *Bee World* has been published by the Bee Research Association, which later became the International Bee Research Association (see IBRA under ASSOCIATIONS FOR BEEKEEPERS). In addition, several journals devoted exclusively to research on honey bees and/or social insects are also being published. These include: *Journal of Apicultural Research, Insectes Sociaux, Honeybee Science,* and *Apidologie*.

Other journals from various countries include: *The New Zealand Beekeeper* – National Beekeeper's Association of NZ; *Bees For Development* – Information for beekeepers in remote areas; *The Beekeepers Quarterly* – Published by Northern Bee Books in the U.K., but is global in scope; *Manitoba Beekeeper* – published quarterly by the Manitoba Beekeepers' Association; *Honey Bee Science* – a Japanese Science and Research journal; *Mellifera* – published twice a year by the Development Foundation of Turkey; *Alberta Bee News* – published monthly by The Alberta Beekeepers Association, serving the interests of beekeepers since 1933; *Teknik Aricilik* – from Turkey; *L'Abeille De France* – the most important of the monthly publications in France, for all beekeepers from the amateurs to the professional; *Apiacta* – an international magazine of technical and economic information on beekeeping, quarterly issues in four versions – English, French, German and Spanish, published by

The first Gleanings In Bee Culture editor was A.I. Root from 1873 - 1921 (see next page for more).

Bee Culture Editors

Amos Ives Root	Jan 1873 - Jan 1921
Ernest Robert Root	Jan 1884 - Jan 1921
H.G. Howe (Managing Editor)	1918 - 1931
George Demuth	Jan 1921 - March 1934
E.R. Root	Mar 1934 - May 1934
Jack Deyell	May 1934 - 1962
Jack Happ	1962-1972
John Root	Oct 1972-1975
Larry Goltz	Jan 1975 - Dec 1983
Mark Brunner	Jan 1984 - Nov 1985
John Root	Nov 1985 - Apr 1986
Kim Flottum	Apr 1986 - present

E. R. Root, Editor of Gleanings In Bee Culture, with A.I. Root January 1884 - 1921 and again on his own from March, 1934 to May, 1934.

George Demuth, Editor of Gleanings in Bee Culture, from January 1921 to March 1934.

Jack Deyell, Editor of Gleanings in Bee Culture, *from May, 1934 to 1962.*

Jack Happ, Editor of Gleanings in Bee Culture, *from 1962 to 1972.*

Larry Goltz was the Gleanings In Bee Culture *editor from January 1975 to December 1983.*

John Root was Editor of Gleanings in Bee Culture *from October, 1972 to 1975 and again from November, 1985 to April 1986.*

Mark Brunner, Gleanings in Bee Culture *editor from January 1984 to November 1985*

The present day editor of Bee Culture *magazine, Kim Flottum has been on the job since April 1986*

American Bee Journal Editors

Samuel Wagner	1861-1872	M.G. Dadant	"	"
W.F. Clark	1872-1874	Roy Grout	"	"
W.F. Clark	1875-1878	Adelaide Fraser	"	"
E.S. Tupper	" "	G.H. Cale	1951-1956	
T.G. Newman	1879-1892	M.G. Dadant	"	"
George W. York	1892-1912	Roy Grout	"	"
C.P. Dadant	1912-1937	Adelaide Fraser	"	"
G.H. Cale	1938-1940	G.H. Cale	1957-1965	
F.C. Pellett	" "	M. G. Dadant	"	"
M.G. Dadant	" "	Roy Grout	"	"
G.H. Cale	1940-1945	G.H. Cale	1965-1966	
F.C. Pellett	" "	M.G. Dadant	"	"
M.G. Dadant	" "	Roy Grout	"	"
J.C. Dadant	" "	Vern Sisson	"	"
G.H. Cale	1945-1947	Vern Sisson	1966-1972	
F.C. Pellett	" "	Roy A. Grout	"	"
M.G. Dadant	" "	M.G. Dadant	"	"
J.D. Dadant	" "	Vern Sisson	1972-1974	
Roy Grout	" "	Joe Graham	1974-1975	
G.H. Cale	1948-1949	William Carlile	"	"
F.C. Pellett	" "	James Scheetz		
M.G. Dadant	" "	Joe Graham	1975-1977	
Roy Grout	" "	William Carlile	"	"
G.H. Cale	1950-1951	Joe Graham	1977-present	
F.C. Pellett	" "			

The American Bee Journal was originally printed in Philadelphia and later in Washington, DC. In 1872 it was moved to Chicago, and in 1912 moved to Hamilton, IL.

Apimondia; *The Australasian Beekeeper* – senior beekeeping journal for the Southern Hemisphere. Complete coverage of all beekeeping topics in one of the world's largest beekeeping countries. Published by Pender Beekeeping Supplies Pty., Australia; *The Australian Bee Journal* – caters to both amateur and commercial apiarists; *Bee Craft* – monthly magazine for beginners and experts alike covering all aspects of beekeeping in Great Britain and Ireland; *Hivelights* – National magazine of the Canadian Honey Council, published quarterly; *The Scottish Beekeeper* – monthly magazine of the Scottish Beekeeper's Association, international in appeal, Scottish in character; *South African Bee Journal* – the official organ of the S.A. Federation of Bee Farmers' Association, published bimonthly in English and Afrikaans, primarily devoted to the African and Cape bee.

BEE LINING – Bee lining is the term describing any method of locating a bee colony using foragers as guides. Most often one is in search of a feral colony with the goal of hiving it.

One first needs to find and collect or attract foraging bees. While historical accounts mention many different baits claimed to attract nectar foragers ranging from heated honey to human urine, none have proven reliable enough to be of use in scientific studies as a quick method of getting bees to visit feeding stations. A more reliable method is to look for foragers when they visit nectar or water sources.

Once foragers are located, one must recognize that foragers will not return to the hive until they have acquired a full load of nectar, pollen, or water. Water foragers tend to take on a full load of water in one visit, while nectar and pollen foragers must visit multiple flowers to acquire a full load.

A bee-lining box allows one to capture foragers and provide the bees with a ready supply of sugar syrup or honey, thereby assuring that bees released from the box will return directly to the hive upon release.

Techniques for locating the hive also vary, but the simplest methods involve the release of bees at various points, noting the flight direction of each bee when released. Upon take-off, bees invariably circle a few times in a six to eight-foot diameter to get their bearings. To avoid dizziness or neck strain, a length of string is used to release each bee from the box so that the circling can be observed from a distance.

Keeping track of the bee as it flies away can be a problem for those with less than perfect eyesight, prompting the use of day-glo chalk dust as a marker in the release chamber of the bee-lining box.

Examples of bee-lining boxes.

A gentle shaking of the box coats the bee with chalk dust, and after cleaning its wings and antennae, enough dust remains on the bee to make it easier to see against the sky.

"Triangulation" of hive locations by releasing bees from widely separated sites allows one to estimate the hive location based solely upon compass bearings of bee flight paths. "Linear" bee releases involve following each bee's flight path, releasing another bee at the point where the prior bee disappeared from sight.

Triangulation has the advantage of finding a general location with fewer captured bees, while a linear approach may be more useful when one is in the general area indicated by triangulation, but cannot see or hear the colony.

BEE LOUSE – (see *BRAULA COECA*)

BEE MILK – (see ROYAL JELLY)

BEE RESEARCH – (see RESEARCH IN APICULTURE)

BEE SPACE – This term describes the walking space, or 'being' space that is

This cross-section through a modern Langstroth moveable frame hive shows the concept of bee space, bee boxes and frames. It also shows where problems arise. Figure 1 shows a bottom board, and two boxes above it, with the inner cover above that. You can see the space – 1/4" - 3/8" (1 cm) – that separates the sides, tops and bottoms of the frames from the insides of the boxes, the top and the bottom.

Figure 2 shows the position a queen excluder has, and now you see why they so easily get clogged with burr comb. The space between the grid of the excluder and the bottom of the frame exceeds that necessary measure, and bees fill it with comb.

Figure 3 shows what will eventually end up a real mess. The dado in the bottom super is so deep a riser is required to keep the top of the top bar the necessary distance from the bottom of the bottom bar above it. But look at the space between the top box and the frame shoulder – too small, and it'll get filled with propolis. And that empty space between the riser and the edge of the box? Too small and it, too will get filled. Violate bee space and the bees are very unforgiving.

left between combs in a natural nest. Bee space is about 1/4 to 3/8 of an inch (six to 10 mm) wide. Bees never make holes through a comb to move from one comb to another, but they walk around the edges. In a natural nest the combs are usually attached to the sides of the cavity, but there are galleries, one "bee space" wide, around the edges through which they may pass. The natural combs built by temperate climate bees may be straight but are more often crooked.

The concept of using bee space in a hive was discovered in 1851 when the Reverend L.L. Langstroth, a hobby beekeeper, observed that if he left a space 1/4 to 3/8 inch (six to 10 mm) wide between a cover and the tops of the wood holding the combs in a hive, the bees would not fill this space with comb or propolis. Bees need and respect this walking space. At the same time, Langstroth realized that if he put the wax combs into wooden frames and left a bee space around them, he could construct a movable frame hive.

We realize today that the ancient Egyptians had understood that such a space existed but no one had sufficient imagination before Langstroth to envision a movable-frame hive that respected bee space. Langstroth built and tested his first hive in 1852, a time when great changes were occurring in other areas of agriculture. The manufacture of the movable-frame hive utilizing the bee space concept and the multi-storied modern beehive changed beekeeping from a cottage industry into a major agricultural industry. By the late 1800s several beekeepers were managing thousands of colonies of honey bees for honey production. Today's beehive is built largely on this simple observation by an insightful individual.

The modern Langstroth box is designed to hold 10 frames. However, in practice most beekeepers use only nine frames in the brood nest and sometimes only eight in a honey storage super. It is best to have ten frames in any hive body when the bees are drawing foundation (making new drawn combs) so that the combs are drawn evenly. When manipulating combs in a brood nest, it is much easier if only nine are used, since more space is available to remove and replace the combs. After a few years, burr and brace comb will accumulate on the more widely spaced frames. For further information about Langstroth, his book (*Langstroth on the Hive and the Honey Bee: A Beekeeper's Manual*), his life (Naile, F. *America's Master of Bee Culture: The Life of L.L. Langstroth*, Cornell University Press, Ithaca, N.Y. 215 pages. 1942, reissued 1976.) and his patent. (See LANGSTROTH, L.L.)

BEE STINGS, FIRST AID – When a worker honey bee stings, the tip of her abdomen is usually perpendicular to the rest of her body and the sting is driven at least half of its length directly into the enemy before the bee pulls her body away. Pain, from both the physical penetration of the sting and the venom, is felt almost immediately. However, most of the pain is a result of certain components, such as proteins, of the venom. The pain of the sting serves its purpose and drives a potential enemy away. Much of the bee's venom in the poison sac flows into the wound immediately. The sting is torn from the bee's body, usually within a second or two. The lancets, which are slender rods with barbed teeth on their outside surface, are alternately driven deeper into the flesh so that soon the entire sting shaft is embedded to its hilt into the

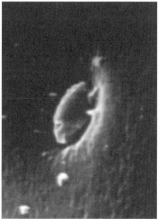  

Sting – left, wound site from sting, center shows the two lancets and the shaft the venom flows through. Right shows entire abdominal end segments with sting chamber opened and sting extended. (from Erickson)

enemy. As the sting shaft buries itself deeper, additional nerve endings in the enemy may be hit and the venom can flow more deeply into the wound.

Because of the way in which the sting works to bury itself in the flesh, one of the first first-aid acts is to remove the sting from the body. Most beekeepers flick the sting out with a hive tool or fingernail. It makes no difference how the sting is removed. Most of the venom flows from the sting rapidly.

Honey bee venom is a complex substance with many components including those that are designed to produce pain, swelling, and irritation over a period of several days. Once the venom is injected it cannot be removed nor are there any substances that may be applied or injected to counteract its effects. First aid for stings means taking steps that may relieve the sensation caused by the venom and preventing any infection because of the wound.

Applying ice to the site of the sting creates a different sensation and probably offers as much relief as any treatment. Rubbing alcohol, which evaporates rapidly and causes a cooling reaction on the surface of the skin, may also be considered a good first aid treatment. No data suggest that other treatments, that range from covering the sting site with mud, baking soda or other materials, do much other than to give some symptomatic and/or psychological relief. It is not uncommon for the area around a sting site to remain swollen and itch for several days in the case of those not accustomed to being stung. Commercial beekeepers, who are routinely stung almost every day, soon build an immunity to the effects of a honey bee sting though they may be as sensitive as anyone else to venom from another species of stinging insect (see REACTIONS TO BEE AND WASP STINGS and STINGS, HOW TO AVOID).

BEESWAX – Many insects secrete wax from glands in their exoskeletons. The wax covers their bodies and protects them from too much water from the outside or drying and loss of water from the inside. In honey bees we find that certain of these glands have developed further so worker bees have four pairs of wax glands on the undersides of their

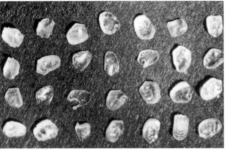

Beeswax is secreted by young worker honey bees from glands on the dorsal side of their abdomens. Liquid and clear when first excreted, the wax cools and turns a pristine white almost immediately. Though similar, wax scales, like snowflakes, all are different.

abdomens that produce wax in great quantity. Bees use beeswax to build honeycomb.

Beeswax is not one material, but a mixture of many long-chain molecules. About 300 components are found in beeswax, the most common of which makes up only eight percent of the wax. It is therefore a complicated substance that clearly will be impossible to synthesize or duplicate. A comprehensive review of beeswax is Coggshall, W.L. and R.A. Morse, *Beeswax, Production, Harvesting, Processing and Products*. (Wicwas Press, New Haven, CT. 192 pages. 1984.)

Beeswax secretion by honey bees – The four pairs of wax glands on worker honey bees (drones and queens do not have wax glands thus do not produce beeswax) are found on the underside of abdominal segments four to eight. Each gland is made up of hundreds of adjacent cells. Opposite, or under, these glands are plates, sometimes called wax mirrors, onto which the wax is secreted as a liquid. The wax solidifies when it comes into contact with the wax plates and air, and appears as scales. If wax is not needed immediately, a worker bee may pile one secretion upon another, and the scale may become very thick and have a laminated appearance. No two wax scales are exactly alike in size or shape.

Wax-producing bees are usually two to three weeks of age. When adult bees just emerge from their cells their wax glands are not yet developed. After bees become foragers and are older than about three weeks, their wax glands degenerate. When bees are producing wax in quantity, bees hang perfectly still and engage in no other activities. Wax secretion is encouraged by a lack of space to store food at a time when it is available in quantity. At such time the bees remain engorged with carbohydrate food (honey), another prerequisite to wax secretion.

Bees that are fed either sugar syrup or honey can continue to produce wax for long periods of time. The quantity of honey needed to produce a pound (one-half kilogram) of wax has never been

vary greatly in their weight, but on average 800,000 wax scales are required to produce a pound (one-half kilogram) of beeswax.

Beeswax bloom – Beeswax foundation or candles will often have a frosty appearance. This white crystalline substance that forms on the surface is called bloom. Beeswax is a complex substance with over 300 components. One or more of the 300 components of beeswax may migrate to the surface of foundation, candles, or blocks of wax where it will look similar to mold but it is not mold. Bloom melts at about 102°F (39.8°C), much lower than the melting point of beeswax itself, about 143°F (62°C). Bloom appears less rapidly on wax that is extruded; it may be seen after a few months on wax that is cast or molded.

Bloom causes no problems either on foundation, where the bees chew it and mix it back with the rest of the comb, or on candles. Bloom may be easily wiped off candles with a cloth. Some candle users feel that bloom gives candles an attractive or perhaps even an antique appearance.

Beeswax bloom as it appears on a (colored for contrast) beeswax candle, and on regular foundation. (Morse photo)

determined precisely, but it is around eight pounds (3.6 kg). No one has been able to force bees to produce beeswax in an artificial hive or chamber. Under ideal conditions a colony of 50,000 bees should be able to produce half a pound (0.23 kg) of wax in a day. Wax scales

Granular and mushy wax and emulsions – Beekeepers sometimes report that the beeswax they render is granular or mushy. When this occurs it is because the wax and the water have emulsified or mixed during rendering. Emulsions are of two types: water in wax and wax in water. Granular wax often occurs when cappings or combs containing honey are melted with too little water and the water is incorporated into the beeswax. The opposite of granular wax is the wax in water emulsion that gives a "mushy mass from which a large volume of solution may be squeezed." The late Professor W. L. Coggshall reports in his thesis on beeswax that with two minutes of stirring melted beeswax and hot honey that he could "incorporate 500 grams of wax into 1500 grams of honey." The two mix quite easily. It was only after adding a considerable amount of water that he could force the granular wax to the surface. Coggshall suggests that honey may contain a natural emulsifying agent since he could also incorporate beeswax into a sucrose solution but it would separate out after about an hour.

In order for emulsions to be permanent an emulsifying agent "must be added to promote the dispersion of one liquid in the other and prevent the coalescence of the droplets." Only a small amount of an impurity is often needed to make permanent emulsions and these are what are used in such products as cold creams in which beeswax has historically been an ingredient.

It is not difficult to break beeswax emulsions whether they are granular or mushy wax. The usual procedure is to re-melt the wax in clean water. Dry heat may be used too, especially to treat granular wax. The beeswax is not harmed in either case. Honey can be moderately or even severely damaged when beeswax is incorporated into it. This occurs rarely but sometimes one can taste a bit of beeswax in a honey.

Harvesting – Old comb, cappings and beeswax refuse can be damaged in storage especially by fermentation, excessive heat, contact with certain metals and the misuse of acids that are sometimes used to clarify the wax. Very few animals can digest beeswax, but wax moths can do so to a very limited extent if they are left unchecked. It is therefore important that beeswax be harvested and rendered as rapidly as possible into solid beeswax blocks that will have a long life. Wax moths, small hive beetles, and other animals that will attack comb and wax refuse will not attack solid blocks of beeswax or foundation even after many years.

Beeswax can provide beekeepers with additional income. The price of beeswax fluctuates. In the United States beekeepers produce one to two pounds (one-half to one kilogram) of beeswax for every 100 pounds of honey produced. In parts of the world, notably east Africa, beeswax is the chief product harvested from the hive, and three to four pounds (1.5 to two kilograms) of beeswax may be obtained per 100 pounds (45 kg) of honey. Methods of harvesting, rendering and storing beeswax have remained much the same for many decades. Unfortunately, many synthetic waxes have been developed, and beeswax does not command the same position in the marketplace that it once did.

Metals, effect on beeswax – Several metals should be avoided when rendering and processing beeswax. Iron, brass, zinc and copper (listed in order of degree) will all discolor beeswax. Monel metal has no effect on beeswax except when water is present, in which case greenish discoloration occurs after long exposure.

Stainless steel is the best metal to use when processing beeswax. Aluminum, nickel, platinum and tin cause no appreciable discoloration. Tinned iron is satisfactory for rendering beeswax but probably should be avoided because cracks, breaks or areas that are worn may expose the wax to the iron.

Rendering beeswax cappings – Although many ways are used to

Two old-style wax presses are still excellent, if labor-intensive ways to remove wax from slum.

separate honey from cappings, old combs and refuse, none is really satisfactory. Much misinformation also exists on the effectiveness and efficiency of various methods. It is well to remember that honey can be dissolved in beeswax and vice versa to the detriment of both; however, honey may be separated from a solid block of beeswax by re-melting, while removing beeswax from honey is not possible. A variety of beeswax melters have been designed and made at home or manufactured and sold commercially.

Brand melters – The Brand cappings melter was designed and first built and sold by a beekeeper, W.T. Brand of Nebraska; it appeared in 1935. The melter became popular with beekeepers almost immediately. The Brand melter has a grid of pipes through which steam flows. Cappings from a power uncapper, or cut from combs by hand, fall onto the grid where they are melted and the honey and wax separated. The melted wax rises to the top and is in constant contact with the heated pipes, whereas the honey stays below and, in theory, is not overheated; however, experience has shown that this is not true. The Brand melter is no longer on the market.

Chemical extraction – Beeswax can be extracted from old comb or slumgum with chemical solvents. However these solvents are only

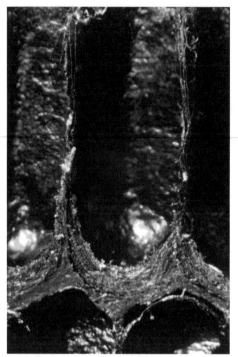

A cross section through an old brood comb shows the many layers of cocoons and propolis laid down after succeeding generations of larvae have emerged. Recovering the wax, it is obvious, will be a challenge. (Jaycox photo)

available to industry, not to individuals. The beeswax obtained from chemical extraction has properties different from that of natural beeswax.

Hot water pressing of cappings and slumgum – Cappings and often old and broken combs may be rendered by melting them in a hot water tank where much of the wax will rise to the surface where it may be dipped directly into molds or allowed to cool and harden. The material left floating or submerged in the hot water is called "slumgum," a rather undesirable term of unknown origin. Because of cocoons and other debris in the slumgum, it is not possible to remove all of the wax from it without pressing.

One way to remove beeswax from slumgum is with a hot water press that makes use of heavy screws or hydraulic cylinders. These serve to compress the slumgum and force the wax to the surface of the hot water. Pressure must be alternately applied and released for the press to be effective. The wax is removed from the press by dipping or allowing it to flow from a spigot or gate. A hot water wax press is effective and will cause the release of all but about one or two percent of the wax only if pressure is alternately applied and released over at least ten hours. There is no alternative to allowing sufficient time for the press to work; increasing pressure is of no additional value and will not shorten the needed pressing time. While hot water pressing of beeswax refuse and slumgum is the most efficient of all of the known methods, the long period required for pressing is a serious disadvantage.

Infrared and other overhead cappings reducers – Several types of cappings reducers have been used. These use no water but work via various types of overhead lamps, especially infrared lamps, which melt the cappings and free the honey. Since the wax is lighter than honey, it rises to the top and the two may be separated. The honey is often overheated and damaged and often has beeswax dissolved in it. The beeswax is usually little harmed, though emulsions may form (see BEESWAX - Granular and mushy wax and emulsions).

The best method of rendering cappings is not a new question. Phillips and Coggshall wrote one of several papers on the subject of melting beeswax in the presence of honey. These researchers used a light clover honey, since light honeys are far more resistant to damage from overheating than most honeys. When beeswax was melted in the

Solar wax melters come in all sizes, shapes and styles. Many are homemade, using doors or windows as the glass top, covering a homemade box and catch pan.

presence of honey, the honey was darkened and the flavor damaged. The conclusion is simply that one should not expect to recover quality honey when honey and beeswax are melted together.

Solar wax melter – A solar wax melter is sometimes also referred to as a solar wax extractor. This is a simple method of recovering high-quality wax from both old and new comb and cappings. Because of size limitations, solar wax melters are used primarily by beekeepers with small volumes of

Some are commercially made and have handles, wheels and fancy catch pans and covers – they all do the same thing. Solar melters recover most of the wax in old combs, but not all. The efficacy of recovering the remaining small percent is questionable, but a wax press will do that job.

Better Way Wax Melter – When the wax has melted and drained off the frame, what remains is the accumulation of old cocoons that comprised the brood area (note the shape). Some wax remains in these, but the cost of extraction generally exceeds the value of the wax received.

beeswax to render. Unfortunately, solar wax melters remove only 50 percent or less of the wax from old combs and 75 percent or less of the wax from new combs. Apparently the cocoons in old combs act much like a sponge, retaining the wax even after several days at high temperatures. (Lesher, C. and R.A. Morse. "The efficiency of solar wax extractors." *American Bee Journal* 122: 820-821. 1982.) An advantage of the solar wax melter is that what is not rendered is reduced in volume and can be easily packaged to send to a rendering plant where steam and pressure may be used to remove the remaining wax.

Solar wax melters are available from bee supply companies. Additionally, plans for building a solar wax melter may be available from a number of sources. Temperatures of 190°F (88°C) and higher can be reached in a solar wax melter. Since beeswax melts at about 145 to 150°F (63 to 65°C) this temperature is sufficiently high.

Most of the efficient solar wax melters are made of wood. The outside is painted black to absorb the heat; the interior is painted white to reflect the heat around the inside; the inside should not be painted black. One pane of glass, or even plastic, is sufficient to cover the box, but two layers or pieces of either substance, about one-quarter of an inch (0.5 cm) apart, are even better and will raise the temperature higher. The glass or plastic is usually built into a cover much like a hive cover that can be easily removed to fill the extractor or remove wax.

An aluminum or galvanized sheet metal pan is placed on the bottom of the box, which is raised on one end so that the melted wax will run into the catch pan. In most solar wax melters the glass cover is about five to six inches (10 to 20 cm) above the floor of the melting pan.

It is difficult to advise how long a batch of combs or cappings to be rendered should be left in the extractor. However, two to four days at a temperature of 190°F (88°C) are usually sufficient to remove all the obtainable wax. One should watch the extractor for several minutes to observe how much wax is being freed and dripping from the main pan into the catch pan.

The beeswax from a solar wax melter is of high quality, probably partly because it has been sun-bleached. Any honey that collects on the bottom of the catch pan is burned and black from overheating and should be discarded.

Another wax melter, long on the market is the 'Better Way Wax Melter.' This is a simple device made of galvanized metal or stainless steel with electric heat coils running inside the walls and bottom. It has accommodation to hold frames, which can be hung inside. The heat generated by this machine is significant, and melts even old, cocoon-filled comb rapidly. Melted wax drips to pan below and exits at the bottom to a catch pan.

Some models of this will heat frames to the point of killing American foulbrood spores. It is suggested that frames have propolis removed before the process begins, so the finished wax does not get propolis in it. High heat can cause wax, propolis and debris to smoke, so operating this unit outside reduces the chance of a smoke-filled room.

This is an efficient and easy-to-use unit for the beekeeper who routinely rotates two to three hundred combs out of brood boxes each season and wants to recover the wax.

Physical and chemical properties of beeswax – Beeswax was our first plastic. It is brittle at low temperatures, but at higher temperatures it is more plastic and malleable. Extruded beeswax, i.e., that produced under pressure, is particularly plastic because its molecular structure is disrupted; this is not the case with cast wax. Broken beeswax presents a dull, granular, non-crystalline type of fracture. Beeswax melts at about 143 to 151°F (62 to 66°C). It is insoluble in water and only partly soluble in most solvents, though it mixes easily with many fats, oils and other waxes. Like most waxes, beeswax shrinks when it cools, losing 9.6 percent of its volume. Beeswax is a remarkably stable material; samples thousands of years old appear to have undergone little change.

The chemistry of beeswax has come under close scrutiny during the past several years especially since the development of the gas chromatograph and other sophisticated chemical apparatus. However, with so many components, and given the variability of natural products, few precise measurements can be made. Much of the concern in studying the chemistry of beeswax has been to detect adulteration. Most forms of adulteration can be detected. Some contaminants, such as microcrystalline wax, are quite easy to detect. The beeswax on the market is remarkably pure given the temptations that exist to dilute a high-priced commodity.

However, due to the fact that beeswax is used in a variety of cosmetics, as foundation for edible comb honey, and as the very foundation for further generations of bee brood, purity has come under close inspection. Many of the pesticides used in a beehive to combat mites are absorbed into the wax during in-hive treatments when the chemicals (usually plastic strips) are in direct contact witht he wax.

Rendering does not remove all or even most of these chemicals. Honey stored in contaminated comb is exposed, as well as larvae, and adult bees that are feeding, cleaning or just walking on the wax.

The only sure way to insure clean wax is to use only wax that has not been exposed to chemicals, and produce your own foundation. Or, purchase plastic foundation, uncoated with wax, and apply wax from a known, clean source.

This problem is slowly being reduced, as organic acid chemical treatments, which leave minimal, harmless residue are being used to treat mites. Also bees that show tolerance and resistance to mites are more commonly used.

Processing and bleaching beeswax – The wax that is secreted by honey bees is white. The beeswax that results after collection, melting and filtering is yellow because it is stained by pollen, propolis and even by the metals of the processing equipment, especially iron. Many of the commercial uses for beeswax call for a wax that is white or a light, bright yellow. Since lighter beeswax usually commands a better price on the market it is sometimes bleached. A variety of ways may be used, some more practical than others.

Most beeswax processors use stainless steel tanks because it does not affect the wax color. When acids are used it is necessary to use wooden tanks since some acids, notably oxalic, will corrode stainless steel (see BEESWAX - Metals, effect on beeswax).

The first step in preparing high quality beeswax is to wash it. Most of the solid and water-soluble impurities may be removed using a tank filled with about one-fourth water and three-fourths beeswax. The water is heated by injecting steam and in turn the wax is liquefied; most of the debris will settle to the bottom of the tank though some remains in the water.

The use of the sun to bleach beeswax is thousands of years old. Beeswax to be sun-bleached may be flaked or exposed as a liquid in shallow stainless steel pans.

Hydrogen peroxide can be used to bleach beeswax. The wax is melted by heating the hydrogen peroxide. The length of time required depends on the initial color of the beeswax.

A word of caution when processing beeswax: bleaches using chlorine must be avoided as the chlorine is absorbed by the wax and will be released as a toxic gas if burned in a candle. For further information on processing and bleaching beeswax see Coggshall, W.L. and R.A. Morse. *Beeswax, Production, Harvesting, Processing and Products.* (Wicwas Press. New Haven, CT. 192 pages. 1984.)

Most wax used commercially, however, is cleaned and lightened by passing it through a pressure filter. Wax is melted, diatomaceous earth and activated charcoal are added and mixed. The resulting hot slurry is passed through a pressure filter. Filters, 20-100, depending on the machine, are made of felt or other cloth, as the wax exerts considerable pressure, plus the building of charcoal and earth exerts considerable pressure on each filter.

The wax is heated to just over 160° so it is well liquified and flows easily. Depending on the amount, even the darkest, dirtiest beeswax can be rendered nearly white by this process.

Filters are often rendered after using to recover even the small amount of wax remaining.

BEESWAX FOUNDATION – The invention of comb foundation in 1857 was one of four events between 1851 and 1872 that changed beekeeping from a backyard hobby into a commercial industry. The others were the discovery of bee space, the invention of the extractor and the invention of the smoker. All four of these items are probably equally important, each for

Beeswax Foundation

Generalized Wax Processing

No matter the scale of your operation, beeswax travels basically the same route from beginning to end. Wax is harvested, as cappings (1) from extracted frames, as wild wax (2) from feral colonies, or as old comb.

Harvested wax is collected (3) and cleaned of large debris and clinging honey by rinsing in clean water or melting. When clean, wax is added to a melting device (4), water added and heat applied. Melted wax escapes, runs into the collection pail where it can be reheated again and even cleaner wax exits via the pipe.

Clean wax, collected in any

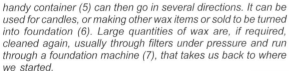

handy container (5) can then go in several directions. It can be used for candles, or making other wax items or sold to be turned into foundation (6). Large quantities of wax are, if required, cleaned again, usually through filters under pressure and run through a foundation machine (7), that takes us back to where we started.

Beeswax Foundation

their own reason. Left to their own devices temperate climate honey bees may sometimes build straight combs. However, all too often the combs are wavy or otherwise distorted. The use of comb foundation makes it possible for a beekeeper to make straight, movable combs.

Comb foundation is exactly what the name suggests. It is the base or midrib of a comb. If one slices through a comb perpendicular to the midrib it will be noted that the foundation is built to give the greatest strength and at the same time keep the larvae in as compact a group as possible. Conservation of heat in the brood nest is an important part of hive biology. At the base of each natural cell we find the bases of three opposite cells.

A German beekeeper, J. Mehring, is credited with making the first comb foundation in 1857. In the early days most of the foundation was made using flat molds or presses, and too often the cell bases and walls were not well-defined. In 1875, A.I. Root, in collaboration with a resourceful mechanic, A. Washburn, made the first practical wringer-type foundation mill that could produce comb foundation in quantity. Root recognized the importance of building foundation with high cell walls. Much later the production of satisfactory plastic foundation was delayed for several years because some of this history had been lost or forgotten and it was not recognized that without good cell walls the bees cannot make comb on plastic any more than they can on a sheet of flat beeswax (see PLASTIC BEEKEEPING EQUIPMENT).

Plain, unreinforced sheets of foundation in a frame are easily broken in an extractor or when colonies are moved. Soon after the first foundation was made it was found that reinforcing it with wooden or metal splints gave the foundation greater strength. Aluminum and plastic core foundation coated with beeswax has been manufactured but was never too satisfactory because the cell walls were imprinted on the beeswax coating but not on the core. If this foundation was put into an active hive during a good honey flow the bees would build on the beeswax cell walls and a good, strong comb could be made. However, if there was no honey flow the bees would chew and remove the wax;

this would expose the plain core that had no cell walls, and the bees would build no cells. The best method for making strong wax foundation is to use wires imbedded into the foundation. One may wire a frame by hand or buy already wired wax foundation. The method used depends upon the time and money available to the beekeeper.

Comb foundation is made in several weights or thicknesses. The foundation used to make comb honey or cut-comb honey is carefully lightened and made extra thin so that those who eat the comb honey will not find a thick midrib that might be objectionable. Foundation that is used to make brood comb and combs for honey storage is barely lightened. For all types of foundation the wax is filtered to remove any debris.

Three steps are usual in the manufacture of comb foundation. The first of these is sheeting. Sheeters are drums with a diameter of from about 10–40 inches (25–101 cm). They are cooled on the inside by flowing water. The revolving drum dips into a pan of liquid wax that solidifies and is scraped off on the opposite side by a knife, passed through a set of smooth rollers that reduces and evens the thickness of the sheet and make it more uniform than when it first comes from the sheeter. The next step involves passing the thin sheet through two precision-made, engraved rolls, which produce the cell walls. After this the sheet is cut to the size desired, wire embedded if required, wire cut, sheets stacked, often with tissue between, and, when enough are cut, boxed. This occurs, in a modern shop, on a machine less than eight feet long and four feet wide.

Making a foundation mill is an

Natural Comb

When bees draw their own comb sheets (1), there can be great variety in the construction. Comb thickness can vary (2) compared to foundation-aided comb, and the orientation of the cells (3) can vary. Compare (4), for uniformity.

Plastic foundation is rapidly replacing wax, especially in large outfits. You can obtain plastic sheets, plastic with wood frames already assembled, or complete, one-piece plastic frame and foundation. (Pierco photo)

exacting process. The two rolls must match perfectly so that the bottom of one cell is precisely a part of the face of three opposite cells. The best foundation mills will produce 50 to nearly 100 sheets of foundation a minute; the mills cannot run too fast or the friction will cause them to overheat and the high temperature may distort the foundation. The history of the manufacture of comb foundation is covered thoroughly in: Coggshall, W.L. and R.A. Morse. *Beeswax, Production, Harvesting, Processing and Products.* (Wicwas Press. New Haven, CT. 192 pages.)

Cell diameter in manufactured foundation – A Belgian professor suggested that a cell diameter of 5.0mm was detrimental to colony vigor and productivity. Manufactured foundation allowed him to launch a series of studies using larger cells to produce larger bees. He later advocated the use of 6.0mm cells and demonstrated that larger cells produce larger bees. However, he incorrectly concluded that increased bee size via the use of larger cells would impose a heritable trait. Even so, the professor's work precipitated a long series of studies conducted in other countries. These studies ultimately led to the variation in cell size now common between manufacturers of commercial foundation. During the past century, some manufacturers have worked to retain a relatively 'natural' cell diameter (~5.1mm) for their foundation, while others used larger cells. The current variance in cell diameter for available manufactured foundation, both wax and non-wax, ranges from 4.8–5.6mm. Beekeepers wishing to explore the use of cell size as a colony management strategy need to determine the optimal size for their region and its availability from suppliers of beekeeping equipment. [(Grout, R.A. 1931. A biometrical study of the influence of size of brood cell upon the size and variability of the honeybee (*Apis mellifera* L.) M.S. Thesis, Iowa State College) (Erickson, E.H., D.A. Lusby, G.D. Hoffman, and E.W. Lusby. On the size of cells; speculations on foundation as a colony management tool. *Gleanings in Bee Culture*, 118: 98-101, 173-174. 1990. (Lusby, D. and E. Lusby. Is smaller better? *Gleanings in Bee Culture*, 126: 25-27. 1998) - EHE

BEESWAX IN ART AND INDUSTRY – Despite the sophistication of most fields of endeavor today, a few areas can be found where an old-fashioned product such as beeswax is still in demand. Also a certain romance is connected with its use, as well as the knowledge that beeswax is a special commodity. Since beeswax is a mixture of substances it cannot be imitated and no product exists that may be substituted for it. Beeswax does vary from one country to another

Batik

but this is because the plant gums and resins, as well as the pollens that may stain and color the wax, vary from one part of the world to another. These differences are well known to beeswax processors, especially those who deal in bleached and refined waxes. As a result, wax from some areas is more desirable and commands a higher price.

The uses for beeswax today are not nearly so many as could be found several decades ago. Still, there is a good demand for beeswax and its sale remains an important part of a beekeeper's income.

Batik – This is a method of making colored designs on cloth. If molten beeswax is poured or pressed onto a piece of cloth, the waxed portion resists dyes that are applied after the wax is cooled. Both sides of the cloth are dyed equally since the wax soaks through the cloth.

The origins of batik are unknown, although some believe it was first made in Egypt. Our present knowledge and use of the process came about with the European occupation of Southeast Asia several centuries ago. Originally only beeswax was available and used in making batik designs, and thus its mention here; today paraffin and other waxes are used in conjunction with beeswax for different effects.

Beeswax figures – For thousands of years beeswax has been used to make life-like, realistic figures. The production manager of Madame Tussaud's famous wax works in London wrote in detail how this is done. (Sargant, J. *Two hundred years of wax modeling: a history of Madame Tussaud's Central Association of Beekeepers*, London. 10 pages. June, 1971.)

To make a beeswax head, or other body parts, one uses three parts beeswax and one part Japan wax (from the fruit pulp of *Rhus succedanea*). A light-colored beeswax is used, and dyes and colored oils are added to reproduce the complexion of the individual. A mold is first made from plaster of Paris. The mold is soaked in warm water for about 30 minutes to make it less porous. Molten wax is poured into the mold and allowed to set for 15 to 30 minutes after which time about half an inch of wax will have solidified against the side of the mold. The remaining liquid wax is poured off, and the solidified wax is allowed to cool to room temperature. Then the plaster of Paris is chipped away. Coloring to accentuate certain features is added along with hair that is inserted one piece at a time. The figure is then ready for display.

Candles – For thousands of years the chief use for beeswax has been the manufacture of high-quality candles both for home and religious purposes.

Making Candles

You can make a multitude of candle styles, the most common using plastic, metal or polyurethane molds, poured containers or dipped candles. Certainly one-of-a-kind individual molds are popular, as are candy molds. Shown here are some of these many styles, and basic instructions on making poured mold candles and candy mold ornaments.

Poured Mold Dinner Candles

The four, six, eight or 12 cavity dinner candle mold is still common and, when used correctly, makes excellent candles.

Clean each cavity with a solvent to remove any clinging wax particles.

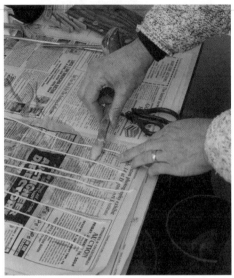

Dip wicks in liquid wax before beginning, and while cooling, straighten using a small, flat stick . . .

. . . pulling the wick underneath.

Cut to length, so the wick has at least a couple of inches above the top and below the bottom.

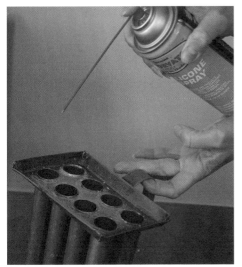

Coat the inside of the mold cavities with a releasing agent. Silicone spray works well.

Insert the wick, pull out and . . .

. . . bend over so it doesn't fall out.

Use a flat, wet sponge to stop wax from flowing out the wick hole.

Properly-made beeswax candles provide a steady and strong flame, produce a good odor when burning and will not drip or smoke if kept out of drafts.

The use of beeswax in candles has declined somewhat in recent years for several reasons, perhaps the most important of which is expense. Good candles can be made from other materials for much less cost. A demand does exist for beeswax candles by people who appreciate the high quality of these candles. The use of beeswax candles in churches continues to be a good market. Beeswax candles made by beekeepers sell very well in craft fairs and shops, and the internet has made these candles available to many people who do not have access to fairs and specialty shops.

A variety of candle-making methods

Secure entire mold, with sponge in place on a stand.

Use pencil or a template to hold wicks in center of cavity.

Fold wicks to secure in center of cavity.

Fill with warm, not too hot wax, let cool, top off so cavity is completely full. Let cool for an hour or so at room temperature then place in freezer if necessary. This is best decided after a few trials – solidifying depends on room temperature, humidity and other factors. If the cavity was cleaned well (this is the big secret to these) when the wax is solid the candle will nearly fall out. Experiment with these for a bit.

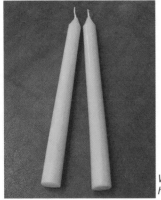

When removed, square off the bottom by rubbing on a hot pan or a hot plate. Polish and use.

include dipping, pouring, molding, rolling, extruding, drawing and pressing. Dipping, molding and rolling are the most practical for the home candlemaker as the other methods require the use of elaborate machinery and methods.

Beeswax candles burn longer than comparatively-sized paraffin candles and also produce a pleasant odor. Beeswax candles are decorative and they make wonderful gifts. For a full discussion on the commercial production of candles see Coggshall, W.L. and R.A. Morse. *Beeswax, Production, Harvesting, Processing and Products.* (Wicwas Press., New Haven, CT 06410. 192 pages. 1984), and Berthold, R. *Beeswax crafting,* (Wicwas Press, New Haven, CT 06410. 125 pages. 1993.)

Melting beeswax – Since beeswax is highly flammable, it should never be heated over an open flame. Extreme care should be taken if not using a double boiler. Experienced candle makers use hot plates without a double boiler. Small quantities of beeswax can be melted in a crockpot or some type of double boiler. A meat thermometer is useful in candle making. For metal molds the wax temperature should be approximately 180° to 185°F (82° to 85°C). As the beeswax hardens it contracts and pulls away from the mold, making removal easy. For polyurethane molds beeswax may be poured as soon as it is melted or at about 150°F (65°C), since contraction is not a key factor in candle removal from this type of mold. Frequent heating of a batch of beeswax will remove some of the aroma and darken the wax.

Preparing wax – A quality beeswax candle will burn with minimum dripping, sputtering, and smoking. It is important that the wax is free of contaminants – honey, pollen, propolis, and dirt.

Wax should be first melted in water to remove honey and microscopic particles of pollen or propolis. Then allow the liquid beeswax to cool and solidify. The bottom of the block can then be scraped to remove any debris there. When the wax is dry it can be again melted and fine-strained to remove any small particles that remain. For straining small quantities of melted beeswax one can use paint strainers available at hardware stores or use sweatshirt material fuzzy-side up.

Candle wicks are available in an incredible array of thicknesses, materials, weaves, shapes and sizes. Moreover, there is little standardization in nomenclature or recommendations of which wick works best in what candle. Certainly, candles, too, sport a variety of sizes, widths and burning challenges. Candle supply catalogs and craft stores generally offer advice on which to use, but the best barometer is trial and error.

Wicks that are too thin for a particular candle will not stay lit, and wicks that are too thick will burn with a large flame, produce smoke (and the resulting soot), and will burn the candle unevenly, leading to safety problems.

Never sell or give away candles you have made until you have found the best wick for that candle.

Cleaning Wax

When making candles or ornaments, the beeswax needs to be finish-clean. After the wax has hardened after its initial cleaning there will probably be some sediment on the bottom of the block. Scrape this off with a hive tool or other tool before remelting.

Remelt the wax in the first container (white in the photo) keeping the temperature as cool as possible to avoid darkening the wax. **CAUTION: NEVER LEAVE HEATING WAX UNATTENDED.** When the wax reaches 175°F to 180°F (yes, use a thermometer) pour it into your second container (black in the photo) filtering it a second time. Use paper towels, sweatshirt material, milk filters or the like as a filter. This wax will make excellent candles or ornaments.

Do not mix wax from cappings and old brood comb when making candles. Old, dark comb, no matter how well strained, will have propolis, and some particulate matter in it which will cause the candle to sputter, smoke and burn unevenly.

Wicking – Wicking cannot be made properly at home. Good candle wicking is braided or plaited and pickled, that is, treated chemically to control the afterglow. Hundreds of different sizes and types of wicking are available, including flat-braided, square-braided, oval and round braided, stranded, twisted, glass fiber, and hollow varieties. It is very important to use the correct size and style of wicking for any candle. Many books on candle making can easily be found.

Only commercially-available wicking should be used when making beeswax candles. It is specially woven and treated to burn properly. Many hobby and craft shops carry wicking for beeswax candles. The type and size needed varies with the diameter of the candle. If the wick is too thin it burns into the candle and extinguishes itself. If it is too thick the candle burns too quickly. As a starting point, 2/0 wick works quite well for standard size beeswax taper candles with a base diameter of about 1" (2.5 cm). Two-inch (five cm) diameter candles generally require 60 ply, or use the guidelines suggested by candle supply manufacturers, and experiment.

Metal molds – Metal candle molds for standard tapers have been used for hundreds of years, and although antiques can sometimes be found they usually have deteriorated past use. Any rust inside the tubes prevents removal of candles. However, reproductions of these antiques are available from a variety of sources.

Polyurethane Mold Candles

These candles reflect the essence of simplicity in production, and excellent quality when complete. The molds are split down two sides. The wick is brought up from the bottom of the mold, which will, in turn, become the top of the finished candle. The wick is centered across the opening with a hair pin. The mold is secured with rubber bands. Wax is poured in and let cool, then poured again to accommodate the shrink that occurs. Depending on the size of the mold you may need to do this more than once.

After the candle cools to room temperature, the rubber bands are removed, the sides peeled back and the candle is complete.

For molds with multiples of two tubes, the wick should be cut long enough for two or more candles, plus an extra two inches (five cm) or so for support pins. Thread the wick in the mold using a thin, stiff wire (like frame wire). The wire should be about twice the length of the mold and then doubled. Twist the free ends together and leave an "eye" in the other end. The twisted end fits through the top of the mold tube, while the wick is inserted through the loop in the wire. The wire is then used to pull the wick through the first tube. The wire with the wick is then threaded into the second tube, from the bottom. Grasp the two free ends of the wick that extend above the top of the mold and pull them taut. Support each with a bobby pin or homemade guide and make sure the wick is centered in each tube.

Seal the spaces where the wick comes through the tips of the tube to avoid wax leaking out. To do this use a piece of one-inch thick household sponge cut to fit inside the candle mold base. Soak the sponge in cold water, and when soft wring out the excess water. If you don't, water gets into the wick, causing the candle to sputter when it burns. On a drip tray, press the base of the mold down on the damp sponge, and fill the mold just up to the top. The wax will solidify in a minute or so and then the mold can be removed from the sponge. As the wax cools in the mold it shrinks and will form a cavity around the wick. You need to top-up each candle two or three times to produce a solid, level base.

When the wax has hardened in the tubes, cut the wick joining the two candles. Place the mold in a freezer for a few hours. This causes the wax to contract and makes the candle easier to remove from the mold. If this does not work try tapping the mold gently on a solid surface. As a last resort grasp the wick extending from the base of the candle with a pair of pliers and steadily pull while holding the mold under hot, running water.

Factors which can make removing candles from metal molds difficult are: 1) dirty molds, 2) using wax not completely cleaned, 3) not putting the mold in the freezer.

Before filling, clean the inside of each tube with a grease solvent to remove any clinging wax, dirt, or other material. Wipe clean with a soft rag. Then, liberally spray the interior of the molds with a candle mold release. This will allow the candles to slip out easily. Candle mold release can be bought at craft shops, or from bee equipment suppliers.

Polyurethane molds – Polyurethane molds are extremely popular. These molds last longer than the older two-piece plastic molds and they do not need to be supported. However the individual standard taper molds will be easier to use if supported in a stand. Best of all, these polyurethane molds give extremely sharp detail and the candles are easily removed.

To thread the wick through polyurethane molds use a long upholstery needle or other long bodkin available in sewing stores. The wick, threaded on a needle, is forced through the mold using pliers.

Wax should be heated to just above its melting point, to about 150°F (65°C) when using these molds. The insides can be sprayed with some type of mold release to make removal easier.

After the candles have been poured, topped-up if necessary, and the wax has hardened, the two- and three- piece molds can be placed in a freezer for a few hours. This hardens the wax and makes the candles less subject to scarring when removed from the mold. However, the one-piece polyurethane molds should not be placed in a freezer because polyurethane contracts more than beeswax, locking the candle in the mold.

Foundation candles – Candles rolled from sheets of beeswax foundation can be quite decorative. They are made from standard non-wired beeswax foundation which has been colored by the manufacturer, from paraffin-blend foundation sheets, or from thin sheets of beeswax.

To roll a candle from a sheet of foundation wax be sure the sheet and the place where it is to be rolled are clean and warm. 80°F (27°C) is best, less than 70°F (20°C) will cause cracking of the foundation. Cut a piece of wick the length of the candle being rolled plus about an inch. The sheets of wax can be cut in various ways to make different sizes and shapes of candles. It is important to press the wick firmly into the foundation to make the first turn. The most difficult step in this process is the first turn. After that, rolling becomes progressively easier as more and more turns are completed. Once mastered, any manner of variations can be tried. Instructions for making foundation candles are available from beekeeping equipment suppliers - RB

Dipped candles – Dipped candles require more effort to make but they are well worth the effort. As the name implies, dipped candles are made by literally dipping a wick in and out of the dipping pot containing liquid beeswax. Usually a frame is used to keep the wick taut for the initial dips, for keeping the dipped candle straight. The wick is dipped in the dipping pot and then hung until the beeswax is solidified, usually about five minutes between dips.

Cosmetics – Beeswax is still used today in the manufacture of a wide range of cosmetics, even though the percentage used in an individual commodity might be small. The manufacturers of cosmetics are aware that beeswax has several attributes: it will not become

Making Cosmetics

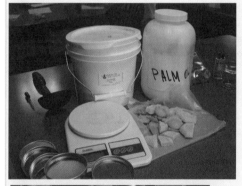

Top left, cosmetic ingredients. Above, melting and combining ingredients. Left, filling containers with finished product.

Jeanne's Hand Cream – A simple formula for an attractive product

2 cups (475 ml) olive oil (the base oil)
1/4 cup (60 ml) palm oil
3/4 cup (175 ml) coconut oil
6 ounces (170 g) beeswax
40-50 drops essential oil(s) (optional)

Combine the oils in a 2-quart (2-l) stainless-steel sauce pan, and stir over medium heat until the oils (and butters, if using) are melted. Add the beeswax to the pan, and stir until melted. Then test the mix by dropping five or six drops onto a sheet of waxed paper, let cool, and test the hardness. If it is too hard, it will be difficult to rub onto your skin, and your mix will need a bit more base oil added to soften it. If it is too soft or greasy, add a bit more beeswax to stiffen it.

When the hand cream has the exact thickness you want, remove the pan from the burner and allow it to cool until the cream begins to harden on the sides of the pan. Then stir in two vitamin E tablets or six drops of vitamin E oil to enhance the healing properties of the hand cream.

For a fragrant hand cream, add any combination of essential oils after removing the pan from heat, or simply rely on the subtle fragrance of beeswax and the oils in the basic recipe. A popular blend that offers a whiff of fragrance without being overpowering contains 15 drops of lavender oil, 15 drops of rosemary oil, and 15 drops of geranium oil. You can use other oils, or vary the proportions of these, to suit your taste.

When well mixed, pour the cooling cream into individual 2-ounce (55 g) containers and allow it to cool, uncovered, overnight. Then cover the containers, label them, and store at room temperature, out of direct sunlight.

---

rancid, is not a skin irritant, holds color well and acts to stabilize other components.

One of the first uses of beeswax in cosmetics was in the manufacture of a cold cream for cleansing the skin; this was first done over 2000 years ago. The beeswax was mixed with olive oil, water and probably some perfume (unfortunately the olive oil could become rancid and thus the early cold creams had a short life). Even as late as 100 years ago cold creams had to be made fresh, in a local pharmacy, each with its own formulation. It was found in about 1890 that adding borax would shorten the manufacturing time, and produced a whiter and more stable product. Mineral oil was also substituted for olive oil, decreasing the likelihood of rancidity.

Those interested in the use of beeswax in cosmetics will find a variety of recipes available in books and magazines specializing in this craft.

Encaustic painting – Painting with hot beeswax, into which colored pigments are mixed, is an ancient art form first practiced by the Egyptians from about 60 to 230 A.D. In the Fayoum area of that country it was popular to paint life-size portraits of the deceased that were placed over the face of the mummy. It is some of the finest artwork in the world. The full face of the person is shown, whereas before this time Egyptian portraits were side-view and showed only half the face.

In ancient times beeswax painters kept the beeswax hot on a sheet of metal

Encaustic painting done by Michael Young from Ireland.

A bronze and jade statue made in China in the third or fourth century cast using the lost wax process.

placed over a pan of hot coals. Those that paint encaustically today use an electrically-heated palette or a warm applicator to melt the wax.

When an encaustig painting is finished, it needs to be subjected to low heat – the sun's rays in the case of the Egyptian painters, and heat lamps today. This heating causes a slight blending or "burning" of the surface; it is from this treatment that the term "encaustic" is derived. It is also reported that the Romans made encaustic paintings. Some of the paintings were unearthed in excellent condition at Pompeii. (MacMillan, J.T. "Through molten lava of Mt. Vesuvius." *American Bee Journal* 77:436-437. 1937.)

Grafting wax – Horticulturists use a grafting wax containing beeswax to cover new plant grafts. Grafting wax will protect the plant wounds from infection and keep the grafts sealed. A good grafting wax must have several qualities. It must be soft and pliable so it can be worked into place but will expand as the cells in the graft grow. It must be non-toxic and it should have a life of at least two months during which time the plant graft will heal.

Lost-wax casting – The oldest method of making intricate and detailed metal castings involves the use of beeswax. The artist first uses beeswax to make the model. The model is next plastered with wet clay and then baked. In the baking process the clay hardens and the wax runs out and is "lost" from the form.

Tree Ornaments

These are easy to make and use little wax. Candy molds and cookie molds work well for these. Begin by making sure the mold is clean. Lay in the hanger by looping a piece of string, ribbon or unused wicking as shown. Pour in wax, let cool, refill if needed. Remove, trim and hang.

Decorative & Functional Uses of Beeswax

Beeswax is an excellent medium for carving, and many people make one-of-a-kind molds for artistic expression.

Exotic glow-candles can be made of pure beeswax. A liquid-filled balloon is dipped in a wax-filled container to make the bowl. The top is evened off with a sharp, hot knife. Decorations can be pressed on to the sides right after the last dip, or previous to that and covered with a thin layer. A beeswax candle is burned inside this globe, and the candle, and the warmed wax in the globe are very aromatic.

Beeswax has many uses besides candles. A solid block should be kept around the house to rub on the edges of sticky drawers, to rub on the threads of screws to ease insertion (and provide a water-proof seal) and on nail shafts for the same reason.

Sewers/quilters run their thread across beeswax to strengthen it and prevent tangling, and the blocks can be sold to crafters, and those making creams, soap and even candles. Clean, good-colored beeswax commands a premium price in the market.

Molten metal is poured into the form and the casting is made. Today, in most countries, less expensive waxes, especially paraffin, are used for lost-wax casting. In either case, the wax is recovered and may be used many times.

Sealing wax – For thousands of years it has been popular to seal personal and private documents so that they could be read only after the seal was broken. Additionally, legal documents often had a wax seal on them to attest to their authenticity. Some drops of melted sealing wax are poured into place and a design is impressed.

Sealing waxes may be made hard or soft. Resin (from plants) is used to make hard seals. Soft sealing waxes contain lard, tallow and/or turpentine. Seals are often colored by adding plant pigments.

BEGINNING WITH BEES – A tremendous amount of literature and other information can be found about bees and beekeeping for beginners. This information includes books and bulletins from state universities, short courses, correspondence courses, movies, videos, and slide sets, and both national and international trade journals. Most important, there are a number of bee associations along with thousands of beekeepers in the United States who own a hive or more of bees who would be most interested in sharing their knowledge.

The best source of information for the beginner is usually a local beekeeping organization that can pass on information on local ordinances about beekeeping and how many colonies an area will support. The single most important factor that limits the number of colonies one may keep in an area is the amount of forage (nectar and pollen plants) available. Usually the local county agricultural agent (Cooperative Extension Service) has information about local beekeepers, clubs and the bulletins that are available from the state universities.

There are basically five ways to start in beekeeping. In order of practicality they are: 1) to buy an established colony(s), 2) a package(s) of bees with queen, 3) a nucleus colony(s), 4) capture a swarm with a bait hive, or 5) remove a colony from a tree or house. All of these methods have advantages and disadvantages.

Buying established or secondhand colonies – The advantage of this method is that an established unit is ready to produce honey, which is the goal of most beekeepers. The best time to buy an established colony is in the spring after the colony has survived the rigors of winter. However, colonies ready to produce in the spring are usually more expensive.

Secondhand colonies should always be examined for diseases, mites and other enemies by a state apiary inspector. In some states it is required that bees be examined for diseases and mites before they are sold or moved; however, not all of the states have active apiary inspectors. It may also be helpful to have an experienced beekeeper evaluate the colony for honey and pollen stores and to determine that the hive has standard equipment that can be interchanged from colony to colony. Frequently you may be able to get some good advice from the previous owner on bee management and apiary locations. (See EQUIPMENT FOR BEEKEEPING)

The most important piece of information you can get from the previous owner is the medication history of the colony. If antibiotics have been routinely used in the past to supress disease you will need to continue that practice, or your new colony definitely will succomb to disease in the first season. What controls have been used to supress *Varroa* mites? Some chemicals are absorbed in the wax combs and can lead to queen and drone problems. The age of the combs themselves is indicative of past management. If, when held up to the sun, you cannot see light through the comb, it is too old, and replacement is called for.

In the overall picture, we recommend that beginner's do not purchase full size colonies until they have had several season's experience or they are working with an experienced and trusted mentor.

The chief disadvantage of buying an established colony is that it takes away the fun of watching a colony develop. And of course manipulating and examining a package colony with fewer bees is much easier than an established full-sized colony.

Package bees – One popular way to begin beekeeping is to buy a package of bees and new or used equipment in which to install them (see PACKAGE BEES, Care of and Installation).

It is easy to start beekeeping with a package of bees because their numbers are relatively small (8000–12,000 bees), have few guard bees and are gentle. Most new beekeepers who install package bees use new equipment and frames with foundation. By starting in this manner, it is easy to study how the hive is put together, comb construction, colony development and other aspects

Package Installation

Your 'package' of honey bees looks like this (1). It is a thin-wood box with screen on the two long sides. A piece of cardboard or wood will be stapled to the top covering the entrance to the inside of the box. Immediately under the cover hangs, or rests, a tin can containing sugar syrup feed for the bees during their trip to you. Hanging next to the feed can by a wire or metal strap will be the small cage that holds the queen. There should be very few dead bees on the bottom of the cage. A floor covered with dead bees has a problem – too long in the cage, overheated, chilled, starved or exposed to a pesticide may be the reason. Consult with the seller if this occurs. Also, the queen should be active and healthy.

Briefly, installation is easy, fast and nearly foolproof. There are several techniques to successfully install that package. Here's one.

1. Examine the package when you first receive it. Look for leaks, dead bees on the bottom and weight (it should weigh about a pound and a half more than the bees – 3 lb. package = 4½ lbs. total, about). If keeping for any amount of time, feed immediately with heavy syrup, using a spray bottle, and store in a cool, dark area.

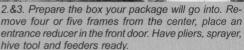

2.&3. Prepare the box your package will go into. Remove four or five frames from the center, place an entrance reducer in the front door. Have pliers, sprayer, hive tool and feeders ready.

4. Spray the bees with syrup, lift and thump down the box (you won't hurt those bees) and knock them to the bottom.

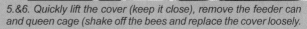

5.&6. Quickly lift the cover (keep it close), remove the feeder can and queen cage (shake off the bees and replace the cover loosely).

7. Dump the bees into the cavity, slowly replace the frames and place the queen cage between the two middle frames, with the screen facing the space between these frames so the bees have access to her. Put on the feeder (not directly over the queen cage), place an empty super over the feeder and close it up. In three days return and remove the cork from the queen cage, and in two-three more days check to make sure the queen is released. Do so if she's not.

Too many dead bees on cage bottom – too long in cage, starvation

of bee biology. Many package-bee shippers have their source colonies inspected for diseases and mites and the packages themselves carry certificates of inspection.

This inspection certificate, however, is not a guarantee that every package is perfect. Nor does it guarantee that the source colonies were inspected recently. When purchasing a package, quiz the seller about inspections, past complaints, prevention techniques to guarantee that *Varroa*, small hive beetle, and American foulbrood are not being shipped. Ask for references. Talk to your State Apiary Inspector about compliance, complaints and customer service from the producer you are considering.

Ask other beekeepers, too, about their experience with this producer, not only in quality, but service, and perhaps more importantly, the quality of the queen that comes with the package. The queen dictates the future of your colony, whereas the bees in the package are only there to get her started.

The disadvantage of starting with packages is the cost. In many locations a colony started with package bees will not produce surplus honey the first year and will need feeding to survive the winter. The queen that accompanied the package may sometimes fail and need to be replaced. The population of bees will decline to about half before starting to expand. However, being aware of these events will preclude problems, and beginning with package bees on new equipment is the safest, and most predictable way to launch this undertaking.

Nucleus colonies – Nucleus colonies can be considered a combination of an established colony and a package of bees. A typical nucleus colony is made with four or five combs, one or more containing brood, pollen and some honey. Bees on the frames and a queen are installed in the nucleus beehive. Then, when the population fills this box, the entire colony is moved to full-sized equipment. Or, more likely, it is moved immediately after purchasing. The advantage of a nucleus colony is that it has a good start with a population of adult bees, an accepted queen, brood, and food stores. Like a package colony, a nucleus colony will probably require feeding to develop winter stores. One disadvantage of purchasing a nucleus colony is that the drawn combs may be a source for diseases and pests, and prior inspection, if not the law, should be insisted on. One precaution when purchasing 'nucs' (the common term for a nucleus colony), is to make sure the equipment your nuc is raised on (usually deeps), fits your equipment, which may

Nucleus Colonies

A nucleus colony, generally called a 'nuc,' is a full fledged, thriving colony with workers, brood and an established, laying queen – all in a four or five frame (instead of eight or 10) colony.

Buying a nuc insures that your queen is healthy and your colony is established. The seller has assumed the risk and time to do this so it will cost more than a package.

Once you obtain your nuc, remove the frames and place in your eight or 10 frame equipment.

The caution here is the frames you obtain. They should be free of disease, no more than two to three years old, and have been inspected. You could be buying someone's old, junky, diseased equipment. Make sure that doesn't happen.

Capturing Swarms

Swarms can be easy to collect, especially when low to the ground and simply hanging on a branch or other object. Sometimes, they're a bit more challenging, as when spread over a wall, in the center of a shrub, or 60 feet in the air.

Easy . . .   *. . . not so easy.*

be different. Equipment that doesn't interchange is worse than useless. Check before you buy.

Capturing swarms with bait hives - Beekeepers have used bait hives to capture swarms for thousands of years. However, only recently have studies been made to determine what kind of nest bees prefer when given a choice. Sufficient information exists to build and position a reasonably successful bait hive to capture a swarm as a means of starting in beekeeping. Information on bait hives can be found in the section on BAIT HIVES. Hobby beekeepers in both rural and urban settings have found that bait hives in their backyard, properly located, can be an easy way to capture swarms.

An advantage of a bait hive is that it often presents an opportunity to observe bees scouting it and perhaps even moving into it. This can be an exciting experience. It is important to check bait hives frequently and move the bees that may occupy it into a movable frame hive as soon as possible. A chief disadvantage in capturing swarms in a bait hive is that they will need feeding to survive the winter and will need to be requeened. Any brood, wax and stores in a bait hive are nearly impossible to recover because the comb is new and so fragile that it is easily damaged.

Capturing swarms – Most swarming in honey bees takes place in the spring. When swarms emerge they usually settle on a tree or bush within 50 to 100 feet (15 to 30 meters) from the parent colony. They may remain in this location, or another temporary location, for a day or sometimes several days. At this time, with the exception of those swarms that settle high in a tree, it is easy to capture the swarm in a hive and to move it, at night, where it is wanted. There are several methods of capturing a swarm.

One easy way to capture a swarm is to hold a cloth bag or a bucket under the hanging swarm, and use one hard, swift stroke to shake the bees into it. The bees are then taken to their new hive and shaken in. A plastic bag should not be used since the bees may suffocate.

A second method is to shake the swarm into a hive body with combs. First staple a bottom board into place under the hive body. The cover is put on the hive after the swarm is shaken into it and put nearby in the shade. Swarms will not always remain in a hive into which they are shaken. They are especially inclined to leave if the sun hits the hive and it becomes too warm. If a frame of brood is placed in the hive the bees are much more inclined to remain. Stray bees, or those lost from the swarm, will find the hive if it is within 15 to 20 feet (four to six meters) of the original clustering site.

Capturing swarms from trees or buildings – The most difficult and time-consuming way to start in beekeeping is to attempt to take a colony alive from a building or tree. Still, from the point of view of learning about bee behavior and the natural nest, it is something everyone should attempt once. Removing bees from a feral nest often results in loss of the queen and some of the bees (see REMOVING FERAL COLONIES). The best time to remove a colony from a nest in a tree or building is early spring when the colony population will be the smallest and the colony will have time to recover and prepare for winter in its new home.

The advantage of capturing a swarm is that the bees are free of charge. Beekeepers sometimes worry that bees in a swarm may carry a disease; usually only healthy colonies swarm but swarms infested with *Varroa* have been captured. Therefore it is a good idea to examine the colony for disease and mites. The chief disadvantage of capturing swarms is that the bees may not have time to gather sufficient food for winter and it may be necessary to feed them. A second disadvantage is that the queen that accompanies the swarm will need to be replaced within 4 to 6 weeks.

Caution: Bees in a swarm are usually well-fed and gentle. However, if the swarm has been away from the parent colony for some time, and has exhausted its food reserve, it becomes a "dry" swarm. Dry swarms can be extremely vicious and nearly impossible to manage. Before one starts to collect swarms, it is best to practice with a local beekeeper.

Number of colonies to buy or manage – Owning one or two colonies of bees is difficult because one is never certain what is normal for colony population and progress. Starting with three or more colonies gives one an opportunity to observe how much variation may exist in colonies. For example, anyone who has worked with bees knows that in a large apiary it is not uncommon for one colony to store twice as much honey as the poorest producer. Of course, in a perfectly-managed apiary such variation would not exist, but in practice such variation is commonplace.

Owning three or more colonies may not be practical for all beginners. A reasonable alternative is to keep in close contact with other beekeepers in the vicinity and to attend outdoor beekeepers' meetings where colonies are opened and their progress assessed. If one's goal is honey production, then a thorough understanding of honey bee biology and the variation in colony growth will be helpful.

BENTON, FRANK (1852–1919) - Around the end of the 19th and beginning of the 20th centuries Benton, an employee of the U.S. Department of Agriculture, spent many years abroad searching for better races of honey bees. He succeeded in bringing queens of several races into the U. S. He also wrote one of the early USDA bulletins on beekeeping, *The Honey Bee*, in 1899. Benton attempted to bring a colony of *Apis dorsata*, then the largest known honey bee species, into the U. S. from Asia but the bees died enroute. The standard wooden queen mailing cage in use for many years was named for Benton.

BERRY, M.C., SR. – In 1902 M.C. Berry, Sr. moving his colonies to the small town of Tyson, Alabama, where the sweet clover and many other honey-producing plants grew rank and wild over the pastures and waysides. These sources yielded heavy crops of light colored honey for many years.

With his increased holdings in apiaries, Mr. Berry began shipping packaged bees and queens to all parts of the United States and Canada. At first, shipment was entirely by rail express. This held shortcomings as purchasers in isolated communities had no express office. Berry decided to find an alternative with Parcel Post. With the assistance of the Postmistress, Mrs.. Sally of Hayneville, Alabama, they searched through postal regulations but could find nothing to prohibit the shipment of bees properly caged. Baby chicks were shipped in this manner.

Benton Queen Cage

The Benton queen cage is the most popular, but not the only cage queens are shipped in. The standard three-hole cage allows cages to be bundled together.

The cage itself is simply a block of wood with three (sometimes two) cavities drilled out. Each end has a hole that has a small cork or other blocking device in it. One of the three cavities is filled with a candy, made from sugar and corn syrup. A screen covers all three holes.

To release the queen remove the cork from the candy end, poke a narrow hole through the candy to assist the hive bees eating in, and the queen to eat out, and place between two frames so the screen is exposed so the hive bees can get to the queen. (*Morse photos*)

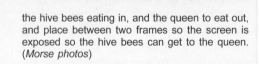

BIRDS – In North America, birds are not serious pests of honey bees. Occasionally people have reported observing a bee-eating bird especially in the vicinity of a queen-rearing apiary. However, the reports are so infrequent that birds are not mentioned as pests in many textbooks on bees and beekeeping. One exception may be the woodpeckers. (see WOODPECKERS)

The eastern kingbird, *Tyrannus tyrannus*, has been occasionally mentioned as a bee pest in the U. S. The western tanager, *Piranga ludoviciana*, has been seen eating live bees in California. Birds will sometimes be seen feeding on dead bees in front of a colony.

Outside of North America, especially in Africa and Asia, several bee-eating birds are pests. Their numbers may be so great that strong measures must be taken to prevent their having an adverse effect on an apiary. In Africa, and to a lesser extent Europe and Australia, over two dozen species of bee-eaters belong to the family Meropidae. These birds appear to favor eating stinging insects. Some are migratory. The birds usually catch the bees on the wing.

Other birds that are sometimes reported to be a problem include some of the tits, shrikes and swifts. In the Philippines, for example, flocks of swifts may visit an apiary almost daily and single birds may eat many bees. The problem has been reported to be so severe at times that beekeepers have countered by capturing the birds. Such birds are especially a problem when queens are being reared.

BOCH, ROLF (1928-1984) – Dr. Rudolph (Rolf) Boch was born in the town of Stiefenhofen, Germany. His father had a typical German Bienenhaus in which he kept a few colonies of bees.

Rolf received his education in German schools, and completed his doctorate under Professor Karl von Frisch, at the University of Munich.

His thesis work was concerned with the relationship between the bee dances and the distance of the nectar sources from the hive. Rolf was among the first to notice the change in direction of the dance with the changing location of the sun through the day.

Rolf came to Canada in 1956 to join the Bee Division under the late Dr. C.A. Jamieson. He married Ana R. Bizetsky who had also studied at Munich with Professor Frisch.

His first studies at Ottawa were on the relation between the size of queens and their productivity, proving what many beekeepers "had always known." He also developed, along with Mr.. Fairburn of the Ottawa Valley Beekeepers Association, a cardboard wintering case for bee colonies. His main career efforts were centered, however, on the effects of specific odors on bee behavior.

Specifically, he and his chemistry associates found that isopentyl acetate (banana oil) found in the worker bee sting gland was mainly responsible for alerting and alarming hive bees for defense of the hive against intruders; that queen substance not only attracted drones to queens in flight, but also attracted workers to queens in swarms and in the hive; this attraction was reinforced by worker scent produced in the scent glands of the workers, themselves. In harmony with these findings were studies showing that acceptances of queens introduced to new colonies was controlled by the genetic makeup of a given queen plus "hive odor."

In recognition of these and other studies, Rolf was the first scientist to

Rolf Boch

receive the James I. Hambleton Award for excellence in bee research, given by the Eastern Apicultural Society.

BONNEY, RICHARD (DICK) (1929-2001) – Born in Newton, Massachusetts, he graduated from Norwood High School in 1946, served in the U.S. Marine Corps until 1951, married Joan Morrton of Wellfleet, Massachusetts in 1953, graduated from U. Mass Amherst with a degree in landscape architecture in 1954; and received a masters in science education from Fairleigh Dickinson University in 1971.

After a successful first career in computer systems management, in 1978 he and Joan moved to Charlemont, Massachusetts to start a second career in beekeeping. There he founded and operated Charlemont Apiaries, wrote numerous books and magazine articles about bees and beekeeping, served as Massachusetts State Extension Apiculturist, and taught beekeeping at U. Mass Amherst until retiring in 1997. He was a columnist for *Bee Culture* for 10 years. He also published beginning and intermediate level beekeeping books.

BOTULISM – (see INFANT BOTULISM)

BOX HIVES – A box hive is just what the name suggests, a colony of honey bees in a box of any size. Despite the abundant literature and information available on bees today, a few beginning beekeepers, or people who find a swarm, actually dump the bees into a box. Often a box is suitable for a nest site, at least as far as the bees are concerned. Once the bees have built comb and started brood rearing, they will not abandon the box except under unusual circumstances. Swarms have also settled in wooden kegs, crates, and packing boxes.

It is illegal to keep bees in boxes because the brood combs cannot be removed for inspection and replacement. Consequently many colonies in box hives perish because they can neither be inspected nor treated for diseases and parasitic mites. Box hives are not ideal units for the production of honey because honey produced in box hives is difficult to remove.

It is usually possible to salvage the bees and the wax from such nests. This is done in a similar manner as removal from a hollow tree. A hole is cut in the top of the box, and a hive body with combs is placed above the box. The box hive must be rigid enough to support the hive body so that the box will not collapse when a portion of the top of the box hive is cut away and the nest exposed.

Another alternative is to turn a box hive upside down and to remove the bottom, instead of the top, and to allow the bees to work upward from their upside-down position. Bees rarely attach comb to the bottom of their nest, and removing the bottom is often much easier than removing the top. If the comb fills, or nearly fills the box, then bees

Braula coeca

Many styles and types of branding irons are available using gas or electricity to heat the brand.

will move from the comb in the box hive to a hive body of combs placed over the upside-down box. Turning a box upside-down will not have an adverse effect on the brood and it will emerge normally. In fact, before she moves up into the comb above, a queen will continue to deposit eggs in cells turned upside-down even though their orientation is slightly different. Cells in a comb are not flat but tilt slightly upward toward the outside; presumably this helps them to retain their contents whether it is brood or food. Yet another method of removing bees from a box hive is by drumming (see DRUMMING).

BRANDING HIVES – Because bee hives are sometimes stolen, beekeepers seek deterrents to theft. One good deterrent is to brand the hive parts, especially the supers. Some beekeepers have branded the tops of frames as a further deterrent.

Electric, gas-heated, and branding irons that must be heated in a fire are available. The deeper the brand can be impressed into the wood, the more permanent and effective it will be. Most beekeepers prefer to use their initials as the brand. Obviously, beekeepers who have practiced branding for several decades, and have bought and sold certain pieces, may end up with many different brands on their equipment.

BRAND MELTER – (see BEESWAX, Brand melter)

BRAULA COECA (Bee Louse) – The braulids are a curious group of pinhead-size insects that infest honey bees. Braula are sometimes called bee lice but are really wingless flies. They do little if any harm other than taking food from the mouths of honey bees. Usually one sees only one or two on a bee at one time, but a few people have reported finding as many as 100 or more on a queen, which must be a nuisance.

Only one species, *Braula coeca*, has been found in the U.S. This species has been reported from only 14 states, all east of the Mississippi River, from New York to Florida; in the latter two states they have been reported only once. The greatest number has been reported from Maryland, especially in Carroll County. *Braula coeca* have never been found in Canada. The limited distribution of these small insects is very strange, especially when we consider how many colonies of bees are moved around the United States.(Smith, I.B. and D.M. Caron. "Distribution of the bee louse, Braula

Braula *on queen, and worker.* (Morse photos)

coeca, in Maryland." *American Bee Journal* 125:294-296. 1985.)

When a braulid feeds, it moves to the bee's mouth, where it clings to hair on the face and claws at the bee's upper lip with its front feet. This action prompts the bee to extend its tongue, and when it does the Braula inserts its mouthparts into those of the bee and sucks up whatever material it finds there. Apparently, female braulids deposit their eggs on cappings over honey, not over brood. The young burrow just under the cappings and their tunnels can give comb honey sections a bad appearance. Since little comb honey is produced today, the braulids are not of much economic consequence. For those who exhibit comb honey in honey shows they can be a nuisance.

Several other species of Braula have been described from Africa but their biologies have not been studied. In Nepal, two species of *Megabraula* have been found recently. These braulids are much larger than the others and so far found only on *Apis laboriosa*, the world's largest honey bee.

A thorough investigation of *Braula coeca* was performed by I. Barton Smith, Chief Apiary Inspector for Maryland (retired), who wrote a thesis on the subject in 1978. *Braula coeca* is about the same size as the Asian mite *Varroa destructor*, and it is possible that the two could be confused. The various chemicals used for control of *Varroa* seem to have diminished the presence of Braula in the United States.

Tunnels under honey cappings made by immature Braula.

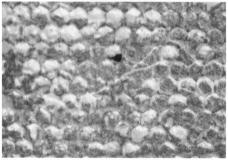

Braula *Adult* (OSU photo)

BREEDING BEES – It is not difficult to breed or select bees for special purposes or characteristics. Approximately 26 natural stocks of honey bees were found imported from Europe, Africa and the near east from the late 1800s until imports were prohibited in 1922. These bees varied in size, color, temperament, honey- and pollen-collecting ability, wax production, tendency to abscond and a host of other characteristics. Even within a stock there is great variation and wherever there is variation one can breed or select for the desired characteristics. At the present time beekeepers are very much concerned about developing bees that are gentle, disease and mite resistant and are good honey producers. (For further information on the selection and breeding of bees see INSTRUMENTAL INSEMINATION, GENETICS OF THE HONEY BEE, QUEEN REARING and DRONE BREEDER COLONIES).

BROOD – This is the collective name given the eggs, larvae, and pupae in a colony regardless of their age or distribution in the hive or combs. Capped brood refers to pupae whose cells are covered with cappings; uncapped brood consists only of eggs and larvae. (See ANATOMY, Average Development Time, for additional information.)

BROOD CHAMBER – This is the hive body or box(es) containing the frames with brood. The term is used to distinguish this chamber from one containing honey or food.

BROOD FOOD – Brood food is a general term that is usually used to describe the

Brood Box

Typically, a full size, 10-frame hive has two brood boxes, in these photos painted white. A small colony will probably start in a single box, with another added when six to eight frames are full of brood, bees and honey. When three to five frames are filled in the second box, honey supers are added, with, or without a queen excluder between.

The beeyard shown here is not only attractive, but educational, with brood boxes painted white, and honey supers yellow.

Typical 10 frame colony with two brood boxes and one honey super.

Beeyard with identifying paint.

secretions from glands in the worker bee's head. Worker bees feed the brood progressively, that is, as the larvae need food it is added into the cells. The larvae are not fed mouth to mouth, but rather the food is deposited on the lower side of the cells where the larvae can consume it as they turn around in their cells (remember, cells lie on their sides, not up and down).

The food fed to worker larvae changes in mid-larval life after initially being fed royal jelly. The older worker larvae, and presumably drone larvae, receive food that contains less of the glandular secretions and some honey. The term "brood food" is used to include all of the above. Royal jelly is the brood food fed to queens, while worker jelly is the brood food fed to workers (see ROYAL JELLY).

BROOD NEST – The brood nest is that section in the hive where the brood is reared; honey bees typically separate their brood and their food, making it easy to distinguish between the brood nest and the honey and pollen storage areas.

The brood nest takes the shape of a round or oblong ball, depending on the shape of the hive or natural nest. In a typical hive the brood nest will cross several frames. Because of the shape of the nest the frames on the outside of the nest will contain much less brood than those in the center.

When colonies are rented for pollination, the beekeeper often rents a colony(s) containing a given number of frames with brood. This is different from saying "so many frames full of brood." This distinction is made because it is easy to measure the size of a brood nest by counting the number of frames with brood, whereas it is nearly impossible to state how many bees may be present. There is a close correlation between the two, however.

We sometimes speak of a compact brood nest as one that is especially orderly and where the brood and food are clearly separated. A compact brood nest should not contain cells filled with pollen and honey. Honey bee colonies vary in the compactness of their brood

Brood Nest

In a typical colony the brood nest begins small, near the center of the center frame. As the season progresses the queen will expand her territory, as a growing population of workers expands to care for more brood. A frame will, generally be filled on the side closest to the center, then the other side. The brood nest radiates outward, toward the edges of the box, and upward, as room and caretakers permit. At the height of the season the brood nest will resemble an oblong ball, thickest and tallest in the center, with decreasing brood toward the edges.

This brood frame is nearly full on both sides, with just a few drone cells near the edge and the bottom.

This brood frame comes from closer to the edge, with only a small amount of brood in the center.

nest, sometimes according to race and sometimes according to the age of the queen. Old queens may produce less compact brood nests. It appears that younger queens are somehow capable of forcing the workers to keep honey and pollen out of the brood nest proper, probably because they produce and deposit eggs as soon as cells are prepared.

BROOD PATTERNS – The best way to judge a queen's egg-laying ability is to observe the pattern of eggs, larvae and pupae in frames in the brood nest. In making an evaluation, keep in mind that a queen with a good brood pattern has brood of the same age adjacent, that is, eggs should be next to eggs, larvae next to larvae of the same age, and pupae next to each other in the same manner. Queens lay eggs in ever-expanding concentric circles. In a comb with a good brood pattern, only a few cells are empty. The brood should be in a compact area so that brood nest temperature is more easily controlled by the bees. Cells within the brood nest should not contain pollen or honey; these should be stored alongside of, but mostly above, the brood. The brood nest in a colony takes the shape of a ball, crossing many frames, and thus the above pattern should exist not only on the side of a comb, but from one frame to the next.

While what is described above is desired, and typical of the pattern of a young queen, one must be cautious in making judgments on brood patterns. A larval disease may cause bees to remove the sick and dying and disrupt the pattern. Old queens may have eggs that fail to hatch or they may have exhausted their sperm, in which case drone brood may appear in worker cells. During an unusually good nectar and/or pollen flow in a space-limited hive, bees may store pollen and honey where they are not wanted. Chilling of the brood will also disrupt the pattern.

One should not use brood patterns alone as a criterion in evaluating queens. Bees that are stimulated to produce brood during a honey flow can be a problem. Because large numbers of adult bees are involved in brood feeding and care, they are lost from gathering and processing nectar. It is important that brood be reared before, not during the honey flow.

In a natural nest the drone brood is usually found in a compact area on the edge of the nest and away from the entrance. However, since man-made frames and the resulting combs vary in terms of the amount of drone comb they contain, one must be very careful when judging a queen's ability based on the quantity or position of the drone brood present.

BROOD REARING – Brood is the collective name given the eggs, larvae and pupae. The process of rearing the young bees is called brood rearing. Honey bees are the only insects, social or otherwise, that rigidly control the temperature of their brood-rearing area. The temperature in a brood nest is remarkably constant, about 93° to 96°F (34° to 35°C). A few records can be found of lower temperatures in the spring when the temperature may drop dramatically or the bees may be rearing more brood than they can cover with clustering bees during such cold snaps. Brood may be chilled and killed if the temperature falls too low. It is also thought that cool temperatures may adversely affect larval and pupal growth and cause abnormal bees (see ABNORMAL BEES).

The control of the brood-rearing

Brood Rearing

Because of the accumulation of pollen, cocoons, propolis, dirt and other material, brood combs darken over time. Additionally, brood combs are exposed to medications on a (sometimes) routine basis, and beeswax will absorb these, too. Thus, after a while the comb that bees walk on, are raised on and store pollen and honey in becomes contaminated.

When a brood comb, held up to the sun is so dark that light does not shine through, the comb is old enough to replace. This normally takes about three years.

Old, dark, contaminated comb contains disease spores (AFB, nosema, chalkbrood), chemicals and dirt. Chemical residue is detrimental to both drone and queen fertility. Replace old comb.

temperature in the brood nest is necessary for queens, workers and drones to mature in more or less precise periods of time. One of the factors that makes queen rearing possible and practical is that we can predict when changes will take place in the growth of queen brood and large numbers of queens can be reared together under conditions that would otherwise be impossible (see LIFE STAGES OF THE HONEY BEE and QUEEN REARING: GRAFTING AND COMMERCIAL QUEEN PRODUCTION).

Feeding occurs only during the larval and adult stages in honey bees. Larval feeding is done by young bees, called house bees or, more specifically, nurse bees. Larvae are fed progressively; that is, food is given to them frequently; their cells are not mass-provisioned all at once as is the case with the larvae of solitary and many subsocial bees. The larvae are not fed directly but the food is deposited in their cells and they consume it by rotating slowly around and around within the cell.

Brood rearing is controlled by several factors including temperature and the number of bees available to hold that temperature, the availability of pollen and nectar, and the number of bees available to feed the brood. Egg laying on the part of the queen, and therefore brood rearing, appears to be stimulated by an increasing day length and discouraged by a decreasing one (see LIFE CYCLE OF THE HONEY BEE).

BROTHER ADAM (1898-1996) – Brother Adam became a lay brother with the order at Buckfast Abbey, in Southwestern England, early in his life. He specialized in beekeeping. During the early part of the 1900s bees in England were apparently devastated by tracheal mites, though the cause of the disorder was not identified until later. Brother Adam became convinced that the solution lay in breeding better bees, especially those resistant to these mites.

Brother Adam soon became known as a master beekeeper and managed the Abbey's 400 colonies for the production of heather honey from the nearby moors. The honey is sold in a specially designed package to help support the work of the Abbey. Starting in 1950, he made a

Buckfast Abbey, Devon, England, looking from the apiary.

Brother Adam in his queenyard in Dartmoor, England.

series of journeys into parts of Europe and the near east and Africa looking for bees that might have special attributes. It was from these, and through a intensive selection and breeding program, that he developed the Buckfast® bee. Brother Adam was a prolific writer whose works have been widely read and cited. A special interest of his was the production of mead. He was a strong supporter of research, a vice president of the International Bee Research Association and received several English and international accolades.

BURLESONS, INC. – T.W. Burleson started keeping bees in 1906 when Flavius Davenport of Nash, Texas, gave him a colony. Burleson brought the bees back to Waxahachie in the back of his buggy. T.W. was District Clerk of Ellis County, but he became so interested in bees that he eventually gave up his political career to devote himself entirely to apiculture.

Business gradually expanded until 1925 when they suffered a $30,000 loss from a bad fire.

Upon graduating from Trinity University in 1929, T.E. Burleson (1904-1996) joined his father, and they started the first commercial honey bottling plant in Texas.

The first honey packing plant was at the T.E. Burleson residence at 1206 West Main Street, but because of the continual increase in business a new bottling plant was built. Thirty years later Burlesons had again outgrown their facilities, and in 1966 they moved into a new 27,000 square foot building they had constructed.

The family-owned business, T.W. Burleson & Sons, remains the oldest and largest honey packer in the southwest.

T.E. (Edward) Burleson served as president of the Texas Beekeepers Association; the National Honey Board; and the National Honey Packers and Dealers Association. He retired in 1974, however Mr. Burleson remained active inthe business until 1995.

BURR AND BRACE COMB – In a natural nest bees rarely build comb to join two combs together because they respect bee space. Beekeepers, however, are often less concerned about bee space when they construct hives so the measurements may not be precise. As a result, combs are bridged and joined by bits and pieces of comb called burr or brace comb. Whenever this type of comb is seen, it is a reminder that beekeepers have carelessly measured hive parts when cutting them or have spaced the combs incorrectly.

Honey bees, of any subspecies, do not build a passageway through a piece of comb. Whenever a person accidentally or purposefully cuts a hole through a comb, or a piece of foundation, it is almost always repaired and replaced with new comb. Around the edges, where the comb is joined to the inside of a tree or building, peripheral galleries or passageways are located. Both queens and workers walk around the edge of the comb as they move from one comb to another.

T.W. Burleson

Forget to replace a frame, or put in one too short and the bees will fill the cavity with comb to fit their needs, not yours.

Burr and brace combs are a nuisance and whenever found, especially between boxes, steps should be taken to correct the matter. This should be done especially when supers are too deep or too shallow and removing them from the hive is difficult because they are stuck together with burr comb.

Most often, the cause of this is the violation of bee space. And the most common violation is when a super with frames is placed upon another super with frames, and the resulting space between the top of the top bars of the bottom box, and the bottom of the bottom bars of the top box is larger than or smaller than normal bee space. If too large, the bees will attach the two with comb. If too small, the bees will attach the two with propolis. Both events make management, at best, difficult.

A too-wide space between frames will be filled with comb.

CALE, G.H., (BUD) JR. (1919-1978) – G.H. was born in Washington, DC on August 29, 1919, the son of G.H. and Elizabeth Chapin Cale. His father was on the staff of the Apiary Department of the USDA. He brought his family to Hamilton, IL on February 1, 1921.

Early in life Bud evidenced an intense interest in entomology, beekeeping in particular, by accompanying his father to outyards. He learned to raise queens as a child of six, and acquired his own colonies at 10. His father and Frank C. Pellett, noted horticulturist and author, played an important role in his early life. Both of these older men, through their editorial positions with the *American Bee Journal*, were able to expose Bud to wide-ranging interests in beekeeping subjects of both practical and scientific vein.

He participated in the early efforts to isolate disease-resistant colonies in the early 1920s and closely watched the progress of the program until it was clearly established that resistance did, in fact, exist and that resistance was inheritable. By 1932 a full yard of resistant colonies had been isolated.

A milestone occurred in 1927 when Dr. Lloyd R. Watson of Alfred University in New York came in Hamilton to demonstrate his technique for artificial insemination of honey bees. Bud was one of those who observed the demonstrations. Future refinements by Dr. Otto Mackenson and Will Roberts finally made the procedure practical.

Bud joined the staff of Dadant & Sons, Inc., immediately after conclusion of his military service, and promptly launched intensive work with queens. He became convinced, from his observations of the results of natural matings, as used in the disease resistant breeding program, that measurable and predictable improvement must necessarily come from a thoroughly controlled scientific genetic breeding program.

Spending summers in Hamilton and winters at Iowa State University, Bud

G.H. (Bud) Cale, Jr.

obtained both a Master's and Doctor's degree in Genetics and Plant Breeding. He strongly felt honey bees and their heredity more closely resembled plants than animals.

The first major impact of Bud's work appeared in 1950 with the release of the Starline hybrid variety. A second hybrid, planned especially with gentleness for hobbyists, was called the Midnite hybrid. It was released in 1958.

Bud developed another line, the Cale 876 hybrid, named in his honor by other members of the Dadant Company. The Cale 876 hybrid was the first three-way hybrid produced on a large scale. Unlike its forefathers, the Starline and Midnite, each Cale 876 queen was instrumentally inseminated to avoid chance free-flight matings.

Some of Bud's other work in the industry included keeping several hundred colonies on a commercial honey producing basis; his research in Mexico for several years and his many scientific papers, articles in the *American Bee Journal* and chapter on "Genetics and Breeding of the Honey Bee" in the *Hive and the Honey Bee* co-authored with Dr. Walter Rothenbuhler, professor of entomology, zoology and genetics at the Ohio State University in Columbus.

Although his career in beekeeping has covered many aspects of the industry, it is for his revolutionary work in honey bee genetics that Dr. G.H. Cale will long be remembered.

CANDLES – (see BEESWAX IN ART AND INDUSTRY)

CANDY, QUEEN CAGE – (see QUEEN REARING)

CAPPING OF CELLS – Bees build wax cappings over cells containing honey and those that contain mature larvae that are about to pupate. The two types of cappings are different – one type is over honey and one type is over larvae. Only after the cells are capped do the larvae spin cocoons (see Cocoons). The chief purpose of the cappings in both cases appears to be keeping the cell contents clean, or at least free of debris that might fall from a bee as it walks over the cells. The cappings over both honey cells and brood cells are not completely airtight. A developing pupa must be able to breathe oxygen through the capping over its cell. As for cells containing honey, honey has the ability to pick up or lose moisture, depending on the humidity, in the area of the hive the honey is in.

'Dry' cappings, shown in the top frame, are simply cells covered with wax, but having an air space between the surface of the honey and the wax. Some comb, chunk and cut-comb honey producers find this improves sales. Below, are 'wet' cappings, where the wax sits directly on the honey. Neither capping affects the quality, flavor or color of the honey in the cell.

Capping of Cells

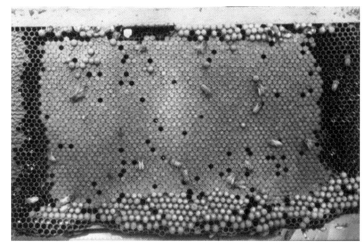

This frame has a large area of capped worker brood in the center, a large area of capped drone brood below that, and off to the left a small area of capped honey. (Calderone photo)

The cappings over honey are almost pure wax. The wax is new in cappings over honey and its production is stimulated by the incoming nectar. Races of bees vary in their methods of capping honey. Bees from northern Europe are more inclined to build their cappings above the honey so that they do not touch the cell's contents (see Comb honey production). This is called "capping the honey white" or "dry cappings" as opposed to "watery cappings." In the latter case, the cappings touch the honey, giving the cappings an altogether different appearance. When bees cap honey, they apparently cap an entire area at once as opposed to capping cells individually as they do with brood cells. Cappings over honey have a wrinkled appearance, and often a single fold or wrinkle will cover several adjacent cells. It is often impossible to determine exactly where the cell edges lie.

Cappings of worker brood are usually convex, or raised slightly; this varies, however, and certain races appear to construct cappings that have a dimple in the middle. Cappings over drone brood, whether in drone or worker cells, are more severely convex, bullet-shaped is a common description, to accommodate the drone pupae, which are larger than worker pupae.

Only a few people have recorded how bees cap brood. Most researchers have

On the left is a frame of worker brood fully capped, probably from the very middle of the brood nest. Often honey is stored in the corners, and a narrow band of stored pollen will separate the brood and honey. (Morse photo) *Right, shows the slightly convex structure of worker brood cappings.*

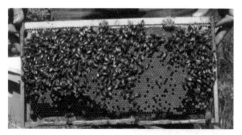

This shows capped worker brood in the center, capped drone brood near the bottom, right hand side, and at the very bottom right, a queen cell.

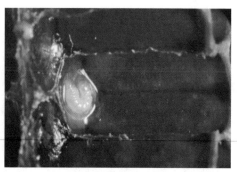

Larva float in a puddle of food. (Jaycox photo)

observed that many workers participate in the capping of a single cell; one researcher, however, observed a bee capping a cell alone (Smith, M. V. *A note on the capping activities of an individual honeybee.* Bee World 40: 153-154. 1959). Smith watched a single worker cap two cells and work on capping others. The material used for the capping was taken largely from the surrounding cell rims. The worker also appeared to insert her head into a cell and take wax from the cell interior. Smith stated, "The edge of the capping on which she was working had a roughened irregular appearance, with little strings of wax and fibers of other material extending from the unfinished margin." Cappings over brood range from nearly red to tan to very dark brown in color and are quite different from those over honey. Wax is recycled from other parts of the comb and combined with new wax to cap brood.

CAPPINGS, TREATMENT AND RENDERING OF – (see BEESWAX - , Rendering beeswax cappings)

CARBOHYDRATES – This is the term given to a wide range of compounds composed of carbon, hydrogen and oxygen. Carbohydrates include the sugars and starches made by green plants. In general terms in beekeeping the carbohydrates are the sugars and are sources of energy for bees. Sugars are, of course, the principle ingredients in honey. Sucrose is the primary sugar in nectar that worker honey bees collect and convert into honey which is primarily glucose and fructose (see NECTAR, CONVERSION TO HONEY).

CASTE – E.O. Wilson, in his work, *The Insect Societies* (Belknap Press. Cambridge, MA. 548 pages. 1976), defined *caste* as any group of individuals of a particular *type* in a colony that specialize on one or a few tasks. More specifically, a *caste* refers to a group of individuals of the same sex that behave similarly to each other but differently from other members of their sex. Behavioral differences among castes are usually accompanied by specific differences in morphology and/or physiology. Castes are thought to have evolved because specialization increases the efficiency of task performance, and colonies with more efficient labor forces have greater reproductive success than colonies with less efficient labor forces. There are no known castes among males of the social Hymenopterans (the ants and the highly social bees and wasps), but male termites (Isoptera) often exhibit caste differentiation. While females are a different *type* of individual than males, the accepted term for this distinction is *sex*.

Division of labor typically exists at two levels. A primary or reproductive division of labor between the queen and worker castes characterizes all advanced societies. Queens lay all or most of the eggs, while workers are reproductively limited, investing their energy in the other tasks required for the growth and maintenance of the colony. In nearly all social insects, this reproductive division of labor is due to environmental factors. In the honey bee, females arise from fertilized eggs that are heterozygous at the sex locus. Such an egg may take one of two paths during development, depending on the nutrition it receives during the larval stage. Larvae fed an abundance of royal jelly emerge as queens with all of the necessary morphological, physiological and behavioral specializations required for life as a reproductive. However, most female larvae are fed a different diet – in both quantity and quality – of worker jelly and emerge as workers with all of the specializations required for the performance of the many non-reproductive tasks that must be performed in order that the colony survives and grows. In other species, other mechanisms of caste differentiation may be involved. For example, in the ant genera *Formica* and *Myrmica*, larvae reared at higher temperatures tend to develop into queens.

Most species also exhibit a secondary division of labor among the members of the worker castes. For example, in the tropical leaf-cutter ant, *Atta cephalotes*, there are three morphological worker castes: minor, media and major workers. Each of these three types of workers is characterized by unique physical traits and each specializes on a different task or set of tasks. In other ant species, one finds only major and minor workers. In most ant species, and in most social insects, including the honey bee, one finds only a single morphological worker caste. Morphological caste differentiation is also due to environmental factors, including larval nutrition; however, this process varies from species to species and is not well understood.

The more common type of division of labor among workers is based on *age* or *temporal castes*. Worker honey bees progress through a series of tasks as they age. T. D. Seeley in *Honeybee Ecology*, (Princeton University Press, Princeton. NJ. 201 pages. 1985) described four age castes in honey bee workers: cell cleaners, brood nest workers, food storage workers and foragers. Workers in each age caste perform a number of related tasks, beginning as cell cleaners, gradually shifting to nursing and then food processing tasks, and finally ending their lives as foragers or scouts. As a worker shifts from one set of tasks to another, her underlying physiology changes in concert, providing her with the exact set of tools she requires to perform her new jobs.

While this framework provides a general description of the honey bee's system of division of labor, there is actually considerable variation among workers with respect to the age at which they engage in most tasks, and some workers skip certain tasks altogether. To emphasize the fact that a worker's age, *per se*, does not determine what task she performs, modern biologists prefer the terms *temporal polyethism*, *temporal castes*, or *temporal division of labor*, rather than *age castes* or *age polyethism*.

Although the association with age is imprecise, the many tasks that workers perform do appear to comprise a sequence, with nest duties performed before field duties. Once a worker shifts from one task group to the next, she generally does not revert but she can. The transition, from one set of tasks to another is thought to be driven in large part by environmental factors, most notably, the needs of the colony. However, other factors including pheromones produced by older workers are also at play. It is not a simple system.
– NWC

CASTE DETERMINATION – In honey bees there are two female castes: the workers and the queens. The drones are not a caste since they are males and appear in one form only.

Queen honey bees lay one type of egg only. If the egg is fertilized a female will be produced. If the egg is not fertilized then a male (drone) is produced. Whether the larva, which hatches from a fertilized egg after about three days, becomes a worker or a queen depends upon the food it receives during larval life and the size of the cell in which it is reared (see LIFE STAGES OF THE HONEY BEE).

This larva is less than two days old. It was transferred (grafted) from a Breeder Queen colony about 12 hours before this photo was taken, and less than an hour after it emerged from its egg. The plastic cell cup will house this queen-to-be until she emerges to mate and head her own colony.

Honey bee larvae are not fed directly by the nurse bees. Rather, the nurse bees deposit the food in the bottom of the cells and the larvae float in a sea of food that they consume as they turn around and around in their cells. Both queen and worker larvae receive the same food during the first day or day and a half of larval life but thereafter queen larvae are fed more lavishly and differently from their worker sisters. A great number of studies have been undertaken to define how the food for the two castes differs. It appears hormones play an important role. A major difference is that a one- to four-day old queen larvae has over 30% sugar in the royal jelly in her cell while the worker jelly has about 12%. In late larval life the sugar content of the food is about the same. As might be expected, in the laboratory intermediates between workers and queens can be grown by adjusting the quality of the food. This situation occurs rarely in a hive though the lack of good quality food may sometimes result in inferior queens. Long ago it was shown that it was possible to rear larger queens, with more ovarioles and presumably a greater egg-laying capacity, by insuring that the queen-cell-building colony is well fed with a bountiful supply of both pollen and honey or sugar syrup.

It has been suggested that the higher sugar content of the queen's larval food stimulates her to eat more and that this in turn has an effect on her hormones that in turn affect her growth and the differential development of her body. Queen breeders select 12- to 24-hour-old larvae to graft (transfer) into man-made queen cups. However, larvae as old as three and a half days may be forced into becoming queen-like. This again indicates that there is a complicated agenda in the developmental stages.

CELL SIZE – The subject of worker cell size, more specifically, worker-brood comb-cell diameter has been the subject of spirited discussion among beekeepers for decades. In the late 1800s and well into the 1900s, scientists were exploring the concept that somewhat larger bees could be produced if colonies had larger brood cells. This indeed proved to be the case. Their presumption was that colonies having larger bees, with longer tongues and larger honey stomachs, would gather more nectar and produce more honey. Subsequent research has demonstrated that selection and breeding to produce larger bees is far more important than cell size. But more importantly, a correlation between larger bees and increased honey production has never been demonstrated.

The variability of cell size and shape can be seen on this single comb, removed from an empty space in a normal colony. (Morse photo)

Wax combs in a honey bee nest have four types of cells; queen cells, worker cells, drone cells and transition cells. The latter are cells built between worker and drone cells, where comb repairs have been made, and around foreign objects. Transition cells are often oddly shaped. Queen cells, when present, are very large and "peanut" shaped. Drone cells are hexagonal and decidedly larger than worker cells, but relatively few in number. Worker cells, also hexagonal, constitute most of the cells in a colony. Normally, honey is stored in combs having worker-size cells; however, drone and transition cells will be used when available space is limited. Indeed, past beekeepers used drone foundation in honey storage supers believing that fewer, larger cells required less wax to produce, increasing honey production.

Recently, some beekeepers have focused on utilizing brood combs with optimal cell diameter for their geographic region. Their goal is to optimize honey bee reproductive rates, reduce susceptibility to disease and parasitic mites, and produce healthier, more vigorous colonies. Although this philosophy remains largely unproven via the scientific method, it seems to have merit in a small body of scientific evidence. It is also strongly endorsed by beekeepers that employ cell size as a management strategy. Other beekeepers have struggled to determine the optimal cell diameter for their region and obtain and use foundation accordingly. These efforts have often been thwarted by the availability of such foundation, and by industry-wide bee-breeding methods where-in there is constant selection pressure for large productive queens. Large queens produce large workers. It has proven difficult to get strains of large bees to accept and use small cell foundation, and vice versa in some instances. Hence, many beekeepers simply ignore the issue of cell size and use foundation and bees that are generally available throughout the beekeeping industry. - EHE

Size of natural cells – The size of naturally constructed cells has been a subject of beekeeper and scientific curiosity since Swammerdam measured them in the 1600s. Numerous subsequent reports from around the world indicate that the diameter of naturally constructed cells ranges from 4.8 to 5.4 mm. Cell diameter varies between geographic areas, but the overall range has not changed from the 1600s to the present time. While the variance may seem relatively slight, it appears to be significant relative to the perceptions of the bees. Broad interpretation of available data seems to suggest that, as a general rule, cells, as well as bees, appear to be smallest at sea level near the equator. Cell diameter, as well as bee size, increases with distance toward cooler climates and higher altitudes. Like man, honey bees evolved as geographic races, each adapted to specific climatic conditions (Note: Over 20 geographic races of honey bees have been identified with more yet being discovered). Each race of bees appears to construct, or exhibit a preference for, a specific cell size.

It has been suggested that the controlling environmental factor is temperature, e.g. climatic "thermal zones." Further research is needed to define this relationship and determine precisely how critical local adaptations are. It is, however, noteworthy that reported cell size for Africanized honey bees averages 4.5–5.1mm, and populations of Africanized bees build up more rapidly than European honey bees. (Grout, R.A. 1931. A biometrical study of the influence of size of brood cell upon the size and variability of the honeybee (*Apis mellifera* L.) M.S. Thesis, Iowa State College) and (Erickson, E.H., Lusby, D.A., Hoffman, G.D., and Lusby, E.W. 1990. On the size of cells; speculations on foundation as a colony management tool. *Gleanings in Bee Culture*, 118: 98-101, 173-174) and (Lusby, D. and Lusby, E. 1998. Is smaller better? *Gleanings in Bee Culture*, 126: 25-27) - EHE

Measuring cell size – Over the past one hundred years or more, a variety of methods have been developed for measuring cell size. Available data were reported accordingly. Most of these methods were the result of scientists trying to calculate the number of cells on both sides of a unit area of comb when estimating rates of brood production, colony growth, and colony size. Such methods are generally viewed as complicated and confusing. The simplest way to accurately determine cell diameter is to measure ten cells linearly across the face of the comb using a metric ruler, and then divide by ten. Because there can be slight variation from cell to cell, experts have further suggested that this be repeated three times and the average calculated. While these measurements can also be made in inches, useful data now being reported industry-wide refer to the diameter of a single cell in millimeters, e.g. 5.1 millimeters. - EHE

CELLAR WINTERING – (see WINTERING, INDOORS)

CELLS, TYPE OF – (see COMB, NATURAL)

CHALKBROOD DISEASE – (see DISEASES OF THE HONEY BEES – Chalkbrood)

CHECKMITE+® – (see COUMAPHOS)

CHILLED BROOD – It is not uncommon to find brood that has died because of exposure to cold; this is called chilled brood. Such brood is most likely found in the early spring at a time when the bees are expanding their brood nest rapidly and cannot generate enough heat on cold nights to protect the brood. Usually the dead brood is found on the outer fringes of the brood-rearing area and not scattered throughout the comb. The bees apparently recognize that they cannot protect all of the brood, and they abandon a portion of it while still protecting the rest. Brood dead from chilling is sometimes found after a pesticide loss kills a portion of the adult bees in a hive. Chilled brood may also be found in the spring when a beekeeper is making splits and fails to add enough adult bees to cover the brood.

Affected larvae and pupae are discolored, usually yellow with some black on the segmental margins. Eventually the brood turns brown or black. The dead larvae may be pasty or watery. Brood killed because of chilling, will be uncapped and removed slowly by the bees. Frequently pupae dead from chilling may mimic the symptoms of bald brood caused by the lesser wax moth. However, lesser wax moths are usually a problem only in tropical and subtropical areas and thus do not usually occur in areas where chilling is common. In extreme cases, even adult bees have been known to decapitate pupae.

CHITIN – (see ANATOMY AND MORPHOLOGY OF THE HONEY BEE)

CHRYSLER, CHESTER E. (1892-1970) – He was the first to introduce wired foundation to Canada, and invented the equipment that automated making comb foundation. He also developed the Chrysler Automatic Radial Honey Extractor, as well as an excluder welding machine. His company, A. Chrysler and Sons, was started by his father, and operated out of Ontario.

CHUNK HONEY – When pieces of comb honey are placed in a jar and surrounded with liquid honey it is called a chunk honey pack. This is not the same as cut-comb honey that is usually sold in plastic trays but not surrounded with honey. Both cut-comb and chunk honey are produced in the same way and both are usually cut from a frame.

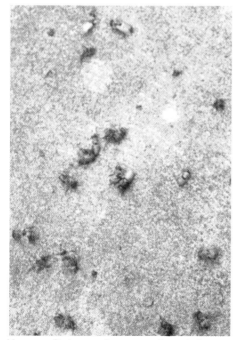
You may find several or many discarded pupae just outside the front entrance, especially if they were nearly mature. They will, generally, be all about the same age.

Chunk Honey

To make chunk honey that is attractive and easy to sell, start the season with unwired thin surplus foundation, so the comb is easy to cut. Put these frames in a colony that regularly produces white, or dry cappings. When the frame is nearly all capped remove it immediately, to avoid getting travel stain on the cappings.

Take this frame and place it on a raised-mesh drain table, so that when you cut the frame into pieces, the honey from the cut cells drains away, rather than puddles beneath the remaining comb. (Also used for Cut-Comb Production.)

Use a template, so you now how big to make the cuts to fit in the jar you use. Make the cuts and let drain overnight.

When drained, place the "chunk" of comb in the jar, and fill with a very light colored honey, so the comb inside shows through. The comb should be attractive on all sides. No dents, even cuts and completely capped. Some producers use two or three "chunks" to fill the jar, using smaller pieces left over from other cuts. This can be even more attractive in the jar, adding additional dimensions to the product.

The origins of the chunk honey pack are obscure but apparently honey was first packed in this manner in the southeastern states where the pack is still popular today. One disadvantage of chunk honey is that some of the liquid honey on the outside of the pieces of chunk honey may crystallize. A crystallized, or partially crystallized chunk honey pack does not have a good appearance and is difficult to use. Only comb and liquid honeys that naturally crystallize slowly should be used so that the pack has a shelf life of three to five months. Even then it is well to place the chunk honey packs in markets where the honey will be moved and consumed rapidly before it starts to crystallize.

CIRCULATORY SYSTEM – (see HEMOLYMPH)

CLARK, W.W. (BILL) (1917-2000) **& BESS** – Bill attended Penn State, and graduated with a degree in agricultural education as a member of the Class of 1939.

After WWII Bill worked with Penn State Extension Service as a beekeeping specialist, retiring in 1974 after three decades of service. Bill and wife, Bess

Bill Clark

moved to Canton, Pennsylvania where they operated an active beekeeping business. During this time Bess was a regular contributor to Gleanings In Bee Culture.

He also was part of the Volunteers in Overseas Cooperative Assistance, to help beekeepers in developing nations such as Bolivia, Gambia, Costa Rica, Egypt, Tunisia, Indonesia and Slovakia. Bess traveled with him, serving as photographer.

COCOONS – Soon after their cells are capped, worker, drone, and queen larvae spin cocoons within their cells. Workers and drones make full cocoons that surround their whole body, while queen cocoons are incomplete and do not cover the bottom of the cell where a supply of royal jelly remains. Queen larvae continue to feed on royal jelly after capping is completed and, in fact, may gain weight after capping.

The way in which the cocoon is spun has been recorded (Jay, S.C. *The cocoon of the honey bee*. The Canadian Entomologist 96: 784-792. 1964). Within their cells, larvae turn 27 to 80 forward "somersaults" over a period of up to two

Bess Clark

or more days. The spinneret, or the organ through which the silk is secreted, is located by the mouth. The silk is produced as a thread or sheet that, in the case of the workers and drones, covers the whole inside of the cell.

Interestingly, the queen cocoon, in addition to not covering the bottom half of the cell, does not touch the inside tip of the cell. Worker bees will often chew away this tip a day or more before the queen emerges. We can only assume that a chemical substance in the cocoon prevents the workers from removing the rest of the cell. Presumably removing the tip of the cell facilitates the queen's emergence, though the entire process is not completely understood. When a queen cell tip is exposed, it can be seen that the cocoon is a solid, protective sheath. While depositing silk the queen larvae add a colorless, pollen-free discharge from the anus that is incorporated into the cocoon. This is followed by light yellow gummy material and then by feces themselves.

It is not until the last stage of worker, queen and drone larval development that the hindgut and the foregut of the larvae are connected and feces can be voided. Most of the fecal matter is deposited in the bottom of the cell and is both covered and incorporated into the cocoon. The accumulation of fecal matter and cocoons spun over a period of time causes cells in old honeycomb to become smaller and, as a result, smaller bees may be produced in the cells. This fact has been of great concern to some beekeepers, especially those from Europe, who wish to grow large worker bees that will presumably gather greater quantities of food. If individual bees are smaller, less energy is used to grow them and the colony as a whole will grow more bees. This is a subject of endless debate. However recent studies show a relationship between worker cell size and increased honey production. (see BEESWAX FOUNDATION).

COGGSHALL, WILLIAM L. – He was born on a farm in central New York not far from Ithaca. After attending grade school in a typical one-room schoolhouse he entered Ithaca High School and graduated in 1931.

During his undergraduate work at Cornell he specialized in beekeeping, and after graduation in 1935 he served as a deputy apiary inspector with the Department of Agriculture and Markets, returning to Cornell to become an assistant to Dr. E.F. Phillips. His summers were spent in commercial beekeeping during his college years.

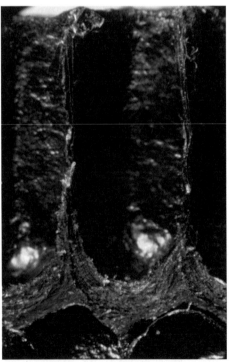

The bottom of a much-used brood cell, showing the layers of cocoons, frass and propolis. Like tree rings, perhaps you can 'date' how many generations have lived here. (Jaycox photo)

William L. Coggshall

After helping organize the Finger Lakes Honey Producers Cooperative at Groton, NY, in 1939, he served six years as its first president. In 1940 he began full-time honey production and was engaged in migratory beekeeping in South Carolina and Florida. Following several years of commercial beekeeping he returned to Cornell, completing the requirements for the Ph.D. degree in 1949.

He was co-author, with Dr. Roger Morse, also of Cornell, of the book *Beeswax*, still the definitive work in the field.

COLONY – In a general sense a colony is a group of animals living and bound together by some kind of mutual tie. In the case of honey bees the term refers to a group of bees and their queen, with or without males, living together in a man-made or natural nest. Dr. Tom Seeley writes "These findings demonstrate that a colony of honey bees, like an individual bee, or a cell within a bee, is an exceedingly intricate piece of biological machinery whose parts cooperate closely for the common good" (see Seeley, T.D. *Honey bee colonies are group-level adaptive units.* The American Naturalist. 150 (Supplement): 22-41. 1997).

The terms nest, hive, and colony are often used interchangeably. Other terms such as skep (usually a round hive made of straw), gum (usually a colony in a section of a tree) and swarm (usually a mass of bees and their queen away from a hive) are rarely used to describe a colony as each has a more specific meaning.

COLONY DEFENSE AT LOW TEMPERATURES – The defense and behavior of colonies at subfreezing temperatures consists of two steps regardless of temperature: When the cover is removed during cold weather, and the colony is disturbed, the bees on the exterior of the cluster react by arching their abdomens upward, protruding their stings, and exposing their sting chambers. In this position the stings are more-or-less perpendicular to the surface of the cluster. In many ways the surface of the cluster is reminiscent of a porcupine, and collectively the protruding stings give the clustered bees excellent protection for if any animal touches the surface it will be stung. Usually a drop of venom forms at the tip of the sting; occasionally a bee will fan her wings. Bees on the surface of the cluster do not fly.

The time colonies remain alert, and bees protrude their stings, varies and it is apparently not a function of temperature but probably of the degree of disturbance and occurs even under

Colony Defense At Low Temperature

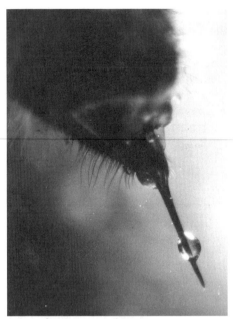

A drop of venom on a sting.

very cold conditions. The greater the disturbance, the shorter the time required for the next step in the reaction to occur, though obviously several seconds must elapse between the two acts.

The next step consists of a rapid expansion of the cluster as warm bees from within come to the surface. This, too, appears to be a function of the degree of disturbance and occurs even

Guards at the front (or any) entrance will challenge intruders, especially during a dearth. They are less likely to challenge a forager returning with nectar or pollen during a honey flow, even if the forager is from another colony. Drones, during the mating season, tend to not arouse much suspicion either.

under very cold conditions. If there is no continued disturbance of the colony, these bees merely mill over the surface and the cluster returns to normal. If a hand is passed over the cluster, or if the observer moves, the bees originally from within the cluster, but now on the outside, will fly at and attack the intruder. If the attacker moves, several may miss their mark and fall onto the snow or ground where they may die. Some may be able to return to the hive but this is a function of temperature, the colder the temperature the less likely they are to return to the hive. These warm, flying bees from the center of the cluster are quite capable of stinging, even at low temperatures, but only for a short period of time until their bodies chill. The number of bees flying at subfreezing temperatures is much smaller than the number that fly from a colony disturbed in a similar manner when the temperature is well above freezing. The number of bees that attack depends upon the degree of continued disturbance. The cold weather defense system used by honey bees is effective and clearly sufficient for their needs.

COLONY ENVIRONMENTAL CONTROL – Honey bees control several aspects of their nest's environment. The way in which this is done is discussed under several headings. See TEMPERATURE CONTROL; HUMIDITY CONTROL IN THE NEST; ANNUAL CYCLE OF HONEY BEE COLONIES; NATURAL NEST OF THE HONEY BEE; NEST HYGIENE; and DEAD BEES, REMOVAL OF.

COLONY ODOR – Honey bees are able to differentiate between bees that are their nestmates and those that are from a different colony. This ability is important to a colony in protecting itself against robbing bees from another colony (see ROBBING BY BEES). Non-nestmate bees are often attacked at the nest entrance by guard bees, though some are admitted, especially if they are bringing in food. Guard bees extend their antennae toward incoming bees and apparently detect intruders because they have a different mix of odors than that which characterizes bees from the guards' nest.

Colony odor also seems to be detected by bees during foraging. Worker bees forage preferentially from feeders used by their nestmates. When foraging at a feeder, workers tolerate nestmates, but sometimes attack bees from a different colony. This kind of defense of food sources is known from stingless bees at rich flowers in the tropics, but it is not known whether it plays a significant role in honey bee foraging.

It has been shown that one component of the colony odor is acquired by the bees from the particular blend of the colony's diet. Discrimination between nestmates and non-nestmates can be experimentally enhanced by feeding colonies with strongly scented food, such as molasses or heather honey. Since bees within a colony exchange food extensively, all the bees in a colony receive a very similar diet at any given time. When a variety of flowers is available, even nearby colonies are unlikely to share the same blend of floral odors, so this environmentally-acquired recognition system provides a reliable means of nestmate discrimination. When large quantities of nectar are being collected by all colonies from the same floral source, the colony odor of different colonies becomes more similar, and the discrimination between bees of different colonies is weaker. However, at just these times robbing is least likely to take place, so the protection provided by colony-odor recognition is less critical.

In addition to the environmentally-acquired odor, colony odor also has a genetic component. Bees genetically related to a given colony, even when reared outside that colony, are more likely to be admitted than unrelated bees reared under the same conditions.

It seems likely that guard bees learn the odor that characterizes their nestmates. The balance between environmentally- and genetically-acquired odors in this process is not fully known, though under most circumstances the environmental sources seem most important. Which bee (herself or nestmates) a guard bee uses as a model of the appropriate colony odor, when she learns it, and how quickly she may change models remain unanswered questions. - PKV

COLONY ORGANIZATION – Not all aspects of the way in which a colony of honey bees is organized are well understood. Good advances have been made in our understanding of some areas such as the dance language and the roles played by pheromones. In addition there is division of labor and certain worker bees specialize on certain tasks. It is clear that the queen is not a "ruler," but secretions produced by her do affect and may even control certain aspects of colony behavior.

Bees patrol within the hive, apparently looking for work, and other bees sometimes appear to guide the activities of the rest. No one has been able to identify certain bees as being bosses or supervisors. However, when bees are seeking a home site the numbers that participate in the activity is small, usually only a few hundred. When bees in a swarm lose their queen it is only a small minority of workers that search for her.

Reinforcement of an activity is an important part of social order and colony organization. For example, a worker bee rarely does anything by herself for very long. A bee exposing her scent gland at a food source, near a lost queen, or at a home site, will not continue to scent for very long unless she is joined in the activity and her actions reinforced by another bee.

Foraging strategy is concerned with how can bees in a colony best use their resources to gather the maximum amount of food in the short period of time that it is available. Honey bees gather half of the total amount of food they use annually during only about three weeks of the year. The other half of the food is gathered over a longer period, but still during a relatively short part of the year. During much of their life honey bees are not fully active but are waiting for a pollen or nectar flow to occur. It is important that they take the greatest advantage of a good food source whenever it is available.

Many studies of bee behavior have been done at a time when little forage is available or in areas where honey plants are not abundant. The classical experiments designed by Professor Karl von Frisch were conducted in the Austrian Alps, a place where it is necessary to feed bee colonies to have them survive the winter. These near-starvation conditions helped von Frisch because his bees readily visited feeding stations for sugar syrup. When a good honey flow is in progress bees prefer natural food sources and it is not possible to force them to visit an artificial one. More observations of bee behavior when a heavy honey flow is in progress need to be made. For example, at a time when a colony might gather 10 to 15 pounds (4.5 to 7 kg) of nectar a day it is possible that worker behavior is markedly different.

Our knowledge of honey bee colony organization is far from complete. An observation hive is still a powerful tool to learn more about what is taking place. In designing experiments to test and learn more about bee behavior it is important that the system be kept simple and that the observations be made in a methodical way that can be repeated both by the original investigator and others who will build on the information that is gained. Several aspects of colony organization and bee behavior are discussed under a variety of headings in this book.

COMB – (see COMB, NATURAL)

COMB FOUNDATION – (see BEESWAX FOUNDATION)

COMB HONEY – This is honey the bees store in new comb, produced on very thin, bleached foundation, and sold in rectangular, square, or round sections. Special supers and foundation are necessary for its production. Comb honey production is one of the finer arts in beekeeping. Colonies used in square-section comb honey production must be crowded. Crowding bees in colonies causes congestion that may lead to swarming. A good system of swarm control is therefore an important part of comb honey production.

COMB HONEY ERA – The period from about 1880 to 1915, when most beekeepers in North America produced comb honey, is called the comb honey era. The reason for producing comb

Sizes of section boxes varied over the years. Only one, 4½" x 4½" exists today.

honey was that before the passage of the first Pure Food and Drug Laws in the United States in 1906, adulteration of many foods, including honey, was common. Corn syrup was added to much of the liquid honey on the market. Most consumers preferred comb honey because they understood it was a pure, natural product.

In addition to the support the pure food laws gave the honey industry, an increased demand for sugar and sweets occurred during World War I. For a short time during World War I sugar was rationed and again during World War II. Because of these pressures, and because it is cheaper and easier to produce liquid honey, less comb honey was produced. As late as 1950 several beekeepers in the United States were still making a full-time living by producing comb honey. Today a rare few beekeepers still produce square-section honey but more produce chunk, cut-comb or round section honey.

COMB HONEY, HOW TO MAKE – The production of marketable (attractive) sections of comb honey is an art that requires attention to detail. Early spring management of the colonies is the same as those for liquid honey production. The bees are given unrestricted space; the queen is given an abundance of room in which to lay eggs, and the brood nest is allowed to grow upward within the brood chambers. Before the honey flow, the colony will occupy three or four brood chambers in the North and probably one less in the South. Comb honey supers are not placed on the colony and no special manipulations are made until the

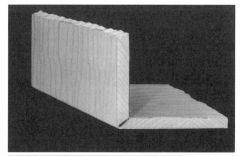
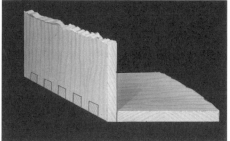

Section box joints. Three joints are "V" shaped grooves, one a box joint.

Assembling Section Comb Boxes

A wooden section super has many pieces – spacers, "M" boards, super springs and more. Shown here is simply the assembly of the boxes, which will be placed in the special sized super.

1. Moisten the "V" joint to soften.

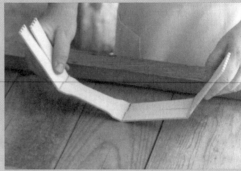

2. Carefully fold the box.

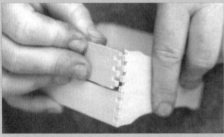

3. Bring the box-jointed sides together.

4. Secure in holder.

5. Insert foundation.

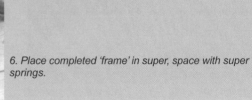

6. Place completed 'frame' in super, space with super springs.

Comb Honey, How To Make

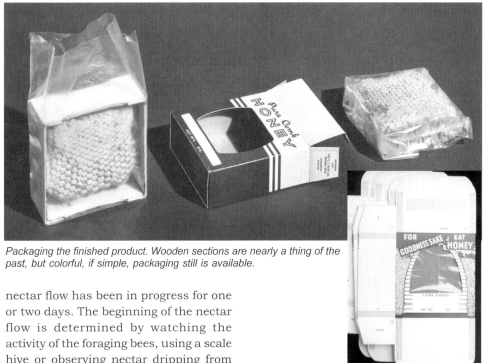

Packaging the finished product. Wooden sections are nearly a thing of the past, but colorful, if simple, packaging still is available.

nectar flow has been in progress for one or two days. The beginning of the nectar flow is determined by watching the activity of the foraging bees, using a scale hive or observing nectar dripping from the combs.

Colonies that are used for comb honey production are given no upper entrance. This is because some of the bees returning to a hive will be carrying pollen or propolis. If one of these substances is deposited, even accidentally, on the new comb honey sections, it will darken them. This darkening is called "travel stain" and results from the bees walking on the comb. Comb honey producers make every attempt to avoid travel stain, since the public prefers clean, white cappings, free of marks or blemishes.

Ventilation may be a problem because of the lack of an upper entrance. This can be avoided by using a deeper than normal bottom board or a screened bottom board. This will give the bees some additional clustering space and also make it easier for them to ventilate the hive and reduce the moisture content of the ripening honey. This bottom board is usually two inches deep (5 cm), and with or without a rack that discourages the building of comb in the space below the brood nest. Although bees are reluctant to build comb in a downward direction, they will sometimes do so if they are severely crowded.

Bees do not care to draw foundation, or build comb, in square comb honey sections.

Foundation and the preparation of sections and supers – Wax foundation expands and contracts as the temperature changes. The thin foundation will warp and buckle with ease. The ideal situation is to install foundation in frames or sections on the day it will be put onto colonies. The round sections are much less likely to cause warping and buckling since they

171

hold the foundation in place more securely. A very few bee supply manufacturers sell split wooden sections that take one long piece of foundation for four sections at one time.

The section foundation, which will later become the midrib of the comb in the section, must be of high quality so it does not detract from the flavor, texture or appearance of the section itself. The foundation that is used to make comb honey is extra thin, which also makes it more difficult to manipulate and place in the sections. It is sheeted in the same way as the foundation that is used in standard frames, but it is rolled thin before being passed through the foundation mill. The beeswax used for section foundation may be sun-bleached or filtered to obtain an almost white wax.

Sections for comb honey production –For many years sections were made only from planed, carefully

The unassembled pieces of a Ross Round frame, the sheet of foundation that goes between the frames two-piece frame, rings that fit in the openings, clear, and opaque covers, and labels. (Spear photo)

As assembled Ross Round super, showing the spacer board (used to keep the frame snug and bee-space correct), and the super spring that keeps tension on the boards. The space also provides excellent ventilation. (Spear photo)

Filled sections, covered and labeled. (Spear photo)

sanded, thin, basswood lumber. Round, plastic sections are now popular with good reason; they are easier to clean and bees fill them better.

Traditional wooden sections – J.S. Harbison, a well-known beekeeper in California, is credited with inventing wooden comb honey sections in 1857. At that time beekeeping equipment was really not standardized and during the next several years sections of varying sizes and shapes appeared. Most beekeepers used pine or basswood lumber 1/8" (3 mm) thick to make their sections. Originally, the four sides of the sections were nailed together, a laborious task that led A.I. Root to manufacture a four-piece section whose pieces dovetailed together. The folding, one-piece section was invented in 1879 by James Forncrook, an employee of one of the larger bee supply manufacturers. His idea was revolutionary and soon all manufacturers of wooden sections were using this type of section with its three V-shaped grooves.

Ever since comb honey sections were first made a debate has raged over the ideal size and shape of a section. The majority of beekeepers apparently favored the 4-1/4" x 4-1/4" (11 X 11 cm) square section. A square section contains about 14 oz (1/2 kg) of honey; round plastic sections weigh about seven or eight oz (1/4 kg).

Round sections – In 1954, Dr. Wladyslaw Zbikowski (1896-1977) of Michigan invented round plastic sections for comb honey and the special equipment to hold them in a super. Today the round sections are available as Ross Rounds®. The round sections have several advantages: all the cells in the comb are usually capped, whereas in sections with square corners the bees are often reluctant to fill the corners with comb and honey.

Bees deposit much less propolis on smooth plastic surfaces and thus round sections are easier to prepare for the market. In the case of round sections plastic covers are placed over each face of the comb and a wraparound label is used to hold the two covers in place. When the two covers are added plus the label the round combs are ready for market. (see ROSS, THOMAS and ROSS ROUNDS)

Removal of filled supers – Removing filled supers of comb honey is a special task and art in itself. Smoked bees will immediately engorge with honey. If a full section-super is smoked, some of the cells will be uncapped as the bees engorge. The holes made in the caps ruin their appearance and may cause them to leak honey. Equally important is that when colonies are smoked it is nearly impossible to prevent bits of charred fuel from being blown out of the smoker and into the hive. These specks usually go unnoticed in the production of liquid honey since the bees remove them quickly. However, in the case of comb honey the bees are abandoning the super and the specks may remain on the sections and give them a bad appearance. It is also important that no bits and pieces of broken burr comb be on the top or bottom of comb honey supers, as this will slow the bees from leaving the supers.

One way to remove filled comb honey supers is to use a bee escape using a minimum amount of smoke. Approach the colony carefully, then remove and examine the supers; empty and partially filled supers are put back into place. The filled super(s) is put on the top of the hive over the bee escape. There must be no holes, cracks or crevices through which bees might enter and rob honey from the comb honey sections. If any such openings are present, they may be temporarily plugged. Porter bee escapes (see REMOVING BEES FROM SUPERS) work quite well and should empty the section super of bees within 24 hours. Bee blowers, or bee repellents, may be used to remove supers of comb honey, and are, generally faster, more efficient, and even more effective at removing bees.

It is rare that all of the sections in a comb honey super will be filled and saleable as first class sections. This is especially true of those along the edge of the super. It is more difficult to fill sections during poor honey flows. Sections that are only partially filled may be separated from the rest and placed back on the colony for finishing. Comb honey producers find that some colonies are apparently better at finishing sections than others, and these should be selected to receive the incomplete sections.

Preparation of comb honey sections for market – Comb honey is a delicate product and should be moved into the marketplace and sold as soon as possible. A special problem with comb honey is that even the slightest damage by wax moths, or the wingless fly, *Braula coeca*, is fully visible to the consumer and may prevent a sale. It is best to place all new comb honey sections in a freezer for 24 to 48 hours to kill any eggs of these insects that may be present. Sections may be kept in a freezer for many months without adverse effects.

When placed in a freezer, round sections should be in a sealed plastic bag. When removed, keep them in the plastic bag until they reach room temperature so the condensed water doesn't leak into the honey.

Since honey is hygroscopic it will pick up moisture at its surface if the humidity is high. If comb honey must be stored, it is best to do so in a warm, dry room. Some beekeepers have used dehumidifiers in comb honey storage rooms to prevent the sections from picking up moisture, weeping, and fermenting.

Another reason for moving comb honey onto the market quickly is that the honey may crystallize. In fact, the danger of crystallization is so great that comb honey should not be made from honey plants whose nectars are high in glucose, since they are more likely to crystallize. There is no way to liquefy granulated comb honey. When the sugar crystals in naturally crystallized comb honey are large, the product loses much of its taste appeal.

Many years ago, when comb honey production was much more popular than it is today, beekeepers selected and used queens whose offspring collected and used much less propolis. This reduced the excessive deposition of propolis on and between the sections, such as one finds in bee hives today. Beekeepers also coated the tops of their wooden sections with paraffin, and then later, easily removed tape. The paraffin and any excess propolis were scraped from the sections before they were packaged or simply removed with the tape. In this way the wood surrounding the section was clean and white. This was a time-consuming effort that may now be avoided by using the Ross Round sections.

Comb honey requires special packaging. Square sections are usually placed in plastic bags and then inside cardboard packages for sale. The plastic bags prevent any leakage and protects against wax moth attack. The Ross Round sections that are in plastic containers require much less preparation for market. Because of the special packaging requirements, many beekeepers prefer to produce cut-comb honey that is packed in solid plastic boxes.

Shook swarming – This quaint expression refers to a procedure sometimes used in producing comb honey. It is a variation of "padgening," a word derived, with misspelling, from the name of the British beekeeper J. W. Pagden who first publicized it in 1870. Padgening consists simply of hiving a prime swarm on the stand of the colony that threw the swarm, meanwhile moving that parent colony off to one side. This strategy permits the augmentation of the population of the swarm by collecting the foraging bees of the original colony and thereby increasing nectar-gathering capabilities.

Shook swarming consists of creating a swarm artificially by shaking most or all of the bees from the combs of a strong colony and inducing them to enter a hive set on the stand of that parent colony. This parent colony meanwhile was moved to another stand and requeened or, if all of the bees have been shaken from the combs, combined with other colonies. In comb honey production, which is the only kind of honey production for which this procedure is suitable, the new hive taken over by the artificially created swarm consists of a shallow super fitted with frames of foundation only or, as some beekeepers prefer, foundation plus one comb containing brood, to prevent the swarm from absconding. Upon this shallow hive are put a queen excluder and two or three comb honey supers. The bees are thus crowded into the supers by the meager size of the hive below the excluder, and abundant crops are sometimes achieved very quickly. The one disadvantage to this method, in addition to the labor involved, is the propensity of the bees to store pollen in the bottom super. -RT

COMB HONEY PRODUCTION, REARING QUEENS FOR – At the peak of the popularity of comb honey, comb honey producers often selected their own breeding stock and produced their own queens. Here are some of the considerations they used for selecting queens for this special purpose. For more information on growing queens, see QUEEN REARING.

Dr. C.C. Miller, a famous comb honey producer in the late 1800s, wrote that among other qualities he sought bees that produced sections whose cappings were "uncommonly white." Both Carl Killion and his son, Eugene, in their respective books on comb honey, noted that they preferred bees that produced clean, white cappings on sections.

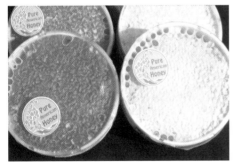

On the left is a round section capped with "wet" cappings. The wax touches the surface of the honey. On the right is a round section capped with "dry" cappings. There is an air space between the top surface of the honey and the underside of the wax cappings. Many comb honey producers prefer the "dry" cappings look, but it has no effect on the quality or the flavor of the honey.

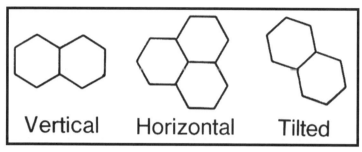

Typical directions of natural comb. (Morse diagram)

These beekeepers were concerned with two factors. First, they wanted bees that collected and used little propolis. Propolis is used by bees in many ways, including the strengthening of comb; bees that use a great deal of propolis will sometimes work it into new wax. Bees collecting propolis may stain cappings as they walk across them with their propolis loads; pollen loads probably cause as much if not more trouble.

A second factor that makes cappings white is the presence of an airspace between the honey and the cappings. The opposite of white cappings is "watery" cappings. Although most books about bees rarely mention the quality of the cappings, examination of score cards from honey shows reveals a criteria that deducts points for watery cappings. Such cappings can become moist when stored in a damp area (see HONEY – Hygroscopicity). Watery cappings are caused when bees do not leave an air space between the honey and the cappings. This space makes a great difference in the physical appearance of the sections. This is especially true with the darker honeys, which will appear quite different if they have white cappings.

Another concern of the comb honey producer was to produce queens that would head colonies that were less inclined to swarm. Since colonies used for comb honey production are more congested than those used for liquid honey production, they are inclined to produce more swarms. There is no question that some strains of bees have a greater tendency to swarm than others.

COMB, NATURAL – Honey bee nests have four types of natural cells: queen cells, worker cells, drone cells and transition cells. The latter are cells built between worker and drone cells; they are often oddly shaped, sometimes with only four or five sides. Transition cells are also built where comb has been repaired or where comb is built around an obstruction. Comb that is wavy and not built in a flat plane will also contain several transition cells. Queen cells are different from the rest of the cells in a hive and play a different role (see QUEEN, Queen cells and QUEEN, Queen cups).

The cells in natural comb are six-sided or hexagonal in shape. In all species of honey bees, the top of the cell is either peaked or flat; the bees build comb with both orientations. In the same natural nest, while the cells within a comb will be the same, each comb may have a different orientation. Apparently, the strength of a comb is not affected by the orientation of its cells. Manufactured comb foundation is usually made with the peak upward; the older bee literature contains much controversy about which orientation is better. Those who argued for one orientation against the other did not ask the bees their preference. One

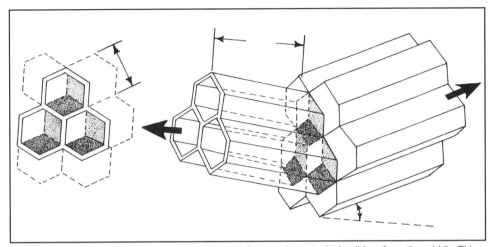

Diagram showing the slight angle (~17°) upward the opening a typical cell has from the midrib. This also shows how the base of one cell is composed of a part of the base of three other cells on the opposite side of the midrib. (Yatcko art)

Englishman who examined natural comb in natural nests found the following:
• Combs with peaked tops and bottoms - 131
• Combs with horizontal top and bottom - 123
• Comb with both orientations - 1
• Combs that were intermediate - 13

Huber, the famous Swiss naturalist of the 1700s, noted that the cells attached to the top of the hive were often pentagons and not hexagons. (For further discussion see Morse, R.A. "Cell orientation and comb strength in honeybee colonies." *Gleanings in Bee Culture* 111:10, 14, 16, 18-19. 1983.)

Interestingly, Charles Darwin devoted 12 pages in his *The Origin of Species* to the subject of honey comb and its origins. Honey comb built by bees is not as perfect as one might imagine. It was once proposed to use honey comb as a natural unit for making standard measurements; however, not only do individual cells vary but the number of cells per unit of distance as well as the sizes of bees of different races also varies. African bees, for example, are about 10% smaller than their European counterparts and thus their cells are smaller (see BEESWAX FOUNDATION). Many races, of course, lie between these two size extremes.

When we study the matter of shape and form, the classical textbook is that written by D.W. Thompson, *On Growth and Form.* (Cambridge University Press, Cambridge, England. 1116 pages. 1942). Thompson devotes 19 pages to "the bee's cell," which he describes as "the most famous of all hexagonal conformations, and one of the most beautiful." He adds that life does not exist without conformity to physical and chemical laws. Thus, the forms living things take "can be explained by physical conditions" in many cases. If, for example, circular pressure is applied to a number of co-equal cylinders, and they are compressed, they turn into hexagons. This is what is found in the honey comb. In this way, the cells are absorbing the space between them that would otherwise be wasted.

If pressure is applied against the ends of the cells on both sides of a comb they take a shape at their bases that is called a trihedral pyramid. It will be noted in examining foundation, or natural comb, that the base of one cell forms the bases of three on the opposite side of the midrib. This makes each individual cell a bit longer and it better accommodates a larva. Separate from this is that while the comb is vertical, the cells themselves slope downwards slightly from their mouth to the bottom of the cell.

Charles Darwin was slightly misled when he wrote that from the point of view of natural selection, "the comb of the hive-bee, as far as we can see, is absolutely perfect in economizing labor and wax." The truth is that while space to contain brood is utilized to the utmost, mathematicians agree that there might be more economy of wax. Thompson does not elaborate further on this question, but the cells are compact so that the larvae might lie as close to one another as is possible. In this way there is greater economy of heat, especially in cold climates where much of honey bee evolution probably took place.

Mathematicians will find that Thompson's section on bees gives greater detail concerning the angles involved. In nature, certain shapes and designs occur again and again simply because they are dictated by the laws of physics. In brief, the cells in a honey comb represent the greatest conservation of space, offer the greatest strength, and, as our knowledge of brood rearing would suggest, offer the greatest economy of heat energy.

COMB STORAGE – It is important to keep stored combs in good condition from one season to the next. Both the bees and the beekeeper invest a great deal of time and effort in making good combs that have a life of several years if given proper care. Estimates vary, but a little over eight pounds of honey are needed to produce a pound of wax. Thus, combs represent a considerable investment so it is important from this point of view that they be properly protected.

One of the first questions asked about comb storage is whether combs should be stored wet or dry. That is, should bees be allowed to rob any remaining honey from the combs after they are extracted, to dry them before they are placed in storage. One viewpoint is that if the honey is left in the combs it may ferment. While that may do no immediate harm to the combs, it does result in a bad odor and probably takes some extra effort on the part of the bees to clean the comb. It has also been stated that when honey is left in stored combs, crystals may form and cause premature crystallization in the honey the following year. (Yet another thought is that robbing may take place when the wet combs are exposed in an apiary just before they are placed on a colony.) Since the small hive beetle prefers comb with honey, storing comb dry is the best.

Some beekeepers stack their supers outdoors and allow bees to rob them free of honey before they are stored. This is not a good idea as the honey could disseminate diseases. It is also possible to dry combs immediately after extraction by stacking them on a strong colony and above an inner cover in which the hole has been left open. The bees will clean the cells carefully and will not damage the comb. However, they may consolidate the honey in random cells throughout the supers.

The greatest problem in storing combs is to protect them against the several animals that would destroy them. Mice, wax moths and more recently the small hive beetles have become the three most

Stacking supers so light and air can get to the frames goes a long way to protect wax from the wax moth.

destructive (see MICE; WAX MOTHS; and SMALL HIVE BEETLES). Pollen mites are sometimes found in stored combs where they feed on any pollen left in the combs. These mites appear to cause little harm though the floor under stored supers may be littered with pollen grain shells that have been worked out of the combs by the mites. Extremes of heat and cold, at least in most climates, appear to do little harm.

In the past many beekeepers have kept stored supers in rooms that could be fumigated and a wide range of fumigants were used. At one time, cyanide and ethylene dibromide were both used to protect stored combs against wax moths. Both materials are extremely dangerous and no longer registered for use in comb storage facilities. Therefore use of either of these two chemicals is illegal. The number of fumigants approved for use to protect food products or items used in their manufacture is fewer each year under guidelines approved by the federal Environmental Protection Agency.

Other control measures include Paradichlorobenzene (PDB) crystals placed on top of a stack of supers with drawn comb. The super edges are taped shut and stand on a cover or other snug device. As the heavier-than-air crystals evaporate, the fumes fill the stack, and are deadly to the larvae and adult stages of the moth. PDB is not a bee, honey, wax or human friendly chemical. Honey and wax will absorb this substance, and supers *MUST* be aired out for several days before placing back on a colony.

Another technique is to store super so maximum air flow and light can reach the inside of the super. Stacking on end, outside under cover and six inches apart allows this, as does criss-crossing supers if stored flat. Racks can be built to accommodate this technique. This is effective, safe and easy to do for a small to medium operation.

Of course if winters are cold – below freezing on occasion – outside or out building storage is fine, if protected from mice.

Fumigation with carbon dioxide is effective, though rare, and requires somewhat sophisticated environmental controls. And aluminum phosphide gas, a restricted use chemical is very effective but can be hazardous to humans.

A small number of beekeepers are using heavily insulated, temperature controlled rooms for comb storage, especially in the southern states. This method has proven to be reasonable in cost. The ideal temperature from the point of view of cost and comb protection is about 48°F (9°C).

Some honey packers that hold comb honey and cut-comb honey for long periods of time have used controlled atmosphere rooms in which the level of carbon dioxide is raised sufficiently high that wax moths, beetles and other animals that might destroy combs cannot live. These rooms have been very successful. Beekeepers and honey packers who might be interested in this type of storage should consult with those who operate controlled atmosphere storage rooms used to hold fruit, especially apples. The technology changes rapidly but this does appear to be an efficient method of storing cut-comb honey.

Some time ago, a *Bacillus thuringiensis* product called Certan® was available in the U.S. This is still available in much of the world, and is used to treat combs before storing. Protection lasted several months, but did not have any adverse effect on honey bees. Re-registration in the U.S. is under review, and a favorable result would be good to reduce the use of toxic chemicals in honey supers.

COMMUNICATION AMONG HONEY BEES - Honey bees are social and a single nest may have as many as 60,000 individuals. It is obvious that some means of communication among bees is necessary if the colony is to function as a social unit; bees can be seen passing their antennae over other bees and exchanging food but the full meaning of all this is not clear. It has also been recognized for many years that there is division of labor in a colony (see CASTES).

The methods that bees use to communicate with one another in the field by way of pheromones, such as the

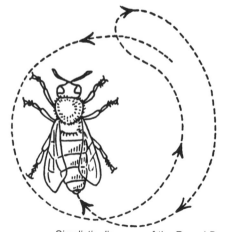 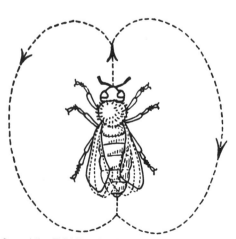

Simplistic diagrams of the Round Dance, left; and the Tail Wagging Dance, right.

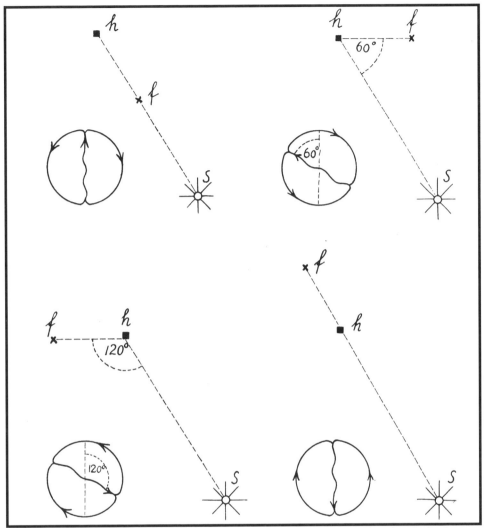

Simple demonstration of how the tail wagging dance communicates the general location of a food source, using the sun as a marker. You would see something similar to this if you watched it in an observation hive.

alarm odor, sex attractant, and other attractant and repellent scents, became understood in the 1960s and the 1970s. At that time chemists developed the sophisticated apparatus necessary to detect the very low levels of these substances that bees use to convey messages. We are aware that pheromones also play an important role within the hive, but apparently the rate at which information is exchanged is slow and complex. However, clearly substances indicate to the whole population that the queen is present or that it is time to swarm. (see PHEROMONES).

A major breakthrough in our understanding of honey bee communication came in 1944 when Professor Karl von Frisch found he could "read" the dance language of the honey bee (see FRISCH, PROFESSOR KARL

VON). During the next several decades, von Frisch and his students undertook a great number of experiments that further defined what information bees could and could not convey with the dance language. During the early 1960s it was reported that a sound was associated with the dance; however, researchers searched in vain for sound receptors, and because they could find none concluded that sound was not an important part of honey bee communication. Recently we have come to understand that this is not true. Honey bees hear through receptors on their antennae. When the antennae are in a receiving mode they are held perpendicular to the face at a 90° angle to each other. Experiments performed leave little doubt that sound is very important (see SENSES OF THE HONEY BEE, Sound perception).

The dance language for food – It can be demonstrated that when a worker honey bee finds a rich source of nectar, pollen or water, she collects some, then returns to the hive and tells her nestmates about it. The information is conveyed by a dance; sound may be a very important component of the dance. Honey bees are successful because this system of communication is rapid and efficient. Honey bees typically gather half of the food they consume during the whole year in a period of about three weeks. Records from some scale hives show that colonies gain weight for only about 10 to 15 weeks of the year, but that in those weeks they store enough for the whole year. The ability of honey bees to share information allows colonies to achieve this high degree of efficiency in foraging and for them to survive long periods of adverse weather.

Two dances are used by scout honey bees to convey information. The chief dance is the waggle dance. Recruits that follow this dance learn the direction and distance of the food from the hive. In the case of nectar they learn the odor either by smelling the floral odor that may cling to the dancer's hair (if the food source is close by and the odor is not lost as the bee flies through the air) or by tasting a bit of nectar the dancer will give if she is begged to do so.

Several factors affect how soon, how long, and how enthusiastically a scout will dance. Scouts may dance after one trip to a rich food source; however, they may require several trips to a moderately good food source to convince themselves that the food is worth collecting and that they should alert the foragers to collect it. Scouts will dance for a longer period of time if the food source is rich. Scouts do not deposit nectar in cells; rather, they give it to house bees that ripen the food further before it is placed into cells. If a hive is full and there is no place to deposit anything, or if so much food is coming in that the house bees are fully occupied, the scouts will find it difficult to unload their nectar and will be discouraged from further dancing and foraging no matter how rich the food source may be.

The direction of the food from the hive is indicated by the scout's orientation during the waggle portion of the dance. When the food is in the direction of the sun (called the sun's azimuth), the scout performs the waggle portion of the dance while moving straight up on the comb. When the food is directly away from the sun the waggle portion is performed as she dances straight down. If the sun is due north of the hive, east and west are indicated by the scout's dancing to the right or left, respectively. The direction danced during the day changes as the sun moves across the sky but performing the waggle portion of the dance straight up always means fly in the current direction of the sun, wherever it may be. The rapidity of the dance, which von Frisch measured as the time for a complete "dance circuit" (that is, a waggle run plus walking back to the starting point to begin another waggle), corresponds to the distance to the food. Bees dance very slowly for food a great distance from the hive.

A second dance that scouts may use is the round dance. This dance indicates that food is nearby, usually less than 100 meters away. However, bees, like people, have different dialects and what one race of bee indicates as a distance of 50 meters may mean 75 meters to a bee of another race. One of von Frisch's students was able to demonstrate this by making a colony that contained bees of two races (for further information and references, see FRISCH, PROFESSOR KARL VON).

The same dance that bees use to indicate food sources is used during the swarming process to indicate the site of a potential new home. A variation of the waggle dance, a very slow dance, is used by African and Africanized bees to indicate that a clustered swarm should become airborne and fly a long distance. Swarming European honey bees apparently fly relatively short distances when establishing a new home but African bees may migrate long distances, though how long an individual swarm may fly, and how many days it may take to migrate, are not known. (see TREMBLE DANCE)

Migration Dance – The "**migration dance**" is a type of waggle dance found in the African honey bee, *Apis mellifera scutellata*, and other species of tropical honey bees. It is performed only in association with seasonal absconding, which is a type of long-distance migratory movement in which colonies

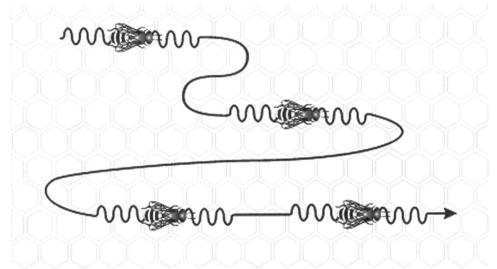

The migration dance of the African honey bee, Apis mellifera scutellata. *The dance differs from the waggle dances used in foraging in that the waggle runs are of very long durations (60 seconds or longer), they lack the "figure 8" pattern of typical dancing, and workers who follow the dances do not leave the hive to investigate the indicated site. The migration dance may help to gradually prepare the entire colony for long-distance travel.*

may travel up to 100 km. Migration dances communicate a consistent direction, which is the direction that a colony travels upon departure. Migration dances also communicate very long distances. However, the distances communicated by consecutive waggle runs are highly variable, suggesting that no particular distance is indicated. Thus, the dance may convey the message to "travel for a long, but unspecified distance in the indicated direction." Because migrating colonies may not preselect a specific destination before departure, the migration dance helps to prepare colonies for this unique type of movement. -SS

Vibration signal – The "**vibration signal**" is one of the most commonly occurring signals in honey bee colonies and is performed on workers of all ages, laying queens, virgin queen and queen cells. Vibration signals performed on workers cause increased activity that enhances brood care, food processing, nest maintenance, foraging, and colony movements during swarming. Most signals are performed by successful foragers, which may help to coordinate many aspects of worker behavior with food availability. Vibration signals are

The vibration signal of the honey bee consists of a worker rapidly vibrating her body dorso-ventrally for one to two seconds, while grasping a recipient with her legs. It acts as a "modulatory signal," because it causes a non-specific increase in activity that enhances many different social activities.

performed on laying queens only during the two to four weeks that precede swarming and may help to prepare queens for flight. Vibration signals are performed on virgin queens during the period when they are battling one another to determine the new laying queen of the colony. Virgin queens who receive more signals survive longer, kill more rival queens and are more likely to become the new queen. Vibration signals performed on queen cells may influence queen development and emergence time. -SS

Odor recruitment to food and the genome analysis – For centuries beekeepers have wondered how hive mates manage to locate food exploited by foragers. Maeterlinck pondered two hypotheses: Bees either "follow" successful foragers or "follow their instructions." He wrote: "Between these two hypotheses, that refer directly to the extent and working of the bee's intellect, there is obviously an enormous difference."

Von Frisch proposed a third, odor-search hypothesis, in the 1930s and insisted that bee dance attendants search only for odor of the target food. However, in the 1940s he instead formed a dance language hypothesis ("recruit use of instructions"), one that soon became "ruling theory."

"Confirmation bias" thereafter controlled events. Instead of testing the available hypotheses against one another, most scientists accepted dance language as fact, comparing honey bee intellect to that of humans. Results of independent experiments during the 1960s supported the earlier von Frisch interpretation, that bee recruitment involved only odor of the target food or locality. In those experiments, searching bees no longer ended up as predicted by the language hypothesis and could not find food without odor. The conflict between those two hypotheses continued for decades. Language advocates embraced only supportive evidence and ignored or dismissed contrary evidence.

The genome sequencing of the honey bee DNA in 2006 provided an opportunity to resolve the controversy that surfaced. Recruitment communication, if an "instinctual signaling system" as claimed, would require the presence of genes not shared with other insects. Researchers found a total of 170 odorant receptor genes (most not shared with other insects), indicating "a remarkable range of odorant capabilities." No genes for "bee language" surfaced. The language hypothesis had thus failed another test. Wenner, A.M. 2002. "The Elusive Honey Bee Dance "Language" Hypothesis." *Journal of Insect Behavior.* 15(6): 859-878.

CONSERVATION, THE IMPORTANCE OF THE HONEY BEE IN – Honey bees, as has been noted several times, are not native to the Americas or Australia. Before European settlers arrived on these continents the insect pollination of the plants that produce the fruits, nuts and seeds so important for the existence of wildlife was done by bumble bees, smaller solitary bees, wasps and flies. A large number of species of these bees exist, nesting in the ground, under rocks and in hollow twigs, but most are very small in size. On average an individual is about 1/20 the weight of a European honey bee. European settlers imported not only honey bees but also many plants. The small, native pollinators are not well-suited to pollinating many of the imported plants.

Birdsfoot trefoil, a legume, is a good example of a plant that requires a large bee for pollination. Birdsfoot trefoil is an important forage plant in pastures and hayfields and in many parts of the U. S. It will grow in heavy, wet or light soils that are not suited to alfalfa. It is the yellow flower that is in blossom throughout much of the summer along roadsides. The birdsfoot trefoil flower must be "tripped" to produce seed; that is, there must be sufficient pressure on the keel that contains the flower's sexual parts to force them out of their protective sheath. Honey bees do this very well but the small, solitary bees are not sufficiently heavy to do so. Bumble bees are good pollinators of birdsfoot trefoil but bumble bees are not very common. Crown vetch, which is commonly planted along roadsides in some states, must also be pollinated by a large bee to produce seed.

The banks along new roads and other disturbed natural areas must be seeded quickly to prevent erosion. Reseeding is also important and the legumes that will reseed themselves if properly pollinated are very much in demand because they require little maintenance. The changing of the forested parts of the Americas to land used for agriculture and recreation has demanded different types of plants from those that once existed in these areas. The honey bee is well-suited and useful to the special needs and changes in our environment created by man. (see ENVIRONMENTAL QUALITY INDICATOR SPECIES – HONEY BEES)

CONTROLLED MATING – The major difficulty in the improvement and maintenance of honey bee stocks has been our inability to control the mating process. Honey bee queens and drones frequently fly several miles to mate on the wing. Queens will mate with 10 to

20 males in Drone Congregation Areas (DCA). While it is relatively easy to select the stock from which the queen is reared it is less easy to control which drones mate with the queen due to the distances flown and the normal population of drones present in any feral or managed colonies in the area.

The presence of *Varroa* mites has reduced the number of unmanaged, feral colonies nearly everywhere. Add to this the ability of beekeepers to supply numerous colonies of known genetic stock to supply a large population of drones to populate the nearby DCAs, and the probability of controlled mating increases. With enough drones in an area, USDA scientists have shown a 90-95% success rate. This is known as 'drone flooding.'

Artificial insemination has overcome even this problem, however, and allows breeders complete control over the insemination of queens. It is an important tool in the breeding of honey bees. However, because artificial insemination is labor intensive, and relatively expensive, it is impractical as a method for obtaining the large number of queens needed to head production colonies. For these reasons it is desirable to obtain queens that have been naturally mated under controlled conditions to head honey-producing colonies.

Numerous attempts have been made over the years to mate queens in small enclosures (see Harbo, J. *Annotated bibliography on attempts at mating honeybees in confinement.* International Bee Research Association Bibliography 12. 1971). While several people have claimed success in mating honey bees in confinement, only Rossignol, Royce and Stringer (*Bee Science* 2:77-81, 1992) have provided photographic proof of success. They built a small chamber about one meter (three feet) high and 60 cm (two feet) in diameter at the base. The top half was a dome that was lined with aluminum. Natural light entered the chamber from the side, reflected up from the aluminum sheet base, and was diffused by the reflective dome. The lighting provided no features for drone orientation and enabled free-flying drones to focus on a tethered queen and mate with her in the chamber.

An alternate approach to obtaining controlled matings, long used in Europe and elsewhere, has been to establish mating yards in areas having no, or relatively few feral colonies of honey bees. By saturating the area with desirable drones varying degrees of control over mating is obtained.

Islands often make good locations for obtaining controlled mating yards because they usually lack feral populations of bees. In the past the USDA, Canada Department of Agriculture and others in the Americas have used islands for this purpose.

While islands offer good isolation they are often expensive or difficult to get to and are not available to the majority of queen breeders. Studies have shown that in North America there are honey bee-free locations present in certain mountainous regions, deserts, and in northern Canada where controlled natural matings can be obtained. While the locations found to date are not perfect from the standpoint of commercial queen rearing they could still be used for this purpose when control over the mating process is an important consideration. These locations may also prove valuable in the future in honey bee breeding programs and in the maintenance of pure European stocks of bees. -MF

COOKING WITH HONEY – (see HONEY COOKERY and HONEY RECIPES)

CORRIGAN, RICHARD (1917-2000) – Richard "Dick" Corrigan made countless contributions to build up both the Essex County Beekeepers (Massachusetts) and the Middlesex County Beekeepers (Massachusetts). He was instrumental in founding the Merrimack Valley Beekeepers (New Hampshire). Dick was also active in the Massachusetts Beekeepers Association.

Dick was early-on involved with EAS especially his work with Dr. Roger Morse and others formulating the rules for the EAS competitive honey show.

Two Checkmite+™ strips. One is lying across the top bars to show color, shape and size. Another is suspended between two frames, using the built-in "holder."

Richard Corrigan

COUMAPHOS – With the advent of widespread fluvalinate resistance by *Varroa*, beekeepers and researchers sought development and registration of an alternative compound to control *Varroa*. Quickly filling this role was the organophosphate compound coumaphos, sold under the product name CheckMite⁺® by Bayer Corporation. As with Apistan®, strips containing 10% coumaphos were hung between brood frames when no nectar was being brought in by bees. Coumaphos initially gave as high as 98% control of *Varroa* (but no control of tracheal mites). Initially the U.S. Environmental Protection Agency allowed only an emergency registration of coumaphos due to concerns about human toxicity issues of organophosphate pesticides. Such an emergency registration is called a Section 18 registration, as opposed to a permanent Section 3 (general use) registration. Eventually, Bayer Corp. received the Section 3 registration of coumaphos for *Varroa* control in hives, primarily through the EPA approval of an acceptable tolerance level of coumaphos for honey and honeycomb. However, *Varroa* resistance to coumaphos has been documented in the U.S., mirroring the difficulties experienced with fluvalinate.

Such resistance by *Varroa* to fluvalinate and coumaphos has necessitated the development of strategies to manage and lessen resistance that can be followed by the beekeeper. One such plan calls for the acaricides to be used in rotation, so that one acaricide is not used exclusively and therefore intensifying resistance to that compound. Resistance management also call for removal of strips promptly after the recommended treatment period, reducing the selection pressure for that compound. Additional resistance management techniques call for treating only when necessary, when a sufficient number of mites are present, so that fewer treatments are made per year and per apiary. - PJE

Dr. Eva Crane

COWAN, THOMAS W. (1840-1926) – Born in Russia, Mr. Cowan lived in England primarily, but also spent time in California. He was one of the founders of the British Beekeepers Association, and President from 1922 to his death. he also served as one of the Editors of the British Bee Journal, and published several books, including *The Honey Bee, It's Natural History and Anatomy*; *British Beekeeper's Guide Book;* and *Waxcraft*. He also improved the workings and efficiency of honey extractors, and worked with the A.I. Root Company in selling his improved models in the U.S.

CRANE, EVA – Dr. Eva Crane began keeping bees in 1942, after receiving degrees in mathematics and physics from King's College in London in 1930-1933, then an M.S. in quantum mechanics and a Ph.D. in nuclear physics. Shortly after receiving her first bees, she became Secretary of The British Beekeepers Association.

In 1948, the Research and Queen Breeding Committee of the BBKA decided a separate research organization should be formed, and in 1949 The Bee Research Association was formed. Dr. Crane was appointed the first Director. The Journals *Bee World* and *Apicultural Abstracts* were published shortly after, edited by Dr. Crane, followed by the *Journal of Apicultural Research*.

Dr. Crane remained Director from 1949 to 1983, when she retired. During her tenure as Director, and continuing after her retirement she traveled the world, eventually visiting 60 countries, exploring the beekeeping practices and customs everywhere she visited.

Dr. Crane authored, co-authored or edited many books, including: *Honey, A Comprehensive Survey*, 1975; *A Book of Honey*, 1981; *Making A Bee Line – My Journeys In 60 Countries*, 2003; *Pollination Directory of World Crops*, 1984; *World History Of Beekeeping & Honey Hunting*, 1999; *Bees & Beekeeping – Science Practices & World Resources*, 1990; *The Rock Art of Honey Hunters*, 2001; *The Archaeology of Beekeeping*, 1983; *Apiculture In Tropical Climates*, 1976; *World Perspectives In Apiculture*, 1985; *The Impact Of Pest Management On Bees and Pollination*, 1983.

She has also written countless other pamphlets, books, directories, dictionaries and other material in the field of Apiculture.

CUT-COMB HONEY – When comb honey is made in frames and cut into pieces, it is called cut-comb honey. Thin surplus foundation manufactured to make section comb honey is used. The honey-filled and capped comb is removed from the frame and cut into pieces. Cut-comb honey is usually packed in one of two ways, either in a plastic box with plastic cover, or in a plastic "clamshell" style container. The advantage of cut-comb honey over square or round section comb honey is that when the combs are only partially filled, parts of the capped comb may still be recovered. When section comb honey is made, partially filled sections are usually not marketable, or are sold at a reduced rate.

Before packaging, the pieces of comb that have been cut must be drained of liquid honey from the cut cells.

The finished product, showing the drained edge and safe, attractive packaging.

When comb is cut and put into a package, it needs to sit overnight on a drain board, so the released honey does not collect in the containers. A "cutter" can be used so the pieces are uniform in size.

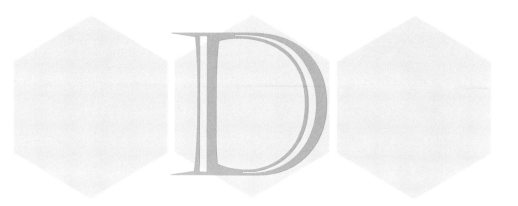

DADANT FAMILY – The Dadant family has been associated with U.S. beekeeping since 1863 when Charles Dadant (1817-1902), together with his wife and his only son, Camille Pierre (1851-1938), came to the U.S. The family settled in a log cabin on a farm just north of Hamilton, Illinois. Charles Dadant had been a beekeeper in his home country of France. Once here he began beekeeping and started to grow grapes for wine making. Apparently the land around Hamilton was not suited to grape growing but the beekeeping operation flourished. Dadant was not pleased with the black German bees that were prevalent at the time and first imported Italian queens in 1868.

In 1885 the Dadants took over publication of *The Hive and The Honey Bee* from L. L. Langstroth, because of his frail health. Charles had been writing for the *American Bee Journal* since 1870, and in 1912, son Camille became editor and publisher and moved their offices to Hamilton, IL. The family has revised their original *Hive and The Honey Bee* several times. Charles even translated it into French and later it appeared in Italian, Russian, Spanish and Polish, giving Langstroth's name even wider publicity. *The American Bee Journal* is still published by the family.

Camille's three sons, Louis (L.C.) (1879-1962), Henry (H.C.) (1882-1966), and Maurice (M.G.) (1886-1972), all entered the candle and bee supply business with their father. Henry invented crimp-wired foundation in 1921, shortly before large migratory beekeeping businesses became popular. This was a boon because it gave combs the extra strength needed in migratory operations.

C.P. Dadant

L.C. Dadant

For many years the Dadants made comb foundation but sold their goods cooperatively with the G.B. Lewis Company, which marketed woodenware under a well-known slogan, "Beware Where You Buy Your Beeware." The two companies published a joint catalogue from 1928 to 1955, when the Dadants purchased the G.B. Lewis bee supply division and incorporated it into their own firm. Acquisition of the A.G. Woodman Company, manufacturers of extractors and other metal goods for beekeeping, was completed in 1971. That consolidated the manufacture and distribution of a complete line of beekeeping equipment by Dadant & Sons, Inc.

Charles C. Dadant (1919-2001), belonging to the family's fourth generation, was president of Dadant & Sons, Inc. Under his leadership with the technical leadership of Dr. G.H. Cale, Jr., the company developed the Starline and Midnite Hybrids. In 1990 he turned over control of the company to his sons Timothy C. Dadant, Nicholas J. Dadant and his nephew, Thomas G. Ross.

Chuck Dadant

DADANT HIVE – (see HIVES, TYPES OF)

DANCE COMMUNICATION BY HONEY BEES – (see COMMUNICATION AMONG HONEY BEES)

DEAD BEES, REMOVAL OF – The removal of corpses of individuals that die within a social insect colony is called necrophoresis. Although most bees die in the field while foraging, a normal full sized colony will lose about 100 bees a day inside the nest. Since these bees may have borne diseases, their corpses pose a threat to the continued health of the colony. The threat is met by a specialized force of "undertaker" bees, that remove the corpses from the nest.

The undertaker bees comprise about one or two percent of the bees in a colony at any one time. No more than about 10 percent of the bees in a colony will serve as undertakers at any point in their lives. Like all labor in bee colonies, necrophoric behavior is performed by bees of a certain age (see DIVISION OF LABOR), in this case about two weeks after emergence. Research also indicates that these bees may be more closely related to each other than to other bees in the colony.

Undertaker bees apparently recognize dead bees by chemical odors that develop in the corpse shortly after death. The corpse is carried away from the nest in the mandibles of an undertaker bee (sometimes more than one), to a distance usually between five and 100 yards (4.5 to 91.0 m); the dead bee is then dropped. Such corpses disappear rapidly, probably eaten by scavenging insects such as ants.

When the colony is less active during cold and rainy weather, corpses are removed more slowly; in winter they may accumulate in large numbers inside the nest and are removed in the spring. The rapid removal of corpses keeps the nest free of dead colony members that encourage pathogens and parasites and interfere with ventilation if not removed. - PKV

DEATH'S-HEAD HAWK MOTH – The Death's-head hawk moth (or sphinx moth) (*Acherontia atropos*) is an interesting insect that does not occur in the U.S. but is found in Europe where it is common in the southern countries along the Mediterranean Sea and even often seen along the warmer, southern coast of England. It is found in the warmer regions of Asia. Interestingly, this live moth is actively sought after by lepidoptera collectors, and trade in live and pinned adults, live pupae and even larvae is international in scope. Several instances of attempted intentional or accidental introduction into the U.S. have been recorded, and stopped.

Because of its behavior it is mentioned in several European textbooks and worthy of discussion here. The adults have a wingspan of up to 4-1/2 inches (115 mm) and it is a pretty moth. The outline of a skull can clearly be seen on the top of the thorax. While this moth usually feeds on fruit, it will sometimes enter hives and eat honey. The moths have short, strong mouthparts and can easily puncture cappings. Apparently not all moths are successful in escaping from hives as they are occasionally found dead inside.

A point of special interest is that the moths produce a squeaking sound when approaching a hive entrance. Perhaps this sound allows them to gain entry. For many years some have said this was proof that honey bees could hear though this was denied by most who have studied bees. However, data indicate that honey bees pick up airborne sound through their antennae (see SENSES OF THE HONEY BEE – Sound perception).

DECONTAMINATION OF HIVE EQUIPMENT – A number of different methods have been proposed to decontaminate hive equipment. Some of the methods merely reduce the contamination level to prevent recurrence of the target disease and in some cases should be combined with antibiotic treatment and follow-up inspection especially in the case of American foulbrood disease. Decontamination of hive equipment has been used for American and European foulbrood, nosema, and chalkbrood disease. Some of the methods that have been used for decontamination include, acetic acid, ethylene oxide, heat, high velocity electron beam, gamma radiation, and hot paraffin. One advantage of these procedures is that they not only reduce the level of the target pathogens but they are also effective against multiple organisms. For instance, when combs are exposed to gamma radiation, the radiation destroys bacteria, viruses, mites, insects, etc.

It is imperative to contact your state authorities to confirm that the procedures are approved by the Environmental Protection Agency and that the proper safety precautions are being observed. For a review on decontamination procedures see Matheson, A.G. and G.M. Reid "Strategies for the prevention and control of American foulbrood." *American Bee Journal* 132: 399-402, 471-473, 534-537. 1992.

DEFENSIVE COLONIES, HOW TO TREAT – It seems that in almost every apiary there are one or two colonies that are more defensive than the others. If one of these defensive colonies happens to be among the first opened by a beekeeper when entering an apiary it is not uncommon for these nasty bees to follow and attempt to sting that person throughout the rest of the time in the apiary. While some beekeepers like defensive bees and think they may be better honey producers, more disease resistant, or have some other valuable trait, no data support such thoughts.

When defensive colonies are identified it is best that they be requeened (see REQUEENING) with new stock. There is no question that defensiveness in honey bees is an inherited trait. Of course, large, populous colonies seem to be more defensive just because they have more bees; however, populous colonies that are easy to manage are common.

On occasion, a defensive colony must be destroyed immediately for the safety of nearby people or animals. If the equipment is to be reused the agent used to dispatch the bees must not leave a residue or must be easily removed. Two simple ideas come to mind. First, with the bees all home, seal the entrance with a reducer or rag, leaving no possibility of escape. Seal any other entrances. Mix up a five gallon (20 l) pail of soapy water. Mix thoroughly, but avoid creating suds. Quickly remove the cover, and inner cover if one is present, and pour the entire contents of the pail between the top bars of the top super. Immediately close the colony. In five minutes or less the great majority of the bees will be dead. Brush them off the combs and rinse with a stream of water to remove soap.

A second technique uses a 10-pound block of dry ice placed on top bars, covered with a super, with all openings sealed.

DEMAREE METHOD OF SWARM CONTROL – (see SWARMING IN HONEY BEES, Demareeing)

DEMUTH, GEORGE S. (1871-1934) – The best-written and most practical treatise on controlling swarming in honey bees was prepared by Demuth in 1921 (*Swarm Control.* USDA Farmers' Bulletin 1198. 48 pages). Probably equally well known is *Commercial Comb Honey Production* (USDA Farmers' Bulletin 1039. 40 pages. 1919). Demuth began keeping bees at an early age in Indiana in the middle of the comb honey era. He attended DePauw University for one year and then became a school

George Demuth

John Mossom (Jack) Deyell

teacher, meanwhile steadily increasing his colony numbers and eventually becoming a commercial beekeeper. In 1911 he joined the USDA and worked for the next nine years with Professor E.F. Phillips, who was then in charge of the federal beekeeping program. They wrote several publications together, including their famous paper on the formation of the winter cluster. Demuth became editor of *Gleanings in Bee Culture* in 1920 where he worked until his death. He was widely known and greatly respected for his practical approach to beekeeping.

DEYELL, JOHN MOSSOM – Better known as Jack, he served as editor of *Gleanings in Bee Culture,* following the death of George S. Demuth, until his retirement in 1962. Jack was first employed by the Root Company when he was 20 years old. He later resigned to obtain a bachelor's degree from Oberlin College. He spent two years keeping bees in Michigan after which time he rejoined the Root Company and became head apiarist and manager for the Root apiaries that consisted of over 1000 colonies. Deyell was a frequent contributor to *Gleanings in Bee Culture* even before he became editor. As editor he was best known for his monthly column, *Talks to Beginners*, a title that had been used earlier by Demuth. Deyell's advice was clear and always based on his own practical experience.

DISAPPEARING DISEASE – (see DISEASES OF THE HONEY BEE, Disappearing disease)

DISEASE DIAGNOSTIC SERVICE – The United States Department of Agriculture provides a bee disease diagnostic service that is available, without charge, to regulatory officials and beekeepers throughout the world. In addition to diagnosing bee diseases this laboratory also can identify mites associated with honey bees. No pesticide analyses are made at this laboratory. Samples to be submitted should be prepared in the following manner.

Brood samples: The sample of comb should be about 4X4 inches (10X10 cm)

A COMPARISON OF SYMPTOMS OF VARIOUS BROOD DISEASES OF HONEY BEES

Symptom	American foulbrood	European foulbrood	Sacbrood	Chalkbrood
Appearance of brood comb	Sealed brood. Discolored, sunken, or punctured cappings.	Unsealed brood. Some sealed brood in advanced cases with discolored, sunken or punctured cappings.	Sealed brood. Scattered cells with punctured cappings.	Sealed and unsealed brood. Affected larvae usually on outer fringes.
Age of dead brood	Usually older sealed larvae or young pupae. Upright in cells.	Usually young unsealed larvae; occasionally older sealed larvae. Typically in coiled stage.	Usually older sealed larvae; occasionally young unsealed larvae. Upright in cells.	Usually older larvae, Upright in cells.
Color of dead brood	Dull white, becoming light brown, coffee brown to dark brown, or almost black.	Dull white, becoming yellowish white to brown, dark brown, or almost black.	Grayish or straw-colored becoming brown, grayish black, or black; head end darker.	Chalk white. Sometimes mottled with black spots.
Consistency of dead brood	Soft, becoming sticky to ropy.	Watery; rarely sticky or ropy. Granular.	Watery and granular; tough skin forms a sac.	Variable, watery and granular
Odor of dead brood	Slight to pronounced characteristic odor of decay.	Slightly to penetratingly sour.	None to slightly sour.	Slight non-objectionable.
Scale characteristic	Uniformly lies flat on lower side of cell. Adheres tightly to cell wall. Fine, threadlike tongue of dead pupae may be present. Head lies flat. Black in color.	Usually twisted in cell. Does not adhere tightly to cell wall. Rubbery. Black in color.	Head prominently curled towards center of cell. Does not adhere tightly to cell wall. Rough texture. Brittle. Black in color.	Does not adhere to cell wall. Brittle chalky white to black in color.

From USDA

and contain a representative sample of the disease. In selecting a sample, look especially for brood that appears to exhibit typical symptoms of the disease to be analyzed. For example, when American foulbrood disease is suspected, select a comb with scattered brood, larva or pupa that exhibit the ropy symptom, and dead or discolored brood. No honey should be in the sample.

Packing brood samples: The comb should be wrapped in newspaper or paper towels. Do not wrap the sample in plastic, waxed paper, or aluminum foil or mail the comb in metal or glass containers. These materials promote the growth of molds and make accurate diagnosis difficult, if not impossible. Mail the sample in a wooden or heavy cardboard box.

Adult diseases: If you suspect the presence of an adult bee disease, collect at least 200 adult bees. Select, if possible, bees that are moribund or have died recently rather than bees that have been dead for an unknown period. Adult bees should be sent in a leak-proof container in 70 percent alcohol or dried, and then sent dried in a wooden or cardboard box. Do not wrap adult bees in plastic, waxed paper, or aluminum foil for the reasons given above.

Mailing samples: All samples should be sent to the following address:

>Bee Disease Diagnosis
>Bee Research Laboratory
>B-476, BARC-East
>Beltsville, MD 20705

The name and address of the sender should be plainly written on the top of the shipping container. Be sure to include, in a letter, any information that may be helpful. If the sample is forwarded by an inspector, that name and address should also be provided.

DISEASES OF THE HONEY BEE (AFB) – **American foulbrood disease** – Until the introduction of *Varroa* and tracheal mites, the bacterial disease, American foulbrood, was the greatest concern to beekeepers in North America. American foulbrood continues to cause consistent losses almost everywhere in the world and beekeepers must not forget its seriousness. The chief problem in attempting to control this disease is that the bacterium forms a spore (resting stage) that remains alive for 70 or more years. Old, stored beekeeping equipment is especially a problem because it may harbor spores. It was the fear of American foulbrood that led to the establishment of apiary inspection programs in departments of agriculture in many states. Over the years, the responsibility of the state apiary inspection programs has expanded to include education, publicity, parasitic mites and other enemies of the honey bees.

Honey bees are the only animals that can contract American foulbrood disease. They become infected in early larval life and usually die in the pupal stage. The symptoms of American foulbrood and European foulbrood disease are quite different and both can usually be easily identified in the field by a trained apiary inspector or

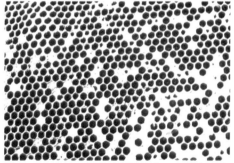

A common symptom of AFB is the spotty brood pattern and many capped cells with small holes.

197

On close examination of an advanced case of AFB notice the "scales," dried and blackened dead larvae that lie on the "bottom" side of cells. They are difficult to remove.

beekeeper. Mixed infections can sometimes occur, making field diagnosis difficult. The names American foulbrood and European foulbrood have nothing to do with the distribution of these two diseases. It is merely that an American, G. F. White, described one and Europeans described the other, both around 1906.

Taxonomic position – The bacterium that causes American foulbrood is called *Paenibacillus larvae* subsp. *larvae*. This bacterium can be grown in pure culture and spores from the culture can be used to re-infect more honey bee larvae. In most cases the field symptoms are sufficient to differentiate American foulbrood disease from the other brood diseases. When the symptoms are confusing, a smear may be taken from a diseased larva and sent to a laboratory for diagnosis (see DISEASE DIAGNOSTIC SERVICE).

Distribution – American foulbrood has been reported in almost all countries where bees are kept. However, the incidence of the disease varies greatly. In the U.S., several states have, through an intense inspection system, kept the incidence of the disease low, sometimes with less than one percent of colonies examined being infected. Methods of controlling the disease have been developed more through experience than by relying on the analysis of data concerning its distribution and spread. There does not appear to be much question that the greatest problem is with beekeepers that have not learned to recognize this and other diseases. Experienced beekeepers are always on the lookout for diseases of all kinds, especially American foulbrood.

Life cycle – American foulbrood disease is spread primarily by beekeepers that re-use equipment from diseased beehives. Other modes of transmission include feeding honey and pollen from unknown sources. Honey transfer by robbing and drifting bees are also possible ways the disease can be transmitted.

Only the spore stage of *Paenibacillus larvae* subsp. *larvae* is capable of inciting the disease. A honey bee larva less than

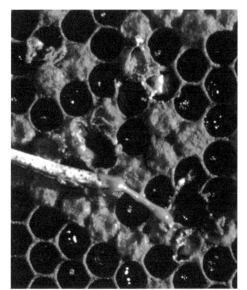

Sometimes, when a larva has been dead only a short while, the "ropey" test works. (File photo)

a day old may become infected if it ingests as few as 10 spores. As the larva becomes older, more spores are required to initiate the disease. Fifty-three hours after egg hatch the larva becomes immune to the disease. Apparently, the older larvae are more tolerant and outgrow the bacterium. Once in the gut of a young-enough larva, the spores germinate within 24 hours; they penetrate the gut wall and multiply in the hemolymph (blood) of the bee.

Death of the developing bee does not take place until after the cell is capped and the larva has spun its cocoon to become a pupa. Since the brood cell is covered, the dead pupa is frequently not detected by the bees or beekeeper. The pupa turns brown and as the disease progresses, an objectionable fish-like or characteristic odor of decay develops. However, odor should not be used as the only means of diagnosis. Eventually the dead pupa dries to form what is called the scale. The scale is rigid and brittle and adheres strongly to the bottom of the cell (as the frame hangs in the hive). An estimated 2.5 billion spores can be produced in one dead pupa. With so many spores being produced, it is apparent that the disease can spread rapidly within a colony.

Detection – American foulbrood, European foulbrood and chalkbrood are examples of diseases that can be diagnosed in the apiary. Decisions to destroy or treat infected colonies are routinely made in the field by trained inspectors or beekeepers. A pupa dead from American foulbrood lies flat on the bottom side of the cell; it is at first brown and then turns black as it dries into a scale. In some cases there may be a pupal "tongue," a narrow projection that points upward from the reclining pupa. This is a symptom typical of no other bee disease. However the pupal tongue is not a frequently encountered sign of the disease.

Application of an antibiotic is one way to control, but not cure the disease. Burning contaminated equipment is still the best way to handle this deadliy, and contagious problem.

Treatment – Apiary inspection programs in several states have shown that American foulbrood disease can be reduced to less than one percent of inspected colonies. This does not necessarily require the inspection of every colony or every apiary. Apiary inspectors use statistical methods to select apiaries and colonies. What is important is early detection followed by appropriate treatment.

Destruction of bees and combs from infected beehives has been shown to be an efficient method of keeping the disease in check. One option exercised by some states is to allow beekeepers to treat diseased beehives with drugs. A few states will permit the decontamination of bee equipment from diseased beehives with gamma radiation, ethylene oxide steam heat, with or without follow-up treatment using drugs.

Drug feeding for the prevention and control of American foulbrood came of age with the discovery of the sulfa drugs in the 1940s and Terramycin® in the 1950s. Sulfa drugs were only effective against American foulbrood while the latter was shown to be effective against both American and European foulbrood diseases. For years beekeepers were highly successful in preventing and controlling American and European foulbrood by a program of inspection and drug feeding. In the mid 1990s resistance of *Paenibacillus larvae* subsp. *larvae* to Terramycin was discovered. This was unfortunate as Terramycin was the only material approved by the U.S. Food and Drug Administration.

The next drug to become available was a soluble form of tylosin, produced by the ElanCo Chemical Co. Tylan, as it is called, is used as a treatment only. The label does not allow this chemical to be used as a preventive, as Terramycin was.

Burning, although a great loss of bees and equipment, is still considered by many the only way to deal with an AFB infestation.

Tylan is water soluble, and as such the opportunity for honey contamination is significantly greater than Terramycin. Great care should be taken to not apply this drug when harvestable honey is on the hive, or within 45 days of adding honey supers.

However the long-range solution to combat American foulbrood disease appears to be the development of resistant bees. It has been shown that bee stocks that display hygienic behavior will go a long way towards the breeding of bees that are resistant to this disease (see DISEASE RESISTANCE).

Amoeba disease – So little is known about this disease that it scarcely seems worth mentioning; however, the name occurs several places in the literature and as we search for better methods of controlling all diseases it is worth

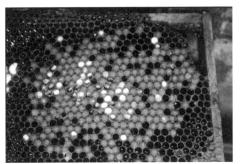

Chalkbrood disease is usually first noticed as white, fluffy material in a brood cell. This is the mycelia stage.

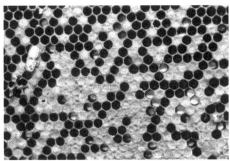

Next, the dead larvae dry down as they are consumed, and the fungus, running out of food, begins its reproduction stage, turning parts of the larvae gray, brown or black.

discussing. The etiologic agent is *Malpighamoeba mellificae,* a microscopic, one-cell organism. This amoeba is found in the malpighian tubules that, in honey bees, are attached to the gut and serve to remove waste products from the blood. Amoeba are said to be most common in the spring and may in some way be related to serious infections of nosema. Where there is nosema, amoeba may also be present but controlling the nosema should take care of both problems.

Bee paralysis – (see DISEASES OF THE HONEY BEE, Virus diseases)

Chalkbrood disease – This disease, named for the chalky appearance of the dead infected bee larvae, is caused by fungi (molds) in the genus *Ascosphaera*. Chalkbrood fungi have only been found in association with bees, and most species infect the larvae of solitary bees, such as the alfalfa leafcutting bee (*Megachile rotundata* Fabr.). At least 21 species of chalkbrood fungi have been found in N. America, but *Ascosphaera apis* is the only one known to infect honey bees. *Ascosphaera aggregata* is the species most common in the alfalfa leafcutting bee. Infections become apparent in the late larval and early pupal stages but may show up in younger larvae if a colony or nest is heavily infected. In honey bees, both worker and drone brood can become infected.

The first sign of the disease in honey

House cleaning bees remove these dry and shrunken larvae and dump them outside.

bees is larvae covered with fungal mycelia (hair-like growth of the fungus). Infected larvae, now called mummies are at first soft and fluffy but soon become hard, white lumps. Later, as the fungus matures and approaches the reproductive stage, the mummies turn black. Thus, in advanced cases of chalkbrood one finds a mixture of white, black, and black speckled mummies. Symptoms in alfalfa leafcutting bees are similar, except the mummies may never become covered in fluffy mycelia; they just turn black.

In a strong honey bee colony, the bees remove their dead quickly so that the infection is not spread to other larvae. However, in heavily diseased colonies, the bees are unable to remove the mummies promptly, and then mummies can be seen in cells, on bottom boards, and in front of the colonies. When mummies accumulate in this manner, they are an ongoing source of infection for the live brood.

The first record of chalkbrood in honey bees is from Germany in 1911. Soon thereafter, chalkbrood was reported throughout much of Europe, but the disease does not appear to have been serious there. In 1968, chalkbrood was seen for the first time in California in the alfalfa leafcutting bee. In 1972, it was reported in honey bees, and within a few years it was found in all states. Initially, it was thought that the same fungus caused chalkbrood in both the leafcutting bees and in honey bees, but it has since been demonstrated that the fungi that cause chalkbrood in the alfalfa leafcutting bee cannot cause chalkbrood in honey bees, and vice versa. Chalkbrood is sometimes a serious problem in honey bees, but it tends to be much more of a problem in alfalfa leafcutting bee management.

Distribution – Chalkbrood has been reported from throughout North America, Europe, Asia, and Australia. *Ascosphaera apis*, the honey bee chalkbrood, was probably introduced into the U.S. from Europe. The source of chalkbrood in leafcutting bees is uncertain, but it – *Ascosphaera aggregata* – likely came from Europe as well. Other chalkbroods that infect this bee may be native to the U.S.

Life cycle – Bee larvae become infected with the disease when they consume fungal spores in their food. Honey bee larvae tend to be more prone to infection when they are chilled. The typical high temperatures in the brood chamber may inhibit the pathogen's growth, and this may also be why honey bees appear to be less susceptible to chalkbrood infections than the alfalfa leafcutting bee. Leafcutting bees are solitary and brood temperatures are not controlled as in a honey bee colony.

Treatment – No effective treatment has been found for chalkbrood in either honey bees or leafcutting bees. In honey bees, managing for strong colonies has proved beneficial in preventing the disease. Management strategies include requeening with a young queen (preferably with hygenic offspring) moving colonies into full sunlight (especially in cool climates), adding supers only as needed to keep the size of the hive to a volume that the bees can keep dry and warm, exchanging wet bottom boards with dry ones, placing hives where they will remain warm, dry and well-ventilated and strengthening weak colonies by merging them with other colonies. And as importantly, removing old, contaminated comb to reduce the available unnocutum.

Control of chalkbrood in the alfalfa leafcutting bee is done mostly through sanitation methods. Bee cells are removed from nesting boards at the end of the season, then tumbled to break the cells apart from each other. Once the cells are removed from the boards, the boards can be disinfected before they are nested in again the next year. Boards can be disinfected by either dipping them in a chlorine solution or baking them at high temperatures. Removing the cells from the boards and breaking apart the nests also reduces the chance that emerging bees will become contaminated with spores. If the nests are not removed from the boards, emerging females must chew their way through siblings who have died from chalkbrood and are blocking the way out of the nest. These dead bees act as a source of contamination. - RRJ

Disappearing disease – There are many references in the beekeeping literature that attempt to explain the sudden reduction of adult populations. The name *disappearing disease* has been used to describe one such condition. In all of these cases, bees died or disappeared in large numbers, and beekeepers sought to understand why this took place. It is important to recognize that there are good beekeepers, careless beekeepers, nutritional deficiencies, toxic pollens and nectars, poor locations, genetic abnormalities, combinations of diseases, and a host of things that can go wrong in honey bee colonies. Disappearing disease has been an attempt by some trying to find a single cause for these sudden population losses.

Many diseases are still little known and need further study. Bees that die do so for a reason. It is the job of science to understand why. Ignorance is no excuse for the multitude of names that have been invented to equate a general symptom with a disease. For instance dysentery is only a symptom that can be nutritional in origin or be caused by nosema disease. Disappearing disease has no standing or credibility.

Dysentery – Dysentery is not a specific disease but a symptom that could be caused by nutrition or disease. (see DYSENTERY)

European foulbrood – This disease affects only the larvae of honey bees which die when they are only four to five days old. The symptoms of European foulbrood disease are sufficiently different that it is usually an easy matter to visually distinguish it from American foulbrood. The disease is most likely to

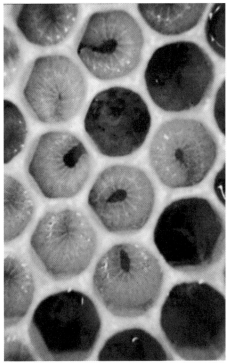

Larvae infected with European foubrood are generally noticeable before the cell is capped.

The larvae are consumed by the disease, the cuticle begins to show rings, and the color continues to darken.

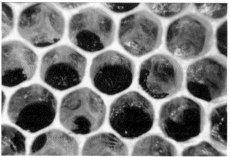

Finally, the larvae are coiled, usually very dark brown or black, lying at the bottom of the cell. The "scale" is usually relatively easy to remove.

be seen when the colony population is building to its maximum strength. Historically, in the U.S., the introduction and subsequent popularity of the Italian race of honey bees resulted in the reduced incidence of European foulbrood disease among its descendants.

Most beekeepers do not consider this a serious disease as European foulbrood usually abates with a good nectar flow. In areas where the disease is endemic, beekeepers have used Terramycin® for its prevention and control.

Taxonomic position – The cause of European foulbrood is the bacterium, *Melissococcus pluton*. This disease is specific to honey bees and affects no other insects or animals. Gross symptoms are usually sufficient for identification although in some cases microscopic followed by cultural tests are necessary for the confirmation of the disease (Shimanuki, H. and D.A. Knox. "Diagnosis of Honey Bee Diseases." *USDA-ARS, Agriculture Handbook* No. 690. 2000).

Distribution – European foulbrood has been found everywhere European and Africanized honey bees are kept. The one major exception is European foulbrood disease has not been found in New Zealand. European foulbrood has also been reported from *Apis cerana* in India. It does appear that European foulbrood is seen more frequently in cooler climates than in warm ones.

Life cycle – The bacterium that cause European foulbrood apparently enter and infect the very young larvae. There is, of course, no feeding during the egg stage and no one has ever reported finding bacteria in honey bee eggs. While larvae of any age can be infected, their death occurs only when the infection starts at an early age. Infected larvae that are not killed have poorly developed silk glands and their cocoons may not fully develop. The death of infected larvae may be hastened by the presence of one or more secondary species of bacteria that have been reported as being associated with European foulbrood disease.

The dried larvae in cells form scales that are soft and pliable and are easily removed by the bees. This is quite different from the situation with American foulbrood where the scales are firm, brittle and cling tightly to the bottom of the cell.

Detection – Diseased larvae, usually in their fourth or fifth day of life, lie curled in the bottoms of their cells. When they die their bodies are flabby, soon turn brown and usually, but not always, have a sour odor. However, the odor is variable depending on the presence of the secondary organisms. The age, position and color of the larvae are frequently used to diagnose European foulbrood and larvae dead from this disease appear quite different from those killed by other diseases.

Treatment – In the case of any disease it is first important to search for the cause. European foulbrood is more common in the spring than in the summer and fall. Both it and sacbrood are likely to be seen several weeks after a colony suffers a loss of adult bees. This suggests that European foulbrood may be called a stress disease, that is a disease that shows itself when a colony is having difficulty holding a normal brood-rearing temperature. If a cause can be identified then steps to correct it should be taken.

One antibiotic, oxytetracycline, sold under the trade name Terramycin®, is approved for the prevention and control of European foulbrood. It is available from any of the bee supply companies. A variety of methods of feeding the antibiotic are used including mixing with sugar syrup (not recommended) and dusting. The U.S. Food and Drug Administration grants approval for the use of antibiotics and methods of application (see ANTIBIOTICS AND OTHER CHEMICALS FOR BEE DISEASE CONTROL). Because the formulations and concentrations in which the antibiotic is sold vary greatly it is important to read and follow the accompanying label directions.

Nematodes – Among the many small creatures that may infest honey bees are the nematodes. Sometimes known as roundworms, nematodes can be both free-living and parasitic. Several people have reported finding nematodes in adult honey bees, but the number of bees infested is so small as to be of little or no consequence. Nevertheless, it is striking to find a two-to-three inch (five-to-seven centimeter) long white worm coiled inside the abdomen of a worker honey bee. We do not know how nematodes infest honey bees, nor do we understand their life cycle or whether or not the honey bee is the primary host for one or more species. Nematodes are usually found by people dissecting honey bees for various reasons, including seeking a disease-causing organism, studying anatomy, or researching a subject such as the sugar concentration of nectar being transported by bees. In one study, only two nematodes were found in over 2,000 bees that were examined, indicating their low incidence. Because the nematodes were found in the older, foraging bees, and since nematode development probably takes a long period of time, we can assume that nematodes have little effect on the lifespan of the bees they infest.

Nosema disease – This disease of honey bees has been a problem for beekeepers, especially in the northern states. It is caused by an organism that can be seen only with a microscope. This organism invades and destroys the epithelial cells in the bees midgut. The result is that the midgut lining is destroyed affecting the bee's nutrition. One of the impacts of the disease is to shorten the lifespan of worker bees. In addition this disease interferes with the food production by worker bees. In

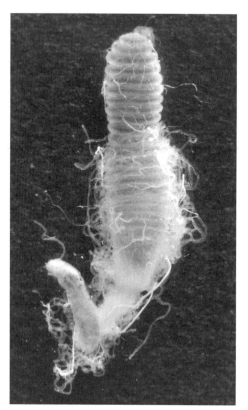

Healthy mid-gut of a honey bee is usually brownish-red to yellow. Circular constrictions are prominent.

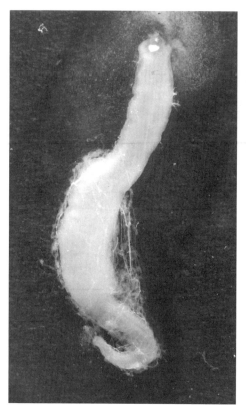

Nosema infected mid-gut. Color is dull gray, to white and constrictions usually cannot be seen.

queens the disease will lead to certain supersedure. As with all diseases, beekeepers can reduce the incidence of Nosema by frequent comb replacement. Feeding pollen substitutes and supplements high in proteins and hive placement to encourage flight are also strategies that may help overcome the impact of the disease.

Taxonomic position – The microbe that causes nosema belongs to a group called the microsporidia; the scientific name is *Nosema apis*. Nosemas are common and widespread among the insects with many species found in the group. A nosema that infects a species of fly cannot infect honey bees and vice versa. Thus, other insects do not serve as a source of infection for honey bees and the nosema in bees cannot infect other insects.

Recently, scientists have discovered a species of nosema affecting *Apis cerana*, the Asian honey bee. It was found that this species, named *Nosema ceranae* could also infect European honey bees. Early tests indicate it is far more virulent than *Nosema apis* in European bees, however more research is needed. Initial tests indicate this new disease is controlled using fumigillin.

Distribution – Nosema disease in honey bees has been found everywhere it has been searched for and presumably is everywhere honey bees are kept. It appears to be present at low levels at all times and becomes a problem only when conditions are right for its growth. Beekeepers in the southern U.S., and in tropical and subtropical areas, do not consider nosema a serious problem. In the north nosema is usually not a problem in the late spring, summer or early fall. Populations of *Nosema apis* appear to build up in the winter and in the early spring the disease may cause dysentery. In severe cases, nosema disease can cause death of colonies especially when weakened by other diseases and pests.

Life cycle – Newly-emerged honey bees are always free of infection by nosema, however, they become infected as soon as they ingest the spores. Adult bees usually become infected when they feed on contaminated pollen and when they clean soiled combs. Once the spores are in the midgut they extrude a filament that attaches and protrudes into a cell and through which the germplasm is injected. The parasites develop within the cell and under normal temperatures new spores are formed after about five days. When the spores have formed, the gut cell wall bursts and releases a new batch of spores. Apparently these may invade other cells or may be passed out of the bee with fecal matter. Researchers have searched other parts of the body of honey bees and while they have found spores that appear similar to those of Nosema it is thought that the microbe grows in the gut only.

Detection – The only clear proof that bees are suffering from nosema can be found by examining the contents of a bee's gut under a microscope. Although not exclusive to nosema disease, heavily infected colonies may have several obviously sick, crawling bees on the ground in front of the hive. The wings may not be connected. These bees may have lost some of their hair. There may be signs of dysentery. Under normal circumstances honey bees do not defecate within a hive or at the hive entrance. Some of these same symptoms are common in the case of other diseases such as tracheal mites and some of the viral diseases. If the head of an infected bee is pinched off, the gut may be pulled from it by grasping the last abdominal segment with the fingernails. The gut of a normal bee is brownish in color while that of an infected be is white and often twice the size of a normal gut.

Treatment – Successful treatment of nosema involves several considerations. Whereas the drug fumagillin works very well to keep nosema under control there are several things that beekeepers may do to aid in the control of the disease. The most important of these is the selection of a good wintering site especially in the north (see APIARY LOCATIONS). The apiary should slope toward the sun and be exposed to full sunlight. Hive entrances should face south to encourage bee flights.

Several methods of fumigating combs and hive furniture that will kill other microbial spores may serve to kill those of nosema. While many fumigants have been tested and recommended, it is necessary that each have approval from the Environmental Protection Agency before they can be used. A state apiary inspector or extension apiculturist at a state university can advise on this matter. Fumigation of stored combs with ethylene oxide, heat, or irradiating hive furniture contaminated with *Nosema apis* can reduce the incidence of nosema disease and other diseases.

Fumagillin is the only antibiotic that is approved to control Nosema. It is remarkably effective and despite the fact that it has been in use for several years, resistance to it has not arisen. The drug is effective only when it is fed in sugar syrup. Fall and early spring feedings are recommended. Feeding fumagillin in dust form or in an extender patty have not proven effective. We emphasize that feeding a drug to control nosema should in no way be considered as a substitute for selecting good apiary sites and other management practices that may be used to control the disease. Label directions should be adhered closely in administering any drug to honey bees.

Parafoulbrood – The causative organism of parafoulbrood was believed to be closely related to the bacteria that causes European foulbrood, thus the name parafoulbrood. This disease remains an enigma, but it was very real to the people who experienced it. Because we still do not know the causative organism(s) and symptoms which make this disease unique, the term parafoulbrood should not be used.

Parasitic mite syndrome, bee – A puzzling array of symptoms has been found in some honey bee colonies that is atypical of any one disease. The one common symptom of this syndrome is the presence of the parasitic mite, *Varroa destructor*. Frequently a reduction in the adult population of the colony can be seen. Most often a spotty brood pattern is seen in the colonies with the syndrome. No typical microflora or odor is associated with the affected brood or adults. (see Shimanuki, H., N.W. Calderone, and D.A. Knox. "Parasitic mite syndrome: the symptoms." *American Bee Journal* 134: 827-828. 1994.)

Sacbrood disease – (see DISEASES OF HONEY BEE, Virus diseases)

Stonebrood – This disease of honey bee larvae, and sometimes pupae, is rare though that is not the statement made in some of the textbooks. It is usually caused by the fungus *Aspergillus flavus*. Larvae and pupae killed by the disease may at first appear white because of the mycelium growing on the surface of their bodies and hence may be confused as chalkbrood. However in the case of stonebrood, the larvae become brownish or yellowish-green and hard, thus the name. It is stated that the fungus may grow in the guts of adult bees that ingest spores but there seems to be some question as to whether or not the fungus will kill normal adult bees. Stonebrood is not considered a serious disease.

***Varroa* disease** – (see MITES ASSOCIATED WITH HONEY BEES, MITES, PARASITIC)

Virus diseases – The viruses are a complex group of living organisms and a much more primitive form of life than are the bacteria. They are, in general terms, the lowest form of life that may reproduce itself but still cannot live independently. Viruses can live, grow

A frame with severe PMS sysmptoms. (USDA photo)

and multiply only within the cells of their hosts. Viruses are genetic material surrounded by a sheath of protein. They feed only on the nutrients in the cells of their host and use the energy they gain to replicate themselves. In the process the viruses destroy the host cell at which time the replicated particles, which are too small to be seen even with a microscope, are released and proceed to infect other cells.

Most, if not all, the bee viruses show a high degree of specificity, that is, they are host specific and do not infect other insects or animals. Thus, virus infections in honey bees pose no threat to humans any more than the virus that causes our common cold threatens honey bees.

Little is known about the cross-infectivity of virus among species of honey bees. A virus known as the Thai sacbrood virus ravaged colonies of *Apis cerana* in Asia. In Nepal, about 90 percent of the colonies died in a matter of two or so years in the early 1980s as a result of this disease; however, bees in some colonies were apparently resistant to the virus and the numbers of colonies of this native Asian species of honey bee have returned to normal in that country. However, the disease continues to result in the death of colonies in India. It is not known if Thai sacbrood virus poses a threat to European or African honey bees.

No cures or drugs are effective in controlling viruses. It is obvious that they have been present for a long time but we cannot describe the circumstances under which the viral diseases may show themselves. Obviously, some bees are resistant to some or most of the viral diseases. As in the case of many other diseases of the honey bee the best defense the beekeeper has is the maintenance of strong colonies with good food reserves. When sick bees are seen it is best to requeen the colony. One should always be on the outlook for especially prosperous colonies that should be used as breeding stock.

A number of viruses have been isolated from honey bees and many of them have not been shown to cause any pathological condition or are of economic importance. Listed below are some of the virus-induced diseases. Some viruses are believed to be transmitted or activated by the mite, *Varroa destructor*. Other viruses may also be found associated with nosema and other protozoan diseases. For more on bee viruses see Ball, B. and L. Bailey. "Viruses" In *Honey Bee Pests, Predators, and Diseases*. Ed. R.A. Morse and K. Flottum, 3rd. edition, (The A.I. Root Co., Medina, OH. 718 pages. 1997.)

Sacbrood – The most common of the viruses that infect honey bees is the one that causes sacbrood. It is a relatively easy disease to diagnose as the dead larvae lie flat on their backs with the head elevated slightly. Brood cells containing larvae dead of sacbrood disease may or may not be capped. Larvae killed by sacbrood virus turn first pale white, then yellow and finally a brown that grows darker with time. The head end of the larvae is frequently darker than the rest of the body. The dead larva is easy to remove from the cell with a pair of forceps. When the larva is removed it hangs from the forceps like a sac. The sac, which is the last unshed larval skin, is filled with watery fluid that flows from it easily when broken. If the larva is not removed by the bees it may dry, shrink, and form a dark brown or black scale on the bottom of the cell that may resemble a larva dead from some other disease. However, the scale does not stick firmly to the bottom of the cell as does the scale formed from a larva killed by American foulbrood.

Sacbrood is usually seasonal and is most common in the spring. However, it may be seen in the summer after a colony has suffered some loss of adult bees, such as from pesticide. Strengthening a colony by giving it additional bees, requeening, or taking steps to improve the environmental conditions in the beeyard are the only recommendations that can be made to control the disease.

Sacbrood has been reported from all countries where bees are kept. It is rarely serious. Some bees are more resistant to the virus than others but there are no data. This virus is similar to the virus that causes Thai sacbrood. The latter virus is found only in Southeast Asia and is a problem only in *Apis cerana*.

Bee paralysis – Beekeepers have written about its existence and strange behavior for over 100 years. Paralysis, as a symptom in honey bees, could be misleading as the symptom actually refers to adult bees that are trembling. This symptom could be the result of viruses or chemical poisoning. The disease occurs rarely, may result in the death of only a small number of colonies in a particular apiary and disappears as rapidly as it appears. Bees that suffer from bee paralysis often lose their hair, their bodies become distended and are shiny and black. The disease is sometimes known as the "hairless black syndrome" or simply as paralysis. Frequently colonies with this disease may show large numbers of bees crawling away from the hive entrance. The cause of this disease is chronic paralysis virus.

At least three paralysis viruses have been described: the classic chronic bee paralysis, the acute bee paralysis and the slow paralysis. The relationship of the acute bee paralysis virus and *Varroa destructor* has yet to be determined.

As with other viruses affecting honey bees no known cure or treatment exists other than requeening, changing the environment, and generally reducing stress. However, as the susceptibility to bee paralysis is heritable, this disease may be the result of inbreeding. Therefore introducing queens from a new source may help overcome the effects of this disease.

Black queen cell virus – This virus kills the developing queen while it is a pupa. The queen cells become brown to black in color. Only developing queens are affected by this virus. It is associated with the presence of Nosema.

Deformed wing virus – The symptoms of adult bees with this virus are easily noted. Adult bees emerge with wings that appear twisted, wrinkled, smaller than usual, and, in a word, deformed. The presence of this virus is associated with *Varroa* mites feeding off

Worker bee with deformed wing virus. (Calderone photo)

of developing worker pupae. When worker pupae are infested, it is thought that the *Varroa* infestation is severe, and finding workers with deformed wing virus is a sign of serious problems.

Adults with this virus cannot fly and do not live long in the hive.

Filamentous virus – This virus causes a disease of adult honey bees, causing the blood to turn milky white and the bee to die prematurely. It is associated with the presence of nosema.

Kashmir bee virus – This virus was first found in *Apis cerana* and now has been found in *Apis mellifera*. Frequently this virus is found in association with colonies infested with *Varroa destructor*. The Kashmir bee virus has caused colony deaths in Canada.

DISEASE RESISTANCE – Any free-living plant or animal species whose potential for reproduction is unchecked would soon overrun earth. However, under many circumstances population explosions are held in check by pests, predators and diseases. The organisms that cause disease are only trying to survive themselves. In a perfect host–parasite or host–predator relationship, the host population is not destroyed by the attacker, for, if this occurs then the parasite or predator loses its food source. Most plants and animals have developed mechanisms that protect them against their diseases, to at least some extent, or have evolved to accomodate some level of infestation by disease or pest, that is some tolerance for the organism. When we develop or encourage disease resistance in the plants and animals we use we are exploiting their ability to do so and following nature's own way of protecting species.

It has been demonstrated repeatedly that some races or groups of honey bees are resistant to certain disease organisms. This resistance can take many forms. An example of behavioral resistance is the ability of the Asian honey bee, *Apis cerana*, to remove, bite and kill *Varroa* mites. Africanized honey bees also show a greater tolerance to the mite. At this writing it is not clear whether this is behavioral, due to a slightly shorter life cycle (Africanized honey bees develop in about 19 days) or if climate has an effect on the interaction of bees and the mites.

The late Professor Walter C. Rothenbuhler, in several papers, demonstrated that some honey bees have an innate genetic resistance to American foulbrood. Two genes are involved. One allows the bees to detect and uncap cells containing larvae killed by the bacteria. The other gene allows the bees to remove the dead larvae. By removing the larvae and cleaning the cells the bees eliminate the source of the infection and the colony does not succumb to the disease. It has also been demonstrated that this hygienic behavior and other desirable characteristics can be incorporated into a bee breeding program including long life, greater hoarding ability and an increased tendency to collect pollen (see Kulincevic, J.M. and W.C. Rothenbuhler. *Selection for length of life in the honey bee*. Apidologie 13:347-352. 1983.) This theme continues to be explored with several strains of bees available that exhibit some level of hygienic behavior, or some level of other resistant behavior. (See VSH BEHAVIOR)

DIVELBISS, CHARLES A. – Charles Divelbiss was an active educator all his life, serving as teacher and principal in

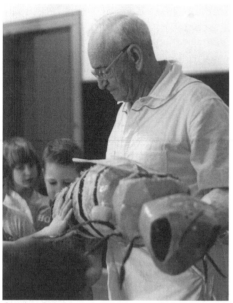

Charles Divelbiss

Mansfield, Ohio. He also was an energetic beekeeping instructor, giving programs to both children and adults, featuring the benefits of bees, beekeeping and pollination. His energy and enthusiasm was rewarded by the Eastern Apicultural Society when they named the 'Educator of The Year' award after him.

DIVIDING COLONIES (SPLITS) – In many parts of the world beekeepers increase their colony numbers by dividing their colonies, or making "splits" early in the season. A split is made so that both parts of the divided hive have brood and food, without paying attention to where the queen may be found. The queenless part of the divided colony is allowed to rear its own queen. The method is not profitable immediately but is quick and easy and the beekeeper need not be very knowledgeable about bee biology or bee behavior. However the queen that the bees raise from the queenless part of the split may not be of top quality, and the development time required to raise a new queen slows the colony, if the queen is lost during mating.

Under certain conditions, this simple method of dividing colonies works quite well. For example, in many parts of the northeastern states the best honey flow is from goldenrod and occurs in late August and early September. If a prosperous two- or three-brood chamber colony is split into two units in mid- or late April the beekeeper will have two producing colonies by August 15. Swarming will probably be prevented but this is not the ideal method for keeping bees because much can go wrong. In a poor year it is possible both colonies may be lost due to insufficient food. In a good year, with more care and a small investment in new queens, it is highly probable that a single strong colony can be split into three or four units that can each produce a surplus in the fall.

The simple division of a strong, two-story colony is done in the following manner. A bottom board is placed next to the colony to be divided. The colony is smoked, cover removed, and top brood chamber placed on the new bottom board to start a new colony. Under normal circumstances in early spring the queen is most likely to be in the top brood chamber, since bees tend to move the brood nest in an upward direction.

The two units are quickly examined and arranged so that essentially equal amounts of open and sealed brood are in each. The number of adult bees needs to be considered for each because the colony that remains in the original location will attract more flying bees, thus have a greater adult population. We recommend hedging toward having more nurse bees in the split that gets moved.

Requeening both splits is also recommended so no time is lost in raising a new queen, and we strongly suggest discarding old comb that is mostly empty, and replacing with newer drawn comb, or foundation.

A hive body of empty combs is then placed on the new colony. If one has only foundation to give such a split it is best to wait until a honey flow occurs, since at least a small honey flow is needed for the bees to draw the foundation. Each split should be fed two or three gallons (eight to 12 liters) of sugar syrup to stimulate the drawing of foundation. An inner cover and an outer cover are added. The colony on the original location, which has now been reduced to one brood chamber, is also given another empty brood chamber, an inner cover and an outer cover.

One must make certain that food – both honey and pollen – and young larvae or eggs are present in each unit.

The two units should be examined in about two weeks to make certain both accepted their queens. If the populations of the two hives are not about equal at the end of the first month their positions may be exchanged in the middle of the day when the bees are flying so that the weaker of the two gains the field force of the stronger of the two. If, after a month, one of the splits is queenless the two splits should be combined using the newspaper method (see UNITING COLONIES AND NUCLEI).

As indicated above, if a colony is sufficiently strong to produce two units with the simple division method it is probably strong enough to make three or four smaller colonies when young queens are given all the queenless portions immediately (see NUCLEUS COLONY). (If one wants to divide a strong colony because it is too defensive see DEFENSIVE COLONIES, HOW TO TREAT).

DIVISION OF LABOR – (see CASTE)

DOOLITTLE, G.M. (1846-1918) – Doolittle's father was a hobby beekeeper who introduced him to bees and beekeeping when he was 10 years old. He lived all of his life in Borodino in central New York and was active in the early days when the commercial beekeeping industry was developing in the U.S. He wrote his first article on honey bees in 1870 and thereafter was a prolific writer and long-time contributor to *Gleanings in Bee Culture*. Doolittle's chief contribution to beekeeping was in the area of queen rearing. His book, *Scientific Queen Rearing*, was first published in 1888 and underwent several printings. He learned that priming queen cups with royal jelly was helpful when grafting larvae. While he did not invent artificial beeswax queen cups he understood their value and encouraged their use. Doolittle wrote extensively about the importance of an abundance of food during the queen-rearing process, a fact that is still one of the most important considerations if one wants to grow good queens.

DRESSING FOR THE APIARY – Several simple rules need to be followed when working in an apiary in order not be stung excessively. Cover as much of the body as is reasonable while maintaining freedom of movement, especially of the hands and arms. Most beekeepers do not wear gloves as they are cumbersome and soon acquire a sweat odor. There is some value in leaving the hands bare as they are a barometer of the bees' defensiveness; when one receives a sting on the hand it is clear that the colony should be given more smoke. The odor of sweat appears to irritate honey bees. Do not wear a wrist watch while working

Dressing For The Apiary

Veils with and without helmets that easily fold when removed . . .

Veils that are stout and square, or can be easily removed from an appropriate hat . . .

Full bee suits with zipper-attached veils are available. The choices are nearly infinite.

The key components of good apiary wear are durability and protection from bees, from honey, wax, propolis and dirt. Well-made seams and enough room to bend, squat and stretch are also necessary. Enough pockets in the right places are invaluable, too. Gloves, if worn, should fit the job. Commercial operators need durable, leather gloves with large gauntlets, while hobby and beginners can use lighter, canvas or plastic gloves, or even dishwashing style hand protection.

a colony; even a metal wrist band will have a sweat odor. Rubber gloves, such as are used to wash dishes, may be used; they may be washed and bees do not sting through them too easily.

Avoid wearing wool, suede, leather or any other type of rough clothing; wear light-colored, smooth-finished clothing. Beekeepers usually wear khaki or white cotton shirts and trousers or coveralls. Dark colors tend to stimulate attacks by bees. Boots that cover the ankles are helpful. Some beekeepers wear white rubber boots. Some beekeepers sew elastic in trouser legs and sleeves to reduce the likelihood of having bees crawl underneath. An alternative is to order bands with loop and hook fasteners from bee supply companies for both trousers and sleeves. Shirts, pants and other clothing with zippers will also exclude crawling bees better than those with buttons.

A wire or stiff plastic veil is better than a flexible cloth veil, as it cannot be blown against the face. The mesh on the face should be painted black on the inside to improve visibility and white or aluminum on the outside. This coloring is especially important when working with Africanized bees. A defensive bee may sting through a veil blown against the ears, face or neck.

One of the most versatile pieces of equipment for beekeepers is a roll of duct tape. The tape can be used to mend clothing and bee veils, repair hive equipment, and prevent bees from crawling under clothing.

DRIFTING – Bees in an apiary will often become confused and enter the wrong hive. This is called drifting. Drones do not seem to care which hive they enter and will do the most drifting. However, drifting among workers is common, and even queens returning from their mating flights may enter the wrong hive. If supersedure takes place in an apiary where colonies are close together and look alike, a queen returning from a mating flight can be lost easily. Worker honey bees carrying food and entering a hive other than their own are said to be more likely to be accepted than are those carrying no food.

Drifting may be reduced by using landmarks in an apiary, painting hives different colors, and by facing entrances in different directions. Since colonies apparently benefit from facing south, entrance direction should not be varied greatly. Painting hive parts different colors has become increasingly popular and is probably the simplest way to reduce drifting. Small trees in an apiary make good landmarks; if pruned frequently they will not shade colonies.

The worst cases of drifting are seen when one installs two or more packages in an apiary at a time when bees are flying. Severe drifting may also take place when colonies are moved from one location to another at a time when bees may be flying in the new location. Under both of these circumstances drifting may soon become rampant. The root of the problem is that the bees become confused and some of them will expose their scent glands at their colony's entrance, or what they presume to be their own colony entrance. Since the scent gland pheromone is the same for all honey bees, those that are flying may be attracted to foreign hives by bees other than their sisters.

An apiary is an artificial situation for honey bees. Honey bees evolved in forests, and their nests were scattered just as the hollow trees that house them are widely distributed. When bees in an apiary drift, they tend to move to the ends of a row of hives rather than to the middle of the row. It is not uncommon for colonies on the ends of rows to have populations many times larger than normal. When drifting is excessive it is possible to pick up hive bodies filled with bees from overpopulated colonies and to place them on other colonies in the apiary. There is a danger of losing or mixing queens in this manner. At times when drifting becomes serious, colonies are usually less defensive than normal, and little fighting occurs between bees when they are moved from one colony to another.

An Italian drone. Note the large eyes, stout abdomen and blunt, tufted rear end.

Drifting is one of the chief ways in which many of the common bee diseases are spread. Apiary inspectors concerned with the spread of the bacterial disease, American foulbrood, have long been of the opinion that the disease is much more likely to spread to adjacent colonies than to those some distance away. American foulbrood is a well-known disease, and anecdotal observations concerning its spread are commonplace and often are corroborated by those with years of experience. Observations on the mites, *Acarapis woodi* and *Varroa destructor,* indicate that they also spread rapidly in an apiary in this manner.

DRONE – A male honey bee is called a drone. Drones serve only one primary function and that is to mate. Drones, however, contribute to the morale of the colony. Drone cells that are destroyed

DRONE-BREEDER COLONIES

The drones with which a queen mates are fully as important as the queen in determining the qualities of the worker bees that are produced by that queen. Beekeepers who actually breed queens attempt to flood the area near their mating apiaries with colonies that have an abundance of drones of their choosing with the qualities wanted in their stock. These are called drone-breeder colonies. They, like the colonies from which larvae for queen rearing are grafted, are selected with great care. Drone-breeder colonies are usually supplied with extra drone comb to increase the drone supply.

Beekeepers who do not rear their own queens, but who depend on natural supersedure for new queens, can influence their stock by encouraging drone production on the part of the better colonies. The easiest way to do this is to add drone comb, or at least to not eliminate drone comb from colonies with desirable traits. Beekeepers may also discourage drone production on the part of other colonies through the elimination or reduction of the amount of drone comb in those with less desirable traits (see MATING OF THE HONEY BEE, Controlling natural queen mating by drone flooding).

Providing drone comb to a colony reduces the drone comb built elsewhere in the colony by the bees. This focuses the *Varroa* population in the colony toward the ample number of available drone pupae in that comb.

Removing this comb specifically for *Varroa* control is at odds with producing a large population of desirable drones for mating. Moreover, treating these colonies with the miticides Apistan® and Checkmite+® affects developing drones, reducing their mating ability.

A queen (yellow circle), many workers and a drone (red circle), for size and shape comparison.

by the beekeeper will be quickly rebuilt during the spring and summer to provide a population of drones for the colony. Drones have a sensory system, including the eyes and antennae, that is much larger and more developed than those of the workers and queens. Drones may feed themselves when they first emerge from their cells but soon learn they can solicit food from workers with ease. Drones do not forage but return to the hive for food after an unsuccessful mating flight. Since the drones serve no useful function in the winter they are slowly starved by the workers in the autumn or during a dearth. When they have become weak they are dragged from the nest and allowed to die outside of the hive. Drones take longer to develop than do the workers and queens (see LIFE STAGES OF THE HONEY BEE and MATING OF THE HONEY BEE).

Using softer chemicals – organic acids or essential oils – is recommended then, to protect the developing drones from the double edge sword of mites, and miticides.

DRONE CELLS – (see COMB, NATURAL)

DRONE COMB FOUNDATION – Foundation, both wax and plastic, where the bases of the larger drone cell size has been made and is available from the bee supply manufacturers. This foundation has four cells per inch versus five per inch (16 per 10 cm versus 20 per 10 cm) for worker cells.

Drone comb foundation is used in extraction supers in areas where the moisture content of the honey is lower than normal, such as Arizona, where beekeepers report that it is easier to extract the dry honey from combs with the larger cells. Drone comb foundation is also used by queen breeders who need to increase drones from selected colonies for mating (see DRONE-BREEDER COLONIES).

Recently, substituting a frame or two of drone comb foundation for worker bee foundation is being advocated as a non-chemical means to reduce the population of *Varroa destructor*. In this case, drone brood production is encouraged as a lure to attract the mite to the combs with drone brood. The combs with drone brood are then removed from the hive and the cells with brood are destroyed before the drones and mites emerge. (see DRONE POPULATION CONTROL)

DRONE CONGREGATION AREAS – (see MATING OF THE HONEY BEE and PHEROMONES, Sex Attractant)

DRONE POPULATION CONTROL – In a comparison of colonies with and without drone combs, Seeley (Seeley, T.D. "The effect of drone comb on a honey bee colony's production of honey." *Apidologie* 33:75-86. 2002) reported that one of the effects of the presence of drone combs was lower drone comb building but the number of drone flights was higher in those colonies with drone combs. He attributed lower honey yields in those colonies with drone comb to the rearing and maintenance of the higher number of drones.

DRONE TRAP – A drone trap is a device that is designed to trap drones and queens leaving the colony. The traps, which are fitted to the hive entrance, consist of two compartments. The bottom compartment is open to the inside of the hive. Exiting bees must crawl through a queen excluder screen to get outside, or through a cone of screening into the upper compartment, which also has a queen excluder screen, allowing worker bees to exit to the outside while retaining queens and drones. Because the screen cones serve as one-way doors, drones and queens that enter the trap become trapped in the upper compartment.

Drone and queen traps were invented for capturing exiting queens in an attempt to stop swarming, and for reducing the number of drones in the colony. Neither of these goals is best achieved with their use, and the traps interfere with both ventilation and the passage of worker bees. Thus, although they are useful for specialized research and breeding purposes, drone traps are not practical for general management.

If, however, a beekeeper wishes to capture drones from a particular colony the following simple trick works well.

Drones leave the colony shortly after mid-day for mating flights. The beekeeper can then set a queen excluder in front of the colony, blocking the entrance from the front, but leaving a gap on either side.

Returning drones, heading straight into the colony, encounter the excluder and, confused, sit on the grid. The beekeeper can then easily pick them off the excluder. Workers, both leaving and returning either go around the grid, or through it, without fuss or confusion. PKV, and Editor

DRUMMING – If one beats rhythmically on the side of a honey bee hive, gum, box hive, or natural nest, the bees will respond by moving upward. The queen and the drones will follow the workers. If a covered, empty super or hive body is placed above an opened hive, the bees will march into it in an orderly fashion in a matter of several minutes. This process, called drumming, is one method commonly used to force bees out of a fixed comb hive. It is important to smoke the bees moderately before starting the beating process, but once the bees start to move further smoking is not necessary. An uncoordinated beating will arouse the guard bees and cause them to attack but a rhythmic one does not. One may drum with bare hands, a stick, or other instrument; a rubber mallet works well.

The origin of drumming is unknown, but the method was widely used in Europe to drive bees out of straw skeps and other contrivances used as hives several hundred years ago. It is curious that bees will abandon brood when drummed, something they are not prone to do under ordinary circumstances.

Some producers of package bees place a queen excluder over a colony and drum the bees up into an empty super. The bees are shaken from the super into a package. The excluder prevents the queen and drones from moving upward and into the empty box. If the drumming is done in the middle of the day, the older bees are out foraging and the package is made up of young bees.

It is not known why bees respond as they do to drumming. Honey bees can detect substrate-borne vibrations, and certain frequencies will cause them to freeze or stand motionless on a comb. The sensory organs involved are apparently located on the feet. How far bees will move when drummed has never been tested.

DUMMY BOARDS – (see FOLLOWER BOARDS)

DUNHAM, WINSTON E. (1903-1980) – Dr. Winston Dunham was the retired professor of Apiculture and extension entomologist at The Ohio State University. He was the extension entomologist for a number of years until his retirement in 1963. After leaving his post at The Ohio State University he founded Deer Creek Honey Farm near London, Ohio, operating apiaries in west central Ohio and an extensive honey packing business selling honey under the Deer Creek Honey Farms label.

Winston Dunham

DYCE, ELTON JAMES (1900-1975) – Early in his career Dyce served as Demonstrator, Lecturer and Professor of Apiculture at Ontario Agricultural College, now Guelph University. He helped found and became the first manager, for two years, of the Fingerlakes Honey Producers Cooperative in Groton, N.Y. He joined the Cornell University faculty in 1942 where he became Professor of Apiculture, retiring in 1966. During his tenure at Cornell Dyce trained more professional apiculturists in the U.S. at the M.S. and Ph.D. level than anyone before him; this included many students from abroad. Dyce visited and worked with beekeepers in most of the major honey-producing areas in the world. He was very active in promoting the International Bee Research Association and was the second American to be elected to honorary membership.

Dyce wrote extensively about honey bee colony management, especially swarm control and prevention. It was during his tenure at Cornell that many of the larger honey packing operations came into being in North America and Dyce worked closely with these firms in honey processing machinery and the packing, processing and marketing of honey. As a graduate student at Cornell he developed and patented a method for controlling honey crystallization. The result was the Dyce process for making crystallized honey that came to be used worldwide. The Dyce Laboratory for Honey Bee Studies at Cornell was named in his honor in 1968.

DYCE PROCESS FOR MAKING CRYSTALLIZED HONEY – E.J. Dyce, then a Canadian citizen, was granted U.S. Patent number 1,987,893, dated January 1935, covering a process for making finely crystallized honey. If made properly the crystallized honey spreads like butter. Its crystals are so small that they cannot be detected with the tongue even when the honey is pushed to the roof of the mouth, a spot that is unusually sensitive to coarse materials.

Dyce came to Cornell in 1928 as a graduate student to study honey fermentation and crystallization. At the time the chief market for Canadian honey was England, where finely crystallized honey was in great demand. Both honey fermentation and crystallization were little understood; beekeepers were suffering some losses to fermentation.

Honey contains two primary sugars, glucose and fructose. We know now that when honey crystallizes only the glucose crystallizes, while the fructose remains dissolved in the liquid portion. If any fructose does crystallize in conjunction with the glucose, the percentage is small. Glucose forms two kinds of crystals. One

Elton James Dyce

Dyce Process For Making Crystallized Honey

is a simple, dry sugar crystal and the other is called a glucose hydrate crystal. The hydrate crystal contains water while the dry crystal does not. However, the glucose hydrate crystal contains only about 9.09% water, which is a lower percentage than the 18% water that is found in honey. Thus, as the honey crystallizes the moisture content of the uncrystallized portion increases and fermentation may occur. Once honey has fermented it cannot be salvaged. Yeasts that are found in honey cannot grow until the moisture content of the honey is slightly above 19%; thus, ripe honey does not normally ferment. Even crystallized honey that is rock hard contains crystals surrounded by liquid in which the yeast cells may grow. When Dyce discovered this, he realized that the honey to be crystallized must first be pasteurized to kill the yeast cells. This is an important part of the process.

Dyce's second discovery was that seed crystals may be added to honey to encourage crystallization. A seed crystal is a crystal with one or more sharp edges on which the crystal may grow larger. To make crystallized honey, one grinds previously-made Dyce processed honey to fracture the crystals it contains. Any of several types of machines, including a meat grinder, may be used to do this. It is important that the honey be kept at a fairly low temperature during the grinding process or the edges of the crystals will melt. Once rounded, the crystals will not grow larger. Dyce also found, through a long series of experiments, that the optimum temperature for crystallization was 57°F (14°C). If the honey is held within ten degrees of this temperature it will still crystallize in a reasonable period of time, but 57°F (14°C) is best. Ten days to two weeks is usually ample time.

Dyce wrote the following concerning his process: "Honey should be heated until it is totally liquefied and until a temperature of about 150°F (66°C) is reached. It should then be thoroughly strained through two or three thickness of fine nylon, an O.A.C. strainer (Ontario Agricultural College metal strainer) or some other strainer that will remove all noticeable wax particles. The honey should be stirred constantly, yet sufficiently carefully to prevent overheating and the incorporation of air bubbles. The agitation should be done below the surface of the honey. The honey should then be cooled as rapidly as possible to about 75°F (24°C). Here again some form of agitation will have to be used which will not incorporate air and which will remove the cooled viscous honey from the sides of the tank or cooling device.

When the temperature of the honey is between 70°F (21°C) and 80°F (27°C), starter, which consists of fine, creamy, previously-processed honey, is thoroughly mixed with the honey which has been heated, strained and cooled. The amount of starter used should equal 10%, by weight, of the honey to which it is added. For instance, six pounds (2.7 kg) of starter should be added to a full five-gallon (18.9 l) pail, which holds 60 pounds (27.2 kg). The honey used as a starter is thoroughly broken up in a grinder or other machine that will not incorporate air. The seeded honey is left to settle for an hour or two, thoroughly skimmed if necessary, and then run into containers of the size desired for market. It is then stored at a temperature not higher than 57°F (14°C) and not lower than about 45°F (7°C) until completely crystallized. This usually requires about eight days. The reason for leaving the honey to settle for a while before running it into containers is to allow the majority of the large air bubbles to rise on the surface of the honey. This precaution helps to avoid a frothy surface on the honey in the containers which have been filled for market.

"Honey high in water content should be blended with honeys low in water content, so that the honey will not have more than 17.5% or 18% moisture. This precaution will usually result in a spreadable product, which is not too hard or too soft. If the processed honey is too hard for table use, it should be placed at a room temperature of about 80°F (27°C), until it becomes sufficiently soft. Once it becomes soft it will not return again to its original hardness.

This is a brief outline of the complete method of processing, but precautions must be taken to prevent overheating, darkening and impairing the flavor of the honey. If this formula is carefully carried out, the resulting product should be a fine, creamy, crystallized product."

There are several considerations and problems in manufacturing crystallized honey that should be clarified. Dyce stated, "Since dextrose [glucose] crystals are pure white, honeys become lighter in color as the crystallization progresses." This may create a problem if the honey is not thoroughly strained, since any specks of comb, especially dark comb, or other dark specks, are readily visible. Also, honey shrinks slightly upon crystallization and has a tendency to pull away from the side of the jar. If the honey is packed in clear glass the white crystals may appear as mold. Some consumers have rejected crystallized honey for this reason, thinking something was wrong with the product. Dyce advised that if the honey was to be packed in a clear glass jar a label should be wrapped fully around it. Crystallized honey should preferably be packed in an opaque container.

Dyce observed that if the moisture content of the honey was too low, the temperature too cool, or the product was stored in a refrigerator, it would not spread easily. Dyce-processed honey made for use in the southern states, or the northern states and Canada in the summer, should contain about 17.5% water. Crystallized honey used in the cooler months should contain about 18% water. This small difference in moisture content will have a profound effect on spreadability at different temperatures. As indicated above, the best way to adjust the moisture content is to blend honeys of different moisture percentages.

A serious problem with crystallized honey is that air bubbles may be incorporated into the honey when it is cooled and/or as seed crystals are added. This air may rise to the surface as the honey cools and before it becomes firm. Foam on the top of a package of crystallized honey gives it a bad appearance and may cause a consumer to reject it. Some found that if they inverted the containers of freshly seeded honey, the foam would form on what would eventually be the bottom of the container. In this position it was not visible to the consumer. More recently, several firms have begun to make crystallized honey in bulk containers, and after the product has become more or less firm, it is homogenized and then put into the final package. This gives the final product a uniform appearance.

Dyce recommended that those who made crystallized honey use ten percent seed; however, he knew that many firms used only 5%. He found that if a grinder was used to reduce the crystals to tiny fragments, a good crystallized honey could be made using only 1% seed. Doing so required the honey to be less than 70°F (21°C) at the time the seed was added, that the seed be thoroughly mixed and, most important, finely ground. Each manufacturer of crystallized honey must develop its own procedure.

How does one determine the best size crystal for Dyce processed honey? No taste tests have been made recently, but those conducted by Dyce many years ago indicated that the crystals should be too small to be felt by the tongue. Not all beekeepers agree, and many make their crystallized honey with larger crystals.

There is no question that many people feel that Dyce-processed honey has a different flavor; many say it is better. An important feature of crystallized honey is that it does not drip the way liquid honey does; this should be emphasized in promotion programs.

DYSENTERY – Honey bees, like all animals, can have digestive problems. Dysentery is defecation by adult bees within and on the outide of the hive, something that does not occur under normal conditions when bees can have flight. Queens must obviously void fecal

Dysentery by itself isn't a disease, but is the result of, perhaps, several non-disease factors, such as long confinement, overwintering food high in indigestible matter, or, these in combination with a nosema infection. The signs are heavy spotting on the ground, fecal matter running down the side of the colony, and fecal spotting inside the hive.

matter within the hive because they do not fly except to mate and accompany a swarm. This is routinely removed by attendants. However, worker bees, and perhaps drones, normally do not void feces in the hive. Dysentery may vary in its severity and if sufficiently bad the colony may die. Colony death probably comes about because of a combination of odor, the stress of the disease, or physical problem causing the dysentery and a breakdown in the communication system. Dysentery can be recognized by fecal spotting on the combs, top bars, inner cover, side bars or other internal hive parts.

It appears there are many causes of dysentery. However, long periods of confinement and inclement weather bring about the greatest problems. Dysentery is most common in the northern states in late winter, especially February through mid-April. Some of the worst cases have occurred in cellars where bees were being wintered. High humidity in a hive, or in the apiary, also appears to encourage dysentery.

Fecal matter may accumulate in the rectum of a honey bee over a long time. The volume may be very great when bees are confined, diseased (such as with nosema) or have food that contains a large quantity of ash or other indigestible material. It is reported that worker bees may accumulate up to 30 to 40 percent of their weight in fecal matter which shows that bees do have a remarkable ability to resist dysentery, but obviously there are limits.

Most authorities agree that a beekeeper may control or even prevent dysentery by the selection of good wintering sites, control of diseases and by providing the bees with good quality food that is low in materials that the bees cannot digest. A good wintering site is one in which the bees may fly, at least for short periods of time, during even the coldest months for the express purpose of voiding fecal matter. Good wintering sites are also dry (see APIARY LOCATIONS).

Poor food is cited commonly as a cause of dysentery. It is agreed by beekeepers who have successfully wintered bees in cellars, or outdoors in the far north, that bees can survive the winter best when fed pure sucrose (cane sugar). This time-tested fact supports the thought that poor quality food such as dark honey, honeydew, burned honey (as from a solar wax extractor), and fermented honey should be avoided as winter food. High fructose corn syrup is a successful replacement for cane sugar, and is, generally, easier to procure and easier to administer to colonies.

Nosema, which is caused by an ever-present disease organism, will no doubt do much to contribute to dysentery as the microorganism that causes the disease destroys the cells of the gut and therefore interferes with digestion. Feeding a drug that will control nosema is recommended. (see DISEASE OF THE HONEY BEE, Nosema).

Colonies severely affected with dysentery – to the point of death – should be thoroughly cleaned. Frames should be scraped of fecal material and comb removed and replaced with foundation. As a matter of routine, comb should be replaced every three to four years to remove the disease causing organisms of AFB, EFB, nosema, chalkbrood and the nutritional and dietary causes of dysentery.

DZIERZON, H.C.J. (1811-1906) – Dzierzon has a special claim to fame in that he discovered that drones were reared from unfertilized eggs and thus that there could be parthenogenetic development in honey bees (see PARTHENOGENESIS). Dzierzon was born in Silesia, in Poland, and played an important role in the development of beekeeping in central Europe. He wrote extensively and, in effect, was one of the world's first extension apiculturists. He had been educated in a Roman Catholic seminary and was a priest. Quite early in his beekeeping career he became aware of the importance of pollen in colony development and wrote about this in the early bee journals and books. Italian breeding stock was brought into north Europe by Dzierzon who recognized the special value of these bees. Many consider him to be the father of practical beekeeping in central Europe.

He was born on January 16th, 1811, in Lowkowice near Kluczbork Silesia, an integral part of an ancient Polish Kingdom under the domination of Prussia. His peasant parents, Szymon and Maria nee Jantos, were Poles. He started his education at the local elementary school, was sent at the age of 10 to the school at Byczyna where he took up German as his second language. The next year he entered the school at

Jan Dzierzon

the Cathedral in Wroclaw and in 1822 he became a student of the St. Mathew secondary school in Wroclaw. In 1830 he passed the final exams there and started studies at the department of theology of Wroclaw University and in 1834 he was ordained a priest.

In 1835 he moved to Karlowice near Brzeg where he became the parish priest, set up an apiary and started studies on the life and breeding of bees.

Dzierzon said "In 1835 I obtained a few bee colonies from my father. I put them into Christ hives which were considered the best at that time. According to the Christ method it was necessary to cut honeycombs from side walls and take all of them out of the beehive. In order to be able to take one honeycomb at a time, I replaced the Christ grate (grille) in the superhive with single bars. To facilitate taking honeycombs out of beehives I made shutters (movable door) in one of the side walls."

The next winter the majority of bees died, and Dzierzon made his own case beehive with side openings. He also designed a twin beehive: i.e. a dismountable beehive for two colonies, later called the DZIERZON. Its lower part was designed for brood and the superhive for nectar. Dzierzon pointed out that the main advantage of such a beehive was that it enabled a colony that had lost its queen to be joined to a colony with a queen, which considerably influenced the total honey yield. In the wall separating two twin beehives was a blocked hole (used for merging colonies).

The discovery of the so-called "displaceable (movable) comb" all over Europe is considered Dzierzon's greatest contribution to beekeeping. Here also lay the secret of the very rapid development of his modern apiary. In this way he initiated modern, rational and intensive development of commercial apiaries which could become a new source of income.

Dzierzon found that an inseminated queen bee lays two types of eggs. Drones hatch from unfertilized eggs, queens and worker bees from fertilized ones. The parthenogenetic origin of drones was discovered by Dzierzon in 1835 and announced 10 years later. For more than 50 years, though, this discovery was the subject of violent, scientific debates and discussions among scientists. It was not until 1906, i.e. just before Dzierzon's death, that his idea was finally accepted during the Congress of Botany in Marburg.

The last international meeting Dzierzon participated in took place in Vienna in 1903. On this occasion, Franz Joseph, the Emperor of Austria, awarded him the golden cross of merit. Some years earlier, in 1872, Munich University conferred the title of *doctor honoris* causa (honorary doctor) on him for his pioneering studies on bees' parthenogenesis.

In 1873 he left Karlowice and returned to Lowkowice where he ran an apiary and carried on his scientific studies on bees, until his death on October 26th, 1906.

Apimondia, 1987, submitted by Andrew Serafin

ECKERT, JOHN EDWARD (1895-1975) – John Eckert was a UC Davis professor emeritus of entomology and parasitology.

Born in Wooster, Ohio, Mr. Eckert received his bachelor of science degree (1916), master of science (1917), and doctorate in entomology and parasitology (1931) from Ohio State University. Before joining the UCD faculty in 1931, he served as apiary inspector with the Ohio State Department of Agriculture, state nursery inspector in North Carolina, in the Army during World War I, associate professor of entomology at North Carolina State College of Agriculture, and associate apiculturist with the U.S. Department of Agriculture in Washington, DC, and Laramie, Wyoming. He pioneered in the use of antibiotics for control of bee diseases and was one of the originators of the modern pollination service to seed producers and cotton growers. He authored and co-authored numerous articles, pamphlets and books on every subject related to bees. His research earned, among others, a Fullbright Grant and took him around the world as he studied bee diseases and helped set up research facilities.

While with UCD, Mr. Eckert served as chairman of the department of entomology. He also served in an advisory capacity on the Honey marketing Board of the California State Department of Agriculture and Legislative Committees.

Mr. Eckert belonged to the Alpha Gamma Rho Fraternity, Entomological Society of America, California Entomology Club, American Beekeeping Federation, California State Beekeepers' Association and Davis Faculty Club, and served for several years on the Davis School Board.

He was the author of several beekeeping books.

ECOLOGY OF THE HONEY BEE – Ecology is a vast subject. It is the study of the interrelations of plants and animals and their environment including natural cycles, community structure and development, and interactions with other organisms. To study the ecology of an animal such as the honey bee is to do more than make observations on its life history and biology. One textbook on ecology is Seeley, T.D. *Honey bee Ecology*. Princeton University Press, Princeton, NJ. 201 pages. 1985. (see ENVIRONMENTAL QUALITY INDICATOR SPECIES – HONEY BEES)

Some of the chapters in Seeley's book include "Honey bees in Nature, The

Honey bee Societies, The Annual Cycle of Colonies, Reproduction, Nest Building," etc. Many aspects of honey bee ecology are treated in detail in separate sections in this text (see ANNUAL CYCLE OF HONEY BEE COLONIES; HUMIDITY CONTROL IN THE NEST; TEMPERATURE CONTROL). It may be said that successful beekeeping depends to a large extent on understanding the ecology of a colony and its needs.

EGG – (see LIFE STAGES OF THE HONEY BEE)

EMERGENCY QUEEN CELLS – (see QUEEN, Emergency queen cells)

ENEMIES OF THE HONEY BEE – Because a tremendous amount of energy is concentrated in a honey bee colony many animals are attracted to it for various reasons. These enemies include bears, skunks, birds, toads, etc. Some of them feed on bees, honey, pollen, wax, and brood, or they may just find the honey bee nest a warm, dry and protected place in which to live. For this reason the enemies and pests of the honey bees and hive products are each treated individually.

Honey bees have many ways of protecting themselves and their nests. For example, studies of bait hives have shown that bees seek sites that will offer them good protection; it is interesting that they even have the ability to select nests with small entrances that they can better defend (see BAIT HIVES). The importance of propolis in controlling bacteria and other microbes has been recognized. Also honey is protected in a way that is unique among foods (see NECTAR, CONVERSION TO HONEY).

Successful beekeeping involves maximizing the natural defensive mechanisms honey bees possess. However, it may sometimes be necessary to supplement these mechanisms with special equipment, or even drugs and pesticides, especially when bees are under stress. Certainly keeping colonies strong (that is, well-populated with bees) is the beekeeper's and the bees' best defense against enemies. Honey bees will always have pests, predators and diseases that would destroy them if given the opportunity; it is the beekeeper's task to control these enemies. (see DISEASES OF THE HONEY BEE; MITES ASSOCIATED WITH HONEY BEES; SMALL HIVE BEETLE; and WAX MOTHS)

ENVIRONMENTAL QUALITY INDICATOR SPECIES – HONEY BEES – The success or decline of honey bee colonies has long been used as a harbinger of environmental conditions. In medieval times, bees were considered to be predictors of everything from famine to war. Bees can now be used to detect the presence of, and to assess the hazards presented by chemical and microbiological materials, ranging from pesticides to toxic pollutants. Because each colony may forage up to two or three miles (three to five kilometers) from the hive, each colony acts as a sampler of air, soil, water, and vegetation over that area. With modern technology, effects of these materials, other than an adult-bee-kill, can be monitored in real time. There also is a much better appreciation of what happens to all of the materials brought back to the hive and how these substances get into the hive. For example, bees possess electrostatic surface charges and absorb heavy metals, radionuclides, and pathogenic spores directly into their bodies. Volatile chemicals, such as dry-cleaning solvents and components of automobile exhaust, show up in the air inside a beehive. (see ECOLOGY OF THE HONEY BEE)

Fortunately, honey tends to remain relatively free of most contaminants, but the bees themselves, the wax, and sometimes pollens tend to accumulate materials foreign to beehives. The cumulative effects of all of these environmental insults range from undetectable, to gradual, to sudden loss of adult bees; reduced egg viability and loss of brood; sub-lethal effects such as changed flight activity, or a breakdown of temperature control in the brood nest. Queens may abandon colonies when high concentrations of volatile solvents are frequent in the hive atmosphere. Other studies have implicated some pesticides as modifiers of critical neural processes that govern olfactory sense, learning, and memory in honey bees.

The number of colonies that can prosper in an area is altered by land-use patterns. Urban development, monoculture cropping patterns, and loss and fragmentation of natural habitats have reduced carrying capacity in many areas. The introduction and spread of bee parasites and pests such as mites and small hive beetles have reduced or eliminated feral bee colonies from many areas. Yet, the new planting of thousands of acres to crops such as almonds, canola, and wild blueberries has greatly increased the demand for colonies of bees for pollination.

Honey bees are a sentinel species. Factors that harm or alter their presence and success are likely to adversely impact other pollinators as well. Certainly the importance of insect pollinators to cultivated crops and food production is well recognized. However, the role of bees and other insects as critical pollinators of approximately 90% of the world's 240,000 species of native, non-crop plants is often overlooked.

Finally, in the coming years, honey bees may provide distinctively new services with respect to environmental safety and security. Bees have been conditioned or trained to search for, and locate by scent, explosives, bombs, drugs, and other materials. The training takes only a few days. The olfactory discrimination of a bee is similar to that of a dog, and the performance reliability of a bee is as good as or better than that of a well- trained canine. Applications extend from possible use of bees in the detection of land mines to border inspections. Bees may not be able to predict war, but they may play an important role in counter-terrorism and agricultural security. – JJB & CBH

EQUALIZING COLONIES – It is advisable to make as many of the colonies in a location (such as a beeyard) more or less equal before a honey flow begins. Colonies that are weak, that is colonies with small populations, are of little value in honey production because a high percentage of the adults are involved in brood production and cannot be spared as foragers. Strong colonies may swarm, rendering them weak and less effective honey producing units. Thus, equalizing can have several advantages. It is understood, of course, that one should determine why certain colonies are weak and ensure that the problem is not caused by disease or a failing queen. The quantity of food, both honey and pollen, in the weaker unit should be checked closely. A food shortage immediately before a major honey flow is not uncommon due to the demand by a rapidly expanding

 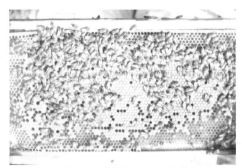

Placing a frame, or frames, of brood in a small colony can stress that colony. Open brood, (left, above) requires a larger population of house bees to care for it (feeding and food, clustering for warmth, sealing when mature), than does a frame of sealed brood (right, above). If adding open brood to a small colony, be certain that you also add sufficient house bees to care for that brood.

population and uncertain food availability early in the season. This food shortage will have a drastic effect on the development rate of a weak honey bee colony.

The easiest and most common way of equalizing is to exchange the positions of strong and weak colonies in the apiary during the day when the bees are flying. This is done by smoking both colonies heavily and just picking them up and exchanging locations. This exchange does not usually prompt fighting since the weak colonies are not especially adept at defending themselves under any circumstances. Field bees from strong colonies appear to adapt to their new hive with little difficulty. Since they are adding nectar and/or pollen they are seldom seriously challenged and can be of great benefit. Usually the stress of a reduced population on the stronger colony is not felt in the brood nest because the colony kept the majority of its house bees. Congestion is relieved in the larger colony because some of the older forager bees are gone. They, now, begin gathering food for their new colony, the one that was weaker.

This method will not work very well if there is a dearth of nectar/pollen. In this situation field bees from the stronger colony may kill the new queen. To prevent this, feed the weaker colony with a 1:1 sugar syrup after moving (see FEEDING).

Another method of strengthening weak colonies is to give them brood from a stronger colony. However, this may place

When equalizing colonies the same principles apply whether you are working with only two colonies, one weaker than the other, or you are manipulating hundreds at a time. When finished, all colonies should have roughly the same amount of sealed brood, open brood, a healthy laying queen, enough of the right-aged house bees to care for all stages of brood present, and enough foragers to gather all the food necessary to keep the colony growing.

too much stress on the weaker unit because it may not have a sufficient number of bees to gather nectar and pollen to feed more brood, enough house bees to feed more brood, or enough bees to cover and keep warm more brood. Sealed brood requires no additional food but enough bees to cover. Moreover, sealed brood will be adult bees sooner. Unsealed brood is the most work for a colony, requiring food and nurse bees.

When adding brood – sealed or open – to a weaker colony remove a brood frame from the chosen stronger colony, give it a moderate downward shake directly over the colony, which will remove between a quarter and a half of the adults on the frame. Generally, these are older bees, which, when moved to a new colony will simply return home. Those bees that hang on are generally house bees, much less inclined to leave a brood nest – old or new. Thus, both brood, and the bees necessary to care for that brood join the weaker colony, reducing significantly the stress this procedure causes both colonies.

When several colonies are being equalized at one time, brood from more than one colony may be added to one weak colony. The same techniques and principles should be employed.

Equipment For Beekeeping – A common arrangement for hobby and sideline beekeepers consists of a hive stand with a slanted landing pad, a bottom board (screened or solid), two deep boxes used for rearing brood, a queen excluder between the brood boxes below and the honey supers (here one medium and one shallow on top), an inner cover (not shown) and a telescoping outer cover.

Switching the position of the weak colony on the far end, with either of those on the close end will greatly aid the weaker colony with an influx of forager-aged bees. Move colonies in the middle of the day when there is much flight activity and, if at all possible, during a strong honey flow.

Parts & Pieces Of A Typical Beehive

- Migratory Outer Cover, or
- Telescoping Outer Cover
- Inner Cover (not used with migratory covers)
- Bee-O-Pac Super (regular medium super with plastic Bee-O-Pac Frames)
- Ross Round Comb Honey Super
- Medium Super
- Deep Super
- Queen Excluder
- Deep Super
- Pollen Trap
- Screened Bottom Board
- Hive Stand With Landing Board

Over the years beekeepers have used a variety of honey super and brood box sizes, and corresponding frame sizes. It's easy to see why standardization was easy to accomplish.

EQUIPMENT FOR BEEKEEPING – In the United States, Canada, and many other countries colonies are kept in 10-frame Langstroth hives of the dimensions chosen by Langstroth in 1851 and 1852. Today's brood chamber is 9-5/8 inches (24.4 cm) deep probably because this was the width of the board Langstroth had when he made his hive. The length and width, 16-1/4 x 20 inches (41.3 x 50.5 cm), outside dimensions, were selected because Langstroth believed this was a convenient size box to carry. Eight-frame hives are used to a lesser degree, and there are even a few hive styles of different dimensions. Both eight-frame and 10-frame hives use frames of the same dimensions. Today, only one size full-depth box is advertised and sold by bee supply companies. Twenty-five years ago three sizes were readily available, and earlier the variety was even greater.

Beekeepers in North America have worked long and hard to standardize equipment for two reasons. First, the resale value of one's equipment is increased if standard dimensions are used. Second, new equipment will always be interchangeable. While full-depth boxes are a standard depth, extracting supers come in two sizes, one called medium, the other called shallow. Some beekeepers use full-depth boxes as supers for honey production, but others feel they are too heavy to lift. Some concern has been expressed that legislation may someday be passed pertaining to the maximum weight a farm worker may lift; this, in turn, could affect the size of supers that will be used in the future.

Many hobbyist beekeepers find that using the medium as a brood chamber as well as a honey super makes lifting full boxes easier. Furthermore, using only one size for both purposes means all equipment is interchangeable and more efficient.

Among commercial beekeepers, especially those who are migratory, a strong effort has been made to simplify and limit the number of pieces of equipment. One beekeeper, with over 30,000 colonies in five states, used only seven pieces of equipment. First, he used a pallet that held four colonies. The pallet was designed to be lifted and moved with a forklift truck. In addition to being a pallet, it was also the bottom board for four colonies. Second, the beekeeper used one or two full-depth boxes for a brood nest. His honey supers were one size, medium. The cover was a piece of plywood with a cleat on either end. He used simple, metal-edged queen

An efficiently designed truck.

Every hive piece is interchangeable with every other hive. Pallets, boxes, frames and covers.

Standardized hives and pallets are designed for easy moving.

Commercial beekeepers have standardized every piece of equipment they use to increase efficiency and reduce labor costs as much as possible. This truck bed is built to accommodate an exact number of pallets. Pipes are located to facilitate securing the load and storage boxes under the bed are all the same size and contain the same tools and equipment on every truck in the fleet.

Hives, too, are standardized. Each forklift-ready pallet serves as a bottom board for the four colonies it holds. Guide strips and clips hold each colony in exactly the same location on each pallet. Each hive has two deep bodies for brood production, and uses deep supers for honey production also. The top is a one-piece migratory cover. In all, a pallet with four double deeps contains: one pallet, 80 frames, eight boxes and four tops – 93 pieces. Four conventional individual colonies contain: four bottom boards, 80 frames, eight boxes, four inner covers, four covers – 100 pieces.

The difference in weight is considerable, but the difference in labor required to move these hives is even more so. With pallets, one person can move four colonies in the same time it takes two people to move one colony.

excluders. In total he used only seven pieces of equipment in his 30,000 colony operation: pallets, full-depth boxes, full-depth frames, medium supers, medium frames, covers, and excluders. If he bought more bees he would keep only those supers and frames that were the same size as those he already owned. All of the owner's trucks were equipped in the same manner: if the hammer was in the front left-hand box under the truck bed on one truck, it was in the same place on all trucks.

Outside of North America, standardization is not always the rule. It is discouraging to realize that many beekeepers try to invent new hives for various reasons. The designers usually believe their hive to be superior. The fact is that honey bees are adaptable and can survive in many different types of hives. The best equipment is that which is most convenient for humans to use and manipulate. (see FRAMES)

Many attempts have been made to introduce plastics, including styrofoam and other materials for boxes, frames, tops, bottoms and inner covers. Plastic foundation has been widely accepted for several reasons (see BEESWAX FOUNDATION), as have one-piece plastic frames and foundation. Management and removal in plastic hive bodies needs attention with increased ventilation because plastic does not absorb, thus remove moisture as wood does. This can be a problem, but it is manageable. Birds have been observed removing pieces of styrofoam from hives to allow themselves better access to the bees inside of the hive, which they then eat. But woodpeckers routinely destroy wooden hives for the same reason. Wax moth will destructively tunnel in styrofoam. The chief reason for using wooden boxes, bottom boards, and covers is that they have a long life and are comparatively inexpensive. This is changing, however, and plastic equipment is becoming common.

EQUIPMENT MAINTENANCE – Hobby beekeeping is an enterprise that may be expanded or contracted with ease, provided the equipment, especially the

Well-constructed and maintained hive boxes, if not mechanically damaged by forklifts, bears, errant trucks or burning, will last as long as most beekeepers. Conversely, poorly-made, and ill-cared for equipment, though inexpensive to make initially, lasts a far shorter period.

Paint isn't the only wood preservative to consider. Food approved stains are equally protective, don't blister or peel, and weather to a nearly invisible gray/brown color. Camouflaging a hive, especially in urban settings has a lot of appeal.

hives and hive parts, is in good condition and interchangeable. Today the major bee supply manufacturers make and sell only Langstroth hives, though the depths of the extracting supers vary. This standardization makes the beekeeper's task easier and, in part, guarantees that the hives will be sellable. However, several considerations for the upkeep of the equipment are important.

If one observes an old, neglected beehive, it is easy enough to see where the weak points are in construction. The top and bottom corners of supers and hive bodies are usually gouged and worn. Tops, especially in the vicinity of the rabbet joints are vulnerable to weathering damage and may warp at these points. While the nailing of the whole box is important, it is at the top corners that special attention is needed.

The holding power of a nail has to do with its length, not its diameter or coating, and thus long nails will hold the equipment together better than will short ones.

One should use box nails, not common nails, when nailing beehives. Box nails have a smaller diameter and are less likely to split the wood.

Many beekeepers use screws to fasten box sides together. These offer the advantage of being easily removable for repairs, can be rust resistant, and with mechanized equipment are faster to install than nails.

The use of wood preservatives to treat beehives has been under heavy scrutiny in recent years and it is important to check the regulations to determine what materials may be used (see WOOD PRESERVATIVES FOR BEEKEEPING EQUIPMENT). The concern is over the use of toxic materials on equipment that may come into contact with a food product, such as honey.

Painting of exterior hive parts with good quality paint is also helpful. With paint, it's almost always true – you get what you pay for. Oil-based paint is longer lived than latex or stain, but costs more initially. Usually, the initial cost of better quality paint is offset by few repaint episodes. Labor is expensive.

Most inner covers are preassembled and made with plywood, usually only one-quarter or three-eighths inch (0.6 to 0.9 cm) thick, or with masonite. These will warp easily in the environment of the hive. Inner covers made of individual boards and interlocked frame are usually more dependable, rigid and long lived.

Bees often stick the tops of frames in one box to the bottoms of frames in the box above with propolis and burr comb. When these frames are separated they may break easily, or the bottom bars may pull loose if they are not carefully nailed. Using a wood glue for joining the parts of the wood frame prevents damage during hive manipulations and in extracting. In fact, a good, waterproof wood glue should be used when joining any wood parts, whether frames, bottoms, boxes, inner or outer covers.

When hive bodies, frames or tops are found with excessive burr comb and propolis it is necessary to determine why such materials have been deposited. Often it is because equipment dimensions have not been properly adhered to and the well-disciplined bee space has been violated (see BEE SPACE). It is more efficient in the long run to eliminate improperly-sized equipment than to tolerate the inconvenience it can cause.

A beekeeper's best protection against the deterioration of equipment is to establish a routine for equipment maintenance. One should plan to repair, paint and/or replace a certain amount of equipment each year. For further information on equipment maintenance see EQUIPMENT FOR BEEKEEPING; FRAMES; HIVES, TYPES OF and HIVE STANDS.

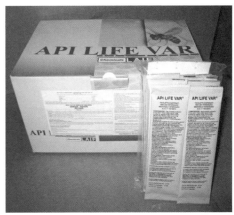

Apilife Var is a thymol-based compound effective in the control of both Varroa and tracheal mites. (Forrest photo)

ApiGuard is a thymol-only gel product effective in controlling Varroa mites. (Flottum photo)

ESSENTIAL OIL BASED COMPOUNDS – Beekeepers in the U.S. enjoy a reputation of pure non-contaminated honey and hive products. Building on this status, the development of natural essential oils for *Varroa* and tracheal mite control has been found to be a worthy endeavor by U.S. researchers. Essential oils are extracted from plant sources, or are synthesized to duplicate natural compounds. Then they are formulated for safe and easy application in the hive. Menthol has had use as an effective compound for tracheal mite control in those areas with appropriate weather. An additional product, a combination of thymol, eucalyptol, camphor, and menthol, known as Api Life Var®, has been used in Europe for years and is registered for use in the U.S. Other essential oil products are under investigation, or are being registered. This is a dynamic process, with new compounds coming on the scene, and older, less effective compounds being discarded. It bears watching. – PJE

EVOLUTION OF THE HONEY BEE - The fossil record of plants and animals that once existed and have gone into extinction is far from complete. It is estimated that something less than 3% of the animals that once lived are preserved as fossils. However, among these are some bees and from these specimens we can construct a general hypothesis concerning the evolution of bees, including the honey bee.

The first bees were solitary ground- and twig-nesters probably closely related to ground-nesting wasps. The bees appeared at the same time as the flowering plants. The time was probably the mid-Cretaceous, that is, about 130 to 65 million years ago. Between 20,000 and 30,000 species of bees have evolved from these original ancestors. Social behavior among the bees probably arose only once, sometime in the Tertiary, a period about 55 to 65 million years ago; however, not all students of evolution agree with this theory and it is possible that the honey bees we know arose from one line while the stingless bees (see STINGLESS BEES) so common in the tropics arose independently from a second line.

Several fossil honey bee-like insects, obviously closely related according to their morphology, have been found. Amber, now a hard material made of once-soft ancient plant gum not unlike that produced by pines today, is the richest source of these fossilized bees. These and other insects became entrapped in the amber while it was still soft.

Amber is especially abundant in Poland around the Baltic sea. The bees found entrapped in Baltic amber apparently arose in the Oligocene period, about 30 million years ago.

Our honey bee species, *Apis mellifera*, came into being in Africa and spread from there through Europe and the Near East. The fossil record that we have indicates there has been little change in *Apis mellifera* during the last two million years.

The study of evolution is complex. Different groups of plants and animals obviously evolved at different rates. Our honey bee, for example, has changed little for millions of generations, but the parasitic mite, *Acarapis woodi*, is apparently of very recent evolutionary origin. Those who wish to pursue the subject of honey bee evolution will find the following article interesting; it contains a number of references pertaining both to fossil honey bees and several aspects of geology and evolution: Culliney, T. W., "Origin and evolutionary history of the honey bees Apis". *Bee World* 64: 29-38. 1983.

EXAMINING COLONIES – Honey bees will defend their nests with as much vigor as their populations, temperature and other environmental conditions allow. Bees in a colony are much more defensive when no incoming food (nectar and pollen) is available to them. During cool weather, or when a rainstorm is

Smoke lightly so bees move down and out of the way, and carefully remove the frame closest to you (if standing on a side) and set aside. Then move adjacent frames toward you, filling in the empty space the original frame occupied until you get to the frame you wish to examine and carefully lift it straight up.

Holding the frame by its ears or lugs, swivel the frame so you can look at the other side without switching hands or setting the frame down.

Stand so the sun is coming over your shoulder and shining directly down into the cells of the frame. This makes eggs, small larvae, any diseases or other problems much easier to see.

pending, the foraging bees are all in the hive and will join the guard bees to quickly defend their home. During a strong nectar flow there are fewer guard bees and colonies do not expend much effort on defense.

There are times, such as periods when queens are reared, that colonies must be examined no matter what the weather. However, on average, beekeepers should select the time they examine their colonies with care. If one opens colonies only at a time when bees are flying, the sun is shining, and there is at least some small amount of nectar available one will be stung much less. Urban beekeepers, and those with close neighbors, must be very careful when they open colonies under adverse weather conditions as the bees may range out from their colonies for up to a hundred or more feet (30 meters) and sting animals and people. Some communities require that colonies of bees be located at least 100 feet (30 meters) from a property boundary line for this reason.

It is possible to reduce the number of stings received by dressing properly (see DRESSING FOR THE APIARY). Most beekeepers do not wear gloves because they accumulate an offensive odor from sweat and stings. Furthermore heavy gloves are clumsy to use and more bees are crushed, thus alarming the colony. Beekeepers often state that they do not mind having a few stings on their hands

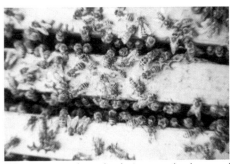

When you see bees begin to come back up and look at you from between the top bars, lightly smoke to keep them occupied while you work. DO NOT oversmoke or it loses its effectiveness. Err on the side of too little smoke until you see how much to apply. Some days will require more smoke – when you are examining a colony later in the day or when inclement weather threatens.

Leaving extra supers setting with exposed honey combs, or frames unprotected, or just having a colony open for too long, especially during a dearth, can allow robbing to begin. This can be disastrous for a weak colony, which can be killed outright by robbing bees. It also causes bees to randomly sting people or pets some distance from your colony. DO NOT let robbing begin.

so they are more alert to a colony's temperament. When one is stung once or twice on the hands it is obviously time for more smoke. Some beekeepers feel they are actually stung less without gloves than they are when wearing them. The thin rubber gloves often used when washing dishes or doctor's exam gloves will give some protection for the hands and can be easily cleaned of offensive odors or discarded for a fresh pair.

When a colony of honey bees is opened, smoked and examined, it is obvious that social order is disrupted. The degree of disruption and length of time a colony is exposed may affect the time it takes for the colony to return to normal. Some activities, such as foraging for pollen and nectar may be resumed in a very short time. Some colonies may become defensive for several hours or up to a day after disruption.

The greatest danger in opening a colony and exposing combs, especially those with honey, is that one may start robbing by the bees when there is a dearth of nectar. Some beekeepers have gone to the trouble of building portable cages that may be used to cover a single colony and the beekeeper when the colony is opened. These cages are usually about four by six foot rectangles (1.5 by 2 meter) and tall enough for a person to stand up inside. Handles on

When examining a colony, stand to the side, so you don't block the entrance with your body. There bees will collect behind, and in front of you, looking for their normal flight path.
Lift the cover and lightly smoke beneath. Replace the cover and wait a minute or so for the bees to engorge. Remove the cover.

the inside make it possible for the cages to be carried from one hive to another. Under some circumstances beekeepers may want to consider this defensive action.

EXHIBITING HONEY – (see HONEY, JUDGING)

EXOSKELETON – Honey bees, like all insects, do not have an internal skeleton, as man and other animals possess. Rather, the insect skeleton is on the outside and thus the term exoskeleton is used. Just as muscles are attached to bones in the interior of a mammal's body so they are attached to the interior of an exoskeleton on both indentations and ridges that are found there. A hard exterior skeleton has both advantages and disadvantages. One advantage is that bees have parts of their bodies formed into strong tools such as antennae cleaners, mandibles for biting and chewing, claws and pollen baskets.

The exoskeleton is made of a material called chitin, a proteinaceous material that is secreted by cells in the interior. The exoskeleton gives the body strength and protection, protects it from both drying out and excess damage by water. A weakness of an exoskeleton is that it may be broken, crushed or abraded. Then the insect has no method of repairing such damage since chitin is not living tissue. If there is too much damage the insect usually dies because the interior of the body dries out.

EXTENDER PATTIES – (see ANTIBIOTICS AND OTHER CHEMICALS FOR BEE DISEASE CONTROL)

EXTENSION APICULTURE – The U.S. Congress created the extension service in the United States in 1914. The purpose was to make new research technology and better methods of farming known to rural America. Beekeeping was a part of the program and several states appointed extension apiculturists that worked out of their state colleges. A number of circulars and bulletins on beekeeping were soon written by these specialists and supplemented the journals and books that had been the chief sources of information. County agricultural agents were appointed for all counties in the country except those in New York City. Today the extension service has become more involved with home grounds, nutrition and has been engaged in social activities. (see ASSOCIATIONS FOR BEEKEEPERS, State and county associations)

EXTRAFLORAL NECTARIES – More than 99 percent of the honey that is produced comes from nectaries associated with flowers. Floral nectaries and the nectar they produce attract

Blueberries, too, have extrafloral nectaries.

Extrafloral Nectaries

Cotton is attractive to bees because of its floral and extrafloral nectaries. (Jaycox photo)

insects that will come into contact with the flower's pollen and move it from one flower to another and thus bring about what we call pollination or fertilization. However, nectaries are also found on other plant parts that have nothing to do with pollination. These are called extrafloral nectaries. They are not common. Their physical structure and chemical activity have been examined and they appear to be the same as floral nectaries. Extrafloral nectaries may be found on stems, leaves and other plant parts.

The function of extrafloral nectaries is not known. Several types of plants have extrafloral nectaries, such as cotton and blueberries. Not all varieties have extrafloral nectaries and in different varieties they may be found on different parts of the plant.

FANNING BY HONEY BEES – Occasionally one will notice worker honey bees standing on all six legs with the abdomen raised fanning their wings. Drones and queens do not fan. Workers fan to disperse Nasonov scent gland pheromone (see PHEROMONES) at a food source, at a hive entrance, or when they are with a swarm. They also fan to move air through their nest to remove moisture from ripening honey, to reduce the internal carbon dioxide level, to cool the hive, or to remove a foul odor such as smoke. No sound is produced when bees fan the wings other than that made by the movement of the wings themselves. The flow of air is from front to rear in the case of our bees; interestingly, in the case of the Indian honey bee, *Apis cerana,* the flow of the air is in the opposite direction when they wing-fan.

Worker bees fanning at a nest entrance to increase air movement through the nest. (Root photo)

When workers are fanning their wings the abdomen is raised slightly. When they are dispersing scent-gland pheromone the tip of the abdomen is bent downward to expose the Nasonov gland; the odor can sometimes be detected when there is little wind if one holds one's nose a few inches behind the scenting bee. The number of bees involved in fanning their wings at any one time depends upon the severity of the problem, the number of bees already fanning or the need to ventilate.

When fanning to disperse Nasonov pheromone, when one bee begins sending out the odor and the activity triggers others nearby so there soon may be many fanning.

FAT BODY – In the spring and summer, honey bees have a small amount of fat stored around the organs in their abdomens but especially under the dorsum (top side) of the abdomen. The situation is different in the autumn; at that time worker bees accumulate a large quantity of fat, which is presumably necessary, or at least helpful, for winter survival. The fat body is an aggregation of cells containing fat globules, glycogen, and some protein. All insects have fat bodies. Fat bodies in honey bees also contain iron particles. A few years ago it

Fat Body

You will, on occasion, see bees fanning their wings, but in different poses. The top photo shows bees standing with the tips of their abdomens pointed down. This position enhances air flow over their backs and creates a steady current of air moving from in front of them to behind them. This is primarily to move air for ventilation. Or, as on the bottom, you will sometimes see worker bees standing with their abdomens raised in the air. Note that the tip is pointed down, exposing the Nasonov gland just in front of the last segment. The air current created by the fanning wings passes over the exposed gland, carrying the scent away from the colony, and creating an odor trail for lost bees to follow home.

was reported that honey bees had magnetic resonance, that is, their bodies responded physically to magnetic fields and that this was important in orientation. It was said that the iron in the fat body was involved. However, it now appears that the iron in fat body cells is an artifact, or iron that may be necessary in ordinary metabolic processes, rather than part of the honey bee's sensory system.

Honey bee larvae also contain large fat bodies that grow larger as the larvae mature. Presumably this fat is required for the development of the pupae, since pupae do not feed, though this has been little studied.

FEEDING BEES – Feeding bees is a critical management practice. This one activity may make the difference in colony survival. Often colonies may need supplemental sugar in the spring and autumn. It is important to recognize that there are differences in both techniques and formulas between spring and autumn feeding. One difference is the concentration of the sugar syrup; it is generally agreed that spring feed should be dilute, usually one part sugar and one part water by weight or volume. In the autumn the syrup should be more concentrated; autumn syrup is usually two parts sugar and one part water. However, in both cases the bees are remarkably adept at utilizing the food in a manner that will best suit their purposes.

Bees should be fed white granulated table sugar mixed with water to form a syrup. High fructose corn syrup (HFCS), in which the sucrose has been converted into a mixture of glucose and fructose, is also good food for bees.

Feeding combs of honey – Feeding sugar syrup is costly and time-consuming. It can be much easier to feed full combs of honey to colonies needing food. Some beekeepers make a point of saving combs of honey to feed their bees in the spring or for emergency feeding. Storage can be a problem since small hive beetle, wax moth and granulation can damage, or destroy stored combs. Keeping honey on strong colonies is recommended, but even there problems can arise. Since new combs on wax foundation break easily when extracted using them for food is often a practical way to use them the first year.

Feeding liquid honey – Liquid (extracted) honey (or even whole frames of honey) should not be fed to bees if the source is unknown. The chief reason is that honey may contain spores of *Paenibacillus larvae* subsp. *larvae,* the causative agent of American foulbrood

Plastic poultry waterers can be used as feeders in an emergency. Small sticks must be placed in the dish to prevent bees drowning. Place on top of inner cover, but not over the hole. Cover with an empty hive body and the cover.

disease. Fermented honey has been reported to cause dysentery in bees though no precise data exist.

Many beekeepers use some of their colonies to clean honey from cappings and whole supers after they have been extracted making it easier to handle both the dry cappings and supers, but it is inconsistent with the recommendation made above. Beekeepers feeding liquid honey, or allowing supers and cappings to be robbed, understand that they are taking this risk. However, feeding supers back to the colony from which they were removed usually poses no problem.

Different methods of feeding syrup – There is no perfect way to feed bees sugar syrup. Below are a few methods that have been used by beekeepers.

Boardman feeders – Using a Boardman feeder is a poor way to feed bees, and is not generally recommended. Such feeders are used more successfully in the South to feed small colonies, but

Bees enter the feeding platform through the opening just inside the hive. They draw syrup through the holes, from below, from the jar that screws into the metal or plastic lid. A leaky feeder can lead to robbing.

even when the temperature is warm this is not a good feeding method. Most Boardman-type feeders hold only a single quart (liter) jar, and a colony that needs feeding usually needs two or more gallons (8 liters or more) of food. The feeder, sitting at the entrance of the colony, is some distance from the brood nest and the bees are not able to take syrup from a Boardman feeder during cool weather when they are clustered. Because it is immediately apparent when the container on a Boardman feeder is empty, since it is exposed to view, these work as a waterer. However, ants will often steal food from a Boardman feeder, and robbing can occur if these are used for small colonies.

Bulk outdoor feeding of syrup – Occasionally articles appear or discussions are heard at meetings by those who feed sugar syrup in bulk from a single exposed feeder in the apiary. A variety of feeders have been used, including barrels and stock tanks. Usually several hundred gallons of syrup are poured into the tank. A wooden or straw float is placed on top of the syrup to give the bees a place to stand and to prevent them from drowning. The distribution of food among the colonies is not equal and weak colonies do not appear to get adequate food. Frequently

A Boardman feeder. Not highly recommended as a feeder to use in cool weather. It does work well in warmer months to provide water for a colony, but a larger container is suggested.

Feeding Bees

Open feeding. It is obvious why this is not a good idea. Robbing, fighting and pest and disease transmission will occur. And weak colonies do not get enough.

there is some fighting among the bees at the feeder. Some beekeepers who have fed sugar in bulk have discontinued this method after a year or two of experience.

Caution: - Extreme care must be taken with this technique so that cattle do not have access to this tank. Even moderate amounts of sugar solution can be lethal, and lawsuits have been filed to replace lost cattle.

Division board feeders – These are plastic containers without tops that are the same depth and length of a normal frame. They have lugs or ears so that they will hang in a standard hive body as a frame does. Most division board feeders are made of plastic an inch and a half or two inches wide (one to two cm wide) and will hold about a gallon (four liters) of sugar syrup. The advantage is that division board feeders will fit into a hive and an extra empty box above is not needed as with jars and pails. Division board feeders must be fitted with a proper float, a ladder or etched sides, so that the bees will not drown.

Division board feeders are refilled by lifting the hive cover (and inner cover if one is used) or even the top hive body and pouring in more syrup from a can or pail. Tanks, pumps and portable hoses with a spout like that found on a gasoline pump, can be used to fill division board feeders. The hose may be pulled through the apiary when feeding is under way and the feeders are filled quickly. Instead of pumps, tanks of compressed air can be used to drive the syrup out of the tank. One advantage is there are no moving parts which could break while in the field. The cost for a compressed-air system is about the same as one with tanks and pumps.

Feeding dry sugar – Beekeepers usually feed their bees dry sugar only

Plastic division board feeders, with easy-to-use ridges on the inside that bees can climb on are efficient and practical.

Feeding Bees

Bulk tanks containing the syrup to fill division or other types of feeders. Syrup is held in a large home-base tank(left), along with water for proper dilution. The two are mixed in proper proportion and pumped to a large tank on the back of a truck. Coupled with this is a large compressed air tank, which pushes the syrup out of the holding tank on the truck, which feeds the hose used to fill individual feeders(center). The advantage is that there are no pumps to break down in the field. Syrup is dispensed with a nozzle(right) attached to the syrup tank, held in place on the back of the truck.

For smaller applications, pumping works well.

placed dry sugar in feeders above the brood nest and some have even fed dry sugar on the bottom board. Feeding dry sugar is not a good method because of the uncertainty of success.

Feeding under the brood nest – In the late autumn, usually October and into November, it is possible to place a Miller-type feeder under a colony (on top

out of desperation. Those who have successfully used this system will admit the bees often carry the dry sugar crystals out of the hive rather than eat them. Bees have the ability to liquefy dry sugar, but are not always inclined to do so. Bees use dry sugar more rapidly when they can fly and collect water with ease.

Dry sugar is usually fed by placing it on top of an inner cover so that the bees have access to the sugar through the inner cover hole. Beekeepers have also

Dry sugar on a sheet of newspaper, placed directly on top bars. This is for emergency feeding only!

of the bottom board), and the bees will carry the syrup upwards and store the food above the brood nest. At this time of year they will empty a feeder under the brood nest more rapidly than they will empty one on top of it.

In autumn one may also place four or five combs of honey with cappings raked lightly or broken with a hive tool into a box placed under the brood nest. The bees will move the honey upwards in a matter of a few days. When four or five combs of honey are placed under the brood area they must be equally spaced. If the super is filled with eight to 10 combs the bees are less inclined to move the honey.

Feeding below the brood nest works well only in late fall. It is not known why. If feeders or frames of honey are placed in hives at other times of the year, in the spring or summer for example, the bees may use the food when nearing starvation; otherwise they will not move the food upwards and store it.

Feeding with jugs and cans through the cover – Beekeepers often use one-gallon metal cans or glass jugs as feeders. A hole is drilled in the hive cover, and the feeder top is inserted into the hole. In such cases the jug or jar is visible above the hive and, in the case of glass jars, the beekeeper can easily observe when they need refilling. If the feeder rests snugly on top of the cover it will not be dislodged by the wind. The covers on the hives in which this system is used are made of wood and have no metal covering. A hole of the proper diameter is drilled in each cover, and a flap, which can be pivoted into place, is nailed on with a single nail. The flap covers the hole when bees are not being fed.

Filling combs with syrup – One may pour four or five pounds of sugar syrup into a standard drawn comb to be placed into a brood nest. In the case of package colonies one may fill several combs for the hive body in which the package is placed. The bees move and ripen the syrup rapidly.

It is possible to fill the cells in a comb only by allowing many fine streams of syrup to flow into the cells, filling one side at a time, because of trapping air in the cells and surface tension on the wax. This is done by using a pail or can with 50 or more holes punched in the bottom or a pressure sprayer on coarse setting. The combs to be filled are placed over a tub, to catch the excess syrup that does not run into cells. Combs may be filled and transported to an apiary without much syrup loss if the excess on each comb is shaken away after the combs are filled.

Fondant – Sugar can be fed to colonies by giving them slabs of fondant, available from bakeries. It is the icing

Fondant, made from table sugar and high-fructose corn syrup, makes an excellent emergency winter food. It can be sliced and put in food storage bags until needed. It has the consistency of medium-hard butter. When placing it on the colony, cut one side of the bag corner to corner in an "X" shape so that the bees can get to it. It is all sugar, so you know how much you fed, and the bees relish it.

Feeding Bees

Cans used for feeding. The cap fits in a hole in the migratory lid cover.

used on confections, donuts and the like. It is made by mixing white table sugar and high fructose corn syrup (see CORN SYRUP) to make a dough-like frosting. This usually comes in 40 or 50 lb. (18 or 23 kg) boxes. It is soft and easily cut with a knife. Slices are cut, put in sealable plastic bags and frozen until needed. When placed on a colony, an 'X' is cut corner to corner in the bag and the bag is placed directly on the top bars, cut side down. The advantage is that the material is soft so the bees are able to move it. Because it is essentially all sugar, the beekeeper knows exactly how much sugar was fed.

Jar and pail feeding – In the past both commercial and hobby beekeepers fed sugar syrup using tin-coated pails

At left – jars used to feed bees.

that hold five, and preferably 10, pounds (two to four kg) of syrup. Today beekeepers now use plastic pails and glass jars.

To feed the bees, containers of sugar syrup are inverted over the brood nest and rest on top of the top bars of the frames. An empty hive body is put around them. It is important to check the colony being fed after about three to five days. If the bees were in need of feeding, they will have exhausted the supply by that time and stored any surplus in empty cells adjacent to the brood nest. If the containers are not checked, and the colony prospers, comb may be built around the feeders and into the empty hive body.

If only one pail or jug of sugar syrup is being fed, it may be satisfactory to have an inner cover over the brood nest and to place the feeder jar or pail immediately over the hole in the inner cover. However, when a colony needs feeding, it usually requires more than one 10-pound (four kg) pail or one-gallon (four *l*) glass jar of food. Feeding one pail at a time is sufficient only if it is refilled as soon as it is empty or nearly so.

Though glass jars are effective as feeders, weight, breakage and rusty lids are troublesome, and plastic pails – one to three gallons (four to 12 *l*) are commonly used. Moreover, glass is less commonly available and even metal cans are difficult to find. Further, plastic pails, with the exact-sized screen outlet, are trouble-free once a vacuum is established, and easy to clean. To use a plastic pail, fill all the way, and turn screenside down before placing on the hive. This lets the vacuum become established without leaking syrup into the colony. When a sugar syrup feeder jar is empty, whether can, jar or pail, bees will close the holes made in the cap with wax and propolis. They will not do so if any sugar syrup is inside. When a beekeeper removes an empty feeder jar and refills it, it must be checked and any closed holes opened with a pin.

Miller-type feeders – Dr. C. C. Miller, of comb honey fame, made a feeder that resembles a half-depth box and fits on top of the hive. These feeders have two pans or open-top trays that can be filled with syrup. A crawl space for the bees is usually left between the two pans; sometimes there is a single crawl space at one end of the feeder.

The principal advantage of the Miller-type feeder is that it holds a large quantity of syrup. The syrup level is checked easily by lifting the cover, and when more syrup is needed it is poured into the trays without disturbing the colony.

The chief disadvantage of this type of feeder is that it is farther from the bees than a pail or jar resting on the top bars of a brood nest. If the weather turns cool bees may cluster. Even if they do not form a firm cluster they may not be inclined to venture far from the brood rearing or clustering area to obtain food.

Miller feeders, like division board feeders, are made of plastic and are nearly leakproof.

Plastic Bags – Some beekeepers, rather than use pails or cans, put sugar syrup in sealable one-gallon (4 l) plastic bags so they are about two thirds full. Then cut an 'X' on one side and place the bag directly on the top bars, 'X' side up. If too full the bag will run out, but when correctly filled the bees can easily feed.

Conclusions – Even beekeepers who try carefully to provide their colonies with adequate food will occasionally have colonies that run short. Colonies that have large quantities of brood when a dearth occurs may break the brood-rearing cycle. Extreme drought or extremely rainy weather conditions can lead to starvation nearly anywhere.

A wood Miller-type feeder, made from a super. The inside is coated with beeswax or paraffin. Bees enter from below by way of the center opening, and floats on the surface of the syrup keep them from drowning. In the winter, insulation can be used to fill the sides, and ventilation is aided by the opening.

A plastic Miller type feeder. Bees can enter from below and take syrup (being filled with the pail), or munch on the Ross Round combs not good enough to sell in the smaller chamber. These feeders allow adding food without opening the colony, a benefit many beekeepers appreciate, especially during cool weather.

Flight Of The Honey Bee

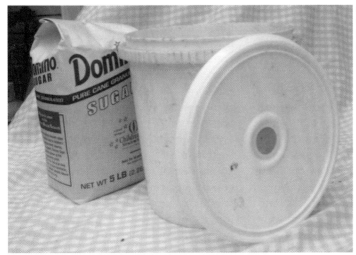

A plastic one-gallon (3.8 l) sugar syrup feeder pail, with screened opening on top, and sugar.

Larger pails can be used also, requiring less labor replacing the syrup.

For these reasons, beekeepers should always be prepared to feed. Only strong, populous colonies will store a surplus of honey for their owner. It is important to keep colonies prosperous at all times, and in some years this means feeding sugar syrup when necessary.

FIRE BLIGHT – Apples, pears and certain other members of the same plant family may suffer from a bacterial disease called fire blight. The cause is a bacterium, *Erwinia amylovora*, which may gain access into a host tree though a wound, openings on leaves, or the nectaries. Honey bees were once thought to be the chief means by which the disease is spread from one blossom to another but now it is understood that wind, rain, many insects and even birds may also spread the fire blight bacteria.

FLIGHT DISTANCE – (see FORAGING DISTANCE)

FLIGHT OF THE HONEY BEE – Honey bees are remarkable fliers. Although they cannot hover, they can fly up and down over a relatively short distance, land on a flower moving in the wind, or land precisely at a hive entrance, all with a load weighing a little more than half of their own weight. Under normal circumstances individual worker bees fly at about 12 miles per hour (19 km/hr); however, they may be able to fly at nearly twice that speed if required to do so. Bees in a migrating swarm appear to fly more slowly, about eight miles per hour (13 km/hr). A wind of about 12 miles per hour (19 km/hr) will usually stop flight from a colony entrance.

While a bee is in flight, the hind and fore wings are coupled together with tiny hooks (hamuli) so that the two wings act as one. The muscles for both walking and flying are contained within the thorax, the middle body segment. Before a bee can fly it must be sufficiently warm so that the flight muscles will function properly. For maintaining steady flight the bee's thoracic temperature must be above 81°F (27°C) but below about 122°F (50°C); a higher temperature would be lethal. It is estimated that at cooler temperatures the movement of the flight muscles causes the temperature to be about 59°F (15°C), which is above the prevailing temperature. Thus, a bee should be able to generate enough heat during flight to fly when the temperature is as low as 54°F (12°C). Some races of bees are known to fly at lower temperatures. They may be able to do so by stopping periodically and warming up their thorax with muscle activity. Cooling the body can apparently be accomplished by regurgitating fluid from the honey stomach and holding a droplet between the folded tongue and the mandibles. As water evaporates from this droplet the head is cooled and the head then acts as a heat sink for the warmer thorax as hemolymph circulates through the body parts.

Eckert, John E., in a classic paper, studied the flight range of honey bees (Journal of Agricultural Research 47: 257-285. 1933). In an experiment conducted on a prairie in Wyoming he found that bees would fly only 1.5 miles (2.4 km) for sugar syrup but at least three miles (4.8 km) for natural pollen and nectar. When colonies were placed at varying distances, colonies within 0.5 to two miles (0.8 to 2.2 km) of food sources made weight gains as great as those located right in the nectar-producing area. Colonies five miles (8 km) from the nectar source lost weight; however, bees were found to fly a maximum of 8.5 miles (14 km) in an effort to collect food (also see FORAGING DISTANCE).

FLOWER FIDELITY – If one watches a foraging worker honey bee it will be noted that she usually flies from a flower of one kind to another of the same kind. This is called flower fidelity. This type of activity is important for both the bee and the flower. In the case of the flower, pollen from an alfalfa flower, for example, will grow and fertilize an embryo on another alfalfa flower only. Pollen from an alfalfa flower cannot be used to fertilize any other plant species any more than sperm from one animal species can be used to fertilize an embryo in another animal species.

In the case of the bee it is important that once she finds a good source of food, whether pollen or nectar or both, that she continue to forage on the same species to obtain a maximum amount of food with a minimum of effort and time. Flowers have distinctive sizes, shapes, designs, colors and odors making them easily identified by foraging bees. Therefore they can practice flower fidelity without making errors.

A small number of bees in a hive are

Bees visiting a particular flower type that is producing nectar and/or pollen will continue visiting that flower type, and only that flower type until it quits producing. This fidelity ensures a steady supply of food, and, in the process, enables like plants to become pollinated.

scouts. The number is usually five to seven percent but may be lower or much higher when there is a dearth. The number of scouts can be determined by examining the color, size and shape of the pollen in the pollen loads of foragers. Under normal circumstances more than 90 percent of the pollen loads will be pollen from a single plant. The scouts that visit a variety of flowers will have pollen loads that are mixed. When a scout finds an excellent food source, and has recruited other bees to it, she ceases to work on this source and searches for something better. Thus, scouts themselves are not great collectors of food but are most important for the ongoing success of the colony. Scouts may redirect the efforts of a whole colony when they find a richer food source.

Flower fidelity has been studied by a number of people. The Greek philosopher and naturalist, Aristotle, over 2000 years ago, observed and recorded that bees flew from a flower of one kind to another of the same. More recently, foraging strategies by animals have come under closer scrutiny. Animals, including honey bees, do not forage randomly or without a plan. One of the keys to the honey bees' success, that allows them to live all over the world in a variety of climates, is their masterful foraging strategy program including flower fidelity.

FLUVALINATE – The compound in longest registration for *Varroa* control in the U.S. is the pyrethroid fluvalinate. Registered soon after *Varros* was discovered in the U.S., fluvalinate has been the acaricide of choice by the majority of U.S. beekeepers in controlling *Varroa*. Fluvalinate is not effective on tracheal mites. Fluvalinate is delivered to *Varroa* via contact of bees in the hive with a plastic strip containing 10% active ingredient, and sold as the product name Apistan®. Strips are hung between frames in the brood nest when no nectar is being brought in, and left to remain in place for 42 - 56 days. Bees are efficient transporters of the acaricide, distributing fluvalinate throughout the hive and causing as high as ca. 98% control of *Varroa*, first for mites on adult bees and then to mites emerging with new bees from infested brood cells. Beekeepers can choose to treat in the spring or fall, or both, as long as no honey crop is being produced at time of treatment. Hazard to the beekeeper in contact with the strips is relatively minor. If used properly, residues of fluvalinate in the honey or wax are not a

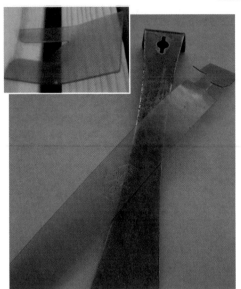
A fluvalinate strip. Inset – strip in place.

problem, as a small, but measurable tolerance for this chemical in honey and beeswax exists. There has been some evidence that fluvalinate may cause sublethal reproductive effects on drones.

In the late 1990s, U.S. beekeepers began experiencing decreased efficacy of fluvalinate in controlling *Varroa*. Fluvalinate resistance by geographically diverse populations of *Varroa* was confirmed in 1998, explaining the cause of *Varroa* population explosions and heavy colony losses. The spread of resistance was likely aided by the migratory nature of beekeeping in the U.S. It is important to note that the use of other acaricides in the pyrethroid class will likely fail, as well. - PJE

FOLLOWER BOARDS – Beekeepers frequently use boards the size of a frame, sometimes with nails for lugs, to reduce the size of a brood nest. Follower boards, sometimes called dummy boards, are especially useful in the spring when it is difficult for bees in small colonies to maintain their brood-rearing temperature. A brood nest can be made with only a few frames, four to six, with a follower board on the side where the rest of the hive is empty. The remaining space may be filled with empty comb.

Honey bees control the temperature in the brood nest only; they do not heat the whole hive interior. Still, weak colonies may not have enough bees to control the temperature easily. Some heat always escapes, and reducing the size or total area of the hive may be very important, especially on cold nights. Follower boards may be made from any kind of lumber or even plywood. The thickness is not important.

FOOD CHAMBER – In a natural nest honey bees store their food, the honey and pollen, above and to a small extent along the side of the brood nest. When beekeepers use the words "food chamber" they are usually referring to the second or top hive body of a two-story colony. Rarely, a food chamber is a third hive body of honey placed on colonies in the fall that do not have sufficient food for winter. During the winter, as the honey is consumed, the brood nest moves upward and eventually occupies some, if not all, of the food chamber.

FOOD OF THE HONEY BEE - Adult honey bees feed exclusively on honey and pollen. Honey is their source of carbohydrate and pollen provides the bees with protein and the small amount of fat they need. The vitamins, minerals and other substances bees need for a balanced diet are contained in both the honey and the pollen though the pollen is richer in these materials than is honey. Pollens vary a great deal in protein content. However, the fact that bees feed on many flowers, and a

Food exchange (trophallaxis) between two workers. (Moody Institute photo)

mixture of pollens is in the hive, makes up for any deficiencies in any one source of pollen.

Young worker larvae and queen larvae are fed a special diet called royal jelly. This is secreted from glands, called hypopharyngeal glands, in the heads of young worker bees. Depending upon the age of the larvae the royal jelly may contain some honey and pollen.

The only other things bees collect in the field besides pollen and nectar are water and propolis. Their specialized feeding habits set the honey bees and other bees aside from all other insects. While many others may also feed on nectar and/or pollen it is rare that it is their exclusive diet or that of their young.

FOOD EXCHANGE BETWEEN ADULT WORKER BEES – If one watches honey bees in an observation hive it will be noted that a great deal of antennal and mouthpart contact occurs between them; much of this activity involves or results in food exchange, called trophallaxis. Food is exchanged under two circumstances. One is the transfer of newly-collected nectar from a forager to house bee (see NECTAR, CONVERSION TO HONEY) and second is the constant exchange of food among bees in the hive. The former is very different from the latter. The constant exchange appears to be an important aspect of social living and is not merely the giving of food for the sake of nutrition.

Some bees beg for food while others offer it apparently unsolicited. A begging bee thrusts her proboscis between the mouthparts of another bee that may or may not respond by regurgitating some food. A bee offering food opens her

259

mandibles where a droplet of food forms as she moves her proboscis downwards and forwards. Queens, drones and workers may beg for food but usually only workers offer it. While exchanging food the two bees are constantly antennating each other.

While it is clear what is being exchanged between foragers and house bees, it is not at all clear what is being exchanged by other bees in the hive. Certainly water, nectar or honey is involved; however, it is generally agreed that secretions (pheromones) that play a role in social control are also an important part of the exchange. The message may be very simple and say nothing more than the queen is present and the colony is normal.

The rate of this exchange between bees was demonstrated in a paper written some years ago. Researchers fed six foragers 20 cubic centimeters of radioactive sugar syrup that the bees carried back to the hive in several trips. After five hours it was found that 27 percent of the workers in the hive were radioactive; after 24 hours 55 percent were radioactive. All of the many unsealed cells in the hive contained radioactive sugar syrup; a large percentage of the larvae being reared in the colony were also radioactive.

FOOTPRINT PHEROMONE – (see PHEROMONES)

FORAGERS – Those bees that search for and collect food (nectar and pollen), or water and propolis, are called foragers. They are the older bees in the hive as opposed to the younger bees, called house bees, that work exclusively within the hive. Foraging is the most dangerous of the tasks bees undertake. The life of an active forager is short, two or three weeks of field duty, because of the demands on the bee's body. A small percentage of foragers are scouts and find the food that the rest of the foragers collect; the rest of the foragers follow the recommendations of the scouts (see CASTES; and JUVENILE HORMONE).

FORAGING BEHAVIOR – "Foraging by honey bees is a social enterprise." So writes T. D. Seeley in his book, *Honeybee Ecology* (Princeton University Press, Princeton. 201 pp. 1985). Seeley suggests we might think of a honey bee colony as a machine that has as its job to "extract energy and other resources from the environment." In many ways it is the success of the whole group, not the individual that is important. Cooperation characterizes the foraging behavior of honey bees. A scout bee, for example, that finds a rich source of nectar, spends a great deal of time communicating this information to others in the hive. The time she spends telling others about what she has found reduces her individual food-collecting efficiency, but her efforts allow the colony to make the greatest gain.

There is a genetic argument that helps explain honey bee cooperation. A worker does not produce female offspring, and rarely produces males. The main way her genes are passed on to the next generation is to make certain that her mother, and the colony, prosper. Thus, colony success is in the self-interest of each forager. The greater the quantity of food collected by the colony the more it will prosper.

To understand the strong selective pressure on a colony of honey bees during the year we will review some of the studies done by Seeley on unmanaged colonies in Connecticut. He found that a feral colony living in a tree will have a minimum population of about 10,000 (late winter) and a maximum of about 40,000 (late spring). During the course of a year the colony will rear about 150,000 individual bees. We have good figures on the amount of pollen and honey required to rear an individual bee. During the year a colony needs about 44 pounds (20 kilograms) of pollen and 132 pounds (60 kilograms) of honey to sustain itself and to grow this number of bees. This food must be gathered in a period of 22 weeks in the area where the study was made.

Feral colonies usually live in small cavities, which become congested easily, and the colonies are thus stimulated to swarm. In fact, a feral colony's continuing success demands that they produce a swarm almost every year. A man-managed colony will be given more space (extra hive bodies) during the year and thus the population will grow larger. A larger population is needed for a beekeeper to produce surplus honey that may be harvested.

One result of the above needs is that we have two groups of bees in a hive: scouts and foragers (see CASTES). Scouts leave the hive each day and search for food. As indicated above, when they find food, they recruit others, the foragers, to collect it. When the foragers are at work the scouts start to search for something better. Of course, they must spend a great deal of time investigating sources that are not rich, but they recruit foragers only to the occasional rich food patch. As an aside, it is important to remember that plants that need bees for cross-pollination are very much in competition with each other and some produce much richer supplies of pollen or nectar, or both, than do others. As a result of their searching behavior, the scouts gather much less total food than the recruited foragers but certainly their exploration efforts contribute to greater colony efficiency.

Foragers, unlike scouts, can be remarkably constant on the plants they visit. In the case of pollen, for example, several studies have shown that 90 to 95 percent of foragers returning to the hive will carry pollen of one type only. Over 2000 years ago the Greek philosopher Aristotle noted this remarkable flower constancy on the part of worker honey bees. One recent study showed that foragers were often so devoted to one source of food that they would overlook a richer source that was in the midst or intermingled with the plant on which they were foraging. One water-collecting bee devoted much or all of her foraging life to this one task (see WATER FOR HONEY BEES). Thus, each forager's behavior is tightly patterned.

Research on foraging behavior and foraging strategy is still taking place. One question that comes to mind at this point is how a forager, whose activities are so closely governed, changes her behavioral or foraging pattern. Apparently, the decisions reflect conditions in the hive as well as in the field. A nectar forager whose load is received quickly and enthusiastically by a "receiver" bee in the hive, is thus encouraged to return rapidly to the field. If there is a delay in finding a receiver bee to accept the nectar, she may cease to forage, or she may seek information from scouts about other food sources. Such a delay may come about when there is little empty space in the hive, so that the receiver bees must spend most of their time searching for empty cells, or if so many foragers are bringing in nectar that the receiver bees are swamped.

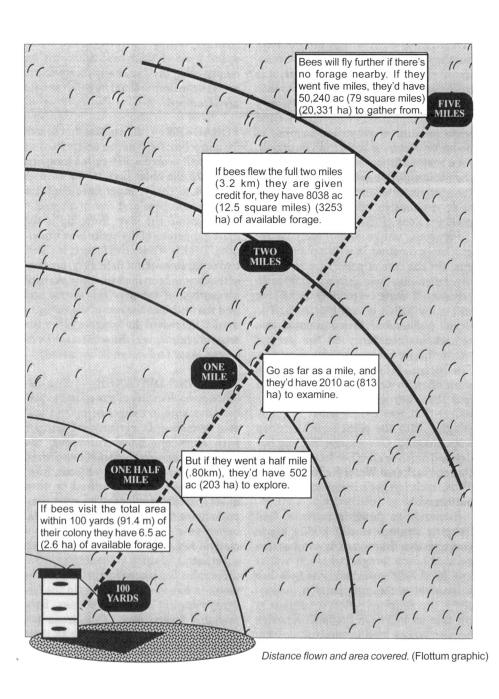

Distance flown and area covered. (Flottum graphic)

FORAGING DISTANCE – The distance over which bees fly to collect food is quite variable, and depends on the available resources. In agricultural settings with abundant floral resources, the median distance can be just a few hundred yards (meters). Under less rich conditions, however, some bees may forage six or more miles (10 kilometers or more) from their colony. The most comprehensive data on colony foraging distances in a forest setting suggest a median foraging of 0.6–1.2 miles (1.0–2.0 kilometers), with substantial foraging extending to about three miles (5 kilometers). Such large foraging distances result in considerable overlap in the foraging range of different colonies. Bees probably forage over large areas because the greater distance increases the choices of foraging patches and allows a colony to increase the average richness of the patches it utilizes.

Bees appear to forage over smaller areas to collect pollen than to collect nectar (1200 and 2000 yards [1100-1829 m], respectively, in a forest in central New York). This may indicate that a colony's pollen needs are more easily met or that there is less of a difference in foraging reward among pollen-yielding forage patches than among those that yield nectar. (Visscher, P. K. and T. D. Seeley. *Foraging strategy of honeybee colonies in a temperate deciduous forest.* Ecology 63: 1790-1801. 1982.)

As the distance from the colony to forage increases, the profitability of foraging falls off due to an increase in the time required for flight and the energy consumed during flight.

Over a season the distance foragers travel is dynamic. Nectar and pollen plants that bloom early in the spring may be near, requiring short foraging flights. Mid-summer crops, however, may be further away, requiring greater foraging investment.

Depending on location and resources some beekeepers move colonies during the season to take advantage of local crops growing too far for their bees to search, or just to reduce the energy required to reach them. -PKV

FORMIC ACID – Because formic acid is a natural compound, found as a normal constituent of honey, the EPA granted approval of this compound for *Varroa* control.

Other forms of formic acid, that is liquid formic acid soaked in absorbent material, contained in plastic bags with evaporative holes, has also been registered. Used extensively in Canada and other countries for years, and costing about the same as the pesticide strips, registration is the U.S. has been approved.

The registered product, MiteAway II is a prepackaged product. It is a pad soaked in a known strength and amount of formic acid, contained in a plastic envelope with holes one side for evaporation.

MiteAway II formic acid pre-soaked pad applied to the top bars of a colony, with an accommodating rim and support blocks added. The acid volatilizes, and, being denser than air, moves down through the hive. Application is made between 50-79°F (10-26°C). As with all chemicals, strict adherence to the label instructions regarding protective gear is mandatory. This is MiteAway II.

Foundation

Plastic frames and plastic foundation sheets make life easier for the beekeeper by reducing the labor needed for construction. One-piece frame/foundation units make labor even less expensive. Plastic foundation sheets are available with, or without beeswax applied. Beeswax on the surface of the embossed plastic sheet hastens acceptance by the bees. Wooden frames with plastic foundation, preassembled, also are available.

FOUNDATION – (see BEESWAX FOUNDATION)

FRAMES – The chief piece of furniture in a beehive is the frame. Properly made, it makes beekeeping possible, practical and simple. Improperly made or finished frames cause honey bees to build excessive burr and brace comb and to use too much propolis.

A frame has four parts: a top bar, a bottom bar and two identical end bars. Bottom bars may be one solid piece, split or grooved. Top bars have either a wedge or a groove to hold the foundation in place.

However the frame is made, one of the most important requirements is that the wood must be carefully sanded and/or planed. Bees use propolis to varnish rough wood within their nest. This is done to eliminate any cracks or crevices in which bacteria, molds, yeasts or other noxious forms of life may live. Propolis is a nuisance for the beekeeper; it makes equipment sticky to handle and frames glued together with propolis are difficult to manipulate and remove. Bees deposit much less propolis on smooth surfaces. Still another question in making frames is the width of the top bar. Many of the frame top bars in use today are 1-1/8 inches (28 mm) wide, which gives them good strength. However, these wide top bars were designed during the comb honey era. At that time wide top bars acted in part as queen excluders and helped to keep queens out of comb honey

Frames in a colony hang on the rabbet, leaving a space on both sides, top and bottom for bees to move throughout the hive.

The lug and rabbet.

How to nail frames is an important question. It is important to nail and wire new frames with great care. Some of these frames may be in service for 50 years. To last, they must be square and the foundation must be properly in place.

It is important to use 10 nails when nailing a frame. Four nails are driven down through the top bar and into the end bars; two are driven through the end bars, under the lugs or frame's ears and into the top bar; and four should be used to nail the bottom bar into place. The nails driven through the end bars into the top bar will prevent the top bar from being pulled off when manipulating the frames. The holding power of a nail is in its length and if a good, soft pine is used to make the frames, long nails will not split the wood.

supers. Unfortunately, they also reduce ventilation and lack of ventilation is one cause of swarming. Many beekeepers today use nine combs in a 10-frame hive body or honey super, and under these circumstances there is no problem. However, when one uses ten frames in a 10-frame hive body or honey super there are difficulties. A top bar 7/8 inches wide (22 mm) would be a better size. Those beekeepers making their own frames may wish to use the smaller dimension.

Soft pine is the favored wood for making frames. Some beekeepers use basswood, also a soft wood. Because soft woods are used, many beekeepers use eyelets in the end bars when frames are wired; if this is not done the wires will not be taut.

Depth of frames – The three most common frame depths used by beekeepers are the full-depth 9-1/8 inch, medium depth 6-1/4 inch (sometimes called the Illinois or half-depth) and the shallow depth 5-3/8 inch. Frequently beekeepers will use the full-depth combs for the brood chamber and either the medium or shallow frames for the honey supers. The medium depth frame is being used increasingly for both the brood chamber and honey supers. The advantage of the medium and shallow depth frames is that they fill more rapidly and the supers are not as heavy when filled with honey.

Hoffman frames – The Hoffman self-spacing frame was the invention of Julius Hoffman (1838-1907) of Canajoharie, New York. Hoffman immigrated to the U. S. from Poland, where he apparently knew and studied with Dr. Dzierzon who became renowned

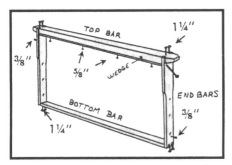

Size and placing of nails when constructing a frame. We recommend using 10 nails for added strength. (Root photo)

Frames

A standard self-spacing Hoffman frame with drawn foundation. This shows the tapered lug for minimal contact with the rabbet; the side bar with spacing shoulders and beespace walkway; pre-drilled holes for wiring; and a two-piece bottom bar, which leave a gap for the bottom edge of the foundation.

If the nail or staple that goes through the side of the end bar into the top bar, directly beneath the lug is left out, the top bar can and will be pulled up and away from the side bar, leaving a difficult mess to repair.

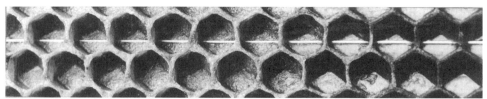
Bees will draw out the cells right over a well-embedded wire.

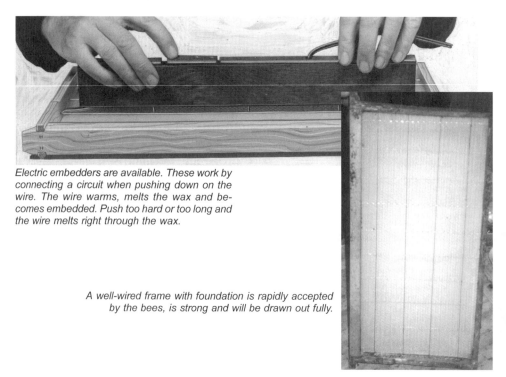
Electric embedders are available. These work by connecting a circuit when pushing down on the wire. The wire warms, melts the wax and becomes embedded. Push too hard or too long and the wire melts right through the wax.

A well-wired frame with foundation is rapidly accepted by the bees, is strong and will be drawn out fully.

Frame Spacers

Wiring a frame that will hold wax foundation is a good idea if the frame is to be used in an extractor. The new wax definitely needs the support. Don't add wax until just before installing the frame in the super.

Inset – For small numbers of frames, a spur embedder works well. Keep the spur in hot water, then roll along the wire, pushing it into the softened wax.

for his research on parthenogenetic development in honey bees.

Hoffman was a full-time beekeeper. He disliked the free-hanging frames that had been designed by Langstroth. Some beekeepers called these "bee smashers" because they could swing, or be pushed against one another, and thus kill or bruise bees. The result of Hoffman's studies was a frame in which the end bars have shoulders that firmly separate them from one another. Others had sought to separate frames with a variety of spacers, but Hoffman's worked best. The shoulders are extended for only about one-third of the length of the end bar. This shoulder length holds the frames in place, but provides ample space for bees to move around the frame's ends. Today some manufacturers of bee supplies still make frames similar to those designed by Hoffman.

Hoffman self-spacing frames touch at the shoulder, but leave walking space toward the bottom.

Frames

For ease of installation use a nail or brad driver when fastening the wedge tightly against the foundation.

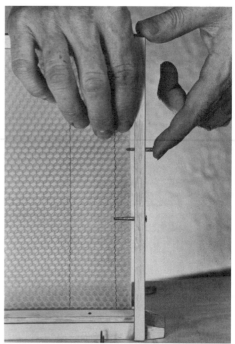

Frames that will be used in the brood nest need not be wired. However support pins are necessary to prevent sagging of the foundation.

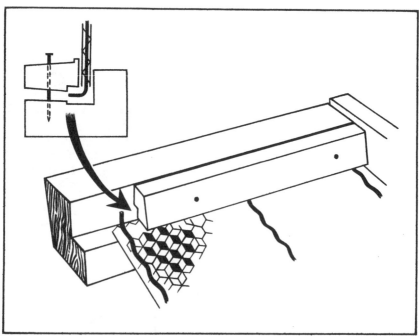

The pre-embedded wires rest in the groove in the frame. Place the wedge bar back in its original position and fasten.

Frame Assembly Pointers – Choose The Right Parts

Split, grooved and solid bottom bars.

Side bar bottoms (l) for solid or grooved bottom bars, (r) for split bottom bars.

Glue all joints with a good wood glue so the nails will not pull out.

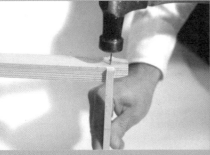

Use nails or staples that are thin enough so the wood parts do not split, and long enough to hold under extreme stress.

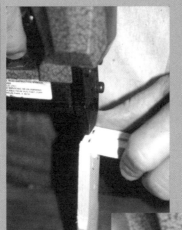

Before the glue sets and the last nail is in, make sure the frame is square.

The most important nail, or staple you put in your frame goes through the end bar and into the top bar to keep them from separating.

Frame Spacers

A sheet of foundation, with both pre-embedded wires and support pins will not sag. The bees will draw out a straight and strong comb.

All plastic frames, with plastic comb (drawn comb) are available, and readily accepted by bees. They are durable and can be used in ultra-high-speed extractors, require essentially no wax production by the bees, and are not damaged by wax moth. One is PermaComb®, another not shown, is HoneySuperCell®.

FRAME SPACERS – An alternative to spacing frames by hand, or using Hoffman self-spacing frames, is to use a metal or plastic frame spacer that fits over a hive rabbet. Several such devices have been invented in Europe but in North America, one spacer, invented and patented by Irwin A. Stoller (1902-1975), has dominated the market for more than 40 years. Based in Ohio, Stoller was a commercial, migratory beekeeper who operated several thousand colonies from Georgia to Ohio and Wisconsin.

Hand held spacers, homemade, or commercially available, avoid bee space problems.

Stoller frame spacers are fastened to a super in the rabbet. Bee space below or above the top of the frame's top bar must be mounted. They come in eight, nine or 10 frame slots.

Watch bee space when using these.

N.E. France

FRANCE, N.E. – N.E. France was the first bee inspector for Wisconsin. In 1891 the province of Ontario, in Canada, passed an act for the suppression of foulbrood, and Wm. McEvoy was appointed inspector. A new figure now came into the limelight, for McEvoy endeavored to rid the apiaries of the disease without destroying the property of the beekeeper. He practiced the shaking treatment already known to well-informed beekeepers and it was widely disseminated under the name, "McEvoy treatment."

In 1897 the state of Wisconsin passed a law patterned after the Ontario statute and N.E. France was appointed inspector in charge. France, like McEvoy, advocated treatment rather than burning, and thus treatment came into popular favor. By this time the details of successful treatment were well understood, and success followed the efforts to eradicate disease in hundreds of apiaries.

France was an active and energetic leader, and Wisconsin soon set the pace for other states to follow. One after another passed laws similar to those of Wisconsin, and no thought was expressed of attempting to eradicate bee diseases by any other method.

FRISCH, PROFESSOR KARL VON (1886-1982) – Professor von Frisch is best known for his discovery in 1944 that honey bees have a language for communicating information about food and nest sites to other workers in the hive. Earlier in his career von Frisch had made notable discoveries concerning the color vision and odor senses of honey bees. He was the first to show conclusively that honey bees could see different colors and that this was important in their orientation to flowers. For his discoveries he was awarded the

Karl von Frisch

Nobel Prize in 1973. During his career, von Frisch served as a professor at several German universities though he was Austrian by birth. Much of his work on honey bees was done at his summer home in the Austrian Alps. His active career spanned a period of more than 70 years. During his lifetime he was able to make a number of contributions to our knowledge of bee biology.

It was the discovery of the language of bees – their dance as a means of communication, associated with the odor of flowers – which brought worldwide fame to von Frisch. At the end of his research, von Frisch concluded that there were two kinds of dances used by the bees for communication among themselves: the round dance and the wag-tail dance. He discovered that the bees remaining in the hive are excited by the dance and the odors of the flowers visited. They then leave the hive to search for the source of nectar.

Professor von Frisch first considered that the round dance was the dance of nectar gatherers, while the wag-tail dance was performed by bees collecting pollen. In the light of later experiments, he modified his interpretation and concluded that when the bees return to the hive from a source of nectar situated less than 100 meters away, they dance the round dance. When the source is more than 100 meters from the hive, the bee which returns to the hive executes the wag-tail dance, which gives more detailed information as to the location of the nectar source.

The direction is indicated by the angle of the sun. That is to say, the angle is found between two imaginary lines extending from the entrance of the hive, corresponding to the direction of the sun, and the other line extending toward the source of nectar. The distance to the source of nectar is revealed by the extent of the distance covered in a straight line during the wag-tail dance.

The research led quite naturally to the training of honey bees. In order to induce the bees to gather nectar or pollen from a specific source, one would give them sugar syrup scented with the odor of the flowers one wished the bees to visit.

The results were beyond expectation; the harvest of forage, food and oil seeds was considerably increased, and the crops of fruit trees were assured to the profit of the growers. Beekeepers saw their yields increased.

He trained a large number of students, several of whom became famous themselves because of their research and studies on honey bees. The best known of his books is *The Dance Language and Orientation of Bees* (English edition: Belknap Press of Harvard University Press, Cambridge. 565 pages. 1967). His work first became well-known in the U.S. as a result of a lecture tour in 1949 and the publication of his little book, *Bees, their Vision, Chemical Senses and Language* (Cornell University Press, Ithaca. 119 pages. 1950. Revised edition. 157 pages. 1971). A book that reviews his life, discoveries and work with students and that makes delightful reading is *A Biologist Remembers* (English edition: Pergamon Press, London. 200 pages. 1967).

FRUIT DAMAGE BY HONEY BEES AND WASPS – In the autumn one will often see wasps, and sometimes honey bees, feeding on ripe grapes, apples and other fruits, especially those that have fallen onto the ground and have been broken as a result. Wasp feeding is much more common than that by honey bees. Honey bees will feed on fruit juices only after all flowers have been killed by frost and nectar is not available to them.

Fruit growers have sometimes accused honey bees of breaking the skins on ripe fruit in order to obtain the sweet juice within. However, honey bee mandibles are rounded and are used primarily for molding and shaping beeswax, not for cutting and shearing. The problem of wasps and bees is rarely serious in an orchard but may become so around fruit stands, especially where overripe and broken fruit may be exposed. The chief culprits are usually wasps, especially yellowjackets, that become especially numerous in the autumn.

FRUIT TREES – (see POLLINATION)

FUMIGATION AND STORAGE OF COMBS – (see COMB STORAGE; DECONTAMINATION OF HIVE EQUIPMENT and WAX MOTHS)

Basil Furgala

FURGALA, BASIL (1932-1996) – Basil Furgala passed away May 11, 1996. He received his B.S. in 1953 and an M.S. in 1954 from the University of Manitoba, and a Ph.D. in entomology from the University of Minnesota in 1959. He was a Research Scientist with Canada Department of Agriculture from 1959 until 1967 when he returned to the University of Minnesota to take a position as associate professor of apiculture.

Dr. Furgala's illustrious career has focused on research designed to aid the beekeeping industries of Minnesota, the nation, and the world. He advised many graduate students including people from Morocco, Indonesia, and Iraq. His research, reported in numerous publications, included the epidemiology and chemotherapy of nosema and viruses afflicting honey bees, sunflower and legume pollination, queen evaluations, and investigations of the honey bee tracheal mite. He was well known for his work on colony management systems including techniques for successful overwintering of honey bee colonies in the upper midwest.

He received the J.I. Hambleton Award from the Eastern Apicultural Society of North America in 1975, a Certificate of Appreciation from the USDA in 1981, the Apiculture Research Award from the Apiary Inspectors of America in 1983, the Outstanding Service to Beekeeping Award from the Western Apicultural Society of North America in 1986, and the Dutch Gold Honey Bear Award.

Basil Furgala also served as a USDA National Research Program Leader and as National Extension Apiculture Program Leader in Washington, DC.

GAMBER, RALPH (1912-2001) – A leader in the honey industry and the area of food safety and quality, Gamber founded Dutch Gold Honey, Inc., in 1946 with a $27 investment in three hives of bees at a farm sale. He, with his wife and three children, built the company into one of the largest honey companies in the world.

In 1957 Dutch Gold Honey was the first to be packaged in plastic honey bear containers, a container he helped develop.

He was also founder of Gamber Container Co., Inc., a distributor of glass and plastic containers.

His professional memberships included National Honey Packers and Dealers where he was past president, American Beekeeping Federation, Lancaster County Beekeepers, Pennsylvania Beekeepers Association, Honey Industry Council.

GARY, NORMAN – E. Norman Gary received his B.S. in Entomology from the University of Florida and his Ph.D. from Cornell in 1959. He was an apiary inspector in New York, a commercial beekeeper in Florida and a Professor of Entomology at Davis, California, from 1962-1994. During his terms as Professor he worked extensively in the film and television industry as a honey bee wrangler in movies, commercials and television shows.

He is best known, as the first to document the reproductive behavior of honey bees on film using a honey bee queen tethered to a balloon and suspended in a drone congregating area. Coupled with this was his discovery and identification of the queen honey bee's sexual attractant pheromone.

Ralph Gamber

Dr. Gary used his beekeeping, and queen pheromone skills at many entertaining public events. The science behind this demonstration, however, is quite serious.

GENETICS OF THE HONEY BEE – Gregor Mendel (1822-1884) spent seven years breeding and studying garden peas and provided us with the foundations of genetics. However, he spent the rest of his life as a beekeeper unsuccessfully trying to produce superior stocks. Mendel failed because he could not control the mating of queens and drones, an essential component of any breeding program. His failure was a consequence of the mating behavior of honey bees, a biological phenomenon that not only fascinates us but has profound consequences for the organization of a honey bee colony.

Mating behavior – Honey bee queens mate while in flight, away from the hive, about six to 12 days after emerging from their cells as adults. During the mating period, a queen may take several flights and mate with 12 to 20 different drones. During mating, each drone deposits six to ten million spermatozoa into the queen's oviducts and dies immediately afterwards. The queen then returns to the hive where the spermatozoa of her many mates migrate from her oviducts into her spermatheca. Queens store only about five to six million of the approximately 170 million spermatozoa that they receive from their mates. These spermatozoa remain viable in the spermatheca throughout the egg-laying life of the queen. Queens never mate again.

Colony structure – Each colony normally consists of a single, inseminated queen that lays all of the eggs that develop into workers, drones and queens. However, workers in a colony come from several different drone fathers. Any two workers that share the same father belong to the same subfamily and are called super-sisters while those that have different fathers belong to different subfamilies and are half-sisters. Therefore, a colony can consist of many different subfamilies.

The term super-sister was coined by H. H. Laidlaw to provide special significance to the genetic relationships of bees as a consequence of their haplo-diploid system of sex determination. Drones arise from eggs that are unfertilized and, therefore, have no fathers. As a consequence, drone honey bees have only a single set of chromosomes (are haploid) compared to two sets in workers and queens (diploids); one set they inherit from their mother and one set from their father. Because they are haploid, the genomes contained in the sperm they produce are identical to each other and have their true origin in the drone mother, not the drone.

As a consequence of the spermatozoa containing identical genomes, workers that belong to the same subfamily are more closely related to each other than are full sisters of diploid species. This is because all super-sisters share all of the same genes that they get from their haploid father and half of the genes they get from their diploid mother, while full sisters share half of their genes in common from each parent.

Stock improvement – The problem of controlled mating of honey bees was not solved until the early 1940s with the development of instrumental insemination technology. Instrumental insemination was pioneered by L. R. Watson while a graduate student at Cornell University, later on the faculty of Alfred University, and W. J. Nolan with the USDA. H. H. Laidlaw, University of California, Davis, and O. Mackensen, USDA, made the major biological and technological discoveries that led to the successful application of artificial insemination to honey bee breeding. With this technology came sophisticated honey bee breeding.

Selection of breeding stocks – The objective of any bee breeding program is to produce stocks that are superior to those that currently exist. This can be accomplished in two ways. First, the environment provided for the bees can be improved. This will result in immediate improvement of the qualities of the bees. Environmental improvements result from better management and improved conditions during queen rearing resulting in the production of superior queens. Good management is the most important aspect of producing better bees.

After the best possible environment is provided, additional improvement can be made by genetic selection. The most important component for a good selective breeding program is a good system of record keeping and clearly-defined stock improvement objectives. Selection of good breeder queens is dependent on being able to accurately measure the traits that are defined for improvement. Good records are necessary so that the queens from superior-performing colonies can be identified and selected as the sources of new queens and drones each generation. Sources of new queens and drones are equally important for selective improvement. Once these superior queens are selected, mating must be controlled to insure against cross-mating with undesirable stocks.

Breeding methods – Several methods of proceeding with a breeding program are available as follows.

Inbred-hybrid breeding – The first use of instrumental insemination in honey bee breeding followed the methods that had been developed, and proven very successful, for corn production—inbred-hybrid breeding. The objective of inbred-hybrid programs is to get superior-performing colonies of bees as a consequence of expressed heterosis or hybrid vigor. The method is to inbreed lines of bees by instrumentally inseminating virgin queens with semen from drones that are related to them. After inbred lines are developed, they are crossed with each other by inseminating virgin queens of one line with drones that belong to another. Comparisons of crosses of different lines are made and stocks are chosen on the basis of either their specific or general combining ability. Specific combining ability is shown when two particular lines produce superior stock when crossed with each other but each may be inferior in combination with other lines. A good general combining line is one that yields superior colonies in combination with several different lines.

The possible mating schemes that can be used in developing inbred lines is limited only by the imagination of the breeder. Some of the most useful ones are:

Maternal mother-daughter mating – A virgin queen daughter is raised and inseminated with the semen of one or more drones (often referred to as her "brother") that originate from eggs laid by her mother. Because the drone genomes originate in the drone mother, it is more appropriate to use genetic, rather than physical pairing terminology. In this case, the physical pairing is a brother-sister mating but since the queen mother is the true source of the male genome, it is genetically a mother-daughter mating. This results in very rapid inbreeding and is the system most often employed in inbred-hybrid breeding programs.

Super-sister–super-sister-mating – (physically an aunt-nephew mating) Daughter virgin queens are raised from a queen that has been inseminated with the semen of a single male. One daughter queen becomes the source of drones that are used to inseminate the other. This system results in even more rapid inbreeding than the mother-daughter mating system but requires more time between generations because of the time necessary for one of the sisters to produce drone progeny. It has the advantage, however, that one of the sister queens, the drone source, can be naturally mated.

Inbred-hybrid breeding programs result in a high degree of uniformity among queens and colonies. Desirable characteristics of different lines can be combined into a single hybrid line for specific purposes or beekeeping conditions. Since lines are maintained by artificial insemination, undesirable traits, such as those expressed by African honey bees, can be excluded from the breeding population.

However, inbred-hybrid breeding systems are labor-intensive and expensive to operate. It takes several generations, often years, to develop good inbred lines that ultimately show severe effects of inbreeding depression and are completely lost. Inbreeding depression is most often observed as a reduction in the viability of worker brood. A condition called "shotgun brood" or "scattered brood" is often observed where there are open, empty cells in areas of capped worker brood. In severe cases more than 50 percent of all worker brood is lost.

Closed population breeding – The severe inbreeding depression and labor costs associated with inbred-hybrid breeding led to a reevaluation of other closed population methods of breeding that had been used by plant breeders. H. H. Laidlaw, R. E. Page, and E. H. Erickson developed a breeding program that is designed to progressively improve the performance of bees by selective breeding while maintaining high brood viability. As with inbred lines, this program requires control over all matings either by artificial insemination or by mating queens in isolated areas that are free of other bees.

Daughters and drones are raised from all queens that constitute the "breeding population." Each daughter queen is mated by artificial insemination to 10 different, randomly selected drones of these queens, or is allowed to mate naturally in isolation. Colonies resulting from these crosses (preferably several daughters from each breeder queen) are then evaluated on the basis of predetermined criteria and the best from among these are selected as the breeder queens that supply the daughter queens and drones of the next generation. With this scheme, a minimum of 50 breeder queens is needed to maintain high brood viability for at least 20 generations (years) of selection. Selection will proceed faster with a greater number of daughters raised from each breeder queen, each generation.

A smaller breeding population can be maintained if each breeder queen has a single daughter represented in the breeding population of the next generation. In this case, the best-performing daughter queen (colony) of each of the queen mothers is selected as a breeder for the next generation. With this selection scheme, a minimum of 25 colonies is needed to maintain good brood viability for at least 20 generations. The response to selection is expected to be slower with this system compared with the one discussed above, but the cost in terms of money and labor is less because fewer breeder queens (colonies) are necessary.

Closed population breeding systems such as these have been implemented as national or commercial programs in several countries including Australia, Brazil, Canada, Egypt, the United States, and West Germany. Successes have been claimed for several commercially important traits including increased honey production, decreased defensive behavior, uniformity of color, and resistance to disease.

Genetic anomalies – Sometimes unusual-looking bees are observed in colonies. Many of these are the result of the more than 40 known single-gene mutations that affect either the color of the integument; the color, shape and occurrence of the eyes; the size and shape of the wings; or the lack of hairs on the bodies of bees. Usually, these are observed only in drones because the mutations are recessive and are not very likely to be homozygous in the diploid worker. Drones, being haploid, are effectively homozygous for all genes and express all of these visible recessive mutations. The most commonly occurring mutation is cordovan. This results in the changing of the normally black color of the integument to brown.

Genetics of the Honey Bee

Geneticists studying honey bees often use the Cordovan mutation as a marker. Cordovan bees are easy to spot in a normal colony because of their color. They can be strikingly beautiful.

A strikingly beautiful bee results when this mutation is expressed in yellow-color stock.

Several eye mutations occur occasionally in colonies of commercially-produced bees. These mutations most often result in red, green, white, brown, or tan eye color. These mutations can be deleterious to colonies when homozygous in workers because they result in blindness. Hairless bees resulting from a single gene mutation are also seen on occasion. However, hairlessness is also a symptom of some diseases. - REP

GIBSON, GLENN (1917-2005) – He began his beekeeping career by working in a honey packing plant in Oklahoma City while in college, but bought the company and 500 colonies and moved to Minco, Oklahoma in 1946. Eventually he reached 1,500 colonies. From 1950 to 1985 Glenn was involved at some level of national beekeeping politics serving as President (twice) and Secretary/Treasurer of the American Beekeeping Federation. In the late 60s he formed the American Honey Producers and served as initial President. He is the only person to hold that position for both groups.

He wrote extensively for *Bee Culture* magazine and *The American Bee Journal*. He was instrumental in obtaining funding for the insecticide indemnity program, the Laramie Research Lab, and funds for USDA bee research including Dr. Johnathan White's honey chemistry studies and grants for the Federal Honey Bee Apicultural Extension Specialist.

GLANDS IN THE HONEY BEE AND THEIR SECRETIONS – Honey bees have two types of glands. One type is called endocrine glands. These produce hormones that work internally in a honey bee to control bodily functions. The second type is the exocrine glands that produce substances externally. The exocrine glands are better known simply because their products, including beeswax, venom, enzymes, and pheromones are more obvious and play definite and obvious roles in bee biology. In this section only those glands about which we have a reasonable amount of information will be discussed. Much research needs to be done before we fully understand the glandular system of the honey bee (also see ANATOMY AND MORPHOLOGY; PHEROMONES, and BEESWAX, Beeswax Secretion by Honey Bees).

Endocrine glands – Three important endocrine glands in larval honey bees have to do with their growth and development. The first of these is the prothoracic gland that, as the name suggests, lies between the first two of the

Glenn Gibson

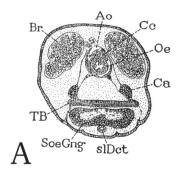
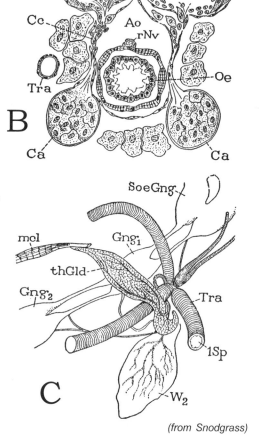

The endocrine organs.
A, section of head of bee embryo, corpora cardiaca cells generated from dorsal wall of oesophagus, corpora allata arising at ends of crossbar of tentorium (from Pflugfelder, 1948). B, corpora cardiaca and allata of honey bee larva, seen in cross section through anterior end of aorta and oesophagus (from L'Hélias, 1950). C, right thoracic gland and associated structures of honey bee larva (from L'Hélias, 1952).

Explanation of Abbreviations
Ao, aorta
Br, brain
Ca, corpus allatum
Cc, corpus cardiacum
Gng_1, Gng_2, first and second thoracic ganglia
mcl, muscle
Oe, oesophagus
rNv, recurrent nerve
slDct, salivary duct
SoeGng, suboesophageal ganglion
1Sp, first spiracle
TB, tentorial bridge
thGld, thoracic gland
Tra, trachea
W_2, rudiment of hind wing

(from Snodgrass)

three thoracic segments. This gland produces a substance called ecdysone that controls moulting in the larva and pupa; this gland is not present in the adult bee.

The second and third endocrine glands are the corpora cardiaca and corpora allata that are connected by nerve fibers to each other and to brain cells. The function of the corpora cardiaca is not clear but the corpora allata produces a substance called juvenile hormone (JH).

This hormone controls development in the larva and pupa. In the adult JH is responsible for workers changing from one role to another as they mature (see JUVENILE HORMONE).

Exocrine glands – These glands are usually large, well-defined, and for many their products are known. Except where noted below, the exocrine glands are found only in workers; only one is known from the drone. The only obvious

Glands In The Honey Bee And Their Secretions

exocrine glands in the queen are the mandibular glands that produce the sex attractant.

Wax glands – Four pair of wax glands are on the underside of the worker honey bee's abdomen. Beeswax is secreted as a clear liquid from the many cells in these glands onto underlying plates called wax mirrors (see ANATOMY). Once on the mirrors, and exposed to the air, the wax solidifies and becomes a bright white. Natural beeswax taken from a hive is yellow because it is colored from contact with pollen and

Wax scales emerging from wax glands of the honey bee. When wax first is extruded it is liquid and clear. When exposed to air it hardens into the familiar flakes often seen on floorboards and turns white. (Morse photo)

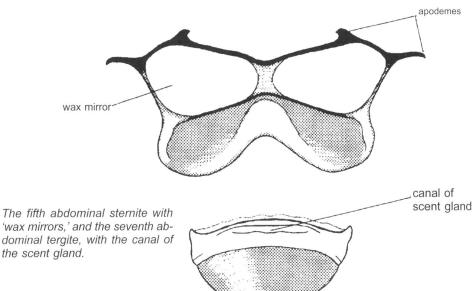

The fifth abdominal sternite with 'wax mirrors,' and the seventh abdominal tergite, with the canal of the scent gland.

(from Dade)

propolis. Honey bees that are not in immediate need of wax may allow several secretions to build up into plates that become laminated on the wax mirrors (see BEESWAX, Beeswax Secretion by Honey Bees).

Sting glands – Despite all that has been written about the sting glands it is still not precisely known how the two glands associated with the sting apparatus, the acid gland and the DuFour's gland (Alkaline), relate and function together (see ANATOMY and VENOM, commercial production of). The acid gland is the source of bee venom. (also see VENOMS OF STINGING INSECTS).

Glands In The Honey Bee And Their Secretions

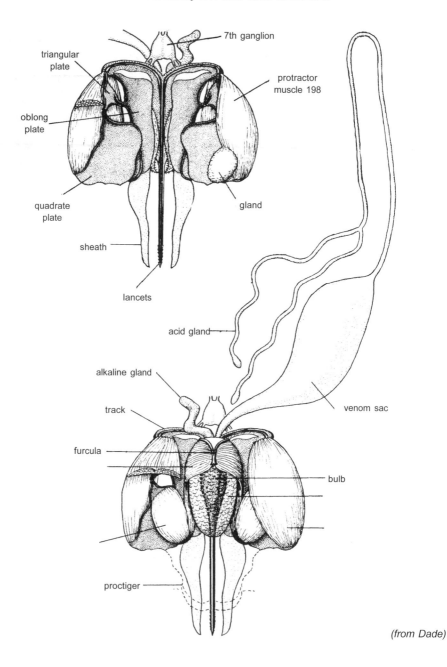

The sting apparatus. Below: dorsal aspect. Above: ventral aspect, revealed when apparatus is lifted out of the bee and turned over.

Glands In The Honey Bee And Their Secretions

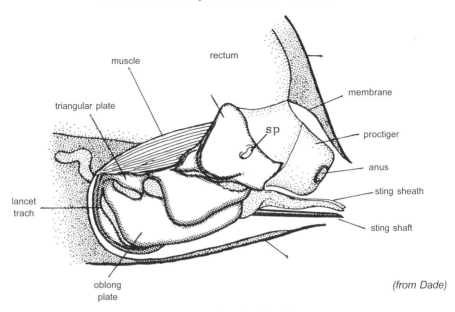

Sting Chamber: lateral dissection of sting chamber, from the left side.

The alarm odor and the sting scent gland – Isopentyl acetate is the principle alarm odor in honey bees. Isopentyl acetate was assumed to be produced by the paired Koschevnikov glands that lie near the base of the sting but that has been questioned by some researchers. When the worker stings, and pulls her body away from the sting leaving it in the enemy, the alarm odor is released for a matter of several minutes from the tissue that is left behind and attached to the sting. It is in this way that the enemy is marked. It is the odor of isopentyl acetate that other worker honey bees follow to chase an enemy where ever it may flee. The alarm odor in honey bees is not part of the venom as it is in the case of yellowjackets and other stinging wasps.

Nasonov gland – One of the more striking and easily seen glands is the Nasonov or scent gland that is found only in worker bees. The gland consists of about 600 cells that lie on the top of the 7th abdominal segment. It is normally covered and the pheromone contained by the 6th abdominal segment or terga. This gland produces four substances (see PHEROMONES) that work in concert to carry the specific message "come here" or "this is it" to signal food, a home site, a lost queen, or an assembly point. The gland is exposed by the bee tipping the end of her abdomen downward. At the same time the worker fans her wings and thus directs the pheromone out and away from her body. It is probably picked up by other workers as much as several feet away when the air is still. One can detect the odor of the Nasonov gland by sniffing near the rear of a fanning bee. This odor is predominantly geraniol, the odor produced by geraniums (see FANNING BY HONEY BEES).

Mandibular glands – These glands are sac-like and attach to the mandibles.

Glands In The Honey Bee And Their Secretions

The Nasonov gland is easily observed and the position of the two bees exposing this gland is typical. The gland is the light area exposed when the segments are separated. Heads are lowered, tails raised and wings fanning to distribute the pheromone.

It is possible to remove the mandibular glands from queens and for the queens to remain alive for a year or more. These glands are the only ones in bees that lend themselves to removal and still keep the bee alive. Thus we have been able to better understand their role in bee biology. The mandibular glands are very large in queens and are the source of the sex attractant. They are very small in drones and may not play any useful role in drone biology. In workers these glands are moderately well-developed. Young worker bees produce 10-hydroxydec-2-enoic acid, the chief lipid or fat component of royal jelly, from their mandibular glands, but older workers produce 2-heptanone. This latter

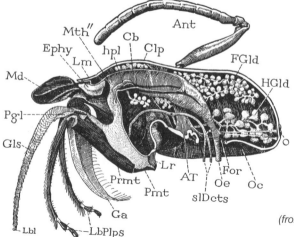

Right half of head of worker bee with mouth parts attached, cut a little to left of median plane, muscles and nerve tissue omitted. See next page for explanation of abbreviations.

(from Snodgrass)

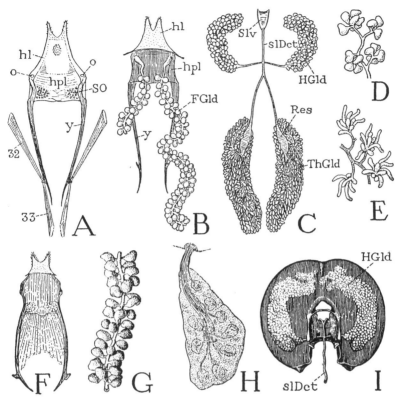

Glands of the head and thorax.
A, hypopharyngeal plate of worker, anterior, showing apertures (*o*) of food glands. B, same, posterior, with proximal parts of food glands. C, labial salivary gland system of worker, including head glands (*HGld*) and thoracic glands (*ThGld*). D, detail of head gland. E, detail of thoracic gland. F, hypopharyngeal plate of drone. G, detail of food gland. H, section of a food-gland lobule (from Beams and King, 1933). I, posterior wall of head of worker, anterior, showing position of head salivary glands.

o, orifice of food-gland duct; *y*, oral arm of hypopharyngeal suspensorium, with protractor (*32*) and retractor (*33*) muscles.

Explanation of Abbreviations
Ant, antenna
AT, anterior tentorial arm
Cb, cibarium
Clp, clypeus
Ephy, epipharynx
FGld, hypopharyngeal food gland
For, occipital foramen
Ga, galea
Gls, glossa (tongue)
HGld, head salivary gland
hl, hypopharyngeal lobe
hpl, hypopharyngeal plate
Lbl, labellum
LbPlp, labial palpus

Lm, labrum
Lr, lorum
Md, mandible
Mth", functional mouth
O, ocellus
Oc, occiput
Oe, oesophagus
Pgl, paraglossa
Pmt, postmentum
Prmt, prementum
Res, reservoir of thoracic salivary gland
slDct, common salivary duct
Slv, salivarium (salivary syringe, *Syr*)
SO, sense organs
ThGld, thoracic salivary gland

(from Snodgrass)

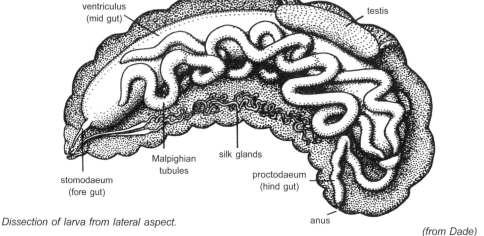

Dissection of larva from lateral aspect.

(from Dade)

material is a mystery. The odor of this material will alarm honey bees as will isopentyl acetate. While both appear to work equally well as alarm odors, it is likely that 2-heptanone plays a role that is not clear to us at present and is not merely a second alarm odor. Recently, this compound has been shown to have negative effects on *Varroa* mites in a colony. Future research may allow scientists to harness this natural compound in controlling this pest.

Silk glands – All larval honey bees spin a cocoon as a last act before becoming pupae and thus the silk glands are well-developed in queen, worker and drone larvae (see COCOONS). The silk glands appear in the same position where the thoracic glands will develop later but about 72 hours after spinning the cocoon the silk glands disappear and no trace of them can be seen in the developing pupae. Soon thereafter the thoracic glands develop from the basement membranes of the silk glands. The thoracic glands are the least well-developed in the drones.

Head and thoracic glands – There are three pair of glands in the worker's head and one in the thorax: labial (sometimes called the postcerebral), hypopharyngeal (or pharyngeal), postgenal, and thoracic. All of these glands are convoluted and are covered with lobules or bulb-shaped organs in which their active ingredients are produced and stored. The shape and design of these glands are such that study of the individual glands is difficult. The enzymes invertase and glucose oxidase that bees need to ripen honey probably originate from the thoracic gland. (see NECTAR, CONVERSION TO HONEY). Part of the royal jelly, the food that is produced by young worker bees and fed to young worker larvae and queen larvae, is also produced in the head glands.

Glands In The Honey Bee And Their Secretions

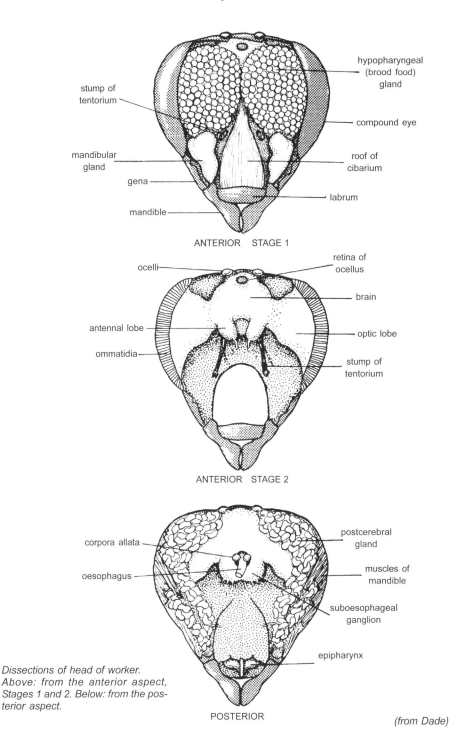

Dissections of head of worker. Above: from the anterior aspect, Stages 1 and 2. Below: from the posterior aspect.

(from Dade)

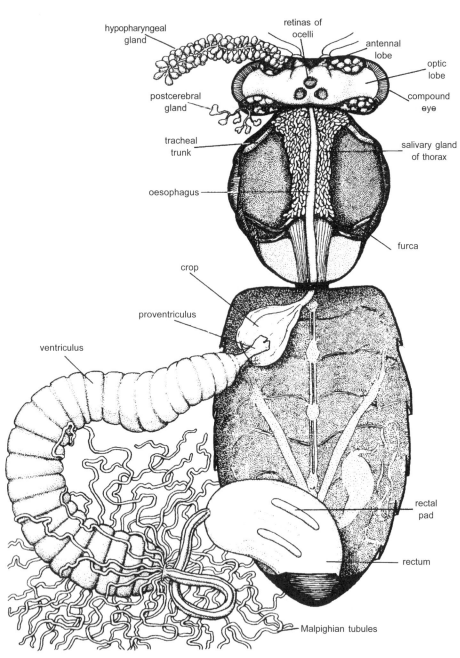

*Dissection of worker, Stage 2.
Glands of head lifted out, indirect flight muscles removed from thorax to expose underlying organs, and alimentary canal displayed.*

(from Dade)

GLUCOSE OXIDASE – (see NECTAR, CONVERSION TO HONEY)

GOOD NEIGHBOR BEEKEEPING – It should go without saying that being a good neighbor is a common practice for beekeepers. Unfortunately, beekeepers occasionally make the assumption that their neighbors share their enthusiasm about the boxes of potentially dangerous insects.

Using the guidelines (see boxes on following pages) will go a long way in reducing close encounters with your neighbors, the local law enforcement and zoning agencies, and others who may have an issue with your honey bees.

We encourage you to be aware of all local zoning ordinances that may impact your bees, and to take all necessary steps to keep your bees, and your neighbors at a safe and comfortable distance.

GRAFTING – (see QUEEN REARING: GRAFTING AND COMMERCIAL QUEEN PRODUCTION)

GRAFTING TOOLS – (see QUEEN REARING, Grafting tools)

GREENHOUSE POLLINATION – (see POLLINATION)

GROUT, ROY A. (1908-1974) – After completing high school in Keokuk, Iowa, Roy entered Iowa State College, in Ames, from which he received both his Bachelor's and Master's degrees for work done on the relationship of the size of the worker bee to the size of the cell in which it was reared.

He married Marjorie Dadant, daughter of Mr. and Mrs. Henry C. Dadant in September 1929. Following graduation they returned to Hamilton, Illinois, and

Roy Grout

Roy began his long-time association with Dadant & Sons (see DADANT'S).

Roy served as president of the American Beekeeping Federation in 1949 and 1950, and was among the group which was instrumental in obtaining a federal price support program for honey.

His initial interests when he came to Dadant & Sons in 1931 centered around plant operation and production, candle manufacturing, and wax refinement. He also cultivated a keen interest in the company's literary programs. From March, 1945, he was on the editorial staff of the American Bee Journal. Roy's major work was the complete revision of C.P. Dadant's *Langstroth on the Hive and the Honeybee* in 1946 under the title *The Hive and the Honeybee*; a second edition with only minor revisions in 1949; and in 1963 he edited an extensively revised edition.

> ## Good Neighbor Practices
>
> **PLACING COLONIES**
> - Locate the colonies away from property lines.
> - If the colonies are close to a property line, put a natural (shrubs) or artificial (fence) barrier at least six feet high between the colonies and the property line.
> - Locate the colonies away from your house, and away from frequently used entrances, walkways, or other areas of heavy pedestrian traffic.
> - Paint colonies with a neutral color (e.g., gray, brown, or military green), rather than white, or stain with a natural-hued wood preservative.
> - Locate the colonies so flight paths and spotting don't cause property damage to hanging laundry, the vinyl siding of homes, or automobiles that are parked in driveways and at dealerships.
> - Don't put colonies in fields or pastures where livestock (e.g., horses, cattle, or sheep) graze unless a fence is installed to prevent animals from scratching on the hives.
> - In temperate climates where the intensity of the midday sun is moderate, locate the colonies in full sunlight.
> - In semi-tropical or tropical climates where the intensity of the midday sun is severe, locate the colonies in sunlight that is tempered with either natural (e.g. trees) or artificial (e.g., fences or buildings) shade.
>
> **PROVIDING WATER**
> - Place colonies near a natural water source that is not frequently used by neighbors or children.
> - Provide a water source (e.g., a tub) near the colonies that contains natural (e.g., branches or wooden boards) or artificial (e.g., packing peanuts) floats that the bees can land on and drink, without drowning.
> - The water source should be changed each week to avoid stagnant water, and breeding mosquitoes.
> - The water source must NEVER be allowed to go dry during water collection months.
>
> **FLIGHT PATTERNS**
> - Place colonies near a tall, natural (hedge of evergreens) or artificial (fence or building) barrier. The barrier will direct or force them to fly high into the air and over the heads of nearby people. The barrier will also force the bees to return at a high altitude, from which they will fly straight down to the hives.

GUARD BEES – When bees are about three weeks old they may become guards for a few days, although this is not obligatory. Guards stand on their hind four feet at the colony's entrance(s) with their front feet raised. As bees enter the hive the guards inspect them to make certain they are residents; however any bee laden with food, and usually most drones, appear to be accepted without difficulty. Guard bees may carry on a lengthy inspection of incoming bees; the latter become submissive and allow the guards to pass their forelegs and antennae over them. Bumble bees, wasps and other insects that attempt to enter the hive are attacked and stung by guard bees. The protrusion of the sting by the guards attracts the attention of other guards, which come to their assistance.

The number of guards present at a hive entrance varies and is greatly affected by the quantity of incoming

Good Neighbor Practices

- Make sure the flight pattern does not direct or force them to fly over public sidewalks or other areas of pedestrian traffic.

HOW MANY?
- Do not locate more than three or four hives on a lot that is less than one acre in size.

WORKING BEES
- Do not work the colonies during bad weather.
- Work colonies between midmorning and midafternoon when the bees are out foraging, fewer are at home, and fewer guard bees are patrolling.
- Make sure there is room behind and to the side of each colony, where the most time is spent working the bees, so you can move freely around the colonies.
- Do not work the colonies when neighbors are out in their yards.
- If a colony is aggressive or mean, requeen it.
- Reduce robbing behavior by not working the bees during a dearth, by limiting exposed honey and by using entrance reducers on colonies that are weak.
- Manage colonies to prevent swarming.

NEIGHBORS
- Give honey to your neighbors every year.
- Before getting bees talk to your neighbors and see how they feel about a nearby beekeeping operation.
- Discuss the symptoms of a temporary reaction to a bee sting – slight swelling, itching, and redness – and the symptoms of a life-threatening allergic or anaphylactic shock reaction.
- Make sure you have an insurance policy that provides coverage for damage caused by bees to any third party.

LEGAL
- Register your bees with the state agricultural official or agency, when required by law.
- Follow all local, state, and federal laws regulating beekeeping.
- Keep good records of all colony management activities, including robbing and swarming prevention techniques. The records may serve as evidence in proceedings involving a beekeeping operation.

nectar. When little or no forage is available, there are many guards, while when food, especially nectar, is available in quantity, there are few if any guards. This suggests that, as with many animals, their own species is the honey bee's greatest enemy.

In the winter cluster, when it is too cold for bees to fly, and those on the outside of the cluster are so cold they can scarcely move, the outside bees can still protrude their sting if molested. This releases alarm odor, and if the disturbance continues, warm bees from the inside of the cluster may attack. In cold weather, they cannot fly very far, but they may arc out from the cluster and will sting anything they contact. Such bees are lost because they chill rapidly and cannot return to the cluster. However, such an attack is often sufficient to deter an enemy from molesting the colony further. (see CASTE)

Guard bees patrol the colony entrance challenging nearly every bee trying to enter. Bees from the guards' home colony will have the correct odor and be allowed to pass. Those without the correct odor will be resisted, especially during a dearth. However, during a honey flow, almost any individual who has a load of nectar or pollen is allowed to pass.

Guard bees also are on the lookout for trespassers already in, or attempting to get into their colony. And there are many that do – yellowjackets, ants, beekeepers, mice, and wax moths are some. When a trespasser is encountered, the guard will attempt to remove it herself, initially. If the struggle proves too much, other guards will be summoned to assist. A small amount of alarm pheromone will hasten other guards to the scene.

GUMS – A gum is a colony of bees in a section of a tree that has been cut from the tree itself, transported to an apiary, and placed there in an upright fashion, resembling its original position in a forest or woods. Often some of the tree trunk is removed so that the comb at the top of the colony is exposed. An ordinary hive cover can be placed over the exposed comb. In addition, if a sufficient amount of comb is exposed one can add a super for honey storage above the gum. The term "gum" has been most popular in the Carolinas and Virginia and originates from the fact that gum wood trees are often hollow and have lent their name to this type of hive.

Gums, like skeps, box hives, and colonies kept in other types of fixed frame hives, have largely disappeared. They cannot be managed and are therefore more a curiosity than anything else. Gums are illegal since the combs cannot be examined for brood diseases. Wherever they are found it is best to transfer the colony to a movable-frame hive (see REMOVING FERAL COLONIES, from trees).

Beekeeping in the mountain forests of western North Carolina has long been a part of the local scene. This old photo, dating from the early teens of the 1900s shows a mountain farm woman and a group of bee-gums, which provided the family's 'sweetening' and some cash income as well. In those days bees, often a milk cow, and a few other livestock and a home garden provided the family with most of its food supply. (Photo courtesy of NC Division of Archives and History)

A close up of a typical gum. The branches/pegs were inserted so the bees had a support for their comb.

This undated photo shows gums of several styles and sizes. Gums are rare now, and in all probability, will be only a museum attraction in the future.

GYNANDROMORPHIC BEES – Honey bees that are said to be gynandromorphic are those that have both male and female body parts. The distribution of the male and female parts can take several forms. For example, a bee may have a female head and a male thorax or abdomen. The gynandromorph may also be bilateral, that is, one side of a bee may be male and the other female. Sometimes patches of male and female parts may appear on the same body part.

There are several causes of gynandromorphism. It may occur when eggs are chilled or overheated. It may also occur when queens are not mated and are producing eggs. The tendency can be inherited. Under normal circumstances bees that are malformed are probably removed from the hive by normal workers. Bees with these characteristics are seen alive only rarely.

In some of the older literature these bees are sometimes called hermaphrodites but this term is not correct as such animals produce both eggs and sperm but gynandromorphs do not. Hermaphrodites are found among the lower animals only, especially certain worms.

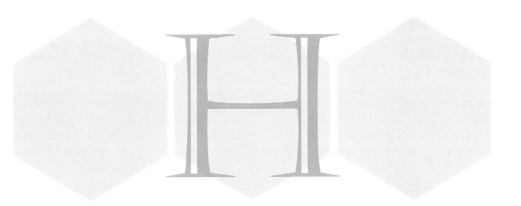

HAMBLETON, JAMES I. (1895–1969) – Though born in Chile, 'Jim' grew up in the Columbus, Ohio area where his father was a teacher. He graduated from Ohio State University in 1917 and took a position at the University of Wisconsin, Madison as an instructor in Entomology and beekeeping.

He served in WWI and stayed on in Paris to study apiculture at the Jardin du Luxembourg. He returned to Wisconsin and stayed until 1921 when he took a position with the USDA. In 1924 he succeeded Dr. E.F. Phillips as Principle Apiculturist, Bee Culture Lab, in Beltsville, Maryland.

During his tenure bee research and the number of USDA bee labs grew significantly. New labs opened in Arizona, California, Louisiana, Ohio, Utah, Wisconsin and Wyoming. He initiated the beekeeping Bibliography and the Bee Library at Beltsville, Maryland.

James Isaac Hambleton

In 1955 Dr. Hambleton attended a successful Tri-State beekeepers meeting held in Rhode Island. He was so impressed with the concept and organization that he assisted in organizing another the following year in Maryland with Dr. George Abrams in charge. That meeting resulted in the formation in 1956 of the Eastern Apicultural Society. The James I. Hambleton award for Academic Excellence was initiated by Mr. A.S. Michael, who financed the award for a time before EAS picked it up.

He retired as Chief of the Apicultural Research Board in 1958, but remained active in the industry attending and speaking at local, regional, National and International conventions. He came out of retirement briefly in 1967 to Chair the Organizing Committee of the International Apicultural Congress (Apimondia), held in Maryland that year. That was the first and only Congress held in the U.S.

HEARING IN HONEY BEES - (see SENSES OF THE HONEY BEE, Sound perception)

HEMOLYMPH – The blood of insects, including honey bees, is called hemolymph and is different from the blood of mammals and other animals. Hemolymph of bees is pale yellow or colorless. It does not carry oxygen as blood does in mammals; oxygen is distributed around a bee's body by a system of breathing tubes or trachea (see Respiratory system). Hemolymph is about 90% water. Insects have no veins or arteries but rather their body organs are bathed in the hemolymph. This blood-like material picks up digested food nutrients as it passes over the gut and makes it available to other parts of the body. As the hemolymph passes over the Malpighian tubules of the bee, which serve much the same function as kidneys, excretory (waste) products it has picked up from various organs are filter out.

Insects have an organ that is heart-like and that serves to move the hemolymph around the interior of the body. The main part of this organ is found in the abdomen and consists of a dorsal vessel and both dorsal and ventral diaphragms. These are driven by muscles that pull the hemolymph into the heart and pump it where it is needed, including up to the thorax and head. The hemolymph also serves to distribute body heat. It is warmed by the action of muscles in the thorax as it moves forward through the thorax to the head and in this way the temperature of the bee's body is controlled.

Insect blood does not have a clotting factor. Thus, if the integrity of an insect's exoskeleton is destroyed the animal's blood flows freely from its body and it dies.

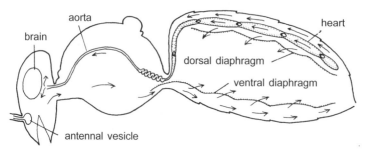

Diagram illustrating the action of the heart and diaphragms. (from Dade)

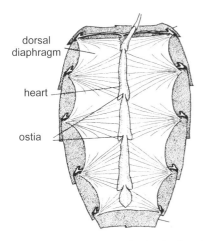

Roof of abdomen inverted, showing heart and dorsal diaphragm attached to tergites. (from Dade)

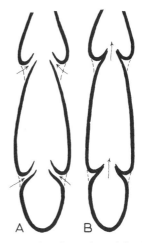

Diagram illustrating the action of the heart, and showing the blind posterior chamber, and the one in front of it. A, the heart dilated, ostia open, blood entering. B, the heart contracted, ostia closed and acting as valves, blood drive forwards. (from Dade)

HEWITT, PHILEMON J. (1917-1980) – Philemon J. Hewitt, Jr., 63, of Richards Road in the Northfield section of Litchfield, Connecticut, died October 4, 1980.

He was a Life Member of the Eastern Apicultural Society of North America, a member of the American Beekeeping Federation for 31 consecutive years, a past president and member of the Connecticut Beekeepers Association and was the editor of the quarterly Connecticut Honeybee newsletter.

He was an EAS Director from Connecticut, the society's Historian and author of the 25-year history of EAS, published in 1980 shortly before his death. This work summarizes much of EAS history and contains records and information on the organization available nowhere else. He was editor of the bi-monthly EAS Journal, a member of Apimondia, International Beekeeping Federation and traveled throughout Europe, Russia, Australia and most of the United States.

His collection of antique beekeeping equipment and many beekeeping books were donated to The Nature Conservatory when he died, and eventually transferred to the University of Connecticut Library.

Philemon J. Hewitt, Jr.

HIGH FRUCTOSE CORN SYRUP (HFCS)
– This is a manufactured sugar syrup derived from the starches in feed corn and called high fructose corn syrup (HFCS). Both HFCS and simple corn syrup consist primarily of the sugars glucose and fructose, plus water. Because of this composition, HFCS has affected the beekeeping industry in three ways: (1) as a consumer item that competes with honey, (2) as a substance that has been used to illegally adulterate honey, and (3) as a winter or stimulative feed that can be given to bees by beekeepers.

HFCS was first manufactured in 1969 and was initially called Isomerose. One method of producing it is by adding bacteria-produced enzymes to starch molecules to create glucose, then converting some of the glucose into fructose by adding another enzyme. HFCS has provided some direct competition with honey for supermarket sales because of its lower cost. In addition, it has indirectly depressed the honey market by diminishing honey's reputation, especially when HFCS is called "imitation honey" (which unfairly uses the word "honey" on the label) or in the few cases where HFCS or HFCS-honey mixtures have been illegally sold as pure honey.

Some key points about feeding High Fructose Corn Syrup to bees are: (1) it is available with either 42% of the sugar as fructose, or 55% fructose. A fructose content of 55% works better; less fructose means more glucose, which means rapid crystallization. (2) Great care must be taken to ensure that no syrup ever gets into harvested honey. Therefore begin fall feeding only after the honey harvest, and cease stimulative spring feeding before the honey flow begins. (3) Any corn syrups that, unlike HFCS, are made by breaking down starch to sugar using acid (rather than enzymes) are unsuitable as bee feed; they contain molecular fragments that are indigestible – and therefore toxic – to bees. -RN

HISTORY OF BEEKEEPING – Honey bees have played an important role in the history of western mankind for several reasons: chief among these is that there was no sugar in Europe and Africa until about 700 years ago when sugar cane was brought from the South Pacific through China to the Mediterranean area. Sugar beets, also a major source of sugar today, were developed only recently. Honey was early man's only sweet other than a few fruits. If people had as much of a sweet tooth

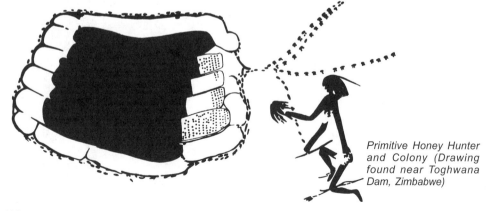

Primitive Honey Hunter and Colony (Drawing found near Toghwana Dam, Zimbabwe)

in ancient times as they do today, then we see why honey was so important in their lives. Honey is mentioned prominently in the Christian Bible, in the Koran and other early literature. Honey, in addition to being important as a sweet and in giving variety to the diet, was also used in many medicines and as a wound dressing.

In addition to honey, beeswax was used by the ancient peoples to make tablets, waterproof cloth, protect iron against rust, give light, and in dozens of other applications. Propolis collected from beehives was used as a wound dressing and in other medicinal applications because of its antibacterial properties (see HIVE PRODUCTS, Propolis). Last, it is obvious that honey bees have been admired by people in many civilizations because of their presumed cleanliness and industriousness. Even today this same admiration can be found; a skep appears on the state seal of Utah and a bank in New York City has long used a skep as its symbol of thrift.

Our first record of ancient people harvesting honey from a wild nest comes from a rock painting from Spain estimated to be over 8000 years old. Other ancient rock paintings showing bees in caves, and the harvesting of honey from them, are found in Africa.

We do not know when the early tribes began to build villages, husband animals and grow crops instead of hunting as nomads; however, this probably occurred in the vicinity of Egypt about 7000 to 10,000 years ago. The first positive records of the Egyptians keeping bees and harvesting honey come from three drawings from ancient tombs that are approximately 4400 and 3450 years old. The hives in one of the scenes looks very much like the mud hives kept in Egypt until recently. More information about beekeeping in ancient Egypt will be found under BEEKEEPING IN VARIOUS PARTS OF THE WORLD. Replicates of the Egyptian scenes and more about the archeology of ancient beekeeping are found in: Crane, E., *The Archeology of Beekeeping.* (Cornell University Press, Ithaca, NY. 360 pages. 1983.)

Following the Egyptians, some of the Greeks and Romans, whose civilizations flourished between 1700 and 2500 years ago, were also beekeepers. The islands of Crete and Cypress have been known as major honey-producing areas, as has the countryside around Athens and Rome. Even today there are large apiaries in all of these areas. While the people in the early Greek and Roman civilizations had among them some philosophers and other educated people, they were not especially interested in science and technology. As a result, medicine, agricultural sciences and other studies were in the hands of the poets, who wrote romantically about honey bees but whose accounts were not factual or based on personal observations.

The Dark Ages began in Europe in 476AD. This was a time when libraries were burned, education discouraged and almost the only people who could read and write were found in monasteries and religious enclaves. Monks and others in these retreats kept bees, harvested honey, made candles, and in northern Europe made honey wine (mead). However the techniques for keeping bees remained primitive.

Probably about 1000 years ago a new type of beekeeping, forest beekeeping, developed in Poland and the surrounding countries. (Barc is the Polish name for this type of beekeeping.)

Beekeepers selected large trees, those more than three feet (about a meter) in diameter, and made cavities in them that were 10 to 60 feet (three to 18 meters) above the ground. A board with a small entrance hole covered the hollowed out area. The inside of the cavity was smeared with beeswax and propolis that made the new cavities attractive and easier for the bees to find because of the odor. The honey was harvested from these barcs once a year. Bears were common in these northern European forests and elaborate methods of preventing bears from climbing the trees and stealing the honey were developed.

When The Middle Ages began beekeepers in northern Europe began to cut trees containing bee nests and to place them upright in an apiary. This was a major advance. During this time boxes of wood, cork and probably straw, skep-like objects, could be placed on top of these upright hives. If an opening was placed between the two the bees would store honey in these added boxes and skeps so that the honey could be harvested without killing the colonies. During this period round hives made from ropes of straw were developed, commonly called skeps. The chief advantage of skep beekeeping is that the hives are easily moved and collected together in an apiary where the beekeeper can capture swarms as they emerge and place them in new hives. To harvest the honey it was common to kill a number of these colonies, usually about half the original number. Burning sulfur was the usual method of killing colonies.

The first book in the English language on honey bees appeared in 1568 (Thomas Hyll, Londoner). The copy of this book in the Cornell University library has been rebound but the original paper, now over 400 years old, is in excellent condition and the text easy to read. The author states at the very beginning that the information given is gathered out of writings by ancient Greek and Roman authors. About this same time a great number of books on bees and beekeeping began to be published including the famous *The Feminine Monarchie* by Charles Butler in 1609 in which it is finally recognized that the queen is a female (the Greeks had called her a king) and that the drones were males. While all this may appear obvious

During the middle ages, and even after, making loud noises was believed to control swarms. It doesn't. (File photo)

History of Beekeeping

Skep beekeeping.

today, this was the beginning of scientific beekeeping. The important point is that at this time people began to determine the truth through experimentation and observation and to write it in a rational matter. Prior to this time authors had merely copied one another and written what they thought might be correct.

No honey bees were present in North or South America, Australia or New Zealand until these areas were settled by Europeans. The exact date and ship on which the first bees were carried to the U. S. is a subject of debate. One reference states that bees were brought to Virginia in 1622. However, by the 1640s and 1650s there are several references to the fact that bees were widespread and prospering along the east coast. Not too many years later swarms of honey bees were spreading west faster than the European settlers. The Indians called honey bees "white men's flies."

Beekeeping remained a cottage industry in North America for over 200

Gum hives remained in use in isolated areas in the U.S. well into the 1970s, when this photo was taken by Charles Divelbiss.

years. One person to write about more practical beekeeping was Moses Quinby (see QUINBY, MOSES) whose book, *Mysteries of Bee-keeping Explained*, appeared in 1853. Quinby began beekeeping in 1828 and he and L.L. Langstroth, who started as a beekeeper in 1837, revolutionized beekeeping everywhere in the world.

Late in 1851 Langstroth popularized the principle of bee space, the fact that the bees needed a walking space around and between combs in their nest. (see LANGSTROTH, LORENZO LORRAINE). His discovery was followed by the discovery of how to manufacture foundation in 1857, the construction of the first extractor in 1865 (see HRUSCHKA, MAJOR F.) and the invention of Quinby's bellows smoker in 1875.

By 1885 many beekeepers were operating several hundred colonies but by about 1890 the champion was W. L. Coggshall of Ithaca, N. Y., who owned and managed about 4000 colonies. Those who worked for Coggshall related that during the active season he and his crew would leave home early Monday morning with a team of horses, a wagon and the men on bicycles. At night they would sleep in farmer's barns. They would return home Saturday night. Each of Coggshall's apiaries contained 40 to 60 colonies. A building in each apiary was used for the storage of equipment. Each building also contained a four-frame reversible, hand-cranked extractor. Bees were brushed from individual combs of honey with the famous Coggshall bee brush made with long bristles, placed in supers on a wheelbarrow and wheeled to the building for extracting. Brushing bees is slow but an effective way of removing combs of honey. Coggshall produced mostly buckwheat honey. The honey was stored in wooden kegs that weighed over 100 pounds (45 kg) each. These were left in the buildings where the extracting was done and collected on bobsleds after the first snow.

Soon after Langstroth's discovery of bee space a great number of journals devoted exclusively to beekeeping appeared. Two that have continued down to the present time are the *American Bee Journal* that was first published in 1861 and is now published by Dadant and Sons, and *Bee Culture* (*Gleanings in Bee Culture*) that was first published by A.I. Root in 1873 and that has been continued by his company since that time.

A good understanding of the natural nest of the honey bee and the way brood, pollen and honey were stored in a hive came about during the second half of the 19th century. Methods of swarm control developed more slowly with contributions from several people (see

W.L. Coggshall

A few of Coggshall's honey houses still exist. Unique in design, these buildings were able to be dismantled in sections small enough to be carried on a wagon. This one was moved from an outyard some distance away, and brought to a home in Ithaca, New York. (Roger Morse photo)

SWARMING IN HONEY BEES, Demareeing). An important fact discovered was that one cause of swarming was congestion of the broodnest that could be relieved by adding a super. It also was understood that one must super in anticipation of the honey flow. The discoveries that young queens produce more eggs and those colonies were less likely to swarm, were also important contributions.

The development of gasoline engines and electric motors changed beekeeping radically. The first central extracting plant, that is, a large building to which supers of honey were carried for extracting, was built in 1937 in New York by N.L. Stevens of Venice Center. Within a few years many beekeepers built central extracting plants. The first automatic uncapping machines were developed in Ontario, Canada, in the 1920s.

The evolution in methods of handling and processing honey is ongoing. The Dakota Gunness uncapper (see HONEY PROCESSING), first advertised in *Gleanings in Bee Culture* in the March, 1988, issue, more than doubled the number of combs that could be uncapped in a day by one person. A three-man crew extracted over 80 barrels of honey in a 10-hour day using one Dakota Gunness uncapper and five extractors, more than double the number of barrels that could be extracted in the same time using the best vibrating knife.

Beekeeping in the U.S. was forever changed with the discoveries of chalkbrood disease in 1968; *Acarapis woodi*, the tracheal mite, in 1984; *Varroa destructor* in 1987; Africanized honey bees in 1990 and the small hive beetle, *Aethina tumida*, in 1998. In the latter half of the 1990s, resistance of *Paenibacillus larvae* subsp. *larvae* to oxytetracycline (Terramycin®) and fluvalinate (Apistan®) for the control of *V. destructor* was found in the U.S. for the first time. Then in the early 2000s resistance of *V. destructor* to coumaphos was demonstrated.

Those who wish to pursue the history of beekeeping further will find that the two books by Crane, E. *The Archeology of Beekeeping.* (Cornell University Press, Ithaca, NY. 360 pages. 1983) and Crane, E. *The World History of Beekeeping and Honey Hunting.* (Routledge, NY. 682 pages. 1999) are the most complete historical documents we have. F.C. Pellett's *History of American Beekeeping* (Collegiate Press, Ames, Iowa. 213 pages. 1938.) has a good record of what had taken place in North America up to that date. Other histories of beekeeping in specific areas around the world are cited in Crane's books.

HIVE CARRIERS – (see MECHANICAL HANDLING OF COLONIES)

HIVE PRODUCTS – Honey and beeswax are usually thought of as the primary products harvested from beehives. However, a number of other products have been collected and sold by beekeepers. In some cases these are sold as medicines or dietary supplements. In the U.S., the Food and Drug Administration requires that a manufacturer substantiate claims for any such products. Therefore, it is important that beekeepers who harvest and sell special products to proceed with caution. For more information on the products of the hive see HONEY; BEESWAX; POLLEN; PROPOLIS; ROYAL JELLY; VENOM.

HIVE STANDS – Hives should be raised several inches off the ground for many reasons. One is to protect the hive parts against rot. Bottom boards should not be treated with wood preservatives and thus the only protection the beekeeper has is to prevent them from contacting the ground (see Wood preservatives). Other reasons for elevating colonies include simply keeping the inside of the bottom boards dry and also raising them sufficiently high that grass does not choke the entrance. In areas where skunks are a pest, raising the hive helps control the damage.

Beekeepers have used a variety of hive stands ranging from bricks and cinder

A plastic version, with legs that can be raised by adding sections, a tray for ant protection and a bottom board can be included that's screened.

blocks to railroad ties and specially constructed wooden or metal hive stands. Cost can be a factor and can dictate what is used.

Those who overwinter in the north, and especially those who pack their colonies for winter, often build special hive stands or use pallets. The goal is to keep the bottom boards protected and dry and to create a dead air space under colonies. The size and shape of such a hive stand depends on the number of colonies in the pack, usually from one to four.

With the increased use of screened bottom boards using a hive stand raised off the ground becomes even more important, as there needs to be a space below the screen where hive debris can accumulate.

Another consideration is the bending and lifting required when working a colony and especially when adding and removing honey supers. Colonies low to the ground will require significant bending to open and examine frames. More agile beekeepers even spend time on their knees while working bees. This can be uncomfortable, or impossible, depending on the physical condition of the beekeeper.

A hive stand that raises the bottom

Hive stand made from treated 2" x 6" boards holds one colony, with room to set removed supers, or two colonies. Inexpensive, easy to make, low to the ground. Moderate life if treated.

Hive Stands

Higher off the ground, better pest protection, easier on the back and knees. Long lasting, more expensive.

hive body off the ground 18" - 24" makes a much more workable situation. The trade-off, however is the eventual height of the stacks of supers. If harvest occurs only once a season the supers may reach too high to comfortably remove when the top box is full. This situation can be solved by multiple harvests.

Commercial operations have perfected the pallet stand, with several types in use. Some are built such that there is a dead air space below to accommodate a screened opening. Others use a solid pallet surface for the bottom boards of the four colonies that sit on the pallet ready to be moved by forklifts..

Treated fence posts keep hives off the ground, as do discarded freight pallets. Both inexpensive, no construction and effective. Low to the ground, short life.

Hive Stands

Ants can be a problem, especially in tropical climates. But an oil-filled barrier will solve that problem.

Hive stand not needed when hives are in a protected location. Working from the rear of the colonies means bending and lifting and disturbance of the bees.

Hive Stands

The ultimate in pest protection. These are the alternatives to electric fences in bear country. Deter theft, too. (photo on right by Mike Hood)

Two ways to move colonies without lifting and straining. The commercial pallet (this one ready-made from Mann Lake Supply) is designed to accommodate four colonies exactly. The trailer can be moved with ease to pollination contracts or nectar flows. Both keep colonies off the ground and easy to move.

High enough to keep pests away, and make working these nucs an easy task was Steve Taber's goal. No kneeling, lifting or stretching required.

Hive Tools

HIVE TOOLS – A variety of tools have been used by beekeepers to pry apart as well as to scrape and clean hive parts and frames. In North America, one hive tool model, sold by all manufacturers, has been a favorite for many years. The tool is made of steel with the last inch of one end bent at a 90° angle to the rest of the piece. Both ends are sharpened. Eight- and 10-inch long hive tools are manufactured but most prefer the 10-inch hive tool. The Maxant® design of hive tool has become popular. Outside of North America a wide variety of hive tools are made and used by beekeepers.

A Teflon-coated hive tool, to reduce sticky wax and propolis from adhering to the surface.

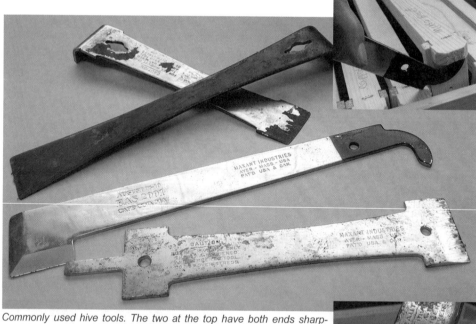

Commonly used hive tools. The two at the top have both ends sharpened for scraping and cleaning hive furniture. One end is bent at 90° to offer a smaller scraping end. This is often used for cleaning the rabbet in hive bodies and supers. This type of hive tool, especially the black tool, came into common use in 1905. Directly below these is the Maxant Frame Lifter model. This tool has one end with two sharpened edges, which makes inserting this end between propolized hive bodies much easier. The other end has what appears to be a hook. This end is used to lift frames rather than pry them up with a conventional hive tool (see inset). An adjacent frame is used as the pressure point, the hook inserted below the end bar of a frame and by pressing down on the handle, the frame is lifted up. The bottom tool, also made by Maxant, is specialized. One end is a sharpened scraper, wider than normal for efficient use. The other end (see insert) is used to clean the space between top bars of wax and propolis.

Examples of pre-moveable-frame hives.

HIVES, DEVELOPMENT OF – In the development of the hive, easier manipulation of the bee-built combs was sought for better bee management. Before the advent of the movable frame hive based on the bee space, comb manipulation was rather limited. With skeps and box hives, honey removal usually required destructive harvesting. Either the colony was killed and the combs cut from the hive, or the colony was heavily smoked and the honey combs were removed leaving a seriously traumatized colony. Some variations of the skep had extensions made of coiled straw. Top extensions were shaped like a cap and bottom ones were shaped like a ring. This arrangement let the colony expand, and by removing just the extension, a less destructive honey harvest was possible.

Some box hives had removable upper chambers where honey was stored, again allowing honey removal without hive destruction. Some wooden hives were even completely subdivided into boxes with each containing several combs. These boxes could be tiered up in a vertical arrangement or placed side-by-side in a horizontal arrangement. These hives allowed not only easier honey removal, but also some limited broodnest management. For example, the beekeeper could divide a colony into two parts by removing some of the lower brood-containing boxes. The queenless colony would rear another queen, resulting in two established colonies. Now the beekeeper could acquire new colonies without relying just on swarming. Each box had a small glass window, giving the beekeeper a limited interior view, enough to tell when it was full with honey or brood comb. It is interesting to note that before the advent of movable frame hives, the parts of the hive that could be moved were the comb-containing boxes. Movable frame hives were a vast improvement over hives that were subdivided into modular boxes.

Early skeps, skep beekeeping on a large scale, and a skep with a honey 'super.'

They allowed inspection and manipulation of individual combs instead of chambers filled with combs that were mostly hidden from view. It is also interesting to note that some hives were not subdivided into modular boxes but rather had properly spaced top-bars. From the top bars the bees built parallel and individually removable combs. These hives were developed independently in Greece and in Vietnam/China (see HIVES, TYPES OF). For an extensive description of the development of the modern hive see Crane, E. *The World History of Beekeeping and Honey Hunting*, (Routledge, NY, 682 pages. 1999.) – WAM

HIVES, TYPES OF – In developing hives for bees many beekeepers and inventors have paid little attention to what bees prefer if given a choice. In fact, it appears that many people were not aware that bees can make choices or that they have preferences. Studies have been done on the preference of bees and their natural nest. Seeley, T.D. *Honeybee Ecology*. (Princeton University Press, Princeton. 201 pages. 1985).

A typical nest in a tree, which is the ancestral home of the European honey bee, is elongate and cylindrical, reflecting the natural shape of a tree (see NATURAL NEST OF THE HONEY BEE). Because certain types of nest arrangements are clearly better than others, modern beekeepers, at least in the major honey producing areas of the world, have almost uniformly adopted the multistoried 10-frame Langstroth hive. Unfortunately, the industry is still, and probably always will be, plagued with those who want to invent a new hive. These people usually claim that their hive is improved or better because it takes into consideration some feature such as safety, ventilation, insulation, mite and pest control, or other factors. If one searches the old bee journals and books that have been published in the past several hundred years it will be discovered that just about every type of hive imaginable has been invented. In this section only some of the more common types of hives will be reviewed. The major American bee supply manufacturers sell standardized hives they sell and all of them offer the 10-frame Langstroth hive. However, the depth of the honey supers used varies and which is best will easily stimulate an argument among beekeepers.

Honey bee colonies prefer to move in an upward direction even though comb building in feral nests is either downward or horizontally. The brood is in a compact area below the pollen and honey stores in spite of the fact that bees prefer to keep forcing the brood nest upwards. Some beekeepers have experimented with having the brood nest on top of the hive and trying to force the bees to store honey under the brood. The thought behind having the brood nest above the honey and pollen was that it would make the brood nest easier to examine. Honey bees will not store honey in this manner. Hives have stopped foraging because there was no room for upward expansion despite the fact that the colonies had several empty boxes below.

At one time those who made hives, both commercially and at home, used the outside dimensions when writing about or making hives. However, lumber sizes, including thickness, have changed and today when describing how to make hives only inside dimensions are used. If these are followed carefully all of the equipment made will be interchangeable. It is important that standard dimensions be used in constructing hives. (see DIAGRAM page 000)

Ten-frame Langstroth hive – It has been stated that the depth of the 10-frame Langstroth hive was determined by the width of the board Langstroth had when he built his first practical bee hive. The other two dimensions originated from what he thought would be a convenient size box for him to carry. From his writings it seems that Langstroth understood the importance of allowing a brood nest to expand in an upward direction. Foundation was invented several years after Langstroth made his first hives and frames. The spacing between combs, and the number of combs per hive were the result of careful measurements made by him.

While many hives of different dimensions have been made, the 10-frame hive has remained the most popular, possibly because of Langstroth's extensive writing and the great influence he had on the development of the industry. Beekeepers in the U.S. and Canada use the 10-frame size. (For a further discussion of the importance of standardization see EQUIPMENT FOR BEEKEEPING.)

Eight-frame hive – Eight frame equipment has been commonly used in migratory pollination because a third more colonies can be placed on a pallet. (A commonly used pallet holds four 10-frame colonies, or six eight-frame colonies.) Thus, a truck load of pallets can haul more colonies in a trip, reducing the cost of transportation.

Hobby and sideline beekeepers, also, enjoy eight-frame colonies because of their lighter weight and reduced storage requirements. Swarming, wintering and honey production management, requires providing enough room for bees and food storage at the right time.

Dadant hive – For a number of years Dadant and Sons made and sold a hive that was deeper, held 11 frames and had a spacing slightly wider than a 10-frame Langstroth hive. Dadant hives were developed after extensive testing in their apiaries and after comments by beekeepers. The wider spacing was thought to give more clustering space to make it easier for the bees to ventilate the hive. Eventually, the Dadant hives were rejected because they were so large that supers full of honey, or even boxes just full of brood and bees, were too heavy to lift.

Deep hive - Some beekeepers have made hives holding frames that are 15 to 20 inches (38 to 50 cm) deep. The thought behind using deep frames is that it might allow the bees to have a more compact brood nest. However no one has ever shown that using two brood chambers with the bee space between top and bottom box has any serious effect on the colony.

Top bar hives or long hives – In these hives, combs are built attached from wooden top bars, lacking the bottom and side supports of standard frames.

An example of a deep hive frame

Nevertheless, individual combs can still be removed from the hive. These hives were described by people traveling through Greece in the 1600s. In that version of the hive, the top-bar combs were housed in a round wicker basket that was wider at the top causing the sides to slope. The slope is thought to reduce attachments between the combs and the inside walls of the hive. Top-bar hives also developed independently in Vietnam/China using upright hollow logs to protect the combs. These hives were used with the Asian hive bee (*Apis cerana*) that does not attach comb to the vertical sides. In the 1960s, top-bar hives were adapted for use in Africa where beekeepers traditionally used horizontal fixed comb hives that greatly limited any bee management. (Hence the hive is sometimes called a Kenya top-bar hive.)

Although the combs are more fragile than those supported by frames, the

Top bar hives can be an interesting addition for study in an apiary, or the main type of hive used. The shape of the comb conforms to the slope of the hive's walls. They remain somewhat fragile, even after some time, but with care can be handled and the colonies managed.

An extractor designed to extract honey from top bar frames. Produced by Swienty Bee Supply Company.

brood combs become stronger with age and can be carefully handled with minimal breakage. Newly-built honeycomb is particularly fragile and must be handled with care. Honey is usually squeezed or pressed from the comb, however, an extractor has been especially developed for these fragile combs so the wax infrastructure can be saved for reuse by the bees.

Today's typical top-bar hive contains about 20 combs in a horizontal arrangement. It is thought that the horizontal expansion of the brood nest slows the colony's development somewhat compared to a vertical expansion encouraged with supers used in a standard hive. This horizontal arrangement makes a full hive awkward to lift, reducing it mainly to non-migratory operations. Top-bar hives have been successfully used in developing countries, especially because their cost is less than modern frame hives, making them affordable to subsistence farmers. Since top-bar combs are removable, bee management is possible, a great improvement over traditional fixed-comb hives. Top-bar hives are simple to build and can be made from locally available materials (see HIVES, DEVELOPMENT OF). -WAM

Warm-way, cold-way construction – An example of human thinking, and not asking the bees what they prefer, is seen when one asks if combs in a hive should be perpendicular or parallel to the hive entrance. Proponents of the warm-way construction say the combs should be parallel to the entrance so that cold winter winds will not blow through the hive; presumably the first comb deters drafts. Proponents of the cold-way say in nature honey bees have no preference and build comb in various directions. Therefore the orientation of the comb to the colony entrance is not an important consideration for honey bee survival in cold climates.

Other than wooden hives – Although wood has been the traditional material for modern hives, in some parts of the world wood is scarce and therefore expensive. Furthermore, in some areas wood is subject to attack from various insects, notably termites, and other animals. In a search for other suitable materials for hives beekeepers have tried various forms of plastics as well as concrete.

Concrete may seem a strange choice but it is a cheap material and not subject to termite attack. Bottom boards of concrete have proven successful, especially for non-migratory beekeeping. Hive bodies can be made of concrete and are not significantly heavier than some of the wood used in tropical countries. However concrete does not "breathe" as wood does and moisture within the hive can be a problem.

Fiberglass hives appeared a number of years ago but these hives proved unsatisfactory. The bottom boards, inner and outer covers warped in a fairly short time. In addition the fiberglass was translucent and permitted dim light in the brood chambers. Therefore the outer frames were not used by the bees and the brood nest was restricted. Like other synthetic materials fiberglass does not "breathe." Fiberglass parts are slippery until coated with some propolis and wax by the bees.

More recently, hives have been made of polystyrene foam. This material is cheap and easily made into hive bodies, tops and bottoms, and even feeders. These hives have become popular in parts of the world where wood is scarce and expensive. These hives are light in weight, but the polystyrene foam is fragile and does not withstand abuse. Therefore the polystyrene foam hives are only suitable for a stationary apiary. A number of bee equipment suppliers are offering the polystyrene foam hives.

Plastic supers, frames, feeders, inner covers, covers, bottom boards and even fully drawn plastic comb equipment are available, inexpensive, durable and in common use. With appropriate management (i.e. applying beeswax to plastic foundation), honey bees successfully use plastic equipment, and beekeepers benefit from the low maintenance, long life and low cost.

Other materials, such as reeds and even corn stalks, have been tried but interest in them has been short-lived. In some countries, particularly Africa, beginning beekeepers have used hives constructed of cardboard covered with mud or a mud-cattle dung plaster. These hives have a short lifetime, usually disintegrating during the rainy season. -AWH

Plastic supers are available, as are many types of all plastic, one-piece frames and foundation, plastic foundation, and feeders. These are typical examples of frames and foundation.

HOARDING BY HONEY BEES – Beekeeping is a practical enterprise for several reasons of which three are most important: Honey bees naturally hoard surplus food, they separate their brood and their food, and they store their food above the brood where the surplus may be easily harvested. While hoarding is a natural instinct, and necessary for the bees' survival during cold or otherwise inclement weather, bees have certain preferences concerning how and when they will hoard a large quantity of food.

Honey bees prefer to store honey in old comb versus new comb (see HONEY, Color). Comb that has been used previously for honey storage, even though it has not contained brood, is preferred over totally new comb. Bees prefer to place honey in worker cells rather than drone cells. The presence of an active, laying queen also stimulates bees to forage and hoard. Most important, having an abundance of storage space encourages bees to hoard more food. It is for this last reason that beekeepers are encouraged to add a number of honey storage supers well in advance of the time they are needed (see SUPERING COLONIES).

HOFFMAN, JULIUS (written by E.R. Root) – Julius Hoffman, the inventor of the frame bearing his name (see FRAMES) was born in Grottkau, province of Silesia, Prussia, 1838, only a few miles from where Dr. Dzierzon spent much of his life. Indeed when a young man he visited the doctor, and from this great bee-master he learned much. No wonder with such a Gamaliel for an instructor the name Hoffman has come to be known in almost every bee-man's home.

In 1862 Mr. Hoffman left his native country and settled in London, England. Four years later he came to America, settling in Brooklyn, New York, where he was employed in the organ and piano business, but bees he had to have, for his heart and soul were in the bee business. As there was hardly room in such a crowded city to keep many bees, he moved to Fort Plain, New York, where he settled in the spring of 1873. In a few years he had increased his stock of bees up to 400 colonies and then to something over 700, all of them on Hoffman frames, and in his quiet way had used them for 10 or 12 years. He could handle two or three times as many

Hoarding

Julius Hoffman

colonies, with the same labor, as he could on the old-style unspaced Langstroth frames, and by way of proof he showed how he could handle such frames in lots of twos, threes, and fours, picking them all up at once; how he could slide them from one side of the hive to the other; how easily he could handle his colonies in halves or quarters instead of one frame at a time.

At this time (1890) nothing but the unspaced Langstroth frame was used by modern beekeepers, except, perhaps, the closed-end Quinby in certain portions of New York; and the idea of a self-spacing frame seemed to be utterly impractical in the mind of the average beekeeper; but so thoroughly impressed was I with the importance and the value of the Hoffman self spacing frame as a labor-saver that I came home and wrote it up for our journal, *Gleanings in Bee Culture*. We soon adopted it in our yards; and my own personal experience with it showed it could be handled rapidly and easily; and that for many operations in the apiary much valuable time could be saved with it over the handling of the old-style unspaced frames. The beekeeping world, not sharing in my enthusiasm took hold of the Hoffman slowly at first; but by 1902, it came to be the leader in nearly all the hives put out by the principal manufacturers in the United States. It should be stated, perhaps, that while the Hoffman frame of today differs in detail from the original frame the main self-spacing feature has been retained. -E.R. Root

HOFFMAN SELF-SPACING FRAMES

– From a very early edition of ABC:
By these are meant frames held apart at certain uniform distances by some sort of spacing device, forming either a part of the frame or a part of the hive. Under Spacing of Frames, Bee Space, and Extracting, the distances that frames should be apart are discussed. Some prefer a spacing distance of 1-1/2 inches from center to center, but the majority prefer 1-3/8 inches.

Self-spacing frames may be defined as those which are spaced automatically either 1-3/8 or 1-1/2 inches from center to center when they are put into the hive. Loose or unspaced frames differ in that they have no spacing device connected with them and are spaced by eye when placed in the hive – or, as some have termed it, "by guesswork". Such spacing results in more or less uneven combs. Beginners, as a rule, make very poor work of it. The users of self-spacing frames get even, perfect combs with comparatively few burr combs and the combs are spaced accurately and equally. Self or automatic spaced frames are always ready for moving either to an out-yard, to and from a cellar, or for ordinary carrying around the apiary. Unspaced frames, on the contrary, while never spaced exactly, can seldom be hauled to an outapiary over rough roads without having some means of holding them in place.

Self-spacing frames can be handled more rapidly. On the other hand, a few who are using the unspaced frame use as an objection that the

Photo at left shows hoarding. The honey bee's tendency to hoard is apparent in this apiary.

Hoffman Self-Spacing Frames

A Hoffman self-spacing style frame. Note that one side of the end bar is flat, while the other is beveled. This reduces the contact between them, and thus the amount of propolis the bees will use on the sides of frames.

self spacers kill the bees. This depends upon the operator, who may kill a good many bees if he is careless. If he uses a little care and patience, applying a whiff or two of smoke between the parts of the frames that come in contact, he will not kill any bees. The myth that self-spacing frames are hard to handle and crush bees is disproved by the fact that they are now in universal use throughout the country.

R.O.B. Manley, in giving at least part of the credit to the inventive American beekeeper for the self-spacing Hoffman frame, hastens to call our attention to an 1879 issue of the *British Bee Journal* which published drawing of frames that had the essential features of the modern Hoffman frame except for the beveled spacing surfaces.

A.I. Root related his experiences in America in an earlier *ABC and XYZ of Bee Culture*: "Knowing that some beekeepers in New York were using self-spacing frames with projecting end bars, in 1890 the author took a trip among the beekeepers of that state. I found, as expected, that all the combs in self-spacing frames were uniform in thickness and spaced exactly right. Contrary to what I had expected, far from being more trouble to handle, they were very much easier and took less time. The hives were always ready to move to and from outyards without fastening the frames.

Of the two self-spacing frames, Quinby and Hoffman, we preferred the latter because with a

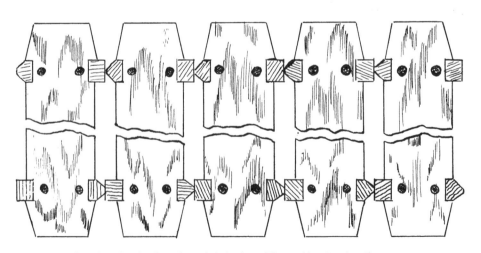

Drawing showing how the pointed edge of the end bar touches the square edge reducing propolis accumulations.

slight modification it could be used in Langstroth hives already in use.

What was more, the Hoffman frame was proof against bad or irregular spacing of frames. The frames could be handled in groups of twos and threes. After the first frame was removed all the rest could be shoved in a hive body or one by one.

When through, all frames could be shoved with one push exactly in place and the right distance apart.

As the result of many tests at our beeyards covering a period of several seasons, we recommended that beekeepers adopt the Hoffman. After some protests and objections the Hoffman frame slowly crept into use. At the present time it is in universal use throughout the country."

The Hoffman is the most extensively used self-spacing frame in all the United States, and there is even a possibility that it is used more generally than any other frame whether spaced or unspaced. All of the hive manufacturers supply it as a part of the regular equipment of their standard hives. (see FRAMES for more information)

HONEY AND HONEY PRODUCTS

Honey– Honey bees make honey from nectar, a plant product. Nectar is a mixture of water, sugars, plant pigments, and related materials that serve to attract bees to flowers. Most nectars have strong odors to aid the bees in finding the flowers. The change from nectar to honey involves the removal of water and the addition of two enzymes that change the major sugars in the nectar (see NECTAR, conversion to honey). For a comprehensive treatise on honey see Crane, Eva. *Honey A Comprehensive Survey*, (William Heinemann, Ltd, 608 pp, 1976.)

The National Honey Board approved a definition of honey. "Honey is the substance made when the nectar and sweet deposits from plants are gathered, modified and stored in the honeycomb by honey bees. The definition of honey stipulates a pure product that does not allow for the addition of any other substance. This includes, but is not limited to, water or other sweeteners." More formal and legally binding descriptions of honey are available, and honey continues to be more closely defined. This is due primarily as protection by the honey industry against products that claim to be honey, or honey-like, and take a very limited sweetener market share. As this work was being prepared, an international standard definition for honey was being prepared. Once this has been completed, honey everywhere will be able to be defined without argument.

Since honey is a natural product, its composition is highly variable. Honey, to be safe from fermentation, should contain less than 18.6 percent moisture. The predominant sugars, which are present in about equal quantities, are glucose and fructose; most honey contains about one percent sucrose, but may contain more. Honey is an acidic food with a pH of about 3.9. The flavor and odor of honey are derived from the plant pigments and other materials that are secreted with the nectar. Honey from each floral source is unique in color, odor and flavor, just as flowers themselves are unique.

Most honeys on the market are blends from several floral sources. Sometimes, when one plant dominates an area either naturally or because of compact plantings, it is possible to harvest a honey that may be almost 99 percent from one source (monofloral honey). An example of monofloral honey is citrus honey, which may be produced in great quantity because of the extensive plantings of citrus in small areas and because the plant produces such great quantities of nectar. However, honey bees do not separate honeys from different sources when they are stored in the hive. Honey packers blend honeys to give the consumer a consistent product relative to color and flavor over time.

Caloric value – A calorie is a unit to measure the amount of energy produced by the food we eat. A tablespoon of honey produces 64 calories.

Chemical and physical properties of honey – Beekeepers in the U.S. usually have not graded their honey for sale at either the wholesale or retail level. When honey moves into the world market more emphasis is placed on honey grading. The European Union has set standards for honey imported into those countries. The grading of extracted honey considers factors such as color, clarity, absence of defects, moisture content, flavor and aroma.

One measurable substance is the quantity of hydroxymethyfurfuraldehyde (HMF) present. This is produced as fructose breaks down. The production of HMF is always taking place but is accelerated by high temperatures. High levels of HMF will be found if the honey is heated too high, or if barrels filled with honey sit in the sun for too long. Some honeys, however, have a relatively high level of HMF produced naturally. Some citrus honeys fall into this category, and exporting citrus honey to European markets carries the additional burden of proving that even the freshest honey was not overheated in processing or storage prior to shipment.

Another test is for diastase, an enzyme that is present in honey and destroyed by high temperatures; a good quality honey is one that is high in diastase.

Acidity of honey – It is not widely understood that honey is an acid. Insofar as its reaction with other products and metals. The pH of honey ranges from 3.42, fairly acid, to 6.10, approaching neutral.

A number of years ago the late Professor E.J. Anderson and a colleague at Pennsylvania State University placed pieces of weighed black iron, galvanized iron, tin, copper, and aluminum in honey for varying periods of time and temperatures. They reported that all of

Storing honey outside in hot weather will overheat the contents of these barrels, and increase the HMF reading of the product.

Honey And Honey Products

Only stainless steel containers should be used for storing honey.

these metals lost weight but they did not publish all of their data. The black iron lost the most weight, more than 200 times the amount lost by aluminum, which lost the least.

Some beekeepers have observed pitting in aluminum containers used for processing honey indicating a reaction between the honey and the aluminum. It is not uncommon to find galvanized extractors that will have some raw iron exposed. Only stainless steel should be used for honey storage and processing equipment.

Color – Honeys can vary in color from water white to dark amber with the lighter colored honeys generally preferred by most consumers. However, this varies greatly from area to area. Dark-colored honeys, such as those produced from buckwheat, are popular in the regions where the plants are grown.

All honeys darken and may become thin and watery in time. Despite popular belief, honey does not have an indefinite life. The proteins in honey break down and the sugars polymerize (join each other) and in the process split out water. How rapidly honeys darken is a function of temperature and light. At room temperature most honeys become very dark after 10 to 20 years of storage. Sunlight will hasten the darkening. Comb honey kept in a freezer will retain its taste and color qualities for a greater length of time than extracted honey. Stories about honey taken from ancient tombs and being palatable are not correct, although the substance is identifiable as honey.

Honey color can also be influenced by the age of the combs. Beekeepers have debated the virtues of using old and new comb for honey storage for decades. Some feel that only new, white comb should be used in honey storage supers. If a piece of old comb is placed in a jar of water at room temperature the color of the water will darken in about an hour from the coloring materials in the comb, such as pollen and propolis. (see HOARDING BY HONEY BEES).

Professor G.F. Townsend undertook an experiment to determine how much honey color was affected by comb color. To perform these tests he cut combs from frames, broke them up and submerged pieces in water and in honey with various moisture contents. He observed that the water and the honeys high in water content darkened considerably. However, "Once the honey has reached full ripeness – approximately 83 percent total solids – relatively little colour is picked up from the comb." (Townsend, G.F. *Absorption of colour by honey solutions from brood comb*. Bee World 55: 26-28. 1974).

Most of the honey marketed in the supermarkets of the U. S. comes from alfalfa and the clovers that yield light-colored honeys with a mild flavor. Varietal honeys, in many colors, have a rising popularity.

Occasionally a beekeeper may come across a honey that has a distinctive color. There are several sources of these unusual colors. Honey in sealed brood comb may sometimes take on a blue-gray soapy color if the comb is exposed to cold temperatures. During periods of dearth bees may collect nectar substitutes. Bees have been seen collecting and storing small amounts of flavored drinks that make honey comb look like a stained glass window for a short period of time. The man-made "nectar" sources are usually avoided by the bees when a real nectar flow resumes. Colored honey can also originate from plant sources. A pale green honey found in much of California is thought to come from morning glories or bindweed flowers (*Convolvulus* sp.). The coastal plain of North Carolina produces a honey which has a very distinctive blue to purplish color. Tests at N. C. State University have shown that this honey is produced by sourwood trees (*Oxydendrum arboreum*) growing in the aluminum-rich, sandy soils of the state's coastal plain. The nectar from the sourwood blossoms is colorless but the processing by the bees can change the color of the stored honey to a distinctively blue to purplish color that is in sharp contrast to the typical water-white color of most sourwood honey. -JTA & RAM

Color grading – Honey is usually sold by color grade, since color often indicates a difference in flavor. Lighter honeys are almost always milder in flavor and are more in demand as table honey, while dark honeys are usually used in the baking trade. One instrument used in the commercial trade to determine honey color was the Pfund grader developed by Dr. A.H. Pfund in 1925. (Fell, R.D. "The color grading of honey." *American Bee Journal*, 118: 782-783, 789. 1978.) The Pfund grader compared the color of a particular honey with that of a colored glass to give a reading ranging from one to 140, as shown below.

Color Grades	Pfund Color Range
water white	8 or less
extra white	9 – 17
white	18 – 34
extra light amber	35 – 50
light amber	51 – 85
amber	86 – 114
dark amber	over 114

Today the Pfund grader has been replaced by two new instruments, the

A Pfund grader.

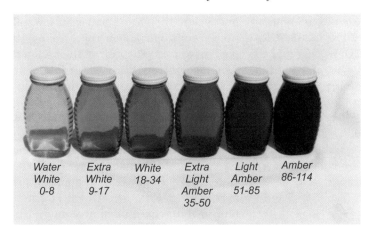

| Water White 0-8 | Extra White 9-17 | White 18-34 | Extra Light Amber 35-50 | Light Amber 51-85 | Amber 86-114 |

Lovibond® grader and the electronic digital Hanna Color Analyzer®. However these instruments use the Pfund scale.

Osmotic pressure of honey – (see NECTAR, COVERSION TO HONEY)

USDA grades – The U.S. Department of Agriculture has grading standards for 155 agricultural products, including extracted honey and comb honey. The grades for extracted honey are contained in an 11-page brochure entitled *United States Standards for Grades of Extracted Honey Effective May 23, 1985,* and available from the Chief, Processed Products Branch, Fruit and Vegetable Division, AMS, USDA, Washington, D.C. 20250. The brochure begins with definitions of terms followed by a table listing the seven color designations for liquid honey. Grades A and B must be honey with 81.4 percent or more of solids (moisture content of 18.6 percent or less). Grade C honey may contain 20 percent or less water. Grade D is substandard and has more moisture. Other considerations include: absence of bits of propolis, beeswax, clumps of pollen; good flavor and aroma; free of a taste of fermentation or burning; and freedom from air bubbles that might cloud the honey. The degree to which a sample might be crystallized is not a consideration though this can have an

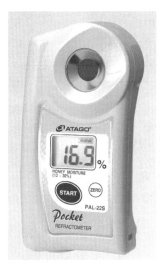

The new digital refractometer. Efficient, accurate and quick (though not perfectly easy to clean).

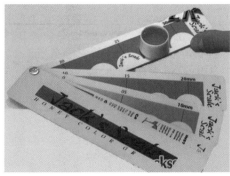

A new color grader. It is easy to use, pocket size and reasonably priced.

effect on the moisture content of the liquid fraction (see DYCE PROCESS FOR MAKING CRYSTALLIZED HONEY). The grades for honey in the comb are contained in a nine page brochure entitled *United States Standards for Grades of Comb Honey Effective May 24, 1967*. These standards are available from the same address as above. There are four color grades for comb honey: White, light amber, amber, and dark amber.

Crystallization – (see DYCE PROCESS FOR MAKING CRYSTALLIZED HONEY)

Fermentation of Honey – All nectar contains microscopic yeast cells that belong to the Genus *Zygosaccharomyces*. These are called osmophilic yeasts and can grow only in rich sugar solutions containing between 30 to 80 percent sugar. They are different from the yeasts that are used in cooking and to make alcoholic beverages. These yeast cells may cause the fermentation of diluted honey but they are inactive in normal honey containing less than about 19 percent water. To protect their stored food bees ripen honey as rapidly as possible thus preventing fermentation. Fermentation may also occur when the glucose fraction of the honey crystallizes leaving the fructose in solution with higher water content.

Heating – Honey can be damaged by overheating. Fructose, which is about half of the sugars in normal honey, is especially delicate and burns easily. When honey is used in baked goods it is fructose that gives the cake, cookie or bread a special golden brown color that enhances its appearance. However, this tendency to burn easily, unlike sucrose and glucose, also indicates that honey is a delicate product.

A second reason that honey is easily damaged by overheating is because of the proteins it contains. This is especially true of the honeys that are dark because they contain more proteins from the plant source.

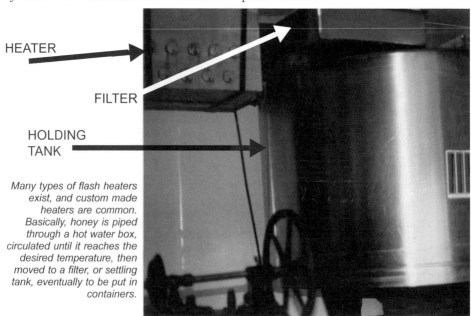

HEATER

FILTER

HOLDING TANK

Many types of flash heaters exist, and custom made heaters are common. Basically, honey is piped through a hot water box, circulated until it reaches the desired temperature, then moved to a filter, or settling tank, eventually to be put in containers.

Honey And Honey Products

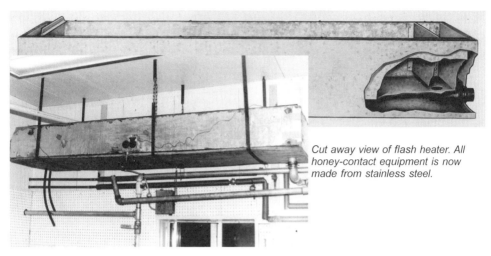

Cut away view of flash heater. All honey-contact equipment is now made from stainless steel.

When honey is heated in any stage of processing, or liquefying crystallized honey, heating should be done rapidly and the honey should be kept at a high temperature for as brief a period of time as is possible (see HONEY, Extracted pasteurization). Various types of flash heaters are available.

In the case of crystallized honey in large containers, the ideal method of liquefying is to heat it in a hot room only enough that it will flow from the containers. The temperature should be raised only long enough to liquefy the crystals and kill the yeasts.

Cautions: Honey should not be heated in a steam kettle. Steam-heated kettles, used to make jams and jellies are not satisfactory. The honey that is in contact with the side of the kettle, which is

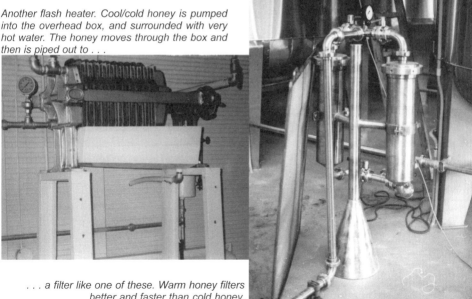

Another flash heater. Cool/cold honey is pumped into the overhead box, and surrounded with very hot water. The honey moves through the box and then is piped out to . . .

. . . a filter like one of these. Warm honey filters better and faster than cold honey.

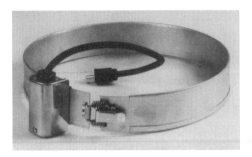

Smaller operations may choose band heaters. Great care must be used with these, as the honey directly inside the pail or barrel, or even extractor, may overheat while the honey in the center of the container remains cool.

surrounded by steam, will burn. Honey may be heated in a hot-water-jacketed tank. The maximum temperature for the water should be 180°F (82° C) (lower in the case of dark honeys). The honey should be stirred carefully while it is being heated.

Honey in drums should not be heated with wrap-around electrically-heated belts. They heat the honey in contact with the sidewalls too much for too long a time, and the honey in close contact with the drum's walls will burn. These wrap-around belts work very well to liquefy crystallized honey in a drum but they can cause damage.

Any heating of honey must be at a controlled temperature for a controlled length of time. Honey is a delicate product that should be handled with great care to deliver to the customer the finest quality product.

Moisture content – Honey with more than about 19 percent water will ferment if it is not pasteurized. Moisture content affects the way in which liquid honey flows and spreads. Crystallized honey must have the correct moisture content to spread properly.

Prior to World War I moisture content was measured in terms of pounds per gallon. Even though this was the official system it was not closely heeded. Following World War I, hydrometers were developed. In the U.S. an 11-inch long Baumé hydrometer was designed. To measure the moisture content about one half-pound (about one-quarter kilogram) of honey was poured into a tall glass cylinder. The hydrometer was placed in the cylinder and allowed to sink to a level depending upon the moisture content of the honey. The moisture content was then read on the stem of the hydrometer.

The Canadian Research Council hired Dr. H. D. Chataway to standardize hydrometer readings with temperature and to investigate the use of honey refractometers (see HONEY, Specific gravity). Her work was published in

Maintaining a constant warm temperature is easier to accomplish if a water jacket surrounds the inner tank containing the honey.

1935. The early refractometers, even though they required only a few drops of honey to make a moisture determination, were costly and bulky. Following World War II Bausch and Lomb Optical Company developed the

A typical refractometer. Many models are available, in a range of prices. Most new models are easy to use, and calibration is either set by temperature scale or simple adjustments.

first hand-held refractometers that are now widely used. Only a drop of honey is required to make a moisture determination. These refractometers have a built-in thermometer for temperature correction. The reading is accurate to 0.1 percent moisture for honeys between about 12 and 27 percent moisture. This history is reviewed by Dyce who was active in urging that standard methods be developed (Dyce, E. J. *Equipment for determining moisture content of honey.* Gleanings in Bee Culture 79: 342-343. 1951.).

Today several types of simple refractometers are available. However, the accuracy of these may be questionable. The electronic digital refractometer is now replacing the previous hand-held instrument. The temperature correction is automatic as well as the calibration.

Humectancy – (see below)

Hygroscopicity – Fructose has a peculiar property in that it readily absorbs moisture when in moderately high humidity. Since honeys vary in the percentage of fructose they contain each one has a different relative humidity at which no gain or loss takes place. An unfortunate feature of honey is that it may pick up moisture at its surface and this water diffuses or moves into the rest of the honey slowly. Thus, at the surface, in an open container, it is possible to have honey with a high moisture content whereas the average water concentration of the whole container may be relatively low.

Just as honey may pick up moisture under high humidity it may lose a considerable amount of it at low humidity. However, at low humidity a dry film usually forms over the top of

A handheld device for reliquifying and extracting heather honey. (Howard photo)

the honey in an open container that slows the loss of water. In actual fact honey loses water far more slowly than it picks it up. For long-term storage it is therefore important that honey be kept in tightly-closed storage containers.

In an open container honey may pick up moisture at the surface and fermentation might take place there. Even honey that has been heated to kill wild yeasts, and is later exposed, may pick up both moisture and yeast cells from the air and as a result may ferment.

Hygroscopicity helps to keep baked products made with honey moist. Even after baking the fructose retains its character of absorbing moisture. As a result honey is often used for some baked products to gain additional shelf life.

A product that is humectant is one that promotes the retention of moisture. In the honey industry the term hygroscopicity is used to describe this feature of honey, but in the food industry the term humectancy is more popular.

Specific gravity – Specific gravity refers to the mass of a given volume of a substance in relation to the same volume of water. The specific gravity of honey containing 18.6 percent water at 68°F (20°C) is 1.4129. At this specific gravity there are 11 pounds, 12 ounces of honey per gallon, in metric terms 1412.9 g per liter. The specific gravity of honey is seldom referred to or used today.

Thixotrophy – A small number of honeys are thixotrophic. This means that in the comb they act like a solid gel but if stirred act like a liquid, but return to the gel form when stirring is stopped. Non-drip house paint is an example of a thixotrophic material. The best known thixotropic honey is heather honey (ling) from *Calluna vulgaris* that is produced in several parts of Europe (see HONEY PLANTS, Heather). Honey from heath (or heather), *Erica sp.*, is not thixotrophic. Honey from Manuka trees honey, from New Zealand, is another thixotrophic honey. Certain proteins give the honeys that quality. Heather and a few other

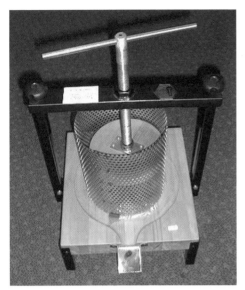

A press used to extract heather honey. (Michael Young photo)

honeys are high in protein (0.2 to 1.86 percent) but most honeys contain only trace amounts of protein.

Extracting thixotropic honeys can be difficult. It is possible to press the honey from the comb, thus destroying it though the wax is salvaged. Another method is to stir the honey in each cell with a gadget called a honey loosener. It is basically a board with many nails spaced the distance of one cell apart. The loosener is inserted into cells and vibrated thus causing the honey to become liquid so that it may be extracted.

HONEY BEE, DEFINITION OF – Honey bees are animals that belong to a group called insects. Like all insects they have three pairs of legs and three body parts – head, thorax, and abdomen. The head carries most of the sensory organs – eyes, antennae and mouthparts. The thorax has a bundle of muscles to which the legs and wings are attached. The abdomen contains the digestive and reproductive systems, and, in the case of the females, the chief defense weapon, the sting.

The members of the colony are the drones (males) and the two female castes, the queen and the workers. Drones are often mistakenly thought of as a caste (see CASTES). The sole function of drones is to mate. The function of the queen is to lay eggs. Queens produce chemical substances, pheromones, by which they are recognized and that affect social order. The queen makes no decisions in the colony other than to determine if a cell is suitable to receive an egg. The worker honey bees, all females, do all of the work and make whatever decisions are necessary for the colony to function.

No honey bee, drone, queen or worker, can live alone. They are social animals and as such cannot function unless a minimum number of bees are present in the unit. Food exchange among members of the community is part of the social system and it appears that information about the status of the colony, including the presence of the queen and the quantity of food available, is conveyed in this way. While a queen and a few hundred bees may be a sufficiently large number to survive under laboratory conditions, or in a queen-mating nucleus in summer, many more bees are required to survive a winter or a prolonged drought. The minimum number is approximately 10,000 to 15,000 worker bees and a queen. Males are not necessary for a colony to survive under harsh conditions. However drones are responsible for the morale of the colony.

No honey bees are native to North or South America, New Zealand, or Australia. The honey bees in North

America come from Europe. Both the European and African honey bees, which are the same species, have the scientific name *Apis mellifera*.

Apis is the name of the genus and is used to designate a group of closely related animals. The second part of the name, *mellifera*, is the specific (species) name. Our system for naming plants and animals calls for the use of two names (binomial nomenclature). The name of the genus is capitalized and the species name is in lower case.

People who name plants and animals agree that those within the same species can interbreed freely and produce normal offspring. For example, the donkey and the horse, animals of different species, can mate, but their offspring, the mule, is sterile and cannot reproduce. Because mules are sterile, the horse and the donkey are considered different species. Most animals from different species show no sexual interest in one another.

The Asian honey bees that belong to the Genus *Apis* are considered to be different species (see *Apis cerana*, *Apis dorsata*, *Apis florea*, and *Apis laboriosa*).

Asian honey bees have not been studied as extensively as European honey bees. While the biology of all *Apis* species is similar, they have different behaviors. In addition, the pests, predators, and diseases that affect each species are often different. (see BEES, DEFINITION OF).

HONEY BEE, HOW TO SPELL – Honey bee may seem to be spelled as two words (honey bee), one word (honeybee), or a hyphenated word (honey-bee). The use of two words for honey bee follows the recommendations of the Committee on Common Names of Insects of the Entomological Society of America. This latter organization has 9000 members and publishes the leading entomological journals in the country. The logic of the committee is that if a common name is used in a way that is systematically correct, then it should be spelled as two words. Since the honey bee is a true bee, two words are used. Bumble bees, sweat bees, leafcutter bees and alkali bees are all true bees and are spelled as two words. Some other correct spellings for insects under this rule are: house fly, face fly, lady beetle, dung beetle, boll weevil, and gypsy moth. Dragonflies and butterflies are not true flies, therefore their names are one word.

HONEY BEE TRACHEAL MITE – (see MITES ASSOCIATED WITH HONEY BEES, MITES, PARASITIC)

HONEY COOKERY – Honey is a versatile cooking ingredient. Sugar imparts only sweetness in a recipe. Honey, on the other hand, not only gives sweetness, but also contributes a subtle flavor that enhances the particular dish.

The mild clover, alfalfa and wildflower honeys are suitable for all recipes. The orange blossom or citrus honeys, with their intense floral aroma and flavor, are particularly delicious with fruits. The stronger, more robust honeies can be used in barbecue sauces and marinades for meats.

Using honey in baked goods has several advantages over using sugar. Honey creams with shortening rapidly, giving a smooth fluffy mixture. Honey produces an attractive, deep brown crust on breads. Since honey absorbs moisture from the air, baked goods remain moist and fresh much longer than when cooked with sugar. Honey cookies are generally soft and chewy. The only disadvantage in using honey in baking is a tendency for easier burning of the crust. However, reducing oven temperatures by 25°F (15°C) from what is called for in recipes will usually prevent scorching.

Although many honey recipe books are available, cooks frequently substitute honey for sugar in favorite recipes. The most important fact to remember is that honey is about 18.6 percent water, the rest being solids. This amount of liquid can change the delicate balance of ingredients in cakes and sweet quick breads. The liquid, which can be milk, water, or fruit juices, in such recipes will need to be reduced, keeping the following proportions:

Substituting Honey For Sugar

If the recipe calls for:	Then reduce the liquid portion by:
8 oz of honey	3 tablespoons
250 ml of honey	45 ml
1,000 g of honey	186 grams (g)

Honey can be substituted for sugar, measure for measure, in meat sauces, such as marinades, barbecue sauces and glazes; in salad dressings; in vegetable sauces, such as sweet-sour sauce and glazes; in all yeast breads; and in beverages of all kinds.

Crystallized or creamed honey can be used interchangeably with liquid honey. However, "whipped honey," if it has air incorporated into it, will need to be liquified over hot water before being measured. –AWH (see also HONEY RECIPES)

HONEY, EXTRACTED – Honey that is pressed, squeezed, or removed from comb(s) with a centrifugal force machine, is called extracted honey. The origins of the term are lost though it probably came into use in the 1860s or 1870s when the first extractors or centrifugal force machines were invented. The honey, unchanged, has been removed – extracted – from a comb. The terms extracted and liquid, to describe honey, are used interchangeably by beekeepers but the term "extracted" is not commonly used on labels. Honey that is sold in the comb can be called comb honey, cut-comb honey or chunk honey, depending on its preparation.

The machine that is used in extraction is called an extractor. Liquid honey may be processed to prevent crystallization, fermentation and to have a long shelf life. Honey in liquid form is easier to ship, to use in making honey products or as a sweetener.

While nothing has a more delicate flavor than fresh comb honey, it cannot be used in cooking. If honey crystallizes in the comb it is impossible to extract and is not suitable for marketing in the comb.

Almost all honey for sale in large outlets is liquid, extracted honey, available in a variety of colors, sizes, container types, from many suppliers, and often many countries.

Loading docks should fit the operation, but there should also be opportunity for expansion. Look, too, at going from muscle power to mechanization. Will the facility handle fork lifts, pallet lifts, or motorized barrel lifts.

Honey houses – A honey house is a place where honey is extracted and processed. The most important consideration in thinking about building a honey house is that it should facilitate a steady flow or movement of supers and honey from start to finish. The following are the essentials of a honey house and the order in which they should be arranged: off-loading dock or room where supers are brought in, hot room for heating and keeping supers warm, and extracting room. At this point the empty supers move in one direction and the newly extracted honey moves in a different direction to the honey processing area. Each of these separate areas and their functions are listed below:

Off-loading dock or room – The entrance to the honey house will depend on the size of the beekeeping operation. For the hobbyist with a small number of supers ease of entering with a handtruck would be the only consideration. For large operations matching the height of the dock and the truck will insure easy and quick off-loading of supers and pallets. Enclosing the off-loading dock will reduce robbing by bees.

It is nearly impossible to harvest honey and to bring the filled supers into a honey house without some adult bees. These are a nuisance and some provision must be made to remove them. A popular technique is to have only one natural light source in the room, usually a window equipped with a bee escape. It

A unique solution to bees in the honey house. Bees are drawn out of the interior to this screened enclosure with a convenient entrance to the hive body in the corner. It contains old comb and fills rapidly, keeping errant bees up and away from workers.

is also useful to have a small vacuum cleaner, with a long hose, to pick up and remove stray bees in the hot and extracting rooms.

Supers containing combs with honey, pollen and bee brood should not be stored for more than three days to prevent damage from small hive beetles. One innovation to control the small hive beetle is the use of forced air using fans directed downward through the stacks of honey supers on specially designed pallets for more efficient air passage. The forced air helps to reduce the relative humidity that appears to enhance the desiccation of small hive beetle eggs. Traps and lures are also used if long term storage or many supers are moved in, and out, of the areas.

Hot rooms – An important first step in honey extraction is to warm the honey so it will flow easily from the comb. Temperature and moisture content influence the rate at which honey will flow. A hot room is a heated, insulated room usually adjacent to the room where the uncapper and extractor are located. Hot rooms are usually held at about 80 to 90°F (26 to 32°C). When honey is removed from hives in the summer it is usually warm enough that extraction is an easy process. However, if cool nights prevail, especially in the autumn, the honey in the storage supers will be cool and difficult to remove from the combs. Beekeepers who have only a few honey supers to extract have often devised small, electrically-heated hot boxes in which to warm the supers; care must be taken to not allow the temperature to go too high in these boxes.

Various methods of forcing warm air through supers have been devised both to speed up the warming process and also to remove water from high moisture honey. The simplest method is to use pallets under stacks of supers through which the air may move. Large circular, ceiling fans are sometimes used to help move air down through the supers.

If warm air is forced through supers filled with honey, even though the cells are capped, it is possible to remove as much as one percent moisture from the honey in 24 hours. The precise quantity of moisture that may be removed varies with the moisture content of the honey, the volume of air and the relative humidity in the room. Beekeepers who live in areas where high moisture honey is common have often taken the time to build special platforms in their hot rooms on which the filled supers are placed so that warm air may be forced through them.

Hot Room Suggestions

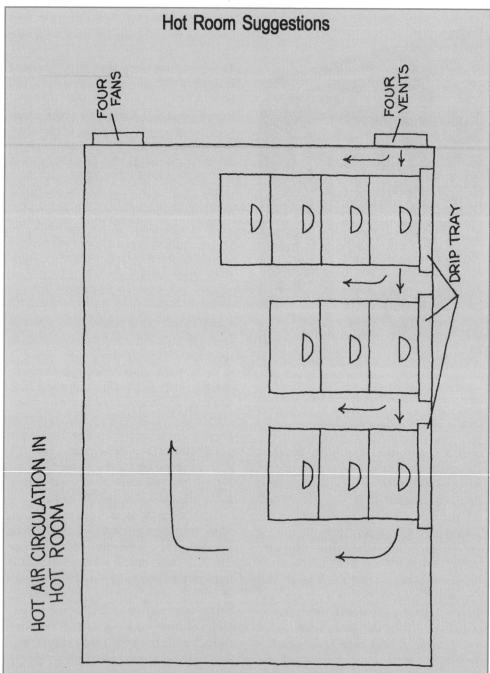

Heat can be supplied by furnaces and forced air, hot water pipes in the floor, or smaller space heaters – all depending on the volume of the room.

Warm, dry air should be blown in, and the moist air removed. Ample space between stacks of supers should be made to allow easy circulation. Some manner of thermostatic cut-off should be employed, bee escapes considered, easy-in and easy-out doors supplied, and washable floor, walls and ceilings installed.

Honey, Extracted

Air movement is increased with a large fan. Note plastic door covering to reduce heat loss.

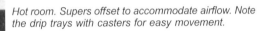

Hot room. Supers offset to accommodate airflow. Note the drip trays with casters for easy movement.

Forced air with some space between stacks of supers. No covers help air flow.

Some honey packers will pay a premium for low moisture honey that they use to blend with high moisture honeys. The ideal moisture content for table honey is 16.5 to 17.5 percent. Honey with this amount of moisture flows easily.

The hobby beekeeper or packer who prepares only a limited amount of honey for market should be aware that today the public seeks many qualities in the foods they buy. One important consideration is consistency from one jar to the next. When a consumer likes a particular product, additional purchases often depend on the quality being the same.

Extracting room – Supers of honey should move only a short distance from the hot room to the uncapping machine. The empty supers should move easily to the area occupied by the extractors. Many beekeepers clean and scrape their supers between these two points. After the extracted combs are put back into the supers they should move to a temporary or permanent storage area without interfering with the flow of people and full supers in the extracting room.

Uncappers – The process of removing the wax cappings over cells filled with honey is called uncapping. When beekeepers first began to extract honey they used cold knives that looked similar to butcher knives. They soon learned that if the knives were kept in a pan of hot water the blades would become warm and a hot knife did a better job of removing the wax cappings, especially when cutting old, dark comb. Today, in addition to plain knives, one has a choice of knives that are heated electrically or with steam. It is very easy to burn honey with a hot blade but the temperature can be controlled with a thermostat.

A variety of power uncappers are available. One type uses two, power-driven, heated, usually serrated, vibrating knives. The frames are held in place with endless chains that force the combs between the blades at a uniform speed. The depth at which the cells are cut may be varied and thus the quantity of cappings wax produced is controlled.

Hand-powered uncappers. A 'cappings scratcher.' The very sharp tines remove cappings by scratching them off. Very little wax is removed. This tool can be used by itself or, more commonly, to scratch cappings missed by other uncappers. Many models are available.

Honey, Extracted

Hand-held uncapping knives. On the left is a serrated bread-type knife. Using two, with one on a warming element so it is hot when used, is good for a small number of hives. Next, an offset cold uncapping knife. Sharp on both edges, using two, one warming while the other is in use is good advice. Second from right is a heated knife. The simplest of the heated knives, it is 'hot,' with only one temperature. Regulation is by an off/on switch. Right, a thermostatically controlled knife, so the blade does not get too hot. Many varieties of these models are available. Steam-heated models are available, but are much less common than in the past.

Flail type or chain uncappers are also popular. These had their origin in Canada in the 1920s. Today these machines have a shaft with chains attached at right angles to the shaft. When power is applied, the shaft spins and the whirling chains strike the surface removing the cappings.

The chief problem with chain uncappers is that the cappings are ground more finely so they are more difficult to separate from the honey. At the same time these uncappers unfortunately tend to introduce small air bubbles into the honey. Such cappings and air bubbles can be removed if the honey is warmed slightly, to about 90°F (32°C), and several days are allowed for settling. The latest design in uncappers is the Dakota Gunness that allows

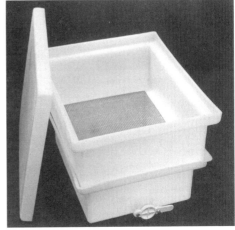

Technique for using a handheld, offset blade knife is to begin at the bottom of the frame, moving upward. The frame is tipped allowing removed cappings to fall away.

A typical small scale uncapping tank. Place a small board across the top to hold the frame while uncapping. Wax falls and is caught in the top container, held by the queen excluder screen.. Honey drains out, and is caught in the tank below.

Honey, Extracted

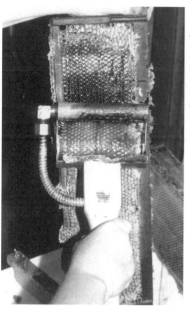

The uncapping plane. The blade is heated, and moved down the surface of the comb, with cappings falling below.

The simplest and earliest of the vibrating knife uncappers. The motor 'vibrates' the knife on a cam as the comb is pushed down. The heated blade cuts back and forth, removing the cappings which fall away.

Clever beekeepers rotated the vibrating knife 90°, covered it so fingers wouldn't get in the way, and provided a place for cappings to fall out of the way. Here, on the left, the knife is being prepared. On the right, the frame is moved from the center to the edge. Fingers, however, are still at risk.

An early version of 'double' knives. Frames are loaded in the descending slot on the top, with the lugs of the frame grasped by endless chains and moved down between the two vibrating knives. The now-uncapped frame eventually was deposited, hanging by the lugs, to the drip tank on the left. This innovation speeded the uncapping process, but tended to have difficulty with frames of varying width. Thin frames would not get uncapped; thicker frames could jam the mechanism.

Advances allowed these uncappers to handle multiple frames at once, to accommodate any width frame, and to do so as rapidly as frames could be loaded.

Modern uncappers are fast, efficient and economical from a labor perspective.

The simplest of the chain 'flail' uncappers. A frame is inserted in the top and manually lowered down, between two sets of steel shafts with chains attached. The chains are adjusted so that they just barely hit the cappings, removing enough wax so that the honey is easily extracted.

Wax cappings fall below.

Motorized models are available, some with wax spinners directly underneath to separate honey and wax. Other models move the frames directly to the extractor, all in one unit.

beekeepers to uncap combs of honey at a great speed.

During uncapping burr and brace comb will be removed as well as cell depths reduced so that the combs will fit back into a super easily. It is difficult to control bee space perfectly and even the best equipment often has extra wax and propolis. Extra care taken during the uncapping process will do much to make the manipulation of combs easier the following year.

Cappings are a useful by-product of honey extraction. The cappings usually yield the lightest colored, and most desirable beeswax. Therefore it is best to keep them separate from beeswax from old combs. (For different methods of processing cappings see BEESWAX, Rendering beeswax cappings).

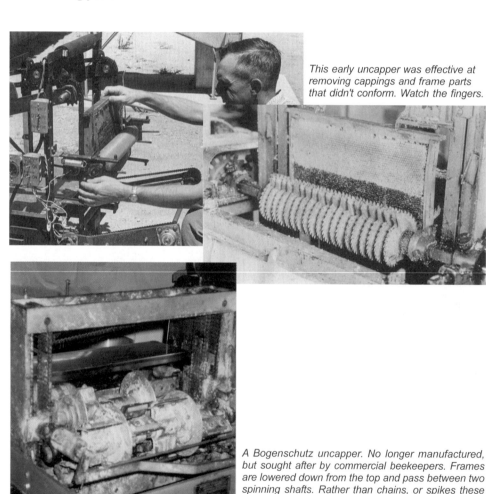

This early uncapper was effective at removing cappings and frame parts that didn't conform. Watch the fingers.

A Bogenschutz uncapper. No longer manufactured, but sought after by commercial beekeepers. Frames are lowered down from the top and pass between two spinning shafts. Rather than chains, or spikes these shafts have 4" blades attached. The frame is guided past the blades by the large wheels shown here. The extreme speed of the rotation, and the strength of the blades make this a very unforgiving machine when considering frame variability.

Honey, Extracted

Shown here the latest design in uncappers. The Dakota Gunness allows beekeepers to uncap combs of honey at a great speed.

Extractors – The principle that honey could be removed from a comb with centrifugal force was discovered by Major F. Hruschka of Italy in 1865. In 1867 L.L. Langstroth built the first practical extractor in the U.S. that was designed to accommodate the frames he built. A number of these machines were sold by Langstroth and Son in that year. Moses Quinby, a famous New York State beekeeper, also designed an extractor using a metal can rather than the wooden tub that was used by Langstroth. In 1868 A.I. Root constructed his famous Novice extractors that were sold by the thousands.

An early version of a dual head uncapper. Frames were dropped down between these two revolving cylinders that had short picks on them. The picks would puncture the cappings on both sides of the frame simultaneously, and honey would escape the cell through the punctured cell cap. This, of course, reduced significantly the amount of cappings wax harvested.

Honey House Guidelines For A Superior Product

It is reasonable to expect that in the future the requirements for beekeeping will parallel other sectors of food production. Examples of such requirements include detailed hive treatments, evidence of proper cleaning and maintenance of the extraction facility, proper training of employees in the safe handling of food product, complete product traceability, and the ability to recall any product that may be determined unsafe. To that end, guidelines have been designed to help producers meet the expectations of the consumer. The general contents of the audit include: building design, construction and maintenance, lighting and ventilation, waste disposal, sanitary facilities, water quality, storage of chemicals, supers, empty drums and containers, processing/extraction from the comb, employee training, sanitation and pest control, recalls, and integrated pest management.

The goal of these guidelines is to establish a workable audit that makes sense for beekeepers, does not incur great cost and gives each facility the flexibility to choose those improvements that make the most sense for their situation. It is not meant to be a "do this or else" proposal. Having participated in an audit allows a beekeeper to have a point of reference – where do you stand in comparison to other beekeepers both in the U.S. and abroad. It is no longer enough to say that we have a quality product, we have to prove it.

1. **Building Exterior: Design, Construction and Maintenance** – Buildings and surrounding areas are designed, constructed and maintained in a manner to prevent conditions which may result in the contamination of food.

2. **Building Interior: Design, Construction and Maintenance** – Building interiors and structures are designed, constructed and maintained in a manner to prevent conditions which may result in the contamination of food.

3. **Building Interior: Lighting and Ventilation** –Adequate natural or artificial lighting and ventilation should be provided throughout the establishment and all artificial lighting in production areas must be of a safety type and protected to prevent the contamination of food.

4. **Building Interior: Waste Disposal** – Sewage, effluent and waste storage and disposal systems are designed, constructed and maintained to prevent contamination.

5. **Sanitary Facilities: Employee Facilities** – Employee facilities are designed, constructed and maintained in such a manner that employees are encouraged to use proper hygiene in order to prevent contamination.

6. **Sanitary Facilities: Bee Control** – Effective bee controls are in place to prevent entry, to eliminate bees in the production area and to prevent the contamination of food.

7. **Premises: Water Quality Records and Program** – The potability of hot and cold water is controlled to prevent contamination. Written records that adequately reflect control of water quality and treatment are available upon request.

8. **Transportation and Storage: Food Carriers/Vehicles** – Carriers used by the establishment are designed, constructed, maintained, cleaned and utilized in a manner to prevent food contamination. This includes both contract carriers as well as vehicles used to transport food, components or equipment.

9. **Transportation and Storage: Chemical Storage** – Non-food chemicals are received and stored in a manner to prevent contamination of food, packing materials and food contact surfaces.

10. **Transportation and Storage: Storage of Honey Supers** – Storage and handling of honey supers is controlled to prevent damage and contamination of the food during both transportation from the beeyard and storage within the extraction facility.

11. **Equipment** – All equipment is designed, constructed, installed and maintained to function as intended, to allow for effective cleaning and sanitation and to prevent contamination of food.

12. **Processing: Handling and Treatment of Supers/Frames** – Supers and frames are handled in a manner designed to prevent contamination from environmental sources (dust, debris, insect parts, soil, etc.), bee removal chemicals, and uncured honey.

13. **Personnel: General Training** – Every food handler is trained in personal hygiene and hygienic handling of food and operation of equipment such that they understand the precautions necessary to prevent the contamination of food.

14. **Personnel: Cleanliness and Conduct** – All persons entering food handling areas maintain an appropriate degree of personal cleanliness and take the appropriate precautions to prevent the contamination of food.

15. **Sanitation and Pest Control** – An effective sanitation program and records for equipment and premises is in place to prevent contamination of food.

16. **Recalls** – The establishment has a written plan to facilitate the complete and rapid recall of any lot of food from the market.

17. **Integrated Pest Management (IPM)** – The establishment is aware of and actively participates in IPM strategies that are designed to reduce costs, inputs and residues while maintaining a viable and profitable crop.

by Eric Wenger, Director of Laboratory Services for Golden Heritage Foods.

History of The Honey Extractor

Some extractor history, from the 1935 ABC

In the olden times the only method of securing honey in liquid form was to crush the combs in some kind of press and strain the honey through cheesecloth. Where there was some brood present in the combs the brood juices mingled with the honey and the product obtained was called "strained honey." This term conveys the impression that the honey itself was separated, not only from the comb but from the dirt, pollen, dead bees and brood.

The modern extractor that takes the honey by means of centrifugal force not only saves the combs, which are worth from 75 to 90 cents each, and can be used over and over, but furnishes a product in point of quality and sanitation that is far superior to the strained honey of old.

In the year 1865 Major D. Hruschka, of Venice, discovered the principle which

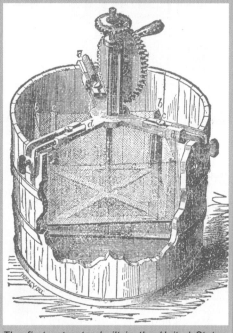

The first extractor built in the United States. Langstroth, the inventor of the hive and frame bearing his name, was the first to build a honey extractor in the United States. With his quick genius for the practical, he early saw the necessity for gearing to increase the speed of the reel. (From *American Bee Journal*, 1868)

led to his invention of the extractor in that year.*

Apparently his discovery and invention did not attract attention in this country until 1867 when L.L. Langstroth, the inventor of the hive and frame bearing his name, built and successfully used an extractor geared back as in the modern machines of today, but instead of a metal can he used a wooden tub to hold the mechanism as shown.

Peabody honey extractor. This is one of the early honey extractors built and sold in this country. This as will be noted had no gearing the whole can revolved. Without gearing it could not do effective work. (photo taken from *ABC*, 1940)

*The legend, oft repeated, that he got the idea of centrifugal force to remove honey from combs from seeing his little boy swing a basket containing an uncapped comb about his head and of the honey flying out is not based on fact. That he did attempt to remove the liquid honey from that partly granulated is true. See *Bee World*, page 118, for the year 1935.

Langstroth's quick genius for the practical and useful in bee culture early saw the value of centrifugal force for removing honey from movable combs. Without his invention of movable combs the discovery of Hruschka would have been of little value. The surprising thing was that Lanstroth immediately used gearing to increase the speed of the reel holding the combs. A number of these machines were listed and sold by Langstroth & Son in 1867.

In 1868 A.I. Root constructed an all-metal honey extractor using the gearing of an old apple paring machine mounted on a wooden cross arm to drive the reel. With this old machine he extracted 285 pounds with the help of an assistant in seven and a half hours. This for 1868 was considered a record-breaking feat. He took in all 1,000 pounds of honey from 20 colonies and increased them to 35. In 1869 he secured over 6,000 pounds of honey from 48 colonies. A.I. Root did not keep his light under a bushel. He told the world about it. Then came a call for information as to how he did it, and immediately a demand sprang up for his machines. He sold literally thousands of them under the name of Novice Honey Extractor.

A.I. Root's improved Novice was so great an improvement over all that had preceded that it found a ready sale at once. The crank was geared so that one revolution made three revolutions of the combs

From The 1880 ABC – That more honey can be obtained by the use of [an extractor] than by having it stored in

The original extractor made and used during his life by Moses Quinby. Note the heavy spur gears, the oak cross-bar, the oak framework underneath, forming a support for the lower bearing. (photo taken from the 1940 edition of ABC)

The modern three-frame non-reversible Novice extractor and capping dryer. (photo taken from the 1940 edition of ABC)

section boxes in the shape of comb honey, all are agreed; but all are not agreed as to *how much* more. If it is nicely sealed over as it should be before being extracted, I do not think more than twice as much will be obtained, on an average, although the amount is placed by many, at a much higher figure. A beginner will be more certain of a crop, than if he relies upon having the bees work in boxes; he will also be much more apt to take away too much, and to cause his bees to starve. This last is a very disagreeable feature attendant upon the use of the implement, especially, where the beekeeper is prone to carelessness and negligence. To secure the best results with the extractor, plenty of empty combs should be provided, that ample room may be given, in case the hives should become full before the honey is ripe enough to remove. If a second story does not give room sufficient, I would add a third for a heavy stock, during a good yield of honey.

Full directions for using extractors are given with the price lists that manufacturers send out; therefore I will not repeat them here.

How To Make An Extractor – Although it will not usually pay to make your own, there are circumstances under which it is very desirable to know how. In places so remote that the shipping rates are very high, it would be well to have some beekeeper of a mechanical turn make them to supply those in his own vicinity. As the manufacture of implements and supplies is getting to be quite a business, the machines can probably be manufactured at many different points. Whoever does the best work will probably get the most orders.

Experiments have been made, almost without number, and the general decision now seems to be in a favor of a machine made entirely of metal, with everything stationary about it except what *must* be revolved. The momentum of heavy, metal, revolving cans, or of honey after it has left the comb, defeats the very object we have in view; and nothing will so effectually convince one of the difference, as an actual trial of the two machines side by side. With the light, all metal machines, the comb is revolved at the speed required almost instantly, and as soon as the honey is out of the comb, the operator is aware of it, by the decrease in the weight as he holds the crank in his hand; but with the heavy, unwieldy machines, the stopping and starting take more time than doing the work. The same objections apply to making machines for emptying four combs at once. They require to be made much larger, and are correspondingly heavy and unwieldy.

A reference to the engraving of the extractor with its inside removed, will enable almost any tin-smith to do the work. The gearing had better be purchased from a dealer in supplies, but if you should have many to make, it may pay you to have them cast, using the sample for a pattern. The shaft of the inside part is made by rolling up a tin tube, double thickness. This is quickly and nicely done with the machine the

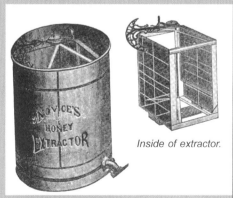

Inside of extractor.

Extractor complete.

tinner uses to make the bead on the edge of eave-spouts. The frame work is made of folded strips of tin.

For a Langstroth frame, we make the shaft the full length of a 14 by 20 sheet of tin. The corner pieces are made of a strip two inches wide, by 14 long, with a seam folded on one edge, and a square fold of 1/4 on the other. The bars that support the wire cloth are six in number, including the top and bottom ones, and are made by folding one inch strips of tin, three times, so as to make a stiff rod of metal. They are 10 inches in length, and our revolving frame is 10 inches one way, and 11 the other. For greater security against sagging, we run a similar rod of metal, up and down across the middle of these bars, and still another lies flatwise across this, to brace the whole something like a truss bridge. This gives a surface very stiff, and yet very light. The wire cloth, which should be tinned like all the other metal work, is made of stiff wire, five meshes to the inch. It may be well to remark here that neither zinc nor galvanized iron should ever be used about honey utensils. The acid principle in the honey quickly acts on all oxidizable metals, and galvanized iron, though bright in appearance, quickly poisons the honey, or even pure water, as has been proven by experiment. Two sheets of wire cloth, 15 inches long by 10 wide, are needed for an L. extractor. They are simply laid inside against the metal bars and tacked with solder. To cover the ragged edges at the top and bottom, we fold a strip of light tin 10 inches long by 1/4 inch wide, at a right angle, so as to make a square trough, as it were; this is soldered on the top rod, so as to cover the upper edge of the wire cloth. A strip of wire cloth, 15 inches long, and four wide, with the edges hemmed by folded strips of tin, is put across the bottom, to support the frames. Two inches from each end, it is bent at right angles, and then 1/4 inch from each end, still again, that it may catch securely over the lower bar of the frame. The frame is completed by the cross pieces at the top and bottom, to hold the two wire cloth frames at the right distance apart. These are strips of heavy tin 1½ inches wide, by 11 inches long. A seam is folded on each edge, so that the bars are left only one inch wide when finished. At each end, a 1/4 inch is folded square, to catch over the outisde of the frame where it is soldered.

Now, to attach this frame to the shaft is a matter somewhat important; for, if we use too much breadth of surface with our arms, they will "blow" like a fanning mill, and we shall have a current of air, that will carry with it a fine spray of honey, over the top of the can. This is a most grievous fault, for who likes to have honey daubed over his clothing? Our first machine was made so that the combs revolved only 1/2 inch below the top of the can, and yet we never had a particle of honey thrown over. This frame was made very light, indeed, and when heavier and stronger machines were made for sale, we were much puzzled to hear an occasional complaint, that the honey was thrown over the top of the can, in a fine spray. I soon found by experiment, that it was caused by the braces being placed flatwise to the line of motion. How to make them strong and stiff, without catching the air, was the problem. We do it nicely, by using 12 braces, made of heavy tin, with a seam folded as just mentioned. The 12 are formed of six pieces. The six pieces are laid across each other in pairs, forming three letter X's. Each letter X has a hole punched at the crossing, large enough for the shaft to be driven through; when it is soldered securely, the ends are bent down and attached to the corners of the frame as shown in the engraving. The lower X also supports the wire cloth that the frames rest on, by being tacked with solder where it passes it.

The gearing is attached to this revolving frame, by driving the small gear wheel into the end of the hollow shaft, and soldering it securely. The casting is first well tinned by a soldering iron, that there may be no slipping loose.

Making The Can – There is nothing difficult about this, except the bottom of the can. It had been, for a long time, quite a problem to get a strong stiff bottom, without some kind of a wooden support; but I struck on the idea, while trying to devise some kind of a bottom that would let the honey all out, the gate or faucet being the lowest part. I will tell you, presently, how I did it. The top edge of the can must be stiff and rigid; more so than we can get it, by any kind of a wire or rod. I found some very stiff hoops that were made for milk cans, and it is these I would advise using. They are so made as to give great stiffness, with but a small amount of metal.

We present a view of a cross-section of the hoop, the concave side, of course being inward. A is the hoop, and B is the tin of which the can is formed. The can is made of four sheets of 14x20, IX tin. For an L. frame, we need a hoop just 17 inches in diameter. For large sizes, we use 20 inch hoops. The two sizes mentioned will accommodate almost any frame used, and we therefore furnish gearing for only these two sizes. After you have made the body of the can, and have your hoop

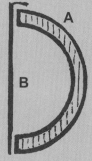

nicely soldered on, you are ready for the bottom.

Lock two of the sheets together, and cut a circular piece 18 inches in diameter. From one side, cut a wedge shaped piece, as shown in the cut below.

How To Make The Bottom To The Can Of The Extractor – The space, A B, should be about two inches in width; and, after cutting it out, you are to fold down the edges of A B, about 1/2 inch. Draw these edges toward each other, and you will make the bottom concave, as shown above in Figure 2. They can be held in this shape for the time, by a slip of tin, temporarily tacked with solder across the gap. Turn over the edge and put this bottom on the can in the usual way. The opening left is for the channel that leads to and holds the honey gate. Cut a piece of tin similar to the wedge shaped piece you took out, but somwhat larger. Fold this up trough shaped, as shown in Figure 3, and fit it over the opening. We are now ready to solder in the gate, but we must have something for our can to stand on. This is fixed by a tin hoop, with a heavy wire at its lower edge, made just large enough to slip closely over the lower part of the can, as seen in Figure 4. This hoop, or band rather, should be about four inches wide, and in one side you are to punch a round hole, just large enough to take in the gate. Solder it securely in place, put in the gate, and then be sure to try your can by pouring in some water to see if it will "hold." We do not want any leaking after we commence extracting honey.

Now, in the centre, C, on the inside,

we solder a piece of steel saw plate; over this, we put a blank iron nut, with a quarter-inch hole drilled in it. This is to hold the bottom pivot, which is made of refined Stub's steel, nicely rounded and polished off on the point. As the bearings for the gearing are all cast steel, our machines should almost run of itself, if everything is made just right. The steel pivot at the bottom is soldered in the end of our tin tube, by rolling some thin tin around it until it will drive in tight.

You should never attempt to use an extractor, and I might almost say *any* piece of machinery, until you have it securely screwed down to the box or platform on which it is to stand. The screw holes are made in the bottom ring just above the heavy wire that rests on the floor. The screws are put in a little slanting. It should also be at a convenient height for easy work. The machine could be made heavy enough to stand still from its own weight, it is true, and it might be made perched on legs, also, to save the trouble of building a box or platform on which to stand, and if you are making them for home use, it may be well to do so; but if making them to ship customers, I would never think of sending them anything that they could procure at home, thus saving heavy shipping expenses. I would say the same in regard to making cans large enough to hold 100 pounds, or more, of honey, below the revolving frame. When the extractor is being used, the honey gate is supposed to be open, and utensils can always be supplied to hold the honey, much cheaper than to have the extractor thus enlarged. Those I have described, can be very conveniently worked over the bung of a barrel, or you can have a tin can made on purpose to set under the honey gate.

The gearing for the extractor, including a tinned honey gate, will cost about $2.00; the materials and labor for the inside should not cost to exceed $2.50; seven sheets of tin for the can, would be 70¢; a half day's work in the making, $1.25; hoop for the top, 50¢; and perhaps the solder and other items, 25¢ this would bring the whole cost up to $7.20. Your own time in "bossing" the tinner, and the liabilities of making mistakes, and doing a bad job on the first one, would probably bring the expense up to about the usual selling price, viz., from $7.50 to $9.00. Machines for different sized frames are made much in the same way; for the American and Gallup frames, we can make a short can, only the height of the width of a sheet of tin, instead of the length. Of course these can be made at a less cost. Where the frames hang in the extractor the same way that they do in the hive, no wire cloth support is needed across the bottom of the comb basket, unless it is preferred for extracting small pieces or bit of comb.

No cover is ever needed over the extractor while at work, for it would be greatly in the way; but after we are through, or only stop temporarily, the machine should be covered to keep out dust and insects. The most convenient thing for this purpose is a circular piece of cheap cloth, with a rubber cord run in the hem. This can be thrown over in an instant and all is secure. When honey is coming in abundantly, it may be safe to carry the machine, located on a suitable platform, around to the hives, especially if the apiary is much scattered about. But if the bees are disposed to rob, all such attempts will come to "grief" very quickly.

Extracting From Broken Pieces Of Comb Or From Section Boxes – As we always use the L. extractor, we have extracted from pieces of comb, by setting them up on the wire cloth at the bottom. The smaller, shallow extractors, for Gallup, Adair, and American frames, have no such attachment; therefore some arrangement is really needed for the purpose. At the same time, it would be very handy for the tall extractors, when any mishap occurs to break a comb down, or when we wish to extract from heavy pieces of comb, in warm weather. Several devices have been described in the journals, but none of them suit me so well as the one figured below, which was sent me by J.D. Slack, of Plaquemine, LA.

He uses it for extracting from section boxes also, but I think I should prefer to do this in the broad frames that hold them, thus doing a full set of eight at one time. With this machine, only one could be extracted at once.

At C are a pair of hinges, that the machine may be opened the more readily to receive a heavy, soft comb. The wires, B, are of one piece, and are also made to turn that they may be hooked in to A, when the comb is properly in place. The hooks, A, are to hook over the top bar of the inside of the revolving frame of the extractor.

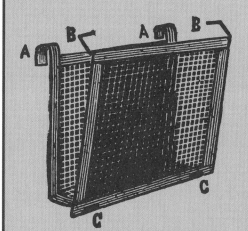

Extractor for pieces of comb. (photo taken from 1880 *ABC*)

From The 1930 ABC
Cowan Reversible Extractor – When the honey from one side of the combs was extracted in the Novice machine the combs had to be lifted out and turned around in order to throw the honey out of the other side.

About the time A.I. Root was experimenting along this line Thomas William Cowan, then editor of the "British Bee Journal," constructed what was called the Cowan reversible extractor. Several "baskets" holding the combs and hung on hinges like a door, could be swung from one side to the other, and either side of the comb could be next to the outside. The first side could be extracted, and then the baskets swung around so that the honey could be thrown from the other side without taking out the comb and reversing it.

The Root Multiple Reversing Extractor
To reverse the Cowan extractor, it is necessary to stop the machine, and with the hand catch hold of the pockets and swing them around to the other position. The multiple reversing extractor reverses the pockets simultaneously when the brake is applied. The lever acts as a brake until the extractor has been reduced in speed to a certain point, when the hub of the reel is held stationary by the brake, and the reel, which continues to turn, accomplishes the reversing of the pockets by means of the reversing levers located on the top of the reel. The strain of reversing is borne entirely by the brake, thus relieving the driving mechanism of all stress.

Control Pivot Reversing Extractor
– All reversible honey extractors on the market made use of one of two principles for changing the sides of the combs. The

Automatic Reversible Root Extractor – This shows the principle of reversing the extractor. The pockets at the top and bottom are hinged on one side. The levers here shown connect each pocket with the reversing drum, which, when temporarily slowed down and then stopped, causes the levers to shift from one position to the other, reversing the pockets to the other position.

This is a top view looking down into the eight-frame Buckeye extractor, the pockets of which are reversed on a central pivot. As will be noted, it is perfectly easy to insert and remove the combs. The tops of the pockets are firmly held in place, no matter how severe a strain may be placed on them. The act of reversing is accomplished by means of sprocket wheels that are made integral with the pinions meshing with the internal gear or rims at the top of each pockets. Each of these sprockets is actuated by a chain driven from a sprocket mounted on a hollow shaft loosely journaled on the main shaft from which power is received.

first one, that of baskets swinging from hinges on one side like a common door, has been used for the last 30 years, and it has given good satisfaction; but it has its limitations. The other one, perhaps just as old but newer in its application, at one time attracted some attention. In the older type the reversing was accomplished by swinging the pockets on their hinges from one side clear to the other. This principle necessitated the stopping of the machine, or nearly so, before reversing could be accomplished. Even at slow speed the centrifugal force tended to throw the baskets over to the reverse side with a bang unless care were used. With new or unwired combs, there was a little breakage, especially when careless help does the work.

In modern practice it is the almost universal custom to start throwing out most of the honey on one side at a comparatively slow speed to reduce the weight of the comb. It is then reversed, when the other side is extracted clean. The first side is then returned to its first position and extracted again. This makes two reversings, and each time the machine must be slowed down, or stopped and started again.

In the other method, although it is as old as the first, the baskets, instead of being hinged on the side and swinging like a door, are pivoted in the center. Of course, it is impossible to have a shaft go through the comb; but the basket can be pivoted at the top and bottom, thus in effect reverse the comb on its center line.

This type of machine requires a much larger can and heavier reel for the same number of combs and is therefore more expensive. There are some who prefer it especially in the West where the honey is thicker.

Extracting Without Reversing – About 1920 a new interest was revived in an old principle that had been exploited some 50 years before (see *L'Apicultur* for that year) by the author and by Hamet in 1867, namely, the possibility of extracting the honey from the combs without reversing. The combs are placed with the end-bars pointing toward the center like the spokes of a wheel. The centrifugal force is applied along the midrib of the comb, thus causing a pressure toward the top-bar of the frame. Such a pressure forces the honey out of the cells on both sides of the comb at the same time. It then climbs over the surface till it reaches the top-bar, whence it flies to the side of the extractor. There are two ways of accomplishing this: (1) Placing the combs in a plane at right angles to the center of revolution; (2) placing the combs on a plane with the center shaft like the spokes of a wheel.

In the September issue of *Gleanings in Bee Culture* for 1888, page 773, the author illustrated and described the two methods. The first is shown as the Bohn's Extractor, reproduced from that number of *Gleanings*. The author demonstrated then (1888) that it was perfectly possible to extract the honey from both sides of the comb at the same time without reversing, but it took from three to four times as long to get the honey out as when an equal number of combs were placed in a machine like those already described in these pages. At that time no attempt was made to increase the number of combs in order to offset the time limit. It would have done no good because this was long before the days of small electric motors or small gasoline engines. It was likewise before the days of commercial beekeeping, when small hand-driven extractors were quite able to do all the work of taking the honey. There were few or no outyards and of course very few beekeepers who produced honey on a large scale. The hand-driven machines requiring the reversal of the combs would take the honey out in from two and one-half to three minutes. The other principle, by which combs were arranged like the spokes of a wheel, required from eight to 15 minutes to do the work. The idea was, therefore, abandoned as impracticable at that time.

The principle was revived in 1915 and 1916. See United States patent issued to Jacquet, No. 1,176,562, March 21, 1916. In 1916 M. Bernard in *L'Apicultur* in the March and April issues, gives particulars of his bilateral extractor. See also in June, 1926, number of the same journal, for a reproduction of the Bernard Extractor. Another U.S. patent was granted G.S. Baird, No. 1,334,585, on March 20, 1920. A French patent

An early extractor in which the combs are whirled vertically. Reproduced from Gleanings in Bee Culture, *November 1, 1893.*

showing this (the radial principle) was issued to M.Sicot, No. 526;342, published October 1, 1921. The French Sicot patent and the descriptions of the same general principles of placing the combs radially as shown in *Gleanings in Bee Culture* for 1888 and in various European journals at that time antedate subsequent patents in the United States for non-reversible extractor having the end-bars of the combs placed like the spokes of a wheel.

In 1921 Herr R. Reinarz, the editor of *Die Deutsche Biene*, published details of his wheel extractor.

In view of the apparent interest in Europe in this principle of taking the honey from the combs, H.H. Root and Geo. S. Demuth in 1921 again tried out the plan, which could be put to the test very easily in the Buckeye extractor. The pockets were reversed to a point where the combs would stand like the spokes of a wheel. The principle was tested very carefully, using an electric motor to drive the machine. It was found that it would extract most of the honey in about three minutes, but it would leave about two and one-half ounces of honey to the comb. Because of this residue the idea was again given up for the time being.

A short time later, 1923, Arthur Hodgson of Jarvis, Ontario, Canada, tried the principle of extracting honey as shown in Bohn's honey-extractor. He discovered that by running the machine longer, say from 10 to 15 minutes, all the honey could be taken. He then built a machine to take 48 combs.

To Mr. Arthur Hodgson and M. Sicot belong the credit of being the first to eliminate the time element by increasing the number of combs. Mr. Hodgson in 1923 built the first practical machine that would throw the honey out in a commercial way without reversing, and at the same time reduce the time limit per comb below the time usually taken per comb in the ordinary reversible eight-frame extractors.

Mr. H.H. Root, who witnessed an early test of the Hodgson machine in 1924, suggested that a cheaper machine holding a like number of combs could be built on the principle as used in the original machine, with the comb end-bars placed like the spokes of a wheel. A machine was built to take 45 combs. It was proven conclusively that this would extract the honey just as efficiently and thoroughly as the Hodgson machine at a much lower cost because of the smaller diameter thus possible.

From the radial principle of extracting without reversing it might appear that one side of the comb would be cleaner than the other, on the theory that the cells preceding the direction of motion would not be as clean as those following the direction. Very extended experience, however, shows that there is no difference. The combs are so close together that the air between them travels with them, with the result that there is no more pressure on one side than on the other.

With either the Hodgson or the principle shown in the Simplicity extractor, the honey is thrown out on both sides of the cells simultaneously because the centrifugal force or pressure is in a straight line away from the center shaft through the center of the combs toward the circumference or the can surrounding the revolving reel. This centrifugal pressure causes the honey to seek the top of the cells. It then climbs over the cells and finally strikes the can surrounding the revolving reel.

It will be clear that the part of the comb nearest to the center shaft will not have the same pull as that portion of the comb near the outer edge of the can. The combs should always be placed with the top-bar next to the outside and the bottom-bar nearest to the center shaft. Most of the honey in the comb will be near the top and the smallest amount will be near the bottom. But as the pull is the greatest near the top, the two parts of the comb will be emptied in about the same time, provided that the bottom of the comb is far enough away to receive sufficient centrifugal pull. It is clear that the radial principle can not be applied satisfactorily in a hand machine,

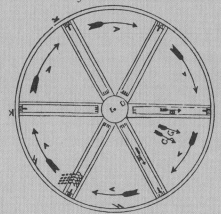

Diagram of radial extractor from Gleanings in Bee Culture, October 1, 1888.

because the bottom of the comb would be too close to the center shaft.

Advantages of the Radial Extractor Over The Reversible Power Extractors – The radial non-reversible extractors in the East and central states are superior to either of the eight-frame reversible extractors, as already described, for the following reasons: (1) On the basis that the eight-frame extractor of the reversible type takes three minutes to extract a load, and that the big machine takes eight minutes to extract 45 or 50 combs, it will be seen that the latter does its work in just about half the time. Or, to put it another way, in a given time it extracts twice as many combs as can be handled in the eight-frame reversible. (2) While the eight-frame reversible requires the constant attention of one man, the big radials are so nearly automatic in the acceleration of speed that they require only about 12 minutes of time per hour. With the reversible it is necessary to extract partially one side, reverse, extract the other side, come back and extract from the first side. All of this takes labor.

With the radial machine, after it is started, no further attention is required from the operator until the combs are ready to remove. It starts at a low speed, gradually increases automatically, throws out three-fourths of the honey at a low speed for about five minutes, then in about three or four minutes more it throws out the residue of the honey at a high speed. It does a cleaner and more thorough job with less breakage of the combs than is done with a power reversible extractor of the old type. During all this time the operator can do other work, such as uncapping, allowing the big machine to spin and finish the job. The only time required is to empty

Root Simplicity radial extractor that does not require reversing. The reel is surrounded with perforated metal to catch particles of comb that would clog the honey pump.

Detail showing how combs are placed in reel.

and refill it, start it, and then forget about it until the combs are extracted. (3) The big radials are very much easier on the combs if they are properly handled. Most of the honey is extracted at a low speed. The speed automatically increases in the Root radial shown, as the honey flies out until high speed is reached. The pressure is all against the top-bar and not against the surface of the combs, as in the old type of machine. This means that scarcely a comb is broken, provided the frames are factory made, well nailed, and the combs wired in the frame. (4) The big radial non-reversible machines have twice the capacity with one-fifth the labor. (5) There is only one moving part – namely the big reel – in the radial machine, while in the reversible type there is not only the revolving reel, but the entire reversing mechanism, reversible pockets and arms, and other parts. (6) The non-reversible 45-comb machine will extract the honey out of the cappings at the end of the day's run, or at the noon hour if preferred (using a cappings basket). Removing the honey from the cappings by the old method of melting cappings and honey or that of draining is very slow and unsatisfactory. When the cappings are melted with the honey, the flavor of the latter after it is separated from the wax is impaired. With the Simplicity extractor the honey comes from the cappings perfectly clear, and the cappings are almost dry. (7) A perforated metal cylinder surrounds the reel of the 45-comb machine. Broken pieces of comb, dead bees, etc., are caught on this screen, thus clarifying the honey to a large extent before it goes out at the honey gate. (8) As the comb surfaces do not come in contact with any part of the machine during extracting, the danger of spreading foulbrood is very much lessened. (9) It is much easier to get the combs out of the non-reversible machine because the pressure or force is against the top-bar that can not crush or stick to the reel. In the reversible machines, especially those using power, the pressure or force is against the surface of the comb. So great is it that new or soft combs are forced against and imbedded into the wire cloth or screen of the basket. When the frame is removed or reversed, there is danger that some of the comb surface will stick to the screen, with the result that the comb will be broken or defaced.

The advantages of the new method of taking the honey without reversing the combs are so apparent that many large producers are adopting it and throwing out their old machines.

How Both Sides Of A Comb Are Extracted At Once – Many do not understand why honey can be extracted from both sides at the same time without reversing. An examination of the arrows in the diagram shows the direction of the force. If the top-bar of each frame is placed to the outside, the honey will come out of one side of the comb as well as the other, the upward tip of the cells favoring the flow of the honey outward and toward the top-bar. The pressure should always be toward the top-bar, in order to clean all of the honey out of the combs. (The combs will also stand more pressure toward the top-bar than against an empty space between the bottom of the comb and the bottom-bar.) If there were pressure on the forward side, and a vacuum on the rear side, the honey should flow from the rear side of the comb more readily than from the other; but practical tests prove that one side of the comb is extracted just as clean as the other. The air on both sides of the comb is moving with the combs, therefore the pull is in a direction outward from the center. This is known as the centrifugal force. It is exactly in line with the midrib of the comb toward the top-bar.

Power vs. Hand Machines – To determine exactly how much honey is left in the cells after extracting, in 1921 the authors made a number of tests with combs that for 2½ minutes had been in an eight-frame Buckeye extractor, speeded up to 350 revolutions per minute. Eight combs were carefully weighed before and after uncapping and extracting, then after these weights were secured the combs were cut out of the frames, melted up, and the honey, thus separated from the wax, was weighed and compared with the original amount of honey extracted from these eight combs. After several tests the amount of honey left in the cells was found to vary from three to 3½% of the original amount in the combs. These combs when taken from the extractor looked perfectly dry – that is, the exact angular

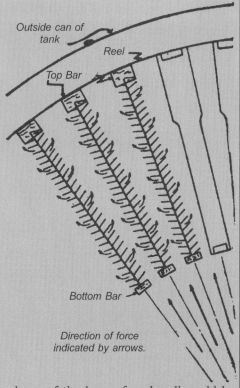

Direction of force indicated by arrows.

shape of the base of each cell could be seen clearly. Where there is enough honey left in the cell so that the angles of the base all run together it is safe to assume that the percentage of honey left is very high, perhaps between 10 and 20 percent.

In the four-frame hand-driven extractors the residue of honey left in the cells is much greater than in any of the power-driven machines, especially the power-driven big radial. The reason for this is plain enough; the hand-power is not sufficient to maintain a high speed. One's hand or arm gets tired except in case of the two or three frame extractor.

Others are applying power to their small machines. When power is applied to small extractors there is no reason why it could not do as clean work as a large machine also power driven. This is because a higher speed can be maintained.

Honey, Extracted

Still one of the best radials, though it hasn't been made in years, is this Hubbard 80-framer.

This parallel/radial extractor has a small footprint and is very efficient.

Hand- and power-driven mid-size radial extractors.

Medium and large parallel/radial extractors, also called 'Merry-Go-Round' extractors. The most sophisticated load and unload automatically. Pushing full frames in from one side pushes an equal number of extracted frames out the other side.

Honey, Extracted

Small and medium sized extractors, made in several countries, are popular in the U.S. Well made, relatively inexpensive, and easy to use, these are favored by small and midsize operations.

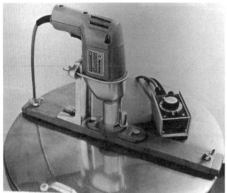

Another way to add power.

A smaller model radial extractor, with a power drive that can be adjusted – slow to start, then faster as the honey is removed.

A typical frame arrangement for an older-model tangential-type extractor. About half the honey is removed from one side of a frame, then the spinning is stopped and the frames reversed. All the honey from the second side is spun out, and the spinning is stopped again, the frame reversed a second time, and the remaining honey removed. This older model had the frames and their individual baskets contained within yet another basket to keep wax and debris out of the honey.

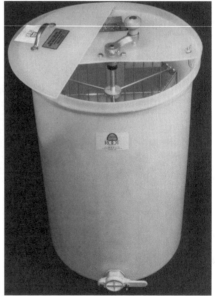

More modern tangential extractors, these without the additional basket. Both are hand driven.

Today, extractors are of three basic types: tangential, radial and the merry-go-round. In the latter the combs are whirled vertically. All three have advantages and disadvantages. The choice depends upon the size of the operation.

It would be difficult to indicate which extractor is more efficient in the removal of honey from the comb. As much as seven percent of the honey, perhaps more, may be left in the comb because of the inefficiency of some extractors. Since warm honey flows from the comb more easily than cold honey the temperature at which the honey is extracted is important.

Honey processing equipment should be made only of stainless steel. However small plastic extractors are useful for the hobbyist. Old galvanized iron extractors may still in use but these should be inspected for wear of the galvanized coating. A food-grade plastic is available to coat the metal parts of these machines that contact the honey.

Super storage area – After the honey has been extracted the supers with comb can be moved into a storage room. Some beekeepers return the supers with wet combs back to the colonies to be cleaned by the bees. Others allow the wet combs to be "robbed" by free-flying bees (a risky practice that could spread diseases and pests) and then store the supers with dry combs. Some beekeepers will store the wet combs in a super storage area.

Wax moths, the small hive beetle (*Aethina tumida*) and other vermin pose a constant threat to stored combs, especially those stored wet. Some beekeepers build their comb storage rooms so that the supers may be fumigated easily. In addition, some beekeepers use refrigerated rooms for comb storage (see WAX MOTHS and SMALL HIVE BEETLES). In the north where the winters are severe, the combs can be stored in unheated buildings after the weather has cooled down.

Other considerations in honey house construction – Several other matters should be considered in thinking about honey house construction. Honey is an acid food. If a concrete floor is used in a honey house it will soon become pitted, rough and difficult to clean. Concrete must have a coating resistant to acids.

Good insulation of the hot room and the settling area is important. Excessive heat from either or both may make work in the extracting room uncomfortable.

Many honey houses have been designed and built in a location where they have no other use and little resale value. For many reasons it is important to build a honey house in such a way and in such a location that it will serve other purposes should the beekeeper choose to move.

Following are three extracting facility layouts we have examined over the years. Seldom is an existing facility perfect, but usually any problems can be repaired, replaced or worked around. These facilities represent small, medium and large operations, and each were moved into an existing situation. When constructing a new facility, take into consideration the advantages and the limitations of each, so you can use the good and avoid the bad. Too, consider growth. The very best 250-colony operations would limit a business that grew to 2000 colonies without room for expansion. Also seriously consider the ever-tightening food safety regulations and allow for improvements in sanitation and production requirements.

Included in this chapter are guidelines for constructing a honey house that meets fundamental requirements for a safe and wholesome product. When constructing a new facility, or improving an existing facility, study these guidelines before you decide on a final design. The five Ps – Prior Planning Prevents Poor Product – will pay you well into the future.

Small Scale Operation Extracting Facility

Uncapped frames are lifted from the blue holding tub to the extractor.

After sitting in the warming area, which is simply a built-in shower area with a curtain to hold in the heat of a small heater, honey supers are moved out, next to the uncapping bench. The warming area is then re-stocked. The uncapping bench holds the cappings tank, and a plastic tub to hold uncapped frames to be put in the extractor.

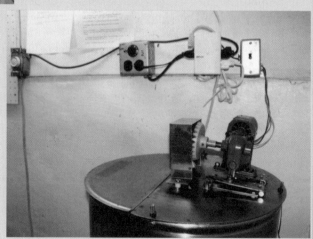

The variable speed extractor. When extracted, frames are unloaded and put into empty supers sitting next to the extractor.

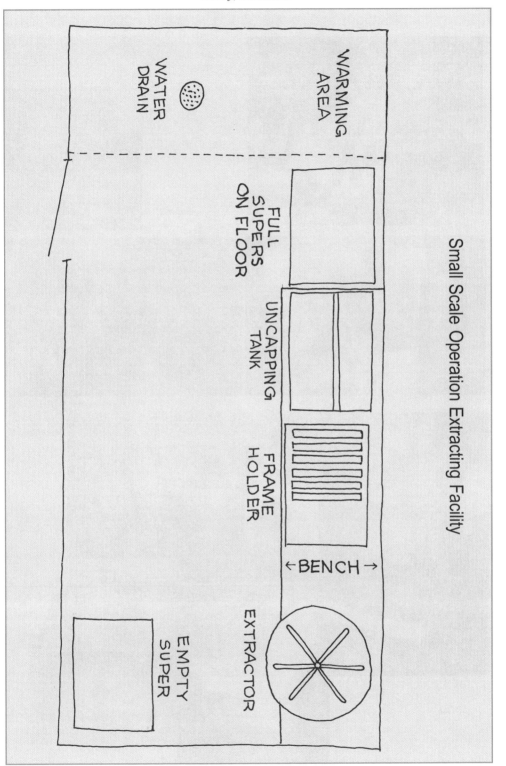

Honey, Extracted

Medium Scale Operation Extracting Facility

Unloading. Note drip board that keeps truck bed . . .

. . . and floor clean.

View from door.

View from wax-melter end of room.

Loading extractor.

Uncapping

Straining into storage tanks.

Wax melter.

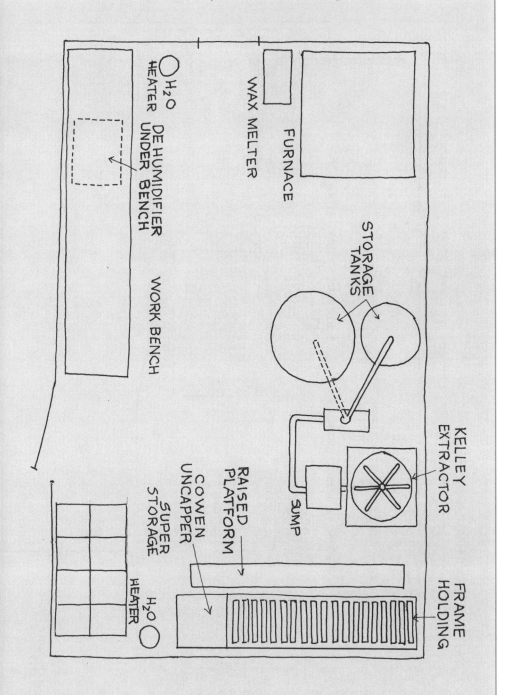

Honey, Extracted

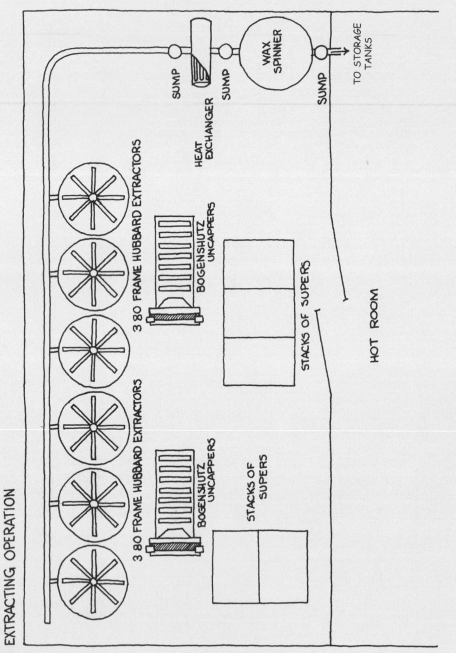

Supers are moved from the hot room to the uncapping area. When supers are empty, they are moved to the ends and the middle of the row of extractors. A crew of two to three people continuously moves supers and load extractors, while each uncapper has an operator.

At the end of the day, the wax spinner is emptied into a 55-gallon drum and melted later. Honey is pumped to 1000-gallon tanks in the next room.

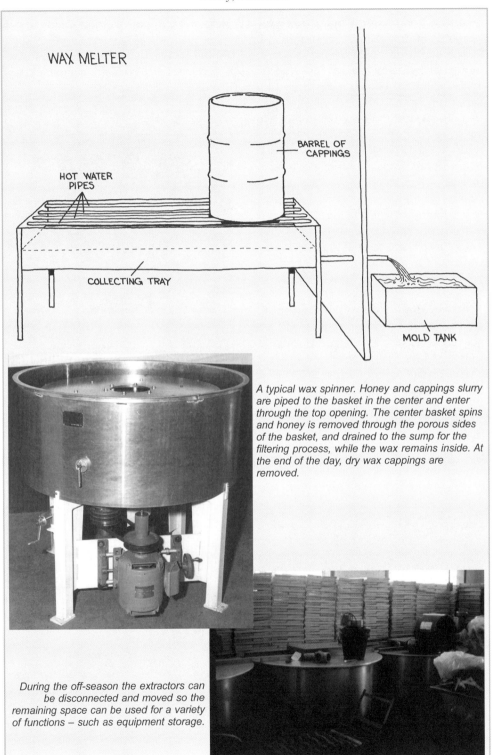

WAX MELTER

BARREL OF CAPPINGS

HOT WATER PIPES

COLLECTING TRAY

MOLD TANK

A typical wax spinner. Honey and cappings slurry are piped to the basket in the center and enter through the top opening. The center basket spins and honey is removed through the porous sides of the basket, and drained to the sump for the filtering process, while the wax remains inside. At the end of the day, dry wax cappings are removed.

During the off-season the extractors can be disconnected and moved so the remaining space can be used for a variety of functions – such as equipment storage.

Honey processing area – This area probably has the most variability in function. Honey can be moved with pumps or by using gravity flow.

It is best to keep the settling area separate, insulated and well-heated, usually about 80 to 90°F (26 to 32°C) to speed up the settling process. Settling tanks for honey should be shallow; a honey depth of two to three feet (60 cm to one meter) is best. A baffle tank in which the honey passes through three to five baffles can also be used. The tank should be covered but allow for inspection. From the settling area honey can either be packaged for retail sale or put in pails or drums for wholesale.

Straining – Honey that flows from an extractor, even though it will be packed in bulk, should pass through a coarse strainer to remove bits of wax, comb, propolis and even wood slivers from a frame. Cheesecloth is linty, disintegrates easily, and threads separate. Do not use it to filter honey. Nylon mesh material has proven to be good material for straining honey. It is strong, lint-free, easily cleaned and has a reasonably long life. Screening and perforated metal are also being used effectively to strain honey. Stainless steel should be used for any metal strainers. A variety of strainers are available from bee equipment suppliers.

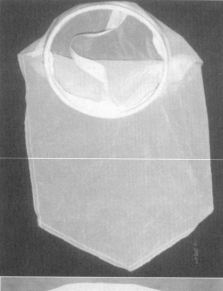

A plastic and metal combination strainer – one for coarse filtering, one for fine.

Hobby-level strainers.

Honey, Extracted

A hobby strainer system. This mesh/screen strainer catches most of the larger pieces of wax and other matter right out of the extractor, then the finer nylon cloth strainer, sitting on the pail, catches the rest.

Hobby and sideliner strainer system. Note the nylon hose used. These should only be those without coloring dye. A dye will contaminate the honey.

Warm honey, under slight pressure is pushed through a series of filters, removing all particulate matter thus insuring long shelf life.

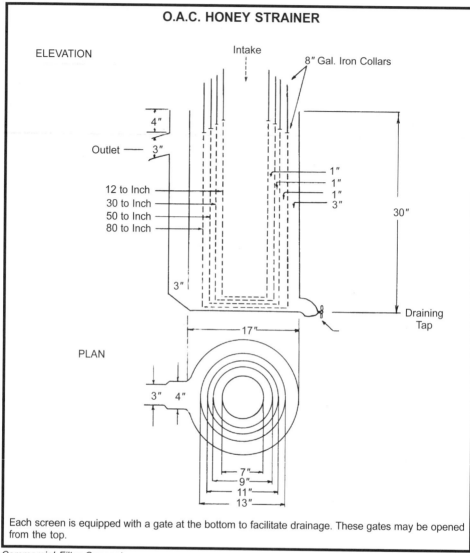

Commercial Filter Concept

Filtering – Most of the honey that is marketed by large commercial firms is filtered. Filtering means a process of pumping the honey through a flash heater then through a filter, followed by flash cooling to bottling temperature. Commercial filters are of various types.

Filtration removes pollen, an integral part of natural honey. In fact bees remove much of the pollen in nectar while it is still in their crop by means of "fingers" on the proventriculus. Much of the pollen that is found in honey finds its way there at the time of extraction when cells, or partial cells, of pollen may be emptied and contaminate the honey. Sometimes the pollen is from a plant species other than that from which the nectar is collected, which can confuse those who use pollen grains in honey to identify its source.

Filtration removes crystals, air bubbles and any pollen and beeswax or comb refuse that might be present. The resulting product is bright and clear and has a long shelf life because the crystal nuclei on which crystals might grow have been removed. Honey must be heated from 140°F (60°C) to 160°F (71°C) to be filtered. Thereafter it is important that it be cooled rapidly to prevent damage. Properly done, filtration has no adverse effect on quality.

Honey gates – Because of the physical nature of honey, the gates (faucets) used on extractors, storage tanks and devices used to fill jars must have a positive cut-off. A shearing-type gate is preferred. Honey must not drip from the gate or faucet and fall onto the floor, filling table or the jars themselves.

When the extracting season is finished, metal gates should be washed,

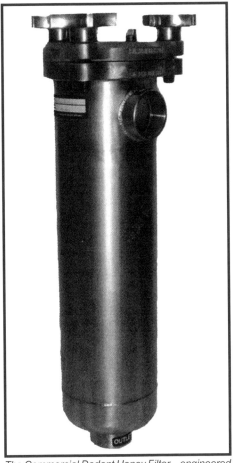

The Commercial Dadant Honey Filter – engineered for honey pump flow rates, normally up to five g.p.m.

Various honey gates.

Other types of honey gates.

dried and coated with a thin layer of food-grade mineral oil. This will prevent the gate from rusting and the following season it will open with ease. Honey gates that are left covered with honey will be stuck shut the following season and may be damaged when they are forcibly opened.

Most honey gates are made of cast iron, brass or stainless steel and have a long life. Plastic gates are popular and work reasonably well though they have a shorter life.

Honey pump – Honey may be pumped with ease, though some beekeepers prefer to have a multi-level honey house that allows honey to flow by gravity from the extractor through strainers, baffle tanks and into filling machines or storage containers. Honey pumps work very well if they are of the proper design and are not run at a high speed. Because honey is such a thick, viscous material, an improperly used honey pump may draw in air that is incorporated into the honey. The air is very difficult to remove. Improperly used honey pumps are often the cause of foam or rings of air bubbles on the top of jars of liquid honey. The lower the temperature the more difficult it is to pump honey and the more likely it is that air will be incorporated.

The most important consideration in using a pump is that the honey must flow naturally into the pump where it is pushed, not pulled, to its final destination. The pump should therefore be positioned below the tank from which the honey is being pumped. It is helpful to use a very large pipe on the inlet side of a honey pump to aid the flow of the honey.

Pumps that are used to move honey should be of the positive displacement type. Centrifugal pumps that are commonly used to pump water should never be used to move honey. A gear-type pump is usually favored though other positive displacement type pumps are available. A gear pump should be run at about 30 to 150 revolutions per minute and the higher figure is used only when the honey is warm. It is usually necessary to use a gear reducer to obtain this slow speed. Even when a honey pump is properly used the gaskets should be checked frequently to make certain that they do not leak and allow air to enter.

Bulk storage of honey – In constructing a honey house, a beekeeper should keep in mind the storage and the moving of containers used for storage of honey. The most popular container for holding and transporting honey today is the 55-gallon steel drum with a food grade liner that holds a little over 600 pounds (about 300 kg) of honey. Five-gallon (28 kg or 60 lb), round plastic buckets are popular containers. The large reinforced plastic tote is used commercially.

Some honey is transported both within the country and internationally in large stainless steel tanks and tank trucks of varying sizes. Most of these bulk tanks are equipped with some type of warming device to aid the flow of the honey at the final destination.

Honey tanks – There are no fixed dimensions for tanks that are used for storing, blending or heating honey. Only stainless steel should be used in honey processing. The reasons for this are that it is practically insoluble in most food products, does not cause off-flavors, cleans easily, and is relatively easy to weld. The first three of these reasons are, of course, the most important. The disadvantages of stainless steel are that it is expensive, sometimes difficult to work into different shapes and can corrode if it is not properly cleaned. So long as stainless steel surfaces are exposed to the atmosphere they remain passive, that is they resist corrosion. The best protection for stainless steel equipment when it is not in use is to wash and dry it thoroughly.

The food industry uses a stainless steel that contains 18 percent chromium and eight percent nickel. The most popular finish is called number four. It does not

Storage tanks of stainless steel should be easily drained.

Plastic 'totes,' held in metal cages for protection are commonly used for both storage and delivery. Many styles and sizes are available in the food industry. They fit conveniently on a pallet and come with a large gate.

show scratches, is easy to clean and has a smooth surface.

Many beekeepers have found that secondhand stainless steel milk storage tanks are excellent for honey. These are often discarded if the insides are scratched or the tanks dented. Such tanks are not suitable for milk storage but they are quite satisfactory for honey. It is usually necessary to enlarge the outflow valve so that the honey will move from the tank rapidly.

One will sometimes find glass or epoxy-lined tanks that are designed for food use. These and food-grade plastic tanks are satisfactory for honey storage but none of these has the long life of a stainless steel tank.

Preparation of honey for market on a small commercial scale – It is important that beekeepers who prepare their own honey for market give the customer the same high quality product that the bees store in the beehive. This section concerns the packing of liquid honey.

Users of honey expect that if they buy liquid honey it will remain liquid for a reasonable period of time. The packing and preparation of crystallized and comb honey are discussed separately. In preparing liquid honey for market there are several considerations including crystallization, fermentation, effect of metals on honey, moisture content, shelf life, and labeling. All of these are discussed separately under the general heading of honey.

A commercial food-packing line consists of an area where the jars are removed from the cardboard boxes, a jet for removing dust from the jars or a washer, a filling machine that is fed from a bulk tank, a capping machine, a labeling machine, and a packing table where the filled jars are placed back into the cases for storage and shipping.

Beekeepers who pack only a small amount of honey will want to proceed with packing in much the same order though they will not use the elaborate equipment used by commercial packers.

A flash heater above, a filter, and a flash cooler, all in one convenient location.

Honey, Extracted

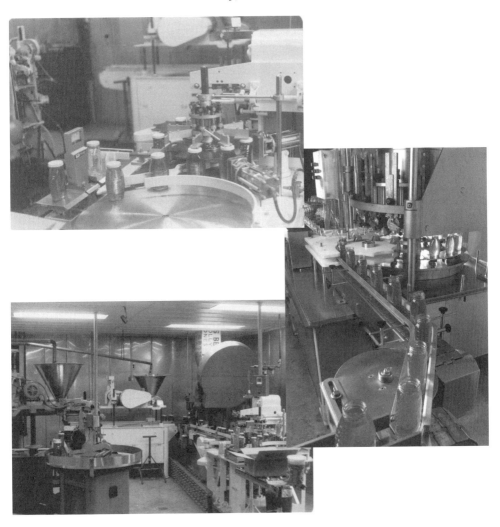

Commercial bottling lines in use. Typically, the bottles are washed or air-blown clean, filled with warm to hot honey, capped, double-labeled, sealed and reboxed in a continuous flow. Speed, cleanliness, accuracy and space are all important in any operation, plus ease of use (simplicity) and amount of use (limited down time) to justify cost.

A bottle filler can be as simple as this. Digitally programmed to fill a specific amount, they are accurate, inexpensive, and uncomplicated. The food industry uses a wide variety of this type of machine.

Equipment for the small producer is available from the equipment suppliers. Automatic, table-top filling equipment speeds up filling the jars. In addition such a device insures the correct weight and also keeps the containers free from spilled or dripped honey. Small producers generally put labels on by hand. The bottling area needs to be planned in the same way as a small commercial operation.

Pasteurization of honey – Wild yeasts are present in honey and can be removed, along with small crystals and other particles that cause crystallization. Flash heating of honey is preferred over bulk heating since the honey is held at a high temperature for much less time and less damage will be done. Heating honey with a flash heater is a simple matter as honey flows more easily as the temperature is increased. However, cooling hot honey is difficult because of the increased viscosity that occurs as it cools. Liquid honey should be packed while warm. (also see DYCE PROCESS FOR MAKING CRYSTALLIZED HONEY).

Stainless bulk milk tanks can make excellent honey storage tanks. A larger drain opening is usually needed.

Honey, Extracted

Though still popular in some parts of the U.S., the common quart and pint canning jars are declining in popularity. These custom made jars were popular in their day, however.

Liquefying honey – To liquefy granulated honey for packing it is important to heat the honey only to the extent necessary. Commercial honey packers liquefy honey in one of two ways. They may place drums or buckets of honey in an upright position in a heated room where they remain for 24 or more hours until they are liquefied. This method is easy but the honey can be damaged to a small extent. A better method is to place drums on their sides in a heated room so that the partially liquefied honey, still containing a number of crystals, flows from the containers as it is warmed. The temperature of this partially liquefied honey will usually be less than 100°F (38°C) when it flows from the hot room and the honey is not harmed. The honey may then be flash heated. Small producers can process small containers of honey using electrically-heated hot boxes that serve the same purpose as a larger hot room.

Blending honey – Commercial packers blend honeys from different areas to offer a uniform product to the public all year. The most popular commercial pack is a clover or clover-alfalfa mixture that is light in color and mild in flavor. This is apparently the type of honey the average American prefers but obviously there are exceptions. Varietal honeys are increasing in popularity. It is not advisable to hold honey for as long as a year before blending because of the dangers of fermentation.

Selecting a jar for packing – There are several considerations when selecting a jar for liquid honey. The queenline-type has been popular for a number of years. The equipment suppliers offer a wide selection of jars in many sizes. The selection of a jar for liquid honey depends on the market chosen.

The jars used for packing chunk honey, pieces of comb honey surrounded with liquid honey, are quite different. The round jar with a slight shoulder is suitable since the piece of comb can be inserted and removed easily.

Cleaning jars – The jars that are used for packing honey should be thoroughly cleaned before they are filled.

Canning jars used today are standard, right-out-of-the-box containers. Labeling can be a problem to accommodate the embossed letters on the jar. Two-part metal, and regular plastic lids are available. Canning jars have the advantage for the end consumer in that they are readily reusable.

Even new jars will contain a certain amount of cardboard dust picked up from the boxes in which they are shipped. Jars also may have a very thin film of mold-release material left from manufacture. Jars that are washed must air-dry upside down to prevent dust from entering.

Selecting a cap – Traditionally the caps for queenline-style honey jars have been round metal caps with a plastic gasket. They have usually been painted white but decorated and colored caps are available. Some beekeepers prefer the plastic caps which do not dent or scratch like the metal ones.

Labeling varietal honey – Beekeepers are often at a disadvantage because the public is not aware that different flowers produce nectars and therefore honeys with different flavors. This problem can be overcome in large part by using a second label that tells the buyer more about the product (see LABELS FOR HONEY JARS). Many beekeepers and honey packers have

This two and a half pound jar was commonly used for chunk honey due to its wide mouth. It is seldom used now.

Honey, Extracted

Plastic containers come in a multitude of sizes, shapes and materials. They are inexpensive to make, light weight and inexpensive to ship, and readily useful for the end consumer.

The 12-ounce bear commands roughly 20% of the U.S. grocery store shelf space (the one pound jar does also). This is because, primarily, ease of use and convenient size.

Honey, Extracted

Honey, Extracted

These plastic containers all have one thing in common – they are a convenient size and easy to use. In the back, left is the 'classic' plastic, next to that are plastic squeeze tubes, and in the back on the right is another commercial and well-made plastic bottle. In front are plastic 'skep' bottles.

In the case of liquid honey, for example, a jar that is partially crystallized is not as attractive to most consumers as is one that is liquid. A crystallized piece of comb honey has little eye or taste appeal. A jar of finely crystallized, Dyce-processed honey

found that explanatory labels are helpful in sales.

Shelf life – The term "shelf life" is used to designate the time that a food product will remain in good condition and continue to have eye appeal for the buyer. Shelf life starts the moment the honey is packed and includes the time during which it is being transported, stored in a warehouse and distributed. Packers of liquid honey seek to have a shelf life of four to six months but this is possible only if the honey is properly filtered. Crystallized, or creamed, honey will liquefy if it is stored in a warehouse at a temperature that is too high.

At left – Colored caps, whether on plastic or glass containers are as important as the container and the label when it comes to attractiveness. They can also denote variety of honey, size of container or other aspects of the product, making it easier for the packer to identify an open case by looking at only the cap.

The one-pound jar, in its many forms is the standard size in the U.S. It commands roughly 20% of all shelf space in grocery stores (the 12 oz squeeze bear and tube also take up that much space). The long time favorite of beekeepers has been the Queenline jar (above right). A plastic, queenline-like jar is also favored (above left). The similar, but easier to use 'classic' jar (bottom), has mostly replaced the original glass queenline jar.

A large variety of exotic and custom containers are available for upscale markets.

should not be partially liquid when opened.

Comb honey has a short shelf life, usually only two to three months, though this depends on the type of honey and how rapidly it crystallizes. Therefore it is important that comb honey be moved into the marketplace and sold as soon as possible.

HONEY FORMS, PACKAGED – Honey has many uses, including in foods, beverages and cosmetics (see HONEY PRODUCTS). However, over 90 per cent of honey used by the bakeries, manufacturers and on home and restaurant tables is in a liquid form.

Surveys of U.S. packers indicate that about 45% of all liquid honey sold in the U.S. in 2005 went into the retail market, 40% into the bulk/industrial market, and 15% into the food services market.

Chunk, comb and cut-comb containers are as varied as liquid honey bottles, and an attractive, leak-proof container goes a long way in making sales.

Though this varies a bit every year depending on market trends, prices, availability and sourcing, these ratios are fairly steady.

In addition to the familiar liquid honey in glass and plastic containers, liquid honey is also available to consumers in natural beeswax combs (see COMB HONEY; CUT-COMB HONEY; and CHUNK HONEY). Another popular form of honey is crystallized or creamed honey (see DYCE PROCESS FOR MAKING CRYSTALLIZED HONEY). Honey is even available as dried honey for specialty markets. A new form, solid 100% honey is becoming available.

HONEY JUDGING – Honey shows, whether local, regional or national, have been popular with beekeepers for many years. A honey show is actually designed to improve the beekeeper's product for the market. Honey bees make a quality product. It is up to the beekeeper to preserve that quality. Having a qualified judge review the product can show the beekeeper where improvement is necessary.

Typically a show can have classes for extracted (liquid) honey grouped by color, for creamed honey, for cut-comb, chunk, square and/or round section, and frames suitable for extraction or as comb honey. Judging criteria for the show will list the classes and the number of points assigned for each item. Judging criteria are not very consistent throughout the United States so beekeepers wishing to enter a show are well-advised to be aware of both the show rules and the judging criteria. Judges must be familiar with the rules of the show to ensure that the competition is fair and equal. However the show supervisor or superintendent controls the actual entries.

Judges have score cards for the various classes. The score cards list the maximum number of points the judge can award. The judges must also make the decision on the number of points to subtract for an item. The score cards usually have room for comments from the judge. In general, items under the beekeeper's control, such as crystals, foam, dirt, wax pieces and lint will be of greater importance than color or flavor. Water content of honey is only partially under the beekeeper's control since the bees cap the honey. Beekeepers who extract uncapped honey do run the risk of high water content.

The value of 18.6% water in liquid honey is a standard maximum number throughout the United States. Honey with over that amount can ferment. Therefore honey in excess of 18.6% water is usually disqualified. Honey of very low water content (12 or 13%) is too thick (viscous) for a marketable product. The acceptable range for water content is usually 15.5 to 18.6%. Judges use an instrument called the refractometer for determining water content. The refractometer should be calibrated frequently and the temperature correction be made for every entry. The water content should be stated on the judge's score card.

Foam can be found on top of both liquid honey and creamed honey. Although beekeepers recognize the source of foam, usually from the extractor or from mixing of creamed honey, foam is not appealing to a customer who may think it is a sign of spoilage. Dirt is always unacceptable, as are parts of bees. The presence of lint could accelerate the formation of crystals. A polariscope (see photo) is used to examine liquid honey for lint and crystals. This device uses polarizing

Honey Judging Tools

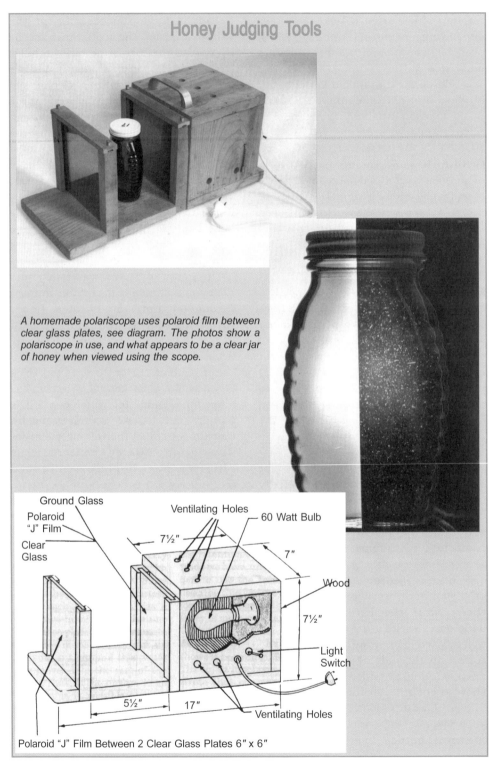

A homemade polariscope uses polaroid film between clear glass plates, see diagram. The photos show a polariscope in use, and what appears to be a clear jar of honey when viewed using the scope.

Diagram labels: Ground Glass; Polaroid "J" Film; Clear Glass; Ventilating Holes; 60 Watt Bulb; Wood; Light Switch; Ventilating Holes; 7½"; 7"; 7½"; 5½"; 17"

Polaroid "J" Film Between 2 Clear Glass Plates 6" x 6"

Honey Judging

Honey moisture is measured using a refractometer. Several models are on the market. They 'measure' solids actually, but the measurement is shown as a "% moisture." Calibration is required when temperature changes. Digital models are available.

Though color, by itself, is not a quality issue in most honey shows, making sure an entry is in the correct color class can be. Shown here are a Lovibond (no longer available) Colorimeter, and a Pfund grader, which measures color in MM. The jars shown have their Pfund color measurement on top. Right, is the newest color grading device which compares honey color to a standard giving final color in MM.

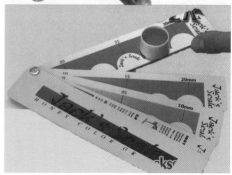

International, National, State and even county fairs once (and some still do) had entries for honey displays. Winning displays were generally architectually complicated, large, and contained lots of varieties of honey products. This from around 1900 or so.

Honey Judging

Eastern Apicultural Society (EAS) Score Cards

SCORE CARD

Exhibitor Number _____

Class Number __C-3__   Class Name __Cake__

MAX POINTS	SCORE	JUDGING CRITERIA	JUDGE'S REMARKS
30		OUTSIDE APPEARANCE: shape color if frosted, distribution, suitability	
40		INSIDE APPEARANCE: texture characteristic of type grain color	
30		EATING QUALITY: taste, odor	
100		AWARD _____	

This card is used for the different food categories – Yeast Bread, Cake, Quick Bread, Muffins, Yeast Bread Fancy, Yeast Rolls, Bars or Brownies, Cookies.

SCORE CARD

Exhibitor Number _____

Class Number __M-1__   Class Name __Mead, Dry__

MAX POINTS	SCORE	JUDGING CRITERIA	JUDGE'S REMARKS
20		CLARITY	
10		COLOR	
20		TASTE	
10		BODY	
20		BOUQUET	
10		BOTTLES	
10		BOTTLE CLOSURE	
100			

AWARD _____

This score card is for the different mead categories – Mead, Dry; Mead, Sweet; Mead With Fruit Juices; Mead, Sparkling.

SCORE CARD

Exhibitor Number _____

Class Number __P-4__   Class Name __Photography, Essay__

MAX POINTS	SCORE	JUDGING CRITERIA	JUDGE'S REMARKS
35		COMPOSITION	
35		TREATMENT OF SUBJECT MATTER	
30		QUALITY AND PRESENTATION	
100		AWARD _____	

This is the Score Card used for the Photography classes – Photography, Essay; Photography, Scenic; Photography, Portrait; Photography, Close-Up.

Eastern Apicultural Society (EAS) Score Cards, continued

SCORE CARD   Exhibitor Number _____

Class Number H-1   Class Name Extracted Honey, WHITE

MAX POINTS	SCORE	JUDGING CRITERIA	JUDGE'S REMARKS
10		DENSITY (water content above 18.6% is disqualified, below 15.5% will be docked points) 15.5-17.0% 10 points 17.1-18.0 9 points 18.1-18.6 7 points	
10		ABSENCE OF CRYSTALS	
40		CLEANLINESS without lint 10 without dirt 10 without wax 10 without foam 10	
10		FLAVOR points reduced if adversely affected by processing disqualified if fermented	
10		CONTAINER APPEARANCE	
20		ACCURACY OF FILLING headroom 1/2" maximum, 3/8" minimum with no visible gap between honey level and cap uniformity of filling	
100			

AWARD _____

This is the Score Card used for the all of the classes of extracted liquid honey – White, Light, Light Amber, Amber, Dark.

SCORE CARD   Exhibitor Number _____

Class Number H-6   Class Name Honey, Square Section Boxes

MAX POINTS	SCORE	JUDGING CRITERIA	JUDGE'S REMARKS
20		UNIFORMITY OF APPEARANCE	
10		ABSENCE OF UNCAPPED CELLS	
15		UNIFORMITY OF COLOR	
10		ABSENCE OF WATERY CAPPINGS	
15		CLEANLINESS ABSENCE OF TRAVEL STAINS	
10		FREEDOM FROM GRANULATION AND POLLEN	
10		UNIFORM WEIGHT OF EACH SECTION	
10		TOTAL WEIGHT OF ENTRY	
100			

Score Card for Honey, Square Section Boxes; Honey Cut-Comb, round sections and the new cassette-type sections.

AWARD _____

Honey Judging

SCORE CARD   Exhibitor Number_____

Class Number A-1   Class Name Gift Arrangement

MAX POINTS	SCORE	JUDGING CRITERIA	JUDGE'S REMARKS
30		GENERAL APPEARANCE	
30		ORIGINALITY	
25		QUALITY OF HONEY AND PRODUCTS	
15		VARIETY OF PRODUCTS	
100			

AWARD _____

This Score Card is used for Gift Arrangements.

SCORE CARD   Exhibitor Number_____

Class Number A-3   Class Name Novelty Beeswax

MAX POINTS	SCORE	JUDGING CRITERIA	JUDGE'S REMARKS
25		ARTISTIC MERIT	
25		ORIGINALITY	
25		SKILL INVOLVED	
25		DESIGN	
100			

AWARD _____

This is the Score Card used for Novelty Beeswax, Miscellaneous Arts & Crafts and Sewing or Needlework Items

SCORE CARD   Exhibitor Number_____

Class Number G-2   Class Name Small Devices

MAX POINTS	SCORE	JUDGING CRITERIA	JUDGE'S REMARKS
25		EXPLANATORY TEXT	
35		PRACTICALITY	
15		EASE OF REPRODUCTION	
10		HELP TO BEEKEEPING	
15		ORIGINALITY	
100			

AWARD _____

This Score Card is for the Gadgets categories – Small Devices and Large Devices

Honey Judging

Eastern Apicultural Society (EAS) Score Cards, continued

SCORE CARD   Exhibitor Number _____

Class Number __C-9__   Class Name __Candy__

MAX POINTS	SCORE	JUDGING CRITERIA	JUDGE'S REMARKS
40		TEXTURE: characteristic of type, free from stickiness	
20		APPEARANCE: uniformity of entry, bite-size pieces	
40		FLAVOR	
100		AWARD _____	

Score Card for judging Candy.

SCORE CARD   Exhibitor Number _____

Class Number __B-4__   Class Name __Candles, Novelty__

MAX POINTS	SCORE	JUDGING CRITERIA	JUDGE'S REMARKS
25		CLEANLINESS AND QUALITY OF WAX	
25		DESIGN AND OVERALL APPEARANCE	
25		FINISHING DETAILS: wick trimmed to 1/2" flat, finished bottom	
25		ORIGINALITY	
100		AWARD _____	

Score Cards for Candles – right, Candles, Novelty and below, Candles, Molded Tapers. Candles, novelty (right), candles, molded tapers (below).

SCORE CARD   Exhibitor Number _____

Class Number __B-3__   Class Name __Candles, Molded Tapers__

MAX POINTS	SCORE	JUDGING CRITERIA	JUDGE'S REMARKS
25		CLEANLINESS, COLOR, QUALITY OF WAX	
25		UNIFORMITY OF APPEARANCE AND SHAPE	
25		UNIFORMITY OF PAIR	
25		FINISHING DETAILS: flat finished bottom, wicks trimmed to 1-2"	
100		AWARD _____	

Honey Judging

SCORE CARD   Exhibitor Number_____

Class Number _H-11_   Class Name _Frame of Honey_

MAX POINTS	SCORE	JUDGING CRITERIA	JUDGE'S REMARKS
25		UNIFORMITY OF APPEARANCE	
20		ABSENCE OF UNCAPPED CELLS	
15		UNIFORMITY OF COLOR	
10		ABSENCE OF WATERY CAPPINGS	
20		CLEANLINESS ABSENCE OF TRAVEL STAINS	
10		FREEDOM FROM GRANULATION AND POLLEN	
100			

AWARD _____

Score Card for Frames of Honey.

SCORE CARD   Exhibitor Number_____

Class Number _H-8_   Class Name _Honey, Round Sections_

MAX POINTS	SCORE	JUDGING CRITERIA	JUDGE'S REMARKS
20		UNIFORMITY OF APPEARANCE	
10		ABSENCE OF UNCAPPED CELLS	
15		UNIFORMITY OF COLOR	
10		ABSENCE OF WATERY CAPPINGS	
15		CLEANLINESS ABSENCE OF TRAVEL STAINS	
10		FREEDOM FROM GRANULATION AND POLLEN	
10		UNIFORM WEIGHT OF EACH SECTION	
10		TOTAL WEIGHT OF ENTRY	
100			

Honey, Round Sections Score Card

AWARD _____

Honey Judging

Eastern Apicultural Society (EAS) Score Cards, continued

SCORE CARD Exhibitor Number_____

Class Number _B-1_   Class Name _Beeswax, Single Piece_

Beeswax, Single Piece Score Card

MAX POINTS	SCORE	JUDGING CRITERIA	JUDGE'S REMARKS
35		CLEANLINESS	
20		UNIFORMITY OF APPEARANCE	
15		COLOR	
15		AROMA	
15		ABSENCE OF CRACKS & SHRINKAGE	
100		AWARD _____	

SCORE CARD Exhibitor Number_____

Class Number _B-2_   Class Name _Candles, Dipped Tapers_

MAX POINTS	SCORE	JUDGING CRITERIA	JUDGE'S REMARKS
25		CLEANLINESS, COLOR, QUALITY OF WAX	
25		UNIFORMITY OF APPEARANCE AND SHAPE	
25		UNIFORMITY OF PAIR	
25		FINISHING DETAILS: last drip left on wicks left joined	
100		AWARD _____	

Score Card for Candles, Dipped Tapers

SCORE CARD Exhibitor Number_____

Class Number _H-9_   Class Name _Creamed Honey_

Creamed Honey

MAX POINTS	SCORE	JUDGING CRITERIA	JUDGE'S REMARKS
30		FINENESS OF CRYSTALS	
25		UNIFORMITY AND FIRMNESS	
20		CLEANLINESS AND FREEDOM FROM FOAM	
15		FLAVOR points reduced if adversely affected by processing disqualified for fermentation	
10		ACCURACY OF FILLING AND UNIFORMITY	
100		AWARD _____	

SCORE CARD

Exhibitor Number _____

Class Number H-10   Class Name Chunk Honey

MAX POINTS	SCORE	JUDGING CRITERIA	JUDGE'S REMARKS
20		NEATNESS AND UNIFORMITY OF CUT upgrade for parallel, 4-sided cuts downgrade for ragged edges	
20		ABSENCE OF WATERY CAPPINGS, UNCAPPED CELLS AND POLLEN	
20		CLEANLINESS ABSENCE OF TRAVEL STAINS, FOAM, WAX FLAKES & CRYSTALLIZATION	
20		UNIFORMITY OF APPEARANCE COLOR, THICKNESS OF COMB, ACCURACY & UNIFORMITY OF FILL	
10		DENSITY AND FLAVOR OF LIQUID A) Density (above 18.6% is disqualified)	
10		B) Flavor (reduced points if adversely affected by processing) C) Disqualified for fermentation	
100			

AWARD _____

Chunk Honey Score Card

filters for the light passing through the jar of honey. Judges can see the presence of lint, dirt, and the amount and size of crystals. Many models are available, though homemade models are often used.

One of the most important items for creamed honey is the fineness of the crystals. The judge is looking for a product as smooth as butter. Another important item is the firmness of the creamed honey. If too liquid, there is no advantage over liquid honey; if too firm it is impossible to spread.

In judging honey in combs, the judges will be looking for a clean comb surface (wax capping), not one with "travel stains" made by the bees walking over the comb surface. The wood frames and plastic rings must also be free of propolis and stain. Judges also look for white cappings although this is actually under the control of the bees and not the beekeeper.

In scoring the flavor of honey, the judges taste for fermentation, smoky flavor from overuse of the smoker and scorching from uncapping knife or overheating the honey during processing. Since honey comes in many flavors, judges are careful not to let personal preference influence the scoring. For information on guidelines for exhibitors and judges see Morse, R.A. and M.L. Morse. *Honey Shows – Guidelines for exhibitors, superintendents, and judges.* (Wicwas Press, New Haven, CT, 35 pages. 1996.) -AWH

The Eastern Apicultural Society of North America produces an annual Conference and Short Course in the

Eastern part of North America each year. A part of that Conference is an Annual Show for attending members. The rules and regulations for this show are straight forward and easy to follow for both Judges and participants. Following are the Annual Show Rules, used at the time of publication. They are the General Rules for each class (Extracted Honey, Class H1-H5, or Photography, Class P1-P4, for example), plus the judging criteria used.

Though individual shows may vary, these guidelines are thorough and can be used nearly anywhere, by any organization. We invite you to do so.

EAS Honey Show Rules

THE PURPOSE OF THE EAS COMPETITIVE SHOWS IS TO PROVIDE A COMPETITIVE FORUM DIRECTED TOWARD IMPROVEMENT OF THE PRODUCTS OF BEEKEEPING. THIS POLICY STATEMENT IS INTENDED TO ESTABLISH GUIDELINES TO CLEARLY DEFINE THE RULES FOR ATTAINMENT OF THIS PURPOSE. As the show is to be a competition, it is fundamental that all participants know in advance the judging criteria for each class and follow the rules herein.

The Honey Show Committee is responsible for drafting policies, rules and judging criteria and for disseminating information. Judges will adhere to the Committee's policies and rules.

The judging criteria are designed to reflect the skill of the participant. For example, taking off seasonal and plant specific honey. These are best illustrated by honey color and flavor. Alteration of either color or flavor in handling or preparation of the entry by the beekeeper will adversely affect the score.

Judges may be professors of apiculture, or students of apiculture under a professor's supervision. Judges may be blue-ribbon winners of local, state and/or regional honey shows, but they may judge only in the area in which they excel. Professionals may judge in their fields, such as sewing or cooking.

Judges must completely fill out each score card unless an entry is disqualified. Then the reason for disqualification must be stated on the score card. Judges should make comments on score cards in order to help the exhibitor improve. Judges must break all contest ties.

The score cards and entries are the property of the exhibitor. Although EAS will exercise all due care in judging and displaying entries, exhibitors enter items at their own risk. If possible, judges may hold an open session to answer exhibitors' questions after the show is opened to the public.

Annually, a Honey Show committee within the host state will be responsible for:
1. Allocating space for the show and arranging entries.
2. Compliance with show rules and judging criteria.
3. Obtaining judges.
4. Cooperating with EAS Treasurer in obtaining awards.
5. Accepting entries and opening show at a predetermined, published time.
6. Within three weeks of the close of the Show, filling out a yearly comparison sheet and making recommendations for future shows.
7. Recording all Blue Ribbon winners for the EAS, and other Journals.

GENERAL SHOW RULES
1. All entrants MUST BE current dues-paying members of EAS. Proof of current paid dues status must be submitted by the entrant at the time of making entries.
2. Only one entry in each class may be made by an individual family, or that individual's family, or that individual's apiary.
3. At the time of entering, the exhibitor may place a small label, with the exhibitor's number, inconspicuously on the entry. The exhibitor must fill out all labels. The labels will be available for the exhibitors' use at the show registration table.
4. Separate section or class rules will apply.
5. Identifying labels on the entries are forbidden. In Arts and Crafts and Gift Arrangements classes, if the exhibitor's name and/or apiary is an integral part of the entry, names are permitted.
6. The exhibitor must choose which classes to place entries in. Judges may adjust classes at their discretion.
7. Entries can only be made during the hours published.
8. Entries must be left intact and on display until released by the Show Chairman.
9. No commercial products or displays are permitted.
10. No EAS entry can be submitted again for three years.
11. The decision of the judges in all cases will be final.
12. Entries will not be accepted by mail.
13. The Show Chairman has the authority to accept, reject and classify entries in accordance with the show policies, rules and judging criteria.
14. Any exhibitor wishing to protest must do so to the Show Chairman within one hour of the public opening of the show.
15. EAS assumes no liability for loss or damage of entries. Although EAS will exercise all due care in judging and displaying entries, exhibitors enter items at their own risk.
16. Entries not claimed by the end of the conference will be disposed of by the Show Chairman.

HONEY SHOW
CLASS DESCRIPTION
H1*   Three 1-lb jars of honey, extracted, white
H2*   Three 1-lb jars of honey, extracted, light
H3*   Three, 1-lb jars of honey, extracted, light amber

Honey Judging

H4* Three, 1-lb jars of honey, extracted, amber
H5* Three, 1-lb jars of honey, extracted, dark
H6 Three section boxes of comb honey
H7 Three packages of cut-comb honey, 4" square
H8 Three circular sections of comb honey
H9 Three 16-oz jars of creamed honey
H10 Three 16-oz jars of chunk honey
H11 One frame of honey, wooden
H12 One frame of honey, plastic

1. Entries in classes H1-H5, marked, *, must be in queenline type jars, and may have plain metal or plastic lids. Canadian exhibitors may enter Classes H1-H5 with 500 gram universal jars and Classes H9 and H10 with 500 gram barrel-type jars.
2. Entries in class H11 must be displayed in bee-proof cases having both sides made of transparent glass or plastic.
3. Entries in classes H6, H7 and H8 must be in the appropriate container: window cartons, round section lids – both transparent, cut-comb box – all sides transparent.
4. Entries in classes H9 and H10 should be in cylindrically uniform, "wigwam" jar OR in the new "shoulder" jar.
5. Honey color classes H1-H5 will be determined by the Show Chairman and the judges. An official honey color grader (i.e. Pfund Color Grader, USDA Honey Comparator, etc.) may be used to make the final decision.
6. All entries must be the product of the entrant's apiary and have been produced since the previous EAS Honey Show.

JUDGING CRITERIA

Extracted Honey
Classes H1 to H5 MAX. POINTS
1. Density 20
(water content above 18.6% will be disqualified & below 15.5% will be docked points)
 15.5 – 17.0% 20 points
 17.1 – 18.0% 15 points
 18.1 – 18.6% 10 points
2. Absence of crystals 10
3. Cleanliness 24
 a. Without lint - 6; b. Without dirt - 6
 c. Without wax - 6; d. Without foam - 6
4. Flavor 8
 a. Points will be reduced for honey flavor that has been adversely affected by processing.
5. Color 8
6. Container appearance 10
7. Accuracy of filling 20
 a. Headroom: ½ inch maximum, 3/8 inch minimum with no visible gap between honey level and cap.
 b. Uniformity of filling
TOTAL 100

Creamed Honey
Class H9 MAX. POINTS
1. Fineness of crystals 30
2. Uniformity and firmness of product 25
3. Cleanliness and freedom from foam 20
4. Flavor 15
 a. Points will be reduced for honey flavor that has been adversely affected by processing.
 b. Disqualified for fermentation
5. Accuracy of filling and uniformity 10
TOTAL 100

Chunk Honey
Class H10 MAX. POINTS
1. Neatness and uniformity of cut 20
 a. Upgrade for parallel & 4-sided cuts
 b. Downgrade for ragged edges
2. Absence of watery cappings, uncapped cells and pollen 20
3. Cleanliness of product (down-grade for travel stains, foreign matter, wax flakes, foam and crystallization) 20
4. Uniformity of appearance in capping structure, color, thickness of chunks and accuracy and uniformity of fill 20
5. Density and flavor of liquid portion of pack
 a. Density (water content above 18.6% will be disqualified) 10
 b. Favor (points will be reduced for honey flavor that has been adversely affected by processing) 10
 c. Disqualified for fermentation
TOTAL 100

Comb Honey
Classes H6, H8 MAX. POINTS
1. Uniformity of appearance 20
2. Absence of uncapped cells 10
3. Uniformity of color 15
4. Absence of watery cappings 10
5. Cleanliness and absence of travel stains 15
6. Freedom from granulation and pollen 10
7. Uniform weight of each section 10
8. Total weight of entry 10
TOTAL 100

Frame of Honey, Wooden Only
Class H11 MAX. POINTS
1. Uniformity of appearance 25
2. Absence of uncapped cells 20
3. Uniformity of color 15
4. Absence of watery cappings 10
5. Cleanliness and absence of travel stains 20
6. Freedom from granulation and pollen 10
TOTAL 100

Frame of Honey, Plastic Only
Class H12 MAX. POINTS
1. Uniformity of appearance 25
2. Absence of uncapped cells 20
3. Uniformity of color 15
4. Absence of watery cappings 10
5. Cleanliness and absence of travel stains 20
6. Freedom from granulation and pollen 10
TOTAL 100

Continued on Next Page

Cut Comb Honey

Class H7	MAX. POINTS
1. Neatness and uniformity of cut, absence of liquid honey	20
2. Absence of watery cappings, uncapped cells and pollen	20
3. Cleanliness of product, absence of travel stains, crushed wax, and crystallization	20
4. Uniformity of appearance (color of honey, capping structure, thickness of comb)	15
5. Uniformity of weight	15
6. Total weight of entry	10
TOTAL	**100**

MEAD SHOW

CLASS	DESCRIPTION
M1	Mead, dry
M2	Mead, sweet
M3	Mead made with fruit juices
M4	Mead, sparkling, made with or without fruit juices

GENERAL RULES: All wines should have been made by the exhibitor by the process of fermentation. A 3" x 5" card should accompany each mead entry. The card should have the exhibitor's number put on at the time of entry. In classes 3 and 4 the type(s) of fruit used must be included.

BOTTLES: Only one bottle will be entered in each class. Still wines should be exhibited in clear (not frosted), colorless (not tinted), wine bottles of approximately 750 ml or 25.4 fluid ounce capacity. Sparkling wines must be exhibited in champagne-type bottles. The domestic (U.S.) Champagne bottle is excellent.

CORKS: Natural cork stoppers are preferred. Screw-top wine bottles or plastic corks may be used in classes 1-3. Corks may be driven straight cork or flanged and hand-applied. Corks are available from wine supply stores or vintners.

LABELS: Entries must not have any identifying labels on the bottles. Small labels with exhibitor's number may be placed inconspicuously if the exhibitor chooses. Labels will be available at the entry desk.

PRESENTATION: Wine bottles should be filled so that when the cork is pushed right home, the air space is between ¾" and 1" in depth. Sparkling wines should have an air space of 1" to 1-1/4".

JUDGING CRITERIA

Mead

Classes 1-4	Still	Sparkling
1. Clarity	20	15
2. Color	10	10
3. Taste	20	15
4. Body	10	10
5. Bouquet	20	15
6. Bottles	10	10
7. Bottle Closure	10	10
8. Carbonation	0	15
TOTAL	**100**	**100**

HONEY COOKERY SHOW

CLASS	DESCRIPTION
C1	Cookies, 1 dozen
C2	Bars or brownies, 1 dozen
C3	Cake, unfrosted or frosted, 1 cake
C4	Yeast bread, 1 loaf
C5	Yeast bread, fancy, 1 item
C6	Yeast rolls, 1 dozen
C7	Quick bread, 1 loaf
C8	Muffins, 1 dozen
C9	Candy 12 pieces

1. Entries must be accompanied by the recipe as used, written on white 3" x 5" cards in duplicate, without name of entrant.
2. EAS reserves the right to publish the recipes.
3. At least 25% of the sweetening agent must be HONEY. Frostings and decorations may be made with 100% sugar.
4. Entries must be made on plain paper or foam plates, in dome-top cake carriers, or on cardboard covered with foil. Plates and covers will not be furnished by the Show Committee.
5. Enter cake, yeast bread loaf and fancy, and quick bread unsliced.

JUDGING CRITERIA

CAKE

Class C3	MAX. POINTS
1. Outside appearance	30
a. Shape and color	
b. If frosted, etc.: distribution and suitability	
2. Inside appearance	40
a. Texture; b. Grain; c. Color	
3. Eating quality	30
a. Taste; b. Odor	
TOTAL	**100**

Cookies, Bars, Brownies

Classes C1 & C2	MAX. POINTS
1. Outside Appearance	30
a. Shape and appropriate size; b. Color	
c. Uniformity of entry	
d. If frosted, etc.: distribution and suitability	
2. Inside Texture	40
a. Texture characteristic of type; b. Grain; c. Color	
3. Eating quality	30
a. Taste	
TOTAL	**100**

Candy

Class C9	MAX. POINTS
1. Texture	40
a. Characteristic of type; b. Free from stickiness	
2. Flavor	40
3. Appearance	20
a. Uniformity of entry; b. Bite-sized pieces	
TOTAL	**100**

Honey Judging

Yeast Breads and Quick Breads

Classes C4 - C8	MAX. POINTS
1. Outside appearance	30
a. Shape characteristic of type; b. Crust or surface	
2. Inside Texture	40
a. Texture; b. Grain; c. Color	
3. Eating quality	30
a. Taste; b. Odor	
TOTAL	100

ARTS AND CRAFT SHOW

Class	Description
A1	Gift Arrangement
A2	Sewing or needlework items
A3	Novelty beeswax with additives permitted
A4	Misc. arts and crafts

1. All items must have a beekeeping theme.
2. Exhibitor must submit estimate of time to make item.
3. Small changes to commercial items or copies of commercial items may be downgraded.

JUDGING CRITERIA

Gift Arrangement

Class A1	MAX. POINTS
1. General appearance	30
2. Originality	30
3. Quality of honey & products	25
4. Variety of products	15
TOTAL	100

Classes 2-4	MAX. POINTS
1. Artistic Merit	25
2. Originality	25
3. Skill involved	25
4. Design	25
TOTAL	100

GADGET SHOW

Class	Description
G1	Large devices (honey extractors, wax equip., etc.)
G2	Small devices

All entries must be accompanied by a typed or written explanation. This is to be used by the judges in scoring.

Classes G1 & G2	MAX. POINTS
1. Explanatory text	25
2. Practicality	35
3. Ease of reproduction	15
4. Help to beekeeping	10
5. Originality	15
TOTAL	100

BEESWAX SHOW

Class	Description
B1	Single piece, pure beeswax, 2 lbs or more
B2	Candles, dipped tapers, one pair, pure beeswax
B3	Candles, molded tapers, one pair, pure beeswax
B4	Candles, novelty, single or coordinated set, containing beeswax

1. All entries in Class B1 must be covered with clean, transparent plastic film.
2. The optimum color for pure beeswax in Classes B1-B3 is light yellow.

JUDGING CRITERIA

Beeswax

Class 1	MAX. POINTS
1. Cleanliness	35
2. Uniformity of appearance	20
3. Color	15
4. Aroma	15
5. Absence of cracks & shrinkage	15
TOTAL	100

Candles, Tapers

Classes 2-3	MAX. POINTS
1. Cleanliness, color, quality of wax	25
2. Uniformity of appearance and shape	25
3. Uniformity of pair	25
4. Finishing details:	25
a. For molded: flat, finished bottom, wicks trimmed to ½"	
b. For dipped: last drip left on, wicks left joined	
TOTAL	100

Novelty Beeswax Candle

Class 4	MAX. POINTS
1. Cleanliness & quality of wax	25
2. Design & overall appearance	25
3. Finishing details wick trimmed to ½", flat, finished bottom	25
4. Originality	25
TOTAL	100

PHOTOGRAPHY SHOW

Class	Description
P1	Close-up, print; Subject must relate to beekeeping
P2	Scenic, print; Apiary subject such as flowers, hives,
P3	Portrait, print; Person or beekeeping procedure in appropriate setting.
P4	Essay, prints; A set of from 4 to 7 pictures depicting a beekeeping story.

1. The photo contest is open to all photographers.
2. Prints must be 5" x 7" inches or larger, mounted on a mounting board that extends at least one inch beyond the print on each side. No frames are permitted. Essay prints may be mounted on one mounting board.
3. Prints may be black & white or colored.
4. Photographs can be entered only once in any EAS show.
5. Each photograph, including the Essay as a set, must be accompanied by a 3" x 5" card giving: photo title, entrant's name, address, city, state, zip or postal code, and telephone. The card must state the class entered.
6. Brief captions may accompany the Essay photographs. The order of Essay photographs must be indicated clearly.
7. Winners must agree to have their photos published.

JUDGING CRITERIA

Classes P1-P4	MAX. POINTS
1. Composition	35
2. Treatment of subject matter	35
3. Quality and presentation	30
TOTAL	100

HONEY PLANTS – To be a successful beekeeper one must live in, or move to, a place where many plants that secrete a large quantity of nectar may be found. It is usually not practical to plant plants for the production of honey, though this concept is changing. In some locations only a few hives of bees can be kept because the quantity of honey plants is limited. In other places one may keep 50 or more hives and still find there is sufficient nectar for all. For these reasons the study of honey plants is an important part of beekeeping.

Generally honey plants are divided into two categories: major and minor. The major honey plants are those that are widespread (or regionally abundant) and responsible for a large portion of the crop that is produced in that particular area. However, in some special locations and situations, a plant that usually contributes little to the crop may become a major yielder of nectar. Minor honey plants are more numerous in the number of species but are usually not present in sufficient numbers to have much effect on the crop. However, minor honey plants, collectively, may be very important in providing the pollen and nectar needed to build colony populations for the major honey flow.

Some plants appear to yield nectar under almost all conditions. However, on average, sunlight, soil type, moisture, temperature and humidity have a great affect on nectar yields. Often the presence or absence of a minor element in the soil is important for nectar secretion.

Plants produce nectar to attract insects,which may then become dusted with pollen in the process of nectar collection. As the insect flies from one flower to another, the pollen is transported and thus cross-pollination is achieved. In the U.S. it is estimated that about half of our honey is produced from weed or free-growing plants and about half from agricultural plantings. Plant breeders have not been concerned with nectar secretion or honey production and thus the plants they have selected may or may not be good producers of nectar. Most varieties of oranges, for example, do not require pollination to set fruit but yet almost all varieties of citrus are excellent nectar producers.

In this section only some of the major honey plants will be described. There are only a few honey plant books. One is out of print but available on the second-hand book market: Pellett, Frank C. "American Honey Plants." *American Bee Journal*, Hamilton, Illinois, later Orange Judd Publishing Company, New York and then Dadant and Sons, Hamilton, Illinois. This book first appeared in 1920. Another book is: Lovell, J. H. *Honey Plants of North America*. (The A. I. Root Co., Medina, Ohio.) This book first appeared in 1926. The original can only be found on the second-hand book market. An abbreviated but updated version was published as follows: Goltz, L.R. *Honey Plants*. (The A.I. Root Co., Medina, Ohio. 96 pages. 1977.) A reprint of the original Lovell book was published by the A.I. Root Co., Medina, Ohio, in 2000 and is still available. The black and white photos in this section are all original Lovell photos from the Root photo library.

Alfalfa – *Medicago sativa*. Alfalfa (called lucerne in many parts of the world) is a legume usually with yellow or purple flowers. It is the major crop grown for hay for dairy cows in the U.S., but also contributes significantly to other livestock feed. Fields are usually

Medicago sativa - Alfalfa

replanted every three to five years. Honey from alfalfa is considered by most users to be a superior table honey. It is light in color and mild in flavor and usually has low moisture content. It is much sought after by honey packers to use in their blends.

Where alfalfa is grown for seed there is usually little honey produced because of the large number of colonies used per acre, typically three to five. The alfalfa blossom has a peculiar pollination mechanism that "explodes" (trips) and hits the visiting bee. Honey bees learn to avoid this mechanism and to obtain nectar from flowers without touching the sexual parts of the flower (see LEARNING IN THE HONEY BEE). As a result of this learning process only about one percent of the alfalfa flowers that are visited by honey bees are pollinated and thus the need for so many colonies.

Where alfalfa is grown for hay, large honey yields are often obtained. Alfalfa

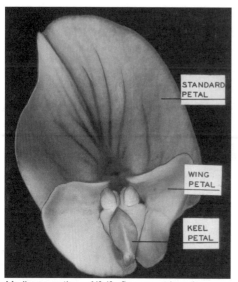

Medicago sativa - Alfalfa flower untripped.

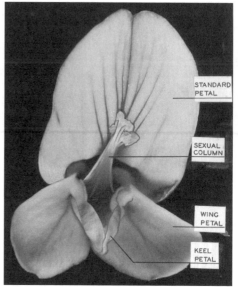

Medicago sativa - Alfalfa flower tripped

is the best plant for hay for dairy cows and is widely used in this way throughout the country. Extension services suggest that the best quality hay with the highest protein content is made from alfalfa that is cut while still in the bud stage, before more than 10% of the flowers have opened. However, for a variety of reasons, including inclement weather, hay is often cut late, after the plants have flowered, and it is under these circumstances that much nectar is gathered. In the eastern states, where little irrigation is used, the second and third cuttings may be delayed in dry years because there is insufficient plant growth. The alfalfa plants may be too small to cut but when they flower they yield nectar copiously. Alfalfa roots may extend eight to 10 feet down. Because of its deep roots water may be available to the alfalfa plants when it is not to more shallow-rooted plants.

Aster – *Aster spp.* An important honey plant in the autumn are the asters. Hundreds of aster species are found throughout much of the world. Aster honey is light in color and strong in flavor. It granulates rapidly and for this reason is thought to be poor winter food for bees. Aster honey is rarely harvested by itself and is frequently found mixed with honey from goldenrod. Many beekeepers have reported that aster honey, as well as a mixture of goldenrod and aster, crystallizes so rapidly as to make extraction difficult and sometimes impossible.

Basswood - *Tilia spp.* A favorite American honey is basswood, produced

Aster spp. - Purple Aster

Aster spp. - Aster

Aster spp. - White Aster

Aster spp. - Pink Aster

Tilia spp. - Basswood

from a tall, stately tree, *Tilia americana*, though several other species of *Tilia* are also good honey plants. The common basswood tree is found from the eastern Canadian provinces south to Texas; however, because it is an attractive shade tree it has been planted as an ornamental in many states, including California. The honey is light in color, much like alfalfa and clover, but basswood honey has a distinctive, almost minty flavor. Some packers avoid using the honey because of its distinctive flavor; however, beekeepers who have packed basswood honey under its own name have found a good market for it.

Close relatives of American basswood include the European lindens, which are also known as excellent honey plants. These have been imported into North America and are widely used as ornamentals. These are also good honey plants. Both American basswoods and European lindens have been widely used by some towns and villages for roadside plantings and where this has been done good honey flows occur.

While the northeastern U.S. was still being settled it is said that basswood was one of the more common honeys, being as abundant as clover. However, the tree is also a desirable source of lumber. It is the favorite wood for wooden comb honey sections. Basswood trees were in part responsible for the rapid settlement of the country as the softwood trees, like basswood, were easily cut and shaped to make log cabins.

An interesting historical footnote concerning basswood trees and the A.I. Root Company: when they established their first commercial queenyard, basswood trees were planted in rows for three reasons. They were to provide summer shade for the queen rearing colonies and, incidentally, the people who worked the queenyard; they were to provide an abundant source of nectar for the colonies; and eventually, as the trees were thinned, lumber for the comb honey wooden sections mass produced by the factory.

Blackberries – *Rubus spp.* Blackberries are native American plants and important sources of both pollen and nectar for bees. Blackberries are closely related to raspberries (see Raspberries), native to Europe. Worldwide, there are over 250 species in the genus *Rubus*. In a few areas a crop of blackberry honey may be obtained.

The blackberry group includes the low-growing dewberry, the loganberry, youngberry, and boysenberry. Generally, true blackberries are tall-growing and may reach a height of eight to 10 feet (about three meters) especially in the Pacific Northwest. Backberries have a cluster of showy flowers in spring that are attractive to a wide variety of insects including honey bees. The wild plants have heavy thorns but thornless varieties are now grown extensively.

Most blackberries have four white petals (some have five) and are usually just a bit more than an inch (2.5 cm) in diameter. The fruit is made up of 50 to 100 one-seeded drupes. There are an equal number of male and female parts (stamens and pistils) since each drupe

Rubus spp. - Blackberries

Rubus spp. - Blackberry blossom

Rubus spp. - Blackberry

(each stigma) must receive a pollen grain for the seed to develop. The druplets are fixed to the receptacles, thus making blackberries firmer fruits. Self-sterility is apparently widespread among the blackberries and cross-pollination is necessary in the case of some varieties.

Blackberries usually have a two-year cane system. In the first year the young canes grow to their full height but flowering and fruit production take place the second year. The canes die in the following winter. For maximum production the dead canes are removed.

Black locust *Robinia pseudoacacia.* – Black locust is one of the finest honey trees. They bloom in spring with fragrant flowers and can produce much nectar. Unfortunately in some areas the early bloom is damaged by rain and wind. Black locust trees covered with bright white flowers are strikingly beautiful trees. They may flower over a period as long as 10 days.

Black locust honey is very light in color, mild in flavor, and very sweet. Black locust flowers are white and look much like other leguminous flowers. They are large, nearly an inch long, with a keel that contains the flower's sexual parts. There is also a large petal with a dull yellow spot in the center that is called the standard petal. Unlike many leguminous flowers, such as the yellow birdsfoot trefoil, there are two additional wing petals that are much smaller than the standard petal but larger than the two petals that cover and make up each side of the keel.

An uneven number of leaves, usually seven to 19, are borne on a single stem. Black locust trees were once planted in large numbers because the wood is heavy, strong, close-grained and resists rot. Most of the plantings were made so

Robinia pseudoacacia - Black Locust

that farmers would have fence posts that would last for a long time. The wood has also been used to make boats and ships, mill cogs, wagon wheels, railroad ties and the pins that were used in making the early American barns.

Black locusts are native American trees. When they were first found, they were transported back to Europe in numbers because of their special qualities. Their range in North America is from New England and the eastern Canadian provinces to California, south to Georgia and west to parts of Texas.

Brazilian pepper *Schinus terebinthifolius.* – Brazilian pepper is one of the honey plants introduced into southern Florida and Hawaii, during the past 50 years. It has caused great and positive changes to beekeeping in these areas. In Bermuda, as an example of a country outside of the United States, the number of honey bee colonies doubled (from approximately 500 to 1000) after Brazilian pepper was firmly established on the islands. It is one of the top three nectar producers in Hawaii. In Brazil, its home, the plant is known to beekeepers but not thought of as a major honey plant.

In the U.S. Brazilian pepper is classed as an exotic, noxious weed. It is a rapid colonizer of habitats that are disturbed and the margins of wetlands. The plant is related to poison ivy and facial swelling, skin irritation, and sneezing are reported as typical reactions to some who come into close contact with the plant. These symptoms may sometimes be severe. The berries may have a short-lived hypnotic effect on birds that eat them. The major way in which the plants are spread is by seeds that pass through the guts of birds and mammals that consume them.

Brazilian pepper was introduced into Florida with enthusiasm as an ornamental in about 1890. One horticulturist wrote in the mid-1920s, before the plant was widespread, that this "ornate berry-bearing shrub" should be in every Florida garden. Another wrote upon finding the plant on Florida's Sanibel Island that it was a "pleasant thing to come upon." Attempts to remove the plant by cutting and the use of herbicides have failed. Biological control using insects that may feed upon the plant is under investigation.

Pepper trees are sometimes called Florida holly or Christmas berry because of the bright reddish berries. Male and female plants occur about equally. There may be 100 to 1200 plants per acre (50 to 500 per hectare). Each plant has many stems. The shallow root system is extensive and main roots may extend well beyond the drip line of the crown; suckers grow from these. The name is apparently associated with the fact that a close relative, *Schinus molle,* also has purplish berries that are ground for pepper; Brazilian pepper is not used in this manner. However, some say the honey has a peppery aftertaste.

Brazilian pepper flowers in the autumn, from late September to about the first of December. The peak of bloom is in late October. Swarming may occur as the nectar flow progresses.

Buckwheat – *Fagopyrum esculentum.* Perhaps one of the most famous honey plants of years past is buckwheat. Buckwheat honey is dark in color, often grading darker than 130 (out of 140 points) on the Pfund scale. It has an equally strong flavor. It is one of those honeys that is greatly liked by some and disliked by others. From about 1880 to 1950 it was one of the major honeys on the market and was well received especially by several ethnic groups in cities in the east. Buckwheat is not grown in abundance today. Small acreages are grown for grain for flour or as a green manure crop.

The great bulk of the buckwheat honey that was produced was from plants grown in the southern tier of central New York and the adjacent area in north central Pennsylvania. In this area the soils are wet, heavy, and acid. Some buckwheat was grown in the New England states, eastern Canada and as far west as Wisconsin. Buckwheat grain can be harvested about 60 days after planting and will yield nectar about 15 to 20 days after planting. When farming was done with horses, and wet and poor soils were present, buckwheat was a salvation. It did not take as long a time to mature as other grains. It could be planted from mid-June until mid-July and still mature. This is much later than other crops.

Buckwheat plants are from one to three feet (30 to 100 cm) tall and the flowers, which are white, are clustered. Curiously, buckwheat secretes nectar in the morning only and beekeepers report that colonies in buckwheat are easy to inspect and manage in the morning, but become defensive in the afternoon when no nectar is available.

Canola – (see HONEY PLANTS, Rape)

Citrus – *Citrus spp.* Citrus is known as one of the best sources of nectar wherever it is grown in the world. Under ideal conditions a foraging honey bee may visit only a few flowers to fill its honey stomach. Almost all citrus honey is sold under the name "orange blossom" whether produced from oranges, grapefruit, tangerines or other citrus varieties.

Citrus spp. - Orange Blossom (note nectar) (Morse photo)

In the U.S., the bulk of the citrus honey is produced in Florida and California with small amounts being produced in Louisiana, Texas and Arizona. In many citrus-growing areas beekeepers do not move their bees into the citrus groves until they flower. Until that time the groves themselves are completely devoid of flowering nectar and pollen-producing plants that may aid in building colony populations. One problem in producing orange honey in the U.S. is that the date the trees may flower and produce nectar may vary as much as a month. In Florida this may mean anytime between mid-February to mid-March or even a little later.

The quality of citrus honey varies from one season to the next, even from the same grove. The honey itself may be light in color or quite dark depending upon the rapidity with which the plants produce nectar. In years when there is a good honey flow the honey will usually be lighter in color and milder in flavor.

Clover – *Melilotus spp. and Trifolium spp.* The clovers, of which there are many species, are important honey plants in North America, as well as in many other parts of the world. Clover honey is generally thought of as the premium table honey; it is light in color and mild in flavor. In discussions about honey quality, clover is generally considered the standard for comparison. Probably over 30 percent of the honey that is harvested and sold is clover honey. Clover is used widely as a forage crop in many areas and is even grown on sandy soils in

Citrus paradisi - Grapefruit

Florida where it yields well. Honey packers often include alfalfa honey in their clover honey packs.

The low-growing clovers belong to the genus *Trifolium*; included in this group are white Dutch and alsike clover. White sweet clover, *Melilotus alba*, is thought by many to be the very best of the clovers for both honey production and honey quality. Yellow sweet clover, *Melilotus officinalis*, is also an excellent honey plant but it blooms earlier than white sweet clover and is not considered as

Trifolium spp. - Alsike Clover

Honey Plants

Trifolium spp. - White Dutch Clover

Trifolium spp. - Alsike Clover

Trifolium spp. - Alsike Clover (above)

Melitotus spp. - Yellow Sweet Clover

Melitotus spp. - White Sweet Clover

Melilotus spp. - Yellow Sweet Clover

Trifolium spp. - Red Clover

good a yielder of nectar. There are about 20 species of *Melilotus* that yield good quantities of nectar but the two mentioned are the most widely distributed.

Cotton – *Gossypium spp.* Cotton, widely grown in the U. S., can be a major honey plant in the irrigated areas of the southwest and California, along with

Trifolium spp. - Burr Clover

parts of the southeast. In these areas many beekeepers migrate to the cotton fields for honey production. There are two important species of cotton and a great number of varieties, with new types still being developed. Pima cotton, *Gossypium barbadense*, the long staple cotton, is the better nectar source. Cotton has both floral and extrafloral nectaries and bees gather nectar from both. (see EXTRAFLORAL NECTARIES). The quality and color of honey from cotton varies greatly; that from pima cotton is light in color and of good quality. Cotton does not require bees for pollination.

Until recently more pesticides used on cotton than any other single crop grown in the U.S. New cultivars, genetically modified, are now immune to the attacks of the boll weevil. Moreover, the boll weevil eradication programs in parts of the country have reduced pesticide use dramatically. These two approaches to pest control have reinstated cotton as a major honey plant almost everywhere it grows.

Arizona cotton field

Gossypium spp. - Cotton blossom

Eucalyptus – *Eucalyptus spp.* Eucalyptus trees and shrubs are natives of Australia. Many species have been transported to other continents where they have been successful in areas with warm climates. Great variation is present in this genus. Some are excellent lumber trees while others have special value as ornamentals. Some dense-growing eucalyptus species have been used in flood control projects in various parts of the world, including the U. S. because of their heavy root system and ability to survive adverse conditions.

Eucalyptus honey is as variable as the plants themselves; it may be light-colored and mild in flavor or dark and strong tasting. Many species of eucalyptus do not flower at regular intervals. Beekeepers in Australia must become good botanists; they may spend long hours studying flower development and growth. Because of this variability in eucalyptus flowering many Australian beekeepers truck their colonies long distances to take advantage of different honey flows. In the U.S. these plants are especially important for beekeepers in the southern states and California.

Fireweed – *Epilobium angustifolium.* In the northern U.S. and Canada this pioneer plant springs up after a forest fire or the clear-cutting of a forest. It is an important nectar plant in the Pacific Northwest.

Fireweed is a tall herb that has attractive, pinkish flowers. It blooms through much of the summer. An area will produce honey for two or three years only, as the growth of trees and shrubs after a fire or clear-cutting is rapid and the new growth soon shades out the fireweed plants. Fireweed honey is water white in color and mild in flavor and is an excellent table honey. Packers will frequently use it in their table blends.

Gallberry – *Ilex glabra.* Gallberry is a major honey plant in the southeastern part of the U.S. It is found from North Carolina through northern Florida. It rarely fails as a honey plant and good crops are harvested from it almost every year. Gallberry blooms in mid- to late May. After citrus has flowered beekeepers will move their bees from the citrus groves to the flatwoods country of north Florida and Georgia for this crop. Gallberry plants thrive on poorer, usually acidic soil; the plants are evergreens, two to six feet (one to two meters) tall. They form a dense growth

Epilobium angustifolium - Fireweed

and choke out most other low-growing vegetation.

Gallberry honey is light in color and mild in flavor. Some gallberry honey is blended with stronger honeys, often orange. Because of its favorable qualities gallberry honey commands a good price

Ilex glabra - Highbush Gallberry

on the market and is much sought after by honey packers.

Goldenrod – *Solidago spp.* One of the great honey plants in North America is the common goldenrod. It is also a producer of a large quantity of golden pollen. Goldenrod is found throughout the country though it is especially important as a honey plant in the eastern states and Canadian provinces from Nova Scotia and New Brunswick through New England and New York to Florida. Westward it is common throughout the northwest and into the plains states. In North America, goldenrod is considered a weed but in Europe it is a common and respected garden flower that may be sold in large bouquets in flower markets. There are a great number of species and hybrids.

Ilex glabra - Lowbush Gallberry

Solidago spp. - Goldenrod

Goldenrod species begin to flower in mid-July in the north and a little later in Florida. Frost kills the flowers in the north but in Florida several species of goldenrods continue to bloom and yield nectar through November. The golden-headed flowers are found atop thin stems that are one to four feet (30 to 130 cm) in height.

Goldenrod honey is golden in color and strong in flavor. Probably the greatest demand for goldenrod honey is in the baking trade; it is especially well-suited for making graham crackers.

Because it flowers in the late summer and fall, goldenrod honey is often used as winter stores by bees. It contains more indigestible materials than does a light

Honey Plants

Solidago spp. - Goldenrod

honey, such as alfalfa or clover. Still, most beekeepers believe goldenrod is suitable for winter food for bees; certainly it is the chief honey used by feral colonies.

Goldenrod honey should be extracted as soon as possible since it crystallizes fairly rapidly, though not as rapid as honey from the asters.

Heather – *Calluna vulgaris*. The heathers are plants native to Europe and Africa though they have been imported into North America where they are often used as garden flowers. In many parts of northern Europe and the United Kingdom, usually on the poor, acid soil found in the moors, lands not suited to agriculture other than for grazing, heather grows in great profusion and is an outstanding producer of nectar. The heather that blooms in August is especially well-known and hillsides in some parts of Europe, notably Scotland, are a rich red as a result of the profusion of flowers. Some heather areas are purposely burned every few years, as the young sprouts of the re-growth are used by wildlife. Apparently, nectar production is especially high in the blooms that grow following the burning.

Heather honey is thixotrophic from its protein content. (see Thixotropy).

It is considered to be a fine table honey and is in demand.

The heath, *Erica spp.,* also produces a fine table honey. This honey is not thixotrophic.

Erica cinerea - Bell Heather
(Young photos)

Calluna vulgaris - Scotch or Ling Heather (Left two Young photos)

Mangrove, black – *Avicennia nitida* Black mangrove is a native American species found from Brazil to Florida and parts of the Mississippi river delta. In Florida, the plant may be found south from St. Augustine on the east coast to Cedar Key on the west coast. Black mangrove is easily killed by cold weather. In its northern range it is a small shrub while on the southern coast of Florida it is a tree that may reach a height of 60 feet (20 meters).

In Florida, the flowering period is May through July. Honey flows from mangrove are variable and in some years there is apparently no nectar produced while in other years large crops are gathered. Black mangrove honey is light in color and mild in flavor and is regarded as a quality table honey. It is often excessively high in moisture and must be blended with other lower-moisture honeys. The plant is much less common in Florida where the coastline is being developed.

Mesquite – *Prosopis spp.* Three species of mesquite are of interest to beekeepers, all native to the southwestern desert areas of the United States and northern Mexico: *Prosopis pubescens* (screwbean mesquite), *Prosopis glandulosa* (glandular mesquite), and the species most famous for it honey, *Prosopis velutina* (velvet or common mesquite). The latter is one of the major honey plants in southwestern United States and northeastern Mexico. All three mesquites overlap to a certain extent in their distribution and all produce fruit that is valuable for wildlife.

Avicennia nitida - Black Mangrove (USGA photo)

Avidennia nitida - Black Mangrove

Prosopis spp. - Mesquite

Mesquite honey varies in color but is generally amber or lighter. It is mild in flavor and considered a premium honey.

The pods of screwbean mesquite are tender and are eaten raw or cooked. A meal may be made from *Prosopis velutina* and in some areas the beans are especially important as cattle feed. The greenish-yellow flowers appear in the spring. The wood from all three species of mesquites is durable and favored for making long-lived fence posts, tool handles, and for firewood.

Purple loosestrife – *Lythrum salicaria*. This plant was brought into the U.S. as an ornamental, however it has spread throughout the country and invaded wetlands. It is found from central China to England and from the Mediterranean basin north to Norway's

Honey Plants

Lythrum salicaria - Purple Loosestrife

North Sea coast. By the 1830s it was well established along the New England seaboard, but it did not become firmly established in the Midwest and West until the 20th century. By 1985 it was found coast to coast. It can be found as far south as South Carolina and as far north as Quebec.

Purple loosestrife creates a mono-specific community, that is, extensive, solid stands. In many wetlands it easily and successfully replaces cattails. While cattails also create a mono-specific community, they are known to be beneficial to several wildlife species. Purple loosestrife is now considered an invasive noxious plant and is being eradicated. (See Thompson, D.Q., R.L. Stuckey and E.B. Thompson. "Spread, Impact and Control of Purple Loosestrife (*Lythrum salicaria*) in North American wetlands." *Fish and Wildlife Research 2, U.S. Department of the Interior.* 1987.)

Beekeepers, especially in the Northeast, have come to recognize purple loosestrife as an outstanding honey plant. It flowers from late June through early September. Since it out-competes other vegetation in an area, one finds huge wetland areas full of these of purple flowers.

Because the honey is produced in the summer it can be low in moisture and heavy-bodied. It is amber in color and has a greenish tinge when held up to the light. It is mild in flavor and, unlike many honeys of amber color, makes a good table honey with a fine flavor. In some of the eastern states where purple loosestrife is especially abundant in wetlands, it is not uncommon to find apiaries that will produce a good surplus.

Rape – *Brassica spp.* Rape (now known as canola) is grown for the seeds which are rich in protein and are used in animal feeds and for oil. The plant itself is often used as a forage crop, especially for swine. The prairie provinces of Canada are major producers of this plant. Rape apparently benefits from cross-pollination. There are two major varieties, one that is planted in the autumn and the other in the spring. Both are excellent sources of pollen and nectar. The chief problem with rape honey is its tendency to crystallize rapidly. As a result beekeepers must harvest two to four times during flowering to prevent crystallization in the combs.

Canola has been genetically modified to make it resistant to certain herbicides and much of this crop is now of that type.

The oil from common rape seed contains a high amount of erucic acid, a material the U.S. Food and Drug Administration considers unhealthy. But in 1988, several varieties were developed

Brassica spp. - Rape (also known as Canola)

in Canada with very low amounts of the acid. To distinguish between them, the developers named these new varieties CANOLA "CAN" for Canada, "O" for Oil and "LA" for Low Acid. Canola is a popular crop in the U.S. -KF

Raspberry – *Rubus spp.* While honey from raspberry is not quite as light as clover honey, it is mild and has a delicate flavor that is sought after by honey packers. Unfortunately, the quantity of raspberry honey that is produced is small, and the raspberry-producing territory is constantly shifting. Beekeepers report that wherever it occurs, raspberry is a good source of nectar and highly attractive to bees.

In many of the northern states, including New York, Michigan and parts of New England, raspberry prospers and is luxuriant following the cutting of hardwood timber. In the past raspberries appeared in quantity where hardwood forests were clear-cut, that is all of the trees removed at one time. Raspberry honey is common in areas of the northwest. Today, it is more common to harvest only the largest trees from forested land every 10 to 15 years. Under these circumstances the forest floor is never exposed to sunlight and the raspberries do not have an opportunity to grow. The result is that much less raspberry honey is produced today than in the past.

Cultivated raspberries are also good sources of nectar but the acreage is small. Raspberries benefit from pollination by honey bees.

Sage – *Salvia spp.* There are several hundred species of sages ranging from ornamental garden plants to the wild sages well-known in California where a few species are major honey plants. The sage honeys are light in color with a mild, but distinct flavor, and low in moisture content. In areas where it is produced, sage honey is usually sold under its own name because of its distinctive characteristics. Sage honey is high in fructose so is very slow to granulate. It is much sought after by packers to improve the quality of their blends.

Salvia spp. - White Sage

Saltcedar – *Tamarix spp.* Saltcedar was introduced from southern Europe and is well established and now considered an invasive noxious plant. However, it is an important honey plant in parts of Texas, New Mexico, Arizona, and California. It is found to a small extent in some adjacent states including the southeast where it is less abundant. Saltcedar is the most hardy and widespread tamarisk in North America. The trees grow to a maximum height of about 30 feet (10 meters). In some areas they are sometimes less tall and thought of as shrubs. It has often been used as an ornamental, as well as in wind and erosion control. It reproduces easily along river banks, in salt marshes and waste areas.

Saltcedar flowers over a long period. The flowers are numerous, pinkish in color and form at the tips of the branchlets. The flowers are complete, that is, they contain both sexes. Saltcedar is an excellent source of pollen as well as nectar. The honey is dark in color and strong in flavor and used primarily in the bakery trade.

Trees in the genus are often difficult to identify and close relatives are difficult to distinguish. Other introduced species in the genus are recognized as good honey plants on a local basis in the warm (largely frost free) parts of the United States and Mexico. Saltcedar is one of several plants marked for eradication by the federal government.

Saw palmetto – *Serenoa repens.* Saw palmetto is a major source of honey in Florida and other parts of the southeast. It is one of the approximately 40 major honey plants in the United States and is North America's most abundant native

Tamarix spp. - Saltcedar

palm. The saw palmetto gets its name from the sharp, saw-like thorns that line the edges of the leaf stalk. Saw palmetto grows as far north as North Carolina and is found west to Texas.

A second palm, *Sabal etonia,* is mixed with saw palmetto but only in the white or yellow sands of upland areas of Florida's central and Atlantic coast ridges. Both flower at much the same time and their honeys are much the same. However, *Sabal* palm is much less common and must be considered a minor source of nectar.

Recent research indicates that the future for both palmettos is bleak in certain areas. The saw palmetto plants may grow only a quarter to less than an inch (five to 20 mm) a year. The stems often lie on top of the ground and may be traced back from the growing crown as much as 12 to 15 feet (four to five meters). Some of this above-ground stem may be dead and in an advanced state of decay but it is still possible to trace it back to its point of origin. Some plants on the Florida ridge are estimated to be 400 to 700 years old. The death rate among mature plants is very low. Of 240 flatwoods and scrubby flatwoods palmettos marked in 1980 not one has died during the 13 annual censuses taken since that time.

Several palmetto areas have been set aside to be protected in central and south Florida. However, the vast acreages that were once available for bee forage are being slowly reduced as land-use patterns change. Housing developments and new agricultural plantings are taking their toll. There are many areas in Florida where palmetto plants in fire lanes cut years ago have never recovered. Once fields of palmettos are chopped to make way for other uses there is almost no re-growth.

Sabal megacarpa - Scrub Palmetto

Serenoa repens - Sawtooth Palmetto

Honey Plants

Sourwood – *Oxydendrum arboreum*. One of the very fine and most highly touted honeys in North America is sourwood. The trees are common from southern Pennsylvania to northern Florida and west through Tennessee, though they are planted as ornamentals in many places. Most of the sourwood honey is produced in the Allegheny and Blue Ridge Mountains. The trees flower from mid- to late June into July. During years when ideal conditions for nectar secretion exist one may shake droplets of nectar from the flowers.

Sourwood honey is light-colored and aromatic, with a distinctive flavor. It is well-known in the areas where it is produced and in such demand that most of it is sold locally.

Glycine max - Soybean

Soybean – *Glycine max*. Many varieties of commercial soybeans are grown in the U.S. and they vary greatly in the quantity of nectar they produce. Since soybeans are self-fertile, honey bees are not necessary for pollination. On a per acre basis, it appears that the best soybean-honey producing area is the lower Mississippi River valley. Soybean honey is light in color, mild in flavor and is often used in blending, especially with stronger flavored honeys.

Spanish needle – *Bidens spp*. The genus *Bidens* is worldwide in distribution with over 200 species, many of which are outstanding nectar and pollen plants. While they are not sufficiently abundant in the U.S. that a honey crop is obtained from them, in many parts of the country they are nevertheless of great importance in helping to build

Oxydendrum arboreum - Sourwood

Honey Plants

Bidens spp. - Spanish Needle

and maintain honey bee populations. A few species are ornamentals but most are thought of as weeds. There are a variety of common names in addition to Spanish needle including stick-tights, beggar ticks, and bur-marigold, however, the common names are often misleading. All of the native American species, except *Bidens pilosa* have yellow flowers.

In the south, Spanish needle may provide some surplus honey but it is dark with a strong flavor, suitable for bakery honey. In other areas it provides winter feed, both pollen and nectar. In some areas it may provide surplus honey.

Sumac – *Rhus spp.* Sumac has worldwide distribution. In the U. S. over a dozen species are known as honey plants. In the east, especially New England and parts of New York State, *Rhus glabra* is especially important. In much of Connecticut it is the chief source of honey. In the northeast, sumac flowers during the last three weeks of July and exceptional yields have been obtained some years. Sumac honey has a good flavor and is often used as a table honey though it is not as mild or light as clover honey.

Sunflower – *Helianthus spp.* Many species of sunflowers are grown. The commercial variety grown for oilseed and as birdfood is *Helianthus annuus*. Commercial sunflowers must be pollinated by insects to produce seed. The plants are also excellent producers

Rhus spp. - Sumac

Rhus spp. - Sumac

Rhus spp. - Sumac

of nectar and thus sunflower growers and beekeepers have close ties. Growers of hybrid sunflower, who produce seed for planting, pay beekeepers for pollination, mostly in California, Texas and Minnesota. Sunflowers produce large quantities of pollen. Sunflower honey has a golden hue. Since it has a tendency to crystallize rapidly it must be extracted as soon as possible. Most sunflower honey is used in the baking trade, though some beekeepers blend it into their table honey. Sunflowers are grown throughout much of the U. S. but the Dakotas and Minnesota are the major producers.

Rhus spp. - Sumac

Helianthus spp. - Sunflower

Tallow tree – *Sapium sebiferum*. The tallow tree, often called the Chinese tallow, may grow 40 feet (15 meters) high. It is one of the major honey plants in the southern states from Texas eastward to Florida. The trees begin to flower in early May and continue to flower for a period of about six weeks. Individual trees are in bloom for about two weeks (Pellet, F.C. *American honey plants*. Dadant & Sons, Hamilton, IL. 467 pages. 1947). The honey is amber in color but mild in flavor. Most tallow honey is used in the baking trade.

Schmitz et al. (*The ecological impact of nonindigenous plants in* Strangers in Paradise, Impact and Management of Non-indigenous species in Florida. Simberloff, D., D.C. Schmitz, and T.C. Brown, Editors. Island Press, Washington, DC.1997) wrote that the tree was introduced into the United States by Benjamin Franklin who sent seeds to a Georgia farmer in 1772. It is described in *Hortus Three* as an ornamental and a "good street tree." A native of China it is used for ornamental plantings. Trees begin blooming in three to four years from seed

Tallow trees are considered an invasive exotic species. The Florida Exotic Pest Plant Council has placed tallow trees in their category I along with Brazilian pepper and melaleuca. These are "species that are invading and disrupting native plant communities" in the state. The tree can now be found in at least 38 of the 67 northern Florida counties

Sapium sebiferum - Chinese Tallow

(Jubinsky, G. *Chinese tallow: in short, this plant is bad news.* Resource Management News (Florida) 6(6) 5. April 1995. It "tolerates a wide variety of soils, moisture levels and salinity gradients." It is typical of many invasive exotic plants in that it grows rapidly, fruits early, produces many seeds, has few pests and because it is attractive has been promoted and distributed by people. In Simberloff, D, D.C. Schmitz and T.C. Brown, Editors, *Strangers in Paradise, Impact and Management of Non-indigenous species in Florida,* (Island Press, Washington, D.C. 1997), all nine reports were negative.

Thyme – *Thymus spp.* Thyme honey is well known from the Mediterranean area and especially from Mt. Hymettus in Greece where it has been a favorite for thousands of years. Over 50 species of *Thymus* are known, several of which are used as garden flowers. Several species of thyme have been introduced into North America and they are now widespread. Since the honey has a reddish color and strong flavor most beekeepers do not consider it a table honey though it does have its advocates. The plant is shallow rooted and requires frequent rains to continue to produce nectar.

Tulip poplar – *Liriodendron tulipifera.* Tulip poplar trees, sometimes called yellow poplar, may grow to a height of over 100 feet (30 meters) and have a diameter of two to five feet (one to two meters). The trees are common from Pennsylvania south to the Gulf states and west to eastern Tennessee. Tulip poplar trees are found only rarely north of there. Everywhere it is found the tulip poplar is considered a dependable nectar source. Tulip poplars bloom early in the season, during May.

Tulip poplar honey is mild in flavor but dark in color with a reddish tinge. Usually we think of the dark honeys as being strong flavored but tulip poplar is an exception.

Tupelo – *Nyssa spp.* There are several species of tupelo that are known for high quality honey. The plants grow from North Carolina south and west to Texas. They flower in the late spring. The most famous tupelo honey-producing area is along the Apalachicola River in west Florida. Hundreds of colonies are moved

Thymus spp. - Thyme

Liriodendron tulipifera - Tulip Poplar

to this area for the nectar flow. Tupelo honey is low in moisture and light in color and has a mild flavor.

Tupelo honey is high in fructose, more so than almost any other honey in North America. The high fructose content means that the honey does not crystallize. Tupelo honey is purchased eagerly by honey packers who blend it with other light honeys to prolong shelf life.

Vetch *Vicia spp.* – Several species of vetch are typical legumes and in some areas important honey plants, especially for colony maintenance. Only rarely are the vetches reported to be present in sufficient quantity to produce surplus honey. Vetch flowers are usually light blue to purple or lavender. The sexual parts are contained in a keel. When a bee lands on the wing petals of a flower and inserts her proboscis to obtain nectar she depresses the keel and the male and female parts are forced out of the keel's tip. Much of the pollen rubs off onto the bee and some pollen from the bee contacts the stigma and pollination is accomplished. The vetches are important both as pollen and nectar plants. Crown vetch, *Corineila varia,* that is commonly used for roadside plantings, is not a *Vicia.*

Other honey plants – There are certainly more plants that contribute to a colony's honey crop than this short list. Some are major contributors, but are found in only a few locations due to environmental constraints. Others are minor, but are so widespread as to be considered important.

Presented here are some of these, primarily from the U.S., with names and comments. Together, these lists still do not comprise a universal honey plant list. When scouting locations for your colonies, use every resource to find the

Nyssa spp. - Tupelo

Vicia spp. - Vetch

Vicia spp. - Hairy Vetch

honey plants that grow in your area, when they bloom, and, how productive they are on average.

One or two moves a year may double, or triple your honey crop, allow you to produce varietal honeys or protect you from seasonal pesticide exposures on some cultivated crops.

American Holly, *Ilex opaca*. Many plants in the genus are good honey producers.

Annise Hyssop, *Agastache foeniculum*. This herb has a stunning reputation as a honey producer, but so many plants are needed to produce a crop it is generally inefficient to use.

Apples, *Malus sp.* Commercial and ornamental apples are attractive to bees world-wide.

Bee Bee Tree, *Evodia danelli*. Sometimes used as a specimen tree. Modest appearing blossoms very attractive.

Birdsfoot Trefoil, *Lotus sp.* A common forage plant in many places, though not as common as in years past.

Century Plant, *Agave sp.* Many in this genus are attractive and productive. This one is special due to its size, and the fact that it blooms after so many years – sometimes approaching 100 years.

Malus sp. - Apples

Dandelion, *Taraxicum officinale*. This common yellow-flowered weed has nearly worldwide distribution, and, where abundant, plays a major role in colony expansion in the spring.

Figwort, *Scrophularia matilandica*. Though extremely attractive, its small size diminishes its value.

Evodia danelli - Bee Bee Tree

Lotus sp. - Birdsfoot Trefoil

Agave sp. - Century Plant

Knapweed (Star Thistle), *Centaurea spp.* Purple Starthistle, sometimes called Spotted Knapweed, generally grows in the eastern U.S. It's cousin, Yellow Starthistle, grows in the west. Yellow Starthistle has a thorny flower head, causing problems for cattle. Both are introduced noxious weeds and controlling pests have been introduced to help.

Taraxicum officinale - Dandelion

Honey Plants

Scrophularia matilandica - Figwort

Lowbush Blueberry, *Vaccinium pennsylvanicum.* Grown only in Maine and the Maritime Provinces, this crop continues to increase in acreage. Extreme pollination, however, has reduced honey production.

Maples, *Acer sp.* Many species in this prolific genus produce both pollen and nectar in the early part of the season and are beneficial to colony survival when they are abundant.

Milkweed, *Asclepias sp.* Several species of milkweed, are relatively attractive to honey bees. They have a unique pollen distribution feature, wherein two pollen 'sacks' are attached with a filament. Bees can sometimes become entangled in this and die on the blossom.

Peaches, Plums, *Prunus sp.* Attractive to bees, but only commercial plantings produce any excess honey.

Centaurea sp. - Knapweed (Star Thistle)

Wild blueberry field, above and right, Vaccinium pennsylvanicum, Lowbush Blueberry

Privet, *Ligustrum spp.* Several species of this shrubby plant are distributed in the U.S. some hardy, others not. Honey is variable, from strong and disagreeable to pleasant and mild.

Redbud, *Cercis canadensis.* This legume tree is an early season nectar producer.

Smartweed, *Polygonum spp.* A wide variety of plants in this genus produce honey colors from nearly blood red to water white. Generally they all are mid to late season, reliable producers.

Prunus sp. - Peach

Honey Plants

Acer rubrum - Red Maple

Acer rubrum - Red Maple

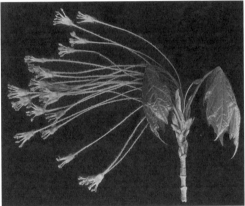

Acer sp. - Sugar Maple

Acer rubrum - Red Maple Blossom

Acer sp. - Maple

Honey Plants

Asclepias spp. - Common Milkweed

Common Milkweed

Swamp Milkweed

Butterfly Bush

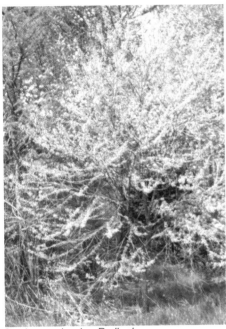

Cercis canadensis - Redbud

Ligustrum sp. - Privet, bottom is Evergreen Privet.

Cercis canadensis - Redbud

Polygonum spp. - Smartweed

Willows, *Salix spp.* Ranging from small shrubs to large specimum trees, these early season bloomers are important for both nectar and pollen.

Yellow Rocket, *Cruciferae family*, This is one of many wild plants in the mustard family. Similar to the commercial crop Canola (rape), it blooms

Salix sp. - Pussywillow

Salix sp. - Willow

Salix sp. - Pussywillow

Salix sp. - Willow

Salix nutallii - Willow

early and is used primarily for spring build up of colonies.

Poisonous Plants – A poisonous plant is one that produces chemical toxins that interfere with metabolic processes of living organisms, either directly or indirectly. A plant can produce nectar or pollen that is toxic to bees and honey made from poisonous nectar can be toxic

Cruciferae family - Yellow Rocket

to humans. Locating the source of poisonous honey presents toxicologists with a number of challenges. Foraging activity is difficult to trace, honey harvest records are sometimes neglected, and moving colonies for pollination may expose them to hazards unknown to their keepers. Furthermore, what is toxic to some organisms is benign to others. It is possible for honey from a thriving colony to be lethal to humans, while nectar and pollen from sources harmless to us can devastate a healthy colony. Making it more difficult still, toxicity is nearly always dose-related. Nectar that can poison a colony in high concentrations may cause minimal effects when diluted with food from other sources. In that case, a poisonous plant may go undetected until bees collect from it in concentrations adequate to cause illness.

However daunting honey toxicology may appear, caution should be used when keeping bees in the proximity of the following plants:

Summer Titi - *Cyrilla racemiflora* – Purple Brood – *Cyrilla racemiflora* is commonly called summer titi or southern leatherwood but is also known as ironwood, and swamp cyrilla. It thrives in the southeastern U.S. as a bushy shrub or small tree and produces abundant, fragrant white flowers May through July.

In 1932, Burnside traced colony poisonings in Florida to *C. racemiflora*, calling the resulting malady "purple brood" for the bluish coloration of affected larvae. Purple brood is still considered a serious problem periodically in Florida, Georgia, Mississippi and South Carolina. The severity of the disease is related to the local abundance of *C. racemiflora* in conjunction with a lack of other food sources.

Purple brood develops suddenly, with the onset of symptoms occurring concurrently throughout all exposed hives. The disease is attributed only to the nectar of *C. racemiflora* as pollen, comb, nor brood from affected colonies can induce illness in those unaffected. Colonies weakened by purple brood can survive, however honey production will be reduced.

Sanford recommends that beekeepers in areas of abundant *C. racemiflora* either move their colonies or feed them sugar syrup to dilute the plant's effect.

Rhododendron – Plants from the heath family (Ericaceae) have long been recognized as sources of toxic honey,

Cyrilla racemiflora - Summer Titi

poisonous to both bees and humans. In fact, the death of the Greek soldiers of Xenophon in 401 B.C. was attributed to the ingestion of honey from *Rhododendron ponticum*. Laboratory tests confirm that several species and hybrids of *Rhododendron* are poisonous to bees, and observations indicate that both the pollen and nectar are suspect. Yet foragers manage to collect both and bring them back to the hive. It is likely that toxicity, in this case, either has a delayed effect or is dose-related. In the later scenario, toxins, which might initially be diluted by the high water content in nectar, become more concentrated as water is removed in the process of making honey.

Mountain Laurel (*Kalmia latifolia*)
– *K. latifolia* is a native evergreen shrub that grows in the moist woodlands of upper elevations of the Appalachian Mountains and blooms prolifically with clusters of predominately white flowers. As with *Rhododendron*, all parts of *K. latifolia* are harmful to humans,

Kalmia latifolia - Mountain Laurel

Kalmia latifolia - Mountain Laurel

Rhododendron maximum - Rosebay

containing an andromedotoxin, which poisons humans. Just a spoonful of honey made from *K. latifolia* can cause numbness and in some cases loss of consciousness. However, *K.latifolia* appears to pose no threat to bees, and no link has been established between this plant and colony poisonings.

California Buckeye (*Aesculus californica*) – *A. californica* occupies more than 15 million acres in California and has long been associated with large losses of honey bee colonies throughout its range. The plant varies in form from shrub-like in the north to a medium-sized (40 ft or 15 meter) tree in the south. White or pink blossoms with bright red pollen produce a showy display on terminal branches.

Vansell has reported symptoms of honey bee poisoning from *A. californica* nectar, honey, pollen, and sap. Field bees, the adult queen and drones were affected as well as the larvae and emerging workers. In mild cases, colonies show a spotty brood pattern and fail to become populous. In severe cases, the entire colony perishes – despite full stores of honey. All that remain are eggs. The absence of uncapped brood – dead or alive – is indicative of a "buckeyed"

Gelsemium sempervirens - Yellow Jessamine

colony, as are piles of light-colored, immature and deformed bees at the hive entrance. Vansell suggested that *A. californica* poisoning affects fertilization after observing that queens which laid only drones in a contaminated colony, resumed laying fertilized worker eggs when moved to a healthy one.

A. californica poisoning correlates to the availability of other food sources. In years of sufficient rainfall, when a variety of sources are present, poisoning is limited. Though it seems counterintuitive for bees to harvest toxic sources of food, they may do so, despite the risks, to prevent starvation.

Yellow Jessamine – *Gelsemium sempervirens* – *G. sempervirens*, commonly known as yellow jessamine or Carolina jessamine, is an evergreen vine with yellow, fragrant, trumpet-shaped flowers. It is native to the Coastal Plain and lower hills from Virginia to Texas, while farther north, it is popular as an ornamental. The toxic alkaloids, gelsamine, gelseminine, and gelsemicine are found in all parts of the vine, but the greatest concentrations are found in the roots and nectar.

Aesculus californica - California Buckeye

Human illnesses and death have been attributed to taking nectar directly from *G.sempervirens* as well as to eating honey made from the plant. Bees foraging on the flowers initially appear intoxicated, then become paralyzed and eventually die. Brood, however, does not seem affected. Huggins suggested that the flower's pollen is the primary source of poison to bees, and studies have found large amounts of *G. sempervirens* pollen in the guts of dead bees during severe colony losses. Laboratory tests to determine bee sensitivity to the plant's nectar have been inconclusive.

The Tutu Mystery – Here is a story that illustrates the challenges inherent in beekeeping toxicology.

In the early 1900s, over 200,000 ha (494,210 ac) of northern New Zealand were closed to beekeeping for fear that honey produced in the region was causing human poisoning. Reports of various symptoms, including abdominal and head pain, vomiting, convulsions, loss of memory, and even coma all implicated honey or honey products as the likely cause.

Initially three plants were suspected of contributing to toxicity: rangiora (*Brachyglottis repanda*), buttercup (*Ranunculus rivularis*), and tutu (*Coriaria arborea*). By 1927, rangiora and buttercup had been eliminated as the first bloomed too early and the second grew in swampy areas too far removed from apiaries. When a group of studies in the 1940s identified tutu as a source of the picrotoxin, tutin, and when local doctors pointed to tutu as the only plant within miles of the suspect hives, beekeepers felt confident they'd found the culprit. However, questions remained. Tutu was not known to produce nectar, and more interestingly, chemists could not isolate tutu's toxin in the toxic honey. Furthermore, tutu pollen fed to rats caused no symptoms.

Beekeeping in the region was abandoned for 12 years, and during that time there were no poisonings. When apiaries moved back in though, symptoms reappeared. As suspected, the offending honey contained pollen from tutu (and other plants). This time the honey was traced to an extraction made near the end of summer. The next spring, new test colonies were established in the region. Curiously though, the test honey was not toxic.

Paterson accidentally found a key piece to the puzzle when, while in the field in late summer, he noticed a sticky substance on the tutu's leaves. This honeydew, a toxic secretion from the passion vine hopper (*Scolypopa australis*), was known to be collected by honey bees in times of nectar dearth, like the end of summer. The bees had been collecting nectar through most of the season, producing non-toxic honey, but when nectar supplies plummeted, they resorted to gathering the poisonous honeydew. Just as toxicology studies had indicated, the tutu nectar itself was not toxic. Nor was its pollen. But, because of a previously unrecognized association with an insect pest, the plant was for years incorrectly pegged as the source of toxic honey.

What to do if colonies have been poisoned – Beekeepers can determine if colony illness is caused by poisonous plants by first eliminating other, more likely, causes. Check for diseases, pests, and pesticide applications. Then, determine if weather conditions have limited preferred nectar sources. Finally, check the flight range for poisonous plants. If adequate evidence of plant poisoning exists, samples of bees, brood, comb, and honey can be sent to local or regional laboratories for confirmation.

The effects of poisonous plants can be reduced or eliminated by moving colonies out of the flight range of the source. If this is not possible, diluting honey with sugar syrup, or pollen with supplements, may also be helpful.

The subject of poisonous nectars and honeys was reviewed in detail by Skinner, J.A. "Poisoning by Plants" in Morse, R.A. and Kim Flottum, Editors. *Honey Bee Pests, Predators and Diseases.* (The A.I. Root Company, Medina, OH. 718 pages. Revised edition, 1997.) Also see Crane, E. *Honey, a Comprehensive Survey.* (Heinemann, London. 608 pages. 1975.) -JS

HONEY PRICE SUPPORT SYSTEM – (see PRICE SUPPORT)

HONEY PURE FOOD AND DRUG LAWS – When commercial beekeeping began in earnest in the late 1800s beekeepers suddenly appeared who owned hundreds and even thousands of colonies. Many of these beekeepers produced liquid honey. At the same time a commercial beekeeping industry was developing, other groups of people learned how to make liquid sugar syrup that was much cheaper than honey. The result was adulteration of honey with this product, in fact so much so that most beekeepers turned to the production of comb honey. At the time most consumers understood that comb honey was a pure, natural and unadulterated product.

The 1906 Food and Drug Act was the first pure food and drug law enacted by U.S. Congress. Prior to that time no restrictions existed for food and drug products or the labels that were on bottles or cans. Drugs such as morphine and heroin were often sold in cough syrups and home remedies. Alcohol was a common ingredient in these products, too, and it was not required that this be mentioned on a label. Adulteration, including the adulteration of honey, was not a crime. Many people and groups of people, including beekeepers, petitioned Congress for pure food and drug laws as early as 1880.

The first federal pure food and drug laws were a great improvement but still weak. Many amendments and changes to strengthen the legislation have been made since that time. States have additional legislation and work through public health agencies to make certain we have a safe food supply. Very slowly the public became aware that liquid honey was no longer adulterated. Sugar was rationed in both the first and second world wars and as a result the demand for honey increased greatly. Since it is easier, and usually more profitable to produce liquid honey, beekeepers changed their ways and equipment. At the end of World War II only a few dozen beekeepers, widely scattered across the country, made a full-time living making only comb honey.

Whenever adulteration has occurred beekeepers have been quick to respond and to call for governmental action. In fact, the beekeeping industry itself has been a very good watchdog and beekeepers have made a strong and conscientious effort to maintain the popular image that honey has as a natural and pure product. Honey maintains a good image today, and since it is naturally free of contaminants, industry leaders have sometimes had difficulty getting the attention of pure drug and food people when their help was needed.

Below are the United States Standards for Grades of Extracted Honey, effective May 23, 1985. (Portions of this document have been deleted due to space limitations. Full descriptions are available from the USDA.)

This is the fifth issue, as amended, of the U.S. Standards for Grades of Extracted Honey published in the **FEDERAL REGISTER** of April 23, 1985 (50 FR 15861) effective May 23, 1985. This issue supersedes the fourth issue, which was effect since April 16, 1951.

Voluntary U.S. grade standards are issued under the authority of the Agricultural Marketing Act of 1946, which provides for the development of official U.S. grades to designate different levels of quality. These grade standards are available for use by producers, suppliers, buyers and consumers. As in the case of other standards for grades of processed fruits and vegetables, these standards are designed to facilitate orderly marketing by providing a convenient basis for buying and selling, for establishing quality control programs, and for determining loan values.

The standards also serve as a basis for the inspection and grading of commodities by the Federal inspection service, the only activity authorized to approve the designation of U.S. grades as referenced in the standards, as provided under the Agricultural Marketing Act of 1946. This service, available as on-line (in-plant) or lot inspection and grading of all processed fruit and vegetable products, is offered to interested parties, upon application, on a fee-for-service basis. The verification of some specific recommendations, requirements, or tolerances contained in the standards can be accomplished only by the use of on-line inspection procedures. In all instances, a grade can be assigned based on final product factors or characteristics.

Copies of standards and grading manuals obtained from: Chief, Processed Products Branch, Fruit and Vegetable Division, AMS, U.S.D.A., P.O. Box 96456, Rm. 0709, So. Bldg., Washington, DC 20090-6456

§52.1391 Product description
Extracted honey (hereinafter referred to as honey) is honey that has been separated from the comb by centrifugal force, gravity, straining, or by other means.

§52.1392 Types.
The type of extracted honey is not incorporated in the grades of the finished product since the type of extracted honey, as such, is dependent upon the method of preparation and processing, and therefore is not a factor of quality for the purpose of these grades. Extracted honey may be prepared and processed as one of the following types:
(a) **Liquid honey.** Liquid honey is honey that is free from visible crystals.
(b) **Crystallized honey.** Crystallized honey is honey that is solidly granulated or crystallized, irrespective of whether candied, fondant, creamed or spread types of crystallized honey.
(c) **Partially crystallized honey.** Partially crystallized honey is honey that is a mixture of liquid honey and crystallized honey.

§52.1393 Styles.
(a) **Filtered.** Filtered honey is honey of any type defined in these standards that has been filtered to the extent that all or most of the fine particles, pollen grains, air bubbles, or other materials normally found in suspension, have been removed.
(b) **Strained.** Strained honey is honey of any type defined in these standards that has been strained to the extent that most of the particles, including comb, propolis, or other defects normally found in honey, have been removed. Grains of pollen, small air bubbles, and very fine particles would not normally be removed.

§52.1394 Definitions of terms.
As used in these U.S. standards, unless otherwise required by the context, the following terms shall be construed, respectively, to mean:
(a) **Absence of defects** means the degree of freedom from particles of comb, propolis, or other defects which may be in suspension or deposited as sediment in the honey. Classifications for the factor of quality, absence of defects, are:
  (1) **Practically free** - the honey contains practically no defects that affect the appearance or edibility of the product.
  (2) **Reasonably free** - the honey may contain defects which do not materially affect the appearance or edibility of the product.
  (3) **Fairly free** - the honey may contain defects which do not seriously affect the appearance or edibility of the product.

(b) **Air bubbles** mean small visible pockets of air in suspension that may be numerous in the honey and contribute to the lack of clarity in filtered style.

(c) **Aroma** means the fragrance or odor of the honey.

(d) **Clarity** means, with respect to filtered style only, the apparent transparency or clearness of honey to the eye and to the degree of freedom from air bubbles, pollen grains, or other fine particles of any material suspended in the product. Classifications for the factor of quality, clarity, are:

(1) **Clear** -the honey may contain air bubbles which do not materially affect the appearance of the product and may contain a trace of pollen grains or other finely divided particles of suspended material which do not affect the appearance of the product.

(2) **Reasonably Clear** - the honey may contain air bubbles, pollen grains, or other finely divided particles of suspended material which do not materially affect the appearance of the product.

(3) **Fairly Clear** - the honey may contain air bubbles, pollen grains, or other finely divided particles of suspended material which do not seriously affect the appearance of the product.

(e) **Comb** means the wax like cellular structure that bees use for retaining their brood or as storage for pollen and honey. Fine particles of comb in suspension are defects and contribute to the lack of clarity in filtered style.

(f) **Crystallization** means honey in which crystals have been formed.

(g) **Flavor and aroma** means the degree of taste excellence and aroma for the predominant floral source. Classifications for the factor of quality, flavor and aroma, are:

(1) **Good Flavor and aroma for the predominant floral source-** the product has a good, normal flavor and aroma for the predominant floral source or, when blended, a good flavor for the blend of floral sources and the honey is free from caramelized flavor or objectionable flavor caused by fermentation, smoke, chemicals, or other causes with the exception of the predominant floral source.

(2) **Reasonably good flavor and aroma for the predominant floral source** -the product has a reasonably good, normal flavor and aroma for the predominant floral source or, when blended, a reasonably good flavor for the blend of floral sources and the honey is practically free from caramelized flavor and is free from objectionable flavor caused by fermentation, smoke, chemicals, or other causes with the exception of the predominant floral source.

(3) **Fairly good flavor and aroma for the predominant floral source** - the product has a fairly good, normal flavor and aroma for the predominant floral source or, when blended, a fairly good flavor for the blend of floral sources and the honey is reasonably free from caramelized flavor and is free from objectionable flavor caused by fermentation, smoke, chemicals, or other causes with the exception of the predominant floral source.

(h) **Floral source** mens the flower from which the bees gather nectar to make honey.

(i) **Granulation** means the initial formation of crystals in the honey.

(j) **Pfund color grader** means a color grading device used by the honey industry. It is not the officially approved device for determining color designation when applying these United States grade standards for the color of honey.

(k) **Pollen grains** mean the granular, dustlike microspores that bees gather from flowers. Pollen grains in suspension contribute to the lack of clarity in filtered style.

(l) **Propolis** means a gum that is gathered by bees from various plants. It may vary in color from light yellow to dark brown. It may cause staining of the comb or frame and may be found in extracted honey.

§52.1400 Grades.

(a) U.S. Grade A is the quality of extracted honey that meets the applicable requirements of Table IV or V, and has a minimum total score of 90 points.

(b) U.S. Grade B is the quality of extracted honey that meets the applicable requirements of Table IV or V, and has a minimum total score of 80 points.

(c) U.S. Grade C is the quality of extracted honey that meets the applicable requirements of Table IV or V, and has a minimum total score of 70 points.

(d) **Substandard** is the quality of extracted honey that fails to meet the requirements of U.S. Grade C.

§52.1401 Determining the grade.

Determining the grade from the factors of quality and analysis.

(a) **For the factor of analysis, the soluble solids content of extracted honey is determined by** means of the refractometer at 20°C (68°F). The refractive indices, corresponding percent soluble solids, and percent moisture are shown in Table III. The moisture content of honey and percent soluble solids may be determined by any other method which gives equivalent results.

(b) **For the factors of quality, the grade of extracted honey is determined by** considering, in conjunction with the requirements of the various grades, the respective ratings for the factors of flavor and aroma, absence of defects, and clarity (except the factor of clarity is excluded for the style of strained).

(c) **The relative importance of each factor is** expressed numerically on the scale of 100. The maximum number of points that may be given each factor is:

Factors	Points
Flavor and aroma	50
Absence of defects	40
Clarity	10
Total Score	100

(d) **The factor of clarity for the style of strained extracted honey is** not based on any detailed requirements and is not scored. The other two factors (flavor and absence of defects) are scored and the total is multiplied by 100 and divided by 90, dropping any fractions to determine the total score.

(e) **Crystallized honey and partially crystallized honey shall be liquified by** heating to approximately 54.4°C (130°F) and cooled to approximately 20°C (68°F) before determining the grade of the product.

§52.1402 Determining the rating for each factor.

The essential variations within each factor are so described that the value may be determined for each factor and expressed numerically. The numerical range for the rating of each factor is inclusive (for example, **37 to 40 points** means 37, 38, 39 or 40 points) and the score points shall be prorated relative to the degree of excellence for each factor.

		TABLE IV - FILTERED STYLE			
Analytical	Factors	Grade A	Grade B	Grade C	Substandard
	% Soluble Solids (Minimum)	81.4	81.4	80.0	Fails Grade C
	Absence of Defects	Practically free- practically none that affect appearance or edibility.	Reasonably Free- do not materially affect the appearance or edibility.	Fairly free - do not seriously affect the appearance or edibility.	Fails Grade C.
	Score Points	37-40	34-36_1/	31-33_1/	0-30_1/
	Flavor & Aroma	Good-free from caramelization, smoke, fermentation, chemicals, and other causes.	Reasonably good- practically free from caramelization; free from smoke, fermentation, chemicals and other causes.	Fairly good- reasonably free from caramelization free from smoke, fermentaion, chemicals and other causes.	Poor-Fails Grade C.
	Score Points	45 - 50	40 - 44 _1/	35 - 39 _1/	0 - 34 _1/
	Clarity	Clear-may contain air bubbles that do not materially affect the appearance; may contain a trace of pollen grains or other finely divided particles in suspension that do not affect appearance.	Reasonably clear-may contain air bubbles, pollen grains, or other finely divided particles in suspension that do not materially affect the appearance.	Fairly Clear-may contain air bubbles, pollen grains, or other finely divided particles in suspension that do not seriously affect the appearance	Fails Grade C.
	Score Points	8 - 10	6 - 7	4 - 5 _1/	0 - 3 _2/

_1/ Limiting rule - sample units with score points that fall in this range shall not be graded regardless of the total score.
_2/ Partial limiting rule - samples units with score points that fall in this range shall not be graded regardless of the total score.

TABLE V - FILTERED STYLE

Analytical Factors	Grade A	Grade B	Grade C	Substandard
% Soluble Solids (Minimum)	81.4	81.4	80.0	Fails Grade C
Absence of Defects	Practically free- practically none that affect appearance or edibility.	Reasonably Free- do not materially affect the appearance or edibility.	Fairly free - do not seriously affect the appearance or edibility.	Fails Grade C.
Score Points	37-40	34-36_1/	31-33_1/	0-30_1/
Flavor & Aroma	Good-free from caramelization, smoke, fermentation, chemicals, and other causes.	Reasonably good- practically free from caramelization; free from smoke, fermentation, chemicals and other causes.	Fairly good- reasonably free from caramelization free from smoke, fermentaion, chemicals and other causes.	Poor-Fails Grade C.
Score Points	45 - 50	40 - 44 _1/	35 - 39 _1/	0 - 34 _1/

_1/ Limiting rule - sample units with score points that fall in this range shall not be graded regardless of the total score.
_2/ Partial limiting rule - samples units with score points that fall in this range shall not be graded regardless of the total score.

HONEY PRODUCTS – Probably 98 percent of the honey that is marketed is sold as table honey or for use in baking. Honey products, that is items where honey is an ingredient or where it is used as a sweetener, other than baked goods, are being marketed. A few honey products have a short shelf life. Honey has been used successfully, for example, as a sweetener in peanut butter but when five to 15 percent peanut butter is added to crystallized honey the product becomes rubbery in time. Many honey products are also made at home as they add variety and interest to the table.

Honey beer – Beer made with honey is a traditional drink in much of Africa. In fact, in East Africa nearly all of the honey produced is used for this purpose.

Native beekeepers sell their honey, which often contains some brood, to the beer manufacturers who use large tanks for the fermentation. Grain is sometimes added but is of little value other than, perhaps, to improve the flavor. The barrels are not cleaned between fermentations so there are always residual yeast cells to start a new fermentation. The fermentation is usually complete within three to four days. About five to six percent alcohol is produced. The beer is sold and drunk immediately as it is neither filtered nor pasteurized.

Honey beer is popular in many areas of the U. S. Many microbreweries feature a honey beer. Home brewers have found they may substitute honey for sugar in beer making with satisfactory results.

Honey butter – Butter and creamed (crystallized) honey may be mixed and made into a delicious spread. One need use only five to 15 percent butter. Honey butter should be consumed within a few weeks of being made as the butter may become rancid. The honey butter that is marketed in the U.S. is made according to a secret process developed in the 1930s. It is kept refrigerated in order to protect its delicate qualities.

Honey candy – It has never been too popular to make commercial candy with honey. One of the chief problems is that honey is too sweet so candy manufacturers use glucose almost exclusively. Glucose is the least sweet of the common sugars and one can eat much more candy if it is made using glucose rather than cane sugar, High Fructose Corn Syrup, or honey. It is, of course, the goal of the candy manufacturer to sell a greater quantity of candy.

Another problem in making honey candy is that steam kettles are often used in candy manufacture. Glucose and sucrose can be heated in a steam kettle without serious damage (burning). Honey is easily burned when it comes into contact with metal that is heated with steam. Both fructose, and the proteins that give honey its distinctive flavor, are easily burned and destroyed. An important consideration when using honey in candy making is to remember that honey contains about 20 percent water and compensation for this must be made in the recipe (see HONEY COOKERY). Recipes for honey candy are seen from time to time in the bee journals and can be found in honey cookbooks.

Flavored honey – The natural properties of honey have made it a favored ingredient in many baked and manufactured foods for centuries. In recent years, food scientists responding both to the honey industry's need for more value-added honey products and the modern consumer's demands for variety have experimented with honey-fruit spreads as an alternative to traditional jams and jellies. The challenge has been to create a spread with proper taste, balance, eye appeal and spreadable texture while maintaining a moisture level low enough to prohibit fermentation.

Cremed honey (see DYCE PROCESS) is mixed with commercially dehydrated fruits until a homogeneous texture and color are produced. Care must be taken in selecting honey that will not overwhelm the taste of the fruit. Generally, clover and alfalfa honey are suitable for this purpose, but any mild flavored honey may be used. Honey-fruit spreads have demonstrated a fine potential for consumer acceptance. They show little deterioration with age, they have expanded the taste parameters normally associated with honey and, when properly prepared, have a texture that many consumers believe superior to commercially prepared preserves. Modern consumers are showing an increased awareness of natural foods and honey-fruit spreads represent an alternative to sugar-based jams and jellies.

The equipment needed to produce a good quality honey-fruit spread is not sophisticated and it is easy to attempt small-scale production. Many beekeepers package part of their honey crop themselves for direct or local sales, and the production of honey-fruit spreads has the potential for them to enhance their income as many consumers are willing to pay a premium price for these value-added products. For further information follow the instructions included with the dehydrated fruits. -SZ

Honey ice cream – Ice cream made with honey is not found on the market too often though several manufacturers have made efforts in this direction. Honey lowers the freezing point of the ice cream if it is used exclusively in place of sugar. As a result, honey ice cream must be stored at slightly lower temperatures to have the proper consistency when served. Because the texture of honey ice cream deteriorates honey ice cream has a shorter shelf life than commercial ice cream.

Honey has been used successfully as an ice cream topping on sundaes. Honey is sometimes variegated, that is introduced in streams or patches, into ice cream with considerable success. The lighter, milder honeys are more satisfactory, but again, this is a matter of taste.

Mead – Mead, honey wine, is considered the first alcoholic beverage made by humans. All the ancient cultures of the world had rituals that involved mead. There were some sweet fruits, but honey was readily available in the wild and the techniques for harvesting it were simple.

Since honey was the only known common sweet in the ancient Egyptian, Greek or Roman world it was also valuable in its own right. As people learned how to make alcohol from grain and selected sweet fruits the use of honey to make alcohol waned. However, in northern Europe, sugar cane would not grow because of the cooler climate and historical records show that from 400 to 1000 years ago honey was the chief ingredient in the alcoholic beverages made in that part of the world.

Mead is the old English word for honey wine; metheglin is the name for spiced honey wine and melomel for mead made with fruit juices. Today mead can be made as a dry table wine, a sweet wine or a sparkling wine. Any number of books are available with instructions for making mead and are highly recommended reading before attempting to make mead. These books also give sources of equipment and wine-making supplies.

It may be necessary to experiment with several batches of mead to select the most suitable honey flavor and the best water. Initially a mild-flavored honey would be the best choice. Water with high chlorine content will not make good mead. Try to use good quality water, either bottled or from a well.

Basic equipment includes gallon jugs, perhaps a five-gallon carboy, a funnel with a long neck, tubing for siphoning, and the essential airlocks. Rubber stoppers, tea strainer, wine bottles and new corks are also necessary. Containers of plastic, glass or stainless steel can be used for making the mead.

Honey contains spores of wild yeasts but these are not suitable for fermentation into mead. Successful mead requires sterilization of all equipment, including the honey and water mixture, called the "must." Sterilization of the honey and water mixture can be done in two ways. One is with the addition of sulfites. The second method is by boiling the must. Boiling also serves another purpose. It destroys the proteins in honey that, if left, cause mead to be cloudy. Unfortunately long boiling can destroy some of the flavors and aroma of the honey.

One of the best yeasts for mead is champagne yeast, although other wine-making yeasts can be used. In addition a yeast energizer (nutrient) is used which helps complete the fermentation. After the yeast is added the mixture should be kept at a constant temperature between 55 to 70°F.

One basic recipe follows:
16 pounds of honey (mild wildflower or clover-type)
4 teaspoons acid blend
6 teaspoons yeast nutrient
1 packet champagne yeast

1-1/2 teaspoon grape tannin
1/4 teaspoon sodium metabisulfite (if not boiling)
water to make five gallons

In a food-grade plastic bucket mix honey and water sufficient to make five gallons of liquid. Add the sodium metabisulfite, stir and let stand, loosely covered for 24 hours. Stir in acid blend, tannin and nutrients. Add the yeast and stir vigorously. Cover loosely. Fermentation will begin. Skim, with tea strainer, the scum and froth daily. When the fermentation slows, siphon the liquid into gallon jugs and fit with rubber stoppers and air locks. You will siphon the liquid from the sediment once a month until fermentation has stopped, anywhere from three weeks to many months. The liquid can then be siphoned into bottles and left to age for a year.

These instructions only cover the basics of mead making. It does require both patience and attention to details.

Some innovative technology was introduced to mead making in the 20th century. Honey is diluted with water and the solution is continuously pumped through an ultrafiltration system. This removes all of the proteins and other products that previously had to settle out of the solution. The end product is then allowed to ferment, with, or without flavorings added. This takes months off the time required from start to finish, and produces an exceptional product.

The process was developed by Robert Kime at Cornell University and is gaining industry acceptance because the end product is predictable, repeatable, inexpensive to produce and sells well.

Soft drinks – Honey has sometimes been used to sweeten soft drinks. Soft drinks manufactured with any sweetener pose special problems with microbes since soft drinks are not pasteurized. Instead, the industry depends on a high degree of sanitation, using ultra clean water that is usually passed through an ion exchanger; a low pH; and, when necessary, a preservative such as sodium benzoate. Most of the colas, for example, will have a pH of about 2.6 while the root beers may be as high as four. The average soft drink has a pH of about three. A medium that has a low pH is inhospitable for most microbial growth. One reason that honey is such a safe food is that its low pH averages about 3.5. Soft drink manufacturers that use honey are seeking to use the fine reputation the product has and are not inclined to use a preservative, thus it behooves them to take special care with sanitation. Micropore filtration, as a last step before bottling, is often used.

Vinegar – Vinegar was discovered more than 10,000 years ago, probably from mead that had turned to vinegar. The word "vinegar" came from the French word meaning sour wine.

Good vinegar is made with good ingredients, and basically starts with good alcohol. In the case of honey it means starting with good honey wine – mead. Vinegar, of course can be made from other sugar-containing liquids such as fruit juices and grains.

Two distinct steps go into the making of vinegar. The first is fermentation with the action of yeasts on sugar to form alcohol. The second step is the changing of the alcohol to acetic acid with the action of bacteria known as Acetobacter or "mother." Vinegar-making supplies can be obtained from wine-making suppliers.

In making any fermentation, all

equipment must be sterilized. Wild yeasts and bacteria cannot be relied on to give a good product. Containers must be glass, enamel, or stainless steel. It is worthwhile to have a simple acid test kit so that the final degree of acidity can be determined. Vinegar should be at least five percent acid for preserving or pickling. Wine should be diluted to about 5-1/2 to seven percent alcohol for converting to vinegar. Steady temperatures between 80 to 85°F are ideal.

The basic recipe for vinegar is simple:
3 measures mead (5-1/2 to 7 percent alcohol)
1 measure vinegar culture containing active bacteria

These are mixed in a container that has a cloth cover or a loose cotton stopper that allows air in but keeps dirt and insects out. The conversion from mead to vinegar will take from four to six months. Testing for acid will indicate when the conversion is complete.

During the process a "mother" will form. This is a leathery, thick, grey film that should not be disturbed. If the film falls to the bottom it should be removed. Another film will form.

The vinegar should be pasteurized to prevent spoilage or conversion to carbon dioxide and water. To pasteurize, heat the vinegar to 150°F for 30 minutes, then bottle. Use plastic caps, since the vinegar will react with metal caps.

Before using allow the vinegar to age about six months at steady temperatures of 50 to 60°F. If sediment forms, siphon the clear liquid off into sterilized bottles. A little of the unpasteurized vinegar can be saved to start another batch since the bacteria are still alive. The mother does not have to be saved.

Times, and techniques have changed.

The following is from an *ABC* published over 100 years ago.

How To Make Vinegar

In the first place it should not be forgotten that vinegar may be made from any liquid containing sugar, provided there is enough sugar to be of any consequence. This includes a number of the fruit juices, of which the apple and grape are the best known examples, and syrups, like honey or molasses.

Vinegar is the product of two absolutely distinct fermentations: first, the vinous, or alcoholic, and second, the acetic, or acid fermentation. The first should be completed before the second is begun; otherwise the first never will be completed and weak vinegar will result. This means, for instance, that the "mother of vinegar," the thing which starts the acetic fermentation, must not be introduced until practically all the sugar in the liquid has been converted to alcohol by the common wine or alcoholic fermentation.

The alcoholic fermentation will usually start spontaneously, but it is far better to insure its starting by the addition of a small quantity of yeast.

Honey vinegar can be made just as cider vinegar is made by diluting the honey with water, then allowing the solution to ferment. Five parts of water to one part of honey by weight is about the right strength to make good vinegar. As the specific gravity of honey varies, it is perhaps well to check up the mixture of honey and water. When the liquid will just support a fresh egg, leaving a spot about as large as a dime above the surface, it is about right. Since such a dilute solution does not contain sufficient mineral matter for the growth of the organisms causing fermentation, it is well to add certain minerals. The Michigan Agricultural College has published a bulletin on honey vinegar in which the following formula is recommended: Extracted honey, 40 to 45 pounds; water, 30 gallons; potassium tartrate, two ounces; ammonium phosphate, two ounces. The bulletin recommends that this solution be boiled for 10 minutes in order to kill all or most of the micro-organisms which would later cause trouble, and to give the vinegar better color. The process of alcoholic fermentation can be greatly hastened by adding a starter prepared from a pure yeast culture. The acetic fermentation ban be hastened by adding "mother of vinegar." The bulletin mentioned above tells how to make these starters.

The tank should not be of metal but of glass or wood. Common jugs are, of course, suitable. To make a quantity use a wooden tank, hogshead or barrel.

The process of fermentation in the first stage can be hastened by having the honey water trickle slowly over cypress shavings so as to expose the liquid to the air as much as possible.

If barrels are used the bung should be left out and the hole covered with cheese-cloth. The barrel or barrels of liquid should be allowed to stand in a warm room and it may take some weeks before fermentation is completed unless cypress shavings are used.

It is only fair to say that unless one has a lot of cheap, off-grade honey that cannot be sold at any price, one cannot make a honey vinegar and compete with cider vinegar. Sweet cider from windfall apples is far cheaper than honey and water. While a honey vinegar when rightly made is superior to cider vinegar, it is not possible to make the public think so when cider vinegar is cheaper.

Yogurt – One of the oldest fermented foods for many centuries has been yogurt, long a favorite in the Near East and the Mediterranean area, now equally popular in Europe and North America, especially as a nutritious, low-calorie food. Flavored yogurt may be made in a variety of ways. One is to layer flavoring ingredients into the yogurt. A second popular method is to incorporate the flavoring into the body of the product.

Honey may be used both as a flavoring agent and as a sweetener. When used as a flavoring agent, a strong-flavored honey such as buckwheat can be used. When used as a sweetener, a milder-flavored, lighter-colored honey is better. A minor problem is that honey contains nearly 20 percent water and this may affect the viscosity of the yogurt. This problem may be corrected by adding small amounts of skim milk solids to the final product. Since both honey and yogurt are natural products, popular demand for the combined product has been high.

HONEY RECIPES – For many years around the turn of the century, the A.I. Root Company was one of the world's leading honey packers. They produced honey under the brand name AirLinE, which stands for A.I. R(oot) Line.

AirLinE honey was served in the fanciest of locations – especially the Grand Hotels of the time, and in the elite Pullman cars that crisscrossed the U.S. day and night.

To promote their fine product, the A.I. Root Company published a variety of recipes featuring honey as the only sweetener. They had recipes for canning and preserving, desserts, main dishes, cool summer drinks, candies and sauces.

This book features several of the best of the AirLinE Honey recipes we could find. You can make everything from hot drinks to puddings, from candy to the world famous Bar-le-duc Preserves.

The rest of the recipes featured here were taken from *A Honey Cookbook*, published by the A.I. Root Company in 1991, many of them gathered from the kitchens and card files of America's best honey cooks.

We've put together two complete meals and hope that you'll try these recipes. You can find thousands more honey recipes by going online.

From A Honey Cookbook, *published by The A.I. Root Company*

GRAPE DRINK

3/4 cup club soda
1/4 cup grape juice
2 tsp. honey (any variety works well)
1 egg

Combine all ingredients into blender. Blend well. Pour into a chilled glass and serve immediately. Single serving.

TANGY COLE SLAW

1 medium head of fresh cabbage
1 cup sour cream
1/4 cup honey (medium to dark)
1 tsp. salt
2 tsp. celery salt
1/4 cup white vinegar

Chop cabbage. Mix all together the rest of the ingredients and pour over chopped cabbage. Marinate for several hours or overnight before serving.

SKILLET BAKED BEANS

1 Tbs. butter
1 onion, chopped
1 green pepper, choped
1/2 cup BBQ sauce or catsup
2 Tbs. brown sugar
1/4 cup honey (light to medium)
Dash garlic powder
2 cans pork and beans

Sauté the onion and pepper in butter. Add the rest of the ingredients and cook on top of stove for 30 minutes or bake in a casserole dish. Serves six.

BBQ BEEF

2 lbs. hamburger (lean)
1-1/2 Tbs. oil (try canola)
1 green pepper, finely chopped
1 cup celery chopped
1 cup tomato sauce
1 onion, chopped
Chili powder – dash (to taste)
1/2 Tbs. salt
1/4 cup honey (dark and flavorful)
1 Tbs. cider vinegar

Sauté the onion, celery, and green pepper in oil. Add hamburger and brown, making sure meat is broken into small pieces. Stir in remaining ingredients and cook over low heat about 45 minutes. Serve on open faced buns. Makes 10-12 servings.

CHERRY PIE

3 cups cooked, unsweetened cherries and juice
4 Tbs cornstarch
1/4 tsp. salt
2 Tbs. butter
3/4 cup honey (light and mild variety)
1/2 tsp. almond extract
1 9-inch unbaked pie shell

Drain the juice from cherries into saucepan and set the cherries aside. In the cherry juice stir cornstarch, salt, honey and almond extract and heat about 10 minutes on low until well blended. Stir in cherries. Pour into pie shell and dot with butter. Cover with criss-cross top for the traditional look. Bake at 400°F for 35-40 minutes.

HOT FRUIT PUNCH

2 cups apple cider
2 cups cranberry juice
¼ cup honey
4 whole cloves
4 slices of lemon
1 cinnamon stick (optional)

Combine all ingredients in saucepan and heat until mixture just boils. Pour into glasses or mugs immediately and serve. Serves eight to 10.

CRANBERRY FRUIT SALAD

1 apple cored and shopped
1 cup seedless grapes, halved
1 banana, sliced
6 seedless oranges, sectioned & cleaned
1 cup cranberry sauce, jellied
¼ cup honey (light and delicate)
1 Tbs. orange juice

Combine oranges, apple, and grapes in a bowl and refrigerate. Heat cranberry sauce in saucepan until melted. Stir in honey and orange juice and cool. Just before serving add the banana to the fruit and pour the cranberry sauce mixture over all before serving. Serves four to six.

BAKED ACORN SQUASH HALVES

2 acorn squash
8 Tbs. honey (light or dark)
½ tsp. cinnamon
¼ tsp. salt
4 Tbs. butter

Cut the acorn squash in half. Clean out the inside by carefully scooping out seeds and pulp. To the cavity in each half add two tablespoons of honey, 1/8 teaspoon of cinnamon, dash of salt and one tablespoon butter. Place upright in bakin dish and add ½ cup hot water to the bottom. Bake 1¼ hours at 350°F. Serves four.

EASY CHICKEN

2½ - 3 lbs. chicken pieces
½ cup butter, melted
2 Tbs. oil
¼ cup lemon juice
1 tsp. ground ginger
½ cup honey (light and mild or dark and flavorful)

Brown chicken quickly in melted butter and oil. Combine rest of ingredients and pour over browned chicken. Cover and cool slowly 30 to 40 minutes until chicken is tender. Don't forget to baste with the sauce occasionally. Serves four to six.

Read this, then try it, at least once. This recipe is taken from the best of the original AirLine Honey recipes.

BAR-LE-DUC PRESERVES

These preserves are believed to be the finest of their kind, and have hitherto been imported at extravagant prices. Other fruits besides currants may be treated in this way, as honey is of itself a preservative. These preserves do not require to be kept absolutely airtight.

Take selected red or white currants of large size. One by one carefully make an incision in the skin ¼ of an inch deep with tiny embroidery scissors. Through this slit remove the seeds with the aid of a sharp needle, preserving the shape of the fruit. Take the weight of the currants in honey, and when this has been heated add the currants. Let it simmer a minute or two, and then seal as for jelly. The currants retain their shape, are of a beautiful color, and melt in the mouth. Care should be exercised not to scorch the honey.

From A Honey Cookbook, *published by The A.I. Root Company*

HONEYDEW – Several species of insects, including white flies, coccids, and treehoppers, but especially aphids, feed on plant saps. These saps contain sugars and provide the insects with the food they need to live. Sap-feeding insects are often not efficient in their use of these materials. As a result they secrete waste products that are high in sugars and are attractive to bees and other insects. The secretions are deposited in droplets on the leaves and needles of the plants on which the insects are feeding. When this honeydew occurs in quantity, as it may in certain parts of the world, and under special circumstances, honey bees may collect it and make it into a honey that is called honeydew honey. It has been reported that trees on which aphids are abundant may glisten in the sunlight from the droplets of honeydew.

In parts of Europe, especially Germany and Switzerland, honeydew honey is called forest honey and is sold as a specialty product, often at a high price. In some areas in Europe, beekeepers move their colonies into forests for the express purpose of gathering honeydew. There are more people per unit of area in Europe and thus greater pressure on growing trees and other crops. As a result the forests are carefully pruned and watched. This means too that forests may consist of only one tree species. This encourages the populations of certain insect pests, such as aphids, and it is under these circumstances that honeydew honey flows occur.

A major problem in producing honeydew honey, especially in quantity, is that there is no pollen accompanying a honeydew flow and colonies may suffer a pollen shortage and be unable to rear brood. At the end of the season the colonies have only old bees. Under such conditions the colonies may perish in the winter that follows because they do not have a pollen reserve and enter the fall and winter seasons not having young bees present.

Honeydew honey is not common in North America, in large part because extensive plantings of single forest species are not common. Dr. J.W. White Jr. and his co-workers (*Composition of American Honeys*, USDA Technical Bulletin No. 1261. 124 pages. 1962.) included 14 samples of honeydew in their analysis of 505 samples of American honeys. The honeydews were quite different in several respects: On average the samples were quite dark, with an average of 85 to 104 on the Pfund scale (see Pfund grader). The following table is patterned after that of White et al.

Composition of Honey (490 samples) and Honeydew (14 samples)

(all figures below are percentages)

	Fructose	Glucose	Sucrose	Und'tmnd.	PH	Ash	Nitrogen
Honey	38.19	31.28	1.31	3.1	3.91	.169	.041
Honeydew	31.80	26.08	0.80	10.1	4.45	.736	.100

It will be noted that honeydew honey has almost the same sugars as normal honey and the fact that it contains nearly as much fructose makes it, like honey itself, markedly different from other sweeteners. The makeup of the undetermined matter in honeydew is not clear. In most honey the undetermined material is usually proteinaceous in nature. Several gums, dextrins and plant pigments are present in honeydew honey giving it its golden to very dark amber color. In general, Americans prefer light-colored honeys for table use. Europeans prefer the darker honeys, such as honeydew honey. Generally, when honeydew honey is found in North

America it is considered inferior for use here because of its dark color and strong flavor and is usually blended with the bakery honeys. It is, however, a highly sought-after product for export to Europe. Honeydew has been associated with aphid outbreaks on alfalfa and occasionally aphid populations on some trees.

In general, honey is not contaminated by molds and bacteria. However, because of its protein and sugar content, exposure on leaf surfaces, and perhaps its higher than normal pH, sooty mold may grow on the honeydew before it is collected by the bees. Once the honeydew is collected and processed it is protected, as is normal honey, against further growth by these microbes but dead mold spores may be present. These are often used to identify honeydew honey and to identify its origin. These molds have not been reported as being harmful for humans.

Rarely, sweet plant fluids may be produced from plant wounds, or extrafloral nectaries, and these may be gathered by bees and the product called honeydew.

HORNETS, YELLOWJACKETS AND OTHER WASPS – Many species of stinging insects commonly known as wasps, hornets and yellowjackets are confused with honey bees. Because of this confusion, beekeepers are often accused of causing problems that are not of their making. The terms wasp, hornet and yellowjacket are general terms and do not refer to any particular species. Yellowjacket refers to several species of yellow and black insects, all of which are social. They build paper nests both above and below ground and are the stinging insects most commonly confused with honey bees.

One particular problem with yellowjackets was the accidental introduction of a European yellowjacket, *Vespula germanica,* into the northeastern United States. This insect is a serious pest at picnic and campgrounds. Under normal circumstances, it feeds on nectar and collects other insects to feed its young, but it will also eat almost all of the picnic foods and drinks that we enjoy. Some people have been stung in the mouth by yellowjackets feeding on the same hamburger or soft drink at the same time. It is sometimes necessary to close campgrounds when these insects become especially numerous. It is important to keep picnic areas sanitary by removing waste food and keeping garbage cans closed in order to discourage foraging by yellowjackets.

Wasps, hornets and yellowjackets are not particularly good pollinators, although they feed on flower nectar as adults and contribute to the pollination of some plants. Wasps have few hairs on their bodies and, unlike bees, their body hairs are not branched and thus pollen does not cling to them so easily. In addition, wasps do not show flower fidelity; that is, they feed on many different species of flowering plants rather than flowers of the same species.

Of the several thousand species of hornets, yellowjackets and wasps, only a few are social and build large populations. Yellowjacket queens mate in the fall and hibernate under rocks and bark; in the spring they start individual nests. The queen constructs a small nest and comb with paper cells. The first brood may consist of ten or fewer individuals. The queen does the foraging, feeding, and brooding of these first young. As soon as they emerge, these young yellowjackets take over the foraging activities and the queen devotes herself to egg-laying. By fall an individual nest may contain a few hundred to a few thousand individuals.

Yellowjackets are usually not defensive until the population is fairly large. Home owners have mowed over ground nests for one or two months without being attacked, but late in the summer several yellowjackets may suddenly attack, causing one to think the nest was recently established.

In the fall, when fruit is ripe, yellowjackets will feed on the juices of fruits with broken skins. They may also bite and break fruit skins, something that honey bees cannot do because their mandibles are not of the biting type. Honey bees also sometimes feed on fruit with broken skins, but only when natural nectar is not available. Fruit growers and open-air market owners, seeing honey bees feeding in this manner, may mistakenly believe the honey bees have broken the fruit skins.

Public relations are important for beekeepers. Coping with the yellowjacket problem, and teaching people the differences between honey bees and yellowjackets is important. It is also well to remember that yellowjackets play an important role in balancing nature. Because they feed on other insects, especially mites, aphids and other insects that are often pests, we need them, too.

The bald-face hornet, or white-face hornet, is not a true hornet but is an aerial-nesting yellowjacket. It follows the same lifestyle of other yellowjackets.

Various wasps of the genus *Polistes* build open comb nests in sheltered places. These insects also follow a lifestyle similar to that of yellowjackets. Mated queens overwinter and establish nests in the spring. The wasps are beneficial insects also since they prey on other insects, some of which are harmful to crops.

No hornets are native to the U.S. However, the imported European hornet, *Vespa crabro,* is established in some areas. It is a cavity-nesting hornet so its paper nests can be found in hollow trees and other enclosed spaces.

HOUSE BEES – In very general terms, bees in a hive are often divided into two groups: house bees and field bees. The field bees are those that do the foraging and home hunting when the latter is necessary. The first task of all worker bees as soon as they emerge is cell cleaning. House bees also feed larvae, secrete wax, build comb, process food, keep the brood warm, help cool the hive, and perform a myriad of lesser tasks. A small number of bees in a hive become guards; this is a task intermediate between those performed by house bees and field bees (see CASTE).

HRUSCHKA, MAJOR F. (1813-1888) – Major Hruschka of Italy, a bee supply dealer and exporter of queens, is credited with inventing the first honey extractor in 1865. It was a tangential type with a wooden tub. It was soon copied worldwide and was one of several inventions following the discovery of bee space in 1851 that made practical beekeeping possible.

HUBBARD APIARIES, INC. – Leland Martin "Lee" Hubbard hived his first swarm of bees at 14 in the orchard on his parent's farm, at Onsted, Michigan. From this small start he built Hubbard Apiaries into the world's largest private packer. More than 30 million pounds of honey were packed annually at Onsted and several million more at Belleview,

Leland "Lee" Hubbard

Richard Hubbard

Florida. Honey was marketed under Great Lakes, Florida Lakes, and Crystal Clear labels nationally, and was exported under the Monarch label. Hubbard custom packed more than 125 labels at Onsted and more than a dozen at Belleview.

Hubbard Apiaries also operated a large bee supply business, and manufactured most of the supplies they sold including hives, frames, foundation and one of the best commercial-sized extractors ever built, as well as making their own plastic containers.

Lee completed the course in beekeeping taught by Professor Russell Kelty at Michigan State University in 1924.

He sold his honey from a roadside stand located on M-50, the highway that runs by their farm near Onsted, then door-to-door, and later to local stores. He also exhibited and sold honey at two county fairs and the Michigan State Fair.

In 1927 he purchased his first truck so he could expand into the national honey markets in Chicago, Cleveland, Cincinnati, Detroit, and Pittsburgh.

In 1928 he purchased part of the Hubbard Farm, cleared brush, and built a small building. The manufacturing of beehives was started in 1929, with Lee designing and modifying the machinery. After he redesigned it, one of these machines could make 10,000 end bars a day.

By 1937 Hubard was keeping more than 1000 colonies in 30 beeyards.

By 1941 they employed 10 men to manufacture and operate their bee supply division.

In 1946 Lee purchased a modern beehive factory from Superior Honey Company in Los Angeles, California and moved the equipment to Springville, Michigan. A filtering system was installed in their honey packing plant in 1948.

At one point Hubbard operated more than 7000 colonies in six states – Michigan, Florida, Wisconsin, Ohio, and North and South Dakota.

In 1967 Mary and Lee Hubbard moved to Belleview, Florida, to retire but by 1979 they had the largest honey processing plant in Florida.

After Hubbard retired the Onsted plant was managed by the Hubbard's two sons, Dick and Jim and their son-in-law, Ernest Groeb, Jr. (who eventually left the company to start his own packing business). – from *Some Beekeepers & Associates* by Joe Moffett.

HUBER, FRANCOIS (1750–1831) – This Swiss naturalist laid the groundwork for the scientific understanding of honey bees. Although blind, he was assisted by his servant, Francois Burnens, an uneducated peasant, who possessed great intelligence and aptitude for natural science. Huber's *Nouvelles Observations sur les Abeilles*, consisting of letters to other naturalists written over many years, was published in 1792. An English translation appeared in 1806. His basic tool was a "leaf hive" of his invention, consisting of twelve combs hinged in such a manner that they could be turned like the pages of a book, and in which every aspect of colony life could be observed. Huber showed, contrary to widespread belief at the time, that beeswax is not made from pollen, but from bees consuming nectar, and he described, for the first time, the manner of its production. He demonstrated that wax combs can be built by bees fed on sugar syrup, and proved that pollen serves primarily as food for developing brood. He described in detail, and added enormously to the knowledge of such things as the function of the antennae, the manner of the fecundation of the queen, laying workers, mutual queen hostility, the origin of propolis, hive ventilation, the conversion of worker larvae to queens, and memory in bees. With respect to the latter he found that bees returned to a familiar artificial source of honey even after the long winter withdrawal. To his contemporaries Huber was known not only for his scientific achievements, but also for his cheerful disposition, which was never dampened by his handicap. - RT

HUMIDITY CONTROL IN THE NEST – The physical nature of the bees' nest, together with the high brood rearing temperature and the low moisture content of honey, serve collectively to help control the humidity in the nest in both winter and summer.

In the summer, at a time when bees are not gathering nectar and evaporating water, the humidity in a hive usually varies between 40 and 50 percent and is lowest in the center of the brood nest where it is the warmest. This is apparently a humidity under which exposed larvae can be reared with ease but is also suitable for both eggs and pupae. When water is being evaporated from nectar the humidity will be much higher but the moist air is usually discharged rapidly. Honey bees gather large quantities of water during hot weather for cooling the interior of the hive.

In the winter cluster the situation is quite different. Surviving long periods of confinement depends upon a delicate balance of humidity. Metabolic water is produced when bees consume honey (see Wintering). A tight winter cluster inhibits the escape of some moisture in winter. However, excess moisture must be removed from the hive. Ventilation of the hive is important throughout the year.

HYBRID BEES – Hybrids are the offspring resulting from the crossing of parents that are of two different sub-species or unrelated lines. The first generation of offspring usually has a greater fitness and overall vitality superior to either of their own parents. This genetic phenomenon is called heterosis or hybrid vigor.

Hybrid vigor has proven to have economic value in agricultural breeding programs. The commercial goal is often a desire for higher production and disease resistance. Impressive successes have been demonstrated in the plant and animal industries. However, the crossing of different species of animals often results in a sterile hybrid, as with the mule.

True hybrids must be distinguished from mongrels that are the result of various interbreeding. The U.S. bee population is largely a mongrelized combination of numerous strains and races because of the random, multiple mating behavior of the honey bee. Genetic control must be accomplished by isolated mating, drone saturation and/or artificial insemination.

To develop a hybrid, inbred lines or different strains of previously tested lines are purposely crossed with the hope that the resulting hybrid will be a superior product. Random mating of future generations is usually disappointing because of genetic dilution.

Daughter queens reared from selected breeders are often naturally mated in areas not adequately saturated with drones of desired parentage. These F_1 (first generation) daughter queens, used for production colonies, often retain most of the selected characteristics though will show a wide range of variability. In each succeeding generation these traits become increasingly diluted by random mating and may be quickly lost.

Viable bee breeding programs require controlled mating and long term commitment to rigorous evaluation and selection over time to ensure maintenance of selected stocks. Beekeeping management practices that do not tolerate the use of supersedure queens are an important part of bee breeding.

In honey bee breeding one of the most successful programs, founded in the 1950s by Dr. G.H. Cale and Dadant & sons, was the establishment of double hybrids, the Starline and the Midnight. Four inbred lines were developed and crossed to produce the superior double hybrids.

In order to control the effects of parasitic mites and diseases various hybrid lines have been developed, such as the hygienic stock and the VSH stock.-SC

IMPORTING BEES – (see LEGISLATION IN THE UNITED STATES RELATING TO HONEY BEES)

INFANT BOTULISM – Spores of *Clostridium botulinum,* that cause the disease known as infant botulism, are found everywhere: in soil, on vegetables, in dust, and in honey. Finding these spores in honey is not surprising since nectar is collected from plants that are exposed to dust from soil.

Scientists have found that the digestive system of a young infant is still immature so that spores of *Clostridium botulinum* are able to germinate and produce toxins in the anaerobic conditions found in an infant's intestine. Since most cases of infant botulism occur within six months of birth, the medical profession has recommended that infants under a year old not be fed any honey. - AWH

INSECTICIDES AND HONEY BEES – (see PESTICIDES, EFFECTS ON HONEY BEES and BEEKEEPER INDEMNITY PAYMENT PROGRAM)

INSTAR – (see LARVA)

INSTRUMENTAL INSEMINATION – Instrumental Insemination is an essential tool for honey bee research and stock improvement. Advances in techniques and equipment designs, combined with an increasing knowledge of bee biology, behavior and genetics, have made it practical and highly successful.

The first recorded attempts to control mating date back to the early 1700s. Over the next 200 years, the challenge of mating queens in confinement proved frustrating and unsuccessful. This led investigators to explore hand mating and the mechanical transfer of semen. Dr. Lloyd Watson performed the first successful technique of instrumental insemination, as he named it, in 1926.

Watson injected a trace amount of semen, not realizing the need for multiple drones or the need to bypass the valve fold. Several critical discoveries greatly improved these results. In the 1940s it became evident that queens multiple mate and semen from several drones was necessary for a complete insemination.

Dr. Harry Laidlaw, in 1944, wrote a detailed description of the valvefold and its function. This is a structure of tissue blocking the passage of semen into the median oviduct. He designed a hook to bypass the valvefold, greatly increasing the efficiency of insemination and the number of spermatozoa entering the spermatheca of the queen. In 1947 Dr. Otto Mackensen showed that repeated treatments of carbon dioxide, used to anesthetize the queen during the procedure, have the beneficial effect of stimulating egg production. It also eliminated the detrimental effects of ether, previously used as an anesthetic.

Instrumental Insemination, in addition to providing a method of complete control of mating, enables the creation of very specific and extreme crosses that do not occur naturally. This capability offers new possibilities for research purposes, mating system designs and methods of stock development and maintenance. [For example; a single drone can be mated to one or several specific queens.] Dr. Walter Rothenbuhler showed that mating single drones to inbred queens allowed one to identify the mode of inheritance of behavior expressed only at the colony level, i.e. hygienic behavior. Single drone inseminations also allow for the isolation and amplification of a specific trait that is not readily expressed in honey bee populations.

A group of queens can be inseminated with a uniform mix of semen from hundreds of genetically diverse drones. For a breeding program this increases the effective breeding population size, viability and fitness, and simplifies stock maintenance. Various degrees of inbreeding can also be created to produce different relationships, including "selfing," the mating of a queen with her own drones.

Another advantage provided is the ability to store and ship honey bee semen. Semen can be shipped worldwide, minimizing the risk of spreading pests and disease. Short term storage has been perfected. Semen can be held for up to two weeks at room temperature and retain high viability and mobility. Long term storage techniques include cryopreservation, or freezing, of semen and embryos.

A high rate of success can be expected, provided inseminated queens are given proper pre- and post-insemination care and sufficient semen dosage. Numerous comparison studies of instrumentally inseminated queens, IIQs, and naturally mated queens, NMQs, dating from 1946 to the present, demonstrate similar performance levels in production of honey and brood, and queen longevity. With the advantage of selection and controlled mating, several studies have shown that IIQs out perform NMQs

The unfounded reputation for poor performance of IIQs can be attributed to poor insemination techniques, inadequate care, insufficient semen dosage, injury and infection. Queens undergo many physiological changes in preparation for egg laying. The care given will affect sperm migration and performance. To optimize results, natural conditions must be maintained as much as possible.

Virgin queens have a short receptive period for mating and should be inseminated when five to 10 days old. However Dr. Ken Tucker successfully inseminated three month old queens during an experiment in Venezuela. Conditions after insemination are critical for sperm migration to the spermatheca. Activity, brood nest temperatures and attendance by worker bees have been shown to aid sperm migration. Virgin queens established in nucleus colonies and given a direct release after insemination will store more sperm and begin egg laying sooner than queens that have been banked in small cages.

Careful planning is also important to assure an adequate supply of select, mature drones. Drones have a high rate of attrition, are vulnerable to stress and are more susceptible to pests and diseases. Their slower development time, 24 days from egg to adult and two weeks post-emergence to mature, must be coordinated with queen rearing. Colonies used for rearing queens and drones should never be treated with miticides, as these may cause fertility problems and comb contamination.

There are a variety of instruments available. Most are a modification of the Mackensen instrument designed in the late 1940s. Today's models are updated with micromanipulators offering precision, a high rate of repeatability and ease in learning the technique. Instrumentation includes a supporting stand, a queen holder assembly, syringe and syringe manipulator, a set of hooks and/or forceps.

Polish. All friction points are coated with teflon making action very smooth. Tips are of glass and easily replaced. An extra base plate is included so that the space between the two feet can be made wider to accommodate microscopes with a wide base. The syringe, moved by turning the wheel activating a friction rubber wheel, gives very smooth action.

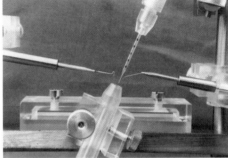

Czech device showing a close-up of queen holder tube, glass tip that is graduated, sting and dorsal hooks.

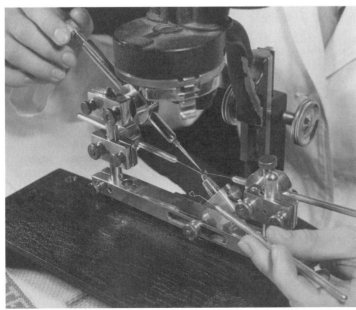

Early insemination device used by USDA as far back as 1947. This is a modification of the original Mackenson instrument.

A trend toward compact and simplified instruments has brought the use of forceps back into vogue, initially used by Watson in the 1920s. A microscope and cool light source are also required. A stereo-zoom microscope with 10x to 20X magnification and working distance of 80 to 120 mm is recommended.

There are choices in techniques of opening the queen's vaginal chamber and lifting of the sting. This can be accomplished with a set of hooks of various designs, or a pair of forceps. Sting-hook designs include the classic spoon shape hook and the perforated hook used to thread the sting. A pair of forceps, used freehand or as mounted pressure-grip forceps, are available.

Most instruments use a piston-type syringe filled with saline solution and glass tips with a capacity of 50 µl. An innovation that has simplified and improved the efficiency of semen handling is the Harbo large capacity syringe. Designed for semen storage, capillary storage tubes and tips are easily detachable. This provides an unlimited capacity and convenience for storing and shipping semen. The flexible latex connection of the tip is also an advantage allowing very fine adjustments for bypassing the valvefold during semen insertion. This syringe is compatible with most instruments on the market.

The size and shape of the tip are important and will interact with differences in queen anatomy and size. Generally, the smaller the tip, the more easily the valvefold is bypassed; a small tip will be less tolerant of mucus, which is carefully avoided during collection. An angled finish of the tip is often preferred over a straight edge finish, providing a fine leading edge to bypass the valvefold and also a larger surface area to skim semen from the mucus layer of the endophallus.

Semen is exposed by eversion of the drone's endophallus, which is a two step process – partial eversion and full eversion. To stimulate the partial eversion, hold the drone dorsally - ventrally and apply pressure on the head and thorax. During this stage, the abdominal muscles will contract and a pair of horn-like structures appear. These are an orange-yellow color in the mature drone and clear in the immature drone.

Mature Drone (Cobey photo)

Immature Drone (Cobey photo)

To obtain semen, eversion of the endophallus is a two-step process, the partial and full eversion. To induce the partial eversion, grasp the head and thorax of the drone between the thumb and forefinger, ventro-dorsally. Roll or crush the thorax between your fingers. The abdomen of mature drones will contract and a pair of horn-like, yellow-orange cornua are exposed. If the abdomen remains soft or the cornua are clear, lacking color, the drone is immature and will not yield semen.

Instrumental Insemination

Exposing Semen

To expose semen, the partial eversion is completed by stimulating further contraction of the abdominal muscles. The buildup of pressure and compression of hemolymph and air sacs force the full eversion of the endophallus. Grasp the base of the dorsal abdomen close to the thorax, with the thumb and forefinger. Apply pressure along the sides of the abdomen, starting at the anterior base and working toward the posterior tip. Squeeze and roll your fingers together in one steady forward motion, forcing the complete eversion. The endophallus will flip up with force, exposing semen. Hold the drone downward to avoid contamination with your fingers. (Cobey photo)

Collecting Semen

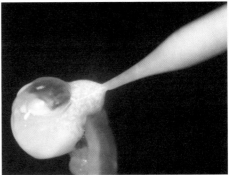

Collect a drop of saline in the tip of the syringe to make contact with the semen on the endophallus. Draw the semen into the syringe, skimming the semen off the mucus layer. To collect semen from the next drone, expel a small drop of semen from the tip on to the semen load of the next everted endophallus and draw this into the syringe. Repeat this process until the desired amount of semen is collected. Maintain sanitary conditions. Contamination can be a major cause of queen mortality. Take care to avoid contamination of semen, critical instrument parts and your hands. (Cobey photo)

The second step of this process, the full eversion, exposes the semen. Once the partial eversion is obtained, apply pressure along the sides of the abdomen with the thumb and forefinger, starting at the anterior base of the abdomen. This pressure forces the endophallus inside out to expose semen. Take care to hold the drone with abdomen downward to prevent the endophallus from flipping back onto your fingers, contaminating the semen.

The exposed semen is a creamy marbled color resting on a layer of white mucus. A drone will yield about one microliter of semen, though the consistency and amount of semen will vary greatly dependent upon age, nutrition and care. The recommended semen dosage per queen is 8 to 10 µl (microliters).

Instrumental Insemination

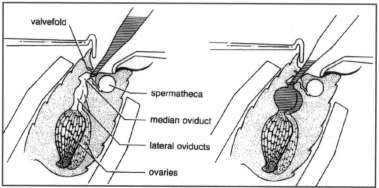

To expose the vaginal orifice of the queen, the abdominal plates are separated using a pair of hooks or forceps. The large sting structure is lifted up and dorsally.

The valvefold, a stretchy flap of tissue covering the median oviduct, is bypassed. Semen is inserted directly into the median oviduct.

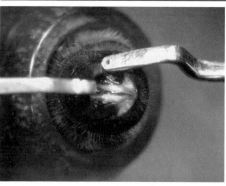

1. The syringe tip is used to bypass the valvefold. The tip is slipped beneath the valvefold and lifted ventrally. The angle of the syringe, and a light "zigzag" movement is used to maneuver around the valvefold.

Position the syringe tip dorsally above the "V", defining the vaginal orifice.

2. Insert the tip about 0.5 mm, slightly forward of the apex of the "V". Positioned correctly, the tip slips easily past the valvefold without resistance. Insert the tip another 0.5 to 1.0 mm into the median oviduct.

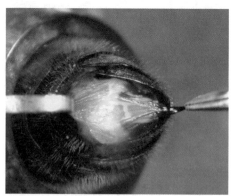

3. Bypassing the valvefold allows passage of the tip into the median oviduct. Test placement of the tip, preceding the insemination with a drop of saline, then insert a measured amount of semen. Eight to 10 µl is the standard dose per queen.

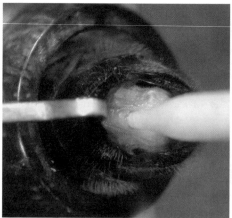

4. With practice, the insertion of semen is performed quickly and precisely, requiring only seconds per queen.

There are a variety of tools to choose from. Use is based on personal preference.

(Cobey photo)

To collect semen into the syringe, skim it off the mucus layer. A small drop of saline, or semen from the previous drone, is used to make contact with the semen. Avoid collection of mucus, which is more viscous and will separate readily from the semen as it enters the tip. Mucus collected with the semen may clog the tip at the critical moment of injection into the queen. Evert the next drone and repeat the process until the desired amount of semen is obtained.

In preparation for II the queen is anesthetized with carbon dioxide and placed in the holding tube of the instrument. The vaginal cavity is opened using a pair of hooks or forceps. The sting structure is lifted to expose the vaginal opening. Properly positioned a "V" of tissue defines the location of the

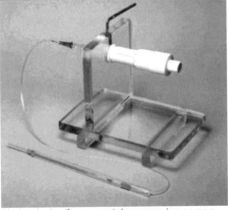

Semen being inserted directly into the median oviduct. (Cobey photo)

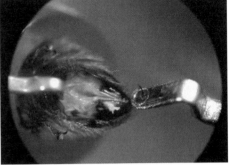

A new tool used in II is this ruby-lined sting hook. The sting is captured within the ruby ring and pulled out of the way. The ruby insert is used because it can be polished smooth, yet is very durable. (Cobey photo)

Harbo device for accurately measuring semen.

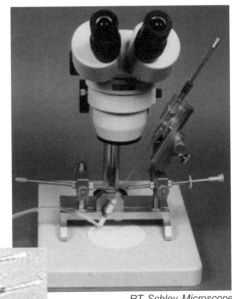

RT Schley Microscope.

Stored semen.

valvefold, which is not readily visible. The valvefold is bypassed using the syringe tip and semen is inserted directly into the median oviduct.

The syringe tip is positioned above and slightly to the right of the "V". Using a slight zig zag motion; insert the tip ½ mm, then move the tip slightly to the left, increasing the incline of the tip. Insert the tip another ½ mm to 1 mm, past the valvefold into the median oviduct. Precede the insemination with a drop of saline. This will test if the valve fold has been bypassed. If the saline freely flows into the queen, the semen is also injected. Otherwise the syringe is removed, the queen repositioned and a second attempt is made. *Sue Cobey*

INTEGRATED PEST MANAGEMENT (IPM) – In times past many growers and farmers have routinely blanketed their crops and animals with pesticides in order to protect them against whatever pest might show itself. That method wastes time and pesticides along with the danger of contaminating the products being grown and the environment. Today, the best method of protecting our plants and animals, as well as ourselves and the public, is to test for pests and to treat only when necessary.

IPM is a decision-making process in the selection, integration and implementation of the best methods of pest control available. It encourages the practitioner to utilize non-chemical control mechanisms including predators, parasites, and diseases and resistant bee stocks that might control the pest or reduce the usage of chemicals. IPM also includes physical methods such as modifying the beehive to reduce the economic impact of pests and diseases.

Two of the most important concepts in IPM are: (1) it is not necessary to eradicate a pest or disease but to remain below the economic threshold and (2) use chemicals only for treatment and not for prevention of diseases.

In the case of honey bees we know that selecting the environmentally correct apiary site will do much to help control diseases and pests such as the small hive beetle, *Varroa destructor,* nosema, and the foulbroods (see APIARY LOCATIONS). The availability and use of mite–resistant and/or tolerant bee stocks make it possible for beekeepers to use fewer acaricides and to enable beekeepers to select materials that may be less effective but are safer to the hive products.

We can expect to hear more about IPM in the future as it is a common sense approach to controlling pests. In addition it forces beekeepers to become familiar with the diseases, pests and predators of honey bees in order to make the best long-range decisions.

INTERNATIONAL BEE RESEARCH ASSOCIATION – (see ASSOCIATIONS FOR BEEKEEPERS)

INVERT SUGAR – When the sugar sucrose is boiled and an acid is added, it breaks into two sugars, glucose and fructose. The resultant product is invert sugar.

Honey bees convert sucrose, which is the chief sugar in nectar, into these same sugars through the use of an enzyme (see NECTAR, CONVERSION TO HONEY). Many years ago there was not sufficiently pure sugar for bee food and one finds reference to this in the older literature. Many queen breeders once used invert sugar to make queen cage candy. (Some beekeepers use honey to make queen cage candy but this is unwise because of the danger of spreading American foulbrood, since the bacterial spores are frequently present in honey.) The subject of invert sugar is of historical interest only, since high fructose corn syrup (HFCS), a mixture of glucose and fructose made from corn sugar, has replaced invert sugar on the market.

INVERTASE – (see NECTAR, CONVERSION TO HONEY)

JARVIS, DR. D.C. (1881-1966) – Author of the best-selling book *Folk Medicine: A Vermont Doctor's Guide to Good Health*, Dr. Jarvis was born in Plattsburgh, New York, on March 15, 1881, the son of George and Abbie Vincent Jarvis. He received an M.D. degree from the University of Vermont in 1904 and, after an internship and residency at Mary Fletcher Hospital in Burlington, Vermont, set up a practice as an otolaryngologist in Barre.

He remained an obscure country doctor until 1958, when he was catapulted to fame by the success of his book. A favorite among health faddists, it sold more than a million copies and was translated into 12 languages. It became familiar to paperback readers as *Vermont Folk Medicine*.

The book extolled the virtues of honey and apple cider vinegar. Although Dr. Jarvis was analytically correct in his description of these commodities, his book conveyed the general impression that they would cure virtually anything.

Dr. Jarvis drew on the knowledge he had gained traveling in New England and listening to the folk tales of its inhabitants.

"Folk medicine reaches very far back in time," Dr. Jarvis wrote in his book. "Nature opened the first drug store. Primitive men and the animals depended on the preventive use of plants and herbs. Vermont folk medicine adapts very old physiological and biochemical laws to the maintenance of health and vigor. In honey we have two sugars, dextrose and levulose. When honey is taken, dextrose passes swiftly into the blood. Levulose maintains a steady level of blood sugar concentration. It keeps honey from raising the blood sugar level too far."

By the end of 1959, Dr. Jarvis had attained international renown. Life magazine commented then that "among medical authorities the value of his cures is, to say the least, controversial."

An accomplished musician, he could play many instruments. He also organized and conducted a junior symphony orchestra in Barre.

Dr. Jarvis helped the local high school's football players improve their passing by making studies of their visual depth perception.

In 1960 his second book, *Arthritis and Folk Medicine*, was published by Holt, Rinehart & Winston.

from American Bee Journal

JAYCOX, ELBERT RALPH, "JAKE" (1923-2004) – Elbert Jaycox, age 80, died March 8, 2004.

Elbert Jaycox became interested in honey bees as an undergraduate at the University of California at Davis in 1947, after serving in the Air Force in WWII. Not long after that Jaycox worked for two commercial beekeepers in California and seriously considered entering commercial beekeeping before deciding to continue his studies with honey bees. He obtained M.S. and Ph.D degrees in entomology/apiculture at U.C. Davis where he studied the use of diluents for honey bee semen and the sexual maturity of drone honey bees.

His first professional appointment was as supervisor of bee inspection for the state of California. After that, he was a research entomologist for the USDA at Logan, Utah and studied the effects of pesticides on bees. He also taught Insect Ecology at Utah State University. From Utah he went to the University of Illinois where he remained for about 18 years, teaching beekeeping and bee behavior, studying bee behavior and the pollination of soybeans, and serving as extension specialist to beekeepers and growers. During that period he spent a year in Switzerland studying the role of juvenile hormone on the behavior of worker honey bees. With Swiss colleagues, he showed that the amount of the hormone in their blood helped to determine the sequence of duties performed by worker bees.

While at Illinois, Jaycox also demonstrated that honey bees visit soybeans in large numbers and that they can be used to pollinate the beans to produce hybrid soybeans. In another project, Jaycox found that swarms of bees of different races have different requirements for living space in a hive or a natural cavity.

Jaycox left the University of Illinois, in 1981, and became adjunct professor of entomology at New Mexico State University, Las Cruces, New Mexico. In Las Cruces he taught two international beekeeping classes and helped to establish a beekeeping program at an agricultural school in the Yemen Arab Republic. He also taught beekeeping and queen rearing to Yemeni students on campus.

Jaycox wrote more than 210 publications of all kinds, including two excellent books, *Beekeeping in the Midwest* and *Beekeeping Tips and Topics*. He authored nine leaflets in the Arabic language which were printed for use by beekeepers in Yemen. From 1975 to 1981, he edited a newsletter, *Bees & Honey*, and produced *The Newsletter on Beekeeping* for several years. He wrote a column for *American Fruit Grower*

Elbert Jaycox

about pollination and the use of bees for fruit production.

From 1986 until 1990 he wrote a monthly column in *Bee Culture* entitled "The Bee Specialist," which explored many beekeeping subjects.

JOHANSSON, TOGE S.K. (1919-2001) – Toge Johansson of East Berne, New York died October 5, 2001. Toge was born in Karlstad, Sweden, and with his parents emigrated to the U.S. in 1923 at the age of four. They settled in Rockford, IL where he attended school. He graduated from Beloit College and earned an M.S. and Ph.D in Zoology from the University of Wisconsin. This was followed by positions at the University of Wisconsin, Grimmell College, Dartmouth College, New York University and Queens College of the City University of New York. After 30 years of teaching at Queens College he retired as Professor Emeritus in 1984.

Toge Johansson

As a child, Toge read everything he could find in the public library about bees but it was not until 1951 when he acquired acreage in the Helderbergs that he had opportunity to actually keep bees. He published over 100 articles on bee behavior, history and management in various journals. His interest in and extensive collection of apiculture literature culminated in a bibliography of Apiculture Literature Published in the U.S. and Canada. The breadth of his knowledge led to a year at the USDA Bee Research Lab in Tucson, another at the University of Guelph and shorter intervals in New Zealand, Mexico and Germany. He left with a stack of to-do items on his desk.

JUVENILE HORMONE – Juvenile hormone (JH) was named for its first known function in insects; its presence in larvae prevents metamorphosis (development to adulthood). JH has since been found to play important roles in many areas of insect physiology and behavior. Its principal functions are the control of metamorphosis, sexual maturation, and reproduction. JH is synthesized by the corpora allata, paired glands in the head lying on either side of the esophagus. The synthesis of JH is thought to be regulated primarily by the brain, via nerve cells and neurohormones.

JH has two special roles in honey bees. Caste determination, the differentiation of queens from workers, occurs during the larval period and is based on the concentration of JH in the blood. A high concentration in a three- to five-day old female larva results in the development of a queen, while a low concentration results in the development of a worker. High concentrations of JH are thought to be a consequence of increased food consumption. Larvae fed royal jelly consume more food than do larvae fed normal brood food because the high sugar content of royal jelly acts as a feeding stimulant.

JH also plays a major role in regulating age-based division of labor among adult workers. Low concentrations of JH are associated with the performance of tasks within the nest early in life, while a rise in JH at about three weeks of age induces foraging. Influences of JH on both pheromone synthesis and the age-dependent development of brood-food-producing glands suggest that this hormone acts as a "pacemaker," coordinating behavioral and physiological maturation in adult workers. JH also contributes to the integration of worker behavior in colonies because environmental factors affect the levels of JH in worker bees. This JH function results in flexibility in the ages at which bees perform certain tasks, enabling colonies to respond effectively to changing environmental and colony conditions. - GER

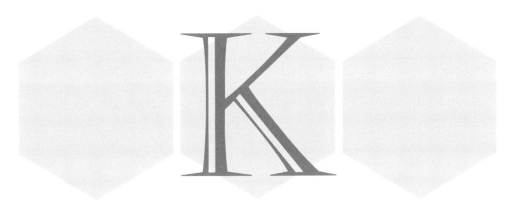

KELLEY, WALTER T. (1897–1986) – Walter T. Kelley, 89, the world's best known Bee Man, passed away Friday, August 22, 1986. Mr. Kelley was born July 30, 1897, in Sturgis, Michigan. He graduated from Michigan State University in 1919.

Between 1919 and 1924 he worked for the USDA in the Barberry Eradication Program and for E.B. Ault in Weslaco, Texas, raising bees and queens. In 1924 he began the bee business full time in Houma, LA, selling bees, queens, honey, and cypress beehives. In 1926 he married his life's mate, Ida Babin.

In 1935 Mr. Kelley moved his business to Paducah, Kentucky, and changed the name of the company to The Walter T. Kelley Company. Mr. Kelley reviewed his company's history in his 1985 catalog: "In 1924, my first year in business, I had a Model T Ford car and a truck. Both had a 22 horsepower engine which had to be cranked by hand neither had a heater nor windshield washer nor other things that we consider bare essentials now and the top speed was little more than 35 miles per hour. This was in South Louisiana and I had several outyards, but life was geared to a much slower speed than we have now. I had started manufacturing Cypress beehives in a very small way as there was a

Walter T. and Ida Kelley

cypress saw mill there, but after 10 years (1934) the saw mill 'sawed out.' These were the panic years when most of the banks went broke and corn was selling for 15¢ a bushel and wheat for 25¢ (when a cash buyer could be found and it was often used for fuel) and we were selling queens at 25¢ each and paying 10¢ a pound for beeswax delivered. Reviewing these conditions, we decided to move to the Central part of the country and to go into the beehive business full time and so we moved to Paducah, Kentucky.

"After a very slow start at Paducah, business picked up and we built a brick house and factory building and World War years came on and we had more business than we could handle, but by 1950 the Federal government decided to build an atomic plant near Paducah and they froze the wages we had been paying. One morning every man quit us for better paying jobs so we started looking for a better location and ultimately moved to Clarkson where we have grown and prospered."

His long-time registered trademark was "Kelley, the beeman" which featured a honey bee with Mr. Kelley's head on the bee body. Although many laughed at the trademark when they first saw it, they didn't forget it and that was Mr. Kelley's reasoning.

Besides selling bee supplies and package bees and queens, Mr. Kelley wrote the book, *How To Keep Bees And Sell Honey*.

Dr. Richard Taylor once wrote of Mr. Kelley: "Like so many beekeepers, Mr. Kelley is a loner. The sole architect of his life and success, with a great quality of individualism. In every sense, he is a self-made man. His bee supply factory is a monument to his industry and his genius. But to Mr. Kelley these are unimportant – modesty, simplicity and a delightful sense of the absurd are the important facts of life. You cannot meet Mr. Kelley without instantly recognizing his integrity. Crusty, outspoken and refreshingly irreverent, you nevertheless see a man of firm principle and the warmth of genuine friendship for everyone he meets."

Walter T. Kelley, early 1980s.

KELTY, RUSSELL – Russell Kelty retired Beekeeping Specialist and Assistant Professor at Michigan State University, died May 8, 1988. He was associated with Michigan State from 1919 until his retirement in 1950 when he became a full time commercial beekeeper with over 2000 colonies. During his active career he was president of the American Honey Association and a Director of the American Honey Producers League, as well as President of the Michigan Beekeeper's Association.

He wrote many beekeeping bulletins and articles while he was Specialist at Michigan State University. The most recognized was his "Seasonal

Management of Commercial Apiaries" a monument to concise and accurate information about beekeeping management.

Russell Kelty

KENYA TOP BAR HIVE – (see HIVES, TYPES OF)

KILLER BEES – This term has been used as a nickname to describe the Africanized honey bees that have moved north from Brazil during the past decades. The term was first used in the United States in an article in *TIME* magazine in the September 24, 1965, issue; however, the term soon became popular, especially with the press that has used it widely ever since. (see AFRICANIZED HONEY BEES)

KILLION, CARL E., SR. (1899-1979) – Carl E. Killion, Sr., world-famous beekeeper and author.

Mr. Killion was born September 2, 1899, near Diamond, IN, the son of Sylvester and Laura Bell Crawley Killion. On September 20, 1920, he was married to Elizabeth Hayes, who preceded him in death in 1964.

Carl was a regular contributor to the *American Bee Journal* and his many articles on comb honey production eventually lead to his writing of the book, *Honey in the Comb*, in 1951. This book was the authority on comb honey production for years. Carl and his son, Eugene, utilized their specialized methods to produce bountiful crops of beautiful comb honey at their Paris, Illinois apiaries.

In 1966 Carl published a second book, *The Covered Bridge*, an autobiography.

In 1970 Carl retired as superintendent of the division of apiary inspection for Illinois, after 32 years of service in that capacity. He served as vice president, secretary and treasurer of the Apiary Inspectors of America. He was elected as vice president of the American Beekeeping Federation. As a member of that organization he also served on

Carl Killion

several important committees, including the honey grades committee, whose recommendations were adopted by the federal government as standards for grading comb honey.

The recipient of the first "Beekeeper of the Year Award" presented by the Illinois Beekeepers Association, Mr. Killion also worked with Kentucky beekeepers in apiculture. For his assistance, he was commissioned a Kentucky Colonel by the state's governor. The Killion Honey exhibits at the Illinois State Fair won countless awards and ribbons and his honey ice cream pleased many Fair visitors.

Carl also authored chapters on comb honey production, care and packaging in the 1975 edition of *The Hive and the Honey Bee* and Eva Crane's *Honey – A Comprehensive Survey*. Mr. Killion previously had written chapters for *The Hive and the Honey Bee* on comb honey production.

Carl initiated a one-man campaign to promote the issuance of a commemorative honey bee stamp. In the early 80s an embossed stamp was produced.

KILLION, ELIZABETH (1902-1964) – Elizabeth Killion, 62, was the wife of Carl E. Killion, Sr., Paris, Illinois.

She was well-known to beekeepers everywhere due to the Killion's production of comb honey and Carl's many years as Chief Apiary Inspector for Illinois.

She was also a president of the Ladies Auxiliary of the American Beekeeping Federation.

KIME, ROBERT W. (1946-2002) – Robert Kime, 56, died May 27, 2002. A dedicated food scientist, beekeeper and outdoorsman, Kime was the operations manager of Cornell University's Fruit and Vegetable Processing Pilot Plant at the New York State Agricultural

Robert W. Kime

Elizabeth Killion

Experiment Station in Geneva, New York. "His expertise was valuable to scientists across the Station and to individuals in the private sector who contracted to use the pilot plant," said associate director Robert Seem. "His knowledge of post-harvest fruit and vegetable processing was well known." Kime helped many fruit and vegetable growers and entrepreneurs develop and refine value-added products, particularly mead, hard cider, and fruit wine.

Kime was an innovative thinker who shared in several patents at the Experiment Station. He and food scientist Cy Lee developed an ultrafiltration method for honey that improved the sensory quality of traditional mead. Ultrafiltration has helped create major new markets for honey producers all over the world. They also obtained a patent on the utilization of honey to clarify fruit juice in processing.

He was also a beekeeper and the owner of Kime Farm Honey. Harvesting wasp venom was a sideline, as was applying bee stings to arthritis sufferers, and manufacturing Kime Skin Cream – a formulation of glycerine and beeswax that he sold locally.

Kime was born in Waterloo, New York, on March 24, 1946. He graduated from SUNY Morrisville in 1966, and received his B.S. in 1968 at the University of Georgia in Athens. First employed at the Station in 1968, he left in 1969 to return to the University of Georgia for graduate studies, and then came back to the Station in 1970 as a research technician. In 1979 he was promoted to research support specialist. In 1995 he was appointed operations manager of the food processing pilot plant.

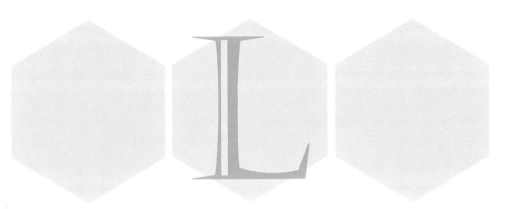

LABELS FOR HONEY JARS – The label is an important and necessary part of packaging honey and other hive products for retail sales. The state laws for labels generally follow the federal laws. It is very easy to comply with label requirements and, in fact, those requirements actually help to sell the product.

For example, to label honey, the word "honey" must be the most visible word on the label. If the producer is certain of the floral source, usually from a pollen analysis or after harvesting from colonies located in isolated locations with a single source of flowers to visit, then the floral source can be stated. Otherwise a term such as "wildflower" is the only appropriate one.

The net weight of the product means the weight excluding the container and any other packaging. The weight must be stated in pounds/ounces (lb, oz) and in metric weight (g). One pound is considered to be 454 grams (g). The weight must appear on the bottom third of the label in a type size easy to read.

Another requirement is called the "contact information." The name and address of the producer or packer must be on the container. A telephone number is optional. The type size for this information must be at least 1/16th inch (1.6 mm) tall. If the package contains honey from other countries, the countries of origin must be on the label, using such words as "Product of...."

The "bar code," actually called the Uniform Product Code (UPC), may be required by certain retail outlets. The producer/packer must apply for a code for each product. To obtain your UPC code contact the Universal Product Code Council, 7887 Washington Village Drive, Suite 300, Dayton, Ohio 43459; telephone 937-435-7317, or apply through the web site: www.uc-council.org.

The appropriate nutritional label, available for hive products, can be obtained from suppliers of beekeeping equipment. The pre-printed nutritional labels include all the required information. If more than 100,000 units of a product are sold during a year, the label is mandatory; if less than 100,000 units, an exemption can be obtained. Contact the Food and Drug Administration's Office of Food Labeling; www.FDA.GOV.

Caution must be used for various descriptive terms. For example if the word "healthy" appears on the main label, the nutritional label must be used. The word "natural" does not require any additional labeling. To use the word "organic" the product must be officially certified as organic. For information on certification contact USDA-AMS-TMP-NOP, Room 4008, South Building, 1400 Independence Avenue, SW, Washington, DC 20250; telephone 202-720-3252, or at www.ams.usda.gov/nop/FactSheets/CertificationE.html.

For a pleasant appearance the shape and size of the label should compliment the shape and size of the container. Logos and decorations add to the interest of the label. However, in the United States a realistic honey bee is often considered threatening rather than identifying; rather, a stylized or cartoon bee should be considered. Colors chosen should also be complimentary to the container and the product within. However, certain colors suggest certain products. For example, blue is frequently found on seafood products.

In general a simple label is the most effective. Labels can be obtained from a variety of beekeeping equipment suppliers, designed on home computers, or by graphic designers specializing in marketing. Local print shops frequently have graphic designers at their disposal who will prepare a very professional label. –AWH

Back or second label – Many beekeepers prepare a second label for their honey jars that describes the contents further (such as description of a distinctive honey) or gives a recipe in which honey may be used. It is useful to tell customers about specialty products such as crystallized honey, especially those that may contain fruit or some other product. If the honey is produced in a limited area, or its production is from a plant that has some historical interest, further information on a second label serves to inform the customer and help to make a sale. Since the annual per capita consumption of honey is a little over a pound per person in the U. S. there are obviously many people in the country that know little about the source, composition, and use.

Further reading – see National Honey Board. *How to: Honey labeling and packaging*, in their National Honey Board Marketing Kit, at www.nhb.org.

LAIDLAW JR., HARRY HYDE (1907 – 2003) – Professor Harry Laidlaw passed away September 19, 2003, at 96 years of age. Professor Laidlaw was recognized worldwide as the father of honey bee genetics and breeding. He was author of many scientific articles and four books on instrumental insemination, breeding, and genetics. He was Professor Emeritus of Entomology at the University of California, Davis, where he served on the faculty from 1947 to 1974.

Professor Laidlaw, "Harry" to everyone who knew him, was born in Houston, Texas on April 12, 1907. He began keeping bees when he was "four or five years old" with his grandfather Charles Quinn, a civil engineer and avid beekeeper who raised queens. Quinn was passionate about controlling the mating of queens in order to breed better bees, a passion that infected young Harry. By the time he was 16 years old, Harry had figured out a way to mate queens by holding the drone in position behind a virgin queen and stimulating him to evert his genitalia into her. This method was affectionately known as the "Quinn-Laidlaw hand mating method." As a teenager, Harry travelled to beekeeping meetings in the Southeast (they were living in Virginia at this time) demonstrating this, the first, method of controlled queen mating.

Harry Laidlaw, Jr.

In 1923 the family moved to Florida. Harry was now 16 years old and worked for a beekeeper in LaBelle named C.C. Cook delivering honey to retail outlets throughout Florida and into Mississippi and Alabama. While in Florida, Harry also was a district apiary inspector and inspected bees in several Florida counties.

The family moved to Louisiana in 1927 and Harry got a job working for the State Department of Agriculture raising queens to resupply beekeepers who lost their colonies in a catastrophic flood. Soon after, he was employed once again as an apiary inspector, for the State of Louisiana. In 1929, he enrolled in the Louisiana State University in Baton Rouge. This is remarkable because Harry never finished grade school! While a student at LSU, Harry worked for the USDA Southern States Bee Culture Field Laboratory where he greatly impressed the resident scientist, Dr. Warren Whitcomb.

Upon completion of his B.S. in Entomology, Harry was encouraged by Whitcomb to work for a M.S. For his research topic, he performed a detailed morphological study of the reproductive tract of the queen honey bee in order to better understand why all attempts at instrumental insemination up to that time had been doomed to failure. Researchers in the USDA had already developed an apparatus to hold the queen and inject semen into her sting chamber, but they had little success in getting queens adequately inseminated. Harry discovered the "valve fold," a tongue-like structure in the median oviduct of the queen that was preventing the proper placement of the semen-filled syringe during insemination.

Harry completed his M.S. in Entomology in 1934 and moved on to the University of Wisconsin to work on his Ph.D. in Entomology and Genetics. His research focused on developing a useful method for instrumentally inseminating queens. He developed a method to bypass the valve fold, allowing the semen to be injected into the appropriate location in the median oviduct, introduced anesthetization of queens (which stopped them from wiggling while trying to insert the syringe), redesigned the hooks developed by W.J. Nolan for opening the sting chamber of the queen, and redesigned the tip to prevent the backflow of semen during injection. The combination of these methods resulted in a successful system of instrumental insemination that enabled controlled mating and selective breeding of bees. This was the first insect for which artificial insemination was developed and has enabled a huge body of research on breeding and behavioral and developmental genetics.

Following the completion of his Ph.D in 1939, Harry worked for the USDA Bee Research Laboratory in Madison, Wisconsin, taught at a small college in Indiana, and served as Apiarist for the State of Alabama. In 1942 he was inducted into the U.S. Army, received a commission, and served as an Army Entomologist for the 1st Service Command in Boston, Massachusetts. This is where he met his wife of 57 years, Ruth Collins Laidlaw.

In 1947, Harry and Ruth moved West where Harry joined the Department of Entomology, University of California, Davis. In 1950, he joined forces with J.E. Eckert and published the first edition of *Queen Rearing*, which became the standard reference worldwide. He published the definitive book on instrumental insemination in 1977. He continued to develop better methods of queen rearing and instrumental insemination while he worked at breeding bees for commercial beekeeping. His queen rearing books were greatly influenced by his observations of successful queen producers.

Retirement came in 1974, but he never really retired. He wrote two books on queen rearing, breeding and genetics. The second was completed and published when he was 90 years old. He also continued to modify the instruments used for instrumental insemination. Harry wanted the instruments to be simple and inexpensive so that they were accessible to everyone. Harry wished to design an instrument that easily could be built from parts available from plumbing and hardware catalogs, for under one hundred dollars. He succeeded and published an article about how to make the instrument, but he still was not happy. So, in collaboration with M.E. Kühnert he modified the queen holder and dorsal hook into a single structure that resulted in a revolutionary new design. They published a description of the new apparatus in 1994; Harry was 87.

Harry served as a consultant with the Egyptian Ministry of Agriculture in the 80s, establishing a closed population breeding program. His breeding methods were adopted internationally by commercial and government programs.

Harry was very fortunate. Not only did he live a long, healthy, and fruitful life, his accomplishments were recognized while he could enjoy the accolades. He was the recipient of many awards. In November, 2000, the UC Davis Honey Bee Research Facility was renamed in his honor, the Harry H. Laidlaw, Jr. Honey Bee Research Facility.

A visit with Harry Laidlaw was on the travel itinerary of bee researchers worldwide. Travellers passed through Davis to see Harry on a regular basis. His door was always open and he greeted everyone with his infectious charm and enthusiasm – always polite, always encouraging, always humble. His contributions to apicultural science will endure, and so will his memory. *Rob Page*

LAND BASED HONEY PRODUCTION – The rather non-descript term refers to a variety of practices that all have one thing in common – a crop is grown with express, or primary purpose of producing nectar to be collected by honey bees to produce a single source honey crop.

This can be as simple as planting a single crop that blooms once and honey is made from that crop. It can be more complicated in that a series of crops can be planted to bloom in sequence so there is available forage all season long.

Some crops, such as sunflowers, can be used for honey and pollen, then harvested later as a marketable agronomic crop. Speciality crops can be used to produce pure varietal honeys – buckwheat or thyme for instance.

The amount of land used will be substantial but will need to be determined for each crop the location the crop is grown, the intensity of the management of the crop produced, and the number of colonies used to harvest the available nectar.

Land-based honey production, in all of its many variables, will become more important as a source for varietal honey as the acreage of nectar-producing crops continues to decline, and, non-agricultural land is consumed by urban development.

L.L. Langstroth

LANGSTROTH, LORENZO LORRAINE

(1810-1895) – Langstroth was the father of practical beekeeping. In 1851 he observed that if a space one-quarter to three-eighths of an inch (1 cm) wide was left between the cover of a hive and the top bars, the bees would not fill this area with comb, propolis or other materials. This is a walking or crawling space for the bees and is known as bee space. Langstroth quickly realized that if he built the same space around and between combs in the hive he would have a truly moveable-frame hive. No one had ever built a hive with bee space before. Langstroth patented his hive in 1852 and published his well-known book, *Langstroth on the Hive and the Honey Bee: A Beekeeper's Manual*, in 1853. Langstroth's patent was infringed upon by many people and he never made any money from it; however, his book continued through many editions and was eventually bought by Dadant and Sons, who issued many more editions and continue to publish revisions under a similar title today.

Langstroth was a Congregational minister who was educated at Yale. He served churches in Massachusetts before moving to Philadelphia where he established and ran a school for young girls. Later he moved to Ohio. The cottage in which Langstroth lived from 1858 to 1885 on the Miami University campus at Oxford, Ohio, was dedicated in 1976 and entered into The National Registry of Historic Places because of the important role Langstroth played in the development of American agriculture. During the time Langstroth lived in Ohio he was a frequent visitor at The A. I. Root Company plant in Medina.

The discovery of bee space by Langstroth triggered the invention of comb foundation, the extractor and the smoker, all within less than 25 years. The Langstroth hive, foundation and these other tools made practical, commercial beekeeping possible. As a result, before the turn of the century, many beekeepers throughout the U. S. were operating a thousand or more colonies and the industry became firmly established. It was all made possible by a simple observation by an observant individual. Langstroth's diary, in which he recorded his discovery of bee space, is a treasured item in the beekeeping library at Cornell University, Ithaca, NY. His life was portrayed in a thorough and interesting book: Naile, F. *America's Master of Bee Culture: The Life of L. L. Langstroth.* (Cornell University Press, Ithaca, N.Y. 215 pages. 1942, reissued 1976.) Additional information is included in the Biography section of this book.

LANGSTROTH HIVES – (see HIVES, TYPES OF)

LARVA (s), LARVAE (pl) – There are four main stages in the life and development of a honey bee: egg, larva, pupa, and adult. The larval stage is short but during this time the developing bees feed heavily and grow rapidly; it is at this time that there is great danger of food contamination and diseases are easily transmitted during the larval stage. For more precise information on development times see LIFE STAGES OF THE HONEY BEE.

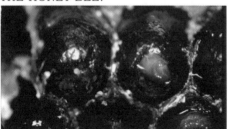
An egg is pictured standing on end, held there by glue produced just for this purpose by the queen. In the cell to the right of the one containing the egg is a first instar larva, already floating in royal jelly fed to it by house bees. (Jaycox photo)

Larvae grow rapidly, going through five instars in six days. Here are three different instars (development stages between molts). (Jaycox photo)

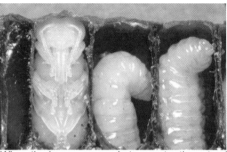
When the larvae are ready to pupate, they stand upright in the cell, stop eating, void their digestive systems into the bottom of the cell, and prepare to spin their cocoons. Noting this change, house bees begin covering the cells with a mixture of beeswax and propolis. Now is the time that female Varroa mites enter the cells to parasitize the larvae. (Jaycox photo)

Several stages of pupating workers are shown here. The wax cappings have been removed to show the developmental stages. At top left is a pupating worker nearing maturity, whose eyes have already developed color. At the center of the bottom row is a worker nearly mature enough to emerge as an adult. (Jaycox photo)

LAWS RELATING TO BEEKEEPING – (see LEGISLATION IN THE UNITED STATES RELATING TO HONEY and LEGISLATION IN THE UNITED STATES RELATING TO HONEY BEES)

LAYING WORKERS – In a normal honey bee colony the ovaries of worker bees do not develop and they do not lay eggs. When the queen is removed or lost and the colony has no brood, or fails in its attempt to rear a queen, the ovaries of some workers will develop and they will lay eggs. For experimental purposes, a queen may be removed and her replacement prevented by destroying any queen cells that the bees build in an attempt to grow a new queen. When this is done the ovaries in about ten to fifteen percent of the worker bees will develop and they will begin to lay eggs in about two weeks.

Diagnosing laying worker colonies is not difficult. A loud "roar" brought about by worker bees fanning their wings is usually noticeable when such colonies are opened. The bees appear "nervous" on the combs. The eggs of laying workers are small and many may be found in a single cell. They may be deposited on the sides as well as on the bottoms of cells. Most of these eggs fail to hatch; those that do hatch develop into drones since workers cannot mate, thus their eggs are not fertilized. Rarely, a normal queen will develop parthenogenetically, that is from an unfertilized egg, into a queen. This happens seldom in European bees but more commonly in one subspecies of honey bees (the Cape bee) from South Africa. (see PARTHENOGENESIS)

It is usually not possible to requeen a laying worker colony, probably because the chemistry of certain of the glandular secretions in the laying workers changes, becoming similar to that of a queen. When beekeepers find a laying worker colony they usually remove the laying worker colony from its position in the apiary, shake the bees out of the hive at the edge of the apiary, then let them drift back into the remaining hives. Those with ovaries in advanced stages of development are killed when they move

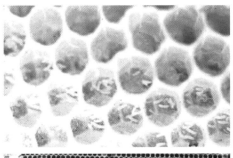

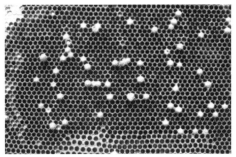

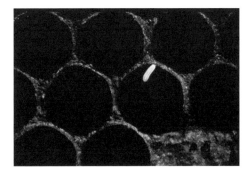

Three signs of possibly having laying workers in your colony. Multiple eggs in cells indicates many layers. Eggs deposited on the side wall instead of the bottom of the cells, and drone brood raised, sporatically, in worker cells. (Morse photos)

into a normal, queenright colony, or, if they continue to lay, their eggs are removed by house bees in the queenright colony. Combs with eggs, brood and honey and pollen are dispersed to other colonies to be cleaned and used. There is no known danger in dispersing laying workers and their eggs in this manner.

Sometimes adding several frames of brood in all stages to a laying worker colony will reduce or eliminate the laying capacity of these workers and at the same time give the colony fertilized eggs to turn into a new queen. The brood may produce an inhibitory effect on the laying workers.

LEAFCUTTING BEES – The alfalfa leafcutting bee, *Megachile rotundata,* has become one of the most important pollinators of alfalfa in the world and can also be used to pollinate many other forage legume species. This bee was introduced to the United States from Eurasia around 1930. Wherever the bee is managed efficiently and effectively, alfalfa seed yields increase.

The alfalfa leafcutting bee is solitary by nature, constructing and provisioning its own nest. After mating, each female cuts, transports, and places suitable leaf material in a tunnel to form thimble-shaped cells. She provisions each individual cell with pollen moistened with nectar upon which she lays an egg. She has little interaction with other females of her own generation and will die before her offspring grow to adulthood. Because they are gregarious, many leafcutting bees will make their nests in holes in boards provided by bee managers.

Large numbers of leafcutting bees (50,000 to 75,000 per hectare) (20,000 to 30,000 per acre) are used to pollinate a crop of alfalfa. It is thus important to

Leafcutting bee on alfalfa. (photo by Theresa L. Pitts-Singer, Ph.D, USDA-ARS, Logan, Utah)

synchronize bee emergence with the beginning of the alfalfa flower bloom. Techniques to synchronize the emergence of the bees with flowering have been easier to develop than techniques to control the blooming of the crop. In order to mass-produce and release the leafcutting bee for commercial pollination, boards containing nests of leafcutting bees are removed from fields at the end of the alfalfa-growing season. Some bee managers may remove bee cells from wood laminate or polystyrene boards, and then tumble the loose cells to filter out empty or diseased cells. These loose cells are then placed into cold storage (41°F, 5°C) for several months. Other managers may leave intact nests inside wood boards, which are placed into cold storage. In the spring, the bee cells are subjected to warmer temperatures (86°F, 30°C) inside incubators, where they complete development. Once adult emergence begins, the bees are placed, along with empty boards, in shelters located in or around alfalfa fields. Here the bees continue to emerge and mate so that the cycle begins again. The timing of bee emergence is more easily controlled with the use of incubation facilities than by relying upon field conditions. Therefore the optimum number of bees are placed in the crop at the appropriate time to obtain a high seed set and a return of viable bees for the following year.

Within the management system of leafcutting bees, populations of insects that are natural parasites of the bees (e. g., chalcidoids, cuckoo bees, blister beetles), as well as insects that feed on stored products (e. g., dermestids, dried fruit moth), may infest boards and destroy developing bees. Control of these injurious insects is achieved through different management procedures, including precise construction of nesting boards, use of controlled incubation and light traps, installation of insecticide vapor strips in the incubator, dipping of cells in insecticides, and removal and subsequent tumbling of cells from the nesting boards. Many of the techniques have been directed at reducing the population of emerging adult parasites and preventing parasitism during incubation. Accurately predicting the emergence of adult parasites during incubation and scheduling appropriate control measures are necessary. Another major concern for this bee is the occurrence of chalkbrood, a disease caused by fungal infections of bee larvae in the field. This disease can greatly increase bee mortality (see DISEASES OF THE HONEY BEES, Chalkbrood).

The current management system permits bee managers to take samples of cells to be inspected via x-ray analysis to accurately estimate numbers of intact cocoons, female bees, parasites and disease. These estimates provide a means for assessing and improving management practices. The estimates also provide quality guidelines when the bees are items of commerce. For further information see Richards, K. W. *Alfalfa leaf cutter bee management in western Canada*. Agricultural Canada Publ. 1495. 56 pages. 1984; Peterson, S. S. *et al.* Current Status of the alfalfa leafcutting bee, *Megachile rotundata*, as a pollinator of alfalfa seed. *Bee Science*. 1992. - JHC

LEARNING IN THE HONEY BEE – As far as is known, the first person to submit data showing that honey bees could learn was Reinhardt (Reinhardt, J. F. *Some responses of honey bees to alfalfa flowers*. The American Naturalist 86:257-275. 1952.). He showed that honey bees could learn to avoid the explosive tripping mechanism of an alfalfa flower. Alfalfa evolved in the Near East, probably in the vicinity of what is now Iran. Apparently no honey bees were native to the area and the flowers were pollinated by ground or twig-nesting solitary bees. This presumably led to this unusual flower structure and activity. In alfalfa, which is a legume and has a flower typical of that group of plants, the male and female flower parts are contained in a part called a keel that lies under two wing petals. (See HONEY PLANTS, Alfalfa for graphic display.) However, in the case of alfalfa the keel is held under pressure. When the top center of the keel is touched by a bee, the flower's sexual parts pop out and the head of the bee is caught, at least momentarily, between the sexual parts and the large standard petal, a typical part of a leguminous flower.

Reinhardt once observed a worker bee that took 45 seconds to free herself from an exploded alfalfa flower mechanism, though clearly most bees do so much more rapidly. This explosive mechanism is easily shown and is an excellent classroom demonstration if enough alfalfa flowers are available so that each student may have one or more to observe. One need only touch the keel with a toothpick, or other small object, to trigger the mechanism. A small cloud of pollen can usually be seen in the air. The flowers must not be wilted or dried for the mechanism to work or for the cloud of pollen to be seen. As the bee wrestles to free herself, she unknowingly deposits pollen on the alfalfa flower's female parts. Apparently, being hit under the head by this mechanism disturbs, or perhaps even hurts, the honey bee. Reinhardt found that after a worker honey bee had been hit in this manner several times, she sought to avoid the mechanism; this can be done by her inserting her mouthparts in the direction of the nectary, but alongside of the keel. Reinhardt also observed that solitary bees did not learn to avoid this exploding mechanism, and thus in some ways they are better pollinators. Pollination in alfalfa is not accomplished unless the sexual parts are exposed.

Reinhardt's observations have had important economic consequences and have shown growers and beekeepers how to obtain the greatest profit from the use of honey bees in pollinating alfalfa. Alfalfa is the chief hay crop for dairy cows, and vast acres are grown across the country for this purpose. Alfalfa fields must be reseeded every few years. This demands an abundant seed supply that can be available only if enough bees are available for pollination. Bees that learn to avoid the alfalfa flower's sexual mechanism do not pollinate. Thus, it is important to use either solitary bees for pollination, which is done in many of the northwestern states and western Canadian provinces, or to flood an alfalfa field with colonies of honey bees, as is done in California, the principal alfalfa seed-producing area in the U. S. Flooding the area with bees provides an abundance of naïve bees that are good trippers until they learn otherwise. Also, some experienced honey bees will occasionally make a mistake and accidentally trip a flower. It is common to use three to five colonies of bees per acre (6 to 10 per hectare) for alfalfa pollination. Alfalfa is a good nectar-producing plant, but when such a large concentration of bees is used, no surplus honey is gathered. When there is a shortage of pollen some honey bees will work alfalfa; these bees are, of course, effective pollinators because they must trip the flower to obtain the pollen.

A second place where beekeepers have observed learning in bees is in watching how honey bees find the location of their hive. If a hive is moved a few feet, the bees will first go to the old location using whatever landmarks are available. If the hive is no longer there, they will search the vicinity until they find it. A beekeeper technique to weaken strong colonies and to strengthen or add to the population of a weak hive, is to exchange the positions of the two at a time when the bees are foraging.

This has proven to be an effective method of swarm control. If done at the right time, a colony in the early stages of swarm preparation loses some of the extra bees that might otherwise cause congestion in the brood nest that could lead to swarming.

Yet another time when beekeepers may take advantage of bees having learned the location of their hive is when one is opening a hive and demonstrating the contents to an audience. If the colony is smoked heavily, picked up and moved 50 or so feet (15 meters), the guard bees that may otherwise defend the nest will fly back to the original location. In this way they are not a nuisance during the examination since they seek to defend the location where they believe the hive to be and not where it actually is.

A researcher in Brazil (Pessotti, I. "Discrimination with light stimuli and a lever-pressing response in *Melipona rufiventris.*" *Journal of Apicultural Research* 11:89-93. 1972.) showed that a species of stingless bee could learn to respond to light and press one of two levers to obtain food. The bee was never perfect in her decision making, but her ability to discriminate between the two levers was statistically significant.

Those who care to pursue the question of animal learning will find a great number of examples of the honey bee's ability to think and learn in the studies of Karl von Frisch and his students. This includes the bees' ability to find and report the location of food and nest sites. Seeley has conducted a thorough study of a scout bee's ability to measure the acceptability of a new nest (Seeley, T. D. *How honeybees find a home.* Scientific American 247:158-168. 1982.).

The question of animal awareness and learning lead one to ask what kind of brain a honey bee has. We know it is a small brain, and from this we assume that the bee cannot encode into her brain cells all the possible circumstances she may encounter. From the point of view of economy and efficiency, a bee's brain must contain an area where some thinking and decision making can take place (also see ANIMAL AWARENESS). At the same time we can find examples of things bees cannot do. For example, when Professor von Frisch placed a food source under a high bridge he could train bees down to the food by lowering it gradually down the side of the gorge over which the bridge was built. Even though the scouts flew back and forth between the food and the hive, and danced a wagtail dance that indicated the correct distance, they did not have the ability to indicate that the recruits should fly downward. Recruits, in response to the dance, flew out from the hive, but at the same level, seeking the food. The same thing occurred when the food was placed on top of a high water tower; again the bees could not convey the information to fly in an upward direction.

LEGISLATION IN THE UNITED STATES RELATING TO HONEY – Honey has been defined as the nectar and sweet deposits from plants as gathered, modified and stored in honeycomb by honey bees. The Food and Drug Administration, which has official jurisdiction over food quality and such matters as food adulteration, has given an earlier definition from the original 1906 legislation "advisory status." This advisory status has been sufficient for the Food and Drug Administration to protect the consumer and to prevent the adulteration of honey.

The standard reference for those interested in honey and how it varies is White, J. W. Jr., M. L. Riethof, M. H. Subers, and I. Kushnir. *Composition of American Honeys.* U. S. Department of Agriculture Technical Bulletin 1261. 124 pages. 1962. In this Bulletin, Dr. White and his co-workers examined 505 samples of honey and honeydew from two crop years (1956 and 1957) from 47 states. While methods of chemical analysis have improved greatly in recent years the techniques used by these researchers still give us a good understanding of honey and its constitution.

The Public Health, Security and Bioterrorism Preparedness and Response Act of 2002 (known as the Bioterrorism Act) is designed to protect the public's food supply from any attacks on its safety. Since honey is considered a food, this act pertains to all beekeepers. Since the Act is quite complex, it is best to contact the Food and Drug Administration (FDA) or visit the web site www.cfsan.fda.gov for information. As this work is being completed a new and more complete definition of honey was being constructed with input from the U.S. honey industry, the National Honey Board and the FDA. Much of this work has been completed by European food regulators and is being adapted to U.S. standards.

LEGISLATION IN THE UNITED STATES RELATING TO HONEY BEES

– No federal regulations relate to honey bee diseases or the movement of honey bees in the U. S. However, many states regulate the beekeeping industry within their boundaries through their department of agriculture. Some states require that colonies of honey bees be registered but because of the high turnover in hobby beekeepers such regulations are difficult, if not impossible, to enforce. A few states require that commercial beekeepers register their apiary locations. This also guarantees that no other commercial beekeeper may move colonies closer than a prescribed limit, usually two to three miles (5 km). How infectious diseases of honey bees are treated varies from state to state; this makes writing about bee disease control difficult because some states advocate the use of drugs for control of certain diseases, especially American foulbrood, others may destroy colonies infected with the same diseases, while some states have no regulations at all.

States where queen honey bees and packages of bees are produced generally inspect the source colonies and issue certificates of inspection, which are attached by the shippers to the queen cages and packages, but each state varies in how rigorous these inspections are. Most states that receive package bees and queens have legislation or rules requiring that certificates accompany the bees, and these, too, vary in their rigor. In most instances migratory beekeepers have their colonies inspected and are issued certificates of inspection before moving their colonies to another state; only a few states do not provide this service. Most states, except for those with no regulations, have a circular or brochure that outlines their bee disease laws and regulations; these may be obtained by writing the state department of agriculture.

In 1922, the federal government enacted legislation restricting the importation of honey bees into the U. S. except from Canada. The Honeybee Act, as it is popularly known, was passed because of the discovery in Great Britain of *Acarapis woodi*, the tracheal mite (see MITES AFFECTING HONEY BEES). The original act has been amended several times and was changed significantly in 2004 when Australia, New Zealand and Canada were allowed to send bees to the U.S. The new revisions, outlined and analyzed in detail, are available from the U.S. Codes, and in the book, *Honey Bee Law: Principles and Practices*, published by the A.I. Root Company, in 2005, written by Sylvia Ezenwa, J.D.

A number of legal importations of queen honey bees had been made directly into the U. S. from abroad (see MITES AFFECTING HONEY BEES), previous to the amendment, and some beekeepers and researchers questioned whether the federal government should restrict the importation of honey bee stock. It is probably well to continue with some import restrictions since we have not surveyed all of the world's beekeeping regions for honey bee pests, predators and diseases. The fact that we cannot answer such a simple question as how many species of honey bees may be found in Asia gives further support to having some restrictions.

Further details on federal and state legislation regarding honey bee diseases may be found in Morse, R. A. and K. Flottum. *Honey Bee Pests, Predators and Diseases.* (A. I. Root Co. Medina, OH. Third edition. 1997), and *Honey Bee Law*, previously mentioned.

LENGTH OF LIFE OF THE HONEY BEE

– How long a worker honey bee lives depends upon how much work she does. The life of a worker bee is divided into two parts: hive life (house bees) and foraging life. Foraging takes much more energy than does in-hive work. The length of time spent in the hive depends on several factors: how much brood to feed and the number of bees available to feed it, how much storage space in the hive, and how much food is available to be gathered. If no storage space is available then the bees will not forage. If the scouts indicate there is nothing to be gathered, then again the bulk of the foraging force remains in the hive doing nothing except perhaps to aid in an emergency or difficult situation such as temperature control, nectar ripening or an attack.

Winter bees, those that are produced in the fall and that do not forage, have long lives because they do little work and are physiologically different from the spring and summer honey bees. The least amount of brood and almost no foraging in colonies occur during the late autumn and early winter months.

The cells in the bodies of adult insects, including honey bees, are usually not replaced except for those in the midgut. Thus, when the body cells are worn out the insect dies. Wounds, if not too extensive, may be healed by multiplication of the epidermal cells around them.

Field work (flying) requires much more energy that burns up the cells more rapidly than work in the hive. In the case of honey bees it has been determined that on average a worker will die after flying about 800 kilometers (480 miles). At this point the cells are worn out and have accumulated a sufficient quantity of waste material that they cannot function properly. It is interesting that unlike aging mammals there is no drop in work performance as the worker ages until the very end of its life. The life of a worker bee may be only four to six weeks during the active season and four to five, rarely six, months in the winter.

A disease may also affect how long worker bees live because it may be sufficiently debilitating to reduce the length of life. It has been said that a honey flow will do much to cure or improve a disease situation in a honey bee colony. If there is an active honey flow, adult bees infested with mites, or infected with a disease such as nosema, live a much shorter period of time because of the combined effects of the work they do when they forage and the disease. When diseased bees die early they are not present to infect or infest others; therefore a honey flow appears to improve a disease situation. The time of year that we have the greatest difficulty with nosema and tracheal mites is in the early spring. At this time the largest number of older bees are in the hive. Because they are older the parasites they may carry have had time to mature and multiply and may be easily transmitted to others. From the point of view of disease control it is important to get rid of these old, sick bees. An article that discusses longevity in worker bees is: Angelika, N. *Dependence of the life span of the honeybee upon flight performance and energy consumption.* Journal of Comparative Physiology B 146: 35-40. 1982. (see CASTES).

Drone honey bees do no work in the hive. They fly for only a few hours on warm sunny afternoons. Thus, if they find no mate some drones may live several months. However, the life of a drone may be cut short by workers that refuse to feed them during a dearth or in the autumn. Queen honey bees have been known to live for several years. Apparently egg-laying does not produce the strain on body tissues that flying exerts.

LEVIN, MARSHALL D. (1922-1994) – Marshall Levin, former director of the Carl Hayden Bee Research Center, Tucson, Arizona, died October 22, 1994. His career with the USDA-ARS made him a world-renowned scientist, educator and administrator with numerous contributions to the field of apicultural research.

He was born May 18, 1922, in Brownwood, Texas, but spent most of his boyhood in Connecticut where he developed a long-standing interest in beekeeping. First introduced to honey bees by his seventh-grade science teacher, Marshall turned his new-found passion into a high school avocation, then into a career of public service.

As a young man in the military Marshall visited beekeepers at each place he was stationed in England and the United States. Following World War II, he completed a B.A. Degree in Entomology in 1947 at the University of Connecticut. He then obtained an M.S. Degree in 1949, and later (1956) his Ph.D. from the University of Minnesota under the late Dr. Mikola Haydak.

In 1950 Marshall joined the USDA Legume Seed Research Laboratory at Logan, Utah where he studied the foraging behavior of honey bees on alfalfa and other crops. Twelve years later he was transferred to the USDA Bee Lab in Tucson and in 1965 became its leader, a position he held until 1969 when he was transferred to Beltsville, Maryland, to serve as Chief of the Apiculture Research Branch of the Agricultural Research Service. While at Tucson he guided the construction and staffing of the facility.

At Beltsville, Dr. Levin occupied several positions of increasing responsibility. In 1972, when a reorganization of ARS created the National Program Staff, he became the National Research Program Leader for Crop Pollination and Bees. In 1975 he was appointed Deputy Assistant Administrator for Plant and Entomological Sciences. Following other reorganizations, he became Chief, Crop Protection Staff, and then Chief, Crop Sciences Staff. During this period he also served as Associate Director of the Agricultural Research Service. A further reorganization of the National Program

Marshall D. Levin

Staff in June 1982 eliminated the Crop Sciences Staff and provided an opportunity for Marshall to return to research. He was reappointed director of the research center at Tucson in February, 1983. He retired in 1986 after 36 years of federal service.

Dr. Levin's research contributed to a greater understanding of the role played by the pollination activities of honey bees and other insects in the production of alfalfa, carrot, onion, and safflower seed crops, as well as cantaloupe and eggplant. He uncovered new knowledge regarding the management of honey bee colonies for pollination, factors affecting the pollinating efficacy of colonies and the effects of pesticides and other environmental factors on honey bees. Dr. Levin published his research results in over 100 technical and scientific papers. In 1982 he received a Special Recognition Award for service and support to the beekeeping industry from the American Beekeeping Federation.

Dr. Levin completed numerous special assignments for the USDA, ARS. These include participation in a scientist exchange program with the USSR in 1967 where he visited bee research installations and state farms in the former Soviet Union. He served as a technical representative on numerous committees and participated in many national and international professional meetings. He was a member of the Entomological Society of America and the International Bee Research Association. During his Tucson assignments, Dr. Levin held a dual appointment as Adjunct Professor of Entomology at the University of Arizona where he taught a course on beekeeping and pollination, and served on graduate student committees.

In retirement Marshall retained his interest in the problems of beekeepers and remained active as a Research Collaborator at the Tucson Laboratory. He also volunteered to help elementary school children with their reading, worked as an instructor for Handi-Dogs, Inc., training dogs to assist the disabled, and served as a volunteer naturalist in the Sabino Canyon recreational area. Marshall received the Mayor's Copper

Letter Award in recognition of his services in the Tucson community.

LIFE CYCLE OF THE HONEY BEE – (see ANNUAL CYCLE OF HONEY BEE COLONIES)

LIFE STAGES OF THE HONEY BEE – The honey bee has four distinct stages (egg, larva, pupa and adult) as do other insects with complete metamorphosis (such as ants, wasps, beetles, flies, butterflies and moths). The egg, larval and pupal stages of the bee are collectively called the brood. The adult bees in a colony maintain a constant 95°F (35°C) in the portion of the comb that contains brood. This uniform temperature in the broodnest causes the brood to develop at a constant rate.

Brood cells of a bee are either open (called uncapped brood) or closed with a thin layer of new wax, used wax and propolis (called capped or sealed brood). Cells with younger brood (all the eggs and most of the larvae) are uncapped. About one day before the end of the larval stage, worker bees put wax cappings over the cells. Therefore, capped brood includes this last day of the larval stage, the entire prepupal and pupal stages, and a brief part of the adult stage. The prepupal and pupal stages represent the transformation of larva to adult. Physical changes take place during both stages and they can be collectively called the pupal stage, however the appearance is different. The earlier stage looks like a larva and the latter stage looks more like an adult.

The following table gives the average duration, in days, of each stage of development of honey bees from the time an egg is laid until an adult bee emerges from the cell.

Three days after being laid, the egg shell dissolves, (called eclosion) exposing the larva, which is the eating specialist. The larva finishes eating, spins a cocoon, and then lies motionless in the cell. Spinning the cocoon marks the end of the larval stage, but the bee, which is now a prepupa, still has a larval skin and the appearance of a larva. During the prepupal stage, the bee begins to transform into an adult. The pupal stage begins when the larval skin is shed, exposing a body that has the general features of an adult bee. A pupa gradually changes from white to nearly black, with the eyes showing the first coloration. When the pupal skin is shed, the wings expand and the bee becomes an adult. The adult bee gradually becomes more active, chews away the cell cap and exits.

Development time is variable. The table assigns an average development period that is about one day shorter than that which has been reported in previous beekeeping books. Recent work has consistently shown that the average development time for worker bees is about 20 days, but variation is common. For example, when starting a colony of bees with adult workers and a queen (no brood), it is common to see adult worker bees emerge from brood cells in fewer than 19 days from the time the queen was released to lay eggs. Conversely, that same colony may have some worker bees emerging from brood cells as long as 21 days after the queen had been removed. Thus there is variation *within* a colony, caused mostly by the location of a cell and the care it received.

Variation also occurs among different colonies and among different races of bees. African races of bees may have total development times that average more than one day shorter than the totals reported in the table. However, the

Average Development Time for European Honey Bees in Days						
	egg	larva	prepupa	pupa	adult	total
worker	3	6	2.4	8	0.6	20
queen	3	5	2	4.6	0.4	15
drone	3	8	3	9	1	24

point of the example is that even within a single colony, one could conclude that the development time of a worker bee is either 19 or 21 days, depending on which bees were examined. Both would be correct, but neither would be the colony average. - JH (For more information see LARVA)

LINDNER, JOHN V. (1917-1995) – John Vincent Lindner, 78, of Cumberland, Maryland was born February 27, 1917, in Cumberland, he was the son of the late John H. Lindner, a pioneer in Maryland beekeeping, and Teresa L. Lindner.

John started keeping bees early in life and continued working with them for almost 70 years. At one time he maintained over 1000 colonies for honey production and pollination.

Mr. Lindner worked as an apiary inspector 30 years. In addition to keeping bees commercially, he began inspecting bees as a regional part-time apiary inspector for the state of Maryland in 1949, and sold most of his colonies in 1967 when he became Maryland's first full-time apiary inspector. Mr. Lindner served Maryland as chief inspector for 12 years until his retirement in 1979. During his tenure, Mr. Lindner instituted an annual apiary inspector's workshop where inspectors from Maryland and surrounding states received training in all phases of bee diseases and the beekeeping industry. He also assisted researchers at the USDA Bee Research Laboratory on various projects. In 1973, Maryland beekeepers began saving diseased bee equipment by having contaminated items fumigated with ethylene oxide. This continued on a limited basis until 1978 when the Maryland Department of Agriculture, through Mr. Lindner's efforts, purchased a portable ethylene oxide fumigation chamber.

Lindner was a member and director of the Eastern Apicultural Society, a member of the American Beekeeping Federation, and active in the Eastern Beekeepers Pollination Association which he helped form in 1969.

John V. Lindner

LIQUID HONEY – The term liquid honey is used to designate honey that has been removed from the comb with an extractor and that has not yet crystallized. Liquid honey is sometimes obtained by crushing the comb and straining the liquid portion away from the rest. However, this method is usually not a practical way to obtain liquid honey. At present it is preferred to use the term "liquid" over the confusing terms "extracted" or "clear" that have often been used to describe such honey. Honey is also sold and/or moved in the marketplace as comb honey, cut-comb honey, chunk honey and creamed honey.

LITTLEFIELD, HOOD – Hood Littlefield was president of the American Beekeeping Federation for three years (1970-1973), chairman of the Honey Industry Council of America, 1964-1965, and also served several years on the executive committee of the Federation.

He was a member of the California Honey Advisory Board from 1954 to 1968 and was chairman for 12 years.

Hood Littlefield

Hood helped found the Los Angeles County Beekeepers Association in 1948, was a member of the California Beekeepers Association since 1950, and joined the Federation in 1953. His wife, Thelma, was active in the Ladies Auxiliary of the American Beekeeping Federation and was president of the auxiliary in 1958 and 1959.

Littlefield was born near May, Texas, where his grandfather had some bees and cut bee trees. In 1927 Hood moved to California and worked for a meat packing company. In 1929 he went to work for Gilmore Oil Company in Imperial where he married Thelma in 1934. In January, 1942, the Littlefields moved to Pasadena. In 1944 Hood started to keep bees as a hobby and had 700 colonies by 1947. Then in 1952 Littlefield left Mobile Oil Company (who had acquired Gilmore Oil Company) to keep bees full time. He made a 250-pound average on his 1000 colonies that year, which also was the best year California has ever had with 50,000,000 pounds of honey being produced.

Littlefield spoke on "Management for Honey in California" at the Apimondia Symposium on Management which preceded the 1977 Federation Convention at San Antonio, Texas.

LOVELL, HARVEY BULFINCH (1903-1969) – Dr. Harvey Bulfinch Lovell, professor of biology at the University of Louisville and a well known writer and lecturer on bird and plant lore, died November 23, 1969, at his home in Louisville, Kentucky. He was 66.

Louisvillians became familiar with Lovell's interest in Kentucky birds, animals, and flowers through his many articles in *The Courier-Journal* Sunday magazine. For years he edited a quarterly journal on birds, *The Kentucky Warbler*.

Harvey Bulfinch Lovell

He had been president of the Kentucky Ornithological Society, the Kentucky Society of Natural History, and the Kentucky Academy of Science.

He often made talks to schools and clubs about natural history and illustrated them with photographs he had taken of Kentucky birds and wild life.

For a long time he operated a government bird-banding station at his home.

Lovell was a native of Maine and a descendant of the Bulfinch family, prominent in the history of New England.

He was graduated cum laude from Bowdoin College in 1924, with high honors in biology. He was elected to Phi Beta Kappa at the Maine college. He received both his Master's degree in 1927 and his doctorate in 1933 from Harvard University.

He came to the University of Louisville in 1929 as an assistant professor of biology. He had been a student assistant to professors at Bowdoin and was a teaching fellow at Harvard.

He belonged to several national organizations in the natural history field, among them the American Ecological Society, the American Ornithological Union, the Wilson Ornithological Club, the American Society of Mammalogists, the American Society of Zoologists, and the Inland Bird-Banding Association.

Among his hobbies were mountain climbing and photography, and his writings included articles on high altitude plants which he studied on field trips in Kentucky, Tennessee, and Maine. He also was an authority on honey plants and pollination.

On November 25, 1933, he was married to the former Ethel Weeter of Louisville. She was a teacher of biology at Durrett High School.

They had a son, John Harvey Lovell, II, of New York, a commercial pilot, and a daughter, Mrs. John Irvin from Pennsylvania.

Natural history was a family tradition. Lovell's father, John Harvey Lovell, was the author of thousands of articles and several books on plants, bees, pollination and a host of subjects. Mrs. Lovell and their children shared Lovell's interests.

Lovell was stricken with a mental disorder in November 1959 that left his eyesight badly impaired. He returned to his home and never returned to his faculty post.

He kept up his interest in the bird and animal life in the woods surrounding his home. Friends often trained their field glasses on the trees, then reported to the ailing biologist on the bird populations of the woods outside.

Dr. Lovell authored the column "Let's Talk About Honey Plants," in *Gleanings*

In Bee Culture from 1954 to 1970. He was also the author of the Honey Plants Manual, a popular book on important honey plants of the United States published by the Root Company.

Mrs. Lovell passed away only two months after her husband.

LOVELL, JOHN HARVEY (1861-1939) – Beginning in 1913, John H. Lovell wrote for the American Bee Journal, and Gleanings in Bee Culture for over 20 years.

While a student at Amherst College in Massachusetts young Lovell became much interested in science. He made a collection of the plants of western Massachusetts which is still preserved and a collection of minerals from various parts of New England. Due to his excellent record he was selected to represent the Biology Department in a contest and was elected to Phi Beta Kappa. He was also offered a position in the United States Weather Bureau which he did not accept.

Upon his graduation at the age of 21, he took up teaching for several years. This was interrupted by the illness and subsequent death of his father, Harvey H. Lovell, a retired sea captain. At the age of 28 he now found himself the possessor of a large home and an income which in those days was ample for his needs.

Several years later after reading Mueller's book, The Fertilization of Flowers, his attention was directed to the study of pollination. He now took up the study of flowers and their insect visitors and pursued it with boundless enthusiasm for the rest of his life.

In 1899 he married Lottie Magune. Her sympathetic interest in his work proved a big stimulus to his studies. She assisted him in collecting, labeling and cataloguing his insects. She also typed nearly all his notes and manuscripts. They had two sons, Harvey Bulfinch and Ralph Marston Lovell.

Based upon original observations, Lovell had begun in 1897 a series of articles in botanical journals upon the pollination of flowers in which he described for the first time the intricate floral mechanisms and life histories of many Maine flowers. Finding it difficult to get the wild bees accurately identified, he began to publish in 1901 a series of articles in Entomological Journals upon the wild bees of the eastern states in which many species new to science were described and the known range of many other species greatly extended. This work naturally led to a great interest in the honey bee and nectar collection. He firmly believed that the pollination of flowers is the most important phase of agriculture. Spurred on by the curious beliefs of Felix Plateau, the Belgian scientist, that honey bees do not see color and that colors are of no advantage in attracting bees, Lovell published a series of articles upon the sense of vision and flower-visiting habits of the honey bee in which he presented experimental proof that bees can distinguish colors. He trained bees to visit a certain colored slide on which odorless sugar syrup had been placed. This slide was then transposed with a variety of other colored slides lacking sweets. The honey bees were found to ignore all colors except the one to which they were trained, no matter how much he mixed up the colors. Blue and yellow were found to be the colors to which the bees showed the greatest fidelity. He had previously found that the removal of the petals from a flower greatly decreased the number of insect visits even though the nectar was still present.

In 1913, at the recommendation of Dr. E.F. Phillips, of the Bureau of Entomology, at Washington, he was selected to write the biological articles for *ABC and XYZ of Bee Culture*. He continued to be the biological editor for several successive editions, contributing some 78 articles comprising about one fifth of the volume.

In 1918 he published his first book, *The Flower and the Bee, Plant Life and Pollination*, in which he assembled in popular form the observations of 25 years of field work. The book was illustrated by 120 photographs of flowers which Lovell had taken natural size. The pictures have been frequently praised for their wealth of detail, the parts of the flower being shown with almost diagrammatic clearness. These pictures were the result of many experiments with various kinds of photographic plates and with various methods of photographing the flowers. He always did all his own developing and printing, believing that no one unfamiliar with the ecology of the flower could be able to bring out the exact details wanted.

In 1926 appeared *The Honey Plants of North America*, published by the A.I. Root Company, a book describing the honey plants and conditions of beekeeping in all parts of the United States. This book was the result of three years continuous work, all data in the book being carefully checked by personal observation and by correspondence with beekeepers in all parts of the country.

In the spring of 1926 he began the publication of a daily series of illustrated articles in a large Boston paper, a series which was continued through many successive summers until the total number of articles reached nearly a thousand. Similar series were published in newspapers in Maine and Louisville,

John Harvey Lovell

Kentucky. The articles consisted of a description of the flower, its life history, insect visitors, economic uses, and curious stories, myths and folklore connected with it. Each article was accompanied by one of his photographs.

In his second article in the *American Bee Journal* entitled "Flowers from which Honeybees cannot Obtain Nectar" (1913), he discussed bumble bee flowers and butterfly flowers in which the nectar is so deeply concealed that honey bees cannot reach it, and wind-pollinated flowers such as corn which do not secrete nectar. This article was followed in quick succession by articles on "The Signals of White Clover" (November 1913) and "Our Wild Bees" (June, 1914).

In his final article written in collaboration with his son, Lovell describes his experiments on the lowbush blueberry. When plants were screened from insects, no fruit was set, indicating that they have to be cross-pollinated by insects. Solitary bees, bumble bees, and honey bees were found to be the chief pollinators.

In 1921, Frank C. Pellett, then Associate Editor of the *American Bee Journal*, visited Lovell in Waldoboro and wrote an account of his visit entitled "The Maine Naturalist" (February, 1922). He emphasized the ideal surroundings of the Lovell home, which is located on the banks of an attractive stream only a few minutes' walk from woodland, pond, swamp, and marine habitats. He described Lovell as one of the last of the field naturalists, men who were interested in how animals and plants lived and behaved in their natural surroundings.

Such visits were very stimulating to Lovell who lived much of his life isolated on the Maine coast hundreds of miles from his colleagues. He kept in touch with them chiefly through correspondence. He subscribed to most of the entomological and botanical journals and was the recipient of thousands of reprints and bulletins from scientific workers both in America and in Europe.

In spite of his many scientific and literary interests, Lovell managed to find time for many hobbies and diversions. In his younger days he was a devoted cyclist beginning with the old highwheel model. He took up tennis at the turn of the century and soon became very proficient with the racquet. He continued to play until his 74th year at which time he was still able to defeat his daughter-in-law.

He owned and operated a motor boat for several years and greatly enjoyed cruising among the islands which dot the Maine coastline. He also liked swimming and rarely missed his daily dips in the river back of his house, often continuing until the first ice began to form along the shore.

Although suffering from high blood pressure during the last years of life, Lovell still continued to work on bees and pollination with youthful enthusiasm and was full of plans for the future. His last technical article, "Pollination of Verbena Hastata," and a newspaper article appeared in print after he was confined to the hospital in his last illness.

Written by Frank Pellet, published in the December, 1939, American Bee Journal

MACKENSEN, OTTO (1904-1995) – Dr. Otto Mackensen, former research geneticist at the USDA-ARS Honey Bee Breeding, Genetics & Physiology Laboratory in Baton Rouge, Louisiana, died October 27, 1995 in Salida, Colorado.

Otto Mackensen was born in San Antonio, Texas, on July 30, 1904. He received a B.S. degree in 1928 from Texas A&M University and a Ph.D in Entomology from the University of Texas in 1935.

During his 36-year research career in Baton Rouge (1935-1971), Dr. Mackensen fulfilled his assignment to bring scientific breeding to the honey bee industry.

In 1944 Dr. Mackensen discovered that virgin queens or instrumentally inseminated queens would begin to lay eggs at a younger age if they were treated with carbon dioxide gas. He then standardized the practice of using carbon dioxide to immobilize a queen during insemination as well as to stimulate oviposition in queens.

By 1948 he had advanced the technology of instrumental insemination of queen bees to make it a practical technique to use in honey bee breeding. The equipment that he designed and the procedures that he developed in the late 1940s are essentially the same as those used at the present time. He published his first manual for the instrumental insemination of queens in 1948 and his second in 1970, one year before his retirement.

In 1951, Dr. Mackensen found that sex determination of honey bees is similar to that previously described for the parasitic wasp (*Bracon hebetor*). He discovered that diploid males do exist in colonies of honey bees, but that they

Otto Macksen

exist only as eggs and very young larvae. He demonstrated that the poor brood viability associated with inbreeding in honey bees is caused by the removal of these diploid males.

Together with ARS scientist William Nye, Dr. Mackensen produced a strain of bees that had a strong tendency to collect alfalfa pollen. This work demonstrated that selective breeding can produce bees that possess very specific characteristics with practical application for the beekeeping industry.

written by John Harbo
1996 Bee Culture

MACLEOD, HUGH JOHN (1914-1980) – Hugh attended his first Eastern Apicultural Society conference at Guelph, Ontario, in 1963. He became a director in 1969, president in 1974 and vice-president in 1979 for the Ottawa conference.

Hugh John Macleod

In addition, he founded the thriving Toronto District Beekeepers' Association in 1975 and served as president in 1978.

He was the superintendent of the Honey and Maple Syrup Exhibits at the national Royal Winter Fair and helped establish the first colonies of bees to be displayed at the Ontario Science Center.

MAGNETIC AND ELECTRICAL FIELDS, EFFECTS ON BEES – Behavioral evidence has demonstrated that honey bees, as well as other organisms (bacteria, skates and homing pigeons) are sensitive to fluctuations in the Earth's magnetic field. It is suggested that honey bees may use this information as an aid in orientation. Five studies have shown the effects of magnetic fields on orientation in adult bees.

Misdirections – It was observed that the waggle dance (performed by scout bees) could be altered by external magnetic fields around the comb. Data were collected on the misdirections (*missweissungs*) of the bees' dances over a number of years. It was discovered that these errors indicated a response to earth-strength magnetic fluctuations.

Circadian rhythm – In the absence of other cues, bees can set their Circadian rhythms to the periodic fluctuations of the geomagnetic field.

Horizontal dancing – Bees can be trained to dance on horizontal combs (no gravity component) and will continue to perform recruitment dances without any visual stimuli. The bees will gradually dance in directions close to the cardinal octomodal points of the compass.

Comb building – Some data show that swarms will build new comb in the same magnetic directions as the parent hive. Swarms that are disturbed shortly after initiating new comb will rebuild in the same planar direction, and these effects can be manipulated experimentally.

Conditioning – Individual bees were trained to fly to feeding stations with sucrose solutions in varied magnetic fields and successfully learned to discriminate between fields that had offered them greater rewards. Similar experiments were performed using shock treatment and modified fields.

Conclusions – It stands to reason that magnetic field reception would be useful as a back-up mechanism to aid in orientation and navigation along with odor and light (polarized, ultra-violet) reception, but no such sensory system has been found. The big question still unanswered here is, "How are bees able to sense magnetism?"

There was much excitement when it was reported in a leading scientific journal in 1978 that bees have magnetic remanence. Physical experiments using sensitive instruments had been performed on honey bees to identify and locate particles that were believed to be magnetite, a highly magnetic form of iron. The preliminary remanence experiments suggested magnetite in the abdomen. The presence of actual magnetic particles would have been a significant breakthrough, but they were never found. Instead, iron particles of a very different nature were found to exist in the bee's fat body. The iron granules in the trophocytes (originally identified as oenocytes) appear to grow in number and volume as a bee ages, with fewer in newly adult bees and drones. They are non-crystalline, unlike magnetite, and are not associated with any cellular organelle. The iron is now believed to be accumulated in response to the high dietary iron found in pollen and doesn't answer the intriguing questions surrounding magnetism and honey bees. References on the subject are as follows: Gould, J.L., J.L. Kirschvink and K.S. Deffeyes. *Bees have magnetic remanence.* Science 201: 1026-1028. 1978. Kuterbach, D.A., B. Walcott, R.J. Reeder and R.B. Frankel. *Iron-containing cells in the honey bee.* Science 218: 695-607. 1982. - JS

MASON BEES – Mason bees are solitary bees in the genus *Osmia* (Family Megachilidae) that use mud to build their nests. Several species of mason bees have been developed as orchard pollinators, including the blue orchard bee or blue mason bee, *Osmia lignaria*, in the U.S.; the hornfaced bee, *Osmia cornifrons*, in Japan; and the horned bee, *Osmia cornuta*, and the red mason bee, *Osmia rufa*, in Europe. These species range in size from slightly smaller than a honey bee to slightly larger. Females of all four species have small horn-like prongs on their face. Mason bee males are smaller than females, have longer antennae, and display abundant white hair on their face.

Mason bees winter as adults and have a single generation per year. They emerge early in the spring and are active for approximately a month. After mating, females select a pre-established nesting cavity (a beetle burrow in dead timber or an abandoned solitary bee nest in a clay embankment) and begin nesting activities. Nests consist of linear series of cells delimited by mud partitions. Each cell is provisioned with a loaf of pollen mixed with nectar. Pollen is

Cross-section view of nesting tube used by Osmia lignaria *that shows the pollen and nectar plug with attached bee egg (left) and mud cap that separates each egg chamber.* (Tim McCabe, USDA photo)

carried in the scopa, a brush of long hairs located under the female's abdomen. Nectar is carried in the crop and regurgitated on the provision. After a provision is completed the female lays an egg on it and starts building and provisioning a new cell. After building a series of cells, females build a mud plug to seal the nest entrance and start searching for a new nesting cavity. Development from egg to adult is completed by autumn, but adults do not emerge out of their cocoon until the following spring.

Mason bees are excellent fruit tree pollinators because they always land on the reproductive organs of the flower and because they are active under cloudy skies and at lower temperatures than most other bees. At least three of the four species mentioned above, *O. lignaria, O. cornifrons* and *O. cornuta*, show a strong preference to collect pollen from fruit trees. Only 250–300 nesting females are enough to pollinate an acre of fruit trees.

Due to their short activity period, their ability to accept man-made nesting materials, their short foraging range, about 300 feet (100 m), and the low densities required for optimal pollination, mason bees are easy to rear and manage. Prior to bloom, nests containing wintered bees are set up in nesting shelters provided with a mud source and man-made nesting materials, such as drilled wood blocks, boxes with paper straws or reed sections. Bees emerge as temperatures increase, mate and start foraging and nesting. After orchard petal fall, nesting materials containing newly-build nests are collected and stored at 70-79°F (22-26°C) or at close-to-ambient temperatures during the summer. By autumn, when all progeny has reached the adult stage, nests can be left at ambient temperature or wintered in a refrigerator at 37-41°F (3-5°C). Rearing and wintering at artificial temperatures allows for better adjustment between bee

Osmia ribifloris. (USDA photo)

Osmia cornifrons. (USDA photo)

emergence and orchard bloom the following spring. If reared at adequate temperature regimes and if well-timed with orchard bloom, populations recovered at the end of the blooming period exceed populations released.

Osmia cornifrons is the most widely used apple pollinator in Japan. Due to their capacity to fly under sub-optimal weather, commercial yields can be obtained in orchards pollinated with mason bees even in years with marginal weather conditions during bloom. Besides fruit trees (almonds, apricots, plums, cherries, pears, apples) mason bees can be used to pollinate other spring-flowering crops, in the open or in confinement. For further information see Bosch J. and Kemp W.P. "How to manage the blue orchard bee as an orchard pollinator," *Sustainable Agriculture Network, Handbook Series Book 5*, Beltsville, MD. 88 pages. 2001. - JB

MATING OF THE HONEY BEE – Before 1961 mating in honey bees had not been seen or accurately described. It was known that mating never took place in the hive, and a few speculations, most incorrect, had been made about where queens mated. Queens and drones had been observed taking flight on warm, sunny afternoons only; their flights were relatively brief, lasting about 30 minutes. Queens may or may not take a short orientation flight before a major mating flight. No one has ever found a normal queen or drone resting in the field; when they have exhausted the food they are carrying they return to the hive. Queens and drones never forage. Probably it is best that they neither rest nor forage since their large size would make them tasty prey for a bird or insect.

Before 1961 a few people had seen masses of drones pursuing queens fairly high in the air; they had presumed mating was taking place, but because the flight was so rapid, the descriptions were brief and not always accurate.

In 1961 one of the components of the secretion from the queen's mandibular gland was identified and synthesized in England though its function was not immediately known. Dr. N.E. Gary, then at Cornell University, soon discovered that the newly synthesized material was the sex attractant in the honey bee and was the pheromone that drones used to

Dr. Norman Gary was awarded the Eastern Apicultural Society's James Hambleton award, in part for his research on queen pheromones.

find queens. When Gary elevated a queen, or an inanimate object coated with the sex attractant, using a helium-filled balloon, drones would be attracted to it. In only one case did one of these tethered queens mate; apparently queens must be free-flying to do so.

The following year Dr. C. Zmarlicki discovered that queens and drones fly to special sites, which he termed Drone Congregation Areas (DCA), to mate. These areas are well-defined and usually smaller than an acre (one-half hectare) in size.

Drone Congregation Areas remain the same year after year. Data indicate that unless buildings have been constructed on the sites, these areas have remained the same for over 25 years. Since drones live only a short time, probably between six and eight weeks, and queens mate only when they are young, it is clear that memory plays no role in the constancy of these areas. Some feature in the surrounding geographical area must be responsible for the location of Drone Congregation Areas, but attempts to discover what these are have not been exactly determined. Drone Congregation Areas have been found in valleys, on hilltops and on flat plains. Despite this knowledge, we still have no method of controlling natural mating other than to use an island or similarly isolated area.

Whenever Africanized and European honey bees are kept together in tropical and subtropical areas, it appears that European drones rarely, if ever, catch a queen of either group but Africanized drones frequently succeed. As a result, all of the colonies in warm climates where Africanized drones are present soon become Africanized.

Mating takes place about 20 to 50 feet (six-15 meters) in the air, well above the normal flight path of worker honey bees, which is about eight feet (2.4 meters) above the ground. Workers are antagonistic toward all foreign queens and will attack and ball any queens they encounter outside of their hive. Strong winds will force queens and drones as well as worker bees to fly closer to the ground. Drones fly from one congregation area to another searching for virgin queens, and will take more than one flight a day in an attempt to find a queen.

To mate, a drone approaches a queen from behind and grasps her abdomen with its legs. The mating act itself is brief. Drones possess genitalia that are larger in proportion to their body size than all other animal species except a few species of fleas. The drone genitalia are contained in the abdomen. When they are everted, a popping sound can be heard. The shock of everting the genitalia results in the death of the male, which falls over backwards and to the ground.

Eversion of a mature drone. (Cobey photo)

Eversion of an immature drone.(Cobey photo)

At this stage the genitalia may be separated from the male and remain in the queen, but usually only briefly. Apparently the queen removes them herself and continues to mate with other drones. Queens usually take two or three mating flights; one study indicates that queens will continue to search for males until they receive sufficient sperm to fill their spermathecae.

Controlling natural queen mating by drone flooding – Each virgin queen mates with approximately 20 drones in flight. Consequently, controlling queen mating presents unique challenges. Queens may be artificially inseminated, but that process is laborious and time-consuming. Furthermore, most artificially inseminated queens perform poorly compared to their naturally-mated sisters and they are more difficult to successfully introduce into new hives. Successful natural matings have been reported in cages but they are rare and the technique has not been perfected. For now, those interested in controlling queen mating on a large scale should employ techniques used by queen breeders for many years. Collectively, those techniques have become known as drone flooding or drone saturation.

Conceptually, drone saturation is simple. If one is attempting to propagate a specific trait that exists at a high frequency in the beekeeper's managed stock but which may not exist in the wild or feral stock then one should attempt to maximize the number of managed drones that mate with the virgin queens produced by the beekeeper. If one assumes that feral drones and managed drones are equally fit, and that feral drones have no mating advantage by reason of time of flight or other factors, then the numbers of wild and managed drones that mate with managed queens should be proportional to their relative population sizes. Therefore one may control open- or naturally-mated queens by minimizing the population of wild drones and maximizing the population of managed drones. Since the wild population may be difficult to control, one floods the area in which queens will mate with managed drones. In short, the beekeeper saturates the queen mating area with drones from his stock.

Implementation of drone flooding can be difficult in practice. First, one must know how many queens are likely to mate during a particular period and plan to provide approximately 20-25 drones per virgin queen throughout the duration of the mating interval. If one is expecting 1000 queen cells to hatch per day in a particular location then one must have enough colonies in that location to produce 20,000–25,000 sexually mature drones per day as well. Of course, not all colonies will be equally strong or productive and drone strength in any colony will vary with weather, pollen flow, season, queen age and a multitude of other factors. Continuous assessment of drone populations and drone population dynamics is critical to success. For instance, suppose the queen in one hive begins laying drone eggs on February 12 and lays 3000 drone eggs over three days, filling all available drone brood cells in the process. Those drones will not emerge until March 8–10 and will not be sexually mature until March 22–24. Any queen emerging before March 15 probably will not encounter drones from that hive during her mating flight. Equally important, that hive will not produce another cohort of drones until April 1 assuming that the queen immediately begins laying drone eggs as the adult drones emerge.

In summary, the hypothetical hive could provide drones to mate with as many as 150 queens that begin mating flights on March 22. But if those 3000 drones are expended on 150 virgins that fly March 22–24, then that hive will not contribute drones to the drone pool again before April 15. Therefore, the hypothetical hive could produce drones for 150 queen cells grafted on March 4 but would not provide drones for another graft before March 28. All queen cells grafted on or about March 5–27 will need other drone-source colonies.

The real world is even more complicated than the illustration, however, as no provisions were included for variables such as honey or pollen flows (which may result in drone cells filled with nectar, honey or pollen before the queen refills the cells with eggs) or dearth periods (which may discourage drone-egg laying) or inclement weather (which may diminish drone survival and mating success) or predators, parasites and disease (spiders, tracheal and *Varroa* mites, and chalkbrood all affect drone production and survival) or queen supersedure (the process of supersedure will result in a sustained period of no drone production) or colony strength (strong hives will produce many more drones and begin producing drones much earlier than weak colonies).

Drone flooding is all about producing desirable drones in great excess so the chances of a virgin queen mating with feral drones are negligible. Production of drones in quantities much larger than the bare minimum needed for adequate mating is therefore essential. Drone flooding demands large numbers of very strong colonies with adequate drone combs.

Of paramount importance today is *Varroa* mite infestation. *Varroa* reproduce preferentially in drone brood with devastating effects on drone populations. Hives infested with *Varroa* mites produce fewer drones with dramatically reduced life spans and a variety of defects that significantly impair the reproductive fitness of the affected drones. Controlling *Varroa* mite infestations is an essential element in drone flooding today. Simply treating a hive with a large *Varroa* population will not have an immediate impact on the sexually mature drone population. Replacing a seriously infested hive with a healthy drone-productive hive is the only "treatment" that will offer an immediate solution to the drone population deficit that *Varroa* infestation brings.

Drone flooding is an important aspect of controlling queen mating for a variety of purposes including commercial queen rearing and controlling the process of Africanization of managed apiaries. *Varroa* infestation is a serious impediment to successful drone flooding. In fact, reduction of drone populations by *Varroa* may be one reason why some beekeepers have reported more premature queen failure and a higher incidence of drone-laying queens in recent years. - DW

Distances queens and drones fly for mating – Precise control of drone populations is also possible by moving drone-producing colonies and virgin queens in their nucleus colonies, to an isolated island or another area where there are no other honey bees. It is generally agreed that isolated areas should be at least 12 miles (20 km) from a location where there are other colonies.

A paper by Dr. D.F. Peer is cited most frequently when we are questioned about how far queens and drones will fly and how much isolation is needed to obtain pure matings. An area in northern Ontario was used for the study. Here apiary registration records indicated there were no honey bees within 35 miles (56 km). It was found that when queens and drones, in their respective colonies were 3.8 and 6.1 miles (six and 10 km) apart, that the queens would start to lay eggs at approximately the same time. This indicates that these distances pose no flight problems for the queens and drones. Apparently, it is normal for them to fly this distance to mate. When the two sexes were located 8.0 miles (13 km) apart, egg-laying by the queen was delayed; it was delayed still further when the two were 10.1 miles (16 km) apart. No mating occurred when queens and drones were more than 10.1 miles (16 km) apart though distances of up to 14 miles (22.6 km) were tested. (Peer, D.F. "Further studies on the mating range of the honey bee." *The Canadian Entomologist* 89; 108-110. 1957.) The data indicate that apiaries that are isolated from all other honey bees by 12 miles (19 km) will have pure matings.

Most mating takes place within one or two miles (1.6 to 3.2 km) of the apiary where the virgin queens are located. From a practical point of view, queen breeders place some drone-producing colonies in the same apiary and at four points, in different directions, about one to two miles (1.6 to 3.2 km) away. An effort is also made to eliminate other feral and man-kept colonies in the vicinity that might add drones to the pool. In this way the queen breeder can produce queens with a higher percentage of the desired traits.

Mating frequency – All of the older textbooks and papers on honey bees state that queens mate once, rarely twice. In 1954, Steve Taber III found that queen honey bees mate several times; he believed this number was seven or eight, but more recent studies have shown that the average queen probably mates with 15 to 20 drones within one to two days when she is three to five days old. Apparently queens are unable to mate after the age of about 30 days; some references indicate this may be closer to 20 days. Drones are not mature and capable of mating until about 12 days of age.

In the process of mating, a queen collects about five million sperm that are packed into a pouch in her abdomen called the spermatheca. This is a curiosity since a drone produces about twice that number of sperm. As an egg passes down from the ovaries to be deposited in a cell, a few sperm are released and one enters the egg and fertilizes it. These eggs develop into worker bees and rarely queens. Drone eggs are not fertilized, and it is not known how the queen controls this. The supply of sperm obtained when the queen is young usually lasts the rest of her life and she never mates again.

In rare cases a queen will run out of sperm and become a drone layer, but still, a drone layer never mates again. Drone-laying queens are rare, but beekeepers must be aware of the fact that they may occur. A drone-laying queen may continue to produce drone eggs for a long period of time but may also produce chemical secretions (pheromones) that cause the bees in the colony to believe that they have a normal queen. Thus, the colony may be doomed. From the point of view of the bees in the colony not everything is lost since the drones produced contribute to the gene pool, at least for a short period of time. The only other time a queen will leave a hive, other than to mate, is to accompany a swarm to a new home site.

Mating in confinement – Several people have attempted to mate honey bees in confinement using greenhouses or large cloth or screen-covered tents. A few people have claimed success, but no one has been able to duplicate their results. Dr. John Harbo prepared an annotated bibliography listing the papers that had been written on this subject; this is available from IBRA (see ASSOCIATIONS FOR BEEKEEPERS, International Bee Research Association). (see CONTROLLED MATING.)

MATING NUCLEUS – (see QUEEN REARING, Equipment for queen rearing)

MAXANT, WILLIAM T. (1908-2005) – Bill Maxant, a longtime Harvard, Massachusetts resident who owned three prominent local businesses, died November 20, 2005 at his home. He was the husband of the late Helen (Korpacy) Maxant.

He was born in Pittsburgh, December 31, 1908, son of the late Frank and Hedwig (Wiebeck) Maxant. He graduated from Ayer High School and Burdett Business School in Boston, AND lived in Harvard for 85 years.

Mr. Maxant was the co-owner of Chandler Machine Co., which manufactured textiles and sewing machines. He also owned Murphy Knife in Ayer and Maxant Industries, which manufactured honey processing equipment.

Bill was a Life Member of the Eastern Apicultural Society, the Middlesex Bee

Bill Maxant

Club, and was instrumental in founding the Western Apiculture Society.

He was passionate about beekeeping, admiring covered bridges and flying his Beechcraft.

MAY, GEORGE W. – George May of Chicago, Illinois, founder of the commercial apiary called George W. May and Son Honey Farms of Marengo in north-central Illinois, died January 31, 1982.

He became interested in beekeeping in 1948 when he acquired five colonies. He and his son, Phillip G. May, had as many as 1400 colonies in the apiary. They operated the largest commercial apiary in the state of Illinois.

May was a member of the American Beekeeping Federation, the Illinois State Beekeepers' Association, and the Cook-DuPage Beekeepers' Association in the Chicago metropolitan area.

McEVOY, WILLIAM (1844-1912) – Beekeepers are generally no longer familiar with the name of William McEvoy, who gave first prominence to the so-called shaking treatment for American foulbrood which for a long time was the accepted method for getting rid of this disease. He was born of Irish parentage in the county of Halton, Ontario, Canada, March 26, 1844. Soon the family moved to the little village of Woodburn in the county of Wentworth where they lived the rest of their days.

Here William began his schooling which was destined to be short due to the death of his father. At a youthful age he was engaged as a farm hand with William McWatters, successful neighboring farmer, and the influence of this rigid disciplinarian was most wholesome. It impressed William with the great importance of doing well whatever he attempted, and laid the foundation for the thoroughness and neatness which was a most decided trait all through life.

In 1861 he purchased two box hives, transporting them home by the primitive method of suspending the colonies from a long pole, the ends of which rested on the shoulders of the two carriers. The bees were paid for by piling and cutting 20 cords of wood. His mother protested vigorously about his bringing home those "wasps." However, the "wasps" stayed and this event formed a waymark in his life. The fascination of the study of the bees and their habits led him into a business which was to bring him world renown.

As his apiary grew, the necessity of producing extracted honey grew too, and hearing of an extractor at Kilbride, 30 miles away, he made the journey both ways on foot to ascertain the value of the process. Coming home with his coveted information, he soon after became the proud owner of a very crude type of extractor. Standard equipment was unknown in those days and the hive he finally developed had about the same cubic space as today's standard hive, but in contrast was deeper, his theory being that in cold weather bees would move up to stores when they would not move lengthwise.

In 1868 he began to exhibit honey and wax successfully, capping a long list of winnings with the first prize on extracted honey at the Philadelphia Centennial in 1876. The honey was gathered from Canadian thistle and topped 52 other exhibits from various parts of the world.

In 1875 he had his first experience with American foulbrood. After firmly establishing, by experiments, that disease was carried in the honey, he developed the system of shaking the bees of a diseased colony onto new foundation and destroying the old comb and honey. So successful was he that in 1890 he was appointed provincial inspector by the Ontario government, and was reappointed to the position for 19 years. He successfully treated thousands of colonies in all parts of the province and the shaking treatment became known all over the world as the McEvoy treatment.

In 1892 while attending the convention at London, he drew attention to the damage done to bees by the needless spraying of fruit trees in full bloom. He later headed a delegation which called on the Ontario Legislature and convinced them of the need of a bill prohibiting such spraying.

On his retirement from public life in 1909, he was presented with a purse of gold by the Ontario Association, the members of which expressed themselves as "being under great obligations for the magnificent services rendered by Mr. William McEvoy." He died in the fall of 1912. *Ewart McEvoy, NY, from American Bee Journal, 1944.*

McGINNIS, DAVID K. (1915-1999) – David McGinnis, 83, New Smyrna Beach, Florida, died May 18, 1999.

Born August 2, 1915, in Wellsburg, West Virginia, he moved to Florida in 1937 and established Tropical Blossom Honey Co., Inc., Edgewater, Florida, in 1940.

He invented the globe honey jar that is used widely in the honey and gourmet food industry today. McGinnis was a pioneer in developing international markets for Florida honey. He served as secretary for the National Honey Packers and Dealers Association and was active in state and national beekeeping organizations.

McGREGOR, SAMUEL E. (1906–1980) – McGregor, called "Mac" by his associates and friends, died February 4, 1980. He was 74 years of age and still at work four weeks prior to his death as a Collaborator at the U.S. Department of Agriculture's Carl Hayden Bee Research Center, Tucson, Arizona.

Mr. McGregor was born in Milano, Texas, January 3, 1906. During his early boyhood, on his father's farm, he helped manage honey bees that were maintained for honey production and crop pollination. He pursued his interest in beekeeping at Texas A&M College from which he received a B.S. degree in 1931. He was awarded the M.S. degree by Louisiana State University in 1936.

His first professional job in the field of apiculture was in 1925, when he became a State Apiary Inspector in Texas. In 1930 he moved to New York State, where he continued in inspection work, returning to Texas a year later as a State

Samuel E. McGregor

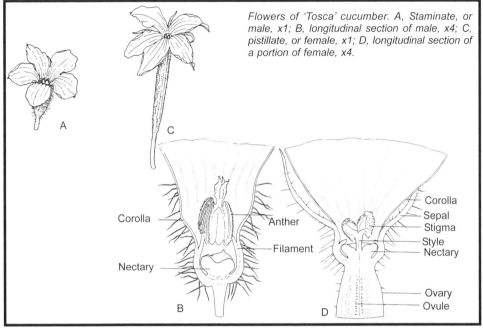

Sample drawing from Insect Pollination of Cultivated Crop Plants. Not to scale. (USDA, 1976)

Apiary Inspector. For two years, he kept bees commercially in Texas before he began to work for the U.S. Department of Agriculture in 1934. His first assignment was at Baton Rouge, Louisiana, where he remained for five years, followed by five years in Arkansas, one year in Wisconsin, and one year in Texas. In 1949 he was assigned to the Bee Research laboratory at Tucson, Arizona, to work with the late Frank Todd on research on pesticide problems of honey bees. Later the emphasis of his work shifted to pollination of melons, alfalfa, citrus and cotton. In 1961 he was appointed to head the Bee Research Laboratory at Tucson, Arizona. From 1965-69 he served as Chief of the Apiculture Research Branch, and directed the Bee Research Program of the Entomology Research Division, Agricultural Research Service at Beltsville, Maryland. In 1970 he returned to Tucson to utilize the library facilities to write a book entitled *Insect Pollination of Cultivated Crop Plants*. This treatise on plant pollination by insects was not only of immediate value to agriculturists but clearly showed the value of insect pollinators to mankind's total environment.

In June 1973 Mr. McGregor retired from his position in the United States Department of Agriculture but remained active as a Collaborator at the Tucson Bee Research Laboratory.

Mr. McGregor's entire professional career was in apiculture. He conducted research on almost every aspect of apiculture, but he is best known for his research on honey bee pollination of cultivated crop plants. His research on crop pollination provided many significant contributions, and he was recognized as a world authority on insect pollination of crop plants.

Mr. McGregor's list of bee research and pollination publications, the result of some 45 years of work with bees, is lengthy. Among the most important are those that appeared in *Annual Review of Entomology, USDA's Yearbook of Agriculture* and the *Encyclopedia Britannica*.

MECHANICAL HANDLING OF COLONIES – A variety of colony and honey-super lifters have been invented to make both work in the apiary, and in moving colonies, easier. Some of these devices are homemade but a few, such as booms and forklift trucks for lifting colonies, or groups of colonies, off and onto trucks, are available commercially. Gas-powered, two-wheel hand trucks for moving stacks of four to eight supers within the apiary and honey house for the honey harvest have also been developed.

One problem in using hive lifters and mechanical devices in an apiary is the rough terrain one usually encounters.

A simple way to lift and move a colony. Do not forget the bottom board.

Whenever colonies are going to be moved in quantity, ventilation can become a serious and deadly problem. Bees can overheat if confined and an entire colony can expire in a matter of minutes. A variety of commercial moving/ventilation screens are available, or you can make your own. The photo shows a simple, practical and useful screen.

Mechanical Handling Of Colonies

Simple two-wheel carts or a motorized cart make moving hives and supers more efficient. (Morse photo above)

Trucks with motorized tailgates – sometimes called Tommy lifts – make moving easier.

On these two pages – mechanical booms, operated by one person have been used for years. They range in size and capability from being able to handle a single colony to a four-colony pallet. They replace a second person, work when muddy conditions exist, and on somewhat uneven ground. They have been available in many boom lengths.

Mechanical Handling Of Colonies

Forklifts and Pallets

A wide range of forklift-style machines have been used over the years. Models range from simple non-motorized hydraulic floor jacks to attenuated, specially designed machines made for moving pallets of honey bee colonies.

Basic warehouse-style lifts work well on level ground with lots of room. Fuel can be gasoline, propane, battery or even diesel fuel.

Large tires, good visibility, enough power, turning radius, long life, fuel source, comfort, noise and ease of repair are all considerations when purchasing these very expensive machines.

Some of this problem can be solved by using large, bicycle tires in place of small wheels. Since most apiary sites are used for a number of years some time and effort put into leveling locations is worthwhile.

In the U.S., beekeepers moving long distances prefer to use pallets that hold four colonies. In Australia, where beekeepers also migrate long distances, it has been popular to place four, five or six colonies on a long pallet so that they are all side by side. Whatever type of pallet is adopted it is important to stay within the width limits required of trucks. Moving colonies long distances cause the bees no hardship provided they have adequate ventilation and the frames are fixed into place so that they will not sway and crush bees and possibly the queen. (see MIGRATORY BEEKEEPING)

Moving hives the old way. Packages (in the back of the automobile) were lighter, but the wind stress must have been a problem. Horses sometimes had to have blankets covering them to avoid being stung.

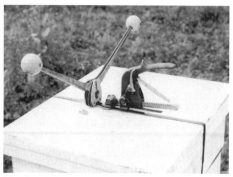

Metal and plastic banding are still used, but easier and more efficient are the ratchet straps that are so inexpensive as to be nearly disposable.

Once a colony is strapped, seal any leaks with tape. Close the front with window screen or a moving screen, and maintain good ventilation.

A colony ready to move. Secured by a durable cinch strap, a ventilated inner cover fits under the migratory cover, and a front moving screen is in place.

Rather than move colonies individually, placing them all on a small trailer and moving them from honey source to honey source, or pollination crop to pollination crop is much easier. (Richard Taylor photo)

Larger loads, of course, require better trailers and secure fastenings. Cinch straps hold each pallet to the trailer and a band holds the colonies to the pallet.

Mechanical Handling Of Colonies

Midsize trucks for short hauls generally come equipped with rear bed grates (headache fences), equipment boxes and rope hooks or rails underneath the smooth bed, and are the length needed to hold appropriate weight. Many operations standardize everything, so every truck has the same material (smoker fuel, smokers, hive tools) in the same location. Workers do not spend time looking when they change trucks.

Securing a load is critical and corner/edge supports, lots of rope or straps are needed.

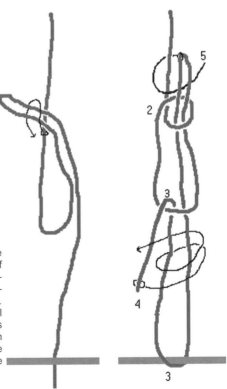

When hauling boxes, full or empty, a rope over every stack is needed. Holding them in place is a hay hitch knot. Self-tightening, but easily undone. It is the universal trucker's knot.

HAY HITCH KNOT
Allow the free end of the rope to drop down behind the anchor point (e.g. trailer rail). 1) Take a double length of rope and place it over the standing part. 2) Use the standing part to throw a hitch or loop around the double section. 3) Bring the free end up around the anchor (e.g. trailer rail) and through the lower loop or rope. 4) Pull the end down tight and throw a couple of half-hitches around the lower section. Alternatively tie a round turn and two half-hitches around the rail. 5) The hitch can be made more secure by throwing a half-hitch around the upper standing part with the doubled rope.

METAMORPHOSIS – Honey bees are among the more advanced insects that go through four stages in their development: egg, larva, pupa and adult. Less advanced insects, such as grasshoppers and crickets, have three developmental stages: egg, nymph and adult. In insects the change that takes place when the larva transforms into an adult is called metamorphosis.

Honey bees feed in the larval and adult stages only. During the larval stage the insect builds a large fat body and stores the energy that is needed for the transformation into an adult. During metamorphosis both the internal and external anatomy change slowly but in a uniform and organized fashion. The organs of the larva, including the digestive tract and nervous system, are discarded and the legs, wings, antennae and other external features develop. Because the honey bee nest temperature is controlled, metamorphosis proceeds in a way that can be measured and observed on a day to day basis (see LIFE STAGES OF THE HONEY BEE).

A. H. Meyer

MEYER, A.H. & SONS – A.H. Meyer immigrated to the U.S. in 1906 when he was 12. He worked for a time for Nephi Miller, in Providence, Utah. He operated bees with his brother, but shortly struck out on his own. Wax processing became a big part of his business, and his son Jack, Sr., operated one of the largest rendering businesses in the U.S. Jack, Jr., operates the business now, and has about 12,000 colonies and is the largest beeswax refining company in the U.S.

MICE – These small animals are the leading rodent pest of the beekeeping industry. Mice can be a serious problem during the cooler months of the year when they seek shelter and warmth inside beehives, stored equipment, or swarm bait hives. Their activities can lead to colony disruption and may seriously weaken even a strong colony when the hive cluster is disturbed during cold weather.

Mice rarely cause problems for the beekeeper in summer, but enter colonies in autumn in preparation for overwintering. They move freely in and out of colonies during cool nights when the bees are clustered. Nests are usually constructed in the peripheral area of the beehive where comb from adjacent frames are destroyed. Sometimes mice will chew frames and other wooden parts of the hive. The beekeeper should immediately remove a mouse nest with minimum disturbance to the colony cluster. The affected frames should be removed when appropriate and replaced with new frames to eliminate all signs of the nest.

Mice feed mainly on pollen and honey in the hive, but have been known to eat freshly dead bees. Other detrimental effects caused by mice include the foul odor in the hive created from their urine

Mice

Mice will build nests in the bottom box of your hive. They move in during fall and set up housekeeping. Be sure when you put on your mouse guards you are not locking mice in.

To make room for the nest they will chew holes in adjacent frames. They will eat dead bees, even live bees, brood, comb, pollen and wax.

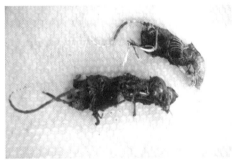

Of course sometimes the bees win. Dead mice are too large for them to move, and a decaying mouse can be unhealthy, so often a carcass will be propolized.

The mice will, if opportunity provides, raise an early brood in this nest, before spring returns and they can leave – or you remove the mouse guard.

and waste. Colonies packed for winter and stored supers of comb are especially vulnerable to mice damage as long as there is an opening large enough to permit passage.

Exclusion by reducing the size of the hive entrance prior to onset of cooler weather is the primary method of mice control. Although some use the wooden entrance reducers, mice can push these

Mice

A wooden entrance reducer is not a mouse guard. Do not use it as one. Mice can, and will chew the wood to gain entrance.

aside or chew the wood to create an entrance. An entrance screen should be made of any material that cannot be destroyed by mice. Small mice can enter openings that are 3/8 inch (one cm) wide or more. Rodent screens can be made of a piece of screen cut to the size of the hive entrance and placed securely over the entrance or bent to shape and placed inside the entrance.

The beekeeper must be careful to use a mouse exclusion method that allows free exit of bees for winter cleansing flight and debris removal. A queen excluder may be placed between the bottom board and brood chamber to exclude mice, but this method may impede colony housecleaning and hive ventilation. As an added protection against entrance blockage, some beekeepers prefer to add an upper entrance.

Exterior hive surfaces should be inspected in late summer for small openings and weak areas that might allow mice entry. A piece of metal or hardware cloth may be cut to size and secured over the vulnerable area as a temporary measure. Other preventive mouse control measures include good yard maintenance such as mowing grass

Welded wire, with openings no larger than 3/8" (no. 5) folded in a "V" shape and stuffed in the front door will keep mice out, yet allow good ventilation and easy passage for the bees.

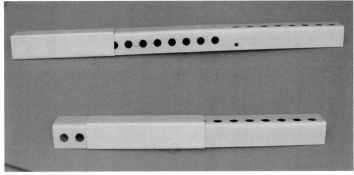

Mice cannot chew through an expandable, fits-all-hives guard such as this one. It will fit large or small hives.

and removal of debris that might attract rodent activities. Equipment storage building doors and other openings should be constructed to minimize rodent entry. Openings larger than 1/4 inch (.65 cm) should be closed with proper materials. Rodent traps have been used successfully in honey houses and storage buildings.

Commercial rodent poison baits are available, but should be used with extreme caution to prevent harm to children and non-target animals. Rodent baits are not very effective outside because of the degradation of the product from weather. - MH

MICROENCAPSULATION OF PESTICIDES – In 1972, the Pennwalt Corporation field-tested the first microencapsulated insecticide, Penncap-M®. It was widely marketed in 1974 and immediately caused problems for beekeepers. The insecticide was methyl parathion; it was contained (encapsulated) in a nylon capsule. Users of the pesticide find that microencapsulation has two advantages: The pesticide is safer for applicators to handle since it is contained, and since it is released slowly through the nylon capsule it gives control of the target insects over a longer period of time.

The chief problem with Penncap-M® for beekeepers is that the capsules are about the same size as pollen grains. When the insecticide is sprayed onto flowers the capsules may contaminate the pollen that is subsequently gathered by pollen-collecting bees. While the Penncap-M® is in storage it is kept in barrels in water that prevents the insecticide from leaking from the capsules. When honey bees collect pollen, they moisten it with nectar, pack it into balls and place it into cells in the comb. Thus, when Penncap-M® is collected with pollen it is much like putting it back into a barrel of water. The insecticide does not leak from the capsule and when the pollen is consumed by bees, even months later, the bees can be killed.

In the early 70s it still was not clearly understood that flowers contaminated with pesticides were killing foraging bees. An airplane or ground air sprayer that deposits an insecticide over the tops of colonies does the bees inside little harm.

Unfortunately, the question of the use of Penncap-M® became emotional with the manufacturers failing to recognize the circumstances under which their product, which was otherwise a boon to agriculture, was causing difficulty; thus, there was little initial effort on their part to cooperate with beekeepers. Several restrictions by governmental agencies have brought about better labeling of pesticides and more protection for beekeepers. As a result, honey bee-pesticide losses are slowly decreasing. (see PESTICIDES, EFFECTS ON HONEY BEES).

MIGRATORY BEEKEEPING – It is difficult to determine when the first migratory beekeeping was practiced, and by whom, but certainly moving bees long distances is an ancient art.

In North America today it is not uncommon to transport honey bee colonies a thousand or more miles at a time. By taking a few precautions this can be done without any damage to the hives, hive furniture, bees, brood or anything else the hive may contain. The greatest danger in migratory beekeeping is overheating, that can result in suffocation of the bees and melting down of the combs. Refrigerated trucks are used by some beekeepers. Honey bees that are moved to a new location reorient rapidly. In a study done in a pear orchard in Ontario some years ago a pollen-laden bee returned to her hive 13 minutes after the colony was placed in the new location and the entrance opened. During the 14th minute six pollen-laden bees returned from their foraging trips and thereafter the number of successful foragers grew dramatically. (Free, J.B. and M.V. Smith. "The foraging behavior of honeybees moved into a pear orchard in full flower." *Bee World* 42: 11-12. 1961.)

When migratory beekeeping started on a large scale in the early 1930s in the U.S. the colonies were treated as individual units. The entrances were closed, usually with a screen, and each colony had a top screen that provided clustering space and adequate ventilation. One beekeeper used a bottom screen but placed the colonies on end while they were on the truck. This beekeeper thought this provided better ventilation for the colonies since there was no closed space where heat could accumulate. A small number of

Pallets are generally custom-made. Some provide space for the bottom board; some are the bottom board, some are screened; some have ridges so the colony is raised; some sit flat. Most hold four 10-frame colonies; some hold six; others are made for six eight-frames. Many have openings both ways for forklifts; others, like these, only one way. Pallets are, by design, useful but generally most useful for the one who builds and uses them. Custom and general design beehive pallets are commercially available, also.

Most long distance transportation of colonies for pollination or other reasons is done by leasing companies or trucking companies who specialize in moving bees. Trucks have air-ride suspensions so colonies are not jostled very much. Drivers are often skilled in loading and unloading, in applying and removing nets and straps, and in negotiating tricky orchard lanes at night.

beekeepers used 24-inch-long (60 cm) bottom boards; each projected beyond the hive entrance four inches so it could accommodate a porch screen to give the bees additional clustering space. Beekeepers have used both ordinary metal and plastic window screening for moving screens.

Keeping colonies well-spaced on a truck is important and aids in keeping them cool. Staples or straps are used to hold the colonies together, although putting the staples in place and removing them at the end of the journey are time-consuming steps. One or two men would place each colony on the truck. In the early 1950s, several companies marketed truck-mounted booms to lift an individual colony on and off a truck, thus making it possible for one man to work alone when moving a truckload of bees.

Another way to move colonies is to use a pallet that holds four colonies, either with two facing one direction and two the opposite direction or with each facing a different direction. U-shaped pieces of iron (sometimes called pallet clips) that project upward about one and a half or two inches (four to five cm) are nailed onto the top of the pallet. The irons fit below two adjacent hive bodies at the same time so that they cannot move. Four of these U-shaped pieces, in a straight line in the middle of the pallet, will hold four colonies in place. It is only necessary to put the supers into place and the time-consuming task of stapling colonies is eliminated. The colonies are taken on and off the truck with forklifts. If the colonies are in two brood chambers the top one is not stapled but the whole load is strapped in such a way that the straps and the propolis that is between the supers hold the colonies together.

The greatest change in modern migratory operations is that the colony entrances are left open so that bees may move in and out of the hives at will. The whole load is covered with a stout nylon net that is securely roped into place. The netting is sufficiently fine to contain the bees. Today, as in the past, migratory beekeepers (except for those using refrigerated trucks) make very few stops and always short ones during daylight hours. A constant flow of air over the colonies will keep them cool. In warm weather the truckers may stop periodically and wet the load with water. As the truck moves and this water evaporates, the load is cooled.

The ability of beekeepers to move large numbers of colonies quickly and easily has made it possible for growers of crops such as almonds, apples, blueberries, alfalfa seed and others that demand pollination to concentrate their plantings in large fields with the knowledge that they can obtain an adequate number of bees for pollination. Unfortunately, the movement of a large number of colonies of bees around the country has meant that tracheal mites, *Varroa* mites, and the small hive beetles have been spread rapidly, too.

As might be expected, a number of accidents have occurred while bees are being trucked. Law enforcement agencies and others are not accustomed to coping with honey bees and this often causes problems. During the active migratory season beekeepers that move their bees usually travel the same route in the event of a truck breakdown or other problems. In this way they may be in a position to help one another. Some beekeepers' associations have organized to help cope with emergencies in their area and have let police agencies know they are available. In warm weather it may be necessary to offload a truckload of bees when a truck breaks down to prevent the loss of colonies through overheating.

Refrigerated trucks for moving bees –Some beekeepers use refrigerated trucks to move bees. One beekeeper who has done so is William Perry of Arcadia, Florida, and Dallas, Pennsylvania. Each year, he takes truckloads of bees, with over 600 colonies per load, north in the spring and the same number south in the fall. His 18-wheel tractor-trailer is a standard model, modified in only a few ways for hauling bees. The truck has two refrigeration units: one standard unit and another mounted under the trailer bed. Perry said that the second unit is necessary in very warm weather when the first unit is in the defrosting mode. The truck usually runs with one unit set on automatic defrost and the other set on manual defrost. A set of lights on the side of the trailer box, visible to the driver, indicates when the automatic unit is in the defrost mode. After the automatic unit has defrosted the manual

A sure sign of a well-secured load is when a trailer tips over and the colonies, and net, stay in place.

Moving bees by boat, following the bloom, was once far more common than it is now. In northern Florida boats still follow the tupelo bloom and it is still the best way to navigate the swampy regions abundant there.

unit is defrosted. It takes 20 to 30 minutes to defrost each unit. This is done while the truck is in motion without the driver leaving the cab.

Perry seeks to keep the trailer box at between 42 and 50°F (five to 10°C). Maintaining constant cooling is important. If the colonies in a load start to warm the bees become more active, produce more heat, and may also cluster outside the hives where they interfere with the airflow needed for cooling. Therefore, it becomes difficult to get the bees back to the cooler temperature.

A full load of bees produces a great deal of carbon dioxide, and if no provision for air exchange is made brood may die, either from lack of oxygen or from excess carbon dioxide. Ventilation must be limited, however, to maintain low temperatures. Perry solves this problem by providing an exhaust fan in the rear door of the trailer, and three one-inch (2.5 cm) diameter air inlets at the front, arranged with tubing so that the refrigeration unit fans pull air through them, providing for several air changes per hour. When the carbon dioxide level reaches about two to three percent the bees will start to fan their wings to change the air.

Perry moves only single-story colonies, with telescoping covers, both spring and autumn. Depending on the number of bees in the colonies, the hives are stacked four or five wide in the trailer. In the spring, the bees have just come from an orange or palmetto honey flow. Since there is little time between the honey flows in the south and those about to start in the north, the colonies will be overflowing with bees and many cluster outside of the entrance. Under these circumstances wider spacing is used to give more ventilation. The colonies are held apart and pallets are placed on the trailer bed, under the first layer of colonies, to provide good airflow beneath them.

Moving bees in a closed van has several advantages, one being that there are no loose bees. When a truckload of bees is covered, even with a very good net, it is very difficult to prevent a few bees from escaping. Refueling and weighing stations complain quite strongly when any bees escape in their areas. An advantage of a refrigerated truck is that the driver may stop more often to rest without worrying about the load overheating. Chilling the brood has not been a problem because if it becomes too cool the bees will cluster around the brood. Perry says he prefers to use a refrigerated truck because when he arrives at his destination he does not need to hurry to unload his truck. Also, he finds there is less queen loss and drifting of bees.

MILKWEED POLLINIA – One will sometimes see honey bees attached to and trapped by a part of a milkweed flower. Milkweeds (several species) are good nectar producers and attractive to bees. What holds the bee to the flower is a wedge or clip-like mechanism attached to two pollinia, which are capsules containing the pollen. In the normal course of events the pollinia detach and the bees carry them to the next milkweed flower where the pollen contacts the female parts of the flower and cross-pollination is accomplished. However, sometimes bees may become trapped in the flower and die. Sometimes the pollinia detach and become permanently attached to a bee's leg and rarely to the antennae or even the mouthparts. Honey bees that are severely hampered by pollinia are removed from the hive by house cleaners just as any sick or dying bee might be.

C.C. Miller

MILLER, CHARLES C. (1831–1920) – C.C. Miller was a physician who gave up private practice to become one of the most famous producers of comb honey in the country. He was a prolific writer whose final book, *Fifty Years Among the*

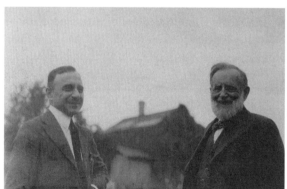

C.C. Miller, right, shortly before his death. George Demuth is on the left.

Bees, became a bible for those concerned with colony management. The final printing of this book contains eulogies by E.F. Phillips and E.R. Root and is considered a collector's item. Miller once wrote that his greatest curiosity was to understand the cause of swarming in honey bees, a question to which we still seek an answer today.

Miller invented no special devices according to his own testimony, but a sugar syrup feeder, which resembles a box without a top and is placed on top of a hive, is named after him. Miller's greatest ability was to select the good management ideas from the bad. For many years he wrote a column for *Gleanings in Bee Culture* entitled *Stray Straws*.

MILLER, NEPHI (1873-1940), **MILLER HONEY COMPANY AND MILLER'S HONEY FARM** – Nephi Miller started keeping bees in 1894 in Utah. In December, 1907, during a visit to California to study techniques of rendering beeswax he noticed bees visiting orange blossoms, while his bees at home in the Wasatch mountains were buried in snow.

Inspired, he purchased 300 colonies in California that winter. He made splits and moved them back to Utah in the spring on the railroad. In the winter of 1909 he moved his entire holdings of 1200 colonies to Colton, California, to enjoy a second season and a second harvest. Migratory beekeeping had begun. Later, he moved bees to various locations to take advantage of seasonal honey crops.

By 1911 Miller had 3000 colonies in Idaho, California and Utah, and in 1912, the business was renamed N.E. Miller and Sons Honey Company.

Woodrow, the youngest of Nephi's

Nephi Miller

sons, started a beekeeping and honey packing business, known as Miller's Honey Company, still in operation in 2006. Woodrow was involved in the design and sale of several unique retail containers – the Hula-Girl squeeze bottle, individual honey serving packs, but most notably the ubiquitous Honey Bear. Woodrow's business included honey packing and marketing, beeswax refining and selling bee supplies.

Earl Miller, Nephi's second son, went into business with his father and was involved with much of its early growth. He purchased the business in 1931 and moved its headquarters to Blackfoot, Idaho. Earl was very involved in the community, serving on the school board and as mayor for 12 years.

Earl's son, Neil, entered the business after school and purchased it in the late 50s. Earl moved back to California and

Neil Miller

worked with Woodrow for a time, summering in Blackfoot, and wintering in milder Colton, California.

When Neil took over the business he had about 1900 colonies, which grew to over 7000 by the late 70s.

Changing agriculture hastened an expansion to Grackle, North Dakota, in 1970. In 1977 Neil's son, John, entered the business, and in the 80s son Jay came on board. Neil, John and Jay have all held positions on the National Honey Board, John serving two terms as Chair.

MITES ASSOCIATED WITH HONEY BEES – Mites are not insects but are closely related to them. Adult mites have four pairs of legs whereas all insects have three pairs. On average, most mites are small and some are microscopic. Although many insects are small, mites are usually much smaller.

Most mites that are found in beehives have little or no effect on honey bees (see Non-parasitic mites). They are there accidentally because they are either scavengers that feed on the debris on the bottom board, or they are predators for the scavenger mites. Still, a few mites are serious pests of honey bees. One species, *Varroa destructor* (formerly referred to as *V. jacobsoni*), is considered the most serious bee parasite of European honey bees. A second mite, *Acarapis woodi*, has caused great problems in European honey bees in North America. A third mite, *Tropilaelaps clareae*, found in Asia, is considered a more serious parasite of European honey bees than *Varroa* mites. Some groups of honey bees seem to have a natural resistance to some of the pestiferous mites. This resistance takes many forms. In some cases the bees have an ability to detect and remove the mites; in other

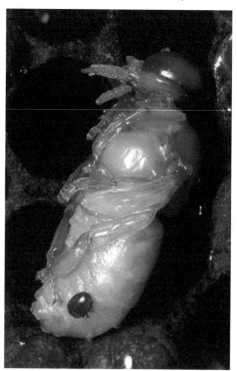

The most devastating honey bee mite, worldwide, is without doubt Varroa destructor. *(USDA photo)*

cases a difference in the life history of the bees, such as a shorter development time, is such that the mites cannot adapt.

The biology of most mites including those that are harmful to honey bees are little or poorly understood. Many mites are very small and a microscope is necessary for diagnosis. Many of them are yet to be named and described. When unknown insects or mites are found in a beehive, it is advisable to put them in a small vial of alcohol and send them to a state or federal laboratory for correct identification. Knowing what pest(s) are in in their colonies helps beekeepers with colony management.

Phoretic mites – If one searches in a beehive for other forms of life a number of species of mites will be found. At least 50 species have been found. None of these are in any way harmful to the bees or their stored food. They are phoretic; they use the honey bees as a method of transport. These mites live in flowers where they may feed on nectar, pollen, flower parts, or even each other. Since they may attach to the bees they frequently end up in a hive where they usually perish. It is not unlikely that if one is surveying a colony for harmful mites, whatever the technique used, these mites may appear. Many species of mites feed on dead animals and debris. These mites are often found on the bottom boards of colonies but they cause no problems.

Pollen mites – Not infrequently, especially in the spring, small piles of pollen dust can be found under stacks of stored hive bodies containing combs. The combs are infested with mites that eat the pollen remaining in the combs after they have been put into storage. These mites are common and do no harm. No effort needs to be made to remove or fumigate these mites. The combs may be put onto colonies and the bees will clean them prior to their being used again.

Mellitiphis alvearius – The mite, *Melittiphis alvearius* is not considered a pest of honey bees. It has been found in New Zealand, Canada, and the U.S. (see Delfinado-Baker, M. *Incidence of Melittiphis alvearius (Berlese), a little known mite of beehives, in the United States.* American Bee Journal 128:214. 1988.

MITES, PARASITIC – The role parasitic mites play in the transmission of bee diseases is still unclear. Some of the viruses pathogenic to honey bees are also found in *V. destructor*. It is not clear whether the viruses are being transmitted by the mite or if the role of the mite is to activate latent viruses.

Acarapis woodi – The popular name of *Acarapis woodi* is the honey bee tracheal mite because this mite infests the tracheae or breathing tubes of adult queens, workers and drones alike. These mites are believed to be the cause of colony losses that occurred on the Isle of Wight (United Kingdom) in 1919. Fear of these mites brought about the enactment of the Honey Bee Act of 1922 by the U.S. Congress. This legislation restricted the importation of new honey bee stocks into the country and served the industry well for many years. However, in July 1984 the first discovery of *A. woodi* was made in U.S. honey bees on the U.S.-Mexican border. Since tracheal mites cause little harm in other parts of the world where *A. woodi* had been established for many years some beekeepers and scientists believed *A. woodi* would not be a problem in North America. To the contrary, U.S. beekeepers lost large numbers of colonies because of *A. woodi*.

Taxonomic position – *A. woodi* belongs to the sub-class Acari, which contains mites and ticks, and belongs to the family Tarsonemidae. Tracheal mites are the only internal parasites of honey bees. These internal mites and the two external *Acarapis* species (see Other parasitic mites) are difficult to distinguish morphologically. The species are usually identified by the location of the mites. Tracheal mites appear to have evolved from one of the external *Acarapis* species probably between 1890 and 1900.

Distribution – The tracheal mites were first found in England in 1921. They probably evolved about 20 to 30 years earlier somewhere in Great Britain. The mites were soon found in other European countries. This suggests that the mites are capable of spreading rapidly. *A. woodi* is now found on all the continents except Australia. In addition, honey bees in Hawaii and New Zealand both are reported to be free of tracheal mites. The tracheal mites infest not only European and Africanized honey bees but also other honey bee species such as *A. cerana indica* in India and Pakistan.

Pathogenicity – Tracheal mites usually infest the large breathing tubes on either side of the first segment of the thorax (prothorax). They are occasionally found in the air sacs of the head, thorax and abdomen. The mites puncture the breathing tubes and feed on the hemolymph (blood) of the adult honey bees. These feeding wounds are instrumental in the occurrence of secondary infections, which may result in early death of infested honey bees. Shorter life span of over-wintering infested bees and an unusually high death rate among infested worker bees in March have been recorded (Bailey, L. and Ball, B. *Honey Bee Pathology*. Academic Press, London. 2nd edition. 103 pages. 1991). Thus, colony mortality increases at the time of the year when young bees are not produced in sufficient number to replace those bees dying early because of mites.

Bailey says that today only about two percent of colonies are lost each year in England and Wales, both of which, until the North American invasion, had the highest infestations in the world. In the northern U.S., beekeepers report that they find many dead and dying colonies, with plenty of stores but no bees in February, March and April. In southern Louisiana dead colonies with high levels of tracheal mite infestation were also found in April.

The damage caused by tracheal mites on *A. cerana indica* is similar to that on *A. mellifera*. In India, high levels of tracheal mites also caused the death of many infested bees and ultimately the loss of colonies.

Detection – The chief problem in dealing with tracheal mites for both beekeepers and researchers is that there is no quick and easy method of detecting their presence in a colony. No outward signs enable someone to make a reliable diagnosis in the field. Many books and articles report the presence of crawling bees with disjointed wings, called K-wings, in front of infested colonies. However, such bees are not always present and bees dying from other

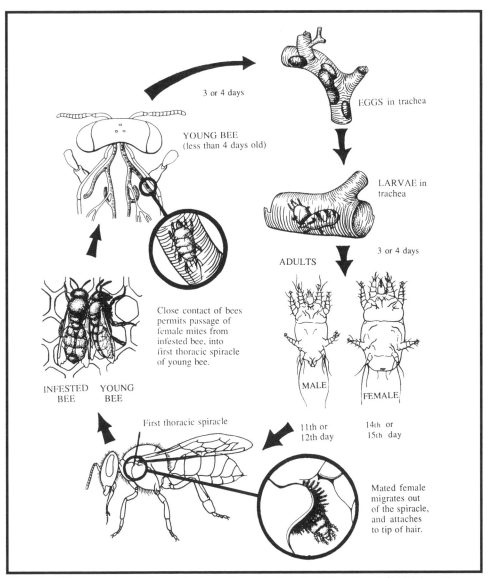

Life cycle of the honey bee tracheal mite, Acarapis woodi. (from B. Alexander and C. Henderson)

causes may also display the same symptoms.

A reliable method of determining the presence of mites is to collect the suspect bees and place them in 70 percent alcohol. In the laboratory, the wings and legs and other body parts are removed and the thoraces are placed in five to 10% sodium hydroxide for 24 hours. This dissolves away the tissue in the thorax and when the prothoracic collar is removed the tracheae are clearly visible. These tracheae may be examined under a microscope to determine whether mites are present.

Another method of diagnosing tracheal mites involves examining a live bee, so the diagnosis can be almost immediate.

Mites, Parasitic

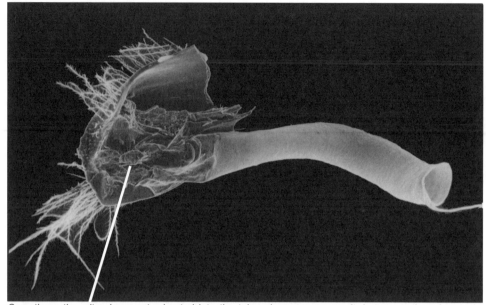

Sometimes the mites have not migrated into the tube when you remove it, but are gathered near the opening.

To do this test, place live bees in a refrigerator to cool them down. After an hour or so, remove a bee and pin her on her side using two or three pins. Remove the abdomen or the sting to eliminate the chance of being stung and, looking through a dissecting microscope (50X), take a fine-pointed forceps and remove the flap that covers the first tracheal opening. Lift the flap up and back to expose the tracheal tube. A spotted or dark-stained tube indicates presence of mites. A creamy white tube probably means absence of mites. Sometimes low infestation of mites does not cause any blotches on the tracheae. To be certain, open tracheae with very small needles. Dissecting trachea is a tedious and cumbersome procedure.

Serological methods, for example an enzyme-linked immunosorbent assay (ELISA), have been developed for the detection of tracheal mites. (Grant, G.A., D.L. Nelson. P.E. Olsen, and W.A. Rice, "The 'ELISA' detection of tracheal mites in whole honey bee samples." *American Bee Journal* 133:652-655. 1993.).

Life cycle – The entire life cycle of tracheal mites is completed within the trachea. Mated female mites move into a trachea usually within 24 hours after a worker bee emerges from her cell. Workers older than about a week cannot be infested. A female mite lays five to seven eggs over a period of several days. The eggs hatch after three to four days. Adult male mites are seen about the 12th day and mature females a few days later. After mating, the females move out of the trachea and attach themselves to the tip of a body hair. In this position they can grasp onto and infest another young passing bee and repeat the life cycle.

The population dynamics of tracheal mites, that is, how populations change during the course of the year, are not clear. As indicated above, mite infestations may cause the death of colonies in the spring. In surviving

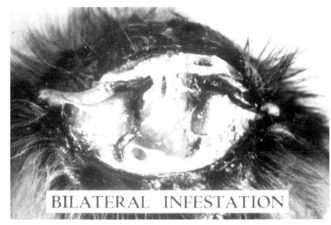

BILATERAL INFESTATION

A field test for tracheal mites. Collect a worker bee from the edge of the brood nest – not too young, but not yet a forager. Place in a jar, then in a freezer until she has expired, but is not rigid. Remove from freezer, hold her with thumb and forefinger on top and bottom of her thorax. With a sharp forceps remove the head by placing the 'neck' between forcep points, then lifting. Discard head. Using forceps, remove the thoracic collar. The inside of the thorax is now visible. On either side you will see the main thoracic tracheae. Pure white probably means no infestation. Black or spotted like this indicates a high infestation.

colonies the death rate of the older bees is also high but as bees begin to raise more replacement bees in the spring the mite population appears to decrease dramatically. During this period, there are increased numbers of new, uninfested hosts since the mites cannot reproduce as the same rate as the rapidly growing bee population. For this reason, the mite population is highest in the autumn when colonies have a higher proportion of older bees and the rate of young bees emerging declines.

A female mite questing on a bee's hair, waiting for a young bee to come close enough that she can transfer from an old bee to a young bee, move into a new trachea and begin another generation of mites. (Webster photo)

Dispersal – Within a hive the spread of mites occurs because of the close proximity of one bee to another. Within an apiary mites are spread from one colony to another by drifting bees, bees that mistake another colony for their own. In the U.S., more than one million colonies are rented for pollination. Bees are often moved long distances for pollination services. For example, the bulk of the over 20,000 colonies that are used for blueberry pollination in Maine are trucked in from Florida. Bees are brought into California from many states for the pollination of almonds. Queen and package bee producers in the southern states routinely ship bees to all states except Hawaii. Finally, colonies are moved for honey production, which may or may not be done in conjunction with moves made for pollination. With such a massive movement of bees around the country it is clear that any newly-introduced pest will soon become widely distributed in the U.S.

Control – Chemical control – Menthol, formic acid and Apilife Var® are the only chemicals registered by the U.S. Environmental Protection Agency (EPA) for the control of tracheal mites in

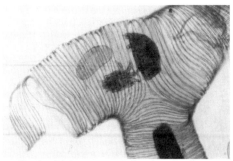
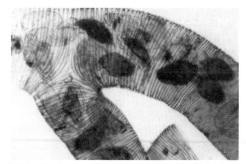

Tracheal tube removed showing egg, larva, and adult tracheal mite. 125X (USDA, Wilson photos)

the U.S. Other chemicals are approved for use in Europe but since the problem is much less severe there, these materials may not be as effective here in the U.S. Other chemicals are under investigations in North America.

It has been shown that placing a solid vegetable oil and sugar patty in a colony can disrupt the questing behavior of young, mated female mites. To make the patty, mix one pound of vegetable shortening with three pounds of table sugar. Blend together the ingredients, and make a patty roughly the size of your open hand. Place the patty on the top of the broodnest. Repeat as needed. Essentially, what happens is that when young bees are exposed to the shortening it masks their odor and a female mite does not recognize them as young workers. The female mite then will continue questing until a young,

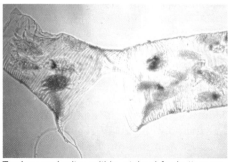

Trachae and mites within, stained for better contrast. (USDA photo)

ungreasy bee comes along, or, failing that, will grab onto an older bee who dies before the mite's life cycle is complete, or failing that, will wait until she starves, or desiccates. In any case, the mite population growth is diminished, and the colony has fewer mites to contend with.

<u>Genetic control</u> – Some races or stocks of honey bees have natural resistance to some of the pestiferous mites. Resistant honey bee stocks are effective in reducing the number of bees being infested with tracheal mites. Several candidate bee stocks have been imported from Great Britain in 1990, Yugoslavia in 1989 and far-eastern Russia in 1997 in hopes of increasing the mite resistant traits in the U.S. bee stocks. There were two importations of the Buckfast bees from southern England. The first Buckfast importation was via Canada and the second was a direct importation in 1990. Various scientific studies as well as reports by several beekeepers showed that Buckfast and Russian honey bees show some resistance to tracheal mites. Colonies of these stocks consistently had low levels of tracheal mite infestations so that little or no chemical treatment is required. Clearly, the use of these resistant stocks is the best approach to controlling tracheal mites in honey bee colonies. All of these

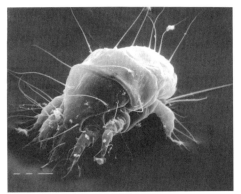

Close-up of tracheal mite adult. Long bar equals 100 microns. (Styer photo)

stocks are now commercially available. Through a serious selection process mite-resistant bees can be developed to reduce or eliminate the need to use chemicals for mite control. HS & LDG

Varroa destructor (formerly referred to as *V. jacobsoni*) is an external parasite of the honey bee (*Apis mellifera*) that feeds on both pupal and adult bees. They were found in the U.S. for the first time in Wisconsin in September 1987. The colonies in which they were found belonged to a migratory beekeeper.

A 'Grease Patty' for controlling tracheal mites is made of one pound of solid vegetable shortening, three pounds of sugar, and a hint of food grade essential oil to provide odor and make the patty attractive. The patty is placed on the top bars of the top super when honey supers are not on the colony. Replace as needed. If the colony is not consuming the patty, replace with one that has a bit more honey added. Some colonies will not consume patties, no matter what.

Given the experience regulatory agencies had with tracheal mites, this finding should have been a clear warning that the mites were already widespread at the time they were first found. It is now commonly accepted that the *Varroa* mites had been in the U.S. for several years before their detection. As discussed below, finding *Varroa* mites is a difficult, time-consuming process.

Taxonomic position – *Varroa* mites also belong to the sub-class Acari, which contain the mites and ticks, and belong to the family Varroidae. There are several species of *Varroa* mites (see MITES ASSOCIATED WITH HONEY BEES, Other parasitic mites). The first identified and described species is *V. jacobsoni*, which infests *A. cerana* in Indonesia. While this mite was discovered and named in 1904, it was not until recently when it was learned that this is not the type of *Varroa* that is causing a lot of devastation to European honey bees worldwide. The *Varroa* mite that is causing all the problems to *A. mellifera* colonies is now named as *V. destructor*. *Varroa* is a relatively large mite (1.05 to 1.09 mm long by 1.5 to 1.58 mm wide), oval with a brown, hardened cuticle. It is about the size of a common pinhead and has a crab-like shape.

Distribution – *V. jacobsoni* was originally found parasitizing the Indian honey bee, *A. cerana*, and reported again in Sumatra in 1912 on the same bee species. It was not until 39 years later that the mite was again mentioned. Data on its distribution and spread after this point are confusing, as numerous reports on the presence of the mite were made, but not necessarily coinciding with the year of its introduction into new locations.

The spread of *Varroa* mites was accelerated by the importation of *A. mellifera* into Asia. The first association of *Varroa* mites and *A. mellifera* probably happened in Japan. *A. mellifera* has been in Japan since 1877 but infestation was not observed until 1957. Subsequently, *Varroa* mites were reported infesting *A. mellifera* in Hong Kong (1962) and the Philippines (1963). Once *A. cerana* and *A. mellifera* inhabited the same region, the range of the mites quickly expanded. Scattered information from 1949 through 1978 indicate its Asian presence in Singapore, the USSR, Japan, Peoples Republic of China, India, Philippines, Hong Kong, Malaya, Vietnam, Korea, Cambodia, Thailand, Taiwan, Iran and Pakistan. It has been suggested that the practice of strengthening *A. mellifera* colonies by supplying them with brood from *A. cerana* was the main cause of infestation of *A. mellifera* colonies. Beekeepers not only moved *A. mellifera* colonies into Southeast Asia, but also moved them back to Europe again. The transport of colonies west is believed to have initiated the spread of the mite to other parts of the world.

The first appearance of the mite in Europe was in Russia in about 1949 followed by Bulgaria in the early 1960s. The natural movement of the mites has been estimated at about six miles (10 km) per year in Europe but migratory beekeepers have speeded up this movement. Today, the mites are found in all European and Mediterranean countries. South America was the next continent to be infested by the mite. In 1971, *Varroa* was accidentally imported into Paraguay on infested bees from Japan, and colonies were then moved to Brazil in 1972 infesting the state of São Paulo. This pest has since spread to Argentina, Uruguay, Bolivia and Peru. There have been no reports of the mite from Guyana, Suriname or French Guiana. In Africa the mite first infested Tunisia in 1975 and was found in Libya in 1976. In 1997, *Varroa* was detected in South Africa. *Varroa* has also been found in Mexico, Canada and New Zealand. Australia is the only continent free of *Varroa* mites.

A beekeeper in Wisconsin found *Varroa* mites in his colonies in September 1987. The mites were soon found in other states and today honey bees infested with these mites have been reported in 49 states. Hawaii is the only state where *Varroa* has not been found.

Pathogenicity – The developing stages and adult females of *Varroa sp.* parasitize immature drone and worker bees within their cells by feeding on their hemolymph (bee blood). In addition, adult female mites suck hemolymph from adult workers and drones. Mites on adult bees can usually be seen on the thorax or between sclerites on the underside of the abdomen. Adult worker bees previously parasitized by only one mite suffer a 50 percent reduction in longevity. *Varroa* mites also cause early drone mortality. Pupae parasitized by five or more mites suffer weight loss and possible deformations.

The overall effects on colonies from *Varroa* mites have been variable in different countries, depending on the degree of infestation and species of bee hosts. In the first two to three years of infestation, no signs of the disease are evident, in spite of the developing mite population. The colony as a whole greatly suffers when infestation levels rise in subsequent years of infestation. For example, in Germany 10-fold increases in infestation occurred every year after the introduction of the mite. During the fourth year, it was common to see newly-emerged deformed bees in infested colonies. In the region of Frankfurt, about 2000 heavily-infested colonies died from July to October 1982. In Tunisia, where beekeeping is mainly done in log-type colonies, the beekeepers lost 90 percent of their colonies from 1978 to 1982. In contrast, Africanized honey bee colonies in Brazil, which have been infested since 1973, have seemingly not been harmed. In Brazil, beekeepers seldom treat for the mites. Since European honey bees are also resistant to *Varroa* in Brazil, the type of *Varroa* mites there are thought to be not too harmful to honey bees. DNA analyses confirmed that Brazilian *Varroa* are genetically different from those *Varroa* commonly found in the U.S., Europe and in some Asian countries.

Continued research in this area has determined that there are, indeed, different races of *Varroa* mites. Moreover, it has been found that there are differences in how these haplotypes affect individual bees and entire colonies. Discriminating these different mites is impossible without sophisticated DNA sampling equipment however.

Additionally, it has been shown that the presence of *Varroa* mites in a colony exacerbates the affects that some debilitating honey bee viruses have on individual honey bees and entire colonies. They have been shown to transmit viruses, stimulate viruses, and carry viruses. All of these activities are additionally detrimental to a colony, besides the fundamental damage mites do while feeding on both larvae and adults.

Detection – Various methods have been used to detect mites in colonies, but none is really considered an outstanding method. State inspectors in the U.S. typically use the ether roll method. This consists of collecting a sample of approximately 50 bees into a small jar. One then quickly sprays inside the jar with ether (car starter fluid) which comes in a spray can. This kills the bees and the mites release from their hosts and stick to the side of the jar. By holding the jar up to the light, the loose mites are easily seen. This is a very quick method, but may not detect very low mite infestations. Confectioners sugar can be substituted for ether. The advantage of the sugar is that it is not harmful to the bees and avoids exposing beekeepers to ether fumes. About two tablespoons of the confectioners sugar is added to the sample of bees in a jar with a five-mesh screen top. After being shaken vigorously the sugar, along with the mites, is dumped out of the jar, through the screen onto a white sheet of paper. The brownish *Varroa* are very visible. The bees, only a bit sticky, are returned to their hive.

Similar to this method is the alcohol-wash technique. Again, one takes a sample of bees (200 to 300) into a jar with 70 percent ethanol, and shakes them vigorously for one minute. Pour the bees with alcohol onto a pan with a screen mesh to filter out the bees but not the mites. The mites are easily seen on the pan. The advantage of this method is that it can be very sensitive. If the sample is infested, there is a high probability that the mites will be found. The disadvantage of the method is that it is slow, requires alcohol and again a sample of 300 bees may not be sufficient to detect low levels of infestation.

Other methods of detection do not require the killing of any bees, and may therefore be preferable. For example, the colony may be tested with an Apistan® treatment strip (see FLUVALINATE) or Coumaphos® if the mites in the colony are not resistant. The falling mites are caught on a sheet of white cardboard or paper placed on the bottom board of the hive. It is best to spray vegetable oil or make a rim of grease or petrolatum all around the edge of the cardboard/paper to avoid moribund mites from crawling away. The sheet should be removed within a day of the introduction to avoid excessive accumulation of debris that makes the search for mites harder. This is a fairly reliable method.

Life cycle – The female mite begins the reproductive cycle by leaving an adult bee and entering an uncapped cell that contains a five- to 5.5-day-old worker or drone larva. More than one adult female may enter the same cell. Once inside the cell, the female mite submerges herself in larval food (the remaining royal jelly) with her ventral (bottom) side oriented toward the entrance of the cell. After capping, the remaining food is consumed by the developing bee thus freeing the mite. The mite then begins to feed on the hemolymph of the developing worker or

Sticky Board Method For Detecting *Varroa* Mites

Sticky boards are placed in a colony such that mites that fall, either because they were knocked off an adult bee, died naturally, or fell, drop to the very bottom of the hive and are captured on a board coated with a sticky substance. The board is located so that bees in the colony cannot reach it to get stuck in the material, or to remove debris or mites that have fallen. A general location is below a screened, five-mesh bottom board, on a removable tray, as shown.

A common test is to leave the board in for three days then remove. The mites are counted, that number divided by three to arrive at an average natural mite fall per day. This number is important.

The number of mites detected will vary with the time of year and the population of the colony. In early spring, when the population is small with little brood, the colony may produce very few mites per day, perhaps only one or two. In late spring to midsummer when the adult population is at its largest, the mite population could be expected to be much higher. In autumn with decreasing bee population and brood, but continuing increase in mite populations due to reproduction in worker cells, the colony may have many mites.

The best IPM methods always call for finding the current pest population, and knowing the economic injury level. Thus, stay current on recommendations made for when to treat in *your* location, given the time of year, the size of the bee population, the race of bees in the colony, and the effectiveness of the control agent(s) used. These recommendations will specify when to treat, and with what chemical or compound. The important part of this is knowing your current mite population.

Removing a sticky board to count mites.

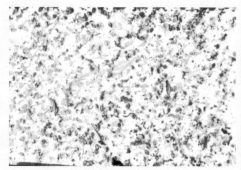

The surface of a sticky board.

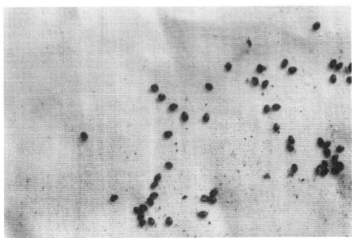

After agitating 200-400 bees in a detergent or alcohol solution, they are poured on a mesh screen and the liquid, containing any mites, passes through. This is further screened to catch and observe mites, which can be easily seen and counted. (Morse photos)

drone. A female mite lays her first egg about 60 hours after the cell is capped, and subsequent ones at 30-hour intervals. The first egg laid will produce a male, and the second and all subsequent eggs are generally females. An infested, capped cell therefore contains one or more adult reproductive females, and offspring in varying stages of development.

After the egg hatches, a mite goes through larval, protonymphal and deutonymphal stages before becoming an adult. Development time from egg to adult is 5.5 to 6.5 days for females and males, respectively. Adult mites mate in the cell in which they are produced. If only one female enters the cell, her female offspring are inseminated by the only male in the cell, their brother. Only if two or more females enter the cell, may out-breeding occur.

Once worker or drone development is completed (20 or 24 days, respectively), the young adult bees emerge. The adult female mites attach to the host bee and leave the cell together, while male mites and all immature mites that did not complete development remain in the cell and die. Close contact between bees

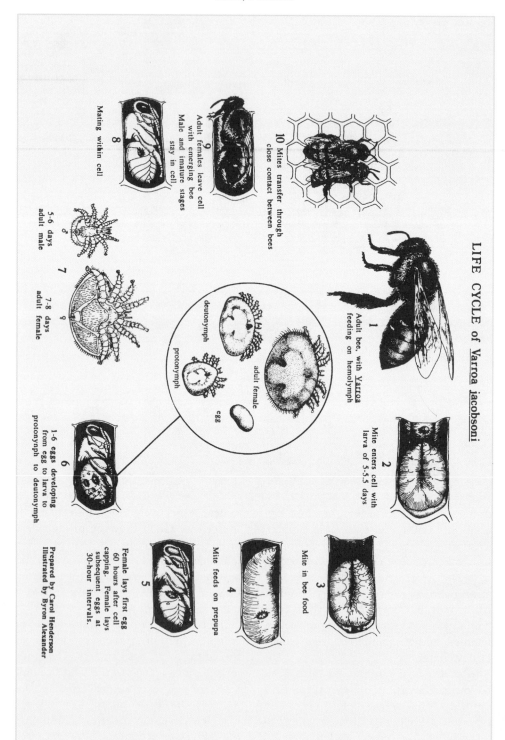

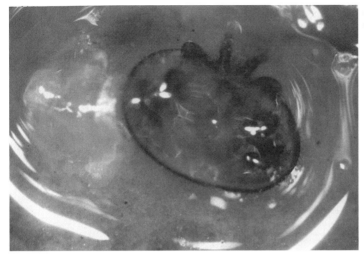

Female Varroa *mite floating in royal jelly.*

within their colony permits the mites to transfer readily from one bee to another, easily infesting new hosts.

Not all mite offspring have time to complete their life cycle and reproduce within the cell. As worker and drone development time is 20 and 24 days, respectively, and female mite development time is 5.5 days, the mother mite must lay her eggs early enough to permit the complete development of her offspring before the emergence of the host honey bee. Any egg laid after day 14 in a worker cell, or after day 16 in a drone cell, will not reach maturity and will die. Mites produce on the average 1.8 adult females in a worker cell, and 2.7 adult females in a drone cell. These relatively low numbers, combined with the fact that only 22 percent of female mites enter and lay a second cell suggests that *Varroa* has a low reproductive rate. This rate seems to be even lower in Africanized honey bees that have a shorter postcapping period.

Varroa destructor cannot reproduce on *A. cerana*. Perhaps the association between *Varroa* and *A. mellifera* is a case of a "new" parasite relationship, where behavioral adaptations on the part of the parasite have not yet occurred.

A female Varroa *mite on a developing pupa.* (USDA photo)

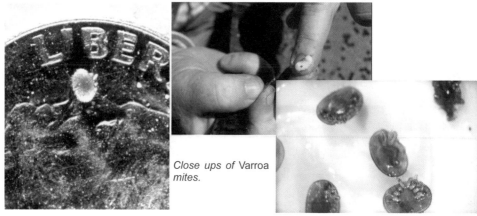

Close ups of Varroa *mites.*

(Morse photo)

Dispersal – Mites disperse through honey bee populations in several ways. Swarming bees can carry mites to new regions, and the drifting of bees (especially drones) can spread mites from colony to colony within a small area. Swarming bees may carry mites to their new nest sites, which are sometimes great distances from the parent hive.

Beekeeping by man dramatically increases the dispersal of the mites. First, colonies are maintained close together, which facilitates drifting. Second, hive manipulations such as brood exchange or combining weak hives can spread mites between colonies. Finally, migratory beekeeping and shipping bees through the mail from one state to another greatly accelerates the long-range spread of mites.

Climates and races of bees in Europe and North America are thought to be sufficiently similar for the disease to take a similar course. The counter argument involves two hypotheses. The first hypothesis refers to the density of colonies in the U.S. versus Europe. West Germany, for example, has considerably more colonies in very close proximity to each other. The number of hobbyists throughout the countryside is large, and dispersal of disease is therefore facilitated. The second hypothesis refers to chemical control methods (see TREATMENT). The U.S. has had the advantage of inheriting the *Varroa* problem at a time when results from efficacy tests of various miticides were already available from Europe. By efficient use of these pesticides, beekeepers may be able to better control their problem.

Control – Chemical control – In Western Europe, where the mite has been present for decades, nearly 10 chemicals have been approved and are

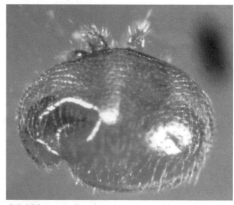

Adult Varroa *mite showing signs of being crushed.*

being used by beekeepers in one or more countries. In the U.S. the only chemicals that have been registered to date are Apistan® (see FLUVALINATE), formic acid, CheckMite+® (see COUMAPHOS), Apilife Var® and ApiGuard®. Coumaphos is registered under an EPA Section 18 emergency use provision. Beekeepers should check with the state apiculturist before using CheckMite+®.

Genetic control – In 1989, the USDA imported queen honey bees from Yugoslavia that were said to be resistant to *Varroa* mites. The Yugoslavian bees, referred to as ARS-Y-C-1, were found to be somewhat tolerant to *Varroa* mites. The stock was released to the industry for its tracheal mite resistance in 1993, but has mostly disappeared.

Certainly if resistance to these mites exists in European bees it is logical to search Russia and the iron curtain countries for resistant bees as the mites have been in Russia for about 40 or more years. Queen honey bees from Russia were imported into the U.S. in 1997. Different lines of Russian honey bees and a recently-developed USDA hygienic line of honey bees have now been released to the industry because of their resistance to *Varroa* mites.

Apis cerana, native host of *V. jacobsoni*, is very resistant to this parasite. This bee species rids itself of this pest through grooming, that is, the physical removal of mites from their bodies. The mites that are found during grooming are crushed with the bee's mandibles and carried outside of the hive where they die. It has been stated that some European bees will do the same but there are no data to indicate which race might be most active. As indicated elsewhere *Varroa* mites can be found in colonies of Africanized bees in Brazil. However, *Varroa* populations never become very great in Brazil and beekeepers in that country seldom treat for them. This low population of *Varroa* in Brazil may be due to the occurrence of different biotypes of *Varroa*. *Varroa* in Brazil are known to be genetically different from the *Varroa* most commonly found in the U.S. and Europe. Further, the *Varroa* mites found on *A. cerana* are a different species than those mites infesting *A. mellifera*.

Until resistant bees with good dispositions and honey-gathering ability are found, it is important that beekeepers pay close attention to selecting good apiary sites and keeping colonies headed with young, vigorous queens. For control beekeepers should follow the bee journals closely as changes in methods and materials for control are made. Beekeepers should also be always watching for colonies that appear to be resistant to or tolerant of *Varroa* mites and other disease-causing organisms.

There are some mechanical methods beekeepers can use to reduce *Varroa* populations in a colony, or to keep them from increasing.

It is well known that *Varroa* females prefer to enter drone cells over worker cells, given a choice. By placing a frame of drone comb in a colony, near the edge of the brood nest, you can focus female mites to enter those cells. Then, when the frame is almost completely capped, the frame can be removed, the larvae and accompanying mites destroyed, and the frame replaced for another generation of mites. This is an effective, non-chemical means of removing many mites.

Making summer splits, with very little brood, open or sealed, and a new queen, reduces significantly the *Varroa* population, and does not provide brood cells to reproduce in for an extended time. This reduces, and does not allow *Varroa* population buildup before fall, and allows the colony to remain relatively *Varroa*-free through the following spring into the summer.

Providing a screened bottom board from early spring to late fall allows any clumsy *Varroa* to fall outside the colony and be removed. This reduces the *Varroa* population a little; research has shown 10-25%, which helps, but more importantly the improved ventilation in the colony reduces stress significantly.

Keeping colonies in full sun, too, has been shown to reduce *Varroa*, and small hive beetle populations significantly.

Other parasitic mites – Two other species belonging to the genus *Acarapis* live externally on adult honey bees. *A. dorsalis* infests the dorsal groove of the thorax, and *A. externus* uses the neck for reproduction. These two external *Acarapis* are found everywhere European honey bees are found. Like tracheal mites, the external *Acarapis* feed on the hemolymph (blood) of the honey bees. However, no one has ever documented if their feeding activities cause any harm or loss of honey bee colonies.

Asian mites – Several mites parasitize different species of honey bees. All of these mites are Asian and at the time of this writing they have not been found outside of their native range. Some of these mites will attack European honey bees when given an opportunity. Very little is known about the biology of these mites. The life cycles of all these parasitic mites are similar to that of *V. jacobsoni*. Of these mites the most devastating for European honey bees is *Tropilaelaps clareae*. The natural host for this mite in Asia is the giant honey bee, *A. dorsata*. The damage and injuries caused by *Tropilaelaps* on honey bees are similar to that brought by *Varroa* mites. This mite has not been found outside of tropical Asia and is presumably restricted to the tropics for reasons that are unknown. Another species of *Tropilaelaps* that parasitize the giant honey bee is *T. koenigerum*.

There are two other *Varroa* species: *V. underwoodi* and *V. rindereri*. *V. underwoodi* has been found infesting different species of honey bees: *A. cerana, A. nuluensis, A. nigrocincta,* and *A. mellifera. Varroa rindereri* is a parasite of *A. koschevnikovi*. Little is known about the biology of the two species. Their

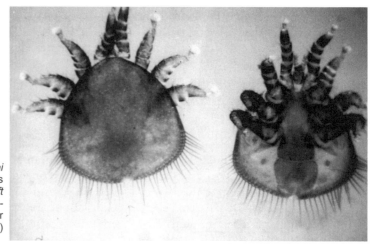

Euvarroa sinhai collected from **Apis florea** in Sri Lanka. Left - dorsal view, right - ventral. (Koeniger photo)

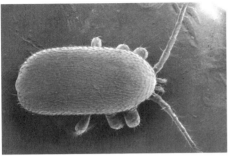 

T. clarae *female – top and side view.*

potential to become a serious threat to *A. mellifera* needs to be recognized. Several more *Varroa* species are likely to be identified.

Other closely related mites are the brood parasites of the dwarf honey bees. *Euvarroa sinhai* is a parasite of *A. florea* while *E. wongsirii* is specific to *A. andreniformis*.

When European honey bees have been taken into tropical Asian countries, they soon become infested with the Asian mites. However, a number of chemical controls have been developed that are effective. No one has yet looked for any European bees that are resistant to *Tropilaelaps clareae*. Our knowledge of Asian and other mites is changing rapidly and readers are advised to follow the bee journals closely as the distributions and methods of control change rapidly. For further information on the biology of mites see Morse, R.A. and K. Flottum. *Honey Bee Pests, Predators and Diseases.* A. I Root Company. Medina, Ohio 1997.

MOELLER, FLOYD E. (1919-1978) – Dr. Floyd E. Moeller, research entomologist at the USDA North Central States Bee Laboratory, University of Wisconsin, Madison, and University of Wisconsin Professor of Entomology, died July 26 1978. He was 59.

He was born July 26, 1919, in Milwaukee, Wisconsin, where as a youth living on the outskirts of town he acquired a strong interest in plants and wildlife and a lasting affection for the out-of-doors. During his high school years he began keeping bees, a practice he continued throughout his life. Upon graduating from high school, he received a university scholarship, and in 1941, he graduated with honors from the University of Wisconsin with a B.S. degree in Plant Science. Shortly after graduation, Dr. Moeller entered military service and after training served in the European Theatre as a navigator in the Army Air Corps. He retired from the Air Force Reserve in 1968 as a Lt. Colonel. Following active military service, Dr. Moeller re-entered the University of Wisconsin and worked as a summer employee of the USDA Bee Culture Laboratory, Madison, Wisconsin, from 1946-1948. On March 29, 1947, Dr. Moeller married the former Eleanor Johnson of Saginaw, Michigan. In 1949, he became a permanent member of the research staff at the Bee Culture Laboratory, a position he held until his death. Dr. Moeller received his Ph.D in Entomology from the University of Wisconsin in 1952. The subject of his dissertation was, "The effect of stock lines upon the honey bee population-production relationship." In 1966, Dr. Moeller was named Research Leader of the Madison Laboratory. He stepped down from this position in April 1978 in

Floyd E. Moeller

order to devote all of his time to teaching and research.

Dr. Moeller's research sought development of improved beekeeping management methods for increased honey production and pollination. He studied disease control, breeding and selection of stocks, colony behavior, application of repellents for honey removal, and protection of honey bee colonies from pesticides. He was author of many articles on bee management and diseases and was a coholder of a patent for bee repellents. Dr. Moeller was a recognized authority on the management of honey bee colonies for honey production.

In 1962 Dr. Moeller was cited for his role in the development of safe bee repellents for use in removal of honey from the colony.

In 1967 he travelled extensively throughout the Soviet Union under a Scientific Exchange Program. He was a member of Gamma Sigma Delta and Sigma XI agricultural fraternities, the Entomological Society of America, the International Bee Research Association, and the Wisconsin Beekeepers' Association.

MOFFETT, JOSEPH O. (1926-) – Joseph O. Moffett was born near Peabody, Kansas, in 1926. After Army service in Europe he completed his B.S. with honors from Kansas State University in 1949 and his M.S., also from Kansas State, in 1950. In 1974 he received a Ph.D from the University of Wyoming.

In 1949 he established the research, teaching, and extension program in Apiculture at Colorado State University. While there he helped develop a cure for European foulbrood and edited *B-Notes*. Upon leaving the University in 1959, Moffett became secretary-treasurer of the American Beekeeping Federation for five years. During this time he edited the Federation newsletter and for three years wrote a monthly column on Federation activities for the *American Bee Journal*.

Starting in 1967 Dr. Moffett was a research entomologist with the United States Department of Agriculture. He and his fellow scientists worked on nosema, American foulbrood, the pollination of citrus and hybrid cotton, bee flight in greenhouses, and the effect of herbicides and insecticides on honey bees. In 1978, after 10 years at the Tucson Bee Culture Laboratory, Moffett transferred to Oklahoma State University to study both the effect of injurious insects and pollinators on alfalfa grown for seed. He retired from that position.

He published more than 90 scientific and popular articles. His book, *Some Beekeepers and Associates*, is the basis for many of the business and personalities presented in this edition of *ABC & XYZ*.

Joseph O. Moffett

MOLD ON COMBS – In the spring, combs in weak or dead colonies will frequently be found covered with mold. The mold is there because the bees cannot control the humidity. The different colors of the molds suggest that several species may be involved. These same combs may contain dead bees or dead brood upon which the mold is often growing. Death may have occurred because of disease, particularly American foulbrood, and combs should be checked for this possibility. Molds will, however, disguise the appearance of diseased brood.

Mold-covered combs do not need to be discarded despite their bad appearance and odor. In fact, the mold *Penicillium waksmanii* has been used as a comb cleaner for combs with scales of American foulbrood. The mold has no adverse effect on the beeswax. It is not necessary to remove the mold from the combs as the bees can clean them, but only when colony populations are strong. When moldy combs are found, including those with dead bees and brood, they should be taken to a warm, dry place where they will dry and mold growth will be stopped. Combs that are stuck together by mold or dead bees should be separated to facilitate drying. When colony populations are growing, these combs, in their original hive bodies, may be placed on top of colonies and the bees will clean and polish them. It is truly remarkable how bees will do this in a few days under the right conditions. The combs may then be reused for brood and/or honey production.

MORSE, ROGER ALFRED (1927-2000) – Roger Alfred Morse was born July 5, 1927, in Saugerties, New York, to Margery and Grant Morse. Grant had grown up in the heart of the Catskill Mountains and as a child worked the fields, harvested hay, hoed potatoes and helped with the bees. Grant Morse attended a one-room school and worked on the school's magazines and newspapers. He earned his M.A. from Columbia University and his Ph.D from New York University. For over 40 years Grant Morse served as an administrator in the public schools of New York State, 37 of them as a school superintendent. He was a prolific writer, making contributions to this and other magazines, and published two books of verse on Catskill Mountain philosophy and wisdom.

Roger's father remained a hobby beekeeper for years, and gave Roger, when he was about 10, a hive of his own. This was more bribe than gift as the elder Morse's intent was to instill an interest in the hobby, and then enlist his son's help with the work the hobby required.

By the time he was in his teens Roger was operating nearly 200 colonies spread around the Hudson Valley and Catskill Mountain areas of New York.

He entered the Army in 1944 and served in Europe until early 1947. Then he enrolled in Cornell and received his B.S. degree in 1950. He continued at Cornell and in 1951 was married to Mary Lou Smith. He received his M.S. degree in 1953 and, still at Cornell, received his Ph.D in 1955 studying mead making under the tutelage of Dr. E.J. Dyce.

After graduation Roger and family moved to the Gainesville, Florida, area where he took up the position of Apiculturist with the State Plant Board. While there he worked in the fruit fly spray program, produced the booklet on Florida Beekeeping and wrote another dozen or so articles on various aspects of beekeeping in that state.

A result of his time there was that Roger and his family spent part of almost every winter at the Archbold Biological Station near Lake Placid, Florida. This allowed Roger to extend his research season and contributed much to his graduate students' education.

Roger spent two years in Florida, did a six-month stint in 1957 as Entomologist at the Waltham Field Station in Amherst, Massachusetts, and in 1957 he returned to Cornell.

Although Roger's first contribution to *Gleanings in Bee Culture* was made in 1953, he started his regular "Research Review" column in the April issue of 1958, shortly after moving back to Cornell.

"Dr. Morse's task will be to keep our readers up to date on the latest research developments in the world of beekeeping," wrote Field Editor and former classmate Walter Barth when introducing Roger to the readers. John Root was managing Editor and M.J. Dyell was Editor.

During the 40-plus years Roger's column ran in *Gleanings in Bee Culture* he also wrote many, many other articles covering almost every area of beekeeping. By far the most were fundamental, how-to management articles – wintering, feeding, harvesting and the like. But his travels and curiosity led him in some nontraditional directions too. Marketing, interviews, many on making mead, nearly a dozen book reviews, quite a few editorials, and the science of beekeeping were some areas explored. Other insects, beekeeping in New York and Florida, pollination, judging honey, Extension and running associations were also covered.

Much of the world's beekeeping was made available to *Gleanings* readers because, when Roger traveled, his readers traveled with him. Egypt, China, Nepal, Burma, Brazil, Moscow, Poland, Africa, The Philippines, Italy, England, Europe, Asia and Costa Rica were all discovered on these pages.

During his tenure Roger had nearly 40 graduate students that published in *Insects Sociaux, Bee World, Journal of Apicultural Research, Environmental Entomology, Hortscience, Scientific*

Roger Morse

American, Apiacta, Journal Of Insect Physiology, Natural History, Economic Entomology, Nature, Farm Research, Science, Florida Entomologist, Proceedings Of The Royal Entomology Society Of London and *The New York State Journal Of Medicine,* to name just some of the journals.

He spent time as a visiting professor at the University of Los Banôs in the Philippines in 1968, at the University of São Paulo in Brazil in 1978, and the University of Helsinki, in Finland. He attended many Apimondia meetings, and also served as President of the International Bee Research Foundation. He received an Honorary Doctorate from Academy Rolnicza in Poland for his work there.

Sponsoring much of these activities, and some of the graduate students, were grants from a variety of sources, including 13 NSF Research and travel grants, the NIH, EPA, USDA and even the U.S. Army. The UN's FAO sponsored 25 or so trips abroad to consult on bee diseases, conduct research and survey beekeeping conditions.

He taught, for his whole career, Cornell's introductory beekeeping course, which, like beekeeping waxed and waned in popularity. It reached, at its peak, a couple hundred students for the semester-long class.

Roger published many books during his career. *The Illustrated Encyclopedia of Beekeeping* with Ted Hooper, *Beeswax,* with W.C. Coggshall, *A Year In The Beeyard, Making Mead, Rearing Queen Honey Bees, Comb Honey Production, Bees and Beekeeping, The Complete Guide To Beekeeping, Judging Honey* with Mary Lou Morse and *Richard Archbold and The Archbold Biological Station.* As Editor, Roger was responsible for three editions of *Honey Bee Pests, Predators and Diseases,* and for the 40th Edition of *The ABC & XYZ of Beekeeping.* He had several book chapters to his credit, and even a patent in mead making.

MOULTING – This is a process whereby an animal sheds an outer covering of hair, skin, horn or feathers in order to grow a larger or a new coat. In honey bees it refers to the growth and changes that take place in larvae and pupae as they mature. Moulting is controlled by hormones, substances that are secreted internally in the bee and that control life processes. Moulting is sometimes spelled **molting** and often the term is used interchangeably with **ecdysis**.

MOVING BEES LONG DISTANCES – (see MIGRATORY BEEKEEPING)

MOVING BEES SHORT DISTANCES – Field bees are those bees that forage. They learn the location of their nest by making use of various landmarks such as trees, buildings, other hives, or any prominent features in the environment. Thus, moving a colony several to a hundred or more yards (meters) is often confusing to the flying bees. If a colony is moved within this distance the flying bees will go back to the hive's original location. If only one colony is in the vicinity it will soon be found. When many colonies are in one location bees may drift into foreign hives. Fighting may occur among the bees in the colonies under these circumstances.

There are two ways to move bees short distances. One is to move the colonies several miles from the original area and then to bring them back after two to three weeks. By this time any field bees alive at the beginning will have perished and no bees will remember the old location. A second choice is to move a colony, or colonies, two to four feet (one meter), every few days until the new location is reached. When several colonies are moved in this manner, it is helpful to have the hive bodies painted different colors. In this way the bees can learn the color of their own hive and use it in orientation.

The physical action of picking up and carrying a hive poses no problems for the bees inside or the social structure of the colony. Bees reorient quickly and have no difficulty learning a new location.

MRAZ, CHARLES (1905-1999) – Charles Mraz, died September 13, 1999.

Charles was born on July 16, 1905, in Woodside, New York City, son of Karl and Maris Mraz. Charles started with bees in 1914 while living in Queens, New York. He then worked for other beekeepers in upstate New York before moving to Middlebury, Vermont, in 1928 where he established Champlain Valley Apiaries. He became a world-renowned beekeeper and maintained New England's largest apiary for over 60 years. At one point he operated over 1000 colonies. He traveled to South and Central America, Europe, Asia and the Middle East as a consultant on beekeeping methods and technology. In 1992 the American Beekeeping Federation recognized Charles among the five most distinguished beekeepers in this country for his advances in commercial beekeeping. He was instrumental in the development of the fume board, at first using carbolic acid. He also developed a line of queens that were remarkably well adapted to his part of New England. He produced and sold thousands over the years. To spread his good ideas he wrote a column for

Charles Mraz

Gleanings In Bee Culture for many years entitled "Siftings."

Charles was also an avid gardener and dedicated to the practice of organic farming. He was a president and board member of the Natural Food and Farming Association, the precursor to the Northern Organic Farming Association.

Charles was recognized in this country as the pioneer of bee venom therapy, the use of bee stings to treat autoimmune diseases. He initiated clinical research with scientists at Sloan-Kettering Institute and Walter Reed Army Institute. He established the standard of purity for dried whole venom for the FDA and was the supplier of venom to pharmaceutical companies throughout the world. He was a founding member and Executive Director of the American Apitherapy Society. Charles earned the gratitude of thousands of people who traveled to his home for bee venom treatments or met him at apitherapy conferences around the country. In 1994 he authored *Health and the Honey Bee*, a recounting of his experience with bee venom therapy.

MUTH, FRED W. (?-1949) – Fred W. Muth of Cincinnati, Ohio passed away on May 7, 1949.

Charles F. Muth, Fred's father, started keeping bees on top of his house in 1858, was successful from the start, persuaded others to keep bees, and sold their honey through his grocery as well as furnishing the new beekeepers with their equipment.

The Fred W. Muth Co. succeeded the business of Charles F. Muth, both as buyers and packers of honey and in the distribution of bee supplies. Charles Muth developed and sold the notable Muth Honey Jar, made in several sizes. The name continued with Clifford F. Muth as president of the company, and associated with him a member of the fourth generation, Clifford, Jr.

Muth Honey Jars

Fred W. Muth

NACHBAUR, ANTON J. (1938-1999) – Andy was born June 27, 1938, in San Bruno, CA. As a young boy his family moved to the Monterey area where he attended school. In his early 20s he became interested in bees and after working for a commercial beekeeper he purchased hives of his own. He moved to Dos Palos and increased the size of his bee operation to 4,000 hives. In the mid 1970s he moved to Los Banos and resided here until his death. Andy retired from the bee business in 1992.

He was a member and past president of the California State Beekeepers Association and a recipient of the "Beekeeper of the Year Award." He was also a member of the American Beekeeping Federation, Delta Bee Club and Sioux Honey Association.

An outsopken member of the beekeeping community, he was involved in many political and educational activities. He was a pioneer on the internet, forming one of the very first discussion groups on beekeeping. His signature sign off – TTUL, the olddrone, became a classic.

NASONOV PHEROMONE – (see PHEROMONES, Scent gland pheromone)

NATIONAL HONEY BOARD – National Honey Board staff are appointed by the U.S. Secretary of Agriculture pursuant to the Honey Research, Promotion, and Consumer Information Order. The Order, approved by a referendum of honey producers and importers in 1986, established a program for generic honey research, advertising and promotion. Since February 1987, the National Honey Board's programs have been funded by an assessment of one cent per pound on honey sold (domestic, imported or exported).

The National Honey Board consists of 12 elected members including seven producers, two handlers, two importers, and one national honey marketing cooperative representative. Board members and their alternates represent the honey industry of the United States and Puerto Rico and serve without compensation for their time. Board members and alternates are nominated by the National Honey Board

Nominations Committee, which consists of not more than one member from each state. The Board's policies are supported and implemented by the National Honey Board's staff.

The Board is committed to serving the honey industry by increasing demand for honey and honey products. Marketing and promotional programs are directed to consumers, foodservice operators, distributors and manufacturers. The Board's scientific research program focuses on the composition, functional properties and health-related applications of honey to provide scientific support for new marketing messages and new uses of honey. The Board also funds practical, applied beekeeping research, primarily aimed at producing honey safely, and ultimately, in keeping honey bees healthy and producing honey.

Industry service programs are geared to improving communications within the industry and to provide methods and tools for industry members to capitalize on the Board's research and promotional programs. Information on the National Honey Board can be found at www.honey.com. - BB

NATURAL NEST OF THE HONEY BEE – European honey bees evolved using nests in hollow trees in natural forests. Probably a few bees lived in caves. European honey bees are almost never found nesting in the ground except in very dry areas. However, it is not unusual to find Africanized honey bees nesting in underground nests, at the base of trees, and in water meter boxes.

We know very little about honey bee evolution, including type of tree, size of the cavity or the number of trees that might have been hollow. It is possible, however, to imagine what conditions might have been like. For example, on land that has been abandoned for agricultural use over time many old, hollow trees are to be found in wood lots. These wooded areas may be quite similar to those where honey bees evolved in Europe.

In a study of such an area, 21 honey bee nests in trees around Ithaca, N.Y., were dissected and various aspects measured. All the nests studied had survived for at least one winter so they were all normal and reasonably healthy colonies. These studies were made before *Varroa* and tracheal mites were found

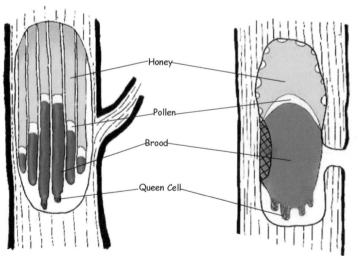

Two views of a typical hollow tree cavity showing the arrangement of honey, pollen, and broodnest. Queen cells are produced on the very bottom of the broodnest, where the wax is the newest. (Yatcko drawing)

in the area. (Seeley, T.D. and R.A. Morse. "The nest of the honey bee." *Insectes Socioux* 23:495-512. 1976.)

A total of 30 bee trees, some of which were not dissected, were examined. While 20 percent of the nests were in oak trees, nests were found in 12 different species of trees. Only 75 percent of the trees were alive, but all nests, even those in dead trees, were in sturdy trees. This indicates that the tree species or type of wood used is not important to the bees. The typical nest site was moderately exposed to the wind, partially shaded, and received moderate rainfall.

All the nests dissected were elongate and cylindrical, conforming to the shape of a tree. The median volume was 45 liters, about the volume of one Langstroth 10-frame box; most nests completely filled the small cavities they occupied. Small nests probably encourage congestion and therefore swarming; from the point of view of perpetuating the species, this is probably good. Of course, in man-kept colonies congestion is exactly what beekeepers try to avoid.

Nest architecture and structure – A typical nest has a small entrance hole near its bottom. Pollen is stored above the brood area and honey above that. Drone comb is found near the bottom rear of the colony. Most of the nests examined (79 percent) had a single entrance but others had up to five entrances. Bees had coated the sides and top, but not the bottom of the nests, with propolis. The nest entrances had been smoothed, probably through washboard behavior (see WASHBOARD BEHAVIOR). None of the nests examined had entrances that were reduced by propolis, a characteristic of Caucasian bees. Most nests were near the ground, but bees prefer to nest high in the air when given a choice. Probably the reason most nests were found close to the ground is that most trees are larger and hollow near the ground. Furthermore a shortage of nesting sites for bees may exist.

NECTAR, COMPOSITION OF – Nectar is a plant secretion that is rich in sugar. The nectar attracts insects, especially bees, to the flower. Nectar is also attractive to a wide range of wasps, flies, beetles, butterflies, other insects, birds and mammals. When insects and other animals are attracted to a flower their bodies become covered with pollen that can be carried to the next flower they visit. Here some is left on the flower's female parts so that pollination may be accomplished. Nectar is attractive to insects, and other animals, because it contains a large quantity of sugar and usually a strong odor. Odor, in addition to being attractive also helps the insect find the nectar. The combination of sugar and odor is usually different from one plant species to another just as the flower's shape, design and color are different. All these differences help to guide the bee from one flower to another of the same kind.

In most nectars the dominant sugar is sucrose. Sucrose is a common plant sugar and is best known as the white table sugar that we use. Its source is sugar cane or sugar beet. The primary sugar in maple syrup, and of course, molasses, is also sucrose. Some nectars contain some glucose and fructose, also common sugars. Glucose is the sugar found in corn and other grass-like plants. Fructose is much less common in nature than the other two sugars.

On a sweetness scale, fructose is the

most sweet of the three sugars followed by sucrose, and then glucose, which is the least sweet. Most candy manufacturers use glucose because it is the least sweet and one can eat much more candy than if it were made of sucrose or fructose, or honey.

The water content of nectar varies greatly. Some nectar contains little sugar. Pear nectar is a good example. Because pear nectar has so little sugar, usually only about 15 percent, it is difficult for pear blossoms to attract bees for pollination. Pear growers must take special precautions to make certain their trees' blossoms have a sufficient number of bee visits to set fruit. Most plants will have nectars that contain 40 to 50 percent sugar and these are obviously more attractive to the bees. Honey bees can measure the difference and will seek out the most attractive nectars.

While water and sugar are the predominate components of nectar it also contains a number of other factors that contribute to its distinctive color and odor. While these are present to the extent of only one to three percent they are the parts of the nectar that make the honeys made from them so distinctive.

NECTAR, CONVERSION TO HONEY – The conversion of nectar into honey involves one physical and two chemical changes. The first of these, a physical change, is a reduction of the nectar's water content while the two chemical changes are brought about by the addition of enzymes from the bee's glands. All of these steps are necessary to reduce the storage space required and to make the honey into a compound with a long life. Many bees work together to bring about these changes.

Foraging bees are older bees and thus their glands are in varying states of deterioration. The foragers do nothing to reduce the moisture content of the nectar; however, they do add enzymes, at least to the best of their ability. In this way they start the process of nectar modification before returning to the hive. Upon returning to the hive, foragers always give the nectar they have collected to a house bee. House bees may be seen standing in the honey storage area processing the honey. Their mouthparts are extended and they may be seen pushing a small drop of the ripening nectar to the tip of their tongue, sucking it back into the mouth and then out again. This process exposes the nectar to air and helps to reduce its moisture content. At the same time the young house bees are adding and mixing their enzymes into the solution. When they are finished manipulating the honey they place it in droplets in cells in the honey storage area. House bees work together over a period of several hours to move a large volume of air through the hive and in this way more moisture is removed through evaporation.

During a honey flow there can be a strong odor in an apiary, especially in the evening, from the most volatile components of the ripening nectar that are also being removed when the moisture is being evaporated. It is possible that not only are the house bees active in moving this large volume of air through the hive in the evening but that the field bees participate in the process as well.

Water reduction does more than just save storage space. It also increases the osmotic pressure of the honey. Osmosis refers to the movement or flow of materials through a membrane such as a cell wall in our bodies or that of any animal. The reason that bacteria, for example, cannot live and grow in honey is that when they are in such a medium the fluids in their cells are forced to pass through their cell walls and into the honey because of the osmotic pressure; no animal has any method of resisting this process. However, when bacteria are in the resting (spore) stage they are not affected by the osmotic pressure because of thick cell walls and very low water content. As a result, any bacteria that are purposefully or accidentally introduced into honey are dehydrated and die. This is the same process we experience when we have our hands in soapy dishwater for a long period of time. The moisture in the cells in our hands moves into the water thus causing our hands to shrink and the skin to wrinkle. After we remove our hands from the dishwater and dry them our bodies restore the normal moisture content of the cells in a matter of minutes. The high osmotic pressure in honey is one of the reasons it is protected from microbial attack and has such a long life. It is also one of the reasons that honey has been used for thousands of years as a wound dressing. The honey keeps the wound moist, kills most infectious organisms present, and protects against attack by microbes.

Glucose oxidase – This enzyme was discovered in honey in the early 1960s. Glucose oxidase had been known from other biological systems earlier. It has been widely known for thousands of years that microbes could not grow in honey. Some literature still refers to "inhibin" that we now know is really the quality imparted to honey by glucose oxidase. The inhibin effect was described and documented scientifically, that is with the appropriately controlled experiments in the 1930s as some unknown, but positively existing quality that was unique to honey. A certain amount of mysticism surrounds the term and the fact that honey has this effect.

Glucose oxidase attacks a very small amount of the sugar glucose in the nectar and ripening honey and converts it into two materials, gluconic acid and hydrogen peroxide. The glucose oxidase enzyme is very sensitive and is easily destroyed by heating. Also, it is active only in dilute honey. As soon as the honey reaches a normal moisture content, 18 to 19 percent, the enzyme ceases to be active. Thus, glucose oxidase starts to protect the freshly collected nectar and unripe honey against microbial attack as soon as it is collected by the foraging worker honey bee that starts to add enzymes immediately. Also, when bees remove honey from cells and dilute it with water to feed to larvae, the system starts to work again.

The gluconic acid that is produced by the action of glucose oxidase is the chief acid present in honey and is responsible for its low pH (high degree of acidity). Honey does not contain much acid but that which is present has a strong effect due to the lack of any buffering agents present. The hydrogen peroxide produced protects the diluted honey. Hydrogen peroxide, which is well known as a common bleach and antiseptic agent, is not a stable compound and rapidly decomposes into water and oxygen. Ripe honey contains no hydrogen peroxide.

Invertase – Several glands in the head and one pair in the thorax of a worker bee empty into the oral cavity. The hypopharyngeal glands produce the invertase that is added to the nectar. The

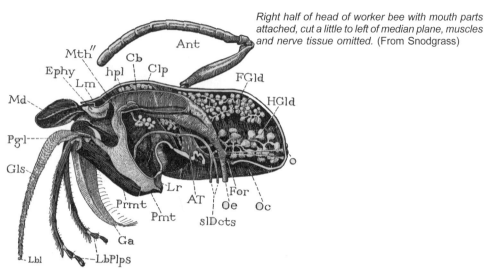

Right half of head of worker bee with mouth parts attached, cut a little to left of median plane, muscles and nerve tissue omitted. (From Snodgrass)

Explanation of Abbreviations
Ant, antenna
AT, anterior tentorial arm
Cb, cibarium
Clp, clypeus
Ephy, epipharynx
FGld, hypopharyngeal food gland
For, occipital foramen
Ga, galea
Gls, glossa (tongue)
HGld, head salivary gland
hpl, hypopharyngeal plate
Lbl, labellum
LbPlp, labial palpus
Lm, labrum
Lr, lorum
Md, mandible
Mth″, functional mouth
O, ocellus
Oc, occiput
Oe, oesophagus
Pgl, paraglossa
Pmt, postmentum
Prmt, prementum
slDct, common salivary duct

function of invertase is simple: it breaks the sucrose sugar molecule, which contains 12 carbon atoms, into two simpler sugars, glucose and fructose, both of which contain six carbon atoms. This process of inversion does several things. It is the first step in the digestion process. It doubles the number of molecules in the honey and thus doubles the osmotic pressure. It also makes glucose available so that a small amount of it may be attacked by the second enzyme, glucose oxidase. The splitting of the sucrose also makes the fructose available and this, as is mentioned before, is the sweetest of the sugars and helps to make honey the distinctive food it is. Like most biological processes the conversion of sucrose into two sugars is never complete and most honey will contain one to three percent sucrose.

Summary – Honey is a safe food and free from growing microbes for several reasons. It has a high osmotic pressure and a low pH when ripe. Ripe means having a full compliment of enzymes and the proper moisture content. Both of these factors make it an inhospitable medium for microbial growth. When the honey is diluted it gains its protection from the hydrogen peroxide that is produced. The hydrogen peroxide system is not fail proof and some yeasts, for example, that are always present in honey, are not deterred from growing in diluted honeys as are most other microbes. Yeasts cannot grow in ripe honey simply because there is insufficient water present. Also, they are inhibited by the high osmotic pressure that does not kill them but that keeps them inactive.

The conversion of sucrose into two sugars poses one problem for the beekeeper and that is that many people think honey is too sweet. This is because of the fructose present. The high fructose level of honey can also be a virtue—if one has a strong sweet tooth a small amount of honey will satisfy one's craving for sweet much more rapidly than will glucose or even sucrose candy.

NECTAR GUIDES – Pollen and nectar are the primary foods of the honey bee, and harvesting them is the bee's chief goal. However, flowers have different goals; they seek to have their pollen transferred to another of the same kind so that pollination may occur. Generally speaking, flowers display their pollen but hide their nectaries. Both the color and odor of the pollen attract honey bees to a flower to collect it. Once they land on a flower the size of the pollen grains attracts the bees. The situation is different with nectar. Because flowers hide their nectar and do not reveal the quantity present, a bee must alight on the flower and probe with its mouthparts to determine both the quality and quantity of the nectar available. In the process the bee usually becomes contaminated with pollen grains, which it subsequently carries to other flowers in its foraging journey.

But where is the nectar? Many flowers ease the task of nectar discovery by using lines and colors on the petals that converge on the nectary and the nectar, which thereby point out their location.

Hybrid sunflowers. Top in normal light. Bottom through a UV filter. Honey bees would see the bright areas highlighting the edge of the flower. (USDA photo)

Sunflower showing UV highlights. Note also the fluorescent nectar in the center of the flower.
(USDA photos)

Nectar Guides

Squash blossom with bright, but lined, petals, but a dark center. (USDA photo)

These lines and colors are called nectar guides. Professor von Frisch found he could fool bees, but only momentarily, by reversing the position of the petals on a daisy-like flower. When the nectar guides pointed to the outer fringes of the flower the bees would probe there, but would soon return to the center of the flower presumably because that was where they had found nectar in other flowers. However, nectar guides may play another more devious role. They are usually arranged on flowers in such a way that the bee straddles the male and female parts of the flower while seeking the nectar. As a result the flower is pollinated at the same time the bee is seeking and/or gathering the nectar. Many flowers have nectar guides that are visible to humans though some can be seen only under ultraviolet light, which is seen by bees. Presumably nectar guides are used by insects other than honey bees for the same purpose.

NECTARIES – From the point of view of beekeeping we normally think of a nectary as a group of specialized cells, or an organ, that lies at the base of a flower. The nectary secretes nectar, a rich sugar syrup, that is attractive to a wide range of insects, especially honey bees. Bees collect nectar and make it into honey. However, the subject of nectaries is not so simple, and as plants vary greatly so do nectaries.

Nectaries that are situated in the region of flowers are called floral nectaries while those that are found on the other parts of plants are called extrafloral nectaries (see EXTRAFLORAL NECTARIES). Floral nectaries may be at the base of the flower, or on sepals, petals, stamens or carpels, all of which are specific parts of a flower. Generally speaking, it is believed that the purpose of floral nectaries is to attract insects or other animals, which accidentally become dusted or covered with pollen while they are collecting the nectar. The pollen is then carried to other flowers of the same species and thus pollination occurs.

Generally we think of flowers as being well-differentiated structures that are sometimes brightly colored. While nectaries are often hidden, forcing the insect to alight to determine if any nectar is present, they are easily seen when the flower is pulled apart. Nectaries take many shapes and may be flat, hollowed, scale- or disk-like. Normally a group of closely-related plants will have nectaries that are similarly situated and look much alike. However, some nectaries look very much like the rest of the plant tissue around them and are not sharply differentiated. Interestingly, nectaries are found on some ferns that have no flowers and no need for pollination. Nectaries clearly have a close connection with the plant's phloem and vascular bundles that carry the sap. Clearly too, plants need an abundant supply of water for the nectary to produce.

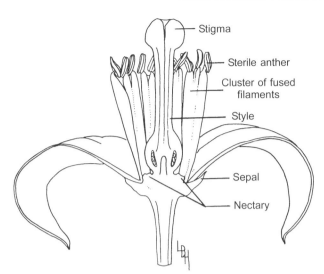

Longitidinal section of Washington Navel Orange blossom. (From McGregor)

NECTAR, POISONOUS – (see HONEY PLANTS, poisonous plants)

NECTAR SECRETION, CONTROL OF – Probably no nectar flow can be accurately predicted since the factors that control nectar secretion by plants are so varied. Soil, water and temperature are the most obvious factors. There are a few general observations on specific plants that may be helpful. For example, in the eastern

A citrus blossom with some parts removed. At the base of the style is the ovule (green) and beneath that is the nectary tissue, here dispensing nectar. As a pollinator pushes down to drink the nectar it forces itself between the anthers and the style, picking up loose pollen grains to be deposited at the next blossom – completing pollination. (Morse photo)

U.S., alfalfa yields best on hot, sunny days; the deep roots of alfalfa find water (to make nectar) when it is too dry for other plants to yield (see HONEY PLANTS, alfalfa). In the case of orange trees, it is nearly impossible to predict when a flow will occur as the buds may form and enlarge but remain unopened for weeks if the weather is not sufficiently warm. Beekeepers claim that irrigated oranges yield better than those that do not receive sufficient moisture. Goldenrod is a plant that is common throughout much of the country and under some circumstances yields large quantities of nectar. However, for a major goldenrod honey flow to take place requires a precise balance of sunshine and moisture, especially during flower bud development.

Beekeepers living in irrigated areas are probably in a better position to predict when nectar flows will occur since moisture levels are carefully controlled by growers. Whereas there has been much discussion of the effect of lime, fertilizers and certain soil elements on nectar secretion by various plants, no one has been able to formulate a program that aids nectar secretion.

However, certain aspects of plant physiology do enter the picture. For optimum pollination a plant must be healthy previous to, and during floral bud formation. This includes adequate sunlight, nutrients, moisture and freedom from significant pest and predator damage or disease. Stress on the plant from too little, or too much of any of these elements can reduce tissue production, or result in foliar, rather than floral development.

As stated, nectar production is generally thought to be associated with the phloem of the plant. Water and dissolved nutrients move through the plant in this aqueous material, some of which moves to both floral and extrafloral nectary tissue. It is the quality of those dissolved nutrients that can affect the amount of nectar produced, and the quality of what is produced. Nutrients, for the most part, are in the form of ions when in water, and these occasionally will affect nectary tissue, cell behavior or reabsorption. This occurs, for the most part, when these ions are in too great, or too small supply than is generally optimal for the plant.

Other factors can affect nectar quality. Already mentioned was adequate moisture. Plants may be subjected to moisture stress during bloom. Stress can be too little water, limiting the phloem flow, and thus available moisture to dedicate to nectar. Reports have indicated that if low moisture stress is accompanied by full sun and adequate temperatures, sugar conversion is enhanced during photosynthesis and though there is less nectar produced, the primary sugar content per bee load is higher. Certainly, some moisture is eventually required or the plant will perish.

Too much moisture can have the opposite affect, wherein the nectar production actually increases in volume, but sugar production remains relatively constant, thus reducing the quality of the bee-gathered product. Excessive moisture, certainly, can kill a plant if it is prolonged and the nutrient absorbing cells in the root hairs do not have access to air in the soil.

Part of successful beekeeping involves moving to those areas where honey plants abound and nectar flows are more predictable due to less environmental influences.

NEMATODES – (see DISEASES OF THE HONEY BEES, Nematodes)

NEPOTISM – Charles Darwin proposed in 1859, in his monumental, and now widely accepted, theory of evolution by natural selection that all living things act in their own self-interest to increase their chances of reproducing. W.D. Hamilton expanded the Darwin theory in 1964 by suggesting that an organism benefits in an evolutionary sense not only by producing its own offspring, but also by helping relatives (kin) reproduce, as an indirect means of propagating its genetic material. According to the theory, the closer the relative, the greater the nepotism.

Honey bees behave in ways that support Hamilton's theory of "kin selection." Bees can distinguish adult worker members of their colony from unrelated workers. The aggressive behavior that is displayed toward unrelated workers under experimental conditions suggests that recognition of nestmates is important as a defense against robbing. Honey bee workers recognize their own queen and show aggressive behavior towards foreign queens, behavior that may have evolved to avoid the usurpation of a colony by a foreign queen. This is the reason why colonies of honey bees often reject queens during requeening attempts by beekeepers.

Nepotism also occurs among the members of a colony. Queens mate with up to 20 drones and use the sperm of at least several drones simultaneously to fertilize their eggs. As a consequence, colonies are continuously composed of numerous subfamilies, each subfamily a group of workers descended from the queen and one of her mates. This relationship is dynamic throughout the life of the queen. Honey bees recognize each other on the basis of the degree to which they are related, and favor closer relatives. In the laboratory, bees are more likely to show aggressive behavior toward members of another subfamily ("half-sisters") than members of their own subfamily ("super-sisters"). Within colonies, individuals groom and feed super-sisters more often than they do half-sisters. Workers prefer to rear queens from female larvae that are their super-sisters rather than their half-sisters. In queenless colonies with egg-laying workers, individuals appear to interfere with the reproductive activities of other workers via aggressive behavior, egg cannibalism, and differential larval care, which may explain why there are differences in reproductive success among workers from different subfamilies.

The findings summarized here must be interpreted cautiously at this point because most studies have been conducted under experimental conditions that may have made it easier for nepotism to occur than in more natural circumstances. However, they point to genetically-based conflicts of interest in the honey bee society, long considered a paragon of cooperation. The levels of conflict in queenless colonies appear to be low relative to the degree of cooperation among workers. Nevertheless, the presence of sophisticated mechanisms for both the production and perception of recognition cues implicate kin recognition and nepotism as important components in the social evolution of honey bees. - GER

NERVOUS SYSTEM OF THE HONEY BEE – Like most animals, the honey bee has a brain located in the head, and a central nerve cord that connects with the

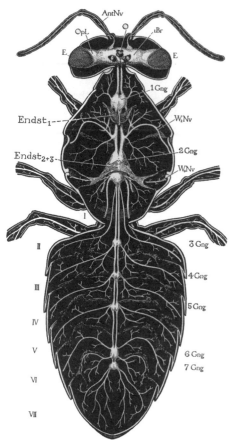

General view of the nervous system of an adult worker bee, dorsal.

Explanation of Abbreviations
AntNv, antennal nerve
Br, brain (1Br, protocerebrum; 2Br, deutocerebrum; 3Br, tritocerebrum)
E, compound eye
Endst, endosternum; $Endst_{2+3}$, composite endosternum of mesothorax and metathorax
Gng, ganglion
Nv, nerve
O, ocellus, or ocellar rudiment in epidermis
OpL, optic lobe of brain
hl, hypopharyngeal lobe
hpl, hypopharyngeal plate
Lbl, labellum
LbPlp, labial palpus

Bees fanning at a colony entrance. Note that the tip of the abdomen is pointed down and the Nasonov gland is exposed.

brain and runs the length of the body. However, this nervous system, which transmits information from one part of the body to another, has several features that are different from that of a mammal.

Honey bees have not one central nerve cord but two, side by side. Enlargements of this cord, called ganglia, are found at two points in the thorax and five in the abdomen. It has been suggested that these might be thought of as sub-brains; places where information concerning local issues is assembled and decisions made.

Professor R.S. Pickard, of the University of Wales in Great Britain, has mapped the bee's brain and identified its various parts. It is a remarkably complex organ (see LEARNING IN THE HONEY BEE and GLANDS IN THE HONEY BEE AND THEIR SECRETIONS).

NEST HYGIENE – The large size and long life of a honey bee colony make it an attractive and apparent resource to a wide range of predators and parasites. The social evolution of honey bees required the concurrent evolution of an arsenal of social and biological defenses against these threats.

The enclosed nests which have allowed *Apis mellifera* to expand its range from the tropics to temperate regions have also posed novel problems. Dead bees and nest debris, which would simply fall to the forest floor from an exposed nest, such as those of *Apis dorsata* or *Apis*

Propolis is used to fill small cracks, such as where the top edge of a super and the inner cover come together. If the weather is warm enough the propolis will be sticky and gummy. If the weather is cool, the propolis becomes brittle and when these hive parts are pried apart, the propolis seal will break, usually with a significant snapping sound.

florea, accumulate in a cavity, where they can harbor disease and pests, or even fill the cavity. Use of the same cavity for many years is made possible by nest cleaning behaviors, in which bees remove corpses (see DEAD BEES, REMOVAL OF) and waste material. In an enclosed nest, the respiration of tens of thousands of bees could quickly deplete available oxygen and build up toxic levels of carbon dioxide. To prevent this, the bees circulate fresh air into the nest by fanning their wings.

Propolis is also used by bees to combat pests and predators. Cracks in the walls of a cavity which are too small for a bee to enter could serve as a refuge for pests such as wax moths, so the bees fill them

Pollen, moistened with nectar, and probably a small amount of enzymes, is brought into the hive in a pellet-like form attached to the corbicula of a foraging bee. The pellet is dislodged by that bee directly into a cell already containing pollen, or an empty cell in the vicinity of the brood nest. House bees pack the pollen tightly into the cell using their heads, until the cell is nearly, but not quite full. The remaining space is filled with honey which preserves the nutritional quality of the pollen for an extended period. The pollen in these cells has not yet been covered with honey.

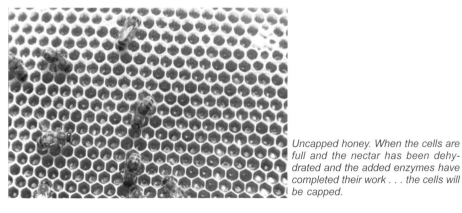

Uncapped honey. When the cells are full and the nectar has been dehydrated and the added enzymes have completed their work . . . the cells will be capped.

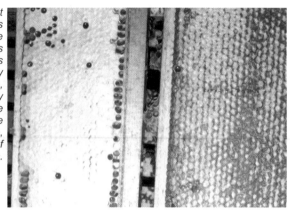

These cells are covered with a thin, yet porus layer of beeswax. These coverings are labeled 'cappings.' If the cappings lie directly on the honey they appear 'wet,' as the photo on the right shows. If there is air space between the top of the honey and the cappings, the wax appears white, and the cappings are referred to as dry cappings. Moisture can, and does move through the cappings, but generally the honey remains at about 16-18% moisture, still low enough to retard the growth of any harmful organisms.

with propolis. The nest entrance is also often reduced with propolis, especially in winter, which restricts the entry of other animals (such as mice), and reduces heat loss. Propolis kills or inhibits the growth of many microorganisms, so its extensive use in the nest probably helps prevent decay of the nest-cavity walls, and may inhibit the spread of diseases within the colony.

The abundant food stores of honey bee colonies are also protected from attack by microorganisms. Honey is protected largely by its high osmotic pressure. The concentration of sugars is so high that honey absorbs water, even from organisms that might otherwise grow in the honey. Also, the action of the enzyme glucose oxidase in honey results in a low pH (acidic), which inhibits bacterial growth. When honey is diluted, for larval food, glucose oxidase produces hydrogen peroxide, a potent antimicrobial agent. Pollen is protected in part by the honey with which it is stored, and in part by the growth of three microorganisms which exclude other harmful organisms, and also enhance the nutritive value of pollen.

These and more varied chemical, physical, and behavioral defenses were a response to the new problems that were posed as increasing social complexity evolved; many such adaptations would be of little consequence to solitary insects. This richness of adaptation against nest contamination thus reflects the complexity of the honey bees' society. - PKV

NITROUS OXIDE – (see SMOKING HONEY BEE COLONIES, Nitrous oxide and laughing gas)

NOSEMA DISEASE – (see DISEASES OF HONEY BEES, Nosema disease)

NUCLEUS COLONY – A nucleus colony is a small colony of honey bees that contains only a few thousand bees and a queen. A nucleus colony has many uses. Such colonies are often referred to as nucs. Baby nucs, sometimes used as mating nucs, are still smaller and may contain only 500 to 1000 bees.

A number of beekeepers keep a few nucleus colonies available for requeening in case a colony loses its queen. For example, a nuc may be placed on top of a queenless colony with a single sheet of newspaper between to slow the mixing of the bees in the two units. A few slits in the newspaper will aid in ventilation. This method is probably the most successful method of

Nucs can be three, four or five frame size. Five are common because of the interchangeability with 10-frame equipment, but four frames for eight-frame are popular too. Three framers are commonly used for mating nucs. (Tew photo)

A commercially available, preassembled nuc. They now come with screened bottom boards, entrance reducers, inner covers and telescoping outer covers or migratory covers.

An entire industry of cardboard, plastic and corrugated plastic materials have enabled nucs to become disposable, used for short term storage, or sold as a one-way unit.

An easy way to use existing equipment to make nucs. If the dividers are removable, individual nucs can be joined if one goes queenless, one can be six frames, one three, or eventually all three can be joined. This is a very versatile unit. (Tew photo)

One problem with a small unit, especially if it sits on the ground, is predators – in this case skunks. Nucs should be on a hivestand at least 24 inches off the ground. Small populations are less able to defend themselves, thus an entrance reducer is also strongly suggested.

A pressed wood or fiber-like material is used here. Somewhat durable this is often used as a bait hive. It comes with frame holders inside and a single entrance.

Feeding with an entrance feeder can be tricky. A small colony will be less able to defend itself if robbing starts. Consider an internal feeder.

Joining two nucs to form a single, large colony is as simple as this. One should be made queenless.

requeening. Since it is many hours before bees in the units mix, little or no fighting occurs. It is important that the nuc be placed above so that bees from the usually larger, queenless unit are slow to come into contact with the new queen.

Nucs may be started from overcrowded colonies. Some colonies build populations too early in the spring and are likely to swarm if not weakened by removing some bees and brood. The entrances of nucleus colonies are usually restricted to reduce the chances of robbing; such small colonies often have difficulty defending themselves.

Nucleus box – Small colonies of bees are often kept in homemade boxes that hold three, four or five standard Langstroth frames and are called nuc boxes. No special designs or considerations are preferred except that it is certainly better to use only boxes that will hold standard frames. Such boxes may have a handle for carrying. A method of reducing the entrance is helpful since small colonies are easily robbed by other bees. Nuc boxes are used for a variety of purposes such as collecting swarms, queen storage, and sometimes as mating nucs. Queen breeders usually use baby nuc boxes for mating boxes since few bees are needed.

Nucleus colony, how to make – The usual way to make a nucleus colony is to place a frame with brood, and sufficient bees to cover it, together with a new queen, in a nucleus box. One may also use a standard brood box and two or three frames, one with honey and pollen, followed with a dummy or follower board (see FOLLOWER BOARD). It is important to have enough bees to cover the brood and to keep it warm on cool nights. There are no absolute rules to follow, but unless one wants a strong colony immediately, usually a frame one-third or half full of brood is sufficient. Inspect the combs carefully to make certain there are no signs of diseases.

Such a unit should be given at least 3000 to 4000 bees, which would probably be about the number shaken from two or three frames. Some of these bees may drift back to the parent colony if the new nucleus colony is kept in the same apiary. If one finds chilled (dead) brood in the nucleus colony at a later date, clearly the number of bees added was not sufficient.

A queen, usually purchased from a queen breeder, is added to the unit. It is preferable to place the queen cage, candy end up, adjacent to the brood. The cage is merely pressed into two adjacent frames and held in place by them so that the screen face of the queen cage is fully exposed. While the queen cage might be placed on the floor of the colony, the bees will be inclined in this case to cluster around both the queen and the brood. This divides their effort and makes it more difficult for them to keep the whole unit warm.

When making a nucleus colony in this manner, a portion of the candy in the queen cage is usually removed. The quantity left should be sufficient, usually about one-third of the original amount, so that the bees can remove it in about 24 hours. If attendants came with the queen they should be removed. When a nucleus colony is made in this manner, the bees in the unit will at first consider the new queen to be foreign; bees in a colony are able to recognize their own queen and will ball (see BALLING) (attack) any queen that is not their own. After about 24 hours the new queen acquires the colony odor (or at least the odor of the old queen is lost) and when the candy is finally eaten away the bees should accept the new queen as their own. As is mentioned above, it is advisable to restrict the entrance of a nucleus colony to reduce the chances of robbing.

A second way to make a nucleus colony is to place a frame with a ripe (nearly mature) queen cell in the nucleus box. The precautions regarding the number of bees remain the same. The number must be sufficient to keep the queen cell warm as well as the brood. Since the queen is so critical to the success of the unit, it is better to err on the side of having too many bees. Queen cells are especially delicate from the time they are capped until about 24 hours before emergence. Use care in transferring the frame to the nuc box.

Certainly a queen cell obtained from an outside source can be added at this time. Many commercial beekeepers make increases this way by removing brood and bees from one or more full size colonies and joining them in a single, new box. A queen cell raised from a grafted larva from a specific breeding line is then added. The entire unit is moved to a mating area so that the virgin queen is exposed to drones of known background.

A queen cell is placed in the center of the brood nest in each nuc two to four days after being queenless.

Once complete, each nuc is plugged with a small piece of screen and left to sit overnight if they were on the pallet, but newly made nucs are hauled to a beeyard that same day.

Commercial nuc production occurs on a much larger scale, and seldom uses boxes smaller than 10 frames. The Adee Honey Farms operation makes nucs in their Mississippi location each year after almond pollination in California. It is instructive to observe the process, which incorporates sound biology, efficient management, and winds up with large, productive colonies.

The process essentially involves dividing or splitting large colonies into smaller units or nucs. Their process has been modified over years of experimenting and has become so well used the process has been named The Mississippi Split.

The goal is to divide colonies so they are all nearly equal in strength. They are generally in good shape when returning from California, loaded with almond pollen and nectar or honey. Since the bees were treated for mites, fed protein and HFCS before moving into the almonds in February, they usually do not need any nutritional attention. Moreover, it is spring in Mississippi and nectar and pollen sources arc available.

The actual process sounds simple enough. Each colony comes to Mississippi as a two-deep unit. The goal is to divide the unit's parts – brood (open and sealed), bees and stores into at least two, or better three, and maybe even four units. The average is three. The standard 'nuc' is one frame of open brood, one frame of sealed brood, one, maybe two frames of bees and one frame of honey plus the remaining frames empty. The queen is found, and killed (about 75 - 80% are found and killed in the time allotted for each colony). Of course it's not as simple as it sounds. There are four colonies on a pallet, and very often all four are open, sometimes more, to get just the right mix for all the nucs that will eventually come from these larger units. That eight-man crew can manage between 320 and 350 California colonies in an eight-hour day which comes to about 10 minutes per unit. An interesting note here. Inclement weather is favored for this activity, rather than warm, sunny conditions. This weather keeps bees at home making these splits is actually easier because fewer bees are out foraging or flying. It is easier to determine the exact population of the colony, plus fewer bees drift or get lost during the confusion.

Once the separations have been made, with brood and stores all in their respective boxes they are reassembled into two story units and left to sit overnight. Nothing is between the boxes as each is queenless. This allows the bees to more or less equally divide themselves to cover the brood in each box. New units, those that were created from the original colonies, are kept as singles and moved that same day to mating yards. Screen is stuck in each entrance overnight to keep all the bees at home.

The next day another eight- or 10-man crew divides the two-deep nucs into singles ("splitting down the middle" they call it) adding tops and bottoms. The nucs are moved to mating yards to be requeened.

These now-queenless nucs are moved to mating yards all around the county. Yards are not yards in the traditional sense. Large pastures are found, and nucs are placed on the outside edge of these, next to the tree line. They are placed on the ground six to eight feet (1.8 - 2.5 meters) apart, with entrances facing in random directions. A large pasture may hold several hundred of these units. They are left there for three to four days, queenless. They are not on pallets, but individual bottom boards. This arrangement means a lot of manual labor loading and unloading each unit, but the configuration leads to nearly no drifting of bees or eventually the queens on mating flights. This method has proven very successful for mating.

By the third or fourth day each unit has come to the conclusion that it is queenless and begins preparations to requeen itself. That is exactly when a queen cell is added. Coordinating the exact number of colonies that need cells in each yard, with the number of cells produced each day requires exact record keeping. You can't afford to miss a yard.

Mating takes place after the queen emerges from the cell and spends a few days getting oriented. There are no beekeepers in the area, so any drones in the area are from Adee's colonies. Drones are fairly common in these colonies, since protein was fed in

Nucs are set along the edges of large pastures, six to eight feet apart facing random directions.

Each colony is checked for a queen. Queenless units are joined with one of the five-framers made earlier so it has a mated queen. Then, each colony is again handled when it is picked up, loaded on a truck and returned to the holding.

These colonies are then loaded on trailers and head out to South Dakota for honey production.

Yet a third way of making a nucleus colony, especially when mated queens are not readily available is to make certain that eggs are present in the comb placed in the nuc box and to allow the new colony to rear its own queen from these. The chief problem with this method of making nucleus colonies is that by the time the queen has been reared from a one- or two-day-old larva, and the time for her to mate is included, nearly four weeks will have elapsed before she lays eggs. It is still a matter of days before these eggs hatch and are ready to be fed by worker bees in the hive. The minimum age of the workers remaining in the hive is then about three and a half weeks, and their head glands, which manufacture the royal jelly fed the larvae, are old and past their prime. Young nurse bees can be shaken into the nucleus. For obvious reasons such a unit is slow to build its population, but in many tropical countries where pollen and nectar flows are long, this method of increasing colony numbers has been used for many years. Just as queen breeders attempt to use their better colonies for a source of the larvae they graft, so the beekeepers who use this method should exercise care in choosing the colonies they select to raise queens in this manner. Furthermore the bees may choose an older larva that will not develop into a satisfactory queen.

California and both pollen and nectar were in abundance in the almond orchards. Thus, drones are from last year's queens, and since breeders are chosen rather carefully, the mating is pretty well controlled. It is a system that has worked and that Adees are comfortable with.

At the same time that these full size units are being made up, several hundred five-frame nucs are also being made. These are treated like the rest, moved, given a queen cell and set in a mating yard. They will be used later.

These units sit in these yards all of April. By the third week of April all the colonies have arrived from California, all the nucs have been made, all the queens have been raised and the next step begins.

NURSE BEES – Those bees that feed larvae are called nurse bees. They are part of the broodnest caste. Very young larvae are fed a special diet of royal jelly that is sccrcted from glands in the heads of nurse bees while older larvae receive the same food but with honey and water added (see CASTES and ROYAL JELLY).

NUTRACEUTICALS – The original definition of a nutraceutical – a single compound, found in a food, that has therapeutic properties – is rapidly changing. Today, the foods themselves containing natural substances that have medicinal or therapeutic value are now being promoted as neutraceuticals. For thousands of years honey and other hive products have been used as medicines for a wide array of diseases and health problems. With modern technology the therapeutic effects and medicinal value in hive products, particularly in honey, are being discovered.

Honey is known to contain trace amounts of vitamins, minerals, enzymes, amino acids and antioxidants. Honey from different plants contains different amounts of these nutrients. Therefore some honeys have more value as a nutraceutical than others. Honey has been shown to be effective in treating both internal and external medical problems. In addition honey as an energy source has been found to have properties superior to other sources. - AWH

NUTRITION OF THE HONEY BEE – Honey bees are no different from other animals on earth; they must have a varied and balanced diet. They require proteins, carbohydrates, lipids, vitamins, and water. Nectar, and rarely honeydew, is their source of carbohydrate and energy. Pollen is the honey bees' source of proteins, vitamins and fat. Most flowers bloom for relatively short periods of time, rarely more than a few weeks, thus a great variety of pollens are consumed by bees during the course of their lives and colonies during a season. This makes up for any deficiency that might be identified with any single pollen source. Water is essential for the nutrition of honey bees and must be accessible or provided by the beekeeper.

The quality and quantity of pollen is critical for the development of honey bee colonies. A number of pollen substitutes are available from bee supply companies for early stimulation of brood rearing and to maintain brood rearing during periods of pollen dearth. Some beekeepers formulate a pollen supplement by adding bee-collected pollen to the pollen substitutes to increase the consumption.

In early 2007, university and government research began again, after decades of neglect, on the nutritional requirements of honey bees, and private companies have conducted or paid for their own. As a result, commercially available protein supplements have improved considerably and meet 100% or more of a honey bee's dietary requirements.

NYE, WILLIAM (BILL) PRESTON (1917-1996) – Bill Nye was born January 10, 1917, in Logan, Utah, the first of two children of Preston William and Lucy Isabella Armstrong Nye. He married Helen Faye Paulsen on August 24, 1945.

He graduated from Logan High School, in 1936 and from Utah State University with a bachelor's degree in 1940 and obtained a master's degree in entomology in 1947.

He began working for the Federal Bureau of Entomology and Plant

William P. Nye

Quarantine, U.S. Legume Seed Research Lab, U.S. Department of Agriculture, in 1947 at Utah State University. He assignment was bee behavior and pollination of agricultural crops. He retired in 1977 after 30 years with the U.S. Department of Agriculture.

As a research apiculturist, he also wrote and taught a class on "the biology of the honey bee," at USU and taught Extension classes at Heber City, Moab and night classes at USU.

He received numerous photo awards from the Entomological Society of America and the Pacific and North Central Branches, Apimondia, and International photo societies; Certificate of Merit Award 1968 from the USDA, ARS; certificate for contribution to research project from the Marion W. Meadows Awards, Horticultural Society; Outstanding Service Award, Western Apicultural Society, 1985; The Hive Tool Award, Utah Beekeepers Association, 1989.

OBSERVATION HIVES – An observation bee hive is usually made by placing a frame or frames in a glass or plastic-sided case with a bee space between the comb and the glass. Observation hives are kept for amusement, educational or research purposes. For demonstration purposes plastic can be substituted for glass, reducing the chances of breakage. One of the chief advantages of using glass to make an observation hive is that it is easier to clean than plastic.

There are no special rules to follow in making an observation hive. The bee space can be made a little wider than normal to make room for more bees and to make it easier for them to cluster to keep the brood warm. However, if one wants to make close observations of the activities of individual bees it is usually best to make a narrow bee space so that the layer of bees over the comb is only one bee thick. When this is done, and the bees are likely to be exposed to cold, the glass should be covered during the times when observations are not being made. Sheets of foam insulation work well and are lightweight if the observation hive is to be carried. It is important to avoid placing the hive where

---

LOCATION • LOCATION • LOCATION

1. Make the runway as short and as straight as possible. It should be wide enough (at least two inches) to allow plenty of bee traffic when the bees are foraging. We prefer rectangular runways instead of glass or plastic tubing.
2. If the runway connects to the middle or upper part of the hive, or if it has a low section, the bees will have difficulty removing dead bees and other debris from the hive. Consequently, this debris will accumulate. Some means of removing this debris with minimal disturbance to the bees would then be desirable.
3. Locate the outdoor entrance away from pedestrian traffic, especially where small children play or walk. The entrance need not be at ground level – a second or higher story will suit the bees just fine.
4. The observation beehive should not be exposed to direct sunlight or temperature extremes. Do not put your hive over radiators or close to heating and air conditioning vents.
5. Be absolutely certain the observation hive is securely attached to a base if it will be on display at any time without direct and constant supervision. We recommend bolting it to a secure and heavy base as a minimum. Tall (three or more frame) units should also be secured at the top with wire or braces to the wall (above the heads of people) or to the ceiling so it will be less likely to tip over.
6. Make it impossible for youngsters to climb in or on the unit. Avoid using a very wide base which will encourage people to lean on the hive. Close or enclose any framework used to suspend the observation hive.
7. Provide adequate space around the observation hive itself to permit unobstructed viewing by the audience. Supplemental display material and handouts should be close to but not interfere with the viewing area itself.

Observation Hives

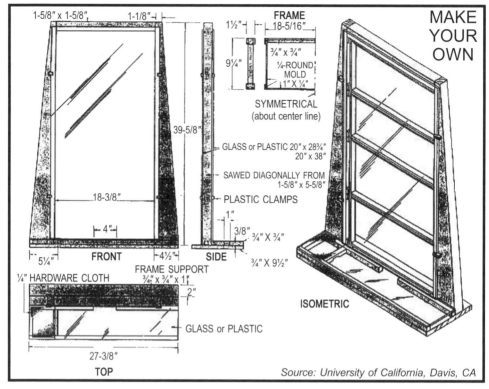

Source: University of California, Davis, CA

it will be exposed to direct sunlight as it may overheat easily.

The hive may be placed next to a window where a tunnel of wood, glass or plastic leads to the outside. Observation hives have been connected to the outdoors by as much as 20 feet (7 meters) of clear plastic tubing through which the bees passed with no difficulty. Screen wire covered holes in the ends or top of the observation hive allow the bees to ventilate the hive as needed. Provisions must be made for feeding. Since an observation hive may contain only a few thousand bees they may have difficulty obtaining sufficient food for themselves so they may need feeding occasionally.

If an observation hive has a normal queen it often becomes over-populated so removing bees from time to time may be necessary. Removing bees can be done in a variety of ways. One way is to close the hives entrance(s) at night and to carry the hive to a remote location several miles away but near another hive. The bees in the observation hive are given free flight the next day and then the hive is closed again at a time when there is good flight. Then it is returned to its original location. In this manner the field bees are left behind; they will join the nearest colony. This removes a portion of the bees from the hive and if the action is repeated as needed the hive can be kept at an appropriate strength. An older queen, two or three years old, makes a good observation hive queen, especially if her egg-laying has declined. It is possible to keep an observation hive in place for many months but only if the population is controlled.

Several simple and interesting experiments may be done with an

> ## Maintenance Tips
>
> 1. Have a standard hive as back-up for your observation hive. This colony should be managed normally. If problems develop, it is usually easier to remove and replace the observation hive frames than attempt to manage it back to the desired condition.
> 2. Examine the observation hive weekly for food supply and population. Populations can be too large as well as too small.
> 3. Plan to clean the runway periodically. Proper hive design can help when debris and dead bees must be removed.
> 4. Be prepared to feed the colony when nectar/pollen sources are not adequate or the population is too low to provision itself adequately. You will need to feed sugar water much of the year.
> 5. Observation beehives are easier to establish and maintain during the active bee foraging season. Outside temperatures above 60°F(15°C) and flowering plant availability are the minimum conditions needed. Fruit tree and dandelion bloom time is recommended.
> 6. Do not overmanage your observation unit. A design that allows for quick removal and permits easy disassembly and reassembly will make your manipulations much easier and reduce the time you need to devote to management. It is best to have the observation hive open for manipulations for as short a time period as possible.
> 7. Always have a *marked queen* in the observation hive. Viewers will be disappointed if they don't see the queen and they generally do not want to spend much time searching.
> 8. Supplement the observation hive with additional display materials, such as drawings, photos and handouts that include visual material but not much text.

observation hive to illustrate various aspects of honey bee behavior and the way in which the colony work force is divided. For example, one may add a red dye to sugar syrup being fed the bees. The movement of the red dyed syrup through the hive is easy to follow. Bees engorged with the red syrup can be seen with ease as the red colored syrup shows through their expanded abdomens and the thin intersegmental membranes. House bees manipulating colored, unripe syrup, forcing it in and out of their mouthparts, can be observed too. Finally the ripened, dyed syrup can be seen in the cells.

One may follow the removal of pieces of grass, or even dead bees, from a hive when they are introduced through a hole in the top. Comb construction may be observed by adding a frame of foundation above the brood nest. It may be necessary to feed the bees heavily to force them to build comb. The process of swarming beginning with congestion; the construction of queen cups and queen cells may also be followed in an observation hive. Usually if a one- or two-frame observation hive becomes congested, the bees may swarm or abscond. However, in large observation hives, with three or more standard frames, the population will usually divide and cast a swarm. There is no question that an observation hive in the living room can be a great conversation piece.

There is no question that an observation hive in the living room is a great conversation piece. Also, an observation hive at a honey stand or farmer's market is an easy way to attract visitors and potentially boost sales. Even people that aren't fond of insects tend to be interested in the activities displayed in an observation hive. It is a good way to remind non-beekeepers of the role

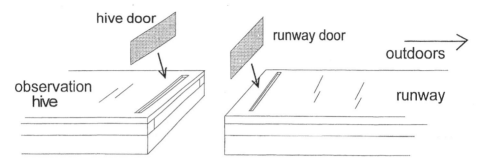

The hive runway has doors that are to be closed at both ends when it is necessary to remove the hive from its stand. Returning bees will accumulate at the outside entrance if the hive is removed during the day. To avoid this, close the hive the previous night, when all the bees are inside.

bees play in our lives and the natural world.

For instructions on how to build, establish, and use an observation hive see Webster T. and D.M. Caron. *Observation Hives*. A. I. Root Co., Medina, Ohio. 1998.

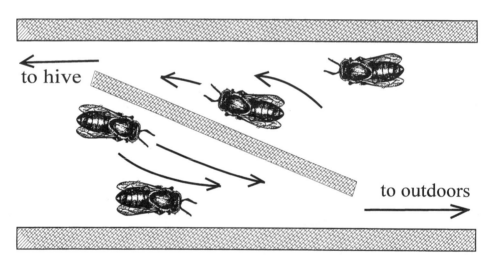

It is difficult to accurately count bees leaving or returning to their hive when flight is extensive. By separating outgoing bees from incoming bees with separate channels, you can get more accurate counts. To do this, you will need a modified runway. With a little training, the foragers will become accustomed to the altered runway, and you can obtain an accurate record of flight activy without having to spend as much time taking data.

Observation Hives

Elaborate, furniture quality hives, such as these two require more work, but are generally large enough to be selfsustaining. The hive on the left is stable and the runway has a clear top so watching bees leave and return is easy. The hive on the right rotates; the runway is below the hive.

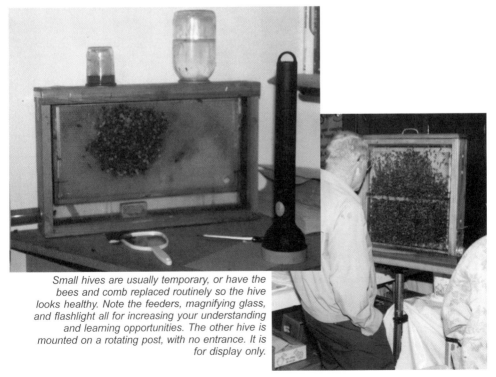

Small hives are usually temporary, or have the bees and comb replaced routinely so the hive looks healthy. Note the feeders, magnifying glass, and flashlight all for increasing your understanding and learning opportunities. The other hive is mounted on a rotating post, with no entrance. It is for display only.

Observation Hives

Safety must be considered when an observation hive is used at a public event: a person in attendance at all times; the hives are firmly mounted; a place for the viewers to lean on; the hives can rotate; the lighting is adequate. Since curious hands can cause problems, physically separating viewers and the hive, as shown on the left, is desirable. Provide educational material nearby.

This style uses a five-frame nuc as the main supply hive. All five frames reside within when not in use in the beeyard. To use, remove one frame, with marked queen, and put in the display area. Fill the empty space with a honey frame or a division board feeder. A queen excluder separates the visible frame from the space below, allowing bees to come and go.

Elegant in its simplicity, this traveling hive, complete with wide base, carrying strap, extra ventilation and very robust design meets all the requirements. This model, created by Ohio State Extension Specialist, Dr. Jim Tew, was designed for abuse, neglect and disaster. It travels well, carries easily and, when stocked with bees, shows everything the way it should.

There are hundreds of ways to draw attention to your observation hive.

OERTEL, EVERETT (1897-1988) – Everett Oertel died October 15, 1988 in Baton Rouge, Louisiana.

He was born on a farm near Prairie du Sac, Wisconsin, May 30, 1897, graduated from public school in Prairie du Sac in 1915, and enlisted in the Army in 1917. He served with the U.S. 32nd Div. in the A.E.F. in France in Word War I. He was wounded while serving as a courier and received the Croix de Guerre with silver star, the Purple Heart, and a Presidential Citation.

Dr. Oertel graduated from the College of Agriculture. University of Wisconsin in 1924. After marriage to Ruth Henika, he attended graduate school at Cornell University, receiving his Ph.D in 1928. His major professor was Dr. E.F. Phillips.

The subject of Oertel's dissertation was "Metamorphosis in the Honey Bee." He was a bee inspector in Wisconsin and New York State during several summers.

After graduating from Cornell, Dr. Oertel joined the Div. of Bee Culture at Baton Rouge, Louisiana (now USDA, ARS, Honey Bee Breeding). He remained with the laboratory until his retirement in 1967. He then spent three summers as a consultant at the Oak Ridge National Laboratory in Tennessee, studying the effect of nuclear radiation on all stages of the honey bee.

Dr. Oertel was the author and co-author of bulletins, scientific papers, and popular articles on bee culture, pollen and nectar plants, and sugar concentrations of some important southern honey plants. His later papers were concerned with nosema disease, effect of radiation on honey bees, and early history of beekeeping in the United States. The bulletin "Beginning with Bees," written in cooperation with E.A. Cancienne and D.K. Pollet, Louisiana Cooperative Extension Service, had a wide distribution. Several editions have been printed.

Before and after retirement, Dr. and Mrs. Oertel traveled in Europe, Central and South America, and in the United States and Canada.

He was a member of Sigma Xi, Entomological Society of America, Louisiana Entomological Society, Louisiana Academy of Science and the Boyd-Ewing Post of the American Legion.

OLFACTORY SENSE – (see SENSES OF THE HONEY BEE, Smell)

ORDINANCES BANNING BEES – Occasionally a city, town or village will attempt to pass an ordinance banning the keeping of honey bees within the boundaries. The moving force behind such ordinances is usually one or more persons who live near a beekeeper and fear bees. Sometimes those who object have a good reason for doing so. Beekeepers who keep their colonies close to people's homes have an obligation to make certain their bees are not a nuisance (see URBAN BEEKEEPING). Some bee associations have drawn up guidelines for beekeepers with apiaries in congested areas.

It should be pointed out that banning beekeeping will not reduce the number of stinging insects in an area. Wasps, bumble bees, as well as honey bees will always be present. As long as food is available, there will be insects to collect it. The only way to eliminate bees from an area is to prohibit the growing of flowering plants. This includes trees, shrubs, annuals, clovers and dandelions in lawns, etc. Obviously this is neither desirable nor practical. Honey bees in cities have little problem finding suitable nesting sites in buildings and hollow trees.

Beekeepers who have fought successfully against ordinances banning bees have pointed out that it is better to have colonies kept under their control. Even with the devastation of colonies from mites, swarms can occur. However, it must be emphasized that without great care and a proper location, keeping bees in congested areas could be a nuisance. (see LEGISLATION IN U.S. RELATING TO HONEY BEES and GOOD NEIGHBOR BEEKEEPING)

For even greater detail, see the book *Honey Bee Law*, by Sylvia Ezenwa, published by The A.I. Root Company.

ORIENTATION BY HONEY BEES – (see SENSES OF THE HONEY BEE and MAGNETIC AND ELECTRICAL FIELDS, EFFECTS ON BEES)

OSMOTIC PRESSURE OF HONEY – (see NECTAR, CONVERSION TO HONEY)

OUT-APIARIES – The term {"out-apiaries" is seldom used or heard today but was an important part of beekeeping jargon. (see OUTYARDS below)

OUTYARDS – Beekeepers use the term outyards to refer to those apiaries other than the home apiary. Beekeepers rarely own the land for outyards, but rent land from a farmer who is glad to have bees on his farm for the free pollination in exchange for allowing the beekeepers access to the location. Beekeepers usually pay rent either in cash or honey.

Selection of locations for outyards should take into consideration the availability of forage and water, slope, exposure to sunlight and quality of the access road. These points are discussed more fully under APIARY LOCATIONS.

OVERCROWDING (OVERSTOCKING) – Perhaps one of the more difficult questions in beekeeping is determining how many colonies a location will support profitably. Commercial beekeepers, those who own a thousand or more colonies and who make their living keeping bees, feel that a location must support 40 or preferably more colonies to be worthwhile. If a smaller number of colonies is kept in each location, the beekeeper spends too much time driving between apiaries. Measuring the suitability of a location and the number of colonies it will sustain usually requires three to five years because of the yearly fluctuations in honey flows.

It should be emphasized that one does not normally plant crops for the sake of producing honey or pollen (see LAND-BASED HONEY PRODUCTION). Plantings around one's home serve to indicate when a nectar or pollen flow may start or end, but for the profitable production of honey one must be in the vicinity of hundreds of acres (hectares) of forage; this includes both major and minor nectar and pollen plants throughout the active season.

In some parts of the world, notably Europe, too many colonies per unit of area can be found. As a result of overcrowding, honey bee colony populations are smaller and honey yields per colony are much lower. To prevent overcrowding of apiaries in the U.S. some states regulate the locations of apiaries.

Even the forests in Europe are manicured because of the need for wood for burning and making lumber. An example of this extreme population pressure is the cultivation of goldenrod. It is strictly a garden flower in Europe and is often sold in bouquets in the flower shops (as it is in some parts of the U.S.), but in many parts of the United States thousands of acres of goldenrod grow wild in unused pastures and fields. In some areas in the eastern U.S., from north to south, goldenrod is so abundant that it is considered a major honey plant.

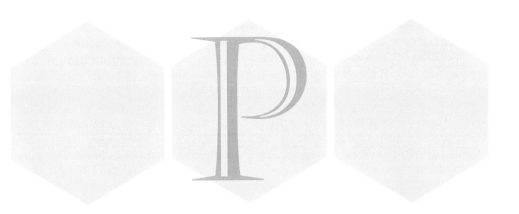

PACKAGE BEES, INSTALLATION AND IMMEDIATE CARE – One way to start beekeeping is to buy a package of bees. Package bees are easier for the beginner to handle as there are fewer bees to contend with and the bees usually are more docile. Another advantage is that the beginning beekeeper can watch the colony develop through the season and thus learn a lot about bee biology.

What are package bees? – A package of bees usually comes in a shipping cage which is a wooden box with wire screen on two opposite sides and partially filled with worker bees and a queen. Males (drones) may or may not be present. Many package bees are produced in Georgia, the Gulf States and California. Most packages are purchased during April, May and June. The best time to start a new colony of bees with a package is when pollen first becomes available in the spring.

Kinds of packages available – Package bees are sold according to the weight of the bees they contain, generally two, three, four or five pounds (one to two kg) and whether they contain a queen. It is better for beekeepers to buy three-pound packages, at least when they first start in beekeeping. There are about 3,000–4,000 worker bees in one pound (see WEIGHT OF THE HONEY BEES).

Temperament of package bees – A swarm, package, or colony of bees is much gentler when it is well fed. Package bees are shipped with a food container, but the container's openings sometimes become blocked. Although bees can carry sufficient food in their stomachs to sustain them for three to four days under normal circumstances, package bees may arrive with their food reserves exhausted. It is prudent to feed package bees before installing them. Even when the package bees arrive with a surplus of food, extra feeding is advisable. (see PACKAGE BEES, INSTALLATION AND IMMEDIATE CARE, Care of package bees before installation)

The temperament of bees also varies with weather conditions. The ideal time to install a package is in the evening when the temperature is above 75°F (24°C), the sun is setting and the air is still.

Another consideration in handling bees properly is the judicious use of smoke. A small quantity of smoke will encourage bees to feed and engorge themselves. This, too, makes the colony gentle. However, it is not advisable to smoke package bees or a swarm unless there is honey available on which they can feed. Too much smoke disrupts the social order within the colony.

Preparation of package bees – Producers of package bees manage their colonies to build up large populations as early as mid-March. The preparation of a queenright package of bees involves providing a suitable shipping cage, removing bees from a parent colony and caging a young mated queen for inclusion in the package.

Standard wooden shipping cages measure about six by nine by sixteen inches (15 by 23 by 40 cm) with wire screen on the long sides for ventilation. A number-three can containing a food supply of 50 percent sugar syrup is held in the cage. A few small holes are made in the bottom of the can through which the bees feed.

Packages are filled directly from strong

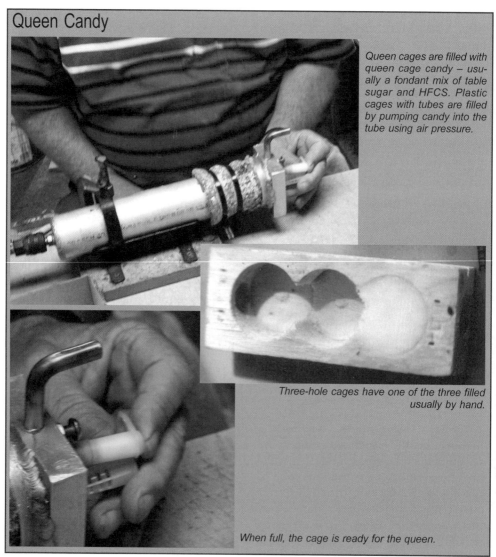

Queen Candy

Queen cages are filled with queen cage candy – usually a fondant mix of table sugar and HFCS. Plastic cages with tubes are filled by pumping candy into the tube using air pressure.

Three-hole cages have one of the three filled usually by hand.

When full, the cage is ready for the queen.

Preparing Queens

Queens that have mated are harvested from the queen mating yard and put in the prepared cages.

Small, queen mating colonies require lots of work. Small hive beetles have been a challenge for these small units.

& Packages In The Field ...

Mating nucs generally have a feeder included, and use heavy cloth for an inner cover.

The package crew begins dismantling colonies to remove queens before shaking bees into the packages. Late season package production means removing honey supers also.

Continued on Next Page

Package Bees

Once the queen is located and replaced in the hive body, frames that had been removed can be shaken.

Finding queens in large population colonies takes a keen eye, and patience.

Frames are shaken so that adhering bees fall into the funnel, and slide into screened box – a 'package.'

Once filled, packages are weighed on a scale tared to read '0' with an empty package. Weight then is only the weight of bees. Most producers try to put 3¼ - 3¾ lbs. in each, depending on time of the day and season. An empty feeder can is placed in the opening temporarily to keep bees inside.

Package Bees

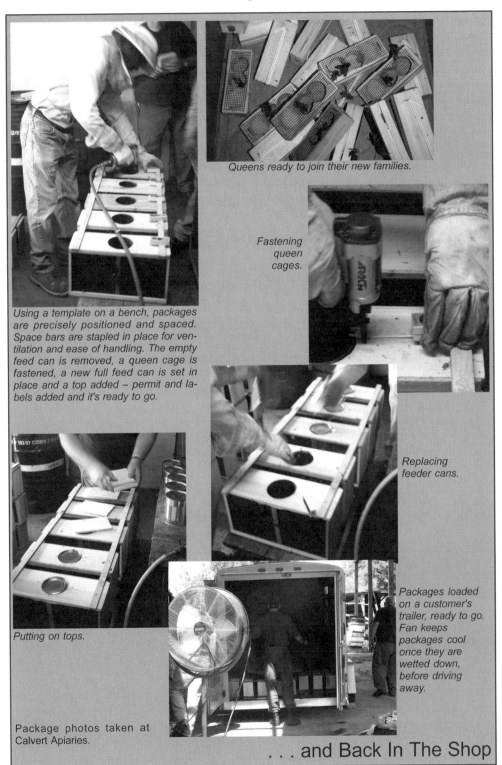

Queens ready to join their new families.

Fastening queen cages.

Using a template on a bench, packages are precisely positioned and spaced. Space bars are stapled in place for ventilation and ease of handling. The empty feed can is removed, a queen cage is fastened, a new full feed can is set in place and a top added – permit and labels added and it's ready to go.

Replacing feeder cans.

Putting on tops.

Packages loaded on a customer's trailer, ready to go. Fan keeps packages cool once they are wetted down, before driving away.

Package photos taken at Calvert Apiaries.

. . . and Back In The Shop

parent colonies. This is usually done in the afternoon when the older foraging bees and many of the drones are in the field, so the packages are composed of mostly young worker bees. As many as possible of the drones and older workers are excluded because they do not aid in rapid colony growth after the package bees have been installed.

Bees adhering on frames from populous colonies are shaken or brushed into packages through a wide-mouthed funnel. The packages are placed on scales to determine their weights. Some producers use the volume of bees instead of weight, especially during a nectar flow. In general, producers prepare overweight packages to compensate for some bee mortality during shipping.

Young, mated queens, reared separately, are placed in small queen cages and added to the packages. Most producers also cage several worker bees with each queen to serve as her attendants. The queen cage is usually supplied with a candy plug that serves as a source of food. This is eaten away by the bees in the colony, thus releasing the queen about a day or two after the package bees are placed in their new home.

Shipping packages – Package bees are sometimes shipped by mail, UPS Next Day, or transported directly to a distribution point by a supply dealer, the package producer, or an independant beekeeper. Generally, a truckload or more of package bees is brought to a local dealer for distribution. Packages are held about a foot apart with wood slats for shipping to protect them from overheating. Upon receipt, package bees should be inspected for unusual numbers of dead bees. Some bee mortality is normal but when dead bees accumulate more than one-half inch (one cm) in the bottom of the shipping cage or when queens are received dead, a damage claim should be filed immediately with the shipping company.

Care of package bees before installation – It is preferable to install packages as soon as they arrive, but installation can be delayed as long as two to four days with little difficulty. Packages should be stored in a cool, dry and preferably dark place, such as a basement or garage. The bees should be fed generously with 50 percent sugar syrup that may be sprayed on the screen sides of the shipping cages with a mister or sprayer. Do not spray the bees so heavily that they are coated with syrup. Heavy spraying can suffocate the bees. A three-pound package of bees may consume a pint or more of syrup in about an hour. Feeding should occur twice a day, morning and evening. If the storage area is warm, misting with water is recommended once or twice a day also. The key is to keep stress to an absolute minimum in the very abnormal environment.

Beekeeping equipment for installing package bees – Before the package is installed, the necessary equipment should be assembled. A hive stand and hive bottom board should be placed in the desired hive location. A hive body with frames of comb or foundation is placed on the bottom board. The choice of frames depends on what is available. If drawn combs from disease-free colonies are available and the combs contain pollen and honey, the colony will start to rear brood faster because the bees do not need to draw out the foundation and make the comb. However if no drawn combs are available, it is recommended that foundation be used in the frames for package bees. Frequently beekeepers will use a mixture of drawn combs and foundation. In such cases, place the drawn combs in the center of what will be the brood nest.

The hive entrance should be reduced with an entrance reducer with an opening about three-eighths inch (one cm) in height and a length of three inches (seven cm). This prevents bees from stronger colonies from robbing the honey stores of the new and relatively weak package bee colony.

A feeder(s), holding about a gallon (four liters) of syrup, must be provided for the colony. It is advisable to use an interior feeder, such as a division board feeder, hive top feeder, or plastic pail feeder protected by an additional hive body to keep animals away.

While rearing brood, colonies try to maintain a temperature of about 93 to 95°F (34 to 35°C). Colonies that cannot stay reasonably close to that temperature will be under stress and may contract certain diseases. The beekeeper can help the colony by keeping the hive off the wet ground, away from wind and in as much sunlight as possible.

Time of day to install packages – Package bees should be installed in late afternoon or early evening when there is little opportunity for flight, and consequently less loss from drifting to other hives. By the following day, when foraging activities begin, the package bees should be familiar with their new abode and drifting will not be a problem. If more than one package is installed at the same time in the same apiary, drifting may be reduced by spacing the colonies apart and facing the entrances in different directions.

Installation procedures – Three basic steps are involved in the installation of package bees. First, the queen in her cage is removed from the package and placed in the hive. Second, the bees are transferred from the shipping cage to the prepared hive. Last, the new colony is fed sugar syrup.

The package-bee shipping cage should be opened adjacent to the prepared hive

In years past there was a thriving business selling packages from the deep south in the U.S. to Canada, where bees generally were not overwintered, but rather were killed after honey harvest.

Different funnels used when shaking bees. Left, a wide funnel with a queen excluder makes sure queens, and drones, do not get into the package. Right, a narrow funnel, wide enough to hold two frames makes shaking fast, once the queen is removed.

by prying off the cage lid (if there is one) with a hive tool and lifting the feeder can from the cage opening. The queen's cage is also removed. The cage lid should be immediately replaced (or a suitable covering previously prepared), without nailing, while the queen cage is being installed in the hive body. It will be necessary to have only nine frames in a 10-frame hive to allow placement of the queen cage.

Package bees do not always accept a queen after she is released from her cage. Therefore, the method of queen release deserves the beekeeper's attention, as it can influence the future of the colony. Keeping the queen in her cage allows the bees to become accustomed to their queen in their new home, while protecting her from initial aggressive worker behavior.

The first step in queen release is to remove the attendant workers from the queen cage if some are present. The worker bees from the package colony will release their queen by eating through the candy plug of her cage. Three or four frames are removed from the hive body the bees are to be shaken into to allow room for the bees. Firmly position the queen's cage, with the candy plug end up, between two frames in the center of the hive. Leave the cork in. The bees should have maximum access to the screen face of the queen's cage for feeding and exposure. Then, simply, quickly, but gently pour the bees out of the package into the open space provided where the frames were removed. Gently shake the cage to dislodge remaining bees, but do not be concerned about getting them all. Then place the cage, with any remaining bees by the front entrance. Stragglers will find their way home.

Package colony management – Three or more days after installation the hive should be opened and the queen cage examined. If the bees are aggressively hanging onto the cage but are rather easily brushed away, remove the cork and reposition the cage. If the bees are aggressive, replace the cage and give the bees a couple more days to acclimate to this queen. Your investment in time will be rewarded with generally higher queen acceptance.

Installing A Package –

The bees cluster around the queen, who is suspended inside next to the feeding can. The package has a cardboard or plywood top keeping the can and the queen in place. If there is more than a half-inch of dead bees contact the shipping company.

Local suppliers drive to the package producer's site and bring bees back for delivery. This reduces delivery time and stress on the bees.

Have all of your tools ready before you begin. You'll need a sprayer and a hive tool and maybe a pair of pliers.

You can put bees in any combination of hive bodies. A 10-frame, single deep, two 10-frame mediums, or two eight-frame mediums. Many combinations work well for the brood nest of your future colony. Here, we use two eight-frame mediums for the brood nest.

In order to remove the queen and her cage you have to first remove the feeding can. Make sure you have a good hold on it.

Continued on Next Page

Package Bees

Once the queen is free of workers place the cage in your pocket for safe keeping and to keep her warm.

After removing a few frames to create a space quickly and carefully dump the bees into the hive. Shake the package to get most of the bees out.

Carefully replace the frames trying not to squeeze too many bees.

Place the queen cage between the bottom bars of the frames in the top box. Make sure the screen side is facing down so the bees can feed and make contact with her.

Put your hive back together and feed, feed, feed.

Once the queen is released remove the cage and replace the frame. About a week after the queen has been released the hive can be opened to look for the presence of eggs. One need not find the queen to make certain she is alive and active – finding eggs is sufficient.

Sometimes when a colony is opened, a mass of worker bees, perhaps as many as 150 or more, form a ball around the queen. This phenomenon is called "balling." It occurs most frequently with young queens. The beekeeper can do little except close the colony immediately and examine it a few days later to determine if the queen is still alive. Directing smoke directly on the 'ball' will sometimes break up the mass of bees. You can also gently pry apart this mass and, if possible, extricate the queen. Always have an unused queen cage in your tool kit, and put her there. Position the cage for good exposure and release her in two or three days. (see BALLING)

If the queen is dead or lost and it is not possible to obtain a new queen immediately, the only practical recourse is to combine the package with another colony. This is done by placing the queenless unit on top of the queenright unit with a single sheet of newspaper between. The newspaper prevents fighting and forces a slow mixing of the bees as they chew through it. Bees that have been queenless for more than a few days will usually not accept a new queen unless they mix slowly with a queenright colony. The ovaries in some workers from queenless colonies develop rapidly and some of them will produce eggs after about two weeks. Eggs from laying workers develop into males. Laying worker colonies eventually perish if the situation is not corrected. (see LAYING WORKERS)

Another technique is to pour the bees on a ramp, or sheet outside the new hive, place the queen inside, and let the bees go inside, leaving dead bees outside.

All things wrong. Inserting a queen cage in this way causes problems. The candy opening is down, allowing any bees inside that perish to block the entrance. The face of the cage is facing the comb, not allowing bees to feed the queen in the cage.

After package installation, the syrup container will need to be refilled every few days until the colony ceases to feed from it. A growing colony of package bees may consume 20 to 30 pounds (10 to 15 kg) of sugar syrup in the first month.

Package Bees

A variation on other techniques. Here, the package is placed in the hive body, the queen positioned nearby. The bees will move from the package to the queen. Frames are replaced in a few days.

Another technique puts an empty super on the bottom board and an opened package put in the super. Above this is another super with frames and a positioned queen. Bees then move up to the queen, rather than over – usually a more reliable technique.

Both, however, assume the bees will move. A turn in the weather for cold and rain, may not allow the bees to move. Then, the queen dies and management becomes difficult, at best.

This method should not be used if the weather is cold. The bees may not move and the queen may die.

Two weeks after installation, the package bee colony should again be examined. At this time the brood combs of the colony should contain many capped brood cells in the central portions of the comb. Around it should be many uncapped brood cells in which bee larvae are visible. If no brood is found at this inspection, the queen has been lost and the package colony should be combined with another.

About one-and-a-half to two months after installation, the package bee colony will require additional space and the next hive body should be placed above the brood chamber. For the next month, the colony may be checked weekly. At no time should the brood nest be divided by the addition of empty comb in the center of it because dividing the brood makes it difficult for the bees to keep the brood warm.

Growth of a package bee colony – During the first 21 to 23 days after installation, the package bee colony experiences about a 35 percent loss in population. Workers require 20 days to develop from eggs. During this time older bees in the package colony die. After this time, the rate of emergence of young workers begins to exceed the rate of death of older bees and the population grows.

About four weeks after installation the population should be at the level it was

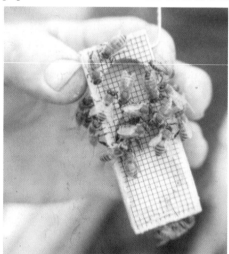

When examining the queen cage three or four days after installation, look for workers clinging tightly to the cage, hanging on with their mandibles, and returning to the cage when brushed off. If this is the case, do not remove the cork on the candy end for two or three more days. Patience will pay off in increased queen acceptance.

at the outset and the real growth of the colony can start. If more space is required and additional hive bodies are given, the population of the package colony will continue to increase steadily.

Due to the reduction in population, some beekeepers give the colony a frame containing some capped brood. This helps to quickly increase the population when the adults emerge and allows the colony to grow faster. Bees should not be given more brood than they can cover and keep warm. One frame, half-full of brood on each side, is usually the most that should be added to a three-pound (1.4 kg) package. A standard Langstroth full-depth frame contains about 6800 cells. Therefore one may estimate the number of bees that will emerge from a frame of brood. The major objection to giving package bees a frame of brood is the possibility of spreading diseases and mites to the package bee colony.

PADDOCK, FLOYD B. (1888-1972) – Mr. Paddock was born November 21, 1888, in Three Oaks, Michigan, and was educated at Colorado State University, Ohio State University and the University of Wisconsin. From 1911 to 1915 he was a professor of entomology at Texas A&M. He became a member of the Iowa State University faculty in 1919 when he became an associate professor of agriculture. He was promoted to professor in 1929.

The author of a number of professional and trade journal articles, Mr. Paddock was a Fellow Emeritus of the American Association for the Advancement of Science, the Entomological Society of America and the Iowa Academy of Science.

He was secretary of the Iowa Beekeepers' Association from 1919 to 1959, president of the Apiary Inspectors of America from 1950 to 1952 and on the executive committee of the National Federation of Beekeepers from 1945 to 1955. He was also a member of Apis international from 1922 to 1959.

Floyd B. Paddock

Floyd Paddock (right) and John Jessup. (photo by O.W. Park, 1935)

PALLETS FOR MOVING BEES – (see MIGRATORY BEEKEEPING)

PARENT COLONY – The parent colony is that from which one makes a nucleus colony or from which a swarm emerges. In the first case the original queen usually remains with the colony, while in the second instance, when a colony swarms, the original queen usually accompanies the new swarm. The parent colony is so-called only because it is the origin of the new unit. A question that biologists have raised is why the old queen should accompany the swarm when one normally thinks of the swarm as the product of reproduction or the reproductive unit. The answer is that the parent colony, since it is already established, has the greatest chance of survival. The old queen, with the swarm, will be superseded in a month or two.

PARK, HOMER E. (1915-2003) – Homer was born in Palo Cedro, California. He started commercial beekeeping in 1942, bought a honey production outfit in Dawson Creek, British Columbia, Canada, in 1955 running 4000 hives. In 1956, he and nephew, Bob Asher,

Lois and Homer Park

bought a 1000 hive business in Brooks, Alberta, Canada selling it in 1973. Homer sold the Dawson Creek business in 1988. Homer sold his Palo Cedro business in 1998.

Homer was instrumental in getting people started in the bee business. J. Park Bee Farms, Tollett Apiaries, Steve Park Apiaries, Wooten's Golden Queens, Woolf Apiaries, Joe Wright Apiaries, Stayer Quality Queens and many more were influenced. Besides shipping queens throughout the U.S. and Canada, Homer shipped queens to Europe, Brazil, Argentina, Israel, Venezuela, Iran, New Zealand, and Australia. His biggest thrill was the year he shipped out 94,000 queen bees.

PARK, OSCAR WALLACE (1889-1954) – Dr. Park was a world renowned authority on the life habits and activities of honey bees, and had contributed

Homer E. Park

greatly to the improvement of honey bee stock through the development of disease-resistant lines. For many years he had initiated and directed the apiculture research program which had placed Iowa State College as a top institution in these fields.

In the February 1986 issue of *Gleanings In Bee Culture* Dr. Walter C. Rothenbuhler, Professor of Apiculture, The Ohio State University, Columbus, published an article offering details and information not available elsewhere, as Dr. Rothenbuhler was one of Dr. Park's students during the exciting days of Park's tenure. Following are excerpts of this article.

"Apiculture has a long history at Iowa State. It antedates entomology because Mrs. Ellen S. Tupper gave lectures on beekeeping here in 1872. Ellen Tupper was prominent in beekeeping circles in Iowa and in the nation. She wrote for at least five periodicals, published one called the *National Bee Journal* in 1873 and 1874, and was an associate editor of the *American Bee Journal* in 1875. Mrs. Tupper was interested in improving stock by importing the Italian race of bees and, along with another lady, formed the Italian Bee Company for this purpose.

"Apparently it was not yet time for an enduring program of apiculture at the college, because it was not until 1914 that C.E. Bartholomew was appointed to carry out beekeeping investigations. A Mr. Atkins succeeded Bartholomew for a short time. Then, O.W. Park succeeded Atkins in 1918 and continued until 1954 with the exception of two short tours of duty elsewhere.

"In those early years, another line of apicultural appointments led to the college. Frank C. Pellett who lived from 1879 to 1951 was appointed the first Iowa bee inspector (for bee disease control) in 1912. It was entirely in keeping with Frank Pellett's personality and philosophy when he concluded that education rather than police power was the better approach to the disease problem. Consequently he recommended to the Iowa Legislature that the bee inspection work be placed under the direction of the Iowa State College Extension Service. This was done and F. Eric Millen was appointed in 1917 to the new position with responsibility for inspection, teaching, and extension. Millen resigned in 1919 and Floyd B. Paddock was appointed to the position which he held until 1959. At this time, the law was changed to put inspection work under the Iowa Department of Agriculture.

"Professor Paddock was a 1911 graduate of Colorado A&M at Fort Collins and received a Master's Degree from The

Oscar Wallace Park, using a an early refractometer in the field.

Dr. and Mrs. Park

Ohio State University in 1916. He had been assistant to Wilmon Newell from 1911 to 1915 at Texas A&M when the two of them were investigating the inheritance of body color in the honey bee. From 1915, when Newell left, to 1919, Paddock was chief of the Division of Entomology in the Texas Experiment Station.

"1922 was a high point for apiculture at the Iowa College of Agriculture. Sixty-five students were registered in apiculture. The staff was composed of five people. Professor Paddock was in charge. John Jessup was an Instructor in Beekeeping. Wallace Park was in charge of research in beekeeping. R.L. Parker was an Assistant in Research; and Newman Lyle was Assistant Professor in Beekeeping Extension. Jessup, Park and Parker were awarded advanced degrees under Paddock's direction.

"And now we come to the O.W. Park story.

"Oscar Wallace Park was born January 14, 1889, near Concordia, Kansas, fifth in a family of six children. His father was a truck farmer, selling produce door to door, and an agent for Watkins Products. As a young man, O.W. lived in Emporia, Kansas, and I recall Dr. Park telling me about the Emporia Gazette and its famous editor William Allen White. Maurine, the older Park daughter told me that when she and her sister, Muriel, were 10 or 12 years old, the family visited Concordia and saw the one-room school where O.W. taught just after finishing high school. They also saw the tiny house, then hidden by tall corn, where he was born. He taught later in high school and was a school principal. In 1914, Wallace Park and Beulah Covert of Lyndon, Kansas, were married. She had also been a rural school teacher but was by this time a graduate of the College of Emporia. In 1914, at the age of 25, Wallace Park enrolled as a freshman at Kansas State. The new Mrs. Park taught German and English at Kansas State which, along with money he had saved, supported the family. The Bachelor of Science degree was granted three years later, in 1917. He remained in Manhattan, Kansas as an assistant in Zoology for one year, studying genetics of grasshoppers under Robert K. Nabours. The help of Wallace Park is acknowledged in one of Nabour's publications. In 1918, Park became an assistant in apiculture at Iowa State with responsibility for research. He was awarded the Master's Degree in 1920 and the Ph.D was the first degree at that level conferred by the Department of Zoology and Entomology.

"As Dr. Oscar Tauber pointed out, in an obituary statement years ago, the changes in Dr. Park's professional interests can be seen in his publications, just under 100 in number. From 1919, when his first paper appeared, until about 1923, his papers reflect an interest in bee management and other practical matters. "From 1923 until 1934, most of his topics centered about the physical

L to R – Park, Paddock, Jones and Pellett

characteristics of nectar and honey, and about the physiology of honey production." Publications on disease resistance appeared first in 1936 and interests in breeding and genetics were dominant through the rest of his life.

"The fifth paper that Park published was entitled "Breed Better Bees." So, as early as this 1920 paper, he was concerned with breeding, and emphasized selecting non-swarming, gentle bees, resistant to European foulbrood. He did not, as yet, carry out any projects along these lines.

"It is of more than passing interest that Park and von Frisch independently and at about the same time, 1923, interpreted the bee dances as a system of communication. Von Frisch published first and Park confirmed his interpretation. Von Frisch credits Park as being the first to observe both round and waggle dances by water collecting bees. At first von Frisch thought nectar collectors danced round dances whereas, pollen collectors danced waggle dances. Park could see no difference in the dances of bees collecting nectar, pollen, or water and he wrote, "So far as I have discovered, the dance performed by nectar carriers and water carriers is identical, and the only way in which that of a pollen carrier differs from the others is that she does not give her load to the other bees in the hive, but deposits it directly in the cell." Subsequent to these early discoveries, von Frisch continued to work on dance bees, and Park turned his attention to other things.

"The other things at this time were studies on sugar concentration in nectars of a wide variety of plants, ripening of honey, and rate of work by bees. He was apparently first to use a refractometer in the study of nectars.

"September 20, 1934 was a pivotal day in apiculture. On that day, Park and Paddock went to Atlantic, Iowa, to discuss with Frank Pellett the possibility of investigating whether or not there was such a thing as natural resistance to the bee disease called American foulbrood. Responsible people believed that once a colony had American foulbrood it was doomed. An occasional odd report claimed otherwise. As a result of this conference at Atlantic, a cooperative agreement was arranged among the Iowa Experiment Station, the Extension Service and the *American Bee Journal*. The project got underway in the summer of 1935 with the assembling and testing of 25 colonies supposedly resistant to foulbrood and six control colonies. From each colony, a piece of comb was cut out and a like-sized piece containing 75 larvae dead of American foulbrood was put in its place. Colonies were inspected regularly for evidence of disease over the next several weeks. By the end of the summer, the six control colonies were heavily diseased but seven of the supposedly resistant colonies were free of all symptoms of disease. The conclusion: Resistance to American foulbrood exists.

"The next question was: Could resistance be transmitted to a new generation? To accomplish more in a year's time, the resistant stock was taken to the Rio Grand Valley of southern Texas. Now, breeding bees presents a problem. Queens and drones mate only while flying so if the queens are to mate with drones of the resistant stock, a mating station, isolated from other bees, must be found since, at that time, artificial insemination of bees was not available. Such a station was found in a 25,000 acre citrus orchard. Park and Pellett reared queens and drones, got them mated and sent stock to Iowa for testing in the summer of 1936. Of 27 second-generation colonies, nine were free of disease at the end of the season. So, resistance was inheritable.

"These results created excitement in the apicultural world. Other states and the USDA, came into the work under a new cooperative agreement. Progress was very fast when compared with some selection programs in other animals.

"The lasting influence of a man is made through several channels. First, of course, is through his personal family. The Parks were parents of two daughters. The entire family went to Texas in the winter of 1937 and spent the winter and spring there while Dr. Park was engaged in rearing a new generation of resistant bees.

"In addition to a personal family, a scholar in academic life has an academic family. Colleagues working together might certainly be considered one segment of this family. Pellett, Park, and Paddock, the three who planned the project on resistance to American foulbrood, were in that family. Also were Newman Lyle, who was on the apiculture staff in 1922; M.G. Dadant, of the *American Bee Journal*; Ed Brown, founder of the Sioux Honey Association; and Park.

"Others, in a photo taken about 1946 by Colin G. Butler, head of the Bee Department, at Rothamsted Experiment Station in England, shows Pellett, Park and Paddock with Glen Jones, who lived in Atlantic and was Secretary of the American Beekeeping Federation for several years.

"In addition to a family of colleagues, in an academic family, a professor is likely to have academic children. Dr. Park had these and they include Roy A. Grout M.S., 1931; E.M. Braman M.S. 1932; J.F. Reinhardt M.S., 1935; T.W. Millen M.S., 1939; Norval Baker M.S., 1942; M.S. Polhemus M.S., 1950; W. Rothenbuhler M.S., 1952; W. Rothenbuhler Ph.D., 1954.

"The first step, taken by me, toward becoming one of his academic children occurred in September of 1945. I was stationed by the U.S. Army in Clarinda, Iowa and was planning my return to school. I came to Ames to see some

About 1950. L to R – Robert Walstrom, Walter Rothenbuhler, Floyd Paddock, Austin Hawes, Martin Polhemus, O.W. Park and G.H. Cale, Jr.

faculty members namely Park, Tauber, Lush, and Gowen. I found Dr. Tauber, who called the Insectary for Dr. Park, and then, since it was noon, called his home. After arranging for me to meet Park, Dr. Tauber took me to the student union and instructed me on where to find the cafeteria line. After lunch I met Dr. Park and saw the beeyards and bee lab. It developed over the next few weeks that Dr. Park wanted someone to introduce artificial insemination into his bee breeding program. Since I had been trained previously in this technique, and was impressed with Iowa State, a mutually satisfactory arrangement was inevitable. I arrived at Iowa State on April Fool's Day, 1946, but was not an April Fool for doing so. I could not have had a better place to go to school, nor stronger support from Dr. Park and the administration during the following years.

"When one has academic children, in due time, academic grandchildren appear. Depending on the requirements for listing as a grandchild, I know of about 25 to 35. Some of those include: Victor Thompson, J.C.M. L'Arrivee, John Bamrick, H. Shimanuki, Harold Borchers, Sharon Mourer, William Wilson, Paul Rab, Roberta Horvath, Nancy Gfeller, Keith Farrell, Nathan Drum, Frank Eischen, Yoshio Hachinohe, Richard Trump, L.F. Lewis, Salah El Din Rashad, Terrell Hoage, Gerald Sutter, Jeanette Momot, Joseph Stewart, Thomas Rinderer, Keith Waddington, Jui-l Ting, Anita Collins, Charles Milne, Susan Grant, Richard Hellmich, Wilhelm Drescher, Jovan Kulincevic and Nicholas Calderone.

"Dr. Park was an avid collector of literature on bees, particularly important bee books of the past. As a material part of the legacy that he left, was this personal library given to the Iowa State University Library. The apicultural holdings of the I.S.U. Library in both books and journals are truly outstanding.

"An immaterial part of the legacy is his worldwide reputation as a careful precise investigator. It is almost impossible to overemphasize the esteem with which his work was and is regarded by apiculturists. His work is a model for his academic children and grandchildren to emulate.

"Looking at his work from a different angle, is there anything to be gained, any legacy to be identified, by noting the areas in which he chose to work many years earlier. These areas were bee behavior, resistance to disease, and bee breeding and genetics. It is behavior of bees that makes them useful. Since certain patterns of behavior vary widely from colony to colony, a genetics of behavior is accessible, and breeding for improved behavior is possible. To be good honey producers, good pollinators, or useful as a hobby, bees must be healthy. Since natural resistance to disease varies from colony to colony, a genetics of resistance to disease is accessible and breeding for higher resistance is possible. Park's work pioneered in these areas. It remains for us to follow, to reap the great practical and theoretical benefits of bee behavior genetics. It remains for us to reap the benefits of breeding for resistance to various diseases.

"And now as a concluding statement, Park's work on certain aspects of bee behavior, on American foulbrood, and on breeding and genetics identified areas of immediate and continuing importance – even to the present day."

Dr. Park was a member of Phi Kappa Phi, Sigma XI, and Gamma Sigma Delta, honorary scholastic, research and agricultural societies, respectively. He was a member of the Iowa Beekeepers' Association, the Iowa Academy of Science, the Entomological Society of America, American Association of University Professors, and American Association for the Advancement of Science.

Aside from his consuming interest in honey bees, Wallace, as he was known to his associates, gave much time to his hobby of photography, and was a skillful nature photographer.

PARKHILL, JOE M. – Professor Parkhill died in January, 1989, after a brief illness.

He taught agriculture and apiculture for over 20 years and was director of the Apiary Department for the state of Arkansas for 18 of those years. He served as President of the Apiary Inspectors of America, and brought the World Wide 9th Pollination Congress to Arkansas. He also served as a Representative to the Apimondia Conference when it was in Russia.

Professor Parkhill was honored to present a paper on pollen at the International Conference on Apitherapy in Herzeiyz-Ti Aviv, Israel, in 1983. He

Joe M. Parkhill

served on the Board of Directors of the American Beekeepers Federation and the American Honey Producers.

Most recently, he was the head of the National Research Educational Institute on Nutrition of Pollen, Inc. He appeared on numerous TV and radio programs promoting the value of pollen, and pollination.

He authored eight books on preventative medicine, nutrition, and cooking.

PARTHENOGENESIS – The development of an egg without fertilization is called parthenogenesis. The eggs of most animals die if they do not receive sperm but in the case of many ants, bees, including honey bees, and wasps unfertilized eggs develop into males. In some situations worker bees develop the ability to lay eggs without mating. Most of the unfertilized eggs laid by the worker bees fail to hatch as they are destroyed by other nurse bees, but those that do hatch develop into males.

Arrhenotoky and Thelytoky Parthenogenesis – In honey bees there are two types of parthenogenesis, one is called arrhenotokous parthenogenesis, the production of drones from unfertilized eggs. The other is thelytokous parthenogenesis, the ability of worker bees to produce diploid workers and normal queens without mating. This latter type of parthenogenesis occurs in all honey bees at a very low rate. It is more pronounced in the Cape honey bees, Apis mellifera capensis (see RACES OF HONEY BEES) where it was first discovered. More usually, the colonies develop parthenogenetic workers and slowly dwindle and die since the workers are unable to produce enough eggs for a viable colony but produce enough queen pheromone to prevent the raising of a new queen.

PATENTS IN BEEKEEPING – As might be imagined, a great number of beekeepers and others interested in apiary equipment and products have obtained patents on their ideas and inventions. There is no current comprehensive list of these patents, but partial lists do exist on various internet sites. A book with a collection of patents on beekeeping, honey processing and packing, honey beverages and beverage powders, honey butters and spreads, other honey foods, and bee and honey health products and medicines was prepared by O.S. North, *Bees and Honey Patents of the World*, *ca*. 1969. Also, a short list of the patents pertaining to beeswax processing and products is found in Appendix 1 in Coggshall, W.L. and R.A. Morse. *Beeswax, Production, Harvesting, Processing and Products*. Wicwas Press, New Haven, Connecticut 192 pages. 1984.

PEER, DON (1927-2000) – Don's connection with beekeeping covered the scientific and commercial sides. He graduated with a Bachelor of Science in Agriculture from the Ontario Agricultural College at Guelph and went on to obtain a Masters degree and a Ph.D from the University of Wisconsin. He also spent periods of time as a researcher for both Agriculture Canada and the U.S. Department of Agriculture. During this period he did significant research on the mating of honey bee queens.

In 1960 Don moved to Nipawin, Saskatchewan and started a commercial beekeeping operation that he maintained until retirement in 1995. Don and a group of commercial beekeepers around

Nipawin made some significant progress in developing methods of wintering honey bees on the Canadian prairies.

Don served as President of the Saskatchewan Beekeepers Association and President of the Canadian Honey Council.

PELLETT, FRANK CHAPMAN (1879-1951) – During the early part of his career Pellett was an attorney and practiced in Missouri; however, his chief interest was bees, beekeeping and especially honey plants. For five years he served as the state apiary inspector in Iowa. For over 40 years he operated and managed a honey-plant test garden near Atlantic, Iowa; plantings included selections from both this country and abroad. The garden was in cooperation with the *American Bee Journal* for whom he served as Field Editor and later Associate Editor. Pellett wrote 13 books, co-authored several others and prepared chapters in the 1946 and 1949 editions of *The Hive and the Honey Bee.* Perhaps his most famous book was *American Honey Plant*s that had four editions, the last one being published by the Orange Judd Company in 1947. In it are listed some 475 plants whose flowers produce either nectar or pollen, together with information on their blooming time, distribution, soil requirements, and the nectar and pollen attraction for bees. Mr. Pellett had in process a revision of this great work for some future date. Another one of his books, *History of American Beekeeping*, Collegiate Press, Ames, IA, 213 pages, published in 1938, has also attracted much attention and is in demand on the second-hand book market. Other books by Pellett include *Our Backdoor Neighbors; Birds of the Wild; Flowers of the Wild; How to Attract Birds; Success with Wild Flowers; Beginner's Bee Book; The Romance of the Hive; A Living with Bees; Productive Beekeeping; Practical Queen Rearing* and with his son, *Practical Tomato Culture.*

Mr. Pellett's recognition of the importance of honey bee pollination to our agricultural economy was far in advance of his time. It was his interest and enthusiasm that initiated the first pollination conferences which have done more than any other factor to bring proper recognition by agriculture to the importance of pollination. On his 67th birthday, he was given special recognition at the pollination conference held at the garden at Atlantic, Iowa. At the Fifth Annual Pollination Conference a banquet was given in honor of his 70th birthday. He long will be remembered for his statement made at this meeting. "The honey bee is the key to our agricultural economy."

In recognition of his outstanding work, Mr. Pellett was made an honorary vice-

Frank C. Pellett

Plaques erected to Frank Pellett.

president of the Apis Club of England and an honorary member of the Bee Kingdom League of Egypt. In 1947, he received the national Skelly award for superior achievement in agriculture, the National Association of State Garden clubs award, and was presented with a special medal by the Iowa Horticultural Society. He was a fellow of the Iowa Horticultural Society, the Iowa Academy of Science, and the Royal Horticultural Society of England. He was also a fellow of the American Association for the Advancement of Science and of the American Association of Economic Entomologists.

PELLETT, MELVIN A. (1907-1987) – Melvin Pellett passed away February 4, 1987, at the age of 80. Melvin, with his wife Elizabeth, had been the proprietors of Pellett Gardens for 50 years. Pellett Gardens was well known to beekeepers throughout the world for sale and distribution of honey plants and seeds. Melvin was the son of Frank C. Pellett, field editor of the *American Bee Journal* for many years and author of *American Honey Plants*. Melvin followed his father's interest in finding honey plants adapted to the many climates and soils of the U.S. He made them available to beekeepers through his annual catalog of trees, shrubs and flowers, especially selected for nectar. He wrote many educational articles for the *American Bee Journal* and *Gleanings In Bee Culture*.

Melvin Pellett

PENDER, W.S. – The *Australasian Beekeeper* in 1931 announced the death of one of its editors, W.S. Pender, a member of the firm of Pender Bros. Mr. Pender was an experienced apiarist and a dealer in bee supplies, besides having been editor of the *Australasian Beekeeper* at West Maitland, New South Wales.

PESTICIDES, EFFECTS ON HONEY BEES – The first recorded death of honey bees as a result of the use of an insecticide occurred in 1881. A beekeeper stated that his 10-year-old plum tree had never given him any fruit. The problem was a weevil called the plum curculio, which punctures the young fruit and deposits an egg through the incision; attack by the curculio causes the fruit to drop. The beekeeper sprinkled the tree, which was still in bloom, with Paris green, an insecticide containing arsenic. The next day he found dead bees in his apiary.

Shaw reviewed the literature on honey bee poisoning through 1940 (Shaw, F.R. *Bee poisoning: a review of the more important literature.* Journal of Economic Entomology 34: 16-21. 1941.). In the 60 years since the 1881 report, several persons had reported problems with London purple and Paris green, both common arsenicals; Lead arsenate had also been a serious problem and was increasingly so, stated Shaw. Several insecticides containing copper were tested in 1930 and caused some bee kills. Rotenone was first reported a problem for honey bees in 1935 as a result of an application made on raspberries. Pyrethrums were first reported as honey bee killers in 1929. However, the reports concerning copper compounds, rotenone and pyrethrums were isolated. Today it is known that

W.S. Pender

these compounds are relatively non-toxic to honey bees under most circumstances. Sulfur compounds were used to control certain apple diseases, but Shaw was quick to point out that sulfur has also been used in an attempt to control some honey bee diseases. Apparently there were no documented cases of honey bee poisoning with sulfur compounds.

One point made in Shaw's review that should have been taken more seriously is: "Poisoned pollen is the principal source of danger to bees and brood." Shaw emphasized that "poisoned cover crops" in flower in apple orchards were a serious problem.

At this point an important aspect of plant biology should be emphasized. Nectar and pollen are the principle sources of food for bees. Most flowers display their pollen and hide their nectaries. Nectar contaminated with an

Large numbers of dead and dying adult bees in front of a hive are one indication of a pesticide poisoning. Taking samples of these bees is important, but too often, if a court case ensues, proof that these bees came from your hive is difficult when the chain of evidence is broken. Photo and video evidence, witnesses, and especially a representative of the State Agricultural Department making and keeping the sample are required. Making collections as soon as possible cannot be overemphasized.

insecticide is rarely a problem for bees or beekeepers. By hiding their nectaries, plants force bees to land on their flowers and search for the nectar. On average, a honey bee spends three to four seconds probing the flower to determine if nectar is present. The bees pollinate the flower whether nectar is present or not. Of course, a plant cannot fool a bee forever and if indeed there is no nectar in any of the blossoms the bees will seek food elsewhere. Plants generally display their pollen so that bees will easily brush against it and bring about cross-pollination. Pollen is attractive to honey bees because of its color (yellow and orange pollens are preferred) and its odor. Black, gray, green and other pollens with unattractive colors have very strong, attractive odors. All this leads to the conclusion that it is pollen contamination that is the cause of most cases of serious bee poisoning.

The arsenicals were used in ever-increasing quantities during the 1940s. This was especially noticeable in the eastern U. S. in apple orchards and as a result the number of bees adversely affected increased. One of the first of what was to become a series of laws, rules and regulations aimed at protecting honey bees was enacted by New York State in 1948. The act stated, "Any person who sprays with or applies in any way poison or poisonous substance to fruit trees or any alfalfa and clovers grown as field crops while in blossom, is guilty of a misdemeanor." It is a curious fact that no one was ever prosecuted under that act but this was because soon after it was passed the use of arsenicals was ended and the DDT era began.

Shaw reported correctly that pesticides in dust form are more toxic to honey bees than are sprays. Dusts are used much less today than in the past.

Combining an insecticide with a repellent – The idea that a honey bee repellent might be combined with an insecticide was first proposed in 1900. It was suggested that carbolic acid might be added to an insecticide to discourage bee foraging. Carbon disulfide, nicotine sulfate, creosote, naphthalene, tar oil and several other compounds were suggested but were not especially successful as bee repellents when bees were visiting flowers. Some of these compounds injured flowers, causing them to wilt, which made them less attractive to honey bees, but that cannot be called repellency. What is understood today, especially as a result of the research of von Frisch, is that if a plant has a good reward for the bees (pollen or nectar or both) the bees cannot be fooled for long and will not be deterred from foraging by a foul odor. It is doubtful if a repellent can be found that can be combined with an insecticide and used in the field to repel bees after an insecticide application.

The DDT era – No documented record exists of any colony of honey bees being killed as a result of the use of DDT. However, it took researchers a number of years of extensive field-testing to determine that this was true. Laboratory tests sometimes indicated that DDT was toxic to honey bees.

DDT was first synthesized in 1873 but it was not until 1939 that Dr. Paul Muller in Switzerland discovered its insecticidal qualities; Muller received the Nobel Prize in 1948 because of the millions of human lives that had been saved as a result of using DDT to control malaria.

Many pesticides have been removed from the market place due to stricter government regulations. Kinder, gentler compounds are slowly replacing them.

The first person to receive DDT in the U.S., under secret cover, was Dr. B.V. Travis who was working for the USDA but contracted an army program to find an insecticide or repellent that would protect U.S. troops against malaria-carrying mosquitoes. The results were outstanding, many lives were saved and Travis's work was widely heralded. Following World War II, attention was given to the control of agricultural pests using DDT and its relatives. Beekeepers and others remained suspicious of DDT for many years but no losses in the field were documented. On January 1, 1973, the Environmental Protection Agency (EPA) canceled all uses of DDT in the U.S. because of adverse effects on wildlife. DDT continues to be an important insecticide in both agriculture and the control of insects affecting human health in many other parts of the world.

The discovery of DDT as an insecticide led to the synthesis of a multitude of compounds, some of which did pose problems for bees and their keepers. It is worth mentioning, however, that two close relatives of DDT, dicofol and chlorobenzilate, which are miticides rather than insecticides, were used to control pestiferous honey bee mites in various parts of the world.

Toxicity of modern pesticides to honey bees – In 1950, Professor L.D. Anderson of the University of California at Riverside began a series of easily-replicated laboratory and field studies on insecticides, with an emphasis on new materials as they were marketed. This work was carried on by his longtime associate, E.L. Atkins. The papers by Atkins and his associates classify pesticides into three categories according to their effects on honey bees: highly toxic, moderately toxic and relatively nontoxic. When questions concerning the toxicity of pesticides to honey bees have arisen almost anywhere in the world, the papers by Atkins and his associates are invariably cited as a baseline. (Atkins, E. L., D. Kellum and K. W. Atkins. *Reducing pesticide hazards to honey bees.* Leaflet 2883. University of California, Riverside. 24 pages. Revised February, 1981.

Paris green mixing plant.

The following paragraph from that publication is especially important. "Most pesticides are not hazardous to bees. In a recent listing of the toxicity of 399 pesticides to honey bees determined from over 30,000 tests, 20 percent are highly toxic; 15 percent, moderately toxic; and 65 percent, relatively nontoxic or nontoxic to honey bees. Of the pesticides most commonly used on agricultural crops, less than 50 percent are highly or moderately hazardous to bees."

One of the few good resources on pesticides and spray problems, and solutions.

A useful publication (Johansen, C.A. and D.F. Mayer. *Pollinator Protection – A bee & pesticide handbook*. Wicwas Press, New Haven, CT, 212 pages. 1990) contains information on toxicity of pesticides. This book also contains symptoms and signs of pesticide poisoning in ground-nesting alkali bee and above-ground-nesting leafcutter bees and honey bees. The authors also include suggestions on reducing bee toxicity for beekeepers, farmers and pesticide applicators.

According to the authors, the toxicity of pesticides to honey bees varies, the following being some of the conditions or factors involved: cool temperatures increase the danger from residues; the use of liquid formulations, especially with stickers, reduces the danger; all fungicides and most herbicides pose little danger; and contamination of pollen constitutes the greatest danger to honey bees.

Night spraying with aircraft – California grows more fruits and vegetables for the nation than any other state. Farms are often large and the most efficient method of pesticide application is with aircraft. In the 1960s it was discovered that when aircraft applied pesticides during the middle of the night, pesticide-caused honey bee losses were reduced by about half. Pilots reported that they often preferred to fly at night because the air was still and they stated they could see aboveground wires with their headlights with ease. In California the humidity is generally low, often less than 20 percent. Pesticides applied while the humidity was low dried rapidly causing them to be less hazardous to honey bees.

Generally, herbicides and fungicides are less toxic than insecticides, but all pesticide applicators should try to avoid spraying blooming plants if bees are present and the label indicates such.

Circumstances under which pesticides kill honey bees – Today's pesticides are chemically very different from those used in the past. However, as in 1940 when Shaw reviewed the subject, most honey bee-pesticide losses occur when pollen on open flowers is contaminated. In the case of several insecticides the foraging bees are killed slowly. Many loads of contaminated pollen may be collected, returned to the hive and packed into cells in the normal manner, before the foragers die or are affected in such a manner that they cannot fly normally. In the hive, when bees (queens, drones and workers) emerge from their cells they first engorge on pollen and nectar. When the pollen is contaminated large numbers of bees may die. In such cases, it is relatively easy to determine what is taking place. Older bees (that is, those with less body hair and with frayed wings) will be found dead at the hive entrance; these are the foragers that did not die until after having made several trips to the field. Also mixed with the older bees will be bees with a full complement of body hair that is often downy in appearance; these are the young adult bees that die as a result of eating contaminated pollen. In some cases, young bees have been killed after eating contaminated pollen that has been stored in a hive for over a year.

Sometimes careful observations are necessary to determine the circumstances under which bees are killed. Beekeepers have reported losing honey bees as a result of the spraying of sweet corn, but when one walks through a field of sweet corn on a warm, sunny day, no bees are seen foraging on the corn. Corn is wind-pollinated, not insect-pollinated, so its pollen is light in weight and easily blown by the wind. Careful observations revealed that bees can collect corn pollen only when it is available, in the early morning. The pollen is usually available for a few hours during the morning before it is completely dispersed by the wind, but during this time large quantities may be gathered by honey bees.

More than 150 pounds (68 kg) of pollen may be produced by an acre of corn and thus it can contribute substantially to a bee colony's resources. If the corn is treated with an insecticide when the pollen is available, in the early morning, a favored time for aerial sprays because of less wind, serious honey bee losses may occur.

Pesticide-contaminated water in puddles near an apiary, especially in an apple orchard, may sometimes cause a loss of adult bees. Honey bees use water to dilute the honey fed to the larvae, to dissolve the crystals in honey, and during warm weather to cool the hive. It is reported that under very hot conditions they may gather as much as two gallons (7.5 l) of water a day. Honey bees will collect water from the nearest source. They prefer water that is warmed, as it might be in a protected puddle exposed to the sun.

Contamination of hives with pesticides – Unpopular as it may sound, an aerial spray over an apiary, even when the pesticide hits the hives, causes little or no losses of honey bees. Under normal circumstances honey bees do not walk over the outside of their hives, except around the entrance where the amount of insecticide deposited is negligible. Some bees may be killed if there is heavy flight to and from an apiary if the bees are caught as the spray descends, but even under these circumstances losses are usually low.

The anthers of sweet, or field corn are released very early in the morning. Later when the humidity drops, the pores on anthers split open and the pollen begins to be released. Honey bees quickly learn what time of day pollen becomes available and begin visiting the field at the same time each day. If fields are sprayed during bloom bees can be killed by direct exposure or by crawling on plant parts exposed to the spray. Corn plants produce pollen and their pollen-receptive silks at the same time. Fields are sprayed during this time to keep corn earworm adults from laying eggs in the silk. It is the time of most bee activity in a corn field, and the time of greatest bee kills.

Present outlook – In 1980, immediately following the end of the federal program that paid indemnification to beekeepers who had lost honey bees because of pesticides (see BEEKEEPER INDEMNITY PAYMENT PROGRAM), some beekeepers continued to report serious losses as a result of pesticide use. However, in more recent years the situation has improved markedly. Discussions with beekeepers indicate that the increasing awareness of the importance of honey bees in pollination by growers has made a major difference in their attitudes toward honey bees, at least in some areas.

Stricter labeling requirements by the Environmental Protection Agency (EPA) on containers of insecticide have made a major difference. Labels state that pesticides should not be applied to crops in bloom, or where flowering weeds might be contaminated. It is now clear that it is the responsibility of the person making the pesticide application to use the chemicals in such a way that flowering plants attractive to honey bees are not contaminated. Unfortunately this is not always strictly adhered to.

IPM programs (see Integrated Pest Management) are being implemented across the country. These programs emphasize that growers should scout fields to determine insect pest infestation levels before treating rather than merely using routine applications of pesticides to control noxious insects. Increasingly, beekeepers are reporting that IPM programs are working and are very much to their benefit as well as that of growers.

As the acreage of individual fields grows larger and the number of farmers decreases, growers will become more aware of problems pesticides can cause beekeepers and honey bees.

Spraying sweet corn in the early to late evening reduces honey bee kill, but does not eliminate it.

Colony recovery following pesticide losses – It is both interesting and remarkable that colonies of honey bees are seldom killed as a result of the application of pesticides. However, colonies may lose most or all of the field bees in severe cases. In instances where pollen is contaminated the losses will include many young bees. Queens are usually not killed, even in severe cases, because the food they receive, royal jelly, is produced and fed to them by workers. Queens do not feed on pollen and honey themselves. The only workers that can feed the queen this food are those that have themselves fed on uncontaminated pollen. When pollen is contaminated the only course of action for beekeepers to follow is to allow bees to continue to eat the pollen and be killed. When the contaminated pollen is exhausted the colony will slowly recover. Most plants flower in the spring or early summer and thus it is at this time of year that pesticide losses are most likely to occur. And, of course, this usually allows time for the colonies to recover. The honey crop is lost during the year the bees are affected because not enough foragers are present to gather the nectar. It may also be necessary for the colonies to be fed to survive the winter. Experience shows that with proper care even seriously damaged colonies can recover, even after severe losses of old and young bees. The financial losses the beekeepers suffer under these circumstances can be serious. And, in some parts of the U.S., pesticide use has been so extreme, beekeeping is no longer possible. This is especially true in parts of the corn belt where field corn is routinely treated during bloom.

Protecting colonies from pesticides – Beekeepers can do some things to protect their colonies when they know a spray is imminent and the crop to be sprayed is the one their bees are currently visiting.

The safest and most expensive method is to move the colonies at least five miles (seven km) from the spray site. Then, after it is safe to return (one to ten days later, depending on the spray and the crop) the colonies can be returned.

The disadvantages of this method are obvious. It can be expensive, time consuming, and will probably reduce honey production. However, eliminating the possibility of contact between your bees and a pesticide removes any chance of poisoning. Care must be taken so that the colonies are not moved to a location with similar problems.

Another technique is to restrict flight just before, during and for a short time after a spray. Although several methods have been devised in the past, the drawbacks have been significant enough to be not practical on a large scale. However, one technique has been found that works well in commercial, sideline or hobby operations in that it is very inexpensive, requires a minimum of time, and is almost 100 percent foolproof.

To protect colonies from sprays in sunflower growing areas, several growers found that two things were required; first, bees needed to stay out of the treated fields for about half a day and second, during their confinement, the colonies required ventilation to stay cool.

To accomplish this, several methods were tried, including placing moving screens on top of the colonies and screen in the front entrances. This accomplished control and ventilation but was expensive.

Finally, it was discovered that by offsetting the top hive body about two to four inches (five to 10 cm), and covering the exposed area with screen, the bees could maintain proper ventilation, when confined, for nearly an entire day, if absolutely required.

Now, when beekeepers are advised of an impending spray, they can go to a yard and simply insert a screened entrance guard into each colony, offset and screen between hive bodies and move on to the next yard. One beekeeper estimated he could close down a 40-colony yard in 10 minutes.

Because the Dakotas have so little rain during the summer, this system of leaving an exposed area between supers works well and can be left for several weeks without causing harm. Areas with potentially greater amounts of rain may have to modify this plan a bit.

Another feature of this technique is that it can be applied any time of the day. Returning foragers will cluster outside the two screened entrances trying to get in to dump their loads and will not forage until they do so.

The cost per colony for this technique is reported to be minimal, and if even only 20 colonies can be protected in 10 minutes, the labor expense is also minimal.

Though not perfect, variations of this method have been successful in areas other than the dry Dakotas and should be considered by any beekeeper whose colonies are threatened by a pesticide spray.- KF

Reporting and preventing further pesticides losses – While it is the responsibility of the federal Environmental Protection Agency (EPA) to approve the labels on pesticide containers, it is the responsibility of the states to enforce the rules stated on the labels. In most states a department of agriculture or conservation undertakes this duty. It is this agency that should be notified when a honey bee loss occurs. The regulators in these agencies will inspect the colonies and take appropriate action. Some state universities have a pesticide coordinator or advisor who works closely with the regulatory agencies. In many counties the county agricultural extension agent will be aware of what pesticides are being used in the area and can also assist in preventing losses.

It is very well understood today that pesticides must be used in such a way that they do not cause honey bees to be killed. The labels on pesticide containers are clear in this regard and there is no need for beekeepers to tolerate loss of their bees. However, as agriculture changes it is always possible that errors will be made. Pesticide losses continue however, due to lack of label enforcement by state agencies, and lack of oversight by the EPA. In almost every situation, beekeepers must pursue a lawsuit to force state and federal agencies to follow their own laws. Beekeepers must be alert to these changes in order to protect their bees. It is, in the final analysis, the beekeeper's responsibility to protect the bees.

PHEROMONES – These are chemical substances secreted by glands to the outside of the body. Pheromones convey a specific message to members of the same species. Many animals produce and use pheromones in their day-to-day existence. The word "pheromone" was coined in 1959. Only a few years before that time people had begun to realize that externally-secreted substances played a role in animal communication. Terms such as "social hormone" and "ectohormone" had appeared and were used for a short while, but these were clearly less definitive than the word "pheromone."

The first and the best-known pheromones identified were sex attractants. Obviously, each species of animal must have its own substance to be able to recognize its own opposite sex. In many animals, especially insects, odors are the only communication needed between the sexes. Encounters are brief, and the males and females come together only once for the sole purpose of mating. In the honey bee, where thousands of animals live together in a colony, a much more elaborate communication system is needed and many pheromones, in addition to a sex attractant, exist.

Honey bee pheromones may be divided into two groups. The first group, commonly called release pheromones, are those that are important and act only outside of the hive, where the need for action is immediate, often because of the danger involved. The second group, commonly called primer pheromones, are those that work inside of the hive where the colony members are relatively well-protected and are more concerned with the long-term maintenance of order. Pheromones are secreted as liquids, but they may be volatile and convey their messages as gases, remain liquids or perhaps, become solids. Those that act externally in the form of gases have low molecular weights so that they will evaporate fast and convey the message rapidly.

Some of the known honey bee pheromones are the sex attractant, alarm odor, and the scent gland substances. Depending upon the circumstances, each of these chemicals may mean something quite different, but in each case the message is clear to the bees.

Our knowledge of honey bee pheromones is expanding rapidly from the use of new technology. Within the hive many pheromones may be present, all of which are necessary to convey the messages. For some early information on pheromones the following book will be useful: Free, J.B. *Pheromones of Social Bees*. Cornell University Press, Ithaca, NY. 218 pages. 1987, but much recent work has been completed and is being compiled.

A bee that strays from home and wanders into a strange colony immediately arouses the guard bees at the entrance. Usually some alarm pheromone is released, recruiting other hive bees.

Pheromones

Alarm pheromone – Perhaps the best-known and simplest of the pheromones, both in chemical structure and to describe, is the alarm odor. When it is released it calls for help as clearly as human cries for assistance in our own languages. Worker bees may release the pheromone by protruding their sting and exposing the paired glands (the Koschevnikov glands) that produce the substance. The glands are a part of the sting apparatus. When a honey bee stings, the barbs on the sting shafts cause the whole sting apparatus, including the Koschevnikov glands, to be torn from the bee's body. The bee soon dies, but the enemy, whether a human, bear, other insect, or another animal, is marked with the alarm pheromone for several minutes no matter where the stung animal goes. Queens and drones produce no alarm pheromone. In many other stinging social insects the alarm odor is not produced in separate glands but is a part of the venom itself.

The chemical in honey bee alarm odor is isopentyl acetate, although several other materials secreted from the vicinity of the sting may reinforce or further increase the action of this chemical. Isopentyl acetate is a simple molecule containing only hydrogen, carbon and oxygen, and thus easily synthesized by bees. The chemical has been known for many years and is commonly found on laboratory shelves; however, no one associated it with honey bees until 1962. Isopentyl acetate smells like bananas.

In general, the alarm pheromone will stimulate a response from other worker bees only when released close to the nest or in a swarm where bees have something to defend. When the alarm pheromone is released near a foraging bee she will usually flee. Young bees produce no alarm pheromone. The greatest quantity is found in worker bees two to three weeks old, which is the time when they may serve as guards, an activity that is engaged in by some but not all bees. As bees continue to age they produce less alarm odor, so that little is found in very old bees.

The mandibular glands of worker honey bees produce 2-heptanone, which will also serve to alarm honey bees. However, isopentyl acetate is more than 20 times as effective as an alarm odor, suggesting that 2-heptanone plays some other role in honey bee biology. Young worker honey bees produce 10-hydroxy-2-decenoic acid in their mandibular glands. This compound is an important component of the food produced by bees of this age for larval consumption. It is not until worker bees are two to three weeks of age, and past the stage of feeding larvae and being nurse bees, that this gland produces 2-heptanone (see CASTE and DIVISION OF LABOR). It is

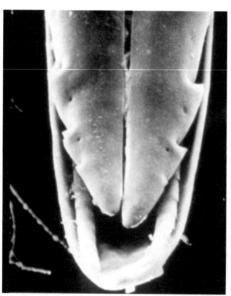

A sting is barbed, causing it and the rest of the sting apparatus to remain in the target. Part of this apparatus are the glands producing alarm pheromone, marking the intruder for more visitors. (From Erickson)

Pheromones

(Jaycox photo)
Each stage of brood, and each caste produces its own message to visitors who come to feed. These differences dictate the diet the larvae need for normal development. Young worker-bound larvae receive a different diet than older larvae. Queen diets are richer in sugar, and drones even different yet. Comb, too produces a message, as do eggs – normal eggs have a different message than abnormal eggs – those to be removed.

possible that 2-heptanone plays a role in the lives of field bees.

When worker bees are confronted with a foreign queen, they will realize she is not their own and will bite and surround her. This is called balling. While balling appears to be an aggressive act, the queen is rarely killed (see BALLING). It is possible that 2-heptanone is used to mark the queen as being foreign. If one places a small amount of isopentyl acetate or 2-heptanone in the vicinity of a group of worker bees scenting around their queen in a swarm, the bees will cease to expose their scent gland and cease to mark the queen as their own. The role of 2-heptanone in honey bee biology is complex. In general it is not efficient for an animal to have different glands that secrete substances that play the same role. While 2-heptanone may act like or mimic an alarm pheromone, it may also have a different function.

Brood and comb pheromones – There is no question that worker honey bees can recognize brood. Since bees feed different food to brood of different ages and also treat drone brood differently from worker brood, their ability to differentiate is obviously complex. One, and probably several, brood pheromones guide the bees. Yet another piece of evidence that drone and worker brood may produce different pheromones is that gravid female varroa prefer drone brood and most likely use an odor to determine which cells they should enter to lay their eggs.

Queen honey bees sometimes produce drone eggs that are diploid, i.e., that have the normal number of

633

chromosomes. These eggs are almost always eaten by workers within six hours of hatching. It has been suggested that it is because their odors (or pheromones) are different and that the bees are able to recognize them.

Brood pheromone definitely has an effect on bees. This is perhaps best seen when hiving a swarm. Swarms will often abandon a hive into which they are placed, even one that has old comb. However, if a comb with a small amount of brood is placed in the hive the swarm will almost never leave.

Old honey comb, like beeswax rendered from it, has a distinctive odor. How much of this is due to the propolis that is mixed with it is not clear, though the propolis has a strong effect. When bees are searching for a new home site they will select a box containing a piece of comb over one with no comb. However, it is also possible that the odor from the comb makes it easier for bees to find such a hive. Dr. J. B. Free of England and Dr. Thomas Rinderer of the USDA Bee Laboratory in Baton Rouge, LA, showed that the presence of old empty comb stimulated bees to store more food. This information strengthened our knowledge that adding many supers to colonies ahead of a honey flow resulted in larger honey crops.

Footprint pheromones – If foraging honey bees are forced to enter their hive through a glass tube, the tube itself picks up an odor and the entrance will be favored by incoming bees. Likewise, if a feeding station is placed on a piece of glass, a residue is left that can be used to attract bees to another feeding station. Since only the feet are in contact with the glass in both instances, the material deposited has been called footprint pheromone although more recently the

Do bees leave a footprint pheromone on flowers they visit – telling other bees this flower has been visited? Or, telling nest mates this is the flower to visit?

term "trail pheromone" has been preferred.

In addition to glass, small pieces of plaster of Paris or wire gauze, over which bees have been forced to walk at the entrance to a hive, also become attractive to bees. It has been shown that glass beads on which bees have been forced to walk can be washed with alcohol and the alcoholic material placed on another object that becomes attractive after the alcohol has evaporated. The chemistry of footprint pheromones will be a topic for research.

Queen substance – At first the term was plural, queen substances, and was meant to include any and all chemical substances secreted by queens that made them attractive and helped to control social order in a colony. This was before any pheromone chemistry was known. In 1960 it was discovered that (E)-9-oxo-2-decenoic acid (9-ODA), a 10-carbon fatty acid, was present in a queen's mandibular glands. For a short while it was thought this was the only substance present, so it was immediately called the queen substance. Virgin queens have almost none of this

Queen pheromone is a mix of many chemicals, in preceise ratios and in the correct amounts.

substance, but it is abundant in a mated, laying queen and continues to be present throughout her life. In 1964, further research showed at least 13 more substances were found in a queen's mandibular glands; even more were found later. Soon thereafter it became clear that 9-ODA, while apparently playing some roles by itself, usually acts in concert together with other chemicals. It remains under study, and new findings continue to occur.

Scent gland pheromone – The scent, or Nasonov gland, lies on top of a worker honey bee's seventh abdominal segment. The gland is normally covered by an overlapping part of the sixth segment. When a bee exposes her scent gland she does so by forcing the tip of the abdomen downward. At the same time she fans her wings so that the air flows back over the abdomen causing the scent gland material to rapidly evaporate and disperse.

Worker honey bees often expose their scent glands when a beekeeper is examining an opened colony. However, such exposure is merely because of confusion since normally bees do not expose their scent glands in a hive. Often when a colony is opened, inexperienced house bees will fly, but since they have had no practice or play flights, they do not have those clues to guide them home. Therefore, a scent trail of this pheromone aids these lost bees in finding their way home. The scent gland is exposed under several circumstances: to mark a hive entrance when a swarm is moving in; to mark a source of food or water in the field; to mark a lost queen or temporary swarm clustering site; to mark an aerial trail as a swarm is moving from its parent colony to a new home site.

Several people have entered into the final chemical identification of the scent gland pheromone. However, the primary work was done by Dr. R. Boch, who patiently wiped the scent glands of hundreds of worker honey bees, and Dr. D. Shearer, who passed the wipings through a gas chromatograph. In 1962, they announced that the pheromone was geraniol, the same substance that gives geraniums their distinctive odor. At that time only a few pheromones had been identified, and researchers did not realize that pheromones usually consisted of complexes of chemicals, not single substances.

When it was discovered that geraniol by itself was not attractive to workers, even though it was the principal component of the scent gland pheromone, the search began for other components. Soon nerolic and geranic acids, as well as citral, were identified.

Worker bees at colony entrances with Nasonov glands exposed, vigorously fanning to propel the very volatile compound out and away from the entrance. Lost bees can hone in on the scent trail and follow it home. When one bee exposes the gland and begins fanning, it signals nearby bees to do the same. When there has been a major colony disruption, such as a thorough exam by a beekeeper, and many bees have taken flight, you may notice many fanning bees on the landing board when all is reassembled.

Today one may buy artificial scent gland lures that will make bait hives more attractive, or that may be used to encourage swarms to settle. The lures are made from a combination of several compounds: (1) (E)-and (Z)-citrals, (2) geraniol, and (3) nerolic and geranic acids. These are usually in a 1:1:1 mixture. The combination, in the solvent hexane, is injected into a small polyethylene vial with a push-in cap. The cap is left in place and the chemicals

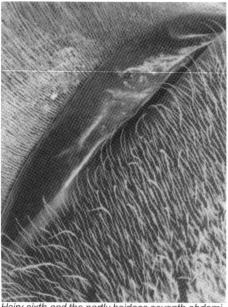

Hairy sixth and the partly hairless seventh abdominal tergites. The glandular area behind the ridge is not exposed. This view complements the one in the right micrograph, showing the entire expanse of this tergite. (From Erickson)

Lateral view of the last (seventh) abdominal tergum (dorsal is left). The smooth hairless area is demarked by the strong submarginal inner ridge; above that is a margin of elevation on the tergum. At the base of this tergum, between terga six and seven is the scent gland of Nasonov. (From Erickson)

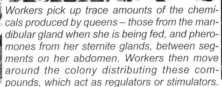

Workers pick up trace amounts of the chemicals produced by queens – those from the mandibular gland when she is being fed, and pheromones from her sternite glands, between segments on her abdomen. Workers then move around the colony distributing these compounds, which act as regulators or stimulators.

One regulator activity is that if the queen is injured, or lost, some workers develop the ability to lay (unfertilized) eggs. They may even develop some pheromone production capability and be treated as queens by other workers. Laying workers deposit eggs in cells rather haphazardly and usually not squarely on the bottom. Moreover, if there is limited room, or several laying workers, several eggs may be deposited in the same cell. In a healthy, queenright colony, eggs laid by workers are detected and removed by house bees. It seems when the queen is gone, policing activity is reduced, or nonexistent.

are absorbed into the plastic and slowly released to the outside. One manufacturer uses 100 microliters of hexane and 10 mg each of the three chemicals that are readily available from chemical supply houses. This quantity of material is believed to contain geraniol equivalent to that from 5000 worker bees

Sex attractant – In 1962 Dr. N.E. Gary discovered that 9-ODA was the sex attractant. When Gary elevated a queen, or a cork with the synthetic material, drones were attracted in large numbers. Before that time it was known that mating took place away from the hive and at higher than visible elevations, but mating in honey bees had never been observed sufficiently to be accurately described. In the following year Dr. C. Zmarlicki discovered that mating took place in well-defined areas half an acre to an acre (0.2 to 0.4 hectare) in size to which both queens and drones flew to mate; he called these Drone Congregation Areas (DCA).

Drone Congregation Areas (DCA) – Many people have tried to determine how DCAs are made or chosen and why they remain the same over many years. Since drones live for only short periods of time, and queens fly only when they are very young, it is clear that memory is not involved. Congregation areas have been found on or near hilltops, in valleys, in both sheltered and exposed areas, and on flat plains.

Some Drone Congregation Areas appear to be more attractive than others, but this may be only because more drones are present in some areas than in others. Drones will fly from one area to another if no virgin queen is present in the first area visited. (see MATING OF THE HONEY BEE)

Wind affects the height at which drones and queens fly for mating. On still days the height appears to be about 20 to 80 feet (six to 24 meters) above the ground, but when there is a strong wind mating may take place at much lower levels. Both the queen and drones fly so fast that they cannot be seen, although they may be heard in an active congregation area. Neither drones nor queens feed outside of the hive, so mating flights last only 20 to 30 minutes.

Worker ovary development – Pheromones clearly play a role in inhibiting the development of ovaries in worker bees. As long as a healthy queen is present, worker ovaries remain small and no eggs are produced.

If a queen is removed from a colony, and the bees are inhibited from rearing a new one, the ovaries in a few workers will develop and after about two weeks they will lay a small number of eggs (see LAYING WORKERS).

It is obvious as one watches a queen in an observation hive that she is attractive to some, but not all of the workers. People who have followed the bees that have licked and attended their queen find that these bees are hyperactive for about 20 minutes. They move around the hive antennating and exchanging food with other workers at twice the normal rate. They appear to be communicating that they have found the queen and that everything is normal. Pheromones play a role in this communication.

PHILLIPS, EVERETT FRANK (1878–1951) – E.F. Phillips became interested in honey bees when he undertook graduate work at the University of Pennsylvania. His thesis work was a study of the compound eye of the honey bee. Because Phillips had no background in beekeeping, and needed to learn about bees, he spent the summer of 1903 in Medina, Ohio, with the Roots. There he became close friends with the Root family and especially E. R. Root, one of two second-generation brothers. The first of his many articles appeared in *Gleanings in Bee Culture* in the September 1, 1903 issue.

In June of 1905, after having completed his Ph.D, Phillips took the position of "Acting in Charge of Apiculture" with the USDA in Washington; soon he was in full charge of the program. Under his direction the causative organism of American foulbrood was discovered and named by G.F. White. A long series of papers on bee diseases followed. R.E. Snodgrass, who later earned the reputation of being the outstanding authority on insect morphology and anatomy, began his work under Phillips. Snodgrass's bulletin, *Anatomy of the Honey Bee,* was published in 1911; it was expanded into book form with the last edition and reprinting, under the same title, appearing in 1984. G.S. Demuth, whose writings on swarm control were well known, also joined the Washington staff. It was during the early part of the 20th century, under Phillips's direction, that

Dr. Everett Franklin Phillips

beekeeping was given its modern, scientific basis.

Throughout his career E.F. Phillips kept up an extensive correspondence with European beekeepers that was of benefit to all beekeepers throughout the world. He visited Europe several times, and Europeans were frequent visitors in his office and laboratory. Phillips published his own book, *Beekeeping*, in 1915 and revised it in 1928. It was the standard classroom textbook on the subject for many years.

Phillips left the USDA in 1924 to join the faculty at Cornell University as Professor of Apiculture. He remained in that position until his retirement in 1949. During the depression, when no money was available for research, and little for graduate student training, Phillips turned his attention to building the nation's largest and most complete library on apiculture. This was funded, at least initially, by financial donations from New York commercial beekeepers, and actual books were donated by many in the industry. Included in the collection is the Langstroth diary in which the discovery of bee space is recorded. The story of Phillips' life and works was written by Mrs. Phillips under the title, *The Bee Man, Life and Letters of Everett Franklin Phillips*. Mrs. Phillips was a professor of home economics at Cornell and herself a noted author and authority on bees and insects.

In 1950, E.R. Root wrote a short biography of Phillips. We reproduce it here.

"Dr. E.F. Phillips is quoted more than any other man in scientific research on the subject of beekeeping. Some four years ago he retired in order to take life a little easier. The life work of such a man should not drop at this point. We can well afford to take a retrospect and I, a contemporary, have assumed the job of telling of his work.

"Dr. Phillips, like Robinson Crusoe, needed a man Friday, and to make a long story short, he selected the late George S. Demuth. Of Demuth he said that he was the best practical beekeeper in America, and one with a scientific approach. Not a better man could have been found. The two worked together for several years with the result that a new epoch in bee culture was born. The work of these two men was outstanding. Perhaps there is not a parallel in all the annals of bee culture which covers a wider scope in scientific and practical beekeeping, unless it is the work of Francis Huber, the Swiss naturalist. Later on, George Demuth became editor of *Gleanings in Bee Culture* and continued as such until the time of his death in 1934.

"While I was at Oberlin College I occasionally wrote articles for *Gleanings*. Later, when my father's health began to fail, he asked me to take over the editorial work. I found many problems unsolved and was looking for some young man who could become familiar with some of these problems and who had back of him some scientific training.

"In 1903 I received a letter from Dr. E.G. Conklin, Professor of Zoology at the University of Pennsylvania, which read as follows:

"Permit me to recommend to your favor Mr. E.F. Phillips, one of my graduate students, and the holder of our Fellowship in Biology for the coming year. Mr. Phillips has undertaken at my suggestion, to go over the whole subject of parthenogenesis in the honey bee in the light of new theories and observations, eliminating, if possible, certain sources of error which are found in the works of older students of this subject and considering many features of the problem from new points of view. Mr. Phillips is a clear-headed, well-trained man, and I consider it highly probable that his work will yield valuable results. He proposes to offer this work, if it should result favorably, as his Ph.D. thesis in which case it will be published in full."

"Here was just the man, and after some correspondence Mr. Phillips came to Medina, and history has proved that, he then only 25, with George Demuth, was the man who played no small part in a new epoch in bee culture.

"During the summer of 1903 when Dr. Phillips had obtained permission to spend some months at Medina, his interest was chiefly in parthenogenetic development of drone bees, the fate of eggs laid by drone-laying queens and, in continuing his work begun in 1902, on the compound eye of the honey bee. He was also interested in methods of queen rearing, having spent the previous summer with E.L. Pratt, known as 'Swarthmore.'

"Dr. Phillips obtained the degree of

Phillips working with injured WWI soldiers.

Doctor of Philosophy from the University of Pennsylvania in June, 1904, and remained there the next year as Research Fellow. He was elected to the Research Fellowship for the second year, starting in September, 1905, but he resigned this to enter the service in the Bureau of Entomology, receiving a temporary appointment when Frank Benton left for his round-the-world trip, hunting for races of bees. He was designated 'Acting in Charge, Apicultural Investigation,' which post he occupied until 1907, at which time he was put in charge of that work.

"The Bee Culture Laboratory at Washington was expanded by the employement of G.F. White, Arthur H. McCray, Burton N. Gates, R.E. Snodgrass, A.P. Sturtevant, and later by a number of others [including D.P. Casteel, G.H. Cale, Eckert (who later updated Phillips' book) and Rece.] Intensive work on causes of bee diseases was begun in 1906, and throughout the progress of this, Dr. Phillips worked on methods of control. He was the first to make a distinction between AFB and EFB, the former a virulent disease, the latter a milder disease. He devised the present plans of control of European foulbrood, and because of these the disease is now inconsequential.

"George S. Demuth, already refered to, entered the Federal Service in 1911 and remained until 1922. Shortly after his appointment, plans were instituted for the two to work together on the wintering problem. After many attempts to devise a suitable program of work in an effort to solve the practical wintering problem, it was concluded that such work was premature and that the proper way to attack the problem was to learn how bees behave in the broodless period. This was done in part by studying the temperatures of the winter cluster in relation to temperatures in the hive and outside the cluster, and outside the hive. A total of over a half-million temperature readings were made the first winter. It was soon evident that there is an intimate inverse relationship between cluster temperatures and the temperature of the air just outside the cluster. From that, after innumerable tests it was found that it was a simple matter to retain the winter cluster at a fairly low temperature merely by insulating the hive properly. Since high cluster temperatures wear out the bees quickly and lead to the development of *Nosema apis*, Dr. Phillips concluded that 'the solution of the practical problem of wintering rests largely on the use of suitable insulation – top, bottom, and sides'. Tests were made of all manner of insulations, and it was found that these were immaterial, just so all parts of the hive are protected.

"During the First World War the Bee Culture Laboratory was entirely changed over to war activities. Extension work in beekeeping was then started. Other offices were induced to take on plans for beekeeping, such as the semi-monthly market news service, work in home economics for the use of honey, special reports on the status of beekeeping annually, and color grading of honey. In fact, under his guidance, commercial honey production increased over 400%.

It was during this time that Phillips sought to assist wounded soldiers returning from the war in learning a new trade – beekeeping. He spent a great deal of time introducing these young men to the science, and the industry of beekeeping.

"In 1924 Dr. Phillips resigned from Federal Service and accepted a professorship at Cornell University. He retired from this post on June 30, 1946. Graduate students working there in this period have since occupied positions in several states, several provinces of Canada, India, China, Ecuador, the Soviet Union, Czechoslovakia, the Union of South Africa, and Mexico."

American Bee Journal, 1951 – At Cornell his great period of teaching flowered; his was a magic personality that expanded the minds of his students; excited them to contribute. The library of beekeeping at Cornell, named for him now rests as one of the largest accumulations of apicultural knowledge in the world, perhaps the center, the pivotal point of his greatness, that future generations shall still feel his ever-remaining influence even though he is no more among us.

PIPING BY QUEENS – (see QUEEN, Queen piping)

PLASTIC BEEKEEPING EQUIPMENT – We live in a world of plastics. Numerous types exist from fragile one-time-use containers to the tough plastic of football helmets. The automobile industry is now using plastics for many auto, truck, SUV and replacement parts. The beekeeping industry has embraced plastics for many items and will probably increase the use of plastics as new types become available.

Plastics have many attributes that make satisfactory products. Many are quite cheap, tough, rigid or flexible as needed, can be colored, made opaque or transparent, easily molded into intricate shapes, and are frequently light in weight. A survey of beekeeping equipment supply catalogs easily demonstrates beekeepers' acceptance of plastics.

Early attempts at plastic hives, using fiberglass, were not satisfactory. Parts warped and hive bodies transmitted light causing bees to make inefficient use of the interior. Today the polystyrene foam hives are popular in a number of countries, especially where wood is almost unavailable and costly. One problem with plastic hives is that they do not "breathe" as wood does so moisture may be trapped inside. Some plastics are fragile and easily damaged.

One of the largest plastic uses today is the plastic foundation and plastic frame-foundation combination. Here again early attempts at plastic foundation did not take into consideration what the bees preferred. Today the manufacturers of plastic foundation realize that cell depth and thin cell edges are the secret to acceptance by the bees. Plastic foundation now comes in several sizes, as well as colors. Queen breeders prefer black since eggs and young larvae are readily visible. However, many beekeepers have found that the bees themselves prefer the black foundation. Since plastic can be colored easily, beekeepers now have drone foundation color-coded green. Plastic foundation is strong, can be cleaned easily and cannot be destroyed by wax moth. As the sales of plastic foundation and frame-foundation have increased, the price has decreased so that the cost advantage outweighs that of wax foundation plus eyelets and wiring. Commercial beekeepers are using plastic foundation exclusively.

Small plastic extractors are suitable for hobby beekeepers. They can combine that extractor with a plastic uncapping tank for inexpensive, lightweight and easily-cleaned honey production equipment. A five-gallon plastic bucket can be fitted with a plastic valve (honey gate) for bottling.

Plastic is also useful in small items such as bee brushes and cappings scratchers; the latter are now made with an extremely strong plastic handle. An inexpensive plastic refractometer is now priced within the reach of small scale honey producers.

Plastics have made feeding bees easier. In-hive feeders such as the division board feeder, as well as plastic pails for top feeding, are cheap and indestructible. Plastic pollen traps, fitting the hive entrance, are light weight and easily cleaned. The plastic propolis trap enables beekeepers to collect clean propolis, free of wood splinters, wax and bee parts.

Comb honey production depends on plastics. The round section inserts for honey supers produce an attractive marketable product, as does the halfcomb cassette and rectangular one-piece comb sections. Cut-comb honey is

presented either in a hard, transparent plastic box or in a transparent clamshell style case.

Queen bee production also involves plastics. The plastic queen cell cups are readily accepted by both the breeder queen and the worker bees. Plastic queen cell protectors and plastic queen cages are now standard supplies. One method of queen rearing involves a plastic box containing plastic queen cups for graftless queen rearing.

Perhaps the largest use of plastics is in containers for liquid honey. The popular squeeze bear now comes in many sizes, either translucent polyethylene or transparent PET. Plastic caps come in a large assortment of colors and in a few different styles. You can choose the classic queenline jar in either glass or plastic.

Larger quantities of honey are contained in a wide selection of plastic jugs and pails. Today the popular five-gallon pail for bulk honey is commonly known as a "60," meaning it holds 60 pounds (27 kg) of liquid honey. Commercial beekeepers are now using storage and shipping tanks of polyethylene held in a wire cage with a capacity of 275 gallons (1040 liters), commonly called a tote.

Plastic is now an integral part of all aspects of beekeeping. As new plastics are developed and costs are reduced, plastics will continue to take an increased role in beekeeping. - AWH

Plastic Containers

Totes

Bottles

Boxes

Round section equipment

Clam shells

Other Plastic Equipment

Honey filters

Pest strips

Valves and gates

Candle molds

Observation hive covering

Shrink wrap for holding equipment

PLAY FLIGHTS – As honey bees age they graduate to more difficult and dangerous tasks. The last job a honey bee does is to become a field bee either as a scout or collector of food. "Play flights" are those taken at the time bees cease to be house bees and before they become foragers. Bees engaging in this activity fly up and down, and back and forth, in front of the colony entrance. Play flights appear to last only about ten or so minutes. Such flights are especially visible on warm days in the spring when colony populations are growing rapidly. At some times a hundred or more bees

Play flights in front of a strong colony. Note that the closest colony has no play flight activity.

may be engaged in this activity at the same time. The purpose is not clear, but play flights probably serve as orientation flights during which bees learn to recognize their hive and the landmarks around it.

POISONOUS HONEY – (see HONEY PLANTS, Poisonous plants)

POLLEN – Pollen is the male germ cells produced by plants. The transfer of pollen to the female germ cells is called pollination. Pollen is the honey bees' source of protein, vitamins and fat. It is an easy matter for a beekeeper to collect pollen in quantity using a trap that is affixed to a colony entrance (see POLLEN TRAPS).

In recent years a number of ads have appeared for pollen sold for human consumption. Pollen is found in pills, pellets and even in snack bars. Pollen, as a food supplement, can be found in health food stores.

Increasingly, microscopic analysis of honey to determine primary floral source is being used by identifying the predominant pollen found in the honey sample. For that reason, being able to identify common pollens is becoming important. Some countries are requiring identify data if claims of varietal honey are being made.

Selling varietal honey offers a marketing advantage over wildflower. Being able to prove variety claims on a label will be necessary.

POLLEN GATHERING BY HONEY BEES – The stimuli for pollen collection by honey bees appears to be precisely regulated in accordance with the colony's needs. A correlation exists between the amount of brood in a colony and the number and proportion of pollen foragers. Doubling the amount of brood in a colony can increase the pollen foraging rate more than twofold in 24 hours. In addition, pollen foraging increases with higher egg-laying rates of the queen. These effects may be mediated by pheromones. Bees can somehow sense the amount of stored pollen in a colony. Adding pollen or pollen supplement reduces the pollen foraging rate, whereas placement of pollen traps stimulates increased foraging. Workers will begin to forage at an earlier age when a colony is deprived of pollen by using pollen traps.

How bees collect and pack pollen – During the course of foraging a bee's body becomes dusted with pollen. The bee then protrudes her tongue and regurgitates nectar that is wiped onto the forelegs, which then brush the head, antennae and front of the thorax free of pollen grains. Pollen on the back of the thorax is combed off by the middle legs, and pollen on the abdomen is gathered by brushing movements of the hind legs. Eventually all the pollen is transferred to the hind legs. Here a specialized rake (rastellum) rakes the moistened pollen off the metatarsal comb of the opposite leg onto the auricle of the metatarsus. This area, at the joint of the metatarsus and tibia, acts as a pollen press when the joint is bent squeezing the pasty mass of pollen up into the corbicula or pollen basket, where the pollen accumulates. The middle legs help mold, pack and shape the pollen load into its characteristic kidney shape.

Photo at right – Typical mixed pollen pellets harvested from a pollen trap.

Depositing the loads in the hive – Upon arrival at the hive, the pollen forager usually grooms herself and then begins to search for an appropriate cell in which to put pollen. The cell may either be an empty one, usually near the brood area, or a cell that already has pollen. When a cell is chosen the bee dips her abdomen and hind legs partially into the cell and uses her middle legs to pry the pollen loads free allowing them to drop into the cell. She will continue to groom off any bits of pollen remaining in her pollen baskets, and then she leaves, allowing another bee to push the pollen loads to the base of the cell and pack them into place with her mandibles, face, and forelegs. After

Commonly Visited Pollen (and nectar) Sources, And The Pollen Grains They Produce

(all pollen photos courtesy of University of Wisconsin Horticultural Department)

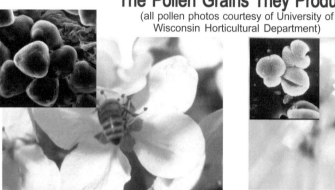

Apple – Malus sp.

Maple – Acer sp.

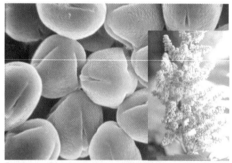

Sumac – Rhus sp.

Black Locust – Robinia sp.

White Sweet Clover – Melilotus sp.

Red Clover – Trifolium sp.

Asters – Aster sp.

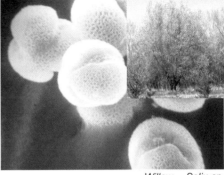
Willow – Salix sp.

Dandelion – Taraxacum sp.

Mustard – Brassica sp.

depositing her load the forager will usually solicit some honey from another bee, and then return to the field to forage. The pollen forager appears to be quite particular about which cell she chooses for depositing, often inspecting several cells before selecting one. However, the factors that determine her choice of a cell are not known. She dips her head into the cell briefly to inspect it, but does not appear to be looking for a cell of the same pollen type. Pollen from different sources is often mixed in a cell.

Once placed in a cell, and packed, enzymes and other substances are added. The pollen undergoes a lactic acid fermentation. A small layer of honey may be placed on top. Pollen thus stored is called 'bee bread.'

Pollen consumption – Young bees eat the greatest amount of pollen, but even older bees that have begun to forage also eat some pollen. Before the third day, larvae are fed a brood food consisting mainly of secretions of the hypopharyngeal and mandibular glands. However, older larvae are also fed pollen and honey. Pollen, as well as honey nearby the brood, are consumed by the colony at a greater rate than pollen and honey stored more peripherally in the comb. - SC

Pollen Gathering By Honey Bees

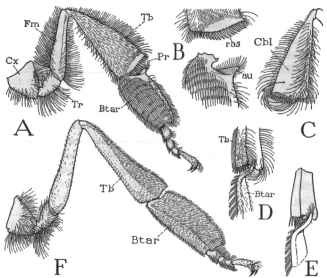

Hind leg of worker and drone and pollen press of worker.
A, right hind leg of worker, posterior (inner) surface. B, ends of tibia and basitarsus of hind leg, separated, posterior. C, pollen basket (corbicula) on outer surface of hind tibia of a worker. D, pollen press between tibia and tarsus of hind leg of worker, dorsal. E, same with tibial hairs removed. F, right hind leg of drone, posterior.

Explanation of Abbreviations
au, auricle
Btar, basitarsus
Cbl, pollen basket, corbicula
Cx, coxal cavity
Fm, femur
ras, rastellum, rake
Tb, tibia
Tr, trochanter

(from Snodgrass)

The hind leg, lateral view. The ascending femur articulates with the descending tibia, the base of which is the basitarsus. The pollen basket is comprised of the curved hairs surrounding the glabrous central area of the tibia. (from Erickson)

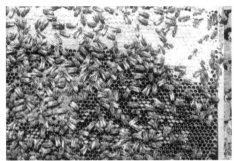
Pollen is usually stored near and around the edge of the brood nest. Honey is often stored-at the edge of the pollen storage area.

A forager returning with a load of pollen to be deposited inside the hive.

Close-up of bees with full 'pollen baskets.'

Individual pollen pellets, all from a single source.

POLLEN MITES – (see MITES ASSOCIATED WITH HONEY BEES, Pollen mites)

POLLEN SUBSTITUTES AND SUPPLEMENTS – A pollen substitute is supposed to replace pollen completely and a pollen supplement is a mixture of a pollen substitute and natural, bee-collected pollen. Supplements usually contain 10 to 20 percent natural pollen that serves primarily to make them attractive and palatable to the bees. Both substitutes and supplements can be made more attractive to the bees by adding sugar syrup or honey. Adding honey always carries some danger of contamination with materials that could cause disease. If the pollen added to supplements is contaminated with chalkbrood mummies, American foulbrood spores or European foulbrood scales these diseases can be spread in this way. In fact, chalkbrood may have entered North America via some contaminated pollen from Europe that was imported and sold as bee food.

Soybean flour and brewer's yeast are both often used in pollen supplements and substitutes. Most beekeepers make their own supplements or substitute but pollen substitutes are available commercially. A popular formula contains three parts soybean flour, one

Which pollen supplement do bees like best? Ask them with a 'supplement sampler.' Put one sample of each on a representative number of hives, and check in one, three and five days. They'll eat the tastiest one first, but usually, they'll eat them all. These samples were from a taste experiment carried out on A.I. Root Company hives 20+ years ago.

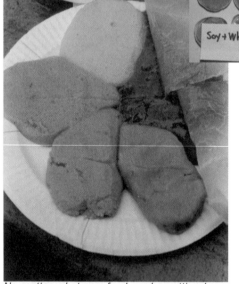

No matter what you feed, make patties large enough that you don't need to refill them too often, but not so large that they dry out before being consumed. Leave the wax paper on top and bottom to keep patties moist. Supplements with lots of sugar are attractive, but be careful on how much you are paying per pound for that sugar in freight and actual cost.

part pollen, two parts white granulated sugar, and one part water. If pollen is not used one may substitute brewer's yeast. The ingredients are blended into a stiff dough that is made into patties that are placed on wax paper or sometimes wire mesh for support. The patties, with the support underneath, are placed on the top bars of the brood chamber in the spring or whenever protein feeding is required. Once one starts feeding a supplement or substitute in the spring it should be continued so that some is always available until the first natural pollen is collected. Disrupting pollen substitute or supplement feeding can be disastrous and even lead to the death of a colony that over-expands its brood nest based on this artificial food. Honey bees reared in the spring do not have a large fat body or body reserves that can be used when

Pollen Substitutes And Supplements

Open feeding dry supplement works quite well when many colonies need feeding. Make sure the powder is well protected from rain and dew and stays dry. Some commercial beekeepers simply pour it on boards, or even the ground every day. Enough is put out to feed every colony, but not so much that there's any left by nightfall or when dew forms.

When open feeding, foragers will walk or even fly right into the powder, covering themselves with the dust. Then, sometimes they'll rise above the mix, comb and gather and pack and return to the powder for more. Sometimes they simply return to the hive covered in dust as it is too dry to pack.

their protein source is no longer available.

Several commercial substitutes are available now. We include a chart from one of the manufacturer's advertisements. You can see there is a vast difference in the composition of these materials.

	Bee-Pro®	Pollen Sub Plus	Brewers Yeast	Feed Bee®	Recommended Amino Acid Profile DeGroot
Crude Protein	48.50%	40.80%	43.00%	36.4%	
Carbohydrate (sugar)	4.90%	10.00%	28.70%	51.8%	
Amino Acids					
Arginine	6.93%	5.54%	4.50%	1.97%	3.00%
Histidine	2.69%	2.15%	2.30%	0.86%	1.50%
Isoleucine	4.94%	3.95%	4.50%	2.10%	4.00%
Leucine	8.36%	6.69%	8.50%	4.62%	4.50%
Lysine	6.28%	5.02%	6.40%	1.87%	3.00%
Methionine	1.59%	1.27%	1.70%	0.91%	1.50%
Phenylalanine	5.26%	4.20%	4.40%	2.15%	2.50%
Threonine	4.24%	3.39%	4.40%	1.63%	3.00%
Tryptophan	1.25%	1.00%	1.20%	0.48%	1.00%
Valine	5.52%	4.42%	5.80%	2.01%	4.00%

POLLEN TRAPS – A pollen trap makes it possible to collect the pollen from the legs of foraging bees as they enter a hive. This device fits onto or replaces the hive entrance so that the bees are forced to walk through it as they enter the hive. On the inside of some traps are one or two pieces of five-mesh hardware cloth (five wires each way per square inch). Other traps use sheets of metal or plastic with precisely punched holes, allowing the same amount of passage space. As the workers pass though the narrow openings their hind legs are pushed to the rear and as they rub against the wire or sides of the holes the pollen pellets are scraped off. The pollen pellets that are removed drop into a tray or box below that is usually covered with eight-mesh hardware cloth. The eight-mesh wire is used so that the bees cannot enter the collecting box and remove the pollen. Various covers are used with pollen traps so that the pollen will not become wet if it rains.

Some bees escape having the pollen removed because these devices are not perfect, and some bees learn to fool the wire grid. A good pollen trap will collect at least 50 percent of the pollen pellets from bees that pass through the grid.

A major problem with pollen traps is that drones are too large to pass through either the five-mesh hardware cloth or the holes. Most pollen traps have special exits for the drones.

Pollen traps place some stress on a colony, especially those that are rearing a large quantity of brood. Colonies respond to pollen traps by having more workers forage for pollen. Honey bees apparently measure the success of pollen foraging by the amount of pollen that is stored in of the hive. Thus, when there is a shortage of pollen in the hive, when a pollen trap is used, more bees forage for pollen. It is usually recommended that pollen traps be removed or inactivated for a few days every week or 10 days to allow normal pollen foraging.

The other obvious problem one encounters in maintaining pollen trap efficiency is to make sure the colony has only one entrance. Cracks, holes or loose covers will be found quickly when the added stress of a pollen trap is added. Bees will avoid a trap if possible. Pollen is a very attractive food for a great number of creatures besides honey bees. You must routinely empty your trap. Too, pollen readily absorbs moisture, and can

Sundance Pollen Trap. (Ross Round photo)

Pollen Traps

Pollen in drawer ready to be collected. (Ross Round photo)

A different approach, with the punched-hole grid vertical, rather than horizontal. This trap opens from the rear of the hive to empty the screened collection tray.

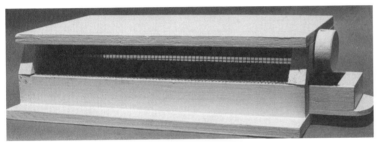

A simple front-door mounted pollen trap. The knob on the side turns the grid to 'close' the trap (the visible grid inside), or open it, allowing unobstructed entry. No drone trap on this model.

When pollen is abundant, a colony will often store nearly entire frames of it. An overwintering colony can use two to three full frames of pollen to feed brood in the spring when brood-rearing starts, before pollen-bearing plants are blooming.

655

become a damp, soggy lump if not removed. The U.S. desert southwest is an ideal location for pollen collection because of the very low humidity. Moist pollen supports the growth of fungus, also. In more humid locations, daily collection is usually necessary.

POLLINATION – The transfer of pollen from the anther, the male part of a flower, to the stigma, the female part, is called pollination. Some plants self-pollinate but in most instances an outside agent, such as wind, water, bird, mammal, or an insect, especially a bee, is responsible for pollen transfer. Some plants would not be pollinated and would produce no seed if it were not for insects. The insects called bees, of which there are many species, are especially well suited to be pollinators (see BEE, DEFINITION OF). The hairs on their bodies are branched (plumose) so pollen is caught in them easily and thus carried from one flower to the next. The honey bee is an especially hairy creature and an important pollinator.

Agricultural crop pollination – In North America the agricultural scene is changing rapidly. Less than two percent of the people in the U.S. produce the food for the rest and that number is growing smaller rapidly. The success of modern agriculture is based on specialization. Farmers use larger and more sophisticated machinery and make plantings in areas where the soil and climate make maximum yields possible. Part of modern agriculture's success is based on renting honey bee colonies in numbers sufficient to pollinate a large crop effectively in the very short time the crop is in bloom.

In one nationwide survey, it was found that more than half of the colonies of

Honey bee branched, or plumose hairs, on abdomen of worker. Pollen is easily caught in these branched hairs and moved from flower to flower. (Erickson photo)

bees in the hands of part and full-time commercial beekeepers are rented for pollination. Beekeepers can rent their bees for two or more different crops during the spring and summer. The demand for hives for pollination is increasing. Honey bees play an important role in the success agriculture has enjoyed. However, in the past when plantings were smaller and widely scattered, beekeepers usually were not paid pollination fees. Often growers made land for apiary sites available, and sometimes the beekeeper paid a small

Common U.S. Commercial Pollination Crops

Apples

Wild Blueberries

Vine crops

Plums

rental fee. The beekeeper's reward was the honey produced.

Today, the situation has changed. A small but growing number of beekeepers devote themselves exclusively to providing bees for pollination; for these beekeepers honey production is a sideline, or even an inconvenience. For example, some beekeepers in Florida use the orange honey flow (which is considered one of the major honey flows in the country) for producing bees for pollination in the north in May, June and early July.

Ninety-five percent of the nearly one and a half million colonies of bees rented for pollination are used on 14 crops in the following order: almonds, apples, melons, alfalfa seed, plums/prunes, avocados, blueberries, cherries, vegetable seeds, cucumbers, pears, sunflowers, cranberries, and kiwifruit. The remaining crops where pollination is demanded, such as squash, pumpkins, birdsfoot trefoil and miscellaneous legumes, use the remaining five percent of the colonies. Some crops, such as sunflowers, where pollination is required to set seed, the honey production is sufficiently great that the beekeepers are paid in terms of the honey they produce. Details concerning the use of bees for pollination may be found in Morse, R.A., and N.W. Calderone. "The value of honey bees as pollinators of U.S. crops in 2000." *Bee Culture* 128. Supplement, 15 pages. 2000. Additional studies are slated to be published spring, 2007.

Almond pollination – The single U.S. crop that requires the most honey bees for pollination at one time is the almond crop in California. Approximately 580,000 acres will bloom during February and March of 2007, and more are coming on line every year. Almond growers tend to rent two colonies per acre (four per hectare). Therefore, nearly 800,000 colonies from out of state need to be moved into California to join the 400,000 "resident" colonies to meet the need. At that time, around 17% of the country's commercial bee population was in California. Following bloom, not enough forage is available to support so many colonies. Beekeepers use some of the bees for pollination of other fruits, a diminishing number to produce orange honey, and most move to other states for pollination, honey production, or move again to be divided, requeened and sent to honey producing areas. - ECM

Pollination contracts – Most arrangements between beekeepers and growers who rent bees are verbal only and concluded with a handshake. Some beekeepers give a contract to growers describing hive size and colony strength. In cases where brokers are involved in renting bees, contracts between the broker and the beekeeper are more common.

When beekeepers indicate what strength colony is being rented they state the number of frames containing brood, not necessarily the number of frames full of brood. Since a broodnest takes the shape of a ball, or football, within the hive, the outer frames will not be full of brood. However, if a brood nest stretches across a given number of frames one has a good estimate of the population of the colony. Bees will not have more brood than they can cover during inclement weather. When colonies of bees are rented for pollination it is assumed that the colony is queenright. Beekeepers usually rent colonies that occupy two full-depth or one full with one half-depth brood chambers.

As stated, pollination agreements have traditionally been settled with a handshake between old friends. However, the trend is changing to corporate ownership of farms and orchards, while beekeepers continue to be self-owned businesses. Dealing with the legal strength of a corporation can be daunting, and expensive if required.

Even if a contract is not used, a review of a good contract by both parties can remind each of their respective responsibilities in the agreement, and reinforce the necessity of cooperation by both to make the venture successful.

For additional legal information and suggestions, refer to *Honey Bee Law: Principles And Practice*, by Sylvia Ezenwa, published by the A.I. Root Company. Numerous sample contracts are available on the internet.

Pollination fees – There are no fixed fees, or methods of fixing fees, for colonies that are rented for pollination. In most parts of the U.S., the rental fee is controlled by the number of colonies available and the willingness of

Pollination Contracts

Salient points to include in a pollination agreement, whether written or verbal.

- Contact information for grower and beekeeper
- Crop and location
- Number of colonies needed
- Payment per colony
- Payment schedule e.g. half on delivery, remaining when removed
- Minimum strength and health of colonies that will be paid for and frames of bees and brood, queen, food
- Average strength to be paid for
- Bonus for colonies exceeding average strength, per frame
- Evaluation procedure and payment
- Delivery notification or % bloom
- Delivery access – roads, damage, assistance
- Removal date or % bloom
- Liability while on location
- Access during bloom for maintenance
- Location and colony count/drop
- Pesticide application warning and procedure
- Additional fees for additional moves
- Water access
- Contract cancellation or transfer

beekeepers and growers to negotiate prices. The number of colonies being rented for pollination for a fee is increasing rapidly. As the demand for bees grows, pollination fees are increasing too. In some areas of the country rental fees have increased dramatically because of a shortage of colonies. As a result some beekeepers are shifting their operations away from honey production and are becoming more concerned with pollination.

When considering what to charge, include the opportunity costs of pollinating a crop –

- Will you lose all or part of honey crop? All or part of a specialty honey crop?
- Do you have adequate transportation and labor?
- Is your equipment in good shape?
- Do you need the income?
- Do you have the time, and can you react on short notice to move bees?
- What resources are available if you don't have enough bees, transportation becomes unavailable, or labor doesn't show up?
- What would it cost you to obtain the correct number of colonies of the right strength from other beekeepers in the event you cannot supply the full compliment contracted for?

Greenhouse pollination – Honey bee colonies have been used somewhat effectively in greenhouses and growth chambers for pollination for many decades. Since greenhouse pollination is usually done in the winter or early spring, the colonies cannot be placed outdoors when finished pollinating, so the colonies usually die. Several things can be done to extend a colony's life and perhaps to allow it to recover after several weeks in a greenhouse.

Only small colonies, usually those with only 5,000 to 10,000 bees and a queen, are wanted or needed in a greenhouse. Beekeepers who rent colonies for commercial greenhouse pollination usually make up the colonies in the fall, measuring both the number of bees present and the amount of food. A laying queen must be present to stimulate pollen collection.

It is a curious fact that the bees themselves regulate the number of bees that will visit a patch of flowers. If, for

Pollinating crops in greenhouses can be tricky. Full sized colonies aren't needed and often just nucleus colonies are used. An often significant problem is older foragers flying to the ceiling or corners, becoming lost, and dying away from the colony. If you are prepared for this it is an acceptable loss, as naive foragers will learn to negotiate windows, ceilings and constrained foraging areas.

example, one has a bouquet of flowers attractive to bees and a certain number of bees visit the bouquet, it is not possible to increase that number greatly without adding flowers or otherwise boosting the nectar content. However, if one doubles the number of flowers in the bouquet then the number of bees visiting it will soon be about double.

The life of a colony in a greenhouse may be extended by feeding it medicated sugar syrup for control of nosema. Providing pollen supplement or pollen substitute is also helpful. Pollen supplement, which contains some natural pollen, is more effective under most circumstances. Since greenhouse plants are watered frequently it may not be necessary to provide water for the bees, but this should be considered, especially if soil-applied pesticides are used, or liquid feeding occurs on a regular basis.

Bumble bees became the primary pollinator of greenhouse tomatoes in the 1980s. The ability to propagate bumble bees year round, methods to ship bumble bees, and an increased demand for greenhouse tomatoes all helped stimulate the interest in bumble bee pollination. Bumble bees work very effectively in greenhouses.

Hand pollination – Most fruit trees, especially apples, produce pollen that will not fertilize flowers of the same variety. That is, pollen from a McIntosh apple tree cannot be used to pollinate another McIntosh apple flower; the pollen must be from a different apple variety. This is nature's way of preventing inbreeding.

Under the old-fashioned system of planting apples there were 27 trees per acre. Every third tree, in every third row (one tree in nine) was a pollinizer, that is, it provided pollen to fertilize the flowers on the other eight trees and they in turn provided pollen so that the pollinizer too would set fruit. In the modern system, where apple trees are planted in tightly-packed rows, growers either plant every third row to a pollinizer or they intersperse pollinizers in the row. Interplanting varieties is a nuisance for several reasons: It is necessary to pick the varieties separately and this slows picking. Different varieties often have different pruning, fertilizing, harvesting times and other requirements. Some varieties may be more susceptible to insect damage than others. For these reasons some growers make solid plantings, though extension services strongly advise against this practice.

It was shown in the 1920s that it was possible to collect pollen from various fruit trees by hand, freeze it and store it for a year. It would remain viable and could be used for pollination the next year. The hand-collected pollen was usually diluted with a non-toxic substance with the same grain size. It was also demonstrated that pollen could be placed in special dispensers on a honey bee colony entrance and, if these were properly designed, the bees would become dusted with the pollen as they left and returned to the hive. Fruit or nut trees in a solid block can be pollinated in this manner. Almond nut production benefits from this techniqueHowever, the pollen must be collected by hand and the dispensers require close care; for both of these reasons the practice is expensive. Some bees that pass through a pollen dispenser stop and clean themselves after leaving their hive; in this way the pollen on their bodies is packed into their pollen baskets and returned to the hive where it is used as bee food.

Honey-bee-collected pollen loses its viability after about two hours. Nectar is added to the pollen as it is collected and this may somehow interfere with the physiology of the pollen. In Japan, where hand pollination is practiced in some areas, it has been discovered that one may collect honey-bee-collected pollen, wash and dry it, and it will retain its viability. When this is done the pollen must be collected rapidly from a colony entrance with a pollen trap before it loses its viability.

For the most economical management of an apple orchard, or other fruit or nut crops where cross-pollination is demanded, there should be proper interplanting of varieties. Where this has not been done, it is best to replant some trees or to graft pollinizers onto existing trees. One must make certain that the pollinizers bloom at the same time as the variety being planted. When interplanting is done there will be a delay of a few years before the new trees or grafts grow large enough to be effective. One method of providing pollinizers under these circumstances is to take large bouquets into the orchard. These are scattered around the orchard so that the flowers on the bouquets will be visited by bees too. The bouquets should be made up of limbs about two inches in diameter, several feet long and are placed in 55-gallon barrels half-full of water.

Some varieties are more effective as pollinizers than are others. For example, it was found that Tangelos (a variety of citrus) required cross-pollination to set fruit; however, the Temple variety of orange was more than twice as effective as a pollinizer of the Orlando Tangelo (that is, more than twice as much fruit was produced) as was the Valencia variety. Extension services are usually in a position to advise what varieties are best used as pollinizers.

Orientation of bees moved for pollination – A study on how rapidly honey bees reorient after being moved to a new location was conducted in a pear orchard in Ontario, Canada. A colony was moved from some distance away into the orchard at 11:00 a.m. at a time when the temperature was 80°F (27°C). It was observed that the first foraging bee returned to her hive carrying pollen 13 minutes after the colony entrance was opened; during the 14th minute another six bees returned from successful foraging trips carrying pollen (see MIGRATORY BEEKEEPING).

When colonies are moved to a new location, placed close together, and the colony entrances opened in midday, some confusion and drifting do occur. However, commercial, migratory

You can imagine the confusion returning foragers may experience when confronted with a home scene like this.

Apple orchard floors should be kept free of flowering plants that attract bees for several reasons. •The flowers on floor plants may compete with apple blossoms for attractiveness, thus reducing visitation, pollination, fruitset, and yield. • Floor plants compete with orchard trees for nutrients and water. • Blooming plants on orchard floors, after the trees are done blooming, continue to attract honey bees and other pollinators. Pesticide applications to these trees will fall onto floor blooms, killing those pollinators visiting floor plant blossoms. Orchard managers should mow orchard floors, apply herbicide to keep plant growth to a minimum, or encourage growth of non-bee-visited plants – grasses, for instance.

beekeepers have never considered this sufficiently serious that special precautions are taken. The drifting of bees from one colony to another, after an apiary is moved, can be reduced by moving the bees into their new location at night.

Plant attractiveness – Plants vary in the quality of the nectar they produce. Pears, for example, have nectar that is usually low in sugar, on average, about 15 percent sugar. Other plants in flower at the same time may have much richer nectar. Apple nectar averages 40 percent sugar. Since honey bees can detect even small differences in nectar quality they will soon find the best sources of food available and abandon plants such as pears that have low quality nectar.

Unfortunately for growers seeking fruit tree pollination, dandelions and yellow rocket, which are in flower about the same time as apples, have nectar that is usually as rich or richer than that of most apple varieties.

It has also been observed that the percentage of sugar in nectar will vary during the day on the same plant. Plants that have more exposed nectaries may produce nectar with only 20 percent

Spraying blooming plants when bees are present – whether on the trees, or the orchard floor plants, is against label instructions, illegal, and incredibly destructive.

sugar in the morning but as the day progresses the combined effects of wind and sunshine may dry the nectar so that it may be twice as rich by mid-afternoon.

In England, it was observed that dandelion is a good producer of both pollen and nectar and attracts honey bees away from apples. Observations indicate the same is true in North America. However, casual observations in an orchard can lead to erroneous conclusions. Dandelions produce most of their pollen in the early morning and by mid-afternoon none is produced. If one observes dandelions in the afternoon one might think dandelions were not offering competition to the apples (Free, J.B. "Dandelion as a competitor to fruit trees for bee visits." *Journal of Applied Ecology* 5:169-178. 1968).

Apple orchards should be mowed while they are in flower to reduce the number of flowering and competing weed plants. In the case of pears it may be necessary to move colonies of bees into the center of a pear orchard when it is in full bloom. Many of the bees from recently-moved colonies will first visit flowers close to the hive before they, and the hive's scouts, find richer sources of food elsewhere.

POLLINATION IN HOME GARDENS – Honey bees play an important role in the pollination of many of the fruits and vegetables grown in the millions of home gardens across North America. Solitary and semisocial ground-and twig-nesting bees, such as bumble bees and leafcutter bees, are also important but are not always adapted to pollinate certain crops where the size and behavior of the insect on the flower is important. In many instances, it is the honey bees owned and managed by hobby beekeepers, often those with only one or a few colonies that provide many of the needed bees. In recent years the parasitic mites have reduced the population of feral bees, therefore the managed colonies are even more important. - RAM and EM

POLLINATION OF CROPS FOR WILDLIFE – Much of our wildlife lives on wild fruit, nuts and seeds. Many of these crops are self-or wind-pollinated but a large number also depend on foraging bees and other insects for pollination. Honey bees have in the past provided a steady and reliable source of pollinators for these crops. However, the recent introduction of parasitic mites, killing feral colonies, endangers this source of bees.

Pollination Shortcuts

Home garden crops require pollination just like commercially-grown crops.

POLLINATION SHORTCUTS – With few exceptions there is no substitute for a good queenright, brood-rearing colony of honey bees to pollinate any agricultural crop. Equally important is the fact that one does not easily fool a honey bee. If a foraging worker does not receive a proper reward for her work she soon loses interest. Equally important, she must receive a proper reward to stimulate others to forage.

An important reason for using honey bees as pollinators in commercial agriculture is that numbers can be measured, their diseases and other weaknesses are understood, and in emergency situations more colonies can be brought into the crop overnight. In other words, there are few pollination shortcuts! However, that has not stopped people from making a variety of suggestions for improving pollination by attracting bees to a crop. These have

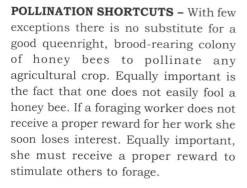

Wildlife at all levels require the food and habitats provided by honey bee-pollinated plants.

ranged from spraying crops with sugar syrup, to using scents and pheromones. These efforts have had mixed results, but in marginal situations may provide enough attraction to obtain fruitset for a crop.

One technique that may have value, but requires additional work, is to feed colonies plain sugar syrup to switch bees away from nectar collection and into pollen collection. This idea appears to have merit but no data support this thought, only what appears to be armchair biology.

What happens when a crop to be pollinated is sprayed with a rich sugar syrup? Years ago researchers wrote that it helped to attract bees to the flowers but the idea never gained momentum suggesting it was ineffective. More research followed and another researcher reported that fewer bees visited the blossoms but many spent time licking the syrup from the leaves. The sugar syrups tested contained between 30 and 50 percent sugar.

Starting in the 1930s, it was advocated that you could use the bees' own language to stimulate foraging on a crop using a scented sugar syrup; Again, the idea flopped but not before at least 27 researchers made tests to determine where the truth lay. What was confirmed was that while you may temporarily stimulate foraging, the bees must receive a proper reward for visiting a crop or they soon lose interest. In other words, if the crop does not produce a rewarding amount of nectar or pollen, or both, the foraging bees look elsewhere for food.

However, feeding sugar syrup to colonies of honey bees may cause them to rear more brood, which, in turn affects the amount of pollen they collect since bees must have pollen (protein) to feed brood. It was also found that when the sugar feeding stopped most of the bees reverted to nectar collection.

Research by several people has shown that feeding sugar syrup also occupies the house bees that process incoming nectar. When foraging bees cannot find house bees to take their loads they may turn their attention to pollen collection.

Some of those who have worked with kiwi fruit pollination have found that the time of day the syrup is fed is important. Feeding at 9:00 a.m. and 11:00 a.m. was

Increasing the brood in a colony, either naturally by increasing egg production by the existing queen; or by replacing a failing queen with a young, vigorous, healthy queen; or, by adding brood from other colonies will stimulate pollen collection in a colony to meet the increased demand for protein.

much more effective than feeding at 5:00 p.m. However, kiwi flowers release their pollen before noon and the method might have different results if the pollen were released in the afternoon or all day.

Other tests on ways to attract honey bees to a crop for pollination purposes are still underway. When you think about any pollination shortcut it is important to remember that honey bees are not truly domesticated animals. Any manipulations we make to change their habits must work within the limits of their biology.

POPULATION DYNAMICS OF HONEY BEE COLONIES – (see ANNUAL CYCLE OF HONEY BEE COLONIES)

POWERS APIARIES, INC. – Irvin Powers (1894-1977) built Powers Apiaries into the largest beekeeping company in the United States. In 1976 the corporation owned 28,000 colonies and produced 3.8 million pounds of honey. The bees were in several states with branches in Arizona, and North Dakota.

Jim Powers wrote "In 1978 we produced 5,000,000 pounds of honey with 29,000 colonies of bees. On January 1, 1979 we purchased 5,000 colonies of bees from Les Walling at Jamestown and Williston, North Dakota. Therefore, in 1979 we ran 33,000 colonies of bees and operated eight extracting plants.

Powers Apiaries at one time had one-half interest in the Kona Queen Company with Weaver Apiaries of Navasota, Texas, the other company involved. Kona produced queens in Hawaii for sale in the continental United States. Powers also had a half interest in the Molokai Honey Company which is located on the Hawaiian Island of Molokai.

Irvin Powers was born in Wallowa, Oregon, but his parents moved near Parma and homesteaded in 1898.

Irvin's father, Francis Ashbury Powers, kept about 100 colonies, invented a pump that is still manufactured, and developed a strain of wheat that was widely grown for many years.

When he was 16, Irvin purchased 100 colonies from his father. Starting in 1916, he operated bees in partnership with Herman Crowther and Howard West. During World War I Irvin Powers served in the Artillery Branch of the Army.

In 1919 he moved his share of the bees to Emmett, Idaho, and ran bees there until returning to Parma in 1932.

Irvin developed a self-spacing frame and extractor in 1927 which made it possible to extract the combs while they are in the supers. This extractor was widely used in the West and Midwest. He produced comb honey until 1927. By 1939 Powers was operating 8000 colonies around Parma. In the same year Irvin, the Bradshaws, and Howard Vanderford started a branch bee operation in Ellensburg, Washington, which was later sold. In 1943, 1200 colonies were purchased in the Lower Colorado Indian Reservation near Parker, Arizona. Later they were expanded to 5000. Powers bought one-third interest in the Cloverdale Honey Company at Manhattan, Montana, in 1945. This interest was sold in 1970.

The Bismarck branch was established in 1955, and between 4000 and 7000 were kept there. The Oakes branch was started in 1961, and between 5000 and 7000 colonies were run there. In 1963

Jim Powers

the Babson Park, Florida, operation was begun. The company had 5000 colonies there and all the bees were moved to North Dakota for the summer.

In 1973 Powers moved to the large island of Hawaii when they purchased 1200 colonies from Woodrow Miller, with 4000 colonies on the island. The same year the Kona Queen Company was started. Later the Molokai Company was established.

In early 1976 and 1977 the colonies in Arizona were severely damaged or killed from spraying Penncap M on alfalfa to control aphids. Therefore, in 1978, 3000 colonies were moved near Bakersfield, California, for $10.50 a colony to pollinate almonds. When the pollination was over these bees were moved into the Arizona desert away from the irrigated areas. The other 1000 Arizona colonies were on river locations which are not close to alfalfa fields. About 2500 of the Parker colonies were rented each year to pollinate cantaloupes.

In 1977 spraying of Supercide to control head moth on sunflowers caused serious damage to thousands of Power's North Dakota colonies. A few of the North Dakota bees were rented to produce hybrid sunflower seed for $20.00 per colony.

Powers Apiaries, Inc. was a small business corporation with James Powers, Irvin's son, serving as president and Blaine Simpson as the general manager. The company was a member of Sioux Honey Association and Jim Powers was a Sioux director.

PRICE SUPPORT – The federal honey support program was established by the Agricultural Act of 1949 to provide for the maintenance of adequate honey bee colonies for pollination of the nation's important seed, fruit, nut and vegetable crops and to provide for market stability for honey producers. After 1950 the program evolved into a combination of a purchase program (where the beekeeper could sell honey to the USDA Commodity Credit Corporation [CCC] at the support price) and a loan program (where the price support could be obtained by taking out loans using honey as collateral).

After 1981 honey support prices exceeded the average domestic market price. Beekeepers found it profitable to default on their loans with the CCC and forfeit their honey to the government. Domestic packers, processors and food manufacturers found it more profitable to import the lower-priced honey on the world market than purchase domestic honey. To correct some of these problems the 1985 Food Security Act established progressively lower support payments for 1986 through 1990 crop years and lowered the loan repayment rate set by the Secretary of Agriculture to minimize the number of loan forfeitures, reduce the costs incurred by the government in storing honey and increase the competitiveness of domestic honey in the world markets. By 1990 the honey forfeitures were zero.

The Food Security Act also initiated a program of loan deficiency payments, which rewarded farmers/producers for not applying for a commodity loan on their production for a particular crop year.

The 1990 Farm Bill further decreased loaned amounts for beekeepers over the next five years and the program was finally terminated in 1996.

In the 1990s increased honey imports and falling global prices put renewed pressure on beekeepers' incomes, even the growing numbers of thosewho were taking their hives on the road for pollination. In 1995, rather than try to resurrect the moribund honey loan program, the two major beekeeping organizations (ABF & AHP) petitioned the International Trade Commission, an arm of the U.S. Commerce Department, to end alleged Chinese honey dumping. The organizations won, and for five years honey imports from China were regulated by a volume quota and a price floor. For a year bulk honey prices spiked to almost US$1.00 a pound but then declined again as other countries filled the import vacuum. By 2000 Argentina was exporting almost 90 million pounds (41 million kilograms) of honey to the U.S. while American production remained flat. In that year, the American Honey Producers Association and the Sioux Honey Association (but not the American Beekeeping Federation) went back to the ITC to complain about both Chinese and Argentine honey dumping. Again, these organizations won.

In 2000 all three beekeeping organizations joined forces to get honey back into the US$180 billion 2002–2007 Farm Bill. Under that legislation the Farm Service Agency would administer a) non-recourse marketing assistance loans and b) loan deficiency payment (LDP) program for the 2002 through 2007 honey crops.

For one year, qualifying beekeepers received loan-deficiency payments of approximately 14 cents a pound on the difference between a USDA-set price of $0.60/lb and the market price. Then the honey price began to rise, reaching as high as $1.50 in mid-2002. (No LDP could be paid if the honey price exceeded $0.60/lb.)

Under the non-recourse market assistance loan program, the producer can put up the crop in approved containers, and then go to the government for a loan for up to nine months using the honey as collateral. The producer can use the money to pay expenses, and then pay the loan back with minimal interest, hoping that the price will rise in the interim. In 2002, Discussion occurred among industry leaders about putting the honey under a crop insurance program, but that was still several years away.

The characteristics and existence of future honey support programs will depend on the administration's commitment to the industry and Congress' willingness to provide market stability for the dwindling number of full-time beekeepers who produce honey and, far more importantly, pollination for over US$15 billion worth of crops. - LSW and BM

PRITCHARD, MELVIN T. (1867-1943) – Pritchard was probably one of the greatest authorities on queen rearing. Few men had the knowledge he had of

Melvin T. Pritchard

all animate life: animals, birds, insects, worms and what not. For this, he went to the greatest resource of all, Nature.

Pritchard did much to improve the Doolittle system of queen rearing. He served the A.I. Root Company for five generations, beginning work for Homer H. Root, father of A.I. Root, as a chore boy. Later he worked for A.I. Root in his gardens for the magnificent sum of six cents an hour.

Later he began to rear queens. He continued to do so until he retired. He said he found that the old body would not respond like it used to. He was 76 years old when he passed away. He was queen breeder for the Roots for 42 years in which time he, himself, reared at least 150,000 queens.

PROPOLIS – Many plants produce gums and resins at wound sites or around buds or newly-emerging leaves. These substances waterproof the area and protect it from attack by potential invaders such as bacteria, molds, yeasts, fungi, insects and other pests. Honey bees often collect these gums and resins and use them inside the beehive where they provide the bees and their nest the same protection the resins give the plant. Beekeepers have given these materials the name "propolis." Just as the plant products vary in color and consistency, so does propolis; however, red- and yellow-colored propolis are probably most common.

Plant gums and resins offer both physical and chemical protection. The sticky nature of these materials serves to entrap many lower forms of life. Most of these gums and resins become hard in time, having a varnish-like appearance and providing further protection. The chemical protection is derived from materials called flavones that are an important part of the plant secretion. Flavones are natural plant products that have a large number of carbon atoms and show a high degree of antibacterial activity.

Inside a beehive or natural nest, honey bees use propolis to varnish the wood (or stone if the colony is in a cave). All cracks and crevices are carefully filled so that no noxious life form may live there and pose a hazard to the bees. Propolis also waterproofs the inside of the nest. When bees nest in hollow trees the propolis may prevent further decay of the tree itself, which is in the best interest of both the bees and the tree. The propolis coating may help lengthen the life of the tree. Interestingly, only the top and sides of a natural nest are coated with propolis. The bottom of the nest, where much debris often collects, is not protected.

Races of honey bees vary greatly in their tendency to collect and use propolis. Caucasian honey bees, from the Caucasus area that lies between the Black and Caspian Seas in Europe, are known for their tendency to collect great quantities of propolis. A colony of Caucasian bees may reduce a colony entrance so that only a few holes that are three-eighths to one-half inch diameter (1 to 1.5 cm) remain for bees to enter and exit.

The indents, small holes and rough spots on these top bars have been filled and polished with propolis. This keeps other organisms from using these small places to hide in or to grow in.

The bees propolized the rough, unplaned section of this board, but left the smoothed section nearly propolis free. (Morse photo)

The bees have used propolis to block a large area of this entrance, making the entrance easier to defend, controlling the environment, and keeping out pests and predators. (Morse photo)

Honey bees may also use propolis to entomb large animals that die or are killed by the bees in the hive. It is not uncommon to find a mouse or small snake dead in the hive. These animals are too large for the bees to remove. In the case of the mouse the bees will usually remove just the body hair and then cover the body with propolis. The propolis will eliminate any foul odor, or at least reduce it to a tolerable level, and at the same time microbial growth is inhibited.

The best-known plant gums and resins are frankincense and myrrh, which have been widely traded for thousands of years from the east coast of Africa, where they are produced, throughout the Mediterranean area. These substances have been incorporated into salves for wound treatment, and are sometimes burned because of the pleasant odor they

Here the bees have closed the space between top bars to reduce exposure. (Morse photo)

Looking up, into a hive body, with the bottom bars of the frames facing. Propolis has been used to close the space between the split bottom bars of one frame, and it is also visible on the edge of the box where it was used to seal the space between the hive body and the bottom board. Most noticeable is the now-deceased snake, who ventured into the hive, but did not leave. The remains are slowly being covered with propolis to keep decaying to a minimum.

When propolis is warm, it has the consistency of chewing gum and will stick to anything it touches.

Propolis will also stick hive parts together when the weather is warm. Another problem with propolis is that when the weather cools, propolis becomes brittle. When separating hive parts such as this, breaking the propolis seals results in a loud snapping sound which will irritate the bees.

produce. The odor that is associated with the burning of beeswax candles comes largely from the propolis in the wax. During the past few decades a small market has appeared for bee-collected propolis that is incorporated into salves, ointments, chewing gum, lozenges, creams, etc., for use in human medicine and for a variety of ills. No data exist that demonstrate that honey bees add anything to the gums and resins that they collect that makes propolis any different from the original material secreted by plants. However, there is a certain mystique about bees and a beehive that encourages this form of apitherapy.

Honey bees have a threshold of acceptance that is variable, that is, when they cannot find a material they seek, such as pollen, nectar or propolis, they may accept a substitute. Bees have been known to collect and use road tar, drying or soft paint, caulking compound and similar items in place of plant gums and resins. In the hive these materials may be distributed and used in the same manner as the gums and resins. This contamination with modern products makes propolis use in any modern medicines questionable. Bees also have been seen gathering propolis from abandoned bee hives.

During the comb honey era, which lasted from about 1880 to 1915, beekeepers searched for and found bees that used very little propolis. They did so because they did not want the propolis to make the wooden comb honey sections sticky, or to give them a bad or stained appearance. Bee-collected propolis has been used as an ingredient in varnish, and some believe it was part of the varnish applied to the famous Stradivarius violins (no good data document that thought).

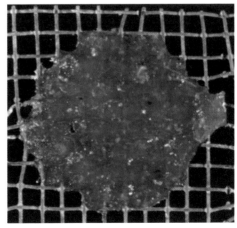

This screen mesh was used to cover a small hole in a super. The bees very neatly closed the opening with propolis.

Bees may be stimulated to collect propolis by placing rough wood in a hive that the bees will then cover with the material. Manufacturers of bee furniture plane hive parts to reduce the amount of propolis bees may be stimulated to collect. Some inventive beekeepers, who wanted to collect propolis for sale, found they could place a piece of wood with many power-saw cuts in a hive in place of a frame. The bees would fill these cuts with propolis that could then be removed with a sharp, pick-like blade or knife. However this propolis contains slivers of wood. Recently, beekeepers have used specially-designed plastic propolis traps placed under the inner cover. After most of the slits are filled, the traps are placed in a freezer. When the propolis is frozen the plastic sheet can be placed inside a plastic bag, then flexed. The propolis (sometimes) falls from the grid. This method produces clean propolis, free from any hive debris. For those who wish to learn more about propolis, a review paper was written by E.L. Ghisalberti, "Propolis: a review." *Bee World* 60: 59-84. 1979. There are several books available on the subject also, generally aimed at harvesting and producing medicinal components.

PUBLIC RELATIONS FOR BEEKEEPERS – Many people view beekeeping as an unusual or rare vocation or avocation. Honey bees sting, unlike other animals kept by people. This makes beekeepers' relationships with the public different and sometimes more difficult, since many people fear stinging insects.

Good public relations for beekeepers involve attempting to calm any fear of bees emphasizing the importance of honey bees in agriculture, pointing out the importance to wildlife of free pollination of crops and our own desire for flowers and flowering trees and shrubs around our homes and in our gardens. Most beekeepers soon realize that a gift of a jar of honey is often a key to good public relations.

A small number of communities have outlawed, or attempted to outlaw, beekeeping within their boundaries. This is not practical. Wherever flowers grow, bees and other stinging insects will also be present. They may live in trees, buildings or man-kept hives. If man-kept hives are absent the bees will find a niche somewhere.

Some beekeepers have found it enjoyable and worthwhile to lecture to civic groups and school children. A closed, plastic-walled observation hive is a good tool for such a venture. Allowing students and others to smell a beeswax candle, see foundation and/or new comb, touch pieces of propolis, and, of course, taste honey can all be features of a lecture or demonstration.

The beekeepers in an urban area can do much to make certain their bees are not a nuisance. This is discussed more fully under URBAN BEEKEEPING, ORDINANCES BANNING BEES AND GOOD NEIGHBOR BEEKEEPING.

PUPA – (see LIFE STAGES OF THE HONEY BEE)

PURE FOOD AND DRUG LAWS – (see HONEY PURE FOOD AND DRUG LAWS)

QUEEN – A colony of honey bees has a single queen under normal circumstances. She is the most important individual in the colony for two primary reasons: (1) She lays all of the eggs; (2) She produces pheromones that inhibit egg production by workers, (see LAYING WORKERS), inhibit her replacement, have a strong effect on colony morale and a multitude of other behaviors on colony inhabitants (see PHEROMONES).

Queens appear to make no decisions other than to determine if a cell is suitable and ready to receive an egg. A queen does not feed herself except for the few hours just after she emerges as an adult after pupation. The food she receives from workers is largely royal jelly, which provides the protein-rich nourishment necessary for her to lay so many eggs.

Beekeepers sometimes find two, and rarely three, queens in a colony. Records that have been kept when multiple queens occur indicate that this is usually a mother-daughter or a grandmother-mother-daughter situation. The older queen may still produce some eggs but her pheromone production – in both amount produced and the ratio of the several pheromones she produces – is still sufficient to inhibit her replacement. In most instances where two queens are found the older queen does not survive for more than a few months.

Many beekeepers do not realize that more than one queen can exist in a colony. When the beekeeper is looking for a queen and finds one the beekeeper assumes that is the only queen. Thus, the second queen is not found.

Once one of the multiple queens is found and removed, the new queen is introduced. Failure of acceptance then occurs. However the beekeeper may not realize the reason for failure.

Even if you wait a day or two, that does not assure acceptance if there is still a second queen in the colony. Did you remove the 'mother' and leave the young, vigorous daughter? You'll still have failure.

Assessing quality of – (see BROOD PATTERNS and QUEEN REARING: GRAFTING AND COMMERCIAL QUEEN PRODUCTION)

Cells – A queen cell is a special cell, larger than any other in the hive, in which a queen is reared. Queen cells hang parallel to the face of the comb, perpendicular to the worker cells, and are attached to the face of the comb or at the bottom of a comb. When queen cells appear on the face of the comb they are grown to replace an old or failing queen, called supersedure, or to replace a queen that was somehow lost, called

Queen cells and cups at the bottom of a comb are often found in colonies that are about to swarm.
(Left Tew photo; right Morse photo)

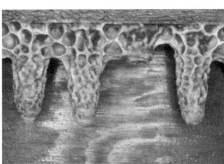

Well-formed, mature queen cells. (Morse photos)

Examining for queen cells – During the swarming season beekeepers conduct frequent (sometimes weekly) examinations of colonies to determine if they might be preparing to swarm and to take steps to prevent or control this. Since queen cells formed at a time when a colony is preparing to

emergency queen cells. Queen cells that are found at the bottom of a comb are found in colonies that are about to swarm; these are sometimes referred to as swarm cells.

The outward physical appearance of a queen cell is unique. Well-formed, mature queen cells have a surface that is wrinkled, much like that of the shell of a peanut. Queen cells with a smooth exterior usually contain inferior queens – that is, queens that weigh less, have fewer ovarioles, and therefore lay fewer eggs. Worker bees will often remove the wax on the tip of a queen cell, exposing the cocoon underneath. It is not clear why this is done but it is common, especially in strong colonies. It has been suggested that it may help the queen to emerge from her cell but this does not appear to be so. The tip of a queen's cocoon does not touch the tip of the wax cell that is the portion removed.

Examining for queen cups and queen cells. When brood is in both the top and bottom box of a two-box brood chamber, a good place to look for queen cups or queen cells during swarm season is on the bottom of frames. Sometimes drone brood will be there also, but careful checking will show the difference.

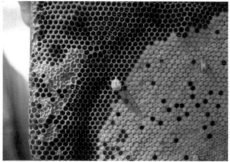

Queen cups on the face of a comb area, near the edge of the brood is most likely the precursor to a supersedure cell. (Morse photo)

swarm appear most frequently at the bottom of a comb, it is not necessary to examine individual combs for cells. During the swarming season colonies should be in two, three or four brood chambers.

To search for queen cells one splits the colony apart where the brood nest occupies parts of two brood chambers. Any queen cups, or queen cells, will be readily visible in the bee space between the two boxes. With a little experience, a beekeeper will be able to determine in a matter of seconds how crowded the colony is. The closer the colony is to being congested, the more queen cups will be present, but a precise value cannot be assigned to the meaning of a given number of cups. Queen cups containing eggs indicate an even more advanced case of congestion, though at this stage it may still be possible to stop swarming. If one finds either uncapped cells with queen larvae lavishly fed with royal jelly, or capped queen cells, swarming is almost certain to occur (see SWARMING HONEY BEES, Swarm control and prevention). The important point is that an assessment of a colony's condition during the swarming season, performed by splitting the brood nest apart, is a simple matter that provides beekeepers considerable information on the population status of the colony.

Queen cups – A natural queen cup in a colony is a round teacup-shaped cell that hangs downwards and is parallel to the face of the comb. Queen cups are the first stage in the construction of a queen cell. Natural queen cups are seldom found in a normal overwintering colony as they are generally removed by the bees in the fall. The first cups appear in colonies at about the time the first spring pollen is available. Their numbers increase as the season progresses and in a small way their presence in large numbers indicates a colony is healthy, strong, essentially normal, and on its way to becoming congested.

Some races and strains of bees tend to produce more cups, and more queen cells that other races, and this should be considered when examining a colony. Strains of bees commonly called Russians (named for the locality where they were first discovered by USDA scientists, and not a true race of bees) are known to expend a great deal of energy making, and then removing queen cups, and then cells. This activity

Commercially available wood bases and wax cell cups. Once common, now seldom used.

Queen

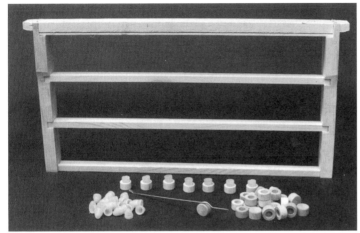

Wax cell cups, wood bases and a cell-holding frame.

is not swarm related, and does not reflect colony population or diminished queen pheromone distribution within the colony, but rather a continual readiness to abscond, or to have ready a new queen if the existing queen is suddenly removed.

Removing queen cups has no effect on preventing or controlling swarming. Beekeepers watch cups in the early part of the season as they are indicative of a colony's condition. The addition of new white wax to cups, called "whitening of cups," is also an indication that congestion is in its early stages. The appearance of the first eggs in the cups is more evidence and suggests swarm prevention should be considered (see Swarm prevention).

Artificial beeswax queen cups are made by bee supply manufacturers for sale to queen producers though many of those who grow queens make their own cups. Examination of natural cups reveals that their inside diameter varies greatly. Those who make their own cups usually use 5/16 inch (0.8 cm) diameter wooden dowels with rounded ends for

Wood bases attached to a frame bar with a drop of wax, and the wax cell cups attached to the inside of the cup. Larvae are being grafted into the cells.

Home-made wooden dowel for dipping and making wax cell cups. (Morse photo)

Homemade wax cell cups work well, and can be deeper than commercial cups.

dipping and making artificial cups for queen rearing. The wood is wetted before being dipped into hot beeswax; usually three dips are made, each being shallower than the first. This leaves a sharp edge on the rim of the cup onto which the bees may add more wax to make the queen cell. Plastic queen cups, however are favored by most queen producers because they are inexpensive, uncontaminated, easy to use, and readily accepted by the bees. Hand-dipped wax cell cups, though still available, are quickly becoming a tool of the past.

Plastic queen cell cups are now the standard in commercial queen production. Easy to use, readily accepted by the bees, uncontaminated, and economical, they are available in several colors and mounting styles.

Drone-laying queens – A drone-laying queen is one that has exhausted her supply of sperm and cannot fertilize the eggs she produces. As a result the eggs she lays are all male. Obviously, colonies with such queens are doomed. However, drone-laying queens may survive for many months. They continue to lay eggs because the worker bees apparently cannot detect the problem and replace their faulty queen.

Queens produce a variety of pheromones by which they are recognized by the worker bees. These pheromones cause the bees to feed and groom the queen differently from the way in which all other bees in the hive are treated. The royal treatment given the queens is a result of these pheromones, which is not related to their ability or inability to fertilize or not fertilize eggs.

Effect of queen age – Excellent data, on a sufficient number of colonies and with several studies extending over three or more years, show that older queens are inferior to young ones. This supports the long-held beekeeper belief, which is mentioned in textbook after textbook,

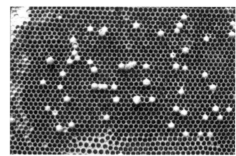

Two views of the result of a drone laying queen. Drone cells – the large, dome capped cells – initially appear randomly on a worker-size-cell frame. Eventually, all eggs laid will be drones, and without repair, the colony will perish. (Morse photos)

that annual requeening is of great benefit for building large colony populations. Large populations mean large crops will be harvested.

In a study conducted in cooperation with a commercial beekeeper in England it was shown that queens that had survived two winters were more than three times as likely to head colonies that would attempt to swarm than would colonies headed by queens that had survived one winter (Simpson, J. "The age of queen honeybees and the tendency of their colonies to swarm." *Journal of Agricultural Science* 54: 195. 1960.) In another study conducted in Connecticut, colonies headed by queens less than a year old had nearly twice as much brood in the spring as did queens more than a year old (Avitabile, A. "Brood rearing in honeybee colonies from late autumn to early spring." *Journal of Apicultural Research* 17: 69-73, 1978.)

Egg laying rate of a queen – The number of eggs a queen honey bee lays in a day is controlled by many factors. The most important of these is the number of adult bees in the colony. Queens will not lay more eggs or have more brood than the bees can keep warm in cool weather. Food is also a controlling factor since bees without pollen will discard larvae and eat eggs. A queen is also limited in the number of eggs she may produce by the number of ovarioles she has in each of her two ovaries. Data indicate that the size of a queen and the number of her ovarioles are controlled in large part by the amount and quality of the food she receives during her larval life. Obviously, too, there are age and genetic differences that influence egg laying.

We find varying estimates of the maximum number of eggs a queen can lay in a 24-hour period; figures as high as 6000 per day have been published but these are probably three to four times too high. Several researchers have indicated that worker bees eat some eggs, though whether they do so to control the population or for some other reason is not clear. It appears that egg-eating by workers occurs most frequently in the spring.

Many decades ago precise measurements of the amount of brood present throughout the year were made in colonies in Maryland. The brood frames were photographed and cells of capped brood were counted at weekly intervals. The figures were divided by 11 since this is the number of days worker bees are in capped cells. Based on these figures, the maximum number of eggs that were laid and developed into pupae was 1587 per day. This high figure was

Queen

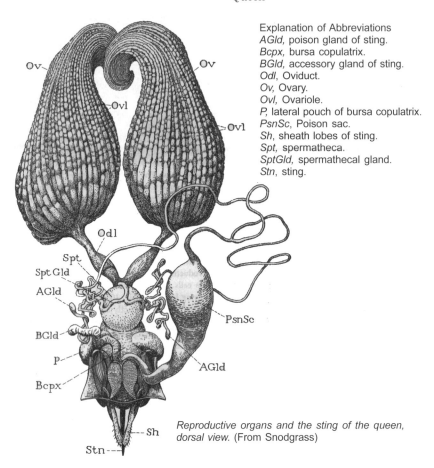

Explanation of Abbreviations
AGld, poison gland of sting.
Bcpx, bursa copulatrix.
BGld, accessory gland of sting.
Odl, Oviduct.
Ov, Ovary.
Ovl, Ovariole.
P, lateral pouch of bursa copulatrix.
PsnSc, Poison sac.
Sh, sheath lobes of sting.
Spt, spermatheca.
SptGld, spermathecal gland.
Stn, sting.

Reproductive organs and the sting of the queen, dorsal view. (From Snodgrass)

attained by only one of the 53 colonies studied. The queens in most colonies produced fewer, but averages of 1000 to 1200 eggs per day for a 12-day period were common. These figures were recorded only during the time of the year when climatic conditions were optimal for brood rearing and did not continue through the whole year (Nolan, W.J. "The brood-rearing cycle of the honeybee." *USDA Department Bulletin No. 1349.* 56 pages. 1925.)

Emergency queen cells – An emergency queen cell is one that is built after a queen has died or somehow been lost. When bees recognize they are queenless they select a worker larva about a day old, sometimes older, enlarge the cell around the larva and feed it royal jelly lavishly. Generally speaking, queen cells produced under the swarming impulse are found near or at the bottom of the nest. Emergency

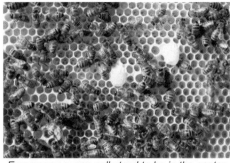

Emergency queen cells tend to be in the center of the brood nest on the face of the comb.

How many eggs can a queen lay in a 24-hour period? One study gave averages of 1000 to 1200 per day during optimal conditions.

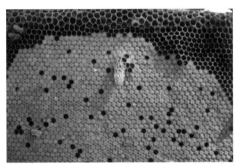

As the emergency cell matures, it is enlarged, and takes the position perpendicular to the comb face. (Morse photo)

queen cells are found nearer the center of the nest on the face of the comb and hang parallel to the face of the comb.

Excluders – The origin of the queen excluder is unknown but probably lies in the idea of using small diameter holes in a board through which workers but not queens and drones might pass. When a board with many such holes was placed above a brood nest the worker bees would move through and store honey, without brood, above the excluder board. The metal queen excluder, made by punching rectangular holes in a piece of metal, was probably first made in France, sometime in the 1860s. At about the same time beekeepers in England made metal queen excluders with round holes. These had the disadvantage that if the bee was carrying pollen it would be scraped off the legs as the bee moved through the holes. In fact, many pollen traps use wire screening with square holes, or perforated metal with round holes, to force the pollen off of the legs of the bees as they drag their hind legs behind them and move though the metal screen.

The first practical queen excluders were made using perforated zinc sheets. However, some beekeepers thought that the rough edges on the perforations might cut the wings of the bees, which is not supported by any data. In fact, if one observes a worker bee under a microscope while she is moving though a perforated zinc queen excluder it will be seen that she does so with ease without tearing her wings.

Because the flat metal zinc excluders were rejected by some beekeepers queen excluders made with stiff iron wires were invented. Presumably they had no rough edges on which a bee might catch her wings. The wires were separated in a variety of ways. The correct distance between the wires, or the width of a perforated hole in an excluder, is 165/1000 of an inch (0.165 inches or 4.2 mm). A normal queen, because her thorax is wider than that of a worker, cannot move through a queen excluder with the proper spacing. Undersize and unmated queens will sometimes not be

Various Queen Excluders

Wooden slatted excluder on hive and up close.

Plastic excluder.

Wood rimmed.

restricted by a queen excluder. Drones are also trapped below an excluder.

Flat plastic queen excluders are popular. An advantage of the wood-bound queen excluders is that the wood may be painted and thus color-coded.

Whether or not a beekeeper should use queen excluders is a question that has prompted much debate. The opponents state that queen excluders affect ventilation, which is possible. Some beekeepers have called them honey excluders. However, if queen excluders are placed on colonies three weeks before the honey is harvested any brood above the excluder will emerge. Those who support the use of queen excluders point out that it is possible to remove the honey much more rapidly since the beekeeper need not check to determine if brood is present in the honey storage supers or not.

Length of the life of a queen – Under natural conditions a queen honey bee may live five years or even more. From the point of view of a practical, honey-producing beekeeper, colonies should be requeened every year for several reasons. It has been shown repeatedly that young queens lay more eggs, produce more of the pheromones by which they are recognized as being queens, and, most important, are less likely to head colonies that will swarm. The length of life of a worker bee is short, apparently because of the hard work she does foraging. Egg laying does not cause such a drain on a queen and they may live for long periods of time (see LENGTH OF LIFE OF THE HONEY BEE).

Locating queens – A queen honey bee always remains in the brood nest area except when she is moving from one side of a comb to another. However, heavy smoking of a colony or much disturbance will cause a queen to move about in the brood chamber. Thus, to find a queen one removes any honey storage supers and goes directly to the brood nest. In searching for the queen it is best to start on one side and to move across the brood chamber examining one comb after the other. The first comb removed is usually one without any brood. It is placed outside of the hive usually leaning up against the hive. This gives the examiner space in which to work. The comb that has been removed will be placed back in its original position after the brood chamber has been examined. The queen can, of course, be on any comb. If there is brood in two brood chambers it may be helpful to set the top one off onto an upside-down hive cover. This prevents the queen from running off a comb in the upper chamber onto one in the bottom.

Several factors affect how rapidly one may find a queen. Light-colored queens

When finished examining a frame, set it down, leaning against the hive body, and examine the next frame. One trick is to quickly scan the frame next to the one you are lifting before it is lifted. As you move that frame away, light penetrates between the frames, and if the queen is on the frame not moved, she will scurry away to the dark. Look fast.

Set the top box of brood combs off the hive onto an upside down hive cover. Remove an end frame first and check for brood, and queen.

Queen

Continue examining each frame, removing it when done. Hold the frame over the box so if the queen falls off the frame, she falls back into the colony.

are easier to find than those that are dark. Some queens run on the comb and appear to hide. Old queens are less likely to run. Some races of bees have a tendency to run and form a ball or mass of bees on one end of the frame. If a queen joins such a group it is nearly impossible to find her. Having marked queens, using the International Color Code, makes the queen easier to find.

It is advisable to use as little smoke as possible when searching for a queen. While smoke causes worker bees to engorge and has a calming effect on

Another way to locate a queen is to fasten an excluder to a box, then place this box over another one with frames. Shake the bees off the frames from the colony from which you want to find the queen on to the excluder. Workers will go below, into the prepared box. The queen(s) and drones remain on the excluder.

When all frames in the top box have been examined, check those in the bottom box, leaving it right on the bottom board, in case she falls.

them, it may cause queens to run and move over the comb rapidly. This makes queen-finding difficult.

Whatever the purpose for finding a queen, whether to clip her wings or mark her, it should be done as soon as she is found. This means having the scissors or marking materials handy and ready to use. While we are aware that colonies can return to normal within about 30 minutes of being opened and smoked it is best to disturb colonies and queens as little as possible.

Queen introduction techniques – (see REQUEENING)

Queen piping – Communication among bees takes place in different ways. One is called queen piping. The queen vibrates her wing muscles while pressing her thorax onto the comb. She makes a "tooting" sound for about one second (frequency of 300-500Hz) followed by quarter-second pulses separated by quarter-second intervals. Workers close to the queen remain motionless perhaps reducing any aggression by the workers. Another result of the tooting is the "quacking" sound from mature queens that are about to emerge from their cells. The quacking sound is composed of only the quarter-second pulses. Queen piping also causes the workers to prevent the emergence of queens during swarming. Piping communicates the number and status of queens and influences queen emergence.

Queen substance – (see PHEROMONES)

Virgin queen – A queen that is not mated is called a virgin queen (see MATING OF THE HONEY BEE for a discussion of virgin queen behavior). Virgin queens tend to be smaller than mated queens. As a result they may be more difficult to locate and will occasionally slip through queen excluders.

QUEEN REARING: GRAFTING AND COMMERCIAL QUEEN PRODUCTION – What follows is a brief overview of a complex matter that has been the subject of several books. One important book is by H.H. Laidlaw, Jr. and R.E. Page. *Queen Rearing and Bee Breeding*. Wicwas Press, New Haven, CT. 224 pages. 1997. Anyone thinking seriously of grafting cells in large numbers should consult one or more books in addition to talking to some knowledgeable queen breeders.

To graft larvae is merely to pick up worker larvae 12 to 24 hours old with a small metal or wooden spoon or hook or tool and transfer them to specially prepared plastic or beeswax queen cups. These cups are then placed in queenless starter colonies where they are held during their first 24 hours of development. The bees in the starter colonies are young, have well developed head glands, are well-fed and have no other brood to feed. Under their special care the larvae are off to a royal start. Following this period of time, the young cells are moved to queenright finishing colonies that feed the larvae through to the capping stage. At this point, the capped cells may be held in an incubator or in a holding colony until it is time to place them into mating nucs. The process sounds simple but there are several important steps that must be done carefully to grow good queens.

Bank colonies – A queen breeder will often have more mated queens than can be sold or shipped at a given time. These may be placed individually in queen cages without attendants or sugar candy, together with other similarly caged queens, in a holding or bank colony. The bank colony is usually queenless though if a queen is present she is confined below a queen excluder with the young queens above. The secret to successful banking is to place a frame of emerging brood in the colony every five days.

A metal grafting tool with an offset hook on one end and a spoon on the other. These are available from suppliers or can be made from wire. Fine sanding of the ends is required to eliminate spurs or cracks that could harm tender larvae.

Queen Rearing: Grafting And Commercial Queen Production

Larva with royal jelly in plastic cup.

Frame with very young larvae is selected from breeder colony.

Larvae are grafted from frame to plastic cell cup.

Frames with cell cups and larvae are ready to put in queenless starter colony.

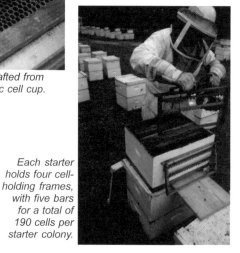

Each starter holds four cell-holding frames, with five bars for a total of 190 cells per starter colony.

After 36 hours the cells are moved from starter to queenright finisher colony.

Each finisher colony gets two bars, or 38 cells, and each bar is at the bottom of the frame, closest to warmth and nurse bees.

Queen Rearing: Grafting And Commercial Queen Production

Cells are removed from the bar and kept covered with a warm towel in a heated truck.

After 8+ days the now-finished cells are harvested.

Older bees are antagonistic toward queens not their own and will ball them. Balling consists of biting and clamping on the queen's antennae, mouthparts, legs or other parts. Any of these appendages can suffer permanent damage as a result. Young bees will feed all of the queens. When young bees are present the older bees will leave the area where the queens are confined and pose no problem. It is possible to hold mated queens in a bank colony for several weeks or longer without any harm. Several people have attempted to hold mated queens over the winter in bank colonies but almost always without success since they could not provide the young bees needed on a frequent basis.

Cells are put into queenless colonies . . .

. . . and virgin queens emerge in about a day, to mate and head the colony.

689

Breeder colonies – A breeder colony is one that has been selected by the beekeeper as being outstanding for one or more of a variety of reasons. Larvae from eggs of the queen in this colony are used in producing new queens. Queen producers usually use several breeder colonies in case one harbors an unknown trait that may be undesirable but is hidden at the time. Most queen producers are concerned first with honey production and those colonies that are selected for breeders are those that did well in production the previous season. However, since a queen may mate with 12 to 20 drones it is easy enough for some undesirable quality, such as defensiveness, to be hidden in the sperm of one of these.

The importance of selecting breeder queens that appear, or can be shown to have resistance to mites and the common diseases, is becoming increasingly important especially since the introduction of the parasitic mites into North America. Some queen breeders are using modifications of an assay for resistance to the tracheal mites (*Acarapis woodi*) that was developed by Page, R.E. and N.E. Gary. *Genotypic variation in susceptibility of honey bees,* ***Apis mellifera****, to infestation by tracheal mites,* ***Acarapis woodi***. Experimental and Applied Acarology 8: 275-283. 1990. Queens that have shown traits such as hygienic behavior and other forms of mite resistance are now available for beekeepers.

Gentleness is usually a consideration when growing queens for hobbyists but is of less interest to commercial beekeepers. One can make a long list of other desirable traits including using less propolis, better wintering ability, cessation of brood rearing at the end of a honey flow and many others.

An important function of breeder colonies is to help in supplying an area with a bountiful supply of drones. Queen breeders who produce selected stocks of bees will also produce sufficient drones for mating.

Cell incubators – Colonies that are used as finishing colonies will hold and incubate cells until they are ripe and it is time for the ripe (ready to emerge) cells to be placed into individual mating nucs. However, some beekeepers use modified refrigerators as incubators, either for the last 24 hours the cells are held or for their whole development period as pupae. Since honey bees control the humidity and temperature of their colony closely it is recommended to hold developing cells in a colony as a first choice. If one builds a home-made incubator for cell storage it is important to know how to control the temperature and humidity within the necessary limits. Also, pests such as ants and wax moths can cause problems.

However, modern, sophisticated incubators are quite common, and not very expensive. These can be programmed exactly for temperature, humidity and excellent air circulation. They are pest-proof and safe, and take

A commercial queen production yard in southern Mississippi. Breeder colonies are closest, on the left, grouped together. Starter and finisher colonies are farther away, situated on both sides of the truck path in the center.

Queen Rearing: Grafting And Commercial Queen Production

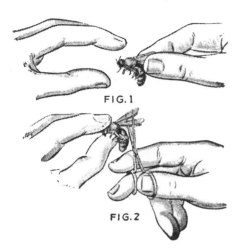

Clipping a queen's wing.

most (but never all) of the problems out of finishing cells. Commercial incubators are available from scientific supply companies and come in sizes ranging from holding only a few cells to walk-in compartments.

Clipping queen's wings – Beekeepers often clip one or both of a queen's wings either to mark her (left wing in odd years and right wing in even years) or to prevent swarming. While a queen with clipped wings can not fly and accompany a swarm, all of the other adverse consequences of swarming, including the cessation of food gathering, occur and the wing clipping really accomplishes little or nothing as a swarm control.

The best way to clip a queen's wings is to grasp her firmly between two fingers (either on both sides of the thorx, or on top and bottom of the thorax) and to cut the wing(s) where desired. The chief problem in clipping wings is to keep fingers clean of propolis. The scissors must be sharp. It is probably best to not cut the wings too close to the base.

However, beekeepers have been clipping queen's wings for many decades without any known adverse effects. Queens rarely attempt to sting when grasped and most beekeepers who clip their queen's wings do not even think about being stung. Make certain that you do not include an errant leg when clipping.

Commercial queen rearing areas – While it is true one can grow a few queens almost anywhere the same is not true of commercial operations. To grow large numbers of quality queens requires being in a location with an abundance of pollen- and nectar-producing plants, especially pollen plants. One can supplement a poor nectar-producing area by feeding sugar syrup; however, there is no good substitute for an abundance of pollen plants. The pollen substitutes and pollen supplements that are available are good but they are not perfect and they do not stimulate colonies in the same way natural pollen does. Furthermore, and just as important, a good natural pollen and nectar flow is much cheaper and more practical than is feeding bees.

Many large-scale queen producers, especially in the southern states, seek out apiary locations near large swamps that have an abundance of flowering plants, especially in the spring. It is a curiosity that many of the good areas that are used for queen rearing in the U.S. are not major honey-producing areas. Often it is the early spring plants needed to produce early spring bees that are most important. Just as in honey production itself, the successful queen producers move to those areas where the plants they want and need are plentiful; there is no alternative. One cannot change or remake natural areas but one rather takes advantage of those that do exist. This guideline applies not just to beekeeping but to the whole of agriculture.

Another concern to commercial queen producers is pests and predators. A Florida queen producer once reported that he had to give up growing queens for two weeks each spring because the dragonflies were so plentiful. Dragonflies, which patrol an area in their search for food, seek out the drones and queens because they are larger. At times, it was reported, the predation would be so great that many airborne mating queens would be lost. In other parts of the country one must deal with skunks, bears, and birds. All these add to the problems that must be faced by a commercial queen producer.

Double grafting – Double grafting involves grafting in the normal manner but replacing a recently grafted two-day-old larvae with one-day-old larvae. The thought behind double grafting is that one is immediately providing the second larva, the one that is to become a queen, with extra food. Some producers believe that superior queens may be made in this manner but the data indicate that this is true only if the starter colony is not as good as it should be. If one delays the second graft more than 24 hours some harm may be done because the jelly in a cell with an older larva is different and may not be suitable for a one-day larva. It is more important to take all of the steps in queen rearing seriously than it is to go to the trouble of double grafting.

Equipment for queen rearing – Since growing queens is a specialized business it requires some special equipment; however, what is needed can usually be made by the beekeeper. One item that

Frames with modified end bars. Notches are made to accommodate queen-cup-holding bars that can be removed easily.

is quite different is the mating nuc. As with other pieces of equipment, it is advisable to use standard items as much as possible to reduce costs and at the same time to make items interchangeable and retain their value in the event of resale.

The brood chambers and equipment used to hold breeder queens and for starter and finishing colonies are the same as those used in other operations. The frames used to hold cell bars are also the same, being modified only to hold the bars.

There are two basic ways to make mating nuc boxes. One is to divide a standard full-depth hive body into four compartments using quarter-inch (0.6 cm) thick plywood as dividers. Each section must have its own small inner cover so that bees cannot move from one

Queen Rearing: Grafting And Commercial Queen Production

A regular-sized hive body can be divided to make two smaller colonies and placed on a bottom board to make separate entrances. (Tew photos)

small colony to another when one is being examined. When a standard box is used it has an exit on each of its four sides. Many queen producers who grow large numbers of queens prefer to use nuc boxes with half-depth frames that are half as long as normal. The nuc box itself holds only one or two nucs. If the frames are half the normal length they can be put end to end into standard hive bodies when the foundation is drawn, or when some are being filled with pollen and/or honey for use by the nucleus colony. How to make good combs is discussed at length elsewhere but it is important here to emphasize that only strong, populous colonies draw foundation properly and can make good combs. Certainly the small nucleus colonies into which one puts ripe queen cells are not capable of making good combs from foundation. How to feed nucleus colonies is also a problem.

A typical small mating nuc-type box. This box holds 10 individual mating boxes, each with its own entrance, a honey frame for food, and a frame of comb for the new queen to lay in.

Individual queen mating nucs. These made of plastic foam offer good insulation for such a small colony.

Small division board feeders are popular with many queen producers. In general Boardman feeders are not considered a good way to feed bees since bees will not take food from them in cool or inclement weather as they are too far removed from the brood nest, and a leaky Boardman feeder can quickly lead to robbing. Too, they can be attractive to small hive beetles.

Queen producers use mostly plastic queen cups but also molded cups made of beeswax, both equally successful.

Finishing colonies – The colonies that are used for finishing cells, that is growing them up to the time they are capped, are usually queenright but the queen is below a queen excluder and the new cells above. One should watch for the construction of queen cells in the queenright part of the colony. Swarming is sometimes a problem with finishing colonies and they must be watched carefully. If the queen's wings are clipped the swarm cannot leave but swarm preparations are in place. A good finishing colony can care for 40 cells every three to four days; this is fewer than are given starter colonies. As soon as the cells are capped they are removed and placed in a mating nuc.

Grafting tools – Good grafting tools can be made from toothpicks. However, for more permanent operations, and the grafting of large numbers of larvae, the metal tools are generally considered the best. Most of those used by commercial queen producers are homemade or shaped to suit the grafter. Sterilize grafting spoons frequently by placing the spoon in alcohol or boiling water. It is during the larval stage that many bee diseases are transmitted.

Marking queens – Queens may be marked in a variety of ways. Marking is done for many reasons. It is valuable to mark queens to be used as breeders and many beekeepers mark their queens to indicate their age. A favorite marker among researchers are the plastic, numbered (1–99), color-coded (five colors), shaped discs that fit neatly on the top center of the thorax. They are held in place with quick-drying glue. Discs have been on queens for two and

Queen Rearing: Grafting And Commercial Queen Production

Holding, and marking a queen. Gently hold the legs as the queen will not pull so hard as to harm herself. Position her to put the paint on her thorax, mark her, and hold, and blow until the paint dries.

more years. These discs are available from some of the bee supply companies.

One may also mark queens with colored paint pens, available from hobby shops or bee equipment suppliers. These have proven to be very reliable and easy to use. Fingernail polish is satisfactory and can be applied with the brush that is usually found in the polish. However some of the brush hairs need to be removed to make a finer brush. The paint mark is applied to the top center of the thorax. It is important that the head, especially the eyes not have any paint. Excess paint must not clog the breathing spiracles on the sides of the thorax and on the abdomen.

The International Color Code for queens is for the years ending in:

0 & 5	blue
1 & 6	white
2 & 7	yellow
3 & 8	red
4 & 9	green

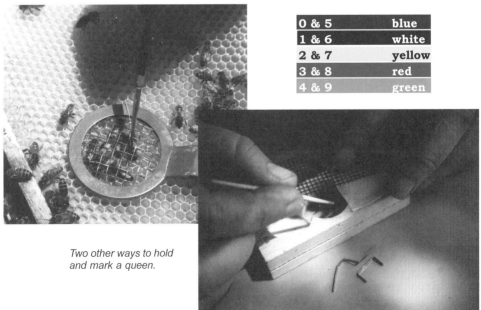

Two other ways to hold and mark a queen.

A queen with a numbered tag glued to her thorax.

Queens with a dab of bright colored paint are easier to find.

But even with paint, dark queens can be hard to spot.

Queen cages and queen cage candy – The first practical cages for shipping queens were designed by F. Benton of the USDA around the end of the 1800s. Benton's name is still often used in describing these cages. Benton's cages were made from a small block of wood with most of the interior cut away from one exposed face that is covered with wire window screening. One end of the wooden cage is filled with sugar candy that serves as food while the queen is in transit.

Queen honey bees are held or mailed in these small Benton cages together with five or six worker bees. The bees are preferably young bees. Under normal circumstances the queen and the bees will remain alive without difficulty for several weeks. While wooden Benton cages have been standard in the industry for many decades there is no question that many queen producers today prefer to use plastic cages that are cheaper. Queen cage candy is made by mixing finely ground (10X) confectioner's sugar with honey or syrup. The two are blended until the candy is firm or like a stiff dough. In a high humidity such candy will pick up moisture but in a queen cage with a few bees this surface water is removed promptly. However, it should be emphasized that the candy must be sufficiently stiff so that it will not run. While most queen shippers state that queen cage candy made with honey is best, most use high fructose corn syrup because honey may contain some pathogens.

Queen Rearing: Grafting And Commercial Queen Production

A typical three-hole queen cage. Each end has an entrance hole through it into the cavity inside. One of the three cavities is filled with candy, the other two are for the queen and her attendants. To release the queen slowly, place the cage in the center of the brood nest for two or three days – without removing the cork protecting the entrance on the candy end. Then check to see how aggressive the colony bees are to the cage. If gentle, remove the cork and let the bees remove the candy, releasing the queen in two or three more days. If aggressive, clinging tightly, wait two or three more days, check again, and remove only when the bees have accepted the new queen. If they remain aggressive, check for an existing queen.

Shipping queens – Most states where queens are reared have state apiary inspectors who inspect the apiaries where the queens are being produced and certify them as being disease free. Many states require that inspection certificates accompany the queens.

When queens are received, a small drop of water should be gently placed on the surface of the cage; if this water is consumed rapidly the bees should be given a second or third drop. It is best to hold queens in a dark area until they are put into colonies. Queens should be placed in a colony as soon as possible.

Some queen producers use the "battery box" method for shipping large orders of queens. As many as 100

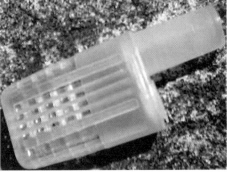

This recent design has become popular when queens are shipped without attendants actually in the cage with the queen. Rather, several cages are carefully held in cardboard boxes, usually double-walled and screened. Worker bees are put in the box loose to care for the many queens in the box. Water, in a sponge or other slow-release device, and food, as fondant, are provided in the box also. This technique provides the queens and the bees, and the shipping people a safe, effective way to travel. (JzBz photo)

A small battery box holding three-hole cages.

A battery box that holds many queens, and attendants are loose inside the box.

697

Another way to pack many three-hole cages. Attendants are inside these cages.

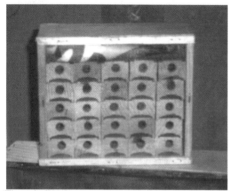

Three-hole cages screened in. Attendants are inside the cages.

queens can be individually caged and held in a battery box for shipment. Attendant bees are then shaken into the battery box to care for the queens.

Starter colonies – In commercial operations, newly grafted cups with larvae spend their first 24 hours in specially prepared queenless starter colonies. These have a large number of young bees with well-developed head glands. Starter colonies are made in different ways by different queen producers. One method is as follows: A strong two-or three-story colony is selected and fed heavily for three days. At the end of this time the colony is reduced to one or one-and-a-half boxes. The queen and the brood are removed and the colony given two or three frames of hatching eggs or larvae less than 24 hours old. One of the frames of mature eggs can be used for grafting the next day. A frame full of fresh pollen should be added to the colony. After 24 hours the frames of young brood are removed and an hour later the newly grafted cups are added. A good starter colony should be able to process 100 to 120 cups at one time. The frame of newly-grafted cups remains in the starter colony for one day only. A good starter colony may usually be used for four successive days but it is best to check the quality of the cells being started on the third day. One continues to feed the starter colony as long as it has new cells. New starter colonies should be made after four days as the young bees used to produce royal jelly will have aged and been under

These beekeepers are moving cells from a starter, in the back, to the now-open finisher, in the top box. A queen excluder separates the two finisher boxes to keep the queen below from venturing up and destroying the maturing queen cells.

Queen producers typically have many small mating nucs in a mating yard. These yards are surrounded by drone sources chosen by the queen producer to enhance the probability that his queens will mate with his drones, giving the customers an open-mated queen that produces bees with the desired qualities.

Many producers use two-way, or even four-way nucs with small frames and few bees to support the mating queen. Enough food and young bees, too many bees, honey storage, small hive beetles and Varroa are all challenges to such small units. Many producers are now wisely making their mating nucs larger with more space and bees to produce less stressed queens.

considerable stress. Remember that worker bees age rapidly and live for only a short period of time.

Stocking mating nucs – Most queen producers use baby nucs for mating queens. These are small colonies with two or three half-length, half-depth frames and perhaps 1000 bees. Such units often abscond and are easily robbed and/or preyed upon by wasps. Keeping baby nucs populated with the right number of bees is difficult. The stronger baby nucs may attract other bees and become overpopulated. The nuc must have young bees since it is young bees that feed and attend a queen.

The newly-mated queens are left in their baby nucs for several days after to check their egg-laying habits including the centering of the eggs in the cell and the size of the eggs. The new queen usually provides the baby nuc with a small amount of new brood, though with such a small unit and so few bees, brood chilling can be a problem and the brood may not emerge. It is usually necessary to add young bees to baby nucs frequently.

Wet grafting – Some beekeepers moisten the bottoms of the queen cups with royal jelly before grafting. This is called wet grafting. One thought is that adding royal jelly in this manner may help the young larvae with their need for food; however, since the royal jelly used for such purposes is taken from more mature queen cells, usually those more than three days old, its chemical makeup is different from that needed by a one-day-old larva and may do more harm than good. Those who prime their cups with royal jelly usually dilute the jelly slightly with water. Experiments have been conducted to test dry versus wet grafting and no data show that wet grafting has any advantages.

Having a small amount of royal jelly in the bottom of a queen cup may make it easier to slide the larvae off of the grafting spoon and into the cup. It is important to not turn the larvae upside-down when grafting. Experience is the best teacher when it comes to grafting and in time one will be able to dry graft (transferring larvae to cell with no royal jelly) as easily as wet graft. In any case, freshly collected royal jelly is desirable. Some grafting tools are more scoop-like than hook-like, and are able to take some royal jelly with the larva in the cell. This has a benefit of ease of removing the larva, and supplying exactly the right food.

Developing larva in a queen cell live in a moist atmosphere. It is important that the larvae being grafted not dry in

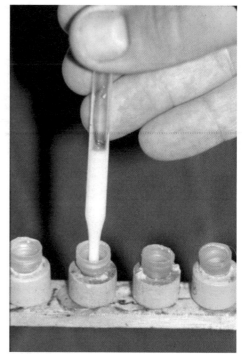

Adding Royal Jelly to a cup before placing a larva in the cup.

the process especially when one is grafting from a large frame full of day-old larvae. Some queen producers keep part of the comb from which they are taking larvae covered with a damp cloth to protect the larvae while grafting; cups into which grafts have just been made may be covered in the same way. Most queen breeders use small buildings just large enough for one person and the necessary equipment that they use for grafting; these are in the apiary close to the center of the operation. Since these must be wired to have a good light present it is just as easy to have a hot plate and to boil water in the building to raise the humidity. This can be especially helpful during the warm part of the day if the natural humidity drops.

QUEEN REARING: RAISING A FEW QUEENS

QUEEN REARING: RAISING A FEW QUEENS – The queen is the most important individual in the colony. We know, from a great number of studies that have been conducted, that the size of the queen and the number of ovarioles present in each of her two ovaries have a strong effect on both her ability to produce eggs and their number. Better-developed queens will be better egg producers. The most important time in the development of a queen is her larval stage. There is no feeding during the egg or pupal stages and thus, how large a queen will be, and how well developed her ovaries will be, depends upon how well-fed and how well-cared for when she is a larva. Therefore, select the time of year to grow queens when you are certain an abundance of food is available. It cannot be emphasized too much how important these points are. One reason many northern beekeepers leave queen production to southern producers is that those in the south have better weather earlier in the year. But many queen producers are now present in the northern states and Canada, and there are more every season.

Even though queens can be produced anywhere bees can be kept, it is important to have an abundant source of pollen and nectar available at the time the queens are being reared. This makes queen rearing much easier. In queen production it must also be remembered that half of the genetic material for the worker bees comes from the drones. For this reason it is important to have an abundant source of drones. Only strong, prosperous colonies will produce an abundance of drones. The drone-rearing colonies need to be kept away from the queen rearing yards. Lastly, it is important to have good flying weather for mating to occur.

Crowding colonies to force cell production – A colony may be forced to produce queen cells by crowding the brood nest. However the bees do not always choose a larva of the ideal age to rear a good queen.

Splitting colonies – One way to rear a new queen is to split a colony, or to make a small, nucleus colony from a larger colony. This technique has been used for thousands of years. It must be done at a time when the bees can keep the brood warm with ease and have an abundance of food. In making a nucleus colony one first places a frame with some

Splitting colonies to raise queens and reduce swarming is almost as old as beekeeping. Take eggs and brood and enough bees to amply cover the brood, along with some honey (or feed as long as necessary) and let the colony raise a queen from the young brood already in the new colony. Remember that the bees will only be as good as the genetics of the queen that produced the egg and the drones the virgin queen mates with.

The Alley method of producing a few queens. Woodcuts from his 1885 book.

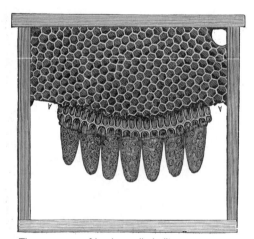

The new way of having cells built.

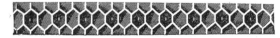

Comb containing eggs in alternate cells.

brood in a box. The quantity is not important as long as there are eggs and larvae present. This frame is usually covered with bees. Two more frames, also covered with bees and containing some honey and fresh pollen, are added. The entrance is restricted so that the bees may better protect themselves against changes in the weather and robber bees. One may reasonably expect, probably in 90 plus percent of cases, that a new queen will be produced from these eggs or larvae. This unit is allowed to grow and soon becomes a producing colony. However, if the queen is grown in the north late in the year, July or August, the new colony will probably need feeding to survive the winter. After the queen in the new colony is laying, one may exchange the place of the new colony with one that is somewhat stronger thereby adding bees from the stronger colony to the weaker one. One disadvantage to this method is the age of the larvae the bees choose.

If the splitting technique is used one should take care to make the split from a colony that has been successful and is thought to be from good genetic stock. Growing new colonies in this manner can be helpful in a swarm prevention program by taking bees from a colony that might otherwise grow too strong and swarm.

The Alley method – Henry Alley of Massachusetts developed a simple method of rearing about a dozen queens at one time. Alley selected and confined his breeder queen in a small hive on a comb about four and one-half by five inches (11 by 12 cm). When the queen was put in the small hive in the evening by the next morning there would be eggs in the comb. In four days there would be day-old larvae in the cells and it would be time to start the queen rearing. New comb was used as it could be cut into strips easily.

The next step was to remove the queen and to smoke and drum a cell-starting colony to cause the bees to engorge. After about ten hours a strip of comb containing the day-old larvae was cut from the larger comb. The cells in this strip were trimmed to about one third their natural height. These were then fastened onto the bottom of a comb, facing downward, in the position they would assume when the bees begin to develop queen cells. Alley said that a good colony would start 25 cells in this manner but he preferred to select a dozen of the best ones and let the bees finish this number rather than more. Since Alley knew the age of these larvae he could remove the ripe cells just before they emerged and place them into mating nucs. The cell-starting colony was to have no other brood and thus the young worker bees would have a maximum of royal jelly and be in excellent condition to rear queen larvae. Alley advised feeding the colony heavily to stimulate the bees.

The Miller method – Dr. C.C. Miller, a famous comb honey producer in the late 1800s, wrote about a simple method, probably modified after that of Alley, of producing a few queens. Miller removed all of the brood except two frames, from a strong colony. A frame containing three or four triangular strips of foundation was placed between these two frames. The bees would draw this foundation and eggs would be deposited in it. At about the time the larvae were to hatch from the eggs along the bottom of the frame Miller would trim the comb and foundation away from the edge and destroy many of the eggs, or hatching

C.C. Miller

larvae, leaving about one in every three or four along the bottom. These were then placed in a queenless colony and the bees would usually select these larvae for cell production. Several queen cells would be produced along the bottom of the frame. Bees apparently prefer to make queen cells from new foundation. When the cells were ripe, that is about to emerge, they would be cut from the frame and placed in mating nucs.

Following is taken from Miller's book *50 Years Among The Bees*, published by The A.I. Root Co. in 1911.

Quality of Queens

The question has been raised whether queens reared in the way I have described are as good as those reared by the latest methods. I think I can judge pretty well as to the character of a queen after watching her work for a year or two; I have kept closely in touch with what improvements have been made in the way of queen-rearing, and have reared queens by the hundred in the latest style; and I do not hesitate to say that the simple method I have given produces queens that can not be surpassed by any other method.

Beginner Improving Stock

I have been asked whether I would advise a beginner with only half a dozen colonies, one of them having a superior queen, to use the plans I have given to rear queens from his best queen. I certainly should, if he intends to give much attention to the business and increase the number of his colonies. The essential steps to be taken are simple enough; and even a beginner can easily follow them. But in a few words, here is what I would advise him:

Take from the colony having your best queen one of its frames, and put in the center of the hive a frame half filled or entirely filled with foundation. If small starters are used in a full colony the bees are likely to fill out with drone-comb. A week later take out this comb, and trim away the edge that contains only eggs. Put this prepared frame in the center of any strong colony after taking away its queen and one of its frames. Ten days later cut out these cells, to be used wherever desired, giving the colony its queen or some other queen.

Now there's nothing very complicated about that, is there?

Using swarm and supersedure cells for increase – Many beekeepers believe one should not use swarm cells from colonies as a source of new queens because the tendency to swarm might be inherited and increased. There is no doubt that some races or groups of bees have a greater inclination to swarm than

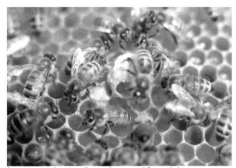

A supersedure cell can be used for raising a queen, as long as the cell, and thus the developing queen inside, is in good condition, and the chosen larva was of the right age. That is usually an unknown, and can be a distinct disadvantage. (Tew Photo)

others. No doubt too, one could select a strain of bees that would be much more tolerant of crowded or congested conditions than others. However, the chief consideration to take into account when thinking about using swarming or supersedure cells to make increase depends upon why the colony produced these cells. In the case of a swarming colony if the bees were congested because of a lack of management then the queens produced should be good queens with no special inclination toward more swarming. The same is true of supersedure cells; if they are produced in a colony with a queen that is poor because of old age or injury, then the cells should be good ones.

There is, however, a more important consideration in selecting the cells to grow queens and that is the physical condition of the queen cells. They should be large and well mottled, that is the exterior surface should be rough and well chewed. Small cells with smooth wax surfaces must be avoided as they will invariably contain small, poorly fed larvae and poorly developed pupae that will develop into inferior queens. The exterior condition of a queen cell, whether selected from swarm and supersedure cells, or grown in any of a variety of ways, is a sure indication of the quality of what is inside the cell.

Queen cells that have just been capped, and about 24 to 36 hours before the queens are about to emerge, must not be jarred or shaken. If this is done the developing queen may fall in the cell where she is growing, hanging head down, and die. When the queens are within one day of emergence they will stand a moderate amount of rough handling. This must be remembered when moving frames with queen cells to be placed in a nuc box. Even tilting the frame during the pre-emergence time may damage the pupa. If a cell is cut from a frame, which is often done when one is selecting swarm or supersedure cells, it must not be tilted in the process. A beekeeper using swarm or supersedure cells is at a distinct disadvantage because the age of the pupae in the cells is usually unknown.

QUEENLESS COLONY – A colony may lose its queen at any time of the year and for a variety of reasons including old age, disease, supersedure or being mauled or mashed while a beekeeper is examining a colony. When a queen is lost, the bees respond within hours by enlarging one or more cells around worker larvae and rearing a new queen from them. However, during the cooler months there may be no brood from which to rear a new queen or drones available for mating. Sometimes a queen rearing effort fails, or, rarely two virgin queens are produced and fight and both may die in the process. In such cases a colony becomes hopelessly queenless and eventually perishes. However, before death finally comes to the colony, laying workers are usually produced and these in turn may produce some drones thus contributing to the gene pool (see LAYING WORKERS and SWARMING AND SUPERSEDURE).

QUINBY, MOSES (1810-1875) – Moses Quinby exerted a strong influence on beekeeping in several ways. His major contribution was to demonstrate, by personal example, that commercial beekeeping could be a viable full-time occupation. Using the primitive equipment and techniques available at the time, Quinby harvested up to 11 tons of cut-comb honey annually in the Mohawk Valley of New York. His successes, publicized in newspapers and in his own book, *Mysteries of Beekeeping Explained* (1853), instructed and inspired many beekeepers, and his hive was used exclusively for some time. Another important contribution was his marked improvement of the smoker: he was the first to attach a bellows to the fire pot. Quinby developed this innovation just before his death and did not patent it. T.F. Bingham quickly made further improvements, took out a patent, and spread the bellows smoker throughout he beekeeping community. -RN

RACES OF HONEY BEES – Recent genome research indicates that the honey bees in North America are African in origin. These bees have migrated to, or been moved to almost everywhere on earth. There are, however, over 20 races (more correctly, subspecies) of the common honey bee; the most popular of these are discussed below.

Probably no race or stock of honey bees anywhere on earth today is pure because humans have been so active in moving bees around the globe. Still, through selective breeding, and in some cases artificial insemination, it is possible to select bees that closely resemble their ancestors. At the same time modern bee breeders have been able to make combinations that are useful in pollination, honey production and resistance to disease and parasitic mites.

African and Africanized honey bees – Africa is a vast continent with many large deserts and several mountain ranges. These have served to isolate groups of bees and the result has been that one may find between 10 to 15 subspecies of honey bees in Africa. Little is known about the African races other than the fact that some in sub-Saharan Africa are generally more defensive and more inclined to sting than are their more northern counterparts.

In 1956 several queens from South

African honey bees will concentrate their defense on dark objects, including black veils, black cameras and black camera straps. Exhaled carbon dioxide is an attractant for our faces. (Wearing a veil that is white outside and black inside significantly reduces that attraction.

Carniolan bees are dark with dark-gray stripes. Queens are very dark, making them somewhat hard to see when among a whole frame of dark workers. Drones, too, have dark to black abdomens, making them easy to distinguish from Italians, and an indication of the mating isolation of his mother.

Colony population increases rapidly in the spring when resources become available and, as a result, they will swarm earlier than most races. They produce very little propolis or burr comb and tend to produce very white cappings. Robbing instinct is low. They tend to forage during marginal weather – rain and cold. Summer brood production is dependent on available forage – little forage reduces brood rearing. They winter well with small clusters and consume the minimum of food.

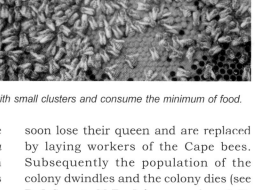

Africa were imported into Brazil. The queens were primarily *A. mellifera scutellata*. Their descendants, which initially hybridized with other subspecies of bees but maintained all of the African traits, are called Africanized honey bees so that we can differentiate between the two groups. The Africanized bees that entered the U. S. from Mexico in 1990 are the descendants of those introduced into Brazil in 1956 from South Africa (see AFRICANIZED HONEY BEES).

Cape bees – The Cape bee, *Apis mellifera capensis*, from the southernmost tip of Africa, is well known because of one aspect of its biology. Colonies that lose their queen become laying-worker colonies. However, after a period of several months a queen will develop from one of the laying worker eggs (see PARTHENOGENESIS). Whereas this may occur rarely in other races of honey bees it is more common in the case of the Cape bee. Cape bees are good honey producers but have no special virtues as pollination and honey production are concerned. They are however of great interest to scientists concerned with evolution and reproduction.

In South Africa, colonies of honey bees that are invaded by the Cape bees soon lose their queen and are replaced by laying workers of the Cape bees. Subsequently the population of the colony dwindles and the colony dies (see D.J Swart, M.F. Johannsmeier, G.D. Tribe, and P. Kryger. "Diseases and Pests of Honeybees." In *Beekeeping in South Africa*. Third Edition, Plant Protection Research Institute Handbook No. 14. 288 pages. 2001.

Carniolan honey bees – *Apis mellifera carnica*, the Carniolan bees, are probably the second most popular race of bees kept in North America. They were somewhat isolated in the Austria, Yugoslavia and the Danube River Valley where they developed their characteristics. They are large bees and usually dark (grayish to black). They are often preferred because they are more gentle than the favored Italian bees. Colonies of Carniolan bees winter well and populations build rapidly in the spring.

Caucasian honey bees – The Caucasian bees, *A. m. caucasica* developed in the Caucasian mountains in Eurasia. Caucasians are dark bees with brownish markings and a good disposition. This latter fact makes them popular with some beekeepers. Some

Caucasian bees are predominantly dark, sometimes with gray or brown mixed in. Their hairs are dark. Drones are dark also. Queens are dark to very dark making them difficult to find on a frame of dark bees. They are gentle, calm and do not run on the combs.

Caucasian bees tend to produce "wet" cappings. Caucasians are slow to build up in the spring, thus swarm late. Brood production depends on available forage. Brood production slows in early autumn. They overwinter on minimal stores.

Known for propolis production, they also tend to be susceptible to nosema. They fly in cooler weather, and in light rain, making them good for pollinating later crops.

Caucasian bees use propolis excessively, which can make management difficult.

German or black bees – *Apis mellifera mellifera,* probably the first bees brought to North America from Europe, were of this subspecies. They are good producers. They were especially desirable for comb honey production because the wax cappings over cells of honey did not touch the honey and were not watery-looking (see HONEY JUDGING). At the same time, these bees used little propolis; both of these qualities made them good bees for comb honey production.

German bees have two faults. First, they have a tendency to run on the combs when colonies are examined. The bees often form large balls of bees on the lowest corner of the comb and these balls may drop off, sometimes onto the ground. It is very difficult to find queens in colonies where the bees run excessively. A second fault with the German bees is that they are more susceptible to the foulbroods, especially European foulbrood, than are Italian bees. When chalkbrood became widespread in the U.S. in the mid-1970s the German bees again proved to be vulnerable and many colonies died at that time. They are uncommon in the U.S., but some lines are still common in Europe.

Italian honey bees – Honey bees from northern Italy, *A.m. ligustica,* were imported in large numbers into the U.S. in the late 1800s and early 1900s. They became very popular because they were more resistant to several diseases. The bees in the colonies are quiet and do not run on the combs; the queens are large, golden in color, and easy to find. Italians are good honey producers. One drawback is that they are more defensive than the races from northern and eastern Europe. Italian bees are the favored bee by most queen producers in North America today.

Other races – Eight of the over 20 known races of *Apis mellifera* have been introduced into the United States according to Sheppard [see W.S. Sheppard. "A history of the introduction of honey bees into the United States."

Races of Honey Bees

Italians, unlike Carniolans or Caucasians, have yellow markings, like their African cousins. Queens are large with gold to leather-colored abdomens and are much easier to find than dark queens on a frame.

They are, generally, short distance foragers, and are quick to rob nearby colonies. They are color oriented, thus tend to drift when colonies are neatly in rows.

They build slower in the spring than Carniolans, thus swarm less, but earlier than Caucasians. They overwinter with large populations. They produce lots of brood, consume large amounts of food, but produce lots of bees. They are good honey producers. The photos show the large queens, and the photo above shows the queen, drones and (with golden abdomen) and workers for comparison.

Yellow Cordovan

Purple Cordovan

Breeders often use a genetic marker to determine if certain characteristics have been passed along. This marker is commonly called the Cordovan gene, and, when used with Italians, the offspring are "yellow" Cordovans, and when used with Carniolans are called "purple" Cordovans. Other than the pigment difference, these bees are similar to other races.

American Bee Journal 129: (9) Pt. 1 617-619; (10) Pt. 2 664-667. 1989]. The introduction of *A. mellifera scutellata* in 1990 increased the number of races in the United States to over nine. As new breeding techniques are developed the number of races will increase.

Russian

Buckfast

There are hybrids – crosses of several races – that are commercially available. Because they are hybrids they tend to have color and marking variability so are not easy to distinguish by visual inspection alone.

REACTIONS TO BEE AND WASP STINGS – The frequency of insect sting allergy in the U.S. population has been estimated at between 0.4 and 4.0 percent. Of that small percentage only a minuscule number will ever develop severe allergic reactions. Only about 40 deaths occur per year in the United States from all types of insect stings. Of these, honey bee stings account for about half the total. Most people will be surprised to learn that allergic reactions to penicillin kill over seven times as many people each year as insect stings and that even so unlikely a cause of death as lightning takes over twice as many lives.

Aside from everyone's initial sensation of pain, reactions to wasp and bee stings are extremely variable and depend on the species of stinging insect, the number of stings received, and factors relating to the victim. Most reactions to bee or wasp stings fall into one of the following categories.

The typical initial reaction to a hymenopteran (the group to which honey bees and the common wasps belong) sting is brief sharp pain, a whitened wheal with a central red spot, an area of surrounding redness, and warmth. In loose tissues, such as the eyelids, there may be considerable swelling. All these symptoms generally subside in minutes to hours. Itching may persist at the site for several days. This typical reaction occurs in the vast majority of people stung, and is called a local reaction. Individuals whose response consists of these symptoms are not allergic to stings.

Next in severity are the large local reactions. Here again, the symptoms involve only the area immediately surrounding the sting. There is more

A typical local reaction to a honey bee sting.

extensive swelling and redness, but the signs are confined to the site of the sting. These reactions may develop slowly, reaching a peak in 24 to 48 hours, then resolve spontaneously over several days. At times the reaction may be so large as to immobilize an entire limb. There is some question whether these local reactions are allergic in origin or whether they are merely an exaggerated normal (i.e., non-immune) response to the venom.

In contrast to local reactions, systemic reactions involve portions of the body and organs distant from the sting site. Also called anaphylactic reactions, they have an allergic origin and can be serious, occasionally even life-threatening. Systemic reactions include generalized urticaria (hives), throat tightness, wheezing, difficulty breathing, hypotension (a drop in blood pressure), dizziness, and loss of consciousness. The most common manifestations are cutaneous: redness, itching, and hives, often on the eyes, lips, tongue, or palms of the hands. Cutaneous symptoms may be accompanied by other manifestations, the most dangerous being respiratory failure and shock. There are two types of life-threatening respiratory events. Upper airway obstruction results from edema (swelling) of the larynx or epiglottis. Death from suffocation can occur rapidly. Diffuse lower airway bronchoconstriction, similar to a severe asthmatic attack, is the other potentially fatal form of respiratory failure. Shock is the most severe form of anaphylactic reaction, which may suddenly develop without any other symptoms and can rapidly lead to death from the sudden drop in blood pressure.

In rare cases, when an individual receives hundreds of stings at a time, a toxic reaction may develop. It is the result of a direct toxic effect of large quantities of venom rather than an allergic reaction. Muscle and red blood cell breakdown, and a drop in blood pressure all contribute to kidney failure requiring lengthy hospitalization for surviving patients. - SC

REFRACTOMETERS – (see HONEY, *Color grading*)

REFRIGERATED TRUCKS FOR MOVING BEES – (see MIGRATORY BEEKEEPING)

REMOVING BEES FROM SUPERS – When the honey flow is finished, and the storage combs are filled with honey, they should be removed from the colony(s) and extracted as soon as possible. A chief concern at this time is that the honey might crystallize in the comb; if this occurs it may be difficult or even impossible to remove it. Crystallization is usually much more likely in the fall when low temperatures may encourage premature granulation.

However, some plants produce nectar that, when converted to honey, will crystallize very rapidly. Canola honey, for example, must be harvested when only 10-25% capped. Waiting until a comb is fully capped may produce a solid chunk of honey.

Removing sections of comb honey as soon as possible is also important to prevent excessive travel stain; this is not a concern in liquid honey production. Still another reason for removing honey at the end of a flow is to keep honey flavors and colors separate, since certain honeys usually command a higher price than do others.

Supers may be removed in a variety of ways. Whatever is done will depend in

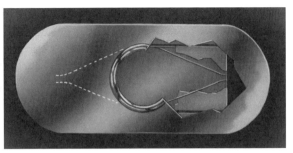

The Porter Bee escape. Bees enter the hole from above, and exit, through the tensioned, but flexible springs, to the super below. They are unable to reenter the above super because they cannot get past the metal springs.

The Porter fits in the oval hole in most inner covers.

Once in place in the inner cover hole, the inner cover is placed, bee entrance up, between the super to be removed, and the supers or hive bodies below. Bees leave the upper super via the Porter escape, and cannot return.

part on the quantity of honey to be removed and the distance to the apiary. The use of a bee escape, for example, requires two trips to the apiary, which might not be practical if it is some distance away. The many methods of removing honey are reviewed below.

Bee escapes – The Porter bee escape was first marketed in 1891 and took much of the labor out of removing honey; its invention was a step forward. The disadvantages of bee escapes are that if the supers above the bee escape are not bee tight, bees may enter and steal the honey. Beekeepers that use bee escapes usually carry duct tape and use it to plug any cracks or holes in and between supers. However, if a bee escape is put into place before midmorning on a day when it is warm enough for bees to fly, the supers should be empty of bees by

Another type bee escape, commonly called a cone escape board. To leave the honey super bees exit via the wide end of the cone and enter the super or hive body below. They are then unable to return through the narrow end of the cone. The board is perforated to allow exchange of colony odor and encourage bees above to leave.

A triangle escape board, under side. Bees above leave the honey super to be vacated, and follow the path to the corners of the two triangles, exiting the above super. They are unable to return due to the very narrow opening at the corners.

the following morning. A bee escape works because it is natural for bees in supers to move downward in the direction of the brood nest sometime each day.

One disadvantage of bee escapes is that two trips to the apiary are required, one to put the escapes into place and one to remove them and the supers of honey. The supers of honey must be removed to install the bee escape, a problem when many supers are on the hive. Some beekeepers leave the escapes in place for several days and then remove the supers full of honey in the early morning before the bees start to fly. At that time of day one does not need, at least under normal circumstances, a veil or smoker. In warm climates, especially during warm nights, the bees are reluctant to move down so the bee escape is less effective. In the opinion of many beekeepers the bee escape is a simple, convenient way to remove supers of honey.

Bees will not abandon supers above a bee escape that contain brood. Those who use bee escapes often prefer to use them in conjunction with queen excluders. When this is done the excluder can be removed at the time the bee escape is put into its place. Then there is no danger of brood being in the honey storage supers.

Blowers – Blowing bees out of supers with bee blowers or leaf blowers is popular with many beekeepers. The bees are blown out towards their own hive. Interestingly, and unlike bees that are

A screened escape board. Similar to using an inner cover, but well ventilated.

An advantage of this system is that it may be used on cool days when repellents may not work.

Repellents – It has been known since the turn of the century that certain chemicals will repel honey bees. Carbolic acid was one such material, but the crude carbolic acid that was used also left a lingering and undesirable odor. In the 1930s it was found that a pure form of carbolic acid (phenol) would repel bees and left no odor. Carbolic acid has been replaced by several better repellents.

All repellents will confuse bees if they are exposed to too much of the chemical. For this reason beekeepers first smoke the supers to be removed to drive the bees down and off the top inch or so of the combs. This starts the bees moving in the right direction. Smoke should not be used in harvesting comb honey.

Fume boards are boxes with the outside dimensions of a super but only one to two inches in depth. They have a top but no bottom. Those that are about two inches deep seem to work best. Usually the fume board has two to three layers of cloth or other absorbant material onto which the repellent is placed. Fume boards are usually painted black so they will absorb the heat from the sun to warm and volatilize the repellent. On hot days less repellent will be needed.

Benzaldehyde – This chemical, better known as artificial oil of almonds and used by bakers to flavor almond cookies, was found to be a good bee repellent in 1963. Usually one or two tablespoonfuls are sprinkled over the fume board cloth at one time. While it will not drive bees as quickly and easily as some other repellents, the fact that it is readily available and registered for use

Using a blower to clear supers of bees. Supers are removed from a colony, using only light, or no smoke. The super, complete with bees, is then placed such that a strong stream of air, pushed through a narrow nozzle, can be directed between each frame, and across the top bars and bottom bars. Bees thus removed are usually disoriented and lost and are unlikely to be defensive. They will, however, often land on the ground and can be easily stepped on.

brushed and shaken, bees that are blown out of supers do not become angry or unduly aggressive. The bees may become confused, and large numbers may come to rest on the hats and backs of the beekeepers blowing out the bees. It may be a good idea to use some sort of ear protection when using a bee blower. It is important to dress carefully in an apiary where honey is being harvested; lost, crawling bees may sting if accidentally squashed or squeezed.

The main disadvantage of this system of removing supers of honey is that the air must be directed between all frames.

Using a fume board. The colony's cover is removed and the bees lightly smoked to get them moving down. Repellent (amount to be applied is on the label) is added to the absorbent pad on the inside of the fume board, and it is placed on top of the super where bees are to be removed. Depending on ambient temperature, number of bees to move, size of super, and whether brood is present, clearing the super will take three to 10 minutes. Bees are very reluctant to leave brood, and may not do so at all.

Once completed, remove the super, smoke the bees in the super below lightly, and place the fume board on the next honey super to be removed.

Bee Go®. A typical fume board sits behind the bottle. A fume board is simply a one-to-two-inch deep rim with a metal top (lined with cloth to absorb the liquid repellent), and no bottom. The top is usually painted black.

as a repellent, has made it popular among beekeepers. Benzaldehyde is not too effective at high temperatures as it volatilizes rapidly and confuses the bees, which then refuse to move in any direction. Most beekeepers state that benzaldehyde works best when the temperature is in the 70s. Benzaldehyde will produce a chemical burn if spilled onto skin.

Butyric anhydride – This chemical is an effective honey bee repellent, but it is not in favor with all beekeepers because of its strong, persistent, and foul odor. This repellent is sold under the trade name Bee Go®. Butyric anhydride is sold by bee supply companies. It is used in the same manner as the other repellents. One manufacturer has prepared a perfumed butyric anhydride which has a diminished odor.

One disadvantage is that if the fume board is left on the hive too long or too much Bee Go® is put on the fume board, the honey will have a smell or taste of it.

Store fume boards as far away from human activity as possible. Do not transport to or from beeyards inside a vehicle, or even in the trunk of an automobile. Do not store inside a house or other dwelling. Do not spill this natural herbs and oils. It has a mild and pleasant odor to humans but is effective in clearing bees from supers. There is no danger of any odors or taste contaminating the honey.

Smoking, shaking and brushing – For several years after commercial beekeeping started in North America, the only method known and used by beekeepers to remove combs of honey was a combination of smoking, shaking,

Sometimes beekeepers will initially apply a fume board askew to get bees moving without overwhelming them with the odor, especially just after applying the chemical. The board will be placed square after several moments to finish clearing the super.

material on your clothes, shoes or beesuit. Read label instructions regarding amount to be used, especially when reapplying.

When too much repellent has been added to a fume board, all of the bees in a colony may vacate the colony, running out the entrance. Do not overdose when using these chemicals.

Fisher's Bee Quick® – This repellent was formulated with

Modern bee brushes have soft, nylon bristles that are kind to bees. To brush bees, move honey super away from colony several feet if possible. Replace cover over half of the top of the remaining honey supers or hive bodies. Remove a frame from the honey super and hold close to the exposed top, or the front door of the colony it came from. Gently shake the frame so that most bees are dislodged, then quickly, briskly, with short strokes brush remaining bees off both sides of the comb. Place the bee-free (or nearly so) comb where robbing bees cannot get to it (an empty super with a tight top and bottom [two covers]). Do not spend lots of time at this, and when complete, close up the colony as quickly as possible and remove the harvested honey.

An old fashioned brush.

and brushing bees off individual combs. The disadvantage is that the method is slow and can anger the bees considerably. Beekeepers who remove honey in this way must dress carefully to avoid excessive stinging.

When a beekeeper has only a few combs or supers to remove, shaking and brushing can still be practical. It should not be used when the colonies are in close proximity to residences where angry bees might be a nuisance. It is a good idea to replace bee brushes periodically as they could be a source of disease transmission.

No simple or perfect way will remove honey bees from supers filled with honey. No matter how it is done, the lifting and removal of supers full of honey is hard work. Harvesting honey may also involve coping with angry bees that are not willing to give up their stores of honey easily.

REMOVING FERAL COLONIES – In many parts of the country a few colonies of honey bees live feral or wild especially where the Africanized honey bees have been established. Many of these feral colonies are swarms from colonies in apiaries that were unattended or where swarm control measures were not practiced. Feral populations have become far less numerous as parasitic mites took their toll.

Removing feral colonies from their natural home can be hazardous (and may require a permit) especially if the bees are Africanized. If you suspect that the bees are Africanized and you are not experienced, call a professional pest exterminator or work with someone who has done it before. Removing established colonies is time-consuming.

Removing bees from buildings – The natural home for a honey bee colony is a cavity such as a hollow tree or a cave. Thus, the side wall of a house or barn would provide suitable protection and make an ideal shelter for bees. Information at present indicates that swarms prefer relatively small cavities for nest sites and the area between two studs in a wall is adequate.

As long as the bees in a cavity remain alive and active they will protect their stores and honey will not leak from the combs and stain the house interior. The bees will also make certain that cockroaches, mice and other vermin that might infest an abandoned nest will not be present. It is when all of the bees in a nest die, usually in winter, that the nest contents are attacked by a host of undesirable organisms making it necessary to take steps to remove the debris. Some colonies in houses, especially ones whose entrance is high in the air so that the bees fly above the

Bees will find a suitable cavity and construct their nest. It may be easier to approach nests like these from the inside of the structures rather than the outside, and it may be less expensive to do that.

heads of people in the vicinity, may not create a nuisance to the homeowners. Bees in a building may harbor diseases or mites which will go undetected since there is no way of inspecting the nest. Because of these potential problems it is best to remove colonies of bees from buildings when it is convenient to do so.

One caution, however. If a colony has died, or worse, the home owner has killed them, the entire contents of the hive must be removed. Wax and honey will melt when the weather warms, and cause many problems. Expediency is recommended, as other bees will be attracted to the odor.

It can be possible to save the bees in a nest in a building if the nest can be made accessible. In some cases it is better to kill the bees before removing the nest. In populated areas removing the nest may anger the bees and cause a stinging incident.

When all of the comb and debris have been removed the area must be cleaned to remove any residual honey. The next most important step is to fill the area with tightly packed insulation to prevent bees from occupying the space again. One can never remove all of the propolis and wax from an old nest, and the odor of these materials will remain attractive for many years to honey bees searching for a new home. It is always necessary

Removing Feral Colonies

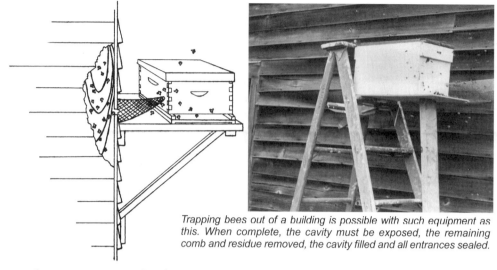

Trapping bees out of a building is possible with such equipment as this. When complete, the cavity must be exposed, the remaining comb and residue removed, the cavity filled and all entrances sealed.

to plug any entrances that bees can use. The area of the nest will remain attractive to bees for many years.

The comb, bees and debris from colonies killed with an insecticide should be disposed of properly. The honey should not be used, and one should not attempt to recover the beeswax. A number of insecticides are approved for killing bees and wasps. However, the technology used to produce these materials changes rapidly. It is important to select material that is labeled for this express purpose and to read the label carefully.

It may be easier, and less expensive to remove a nest from a building, especially a house, from the inside rather than removing boards from the outside of the house. The interior can be easier to repair than exterior boards. However, a great number of precautions must be

Comb can be removed, cut to fit frames, and the entire colony transferred to a standard hive. Bees will often remove the supporting strings or rubber bands.

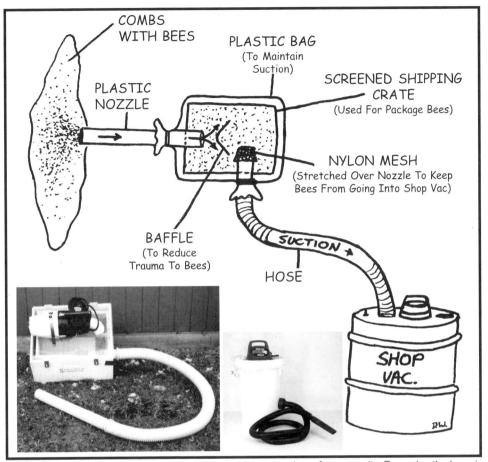

Several devices similar to this are on the market to remove bees from a cavity. Removing the bees is recommended, since it removes the possibility of stings, irritated neighbors, and dying bees.

Homemade 'bee-vacuums' are also common. The concept is simple: provide suction at one end of the system, a hose to connect the vacuum to a box to hold captured bees, a means to keep the bees from entering the vacuum, a cushion to reduce the stress of being sucked out of the nest, and a hose with a collection end to capture the bees.

observed. One is that the furniture should be removed from the inside of the room. All of the bees must be dead or they may crawl into the room and perhaps the rest of the house. When an insecticide is used adequate measures must be taken for ventilation and removal of any toxic residue.

The best time of the year to remove a colony from a building, regardless of the method, is early spring. At this time of year the colony will have the least amount of honey and the smallest number of bees. If one is attempting to save the bees they may have adequate time to store sufficient food for the following winter after the bees have been transferred to a new home.

It is possible to trap, that is to capture, the bees out of a building, but this is a time-consuming task. The first step is to close all but one of the entrances to the nest. A cone, usually several inches long and with a hole at the end sufficiently large that only one bee at a time may exit, is nailed over the entrance. In this way exiting bees cannot return to their home. A weak colony of bees, one containing a frame partly filled with brood and honey and with a small number of bees and a queen, is placed outside of the nest, near the cone, and on a strong platform that will hold the weight of a growing colony. The bees that cannot return to their own nest will first cluster on the outside of the cone but will slowly move into the nearby colony. As brood continues to emerge from the colony in the building, bees will continue to exit. After several weeks most of the bees will have joined the hive on the outside and only a few bees will remain within the building. The queen will not exit the nest but will die without attendant bees. At this time the hive is carried away to a new location. With the bees gone it is much easier to remove the nest and those few bees that remain.

People are often disappointed with the quality of the honey salvaged from combs in a building, or even a tree. Old black comb has a bitter taste as a result of an accumulation of cocoons and especially propolis. The honey must be squeezed from the comb with care. More important is that while honey bees will have a clean, orderly nest in a building or tree, the comb and honey are easily contaminated with bits of wood and debris from within and around a nest when the comb is being removed. This debris can be removed by carefully straining the honey using nylon cloth.

Removing bees from trees – For all of the same reasons that one might care to remove bees from a house or other building, it may be advisable to remove colonies from trees. Trees that have cavities sufficiently large to house a colony of honey bees are usually not very healthy or worth keeping. Bees may be

It is possible to have the bees move into a hive body from this hollowed log. Otherwise, the log will need to be split and combs removed and placed in frames.

removed from a tree in the same manner as they are taken from a house with one exception that is outlined below.

If it is possible to cut down a tree containing a colony, and to put that part of the tree with the colony in an upright position, a brood chamber of combs may be placed above the nest and the bees will slowly move into it. It is necessary to expose the top of the nest. The exposed area should be sufficiently large so that the bees are not inhibited from moving upward. A piece of plywood, of any thickness, but the size of a standard hive body and with a hole as large as that of the hole in the top of tree, is placed on top of the exposed nest. The piece of plywood serves as a bottom board for the hive body. An entrance is cut either out of the plywood or in the hive body. Since bees in a colony move naturally in an upward direction they will soon occupy the brood chamber.

If this is done in the spring, the queen too should move upwards after a few weeks. When the queen begins to lay eggs in the brood chamber, a queen excluder is placed on top of the plywood and under the brood chamber. The queen excluder confines the queen to the brood chamber. The brood below the excluder will all hatch after about three weeks. Then the tree portion of the hive may be removed.

Removing bees from boxes and fixed comb hives – (see BOX HIVES)

REPRODUCTIVE SYSTEM – (see ANATOMY AND MORPHOLOGY OF THE HONEY BEE)

REQUEENING – Replacing a queen in a colony is called requeening. Annual requeening has many advantages. Young queens lay more eggs and as a result head larger (more populous) colonies. Colonies with young queens are less likely to swarm.

Is this larva the best age to be chosen as queen?

It is a good idea to requeen a colony rather than allow the colony to supersede its queen for a variety of reasons. Most obvious is that when a colony is in a situation requiring queen replacement, the bees may choose less-than-ideal larvae to raise queens from because no one to two day old larvae are present, or they choose older larvae; and any queen raised in this manner will mate with unknown drones so the genetics of the offspring is uncertain. Further, there may be too-few drones available, or weather-induced limitations may curtail mating flights.

However, requeening colonies is a laborious, time-consuming task that often fails, even when undertaken by experienced beekeepers. The greater the number of bees in the colony the less likely the bees are to accept a new queen. Small or nucleus colonies, those with 10,000 or fewer bees, are usually requeened with ease. However, it is quite easy to combine queenless and queenright colonies.

Honey bees apparently retain or remember their own queen for about a day. Thus, any method of giving bees a new queen should be done in such a way that the bees being requeened do not have close contact with the new queen for about this period of time. Queens in a queen cage are safe from attack and are gradually accepted by bees, at least under most circumstances.

A number of reports in the beekeeping literature of requeening methods are questionable and probably not advisable. Several of these are discussed below. All of these queen introduction techniques assume the old queen has been found and removed. One common suggestion is that a new queen be dunked in honey and placed in a hive where the bees will lick her clean and accept her. Another recommendation is to smoke the colony heavily so that the sensory receptors of the occupants are deadened and as they recover they will accept a new queen. Yet another method is to drive all of the bees out of a colony with a bee repellent and to confuse the bees in this manner. When the repellent is removed the bees will reenter the hive and will accept a new queen. Yet another suggestion is to feed the bees a rich, heavily scented sugar syrup that may again foul the sensory receptors and allow the slow and safe introduction of a queen. None of the above methods have been thoroughly tested and are not recommended. Bees are much more inclined to accept a queen the same age as the one being removed. During a good honey flow the bees are much more likely to accept a new queen. It is probably this last factor that makes it possible for the above methods to work sometimes.

One method to requeen a small- or medium-size colony follows. One first finds and kills the old queen. It is not advisable to leave her body in the colony as is sometimes suggested. A new, young queen, in a queen cage, without attendants, is placed in the colony. The cage should be placed immediately above, or between frames of brood, especially young brood that is being fed by young bees. Young bees are more attentive to a queen, even one not their own, than are older bees. The cage should be placed so that the screen face is fully exposed.

It is further recommended that when introducing a new queen, she be given as much protection as possible. Most queen cages have a cork covering the cage candy (if not place a piece of duct or other tough tape over the candy opening). Leave that candy hole entrance corked for three days after introducing the cage. Only then remove the cork and let the hive bees release her. That should take two to three additional days.

A new queen needs a protective cage to keep her safe during the 'get-acquainted' stage.

This is what it will look like when the queen's cage is placed correctly between the bottom bars of the frames in the top box, just above the brood nest in the hive body below. Make sure the screen side is facing down and is unobstructed so the bees can feed and make contact with her. This placement is very important.

Many beekeepers keep a few 'nucs,' or small colonies around just in case a production colony goes queenless. A double screen and newspaper are all that's needed to introduce the bees and queen in the nuc to the full size colony.

Also, it is advisable to feed a colony while it is being requeened, and for several days after the queen is released. Use a standard 1:1 sugar:water feed. Do this even if there is a honey flow on (it may end abruptly).

A very satisfactory method of requeening a large colony is to place a small, nucleus colony, with about 5000 bees on top of a double screen separating the two colonies. It is important to place the nucleus colony with a newly-introduced queen on top and provide it with an entrance. When the new queen has been released and is laying, after about a week, the old queen in the lower colony can be killed and removed. The double screen can be removed at this time and replaced with a sheet of newspaper. After about 24 hours the bees in the two units will eat away the paper and the two will slowly become one. It is important to check the two united colonies after about a week to make certain the new queen is accepted.

Many beekeepers keep a few small nucleus colonies in an apiary in the event one of the production colonies should lose its queen. If this occurs the production colony may be requeened using the newspaper technique described above.

REQUEENING WITHOUT DEQUEENING – Many people have searched for a simple way to requeen honey bee colonies. One popular suggestion has been to place a ripe (about to emerge) queen cell in a colony so that the emerging virgin queen would replace the old queen. At least three carefully planned and executed studies have examined this idea. All agree that the method works only in a small percentage of cases and that it is not practical. In one study 919 colonies were used over a period of three years. The ripe queen cell was placed between two top bars in a honey storage super away from the queen where the new queen was less likely to be found and destroyed by the old queen. The results were that 12.7 percent of the colonies were successfully requeened; 53 percent of the colonies retained their own queen and 24 percent reared new queens. The remaining colonies became queenless (Szabo, T.I. "Requeening honey bee colonies with queen cells." *Journal of Apicultural Research* 212: 208-211. 1982.).

Annual requeening of colonies has been advocated for a great number of years for many good reasons: Colonies with young queens are less likely to swarm. Young queens lay more eggs

Placing a queen cell in a colony, without removing the existing queen is just asking for trouble.

than old queens. Young queens will lay later in the fall to provide more young bees as the colony prepares for winter. The best method of requeening requires the laborious effort of finding and removing the old queen.

RESEARCH IN APICULTURE – Research on bees and beekeeping in North America is done by a number of public and private organizations and research stations. The USDA Agricultural Research Service employs in the vicinity of 25 researchers at five research stations in Beltsville, Maryland, Baton Rouge, Louisiana, Weslaco, Texas, Logan, Utah, and Tucson, Arizona. The programs in these laboratories include studies on mites, diseases, pollination, Africanized honey bees and bee genetics. The Canada Department of Agriculture has a honey bee research station at Beaverlodge, Alberta. For a current listing of the federal and state bee research laboratories see the annual listing in Who's Who on the web site of *Bee Culture* magazine, www.beeculture.com.

Several state and provincial universities have full or part-time apiculturists, many of whom have a research program in addition to doing extension and teaching. Not infrequently the extent of the program depends upon how much beekeepers in the state or province are interested in promoting research since programs in apiculture are optional in most states. Only about five state universities have programs to train professional apiculturists. A number of chemists and biologists associated with other public and private colleges and universities may also undertake research on such subjects as genetics, honey products. queen rearing, bee diseases and mites, pesticides, etc.

Bee supply companies and beekeepers will often undertake research projects though they are usually concerned with the more practical aspects of beekeeping. Much of the work that has been done on bee repellents, plastic foundation, honey processing equipment, packing honey and honey promotion, has been done by the private sector.

Funding for bee research programs comes from a variety of sources. The National Science Foundation, The National Institutes of Health and the U. S. Department of Agriculture have funded many of the more fundamental aspects of bee research. Money to study practical beekeeping, especially to search for new and improved bee equipment, is nearly impossible to obtain. However, beekeepers in some states have been able to encourage such research and to prompt local state universities to obtain the necessary funding. Beekeeper's organizations must work closely with the administrators in the state universities.

The sophisticated research equipment needed for some studies is sometimes a deterrent to bee research. However, it is important to remember that Professor Karl von Frisch, who earned the Nobel prize for his research on bee behavior and the dance language, had as his chief research tool a two-frame observation hive. The observation hive is still a useful and widely used tool today for some types of research.

The International Bee Research Association (see IBRA), with headquarters in the United Kingdom, still publishes a refereed research journal specializing in bee-related topics. The most complete libraries devoted to bees and beekeeping in North America are at Cornell University, University of Guelph in Canada and the National

Agricultural Library of the USDA Agricultural Research Service in Beltsville, MD. Only a few endowments support research in apiculture.

RESPIRATORY SYSTEM – (see ANATOMY AND MORPHOLOGY OF THE HONEY BEE)

REVERSING FOR SWARM PREVENTION – (see SWARMING IN HONEY BEES, SWARM CONTROL)

ROADSIDE SALES – Many beekeepers have found that selling honey from roadside stands, usually in front of their homes, or from their homes can be profitable. Direct sales can provide the maximum price for a product. Such pricing is probably the chief incentive for building a honey stand. The chief detraction is that customers will sometimes take up too much time asking questions about bees and beekeeping.

A self-service stand where customers help themselves and leave their money will save the beekeeper much time. Most beekeepers report the honor system is satisfactory and only a small number find that thievery is a problem. Some beekeepers have found it profitable to sell honey to customers who bring their own containers to the honey house.

Building a clientele at a roadside stand takes time and requires a strong advertising program. According to beekeepers who operate stands successfully, repeat sales are most important, and these depend on many factors. The attractiveness of the stand is important. Painting the honey stand each year, and planting attractive flowers around it (but not necessarily those that attract bees) will aid sales. There must be sufficient parking space with easy access. And most important, a large, easily-read sign must be placed far enough away so normal traffic will have

Examples of Roadside Stands –

More honey stands on next page . . .

Roadside Sales

The key element is to have good signage. Signs must be placed far enough away that a motorist, traveling at highway speed, has more than enough time to look for, find, prepare to turn and safely turn off a busy road. Usually two, or even three signs, announcing the remaining, and diminishing distance are required – both directions of traffic.

time to safely react. Jars and packages must be clearly and attractively labeled. Small pamphlets with honey recipes are available from several state and county beekeeping organizations, including the National Honey Board and the American Beekeeping Federation.

The major problem for a beekeeper with a small number of colonies is to be able to provide the customer with a product that is uniform from one year to the next; it is often advisable to buy a small amount of honey for blending to be able to provide a consistent product season after season.

ROBBING BY BEES – It is a curious fact that as long as natural nectar is available to bees in a colony they will not rob from another colony or be interested in exposed honey. However, as soon as a dearth (shortage or lack of) of nectar occurs the scouts turn their attention to thievery, a phenomenon that beekeepers call robbing. Within a colony, bees share their resources equally no matter what the circumstances. However, they never hesitate to take from another colony when the opportunity arises. Robbers appear to seek out weak (less populated) colonies with a vengeance and will not hesitate to kill the bees within in their determination to steal whatever honey is available. Apparently bees do not steal pollen.

Robbing can be especially serious when the crop is removed at the end of a honey flow. It may be necessary to cover supers as fast as they are removed from the hives. If this is not done the whole apiary may be turned into an airborne mass of angry, robbing bees. Some beekeepers have used covered trucks into which they wheel stacks of supers as fast as possible after they are removed from the hives. Robbing in an apiary, whether it be when one is removing honey, or when making routine inspections during a dearth, is frustrating and causes extra work.

Robbing bees exhibit a nervous behavior that is quite unlike that seen under any other circumstances. This behavior may be seen when one exposes a super of combs wet with honey at a time when robbing will occur. The combs will soon be found by the bees. When a number of robbers are active one can pick up the super and shake it. The bees will take wing immediately in a nervous, flighty way.

Beekeepers should take steps to avoid robbing as it can result in the destruction of colonies, especially weak ones. Robbing also places colonies on the alert and makes them more defensive when they are manipulated. Excessive stinging can occur some distance from the colonies, and certainly if people approach too closely to an apiary where robbing is taking place.

When robbing becomes a problem it is difficult to stop, at least that day. It is important to reduce colony entrances immediately, especially those with small populations. Another tactic is to quickly remove the tops from all hives to equally expose all colonies in an apiary, thus making them all vulnerable. It is best not to open and work beehives when robbing is a problem. However, if beekeepers must expose less populous colonies such as queen mating nucs, some beekeepers use portable screen cages, usually about four feet (1.3 meters) square and six feet (2 meters) tall, that can be placed over a colony and themselves, to prevent the robbing of the colony that is being opened. A great deal of effort may be required to move the cage from colony to colony but this may, under some circumstances, be the only way possible to manipulate the hives.

One of the consequences of robbing, aside from the loss of honey, is the transmission of diseases such as American foulbrood and parasitic mites. Robbing occurs not only among managed colonies but can occur between feral and managed colonies. Queen introduction is more difficult when conditions conducive to robbing exist.

Progressive robbing – It is stated in some beekeeping literature that bees will sometimes quietly enter a colony not their own and steal honey that they take back to their own hive. No evidence that this is true has been found and the matter has never received careful scientific study.

ROOT FAMILY HISTORY – The A. I. Root Company, like many successful businesses, had its beginning in a small way. No one else could tell the story of those early years better than its founder, Amos Ives Root.

"About the year 1865, during the month of August, a swarm of bees passed overhead where we were at work and a fellow workman in answer to some of my inquiries respecting their habits, asked what I would give for them. I, not dreaming he could by any means call them down offered him a dollar and he started after them. To my astonishment, he, in a short time, returned with them hived in a rough box he had hastily picked up, and at that moment I commenced learning my *ABC in Bee Culture*. Before night, I had questioned not only the bees, but everyone I knew who could tell me anything about these strange new acquaintances of mine. Our books and papers were overhauled that evening, but the little that I found only puzzled me more and kindled a new desire to explore and follow out this new hobby of mine. Farmers who had kept bees assured me that they once paid when the country was new, but of late years, they were of no profit and everyone was abandoning the business. I had some headstrong ideas on the matter and in a few days I had visited Cleveland ostensibly on other business but I had really little interest in anything until I could visit the book stores and look over the books on bees."

The original of the most commonly used photo of A.I. Root

A rare photo of A.I. Root working bees. When he was much younger, and worked bees routinely, cameras were scarce. When cameras became more common, A.I. seldom worked bees.

"NOVICE'S" Gleanings IN Bee Culture.

1873

Or how to Realize the Most Money with the Smallest Expenditure of Capital and Labor in the Care of Bees, Rationally Considered.

PUBLISHED QUARTERLY.

VOL. I.   MEDINA, O., JAN. 1, 1873.   No. 1.

INTRODUCTORY.

FELLOW NOVICES:—We must confess to a feeling of not being quite as much at home here, just yet, as in the old *American Bee Journal*, but we trust we shall *all*, in time, feel all the liberty here that we have there enjoyed. Remember at all times that Improved Bee Culture is our end and aim, and we trust no one will hesitate to give any facts from experience, because they may tend to overthrow any particular person or "hobby."

If any of *our* especial plans don't work, or if anything we advertise has had its value over-estimated, here in these pages is the place of all others to set the error right. Please don't be hasty or prematurely positive, and when one of our number acknowledges a fault and makes proper reparation, the matter should be overlooked and friendly feelings renewed on both sides, at once and forever.

The advances now being made in Bee Culture, it seems to us, must necessarily bring about *individual* losses often; for instance, one of us may have made up a quantity of hives for sale, and new developments may point out plainly that they are not fully adapted to the present needs of Bee Culture, and when you are satisfied of this, please do not attempt to sell them without telling your customer the *whole truth*, and making the price correspond. The same may be said of Extractors. If necessary to throw them away as old lumber or old metals, do not, we implore you, hesitate an instant.

Our most successful business men of the present day, have discovered it to be a fact that it is more profitable to tell their customers the *bad points* of their wares as well as the good. There are ample opportunities in this world to acquire a competence *honestly*.

One of the most lamentable wrongs in Bee Culture is the custom of taking money for a "right to make and use" a hive, knowing that the buyer could "make and use" a hive so nearly like it as to answer every purpose, without using a SINGLE ONE OF THE PATENTED FEATURES. It will be our especial aim to fully inform the public of all such transactions coming under our observation.

Please give facts all you can without regard to their bearing on individuals, if they are of such a nature as to benefit the masses. Without further moralizing we will try and let our little JOURNAL show for itself what it is; but, dear readers, we hope you have read this carefully for we may refer to it hereafter.

A reprint of the very first page of the very first edition of Gleanings In Bee Culture.

The original A.I. Root Jewelry factory on the square in Medina, Ohio. A fire destroyed many of the buildings some time after A.I. built his new factory. Today, a plaque is located on the original site, now occupied by a newer, larger retail building. Which window did that historic swarm pass?

Late 1800s. Looking east, towards Medina's town square. The building on the right is the original and first addition of the factory on Liberty Street.

At this time, A.I. Root was operating a jewelry manufacturing company on the west side of the public square in Medina, Ohio. The book he chose that day was one by Mr. Langstroth, a Congregational minister credited with popularizing the movable frame hive. The distance between frames was based on the principle of bee space. That is, if bees are given a space of between 1/4" and 3/8" (about 1 cm), they will not build brace comb or deposit propolis in it.

Rev. Langstroth visits the A.I. Root Company. A.I. and L.L. were friends and Langstroth often visited the company. Though undated, this was towards the end of Langstroth's days.

The Root home apiary, showing several additions to the factory since L.L. visited.

A yet later photograph of the Root home apiary. The small building on the left served as an employee cafeteria, a small manufacturing facility, a garage and a print shop over the years. This entire area is now covered by the candle manufacturing plant, retail shops and meetings rooms.

In 1904, A.I. Root traveled to Dayton, Ohio, to visit with the Wright brothers, Orville and Wilbur, with whom he had been corresponding for some time (copies of this correspondance are on record at the Library of Congress). While in Dayton, he witnessed a flight of the Wright brother's airplane as it traveled a good distance, made a 'U' turn, and returned to its original location – a first for air travel.

Attempts to have this story published in prestigious scientific journals were rebuffed, so A.I. published the entire event in his magazine in the 1905 January 1 and January 15 editions of *Gleanings In Bee Culture*.

The above photo was taken about 10 years later, at a demonstration at the Medina County Fair.

Since the bees leave these spaces open, the frames can be removed without damaging the combs. He patented his new hive, but the beekeepers of the time continued the art of beekeeping as in the past by old fashioned methods.

A.I. Root immediately recognized the advantages of this new hive and started his beekeeping with it. The first season he harvested 1,000 pounds (454 kg) of honey from 20 colonies, but he left too little honey on the colonies for the winter and only 11 survived. The next season, he increased to 48 colonies and produced 6,162 pounds (2,800 kg) of honey, an average of 128 pounds (58 kg) per colony. Instead of keeping this a closely guarded business secret, he did everything he could to publicize it to the world. In 1869 he began manufacturing his own version of the Langstroth hive in the building of his jewelry manufacturing company using a foot-powered saw and later windmill power to saw the lumber. The centrifugal honey extractor had been invented in Germany by this time, but he found these machines "heavy and poorly adapted to the purpose." He, therefore, saw no alternative than to manufacture his own honey extractors.

Letters poured in from all over the country when beekeepers learned of A. I. Root's success with this new hive. To answer these he sent out circulars and as he says, "until 1873 all these circulars were sent out gratuitously, but at that time it was deemed best to issue a

The emblem on the front of the original factory. It remains on the original building.

The main office of the A.I. Root Company, built in 1906, sits at 623 West Liberty Street in Medina, Ohio. The exterior of the building remains unchanged after 100 years, though the interior has been extensively remodeled.

quarterly at 25 cents per year for the purpose of answering these inquiries." The name of the new publication was *Gleanings in Bee Culture,* which has been printed continuously through the years and is available to subscribers today. In his early writings which were so popular with the beekeepers of the day, Mr. Root used the nom de plume of "Novice" hoping this would encourage his readers to give their suggestions for improved techniques and ideas freely.

As beekeepers wrote in asking questions about this new modern beekeeping system, A.I. would refer them to back issues of *Gleanings in Bee Culture.* However, in many cases the back issues were not available to the writer for reference. At that point, A. I. realized he needed to publish an encyclopedia of beekeeping. He called that first issue *ABC of Bee Culture.*

The initial sale of the new beehive was so successful it was soon necessary to issue a small price list of "Root's Goods." These early bee supply offerings were beehives, extractors, smokers, and honey. No longer could he depend on wind-powered saws to supply the needs of his customers so he installed a four-horsepower steam engine and more machinery to keep up with the demands

Many years after it was shut down, A.I. Root studies the original windmill that powered the press that printed early editions of Gleanings In Bee Culture.

When belt drive was king in the wood shop.

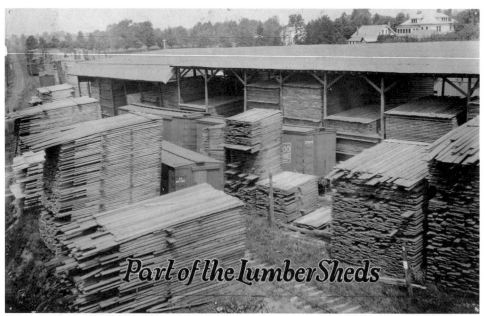
Part of the Lumber Sheds

The amount and size of the lumber storage buildings can be measured by noting the railroad car, nearly buried in the piles. These stacks of lumber were located alongside the property line, next to the railroad.

Part of the beeswax foundation process. Foundation started out, of course, as crude beeswax obtained from beekeepers, and the Root Company's own bees. The wax was melted, filtered, and then went to the 'Sheeter' machine. This was composed of a large drum filled with cold water that rolled through a pan containing melted wax. The wax would cool, stick to the drum as it rolled up and away from the pool. As the drum continued to roll, a large blade scraped the now-solid wax off the drum and collected the 'sheet' of wax on a new roll.

This roll was left to cure for a week or so then run through these machines in front. These 'smooth roll' machines ran the sheeted rolls between two smooth rollers that formed the wax so it was uniformly thick, and the rolls uniformly wide.

These rolls of wax could then immediately be run through the machines at the far end of the row that would emboss the wax with the hexagon design used in foundation, and then cut them to length. Tissue was placed between them, and they could then be boxed and be ready to sell.

If wire was to be embedded, the cut sheets were run through a machine that actually pressed warm wire into the sheet, then cut the wire to fit the sheet.

of the new business. All this time his interest in the manufacture of jewelry was waning and his manufacture of bee supplies increased.

By 1876, he had doubled his work force and had to run his plant at night during the busy season. He did this for a couple of seasons and then in 1878 he sold his building on the public square and purchased the old fairgrounds in the western part of town next to the Medina railroad station. Here he erected a brick structure 40 by 100 feet (12 by 30 meters) with two stories and a basement. In this building he installed a 40-horsepower engine with the latest

For many years the Root Company raised queens. Some of their yards were placed in groves of Basswood trees, planted to give shade, provide nectar and pollen, and eventually end up as section boxes for comb honey.

improved machinery. By 1883 he had to double his capacity again with another wing on this building and as the years went on the floor space of the plant continued to grow until it reached 183,000 square feet (20,000 square meters) of production facilities and storage space, covering some 12 acres (4.8 hectares) of land, not counting manufacturing plants in San Antonio, Texas, and Council Bluffs, Iowa.

During the first 25 years, the business was conducted under the name of A.I. Root. In November of 1894 The A.I. Root Company was incorporated. About that time A.I. Root had been told by his doctor he had less than a year to live because of a serious lung infection. He was advised to turn the responsibilities of his business over to his eldest son Ernest R. Root, and his son-in-law, John T. Calvert. This allowed him to get out into the fresh air and pursue another of his hobbies, gardening. However, the doctor was mistaken by quite a number of years because A.I. Root did not die until 1923.

Perhaps the greatest contribution A.I. Root made to the beekeeping industry is that he publicized the fact that it was possible to make a living by keeping bees. With the new equipment available, such as the Langstroth movable-frame hive and the new honey extractor,

Ernest R. Root (1862-1953). Ernest's favorite activity, when speaking to the public, was to put bees in his hat, and then put the hat on to show that bees were gentle. He traveled many times with a Chautauqua circuit to popularize beekeeping. He visited every state, and every Canadian province.

*Honey packing was a large part of the business early on, and used a significant portion of the manufacturing space. Airline honey was one of the fanciest brands available at the time. Airline was derived from **A.I. Root Line**.*

beekeeping could become a profitable business. He also led the move to standardize manufactured bee supplies and today, rather than having many different sizes and shapes of hives such as they have in Europe, the United States has standardized on the Langstroth type hive. In addition to developing new beekeeping equipment, he made many improvements on the inventions of others.

From the very beginning a large part of the company's sales were in honey. In the 1920s the brand Airline Honey was advertised nationwide, and was known to be the finest brand of table-grade honey available. Late in the 1920s an organization was formed to merge many honey packing interests into one large organization. They offered The A.I. Root Company a good price for their goodwill and equipment and the offer was accepted. This left a void in the production capacity and floor space which was eventually taken up by the entry into a completely new line of

Ernest was awarded an Honorary Dr. of Laws Degree from The Ohio State University in June, 1944, for his work in advancing the knowledge of honey bees and pollination.

Huber Root (1883-1972) examining one of the religious candles manufactured by the company. Huber wrote "Beeswax, Its Properties, Testing, Production and Application" in 1951. It was the industry standard for many years. He also invented the rolled beeswax altar candle, and the glass encased bottled sanctuary light. In 1963 he was honored by Pope John XXIII, who made him a Knight of Saint Gregory.

Alan Root (1905-1991), Ernest's son. During his career Alan worked in the saw room, the honey plant, becoming Plant manager, then Vice President, then President.

business, that of beeswax altar candles. This new business venture was pioneered by Huber H. Root, A.I.'s other son, and developed into a very successful addition to the A.I. Root Company product line, in spite of the fact that it was launched during the height of the Great Depression.

Alan Root, Ernest Root's son, guided the company through the difficult years of World War II. He provided leadership for the beekeeping industry in procuring critical war materials for the manufacture of beekeeping supplies, and later as President of the Honey Industrial Council. Other fourth and fifth generation Roots who have contributed to the ongoing success of the Company are John A. Root and Stuart W. Root, sons of Alan, and John's son, Brad I. Root.

Though A.I. stepped away from the day to day management of his now international company, he didn't step far. He continued to live a stone's throw from the factory, and maintained an office there until his death.

There, he continued to write and make regular contributions to *Gleanings In Bee Culture* in several areas. One column was entitled "Our Homes," that covered a wide variety of topics that he, his family, his employees, friends and others encountered. These included everything from the evaluation of indoor plumbing to bicycles to electric then fuel-driven automobiles.

A.I. was a deeply religious man and wrote about that part of his life in his columns also. He worked with inmates

From left – Huber, Alan (Ernest's son) and Ernest Root.

in the local jail, campaigned continuously against the sale of liquor and cigarettes, taught Sunday School and quoted from, and wrote about the Bible.

From his youngest days A.I. had a strong interest in gardening and farming, and over the years wrote extensively in the magazine about new crop varieties, techniques to use, equipment to try, and results from his own gardening and farming trials.

Occasionally, he ventured back to his roots and examined some new aspect of

The Root Logo Across The Years –

The first Root trademark appeared in the 1890s and had as its focal point the honey bee and a clover leaf representing the relationship between the honey bee and it's prime source of nectar, the legume.

In the early 1920s the script Root trademark was introduced and was used until 1986. In keeping with the cleaner graphics and simpler design of modern marketing the next Root trademark illustrated the Skep, the symbol of beekeeping throughout its long history. The Skep has been used since man began making hives for bees, and to some extent is still in use to-day. Rev. Langstroth's invention of the modern, moveable frame beehive was being largely ignored when A.I. Root encountered his first swarm of bees and caught "bee fever." He was the first to commercially produce the Langstroth movable frame hive and offer it for sale to beekeepers throughout the world. He was also a staunch defender of the Langstroth patent, as later patents attempted to infringe on it. Thus, the Skep symbolized the beekeeping heritage of the 117-year-old firm.

In 2002, the Company again changed the trademark, providing an even cleaner, simple design.

Undated. The bee supply factory and store in San Antonio, Texas.

beekeeping, or wrote about visits to beekeepers during his travels. He never left his beginnings.

A.I. established homes in Michigan and Florida and later in life was away from Medina for extended periods of time. Ernest, Huber, and J.T. Calvert kept the magazine and factory running.

Ernest was involved in the magazine, and traveled extensively to meetings to participate in industry politics and events, and to speak and inform beekeepers of the latest innovations in the industry.

His brother Huber became involved in the candle manufacturing and was instrumental in heading the company in that direction.

Alan, son of Ernest, managed the Company's manufacturing section for several years, then became the President, overseeing the entire operation. He remained active in the beekeeping industry, and business, but steered the Root Company aggressively in the religious candle business, expanding it greatly.

Alan's son, John, led the magazine for several years and he, too, was active in the beekeeping industry. However, as the beekeeping industry continued its slow but steady decline, the church, and then the home decorative candle market expanded greatly, and the Root Company capitalized on its position as a candle-manufacturing company.

In the late 1990s Root ceased production completely of beekeeping supplies, but greatly expanded the candle manufacturing capability at the same factory location. Thee company maintained, and expanded the beekeeping magazine and book publishing business, however, and those continue to inform, educate and entertain the world's beekeepers.

As for the future ... A.I. Root wrote in 1917, "I looked up at the stars and stripes that were floating in the wind from the flag pole over our company and said The A.I. Root Company, God permitting, will last for years after A.I. Root himself is gone."- JAR

John Root, President and Chairman of the Board of the A.I. Root Company. After college, John served in the Air Force as a flight instructor and remained an avid pilot after leaving the service. He served as Editor and Managing Editor of *Gleanings In Bee Culture*, and was editor of several editions of *ABC*.

During the time his father, Alan, was President, John was Vice President, and worked diligently in the beekeeping industry. He served on several National Boards, and was President and the first Chairman of the Board of the Eastern Apicultural Society of North America.

While Alan guided the company during its transition from a major bee supply company toward a church candle manufacturing company, John was instrumental in diversifying those skills toward the home decorative candle market. This aspect of the company expanded tremendously when that direction was chosen. The church market wasn't neglected however, and it, too has grown.

In the early 1990s, all bee supply manufacturing facilities were replaced by new candle manufacturing and warehouse facilities to accommodate this growth.

Publication of the monthly magazine, now renamed *Bee Culture* continued, along with *ABC* and other beekeeping books, under John's direction.

Stuart Root, Alan's son, has been continuously involved in the manufacturing end of the business, specializing in worker safety programs, both inhouse and in the greater manufacturing community, and in Quality Control of the thousands and thousands of products the Root Company manufactures. Stuart has also been very active in the community, like his brother, John, working with and supporting many local groups.

Brad Root, Administrative Vice President of the A.I. Root Company. After completing college, Brad, John's son, returned to the company and worked extensively in nearly every area of the manufacturing, administration, and management of the company. He is well versed in everything from inventory control to information technology to human resources.

Brad is also very active in the local business community, working to ensure local businesses do well and they and their employees thrive.

ROOT, LYMAN C. (1840-1928) – Mr. Lyman C. Root was the son-in-law of Moses Quinby and reviser of Quinby's *Mysteries of Beekeeping* under the title of *Quinby's New Beekeeping*. Mr. L.C. Root was born in St. Lawrence county, New York. He lived in Stamford, Connecticut.

Mr. Root served in the Civil War, was educated at Fairfield Seminary, in New York State, attended the St. Lawrence University and was a graduate of Eastman's Business College, in Poughkeepsie.

He was an active beekeeper and one of the very few remaining noted beekeepers of the old days.

Lyman C. Root

ROSS, THOMAS AND ROSS ROUNDS – The circular section was first suggested in an article in *Gleanings In Bee Culture* in 1888 by a regular contributor known as "The Rambler." His idea was to drill auger holes in a board, line the holes with wood shavings and let the bees build their little combs in them. This idea obviously had no merit whatever. But the following year, there appeared in the same magazine, in an article by Mr. T. Bonner Chambers, what seems to have been a precursor of the modern round section. He suggested getting round glass sections by somehow slicing up glass jars and inserting these "slices in wooden frames, four per frame. His drawing for this idea bears such a striking resemblance to the modern circular section one wonders whether that might be where the modern invention really came from.

In any case, it was a retired physician and hobby beekeeper, Dr. Wladyslaw Zbikowski, who invented the modern circular section in the early 1950s. In 1956, and for the next 20 years I went about proclaiming the merits of these still relatively unknown circular sections. Dr. Zbikowski had revolutionized comb honey production with what he called Cobana Sections.

Sometime in the mid-70s I spoke to a large audience of beekeepers in Ohio and I passed a circular section through the audience. In that audience was Tom Ross, an architect and sideline beekeeper, who expressed keen interest in this round section. Mr. Ross significantly improved on the design of the Cobana Products and had new molds made for the plastic frames, rings and covers. He named his new product Ross Rounds. He sold the business in 1997 to Lloyd Spear, of New York, who continued to produce and sell Ross Rounds, and expanded the product line. (This information was taken from an article by Richard Taylor in the February 1995 issue of *Bee Culture*.) (see COMB HONEY, HOW TO MAKE, round sections)

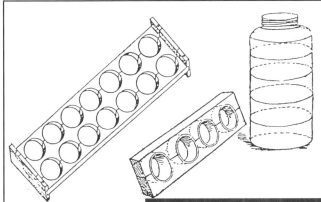

Early forerunners of round sections. Drawing on left, from 1888 Gleanings *(pg. 798) shows "The Rambler's" crude wooden frame. On right is Mr. Chambers' more ingenious wooden frame, into which he inserted glass rings cut from a jar, from the 1889* Gleanings *(pg. 42).*

Left to right – Lloyd Spear, Richard Taylor, Kim Flottum (Editor, Bee Culture *magazine) and Tom Ross. Spear, Taylor and Ross have, over many years, written articles on the efficient production and profitable marketing of Ross Rounds in* Bee Culture *magazine.*

ROSSMAN APIARIES, INC. – Joe Rossman (1902-1983). Joe lived near Moultrie, Georgia, and first became acquainted with bees in 1928 when he helped a neighbor, J.G. Puett, work his bees in the spring.

He soon was operating bees in the south for E.W. Long, a large beekeeper from St. Paris, Ohio. In 1934 Joe went into partnership with Mr. Long. In 1952 he bought out Mr. Long's southern operations.

The Rossman family raised the Island Hybrids (also called Kelley Island or Pele Island Hybrids), for many years. This was the hybrid bee developed by the United States Department of Agriculture, on Kelley Island in Lake Erie, near Sandusky, Ohio.

In 1967 the company incorporated. Two sons, Phil and Fred, were active in the business. Package bee and queen production was their main business for years although some honey was also produced. Most of their packages were sold in the Manitoba and Quebec provinces of Canada and in Wisconsin, New York, Ohio, and other northern states. The business today is located 3½ miles north of Moultrie on Highway 33.

Joe Rossman developed and modified several machines to fit their needs. He built a saw to cut the opening in the top of packages used to ship package bees. He also built another saw that cuts four small strips at a time. These strips are used to fasten the wire on the packages. He also modified a jig that holds frames to be stapled so that the frame also can be stapled on the ends.

The oldest son, Fred, graduated from Auburn in Business Administration. Fred remains in the business, which has expanded to include an active woodenware manufacturing operation and general bee supply company. Bees are sold as packages, queens are produced and honey and pollination are still part of the business.

Joe, Fred, and Phil Rossman all served as presidents of the Georgia Beekeepers Association and held offices in American Bee Breeders Association, and Southern States Beekeeping Federation while it was active. Both Phil and Fred held offices in the American Beekeeping Federation.

ROTHENBUHLER, WALTER C(HRISTOPHER) (1920–2002) – A pioneer in the research on behavior genetics, Walter C. Rothenbuhler obtained his B.S., M.S. and Ph.D degrees at Iowa State University in Ames, Iowa. The American-foulbrood-resistant breeding program was started by O.W. Park. Although Park observed that resistance to American foulbrood was behavioral, it remained for W.C. Rothenbuhler and his students to design the experiments and prove that indeed behavioral resistance was heritable. (for additional information see PARK, O.W.)

In 1962 Rothenbuhler left Iowa State University and went to Ohio State University where he continued his distinguished career. He was an AAAS (American Association for the Advancement of Science) Fellow, member of many professional societies and was the first recipient of the Eastern Apicultural Society J.I. Hambleton Award for excellence in bee research. Rothenbuhler participated in the training of over 40 graduate students and at one point, three of the five USDA bee research laboratories were headed by students who trained under him. He retired in 1985 and in 1989 The Rothenbuhler Honey Bee Research Laboratory was dedicated in Columbus, Ohio.

ROUND SECTIONS – (see COMB HONEY, HOW TO MAKE, *Round sections*)

ROYAL JELLY – Young worker honey bees secrete, from glands in their heads, a special food that is fed young worker larvae, queen larvae and adult queens. This material is called royal jelly (sometimes called bee milk). Older worker larvae are fed a somewhat modified diet that still contains some of the glandular secretion. Workers and queens arise from fertilized eggs and which one develops into a queen depends upon what the larvae receive during their short larval life. Since it is the young worker bees that have active

Walter C. Rothenbuhler

Young worker larvae are fed Royal Jelly for the first day or so as a larva, then they are fed a modified Royal Jelly, different in composition. This larva is less than 24 hours old, and it is floating in a pool of the unmodified Jelly.

brood food glands it is important that a colony of bees contain young as well as medium and older age workers.

Royal jelly is a pale yellowish, milky, viscous liquid. About two-thirds of its weight is water. The remainder is about 90 percent protein and ten percent fat. It may contain a certain amount of sugar (honey), pollen and wax and since larvae are continually growing and molting in it there are always some cast larval skins present. The composition of royal jelly varies slightly during the year depending on what food is available for the adult bees.

Royal jelly, perhaps because of the name and/or the fact that it is the food for a queen, has been touted as a food for man, even an elixir. Many testimonials may be found but no data to show that it has value in human nutrition.

RUTTNER, FRIEDRICH (1914-1998) – Professor Ruttner was born in Eger, Austria. He grew up with a special interest for bees and beekeeping. He became a medical doctor at the University of Innsbruck where he worked for some time in the department of neurology.

As a consequence of a heavy polyarthritis, he had to leave his medical occupation in 1948 and set up a honey bee breeding station in Lunz-am-See, Austria, with the support of the Austrian beekeeper's association. He continued his studies at the University of Vienna where he obtained his Ph.D in biology. In 1965 he became professor zoology at the University of Frankfurt and director of the Bee Research Institute at Oberursel, where he produced remarkable scientific work in the field of genetics, breeding, artificial insemination, taxonomy, mating behavior and bee botanics. In 1969 he founded the scientific journal *Apidologie*, in collaboration with J. Louveaux. After the first occurrence of *Varroa* in Germany, the institute put most of its energy in the control of this important bee pest.

In the International Federation of Beekeepers Association, Apimondia, Professor Ruttner played an important role as president of the standing commission for Bee Biology, where he remained the "past president" until his death.

In 1981 Friedrich Ruttner retired and again took up residence in Lunz, where he produced a number of high-level scientific publications. Throughout his professional occupation he concerned himself with an efficient collaboration between scientific research and beekeeping practice. His best known work remains the *Biogeography and Taxonomy of Honeybees* published by Springer-Verlag, in 1988.

SCALE HIVES – A scale hive is a hive placed on a scale so that a beekeeper may follow its weight as the season progresses. No two colonies of honey bees are alike. However, trends in food consumption and honey and pollen storage can be determined by weighing colonies daily or weekly. Many types of scales are available. Their accuracy depends on their quality and usually their price. Many beekeepers who use scale hives build shelters over the scales and colony to protect both against the ravages of the weather. In order to obtain an accurate measure of what is taking place, the scale hive should be manipulated in the same manner as other colonies. It is exciting to observe a colony gain as much as 10 or more pounds (4.5 kg) a day, which is possible in a good honey flow.

A scale hive in a beeyard can be a good indicator of what is going on with the rest of the colonies. Remember, however, that no two colonies are the same. (McCreary photo)

SECHRIST, EDWARD LLOYD (died 1953) – Sechrist kept bees in Africa, Haiti, Tahiti, Maryland, Ohio and California. In writing the preface for his book *Honey Getting* Sechrist said: "A study of the fundamentals of beekeeping has not hitherto been available in compact and usable form. In the belief that a need exists for such a study, *Honey Getting* is offered. Any originality it may possess lies in the emphasis it places on the Clear Brood

Nest Method of Apiary Management and Colony Balance in Queen Introduction." Those two ideas were a distinct contribution to our thinking."

Long one of Uncle Sam's beekeeping aces in the Office of Bee Culture, USDA, he was the first to study costs in relation to crops, locations and management."

In addition to *Honey Getting*, published in 1944, Sechrist was coauthor of *Scientific Beekeeping* with Dana McFarland, a research engineer, published in 1948; *Applied Thermodynamics in Apiculture*, 1947; *Scientific Beekeeping*, 1946; *Honey Getting*, 1944; plus *Amateur Beekeeping*, *U.S. Standard for Honey*, and *Transferring Bees to Modern Hives*.

SENSES OF THE HONEY BEE – Honey bees use an elaborate and amazing sensory system to take maximum advantage of a short period of time to harvest a maximum amount of food. They are also equipped to respond to emergency situations and to protect the hive against overheating, cold, inclement weather and, of course, attack. Much of the research on the senses used by honey bees was done by von Frisch and is discussed in great detail in his book: von Frisch, Karl. Translated by L. E. Chadwick. *The Dance Language and Orientation of Bees*. Belknap Press of Harvard University Press, Cambridge. 566 pages. 1967.

Taste – Karl von Frisch says that compared to our own sense of taste, a honey bee's is less sensitive to weak sugar solutions. Also, many sugars that taste sweet to us are not attractive to honey bees. When 34 sugars were tested, some of which are rare, it was found that 30 were attractive to humans but only seven to bees. A foraging bee's threshold of acceptance for sugar syrup depends on food availability; when there is nothing else available in the field, bees might accept sugar syrup with as little as two to four percent sugar. Bees will abandon flowers with a low concentration of sugar in the nectar for those that are richer. Bees can detect the difference between a 30 and 35 percent sugar syrup, which is probably a taste difference that is at least borderline for a human to detect; however, if the difference between two sugar syrups is only two and a half percent the bees cannot detect the difference.

Several substances that are bitter to human taste are tasteless to honey bees. Quinine is one example. This is easily demonstrated by adding enough quinine to sugar syrup to give it a disgusting taste for us. The syrup will be enthusiastically consumed by bees.

Sound perception – It has been known for a number of years that sound is produced by foragers that dance to recruit other bees to seek the food they have found (see COMMUNICATION AMONG HONEY BEES). The sound is made by the dancers vibrating their wings. This is different from the sound made by wing movement in flight. Recruits cannot learn the direction and distance of the food from dancers that have had their wings clipped or by genetic mutants with reduced wings. This indicates that the production of sound is an important part of the information given by a dancing bee. Some foragers perform silent dances and these too are ineffective in recruiting.

However, until recently, all of the textbooks on honey bees have stated clearly that they cannot hear airborne sound. The problem, it appears, is that we have been thinking about the wrong kind of sound and looking for the wrong kind of sensory receptors.

The sound with which we are familiar is carried by waves and what we detect with our ears are oscillations in pressure. This type of sound can be carried for long distances. The sound that honey bees hear when following dancers is caused by the movement of air particles. It is a type of sound that can be heard over a short distance only, less than a quarter of an inch (6 mm). Thus, a recruit must be very close to detect the sound made by a dancer.

A recruit that follows a dancer holds her two antennae perpendicular to her face but at an angle of 90° to each other. It is now believed that hairs on the basal antennal segment (the pedicel), called Johnston's organs, are the sound receptors. Interestingly, these organs were described by Johnston in 1855 on mosquitoes. Many insects have such organs but their true function has not been clear, especially in honey bees. These sensory organs have nerve cells that feed into the antennal nerve trunks and the brain.

Experiments to show that airborne sound was received by recruits were done in the following way. Bees taking sugar syrup from a feeder were exposed to a sound of an appropriate frequency and after feeding for four seconds were given a mild electrical shock. This caused them to withdraw from the feeder but at the same time they came to associate the sound with the shock. After repeated trials the bees learned to withdraw from the feeder when they heard the sound, even when there was no shock.

The sound from a loudspeaker was driven through a small glass tube. The tube was directed at the feeding bees. In addition to the above, bees trained to associate sound with an electrical shock were reluctant to land at a feeder when the sound was left on continuously.

When honey bees were exposed to sound and an electrical shock in a large tube, they failed to associate the two. This indicates that sound waves and particle movement cannot be sensed by bees over longer distances. In the research described below a tuned tube was used that created standing waves. In such a tube there are some regions where pressure variations are greatest (humans would hear the sound in these regions) and other regions where pressure did not vary but particle vibrations were great (humans would hear no sound in these regions). When honey bees were tested at these locations they did respond where displacement was great, but not where pressure differences were great.

The findings reported here open a new area of research into how bees obtain

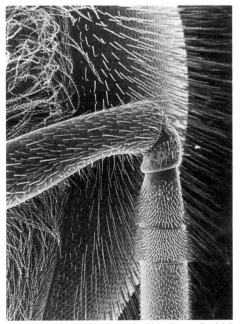

Johnston's organ is situated within the pedicel (p) on the antennae. (From Goodson)

information from a dancer. When we look back through the literature we realize that not everything about honey bees and sound is new. A Russian researcher (E.K. Es'kov) reported in the mid 1970s that bees could hear in this manner but his experiments were apparently not convincing. Also, it has been known for many years that honey bees respond to vibrations in the comb, called substrate-borne sound. For example, walking bees can be caused to freeze in place by some sounds. Also, when one beats rhythmically on a hive ("drumming") the bees respond by marching upwards, even leaving their hive and abandoning brood. The role of substrate-borne sound reception in bee biology remains unknown. For further information see Towne, W.F. and W.H. Kirchner. *Hearing in honey bees: detection of air-particle oscillations.* Science 244: 686-688. 1989.

Color vision – Honey bees use their sense of color to orient in the field to collect nectar, pollen, propolis and probably water. They also use their color sense to return to their home. Beekeepers have long been aware that painting hives different colors will help bees find their own hive and reduce drifting in an apiary.

That bees have and use their color sense was demonstrated clearly by Karl von Frisch in experiments in which bees were trained to a feeder on a table. Under the feeder was a piece of colored cardboard. After the bees had made several round trips to the feeder from their hive, two new feeders were placed on the table. Under each feeder was a piece of colored cardboard, one of which was the same color used in training. The bees went immediately to the feeder atop the same color to which they had been

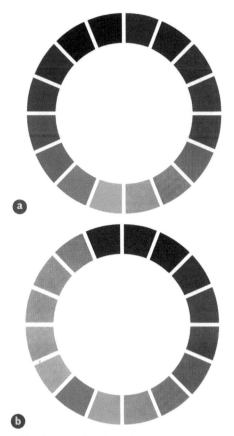

a *The human visual spectrum represented as a circle.* b *The bee visual spectrum can also be represented as a color circle.* (From Goodson)

trained. In the same way Karl von Frisch found that bees can discriminate among different forms and patterns though their ability to do so is not nearly so acute as that of humans. For example, honey bees cannot tell the difference between a solid square and a solid circle of the same area. However, designs that differ in "brokenness," such as an X and a solid circle, are distinguished by bees. Of course, in finding flowers the bees utilize the senses of odor and vision simultaneously.

The bee's field of color vision extends into the color red but not as far as that of humans. Red colors in that far end of the spectrum of vision are invisible to bees. However bees can see into the ultraviolet part of the spectrum whereas humans cannot. Therefore the nectar guides of some plants are visible to bees but not to humans. (see NECTAR GUIDES)

Odor perception – Karl von Frisch reported that honey bees have odor receptors on their antennae and that in some ways the bee's sense of smell is more sophisticated than that of man. In his early experiments he offered bees scented sugar syrups inside boxes with small entrance holes. After the bees had learned to associate food with a box they would enter boxes with the proper scent even though there was no food inside. When bees were trained to colored, scented boxes with food, it was observed that color was used to pick out a box from a distance of one to several feet (30 cm to one meter). However, when bees came close to the box, they would not enter if the odor was not present but would search up and down a row of boxes for the one with the right odor.

Carbon dioxide detection – While it was shown many years ago that honey bees can detect carbon dioxide, the function of the carbon dioxide detectors, which are on the antennae, was not clear. Seeley has shown that the introduction of carbon dioxide gas into a hive will stimulate wing fanning by worker bees. The wing fanning is not random but the bees work together to remove the excess carbon dioxide out through the hive entrance.

In a small colony the carbon dioxide level varies between one and three percent while larger colonies have more precise control. Since animals consume oxygen and give off carbon dioxide, it is important that levels of the latter not become toxic. In an undisturbed, glass-walled observation hive it was observed that wing fanning by a large number of bees might occur several times during a night in order to prevent the levels of carbon dioxide from becoming too high. Carbon dioxide buildup might be a problem when bees are moved in refrigerated trucks; special precautions should be taken for some air exchange while the bees are in transit. For further information see Seeley, T. D. *Atmospheric carbon dioxide regulation in honey-bee colonies.* Journal of Insect Physiology 20: 2301-2305. 1974.

SEX DETERMINATION IN THE HONEY BEE – (see GENETICS OF THE HONEY BEE)

SHAPAREW, VLADIMER (1915-1989) – Oakville, Ontario. He was a valued contributor to the pages of many beekeeping journals for many years.

Born in Russia and educated in Canada, he was a nuclear scientist with AEC and with Ontario Hydro. He contributed much in beekeeping information and equipment development, including bee escapes, hive ventilators and other colony equipment.

Vladimer Shaparew with his Honey Drying Ventilation hives.

SHERRIFF, B.J. – The Sherriff Bee Suit, that is a jacket or full suit with attached, zip-on hooded veil wasn't invented by the Sherriffs, but it was perfected by that company. Brian Sherriff and wife, Pat, were clothing manufacturers in the U.K. Brian became a hobby beekeeper, and put his clothing design skills to work. He designed a better beesuit. It is widely popular with beekeepers around the world because it is easy to use and reliable.

Brian and Pat Sherriff.

SIGHT IN THE HONEY BEE – (see SENSES OF THE HONEY BEE; Color Vision)

SKEPS – A skep is a round, usually peaked hive, made of straw. A few beekeepers use this term to describe colonies in wooden hives, but this is an incorrect use of the word. Skeps have sometimes been used in symbols and seals; the state of Utah has a skep on its state seal that is intended to represent industry, perseverance and the thriftiness of the honey bee.

Skep beekeeping has been popular in parts of Europe for many centuries. A properly woven skep will shed water with ease. The bees will coat the interior with propolis, further protecting the skep against penetration by water; the propolis also strengthens the hive. Often several sticks are poked through the

Wax combs inside a skep.

straw and stretch from one side to the other to provide additional support for the combs. Skeps are sometimes made with a hole in the top over which one may place a glass jar or another smaller skep for honey storage.

The use of skeps is not legal in the U.S. and many countries in the world because the combs are fixed in place and any brood in them cannot be examined for disease. Skep beekeeping is not practical since there is no good method of swarm control. In areas where skep beekeeping is practiced, the beekeepers capture swarms as they emerge and place them in new skeps. In the autumn the number of colonies is usually much greater than it was in the spring. Some colonies are then killed, often by the use of sulfur fumes, and the honey and wax harvested. For more information see Crane, Eva. *The World History of Beekeeping* and *Honey Hunting*. Routledge. New York. 682 pages. 1999.

Skeps

Two different types of skep construction.

1. Starting at the top.

2. Feeding straw into the narrow bottle end.

3. Securing the tightly packed straw with binding.

Skep Construction – Though modern skep builders may use different equipment (a plastic bottle instead of an animal horn), the technique remains relatively unchanged.

SKUNKS

SKUNKS – Few beekeepers own bees very long before they encounter bee-eating skunks. Skunks feed largely on insects including those that sting. Those beekeepers who have observed skunks feeding on colonies state that a skunk will stand in front of a hive, usually at night or early in the morning, and scratch at the entrance. The scratching arouses the bees within and one or more will investigate the entrance area. When a bee appears it is swatted and then eaten by the skunk. Skunks will move up and down rows of hives seeking those that are not too strong and from which only one bee at a time will emerge as a result of their scratching. It appears that skunks do not stay too long at the entrance of a colony from which many bees emerge and attack at one time. Stings have been found in the mouths and stomachs of skunks that have been feeding on bees, and clearly, being stung does not deter most skunks.

It is easy to spot skunk feeding in an apiary as the grass will be torn up immediately in front of the entrance where the skunk has been rolling and pawing as a result of being stung. The colony entrances are usually muddied

Several species of skunks can be found. They all like honey bees.

as a result of skunks scratching at the entrance. In addition, beekeepers report that colonies in apiaries where skunks are feeding become unusually defensive. Apparently, the routine disturbance of

The best, and safest skunk deterrent is to get colonies up and out of the reach of these pests. (Hood photo)

a colony will cause the bees to remain in a near state of alarm; at the same time there may be a greater than normal number of guard bees. The same phenomenon has been reported as a result of other animals feeding on colonies, or even as a result of colonies being examined daily by experimenters or others.

On rare occasions it has been reported that skunks will move elsewhere for food apparently because of excessive stings. Mother skunks have been seen teaching their young to feed at beehives.

Typical signs of skunk predation. You may find skunk droppings nearby containing numerous bee parts.

Two various types of skunk guards and deterrents.

Catching a skunk in a live trap can be a mixed blessing.

Beekeepers report that skunk feeding has become increasingly a problem in recent years as there is less trapping of skunks for their fur than many years ago and as a result skunk populations have grown. At one time it was popular to poison skunks feeding on bees but in many areas this is both illegal and considered a dangerous practice. It is certainly not advisable to trap skunks in the vicinity of a hive because of the odor that is usually released. Some beekeepers have built fences to keep skunks out of apiaries. Elevating colonies on hive stands will also deter skunk feeding, and probably the best defense a beekeeper can use. Various sticky or pointed devices in front of hives have been advocated to deter skunk feeding, sometimes with success. -MH

SMALL HIVE BEETLES – Two hive beetle species are known to infest honey bee colonies. The large hive beetle, *Hoplostomus fuligineus* (Coleoptera, Cetoniinae) is found in the northern Bushveld areas of South Africa and throughout other parts of Africa. The second is the small hive beetle, *Aethina tumida* (Coleoptera, Nitidulidae). The small hive beetle is also found in South Africa, but has recently been found in other parts of the world.

The small hive beetle was first identified in North America from beetle collections made from managed European honey bee colonies in Florida in June, 1998. No supportive evidence has been reported to indicate when or how the small hive beetle entered the U.S. Although the small hive beetle is considered to be a minor hive pest of the African honey bee, *Apis mellifera scutellata*, in its native host range in South Africa, it has caused considerable damage to European honey bee colonies particularly in the southeastern region of the U.S. This pest has also proven to be a problem in and around honey houses.

The first small hive beetle collection reported in the U.S. was made from a colony of honey bees that had been established from a swarm of bees collected in the city of Charleston, South

Carolina, in the summer of 1996 when a hobby beekeeper who captured the swarm noticed a few adult black beetles in the colony shortly after hiving the bees. The beetle sample remained unidentified until the discovery and subsequent identification of the small hive beetles in Florida in 1998.

Taxonomic position. *Aethina tumida* was first described and named by Andrew Murray in 1867 from two beetle specimens that were collected by Rev. W.C. Thomson from Calabar on the west coast of Africa. The first South African record of small hive beetles was made in 1920 by R.H. Harris, who submitted beetles for identification that were collected in Durban. The first collection of small hive beetles in North America was made by W. Weatherford in Charleston, South Carolina, a major seaport. The small hive beetles are in the family Nitidulidae, known for their attraction to and feeding on souring or fermenting plant materials. Small hive beetles may use fruits as alternative food sources and temporary reservoirs in the absence of honey bee colonies. There are no reports of small hive beetles becoming a pest problem on commercially grown fruits or vegetables.

Adult small hive beetles are about one-third the size of a worker honey bee (5.5 to 5.9 mm long, 3.1 to 3.3 mm wide, and 11.7 to 15.0 mg weight). The mature beetle larvae are about 1.2 cm long and have six fully developed legs on the thorax near the head. A. E. Lundie who conducted the first substantial research on the insect implied that variation in beetle size is dependent on the maturation time of beetle larvae. The faster-maturing larvae developed into larger beetles and the slower-maturing larvae developed into smaller beetles. Mature small hive beetles are black and possess a hard exterior cuticle which is covered with fine pubescence (hairs). The European honey bee is unsuccessful in attempts to rid the colony of small hive beetles by physical removal or stinging.

Adult small hive beetle showing distinguishing antennae and thorax shape. (ODA photo)

Small hive beetle larvae, showing legs. (ODA photo)

Small hive beetle side view (ODA photo).

Adult small hive beetle on window screen. (ODA photo)

Small Hive Beetles

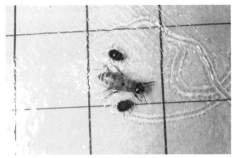

Dorsal and ventral view of the beetle, by honey bee for size comparison. (Hood photo)

Distribution. Prior to 1996, the small hive beetle had not been known to occur outside the African continent. By 2003 the small hive beetle had been confirmed in all regions of the eastern U.S., and by 2007 had been found, to some degree, in nearly every state. This migration was thought to be due to infestation of packages and migratory pollinators. The first discovery of small hive beetles in Australia was made in 2002. The small hive beetles are thought to be capable of flying several miles to infest honey bee colonies. The beetles are attracted to the odors of a bee colony or other volatiles. Other small hive beetle dispersal methods include beekeeper-assisted movement mainly by the transport of beetle-infested bee colonies, empty beekeeping equipment such as supers, package bees or other beetle-infested materials. In the absence of honey bee colonies to invade, small hive beetle adults have been reported to be attracted and feed on cut cantaloupe, pineapple, and bananas.

Detection. A heavily-small-hive-beetle infested bee colony can be detected by the presence of fine pollen and fermented honey at the colony entrance. Upon opening an infested colony, a beekeeper will often observe adult beetles scurrying underneath the hive lid, on top bars, and on the bottom board. A sampling technique includes the removal of a

Top bars (above left), frames, and bottom board of a colony heavily infested with small hive beetle. (USDA photos)

Small hive beetle begging for food, and being fed by worker honey bee. (Hood photo)

suspected beetle-infested super and firmly bouncing the super on an overturned hive top lying on the ground. The vibration causes the adult beetles to dislodge and they can be observed as they scurry about looking for a hiding place. Other detection methods include placement of a container (with openings small enough to exclude bees) of pollen supplement inside the colony. The beetles are attracted to the pollen supplement and can be found hiding, and feeding inside.

Life Cycle. Small hive beetles invade honey bee colonies as individuals but are known to invade a colony in large numbers which will disrupt and kill even strong colonies. Adults feed on pollen, honey, and bee brood. Beetles mate inside bee colonies and the life cycle begins when females lay white banana-shaped eggs inside the hive in cracks and crevices that are inaccessible to bees. Adult female beetles are also reported to make small slits in capped bee brood, insert the ovipositor and lay several eggs just underneath the wax covering. In the absence of patrolling bees, eggs are often found on comb especially in the vicinity of pollen cells or in the bottom of empty cells. A single female beetle can lay up to 1000 eggs during her lifetime. Beetle eggs hatch in one to six days with most hatching in two to three days. Beetle larvae consume bee brood, honey, pollen and wax and leave behind destroyed comb and fermented honey. The larval stage is 10–16 days after which the fully mature larvae exit the hive and enter surrounding soil. Most beetle larvae pupate within 30cm (one foot) of the hive entrance, although some are known to crawl up to 30 m (100 ft) to find a suitable pupation site. The pupal stage lasts three to four weeks. Beetles prefer light sandy soils, although they are known to pupate successfully in heavy clay soils. The fully-developed adults, which are brown in color for the first few days, emerge from the soil to re-infest the same bee colony or fly to a new host colony. Female beetles are sexually mature in two to seven days after emerging from the soil.

The life cycle ranges from 30–60 days depending on food, temperature, and moisture conditions. Hot and humid conditions favor beetle reproduction. In lab tests, female adult beetles lived two weeks to six months. Up to five beetle generations have been reported in South Africa in a single year.

Small hive beetles overwinter as adults in most of North America with few exceptions in the extreme north. The beetles are found in the interior of the bee cluster where there is sufficient food and warmth. The beetles are temperature sensitive and are highly dependent on the bee cluster for cold protection.

Control. In its native range in South Africa, the small hive beetle is usually a minor pest and seldom becomes a problem except in honey houses and on stored combs. Colony control is unnecessary because successful beetle reproduction is limited to weak, stressed, or abandoned colonies. African bee colonies abscond when large numbers of beetles are present, whereas European bee colonies rarely abscond even when heavily infested by the small hive beetles. In European colonies, heavily-infested colonies gradually weaken and die, given favorable beetle reproductive conditions without beekeeper intervention.

Under normal circumstances when a few small hive beetles enter a strong European bee colony, the colony appears to be unaffected by the pests. Apparently, successful beetle reproduction is delayed unless the colony becomes stressed. Any factor that reduces the ratio of the bee population to its comb surface increases a colony's vulnerability to small hive beetles, similar to wax moth invasion.

Worker bees in populous colonies often attempt to sting or grasp and remove adult beetles, but are unsuccessful because of the beetle's hard exoskeleton and the squatting defensive posture by this pest. However, the bee's defensive attempts cause the beetles to seek refuge within the colony. Also, worker bees have been observed to corral and imprison adult beetles in walled-like structures made of propolis. These activities no doubt delay beetle reproduction in populous bee colonies and may play an important role in the natural defense against this pest. Apparently European bees are less effective in their natural defense tactics compared to African bees. However, the selection for hygienic or natural defensive tactics against the beetles in European honey bees may offer promise in controlling this pest. When large numbers of beetles invade a strong European colony, the beetles quickly overcome any natural defenses and the colony will often succumb to the beetle invasion.

Smaller hive entrances allow guard bees to better protect the colony from small hive beetle entry and may play some role in delaying beetle problems in newly beetle-infested areas or in areas of low beetle densities. However the use of reduced size hive entrances offers little control of this pest in a heavily-beetle infested environment and has been shown to limit brood production.

In the honey house, small hive beetles can be a problem when honey supers are stored for more than a few days prior to honey extraction. The use of a dehumidifier in the honey house to lower the relative humidity will help control beetles. Freezing stored honey supers for 24 hours at $-12°C$ ($23°F$) should kill all life stages of the beetle.

At the time of this writing, two

Small hive beetles incarcerated in propolis prison on hive cover. (Hood photos)

pesticide products are available for small hive beetle control in the U.S. Check Mite+® (10% coumaphos) impregnated plastic strips are available for in-hive beetle treatment in many states on an emergency use label. "Gard Star ®" (40% permethrin) is available for soil treatment to kill beetle larvae as they enter the soil to break the pest's life cycle. Currently, in-hive baited traps, and pheromone lures are showing signs of being successful in control. For current small hive beetle treatment recommendations, contact your state extension apiculturist or state apiary inspector. - MH

SMITH, JAY (1871-1953) – Smith was a school teacher, queen breeder, popular lecturer and a prolific writer who especially enjoyed writing tall stories about bees and beekeeping. He was a frequent contributor to *Gleanings in Bee Culture* and other journals. His two books on queen rearing, *Queen Rearing Simplified*, 1923, and *Better Queens*, 1949, are highly regarded and widely read. Both are available on the second-hand market only.

Much of his stock originally came from G.M. Doolittle, of New York, and J.B. Breckwell of Virginia. He used the Modified Alley Method, with no grafting involved.

Smith lived much of his life in Vincennes, Indiana, but the latter part was spent in Fort Myers, Florida which he said was one of the best queen rearing areas in the U.S.

SMOKE, EFFECTS OF – Honey hunters and beekeepers through the ages have been aware that smoke calms the defensive bees in a colony. Certainly the bee's sensory receptors are impaired and they cannot receive, or perceive, alarm odor produced by those bees protecting the hive. However, bees in some colonies are calmed much more easily than are bees in other colonies. Beekeepers that tend colonies of bees in Africa, and colonies of Africanized honey bees in South and Central America, use smokers with fire pots much larger than normal with good results.

Smoke causes many bees in the colony to stop work and engorge on honey. If no honey is uncapped and available from open cells the bees will quickly puncture the cappings to engorge. We do not know why bees behave in this manner; however, it has been suggested that they think there may be a fire and are engorging in preparation for flight to a new home. It is important to not smoke colonies from which one is removing comb honey as the bees may puncture the cappings in the sections and ruin them. Engorged bees are much less inclined to sting than are those that are not engorged.

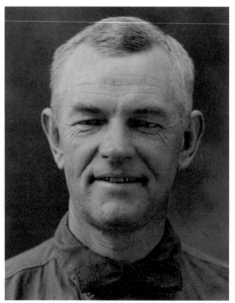

Jay Smith

SMOKERS – For many years beekeepers have used smoldering cow dung and rotten wood. As these materials burned and produced smoke it was blown across the combs. These materials produced only a small quantity of smoke and did not serve well when manipulating irate colonies.

Moses Quinby invented the first practical smoker in 1873, rather late in the history of modern beekeeping considering its importance in keeping bees today. Quinby's smoker consisted of a fire pot and a bellows that would force air through the burning smoker material and cause a blast of smoke that could be directed where needed through a tube at the top of the fire pot. Quinby did not patent his invention, and died

Since smoking fouls the sensory receptors and causes some bees to engorge, it is obvious that smoking a colony disrupts the work force and the work in progress. Several people have made observations of this phenomenon and report that the colony's work is disrupted for a short period of time only. Usually in less than half an hour after the smoking is stopped, even a heavy smoking, foraging and house activities have returned to normal.

From time to time one reads about placing a toxic substance in a smoker and using this to calm, or even to anesthetize, the bees. This can be dangerous as the bees may suffocate or be killed from toxic smoke.

Original Quinby smoker.

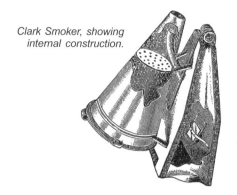

Clark Smoker, showing internal construction.

soon after its discovery. Several practical beekeepers seized upon the idea and improved it so that within a few years the smoker, as we know it today, appeared (see Quinby, Moses).

Today's smokers vary greatly in size but not in design. Many of the larger, and better-constructed smokers have a shield or guard around the fire pot to protect the user from being burned. The largest smokers known are the

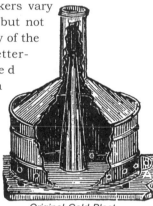

Original Cold Blast

Brazilian smokers that are designed to control the more defensive colonies of Africanized honey bees. For an excellent history of the evolution of the bee smokers see P. Jackson. *Smoking Allowed – A pictorial past of honey bee smokers in the United States.* The A. I. Root Co., Medina, OH. 63 pages. 1995 (out of print).

Bingham Smoker

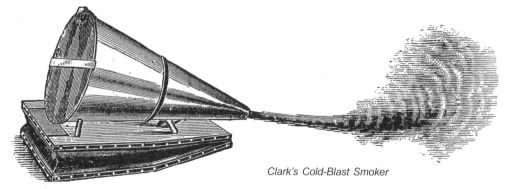

Clark's Cold-Blast Smoker

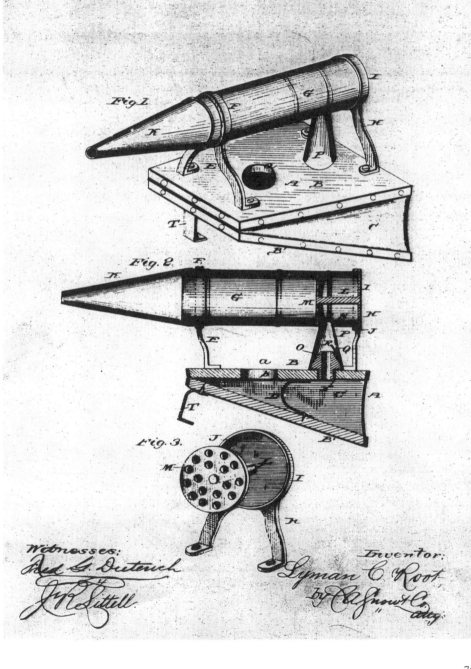

Bingham Smoker, large size.

Assemble your tools: paper, fuel (rotten wood and pine needles) matches, and smoker.

SMOKING HONEY BEE COLONIES –

When opening and working a honey bee colony, it is important to apply smoke and calm bees before they become aroused and defensive. Once the guard bees in a colony detect danger and emit alarm odor it is very difficult to get the colony, and often the colonies around it, under control.

The smoker should be lit and working before entering the apiary. Much time is saved if one goes to the apiary with a bucket of smoker fuel ready so that the smoker may be replenished rapidly and without delay.

It is usually not necessary to smoke all of the colonies in an apiary before starting to work. However, all of the colonies on a single hive stand should be smoked before opening the first colony since any vibrations made with the first colony may arouse bees in the others.

In smoking a colony it should be remembered that there will be guard bees at all entrances. There will be more guard bees when there is a dearth of nectar and fewer when there is a honey flow. Usually two or three puffs of smoke at the entrance will serve to disarm the guards. If there is more than one entrance each should be treated in the same way.

Crumple newspaper into a loose ball that doesn't quite fit into the smoker and light the bottom. Let the paper catch fire and the flame begin to move up without it reaching your hand, push the paper to the bottom of the smoker using your hive tool, and puff the bellows gently two or three times to keep fresh air moving past the burning paper. It will flame up to or just over the edge of the top of the chamber. Puff two or three times more until most of the paper is burning – but not yet nearly consumed – and add a small handful of pine needles and other fuel to the top.

The next step is to lift the cover applying smoke underneath it as it is lifted. Some beekeepers prefer to lift the cover, apply two or three puffs of smoke and then to lower the cover and give the bees a few seconds to adjust and respond to the smoke. Waiting for a minute or two is beneficial when working Africanized bees. If an inner cover is being used it is treated in the same manner.

Smoking Honey Bee Colonies

Puff a couple more times until the flames from the paper reach up and ignite the fuel. Once burning well, push the burning fuel down into the chamber with your hive tool, puffing slowly so air moves through the system. When the first small batch of fuel begins to flame up, add more fuel loosely. Puff several times so you don't smother the fire. This is when most lighting attempts fail because not enough air is coming up from the bottom. The smoldering fuel starves for oxygen and the fire dies.

When the second, or perhaps third, batch of fuel begins to smolder from the bottom, add more durable fuel, if needed, on top of the fuel. Keep puffing slowly. When lots of smoke rises when you puff, and if the fire doesn't quit when you don't pull for a minute or so, close the smoker, still puffing occasionally. The smoker should smolder unattended for many minutes. If it sits idle for a while without use, puff rapidly a couple of times so the smoldering coals flare up a bit, producing lots of cool, white smoke to waft over the bees.

The next step is to identify the first frame to be removed. The first frame should be the one that appears to be the easiest to remove, the one least covered with bees and burr comb. In this way the danger of crushing any bees and making others in the colony defensive is less. Usually the second frame in from the side can be removed first. The ends of the first frame to be removed should be smoked so that there will be no bees in the way when the comb is grasped with the fingers.

After the first comb has been removed, and the colony inspection is underway, one applies smoke only as it is required. One soon becomes accustomed to watching the bees between the combs. When there are five to 15 bee heads peering up from between two top bars it is time to apply smoke. Two or three puffs directed across the top bars are sufficient to cause all of the bees to retreat. If the bees peering from between the frames are not calmed they may fly out and sting. Bees in a colony may appear to coordinate their activities though it is more likely that they are aroused individually.

When not in use, but at a time when it may be needed at any moment in the apiary, a smoker should be left in an upright position. Smokers with burning fuel that are placed on their sides will soon lose their fire.

Burning smokers have been responsible for starting unwanted wood and grass fires inadvertently. If the hot fire pot comes into contact with an easily-lighted fuel or hot sparks are blown from the nozzle dry leaves and grass will easily catch fire. Some beekeepers place a handful of green grass on top of the burning fuel in the fire pot with the thought that this will both cool the smoke and serve to filter out some of the burning sparks and bits of fuel that might otherwise be blown out of the nozzle. Fires have also been started accidentally by beekeepers dumping hot coals from a fire pot onto the ground and not extinguishing the embers before leaving the apiary.

Smoking Your Colony

1. Gently and lightly smoke the front entrance, and any other entrances. Wait a bit.

2. Lift the cover and puff beneath, replace and wait a bit. Remove cover and puff into inner cover hole. If using a migratory lid, puff beneath and replace. Wait a bit.

3. Lift the inner cover and puff beneath. Replace

4. Lift inner cover and waft smoke over the tops of frames. Use any wind to your advantage to move smoke laterally.

5. When bees begin to rise to the top bars, waft smoke as needed. Don't over smoke.

Fuel for smokers – What the best fuel is for a smoker will stimulate a heated debate among beekeepers. Almost any fuel that is easy to ignite and will deliver a large quantity of smoke but not harm or kill bees will serve the purpose.

A chief problem with many smoker fuels is that they do not last too long, that is they burn up too fast, and the smoker must be continually recharged. Beekeepers in the past have used leaves, straw, burlap or other old cotton cloth, worn rope, shavings, tobacco, wood chips and many other products for fuel. Beekeepers today have to be very careful in selecting burlap, jute, bailing twine, cardboard and corn cobs as they may contain plastics or other chemicals that may harm the bees. Dry pine needles and tightly-packed straw are suitable fuels. Some commercial smoker fuels are available from bee equipment suppliers, and one manufacturer produces a commercial quality smoker that burns a chemical that produces smoke. The heat is propane, butane, or electricity (when on a fork lift), and the smoke is safe for bees.

Nitrous oxide and laughing gas – In the early 1950s some beekeepers advised adding one or two teaspoons of ammonium nitrate powder on top of the hot fuel in a smoker to calm the bees. When this smoke was applied to the bees they were anesthetized or at least made groggy. It was also thought that the anesthetized bees would lose their memory and that after such a treatment colonies could be moved short distances without any bees drifting back to the original site. It was stated that nitrous oxide (laughing gas) was produced as a result of burning the ammonium nitrate and that this was responsible for the action. However, some beekeepers reported that large numbers of bees were killed when using this smoke. Further study showed that the treated bees did not lose their memories and that hydrogen cyanide was also produced in the process. The latter was probably the chief reason bees were often killed. Using ammonium nitrate has not been mentioned in the beekeeping literature for a number of years and is certainly not recommended.

Tobacco smoke – Tobacco smoke is not used routinely as a smoker fuel However, it has been used recently for the detection and control of the parasitic mite, *Varroa destructor* (see De Ruijiter, A. Tobacco smoke can kill *Varroa* mites. *Bee World* 63:138. 1982.) However tobacco smoke used in excess can be toxic to the bees.

SMR (SUPPRESSED MITE REPRODUCTION) – (see VSH - *VARROA SENSITIVE HYGIENE*)

SNELGROVE, L.E. – L.E. Snelgrove of Bleadon, England, died November 21, 1965. His books *Queen Rearing, The Introduction of Queen Cells* and *Swarming: Its Control and Prevention* have international acceptance. He was Fellow of the Royal Entomological Society, President of the Somerset Beekeepers' Association, a Vice President of the British Beekeepers' Association, and a member of the Apis Club. He also developed the Snelgrove Board, a double screened divider board with upper and lower entrances.

SNODGRASS, ROBERT E. (1875–1962) – Snodgrass is known as the world's most distinguished insect anatomist. No one before or since has produced such precise observations on anatomy, musculature and, to some degree, the physiology not only of honey bees, but of insects in general.

His first work on honey bees was *The Anatomy of the Honey Bee*, a U.S. Department of Agriculture bulletin published in 1910. This was expanded in his 1925 *Anatomy and Physiology of the Honeybee*, published by McGraw-Hill. In 1956 a third revision was published by the Cornell University Press under the title *Anatomy of the Honey Bee*. It is still a popular reference text, with a fourth printing in 1984.

Most of Snodgrass' career was spent with the U. S. Department of Agriculture. In addition to his reputation as an anatomist, Snodgrass was a popular lecturer, humorist, cartoonist, sculptor and artist. He wrote 79 papers and books during his career and was awarded an honorary doctorate. He was fortunate to work on the insects that fascinated him until the day before he died. His work has been used extensively in this and previous editions of this book.

SOLAR WAX EXTRACTORS – (see BEESWAX, Solar wax melter)

SOUNDS MADE BY HONEY BEES – (see SENSES OF THE HONEY BEE; Sound Perception and QUEEN; Queen piping)

SPERM STORAGE – Storage of spermatozoa in a specialized pouch (the spermatheca) is a normal process for queen honey bees, as it is for most female insects. The spermatheca of a queen bee receives spermatozoa from a natural mating with drones (when a queen is one to two weeks old) or from artificial insemination. About two days after her final mating flight a queen will begin to lay eggs and will never mate again. A queen releases spermatozoa from her spermatheca as she lays eggs, so all spermatozoa are stored for at least two days; some are stored for the entire lifetime of the queen which can be as long as three years, or even longer.

Thus the storage of spermatozoa is standard practice and occurs in every mated queen. Since spermatozoa can be taken from the spermatheca of one or more queens and used to inseminate a virgin queen, a bee breeder has the opportunity to use this natural storage system. Spermatozoa that have been in a spermatheca for over two years are still able to migrate to the spermatheca of a second queen and produce worker progeny.

It is also possible to store honey bee spermatozoa outside a live bee. Nearly all methods use spermatozoa that have been collected as semen from drones. Semen includes cells (spermatozoa) plus a fluid produced by the drone. In most cases glass capillary tubes are used as containers for semen, and the tubes are sealed by various means to minimize the amount of air within the storage chamber.

Short-term storage of spermatozoa (storage for 30 days or less) is best done at nonfreezing temperatures. Temperatures below 50°F (10°C) kill spermatozoa, but spermatozoa survive well between 60 and 77°F (15–25°C). There is good evidence that spermatozoa can survive for at least one week at these temperatures with little or no loss in viability. Therefore, the storage of spermatozoa for less than one week is widely used. This includes semen that is shipped and semen that is collected into syringes or storage tubes and used to inseminate queens the following days.

Long-term storage of spermatozoa has been successful in liquid nitrogen at ultra-low temperatures (-320°F or -

196°C). The procedure requires the use of chemical additives and procedures that protect the cells from freeze damage. Honey bee spermatozoa have been successfully stored at subfreezing temperatures for as long as two years, but survival rates are usually low. Long-term storage is just beginning to be used in the beekeeping industry. Collins, A.M. *Survival of honey bee (Hymenoptera: Apidae) spermatozoa stored at above-freezing temperatures.* Journal of Economic Entomology 93: 568-571. 2000]. - AMC

SPRING NECTAR FLOW, MANAGEMENT FOR - Almost exclusively, the honey from early spring nectar flows is needed and used to produce more bees for the major honey flow that occurs in late spring or early or mid-summer. Spring plants such as apple, blueberry, dandelion and yellow rocket may sometimes yield sufficient quantities of nectar that colonies may store 30 and sometimes as much as 60 pounds (14 to 27 kg) of honey. However, since colony populations are still expanding at this time of year, the honey soon disappears and few beekeepers make any effort to harvest it.

The chief concern in the spring is to encourage colony population growth and to prevent or control swarming. In the spring, colonies should be given supers well in advance of when they are needed. It is especially important to provide space so that the broodnest can expand upward.

STANDIFER, LONNIE NATHANIEL (1926-1996) – Dr. Lonnie Nathaniel Standifer received his early education in Texas, earning a B.S. degree from Kansas University in 1951, and a Ph.D. in Entomology from Cornell University in 1954. He held teaching positions at Tuskegee Institute, Cornell University and Louisiana's Southern University before he was hired in 1956 as an Entomologist by the United States Department of Agriculture, Agricultural Research Service, Tucson, AZ. Specializing in honey bee physiology and nutrition, he was promoted to Research Entomologist in 1960. In 1970 he was appointed Director of The Carl Hayden Bee Research Center at Tucson, a position he held until 1981.

Lonnie Standifer

STARTING IN BEEKEEPING – (see BEGINNING WITH BEES)

STARVATION – Colonies of bees may starve to death and it is the beekeeper's job to make certain this does not occur. Races of bees vary in their ability to conserve food. Some of the lines of queens offered by queen breeders are designed to produce a maximum amount of brood under all conditions; in many areas where honey flows are short this may not be desirable.

Colonies need to be examined in the autumn to make certain they have sufficient food for winter (see WINTERING, Inspecting and weighing colonies for winter). During the spring and summer a colony should have a minimum of 10 to 15 pounds (4.5-6.8 kg) of reserve honey at all times. Drought conditions may lead to starvation at any time. One sign of starvation in the spring is the appearance of discarded larvae in front of a colony. When nearing starvation bees will remove larvae first. When a colony finally exhausts its food, all of the bees starve at once. There is no attempt on the part of one or a few bees to hoard food and bees do not offer their queen any special protection at this time.

When examining a colony that has starved you will often see the remaining dead bees in a small cluster, probably very near the top of the colony – where the last of the honey was located. Many may be head down in empty cells. There may be sealed brood scattered about.

There may also be frames of honey located a short distance from the cluster – too far away for the bees to reach during a prolonged cold spell.

STEPHEN, W.A. – W.A. Stephen was a fellow of the American Academy for the Advancement of Science, Sigma Xi Honorary Fraternity and other professional organizations. He was a veteran of World War II, serving in the Royal Canadian Air Force. Stephen was a graduate of the University of Guelph with a degree in Agricultural Engineering and of the University of Toronto. He did graduate work at the University of Berlin and University of Wisconsin. He was the author of the column "Fundamentals," in *Gleanings*, for many years.

W.A. Stephen

W.A. Stephen served as extension specialist in apiculture in North Carolina from 1947 to 1963, before coming to The Ohio State University in the same capacity in 1963. He was at Ohio State until 1972 from where he retired as Professor Emeritus.

STINGLESS BEES – A group of social bees in the genus *Apis* are related to honey bees. This group has about 500 species. The females cannot sting because their sting is not fully developed. There are several genera of "stingless" bees, but the two most important ones are *Melipona* and *Trigona*. These bees live in perennial colonies formed by (normally) a single mother queen (some exceptions of species have multiple queens), a few hundred to thousands of workers and some dozens of drones, depending on the species. Compared to honey bees, stingless bees are a much more diversified group, both morphologically and behaviorally. The distribution of stingless bees is confined to the tropical areas of the world, with the greatest diversity being found in Mexico, the Caribbean Islands, Central and South America. A few species are also found in Africa, Asia and Australia.

The stingless bees have several ways of defending themselves and their nests.

Some bite with their mandibles and pull hair; others will burrow rapidly in and out of the nose, mouth, eyes and ears of intruders. Some species of *Trigona* secrete a caustic substance from their mandibular glands that they deposit on an enemy causing a burning and itching sensation. All of these defense systems have protected stingless bees from men and animals that would rob their nests.

In spite of these defenses, men have exploited the resources of stingless bees since ancient times, but in only one area of the world has the exploitation of these bees reached a level of domestication and husbandry close to that of honey bees. This area is the Yucatan Peninsula of Mexico, where the indigenous Mayans industriously cultivated colonies of one of the 15 native species: *Melipona beecheii* (the "lady bee" Xunan-kab or Colel-kab in the Mayan language). Meliponiculture (the term used for the domestic exploitation of stingless bees) has been tightly linked to the religion and traditions of the Mayan people. The stingless bee even has a guardian honey-god of its own (Ah-Mucen-kab).

Hundreds of thousands of colonies were kept by the Mayans, and honey from the Xunan-kab was a major trade item across Mesoamerica during colonial times. Meliponiculture was so important in the Yucatan, that the Spaniards decided not to introduce honey bees to this area of Mexico.

Colonies of *M. beecheii* typically have populations in the range of 800 to 2000 individuals. The workers are large (9–11 mm long). Their thorax is densely covered with gray-colored hairs and their abdomen is black with yellow stripes. Workers perform most of the work in the colony and, as in most species of stingless bees, an age-related sequence of tasks similar to that of honey bees occurs. Workers are able to communicate the direction and sometimes the distance of food resources to nestmate bees. In the case of *Melipona*, they vibrate their wings to produce a buzzing sound. The length of

Meliipona spp. cells and nest box. (Guzman photo)

time that the sound lasts indicates the distance from the nest. The bees also may fly towards the food source followed by recruits. Queens are differentiated from workers because of their longer abdomen and smaller head. As in the case of honey bees, the queen's main function is to lay eggs. Drones are large individuals that besides their contribution to reproduction (mating with a queen), they help to ripen honey, secrete wax and use it to help build brood cells and storage pots.

The reproductive biology of stingless bees is somewhat different from that of honey bees. The queen flies only once in her life and mates with one male (though queens of a few species possibly mate with several males). The males of stingless bees usually form swarms near nests, and wait for virgin queens to appear. As in the case of honey bees, drones of stingless bees mate only once and lose their genitalia in the vagina of the queen after mating.

In *Melipona* colonies, queens, drones and workers are reared in cells of the same size. Queens can lay up to 40 eggs per day in the high season. The development of individuals is much slower than for honey bees. For instance in *M. beecheii* workers develop in about 53 days, queens in 51 and drones in 54.

Caste determination in *Melipona* is genetically determined and colonies often rear large numbers of virgin queens. The excess of queens in this genus seems to be an insurance against the possibility of the old queen dying and possibly arises as a consequence of conflict over reproductive interests amongst females. However, virgin queens produced in excess are constantly slaughtered by their sister workers. In *Trigona*, queens are normally larger than workers and are reared in larger queen cells. Caste determination in *Trigona* is dependent only on the amount of food available to the larvae.

Colony multiplication in stingless bees occurs by swarming. But before a colony splits into two sub-units, scout bees select a suitable nest site not too far from the parent colony. Then, the bees carry "cerumen" (a mixture of plant resins, and wax that is produced in the dorsal glands of both workers and drones) and honey from the old nest, and begin to build the new nest. In honey bees, the old queen leaves the mother colony with part of the worker population, but in stingless bees, the old queen is not able to fly due to her distended and heavy abdomen. Thus, a young virgin queen accompanied by a group of workers flies to the new nest after it has been prepared. Soon the virgin queen performs her mating flight and starts laying eggs while the workers continue building the nest and bringing nectar and pollen. There is also a period of time that can last months in which the mother and daughter colonies are connected through foragers that take honey and pollen from the former to provision the latter.

The nest architecture of stingless bees contrasts with that of honey bees. The basic material used by stingless bees to build their nest is cerumen. The cerumen extracted from *M. beecheii* (called "cera de Campeche" or Campeche's wax in the Yucatan Peninsula) was a highly appreciated item for candle making during colonial times. In *M. beecheii* the brood cells are built horizontally in batches forming flat combs at the center of the nest, and, as is the case for other stingless bees, these cells are never used for food storage. Other patterns of comb building are also found in stingless bees. Some species build individual cells in cluster

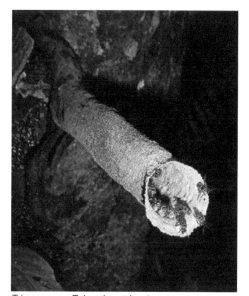
Trigona spp. Tube-shaped entrance.

arrangements, whereas others build irregular, spiral or vertical combs. A unique feature of stingless bees is the pattern of behavior between the queen and workers during egg laying. This pattern is known as the provision and oviposition process. In contrast with honey bees, workers of stingless bees mass provision their brood cells with liquid larval-food before eggs are laid in them. The food provided for the larvae is similar to the royal jelly of honey bees. After the queen lays an egg on top of the food, the cell is sealed by the workers, leaving the individual to develop without contact with adult bees. When the insect reaches the pupal stage, at around day 30, the workers start removing the cerumen on the outside of the cell, leaving only the cocoon with the pupae inside. As a result sections in the brood combs are dark in color containing eggs and larvae and others which are much lighter containing pupae. When the newly-emerged bees leave their cocoons, the workers remove the remains. Combs are constantly built and destroyed. An involucrum, consisting of one to three thin layers of cerumen, surround the brood combs. This structure serves for temperature regulation of the brood nest. The entrance hole of *Melipona* nests is usually surrounded by mud with patterns typical of the species. *Trigona* bees build a narrow tube-shaped entrance, which is covered with a sticky coating of wax and resin to keep out small intruders.

Pollen and honey are stored in separate spherical containers of similar size called pots. Pots are built in the periphery of the brood nest area, outside the involucrum. Colonies collect large amounts of pollen that are tightly packed in the pollen pots. The pollen stored in the pots undergoes a fermenting process that makes it more digestible and that increases its percentage of usable vitamins and minerals.

The honey of *M. beecheii* is thin and more acid than that of *A. mellifera*. Its pH is around 3.5 and its water content is around 25%.

Currently, *M. beecheii* is still cultivated in much the same manner as in pre-Hispanic times. Colonies are lodged in hollow logs called hobones. Hobones are usually made with the trunks of the tree *Vitex gaumeri* (Fiddlewood). The logs are opened at both ends and sealed by means of wooden disks held in place with clay, although there is evidence that stone disks were also used in the past. The length of the hobones ranges between 50–70 cm (20–28 inches) and the diameter of the inner space 10–16 cm (four to six inches). The internal volumes for the nest range between six to 13 liters. An entrance of approximately 1 cm (3/8 inch) width is made in the middle of both extremes, above which it is common to find a carved cross that is used as an indication of the position of the trunk during the honey harvest and as a religious mark, a reminder of the original tree in Mayan mythology (yaax-ché, *Ceiba pentandra*). When a beekeeper has several hobones, they are placed under a hut made of palm leaves (nahil-kab) in two opposite slopes, facing east and west.

Honey is traditionally harvested between March and May. Between one to three harvests are made during that period, yielding a total of around one liter of honey per colony. In spite of the low honey yield of *M. beecheii* colonies, locally its price is in the range of US$20 per liter, which is about 20 times higher than that of *A. mellifera* honey. However, keeping bees in hollow logs represents a major problem for managing the colonies, as well as for extracting and maintaining quality and hygienic standards of honey, which may be contaminated with clay and mud.

The development of modern hives for stingless bees has opened a new possibility for keeping colonies, and has made honey production and reproduction of these bees more efficient. Most modern stingless bee hives use wooden boxes similar in size to wild nests. There are several types of hives that have been developed in Brazil (PNN and Uberlandia models), Angola (Portugal-Araujo models), Costa Rica (UTOB model) and Mexico (Gonzalez-Acereto model).

The basic components of the hives are similar to those of a honey bee hive. There is a floor, a box for the brood with a small entrance hole, a box (or an area inside the same box) for food reserves, and a lid. There are also specific devices in each hive type that are used for ventilation and support of brood combs

(for detailed reviews see Kerr et al. 1996. *Abelha Urucu*. Biologia, Manejo e Conservaçao. Fundaçao Acangaú, Belo Horizonte, and Nogueira-Neto 1997. *Vida e Criaçao de Abelhas Indígenas sem Ferrao*. Edit. Nogueirapis, Sao Paulo, Brazil).

In the rural villages of the Yucatan Peninsula, the cultivation of *M. beecheii* colonies is diminishing, mainly as a result of changes in the economy, ecology and traditions of the region. Only a few hundred *Melipona* beekeepers exist and the majority own between 20–40 colonies. However, since the arrival of Africanized bees (see BEEKEEPING IN VARIOUS PARTS OF THE WORLD; Mexico and AFRICANIZED HONEY BEES) to the Yucatan Peninsula, apiaries of honey bees have been relocated away from villages and human settlements where they were traditionally kept, leaving empty areas where stingless bees could be cultivated again. The use of these spaces, as well as keeping stingless bees in modern hives could help avoid the loss of meliponiculture in the Yucatan Peninsula. Additionally, the University of Yucatan and other government institutions are promoting meliponiculture through the use of modern hives in the tropical areas of Mexico.

The cultivation of stingless bees will hopefully continue with honey as the major product obtained from them. Recent reports support the medicinal properties long attributed to stingless bee honey, pollen and resins. Moreover, the use of stingless bees for crop pollination is another field that has just begun to be explored. The use of modern hives opens the possibility of transporting colonies to distant fields where their services for pollination may be required. There is evidence that stingless bees can efficiently pollinate mango, avocado, tomato, strawberry and coffee. - JQ & EG

STINGS – (see ANATOMY AND MORPHOLOGY OF THE HONEY BEE)

STINGS, FIRST AID FOR – (see BEE STINGS; First Aid)

STINGS, HOW TO AVOID – It is interesting how little attention experienced beekeepers pay to being stung, while those who are not beekeepers worry about it so much. Clearly, all those who keep bees have gone through their first stinging at some time. It is really not a difficult experience. Every beekeeper likes to avoid bee stings. In spite of the best precautions, beekeepers can expect to be stung at sometime or another. Beekeepers will try to reduce the number of stings as much as possible by dressing appropriately and manipulating the hive properly.

Beekeepers are often called upon for advice on how to avoid getting stung. For people who have a fear of being stung, some simple rules can be followed that will at least reduce the number of stinging incidents.

- Wear light-colored, smooth-finished clothing; all stinging insects are much less inclined to sting through this type of garment.
- Avoid wearing hair products, after shaves, perfumes and other sweet-smelling cosmetics that may attract stinging insects. Stinging insects, which feed on sweet, odoriferous flowers, may be attracted to a perfumed person.
- Wear clean clothing and bathe daily; sweat seems to irritate stinging insects. Beekeepers have noted that honey bees will sting more often in the vicinity of a sweaty watchband, even one made of metal.
- Cover the body as much as reasonably possible. Wear ankle-high shoes when walking in fields, and use white, cotton socks. Covering the head and hair with a straw hat or a light-colored kerchief may prevent stinging insects from becoming entangled in the hair.
- Avoid flower gardens and fields where wildflowers are common. Anywhere flowers grow there are likely to be stinging insects because the pollen and nectar produced by the flowers are required by bees and wasps as food.
- Check for ground and aerial social wasp nests by watching for insect flight and activity during the warmer hours, especially in late July, August and September. The social wasps begin their nests in the spring. By late summer, a nest started by a single female might contain several thousand individuals. As populations build, flight activity will increase. One seldom sees the nests in their initial stages. However, when warm weather comes they will have larger populations and should be easily detected and can be destroyed if they become a nuisance.
- Remove refuse and garbage on which wasps may feed. Many social wasps thrive in cities. The adults will feed on soft drinks, milk, jam, beer; they collect bread, hamburger, bits of hotdogs and other refuse for their larvae. Normally the social wasps feed their larvae bits of other insects that they collect. Poor sanitation in parks, around hamburger stands and restaurants and in dumps may encourage social wasp populations to

build more rapidly under man-made conditions than in nature.

- Run if you are attacked by several stinging insects at one time. The only time several wasps or bees would attack simultaneously is when a nest is disturbed or broken. It is best to run indoors. If there is a choice, run through a shaded area rather than an open one. Angry stinging insects have greater trouble following you through a thicket or woods.
- Lie face down on the ground if a single stinging insect appears to be on the offensive. Wasps and bees are likely to sting around the eyes, nose, and ears.
- Stop a car slowly if an offensive insect flies in through a window. A wasp or bee that flies into a moving car is usually more concerned with escaping than stinging. It is best to remain as still as possible while slowing the car and moving to the road shoulder. When the car is stopped, open all of the windows and the insect will soon find a way to escape.
- Capture the insect that stings you so it may be identified. A person who is sensitive to one stinging species may not be allergic to all of them. Killing the bee or wasp with a fly swatter is satisfactory if there is no other way. There are several thousand species of stinging insects, some much more common than others. People who suspect they are sensitive to stinging insects should do their best to capture, keep and identify the insect(s) that gives them trouble.

STOCK – The word "stock" is the preferred term used to describe the various honey bees kept by most beekeepers around the world today. The word can be modified so that it includes certain characteristics, such as Italian stock or Carniolan stock. The original subspecies characteristics have been mixed over the years to create a bee that does show some of the original characteristics of a subspecies but also shows the mixture of genes introduced in the search of an improved honey bee.

STOLLER HONEY – Irvin A. Stoller (1902-1975) started keeping bees in 1920 when he purchased five colonies from his father. In 1975, he and two sons, Darl and Gale, were operating 4,000 colonies in Ohio, Wisconsin, Indiana, and Florida and also were packing honey as Stoller Honey Farms, Inc. Another son, Wayne, was a large honey producer and packer operating as W. Stoller's Honey, Inc.

By 1944 Irvin was operating 3,000 colonies, almost equally divided between Georgia, Indiana, and Ohio.

Stoller started packing honey commercially in 1952. Irvin is known for inventing the Stoller Frame Spacer, which is widely used in the industry.

Wayne Stoller started with 675 colonies in 1952. He started packing

Irvin Stoller

honey in his father's plant in 1952, but built his own bottling plant in 1968.

Eventually, Stoller Honey and W. Stoller Honey joined and became a large honey packing operation in Latty, Ohio, specializing in cremed honey. Dwight Stoller, Wayne's son, was leading the organization when it was purchased by Golden Heritage Foods, of Kansas.

STRACHAN, DONALD J. (1925-2003) **and STRACHAN APIARIES** – In 1953 Don Strachan purchased 600 hives from a retiring beekeeper in Chico, California. In 1954 Don bought "Caucasian Unlimited," a queen raising business, from Tom David. Thus, the beginning of Strachan Apiaries. By that time Don and wife, Alice, had three daughters.

In 1961 Strachan Apiaries moved to its present location in Yuba City, California. In 1980, Don began a new venture with Everett Hastings of Canada and started breeding the Carniolan line of queens and bees. This breed was the backbone of Strachan Apiaries, Inc. Over time the queen breeding stock has evolved and includes the New World Closed Population Breeding Program that was developed by Sue Cobey of Ohio State University. "Time Tested, Industry Proven," is the motto of Strachan Apiaries, Inc. when it comes to queen and package bee production.

Don worked hard, long hours building the business, until the business grew to approximately 10,000 hives, with a busy pollination, package bee and queen breeding business. Don's daughter, Valeri Severson, President of Strachan Apiaries, Inc., worked with her father beginning in 1975. When Don retired in 2001 Valeri took on the task of managing the business.

Don was President of the California State Beekeepers Association during its

Don Strachan receiving the Lifetime Honorary Beekeeper of the CA State Beekeepers Associaiton in 1992.

centennial year in 1989 and an honorary lifetime member. He was a member of the American Beekeeping Federation, the American Honey Producers Association, the California Bee Breeders Association and the Alberta Beekeepers Association.

STRAUB, WALTER AND W.F. STRAUB AND CO. – This large honey packing company got into the honey business when Straubs were awarded 300,000 pounds (136,000 kg) of honey in a bankruptcy settlement in the 1920s. Walter Straub (1897-1964) was manufacturing veterinary medicines when Lake Shore Honey Company asked him to custom pack a drug containing honey. When Lake Shore went broke, Walter was left with the honey. Straub asked a neighbor, who was a buyer for A&P, what to do with the honey. Walter's friend offered to put Lake Shore Honey in A&P supermarkets, and Straub started to pack honey.

Walter Straub

The company also went into beekeeping with 3000 colonies in the 1930s which they increased to 25,000. Apiaries were located in Minncsota, Canada, South Carolina, Georgia, and Mississippi. But it was difficult to manage this far-flung operation and by 1950 Straubs had reduced their colonies to 4000. Only a handful of colonies remained by 1960.

The firm developed and patented a dry honey, Honi-Bake, in the late 1950s that was widely accepted by bakers.

John Straub believed honey improved the texture and increased the shelf life of baked goods. Honi-Bake improved honey's use in baking. It was used in packaged cake mixes, by commercial bakeries, and in some meat products.

Dark strong-flavored honeys were preferred for Honi-Bake since the drum-drying process drives off the strong aromas and also because the additional ingredients added to honey to make Honi-Bake further reduce its flavor. Straub expanded his honey buying in the southeastern United States to meet the demand for the stronger-flavored honeys.

The founder, Walter F. Straub, was prominent in public, social, and business affairs. In 1943-1944 during World War II, he was national director of food rationing. After the War he headed a team investigating nutritional developments in Germany. In 1946 he was appointed Administrator of Emergency Famine Relief. Former President Herbert Hoover and Fiorello LaGuardia, a former mayor of New York City, were in charge of this program.

John W. Straub (1926) became president of the company in 1962 until his retirement. He has a degree in economics from Yale and was a city planner before returning to the company in 1953.

SUBMISSIVE BEHAVIOR – (see GUARD BEES)

SUBSPECIES – This term is a part of the scientific name for the various animals and plants. Subspecies further defines a particular species, one that can be distinguished from other subspecies. In the case of honey bees some of the common names used are actually the subspecies name. For example: *Apis mellifera ligustica*, or the Italian bee. *Ligustica*, written in italics, is the scientific name of a particular honey bee. Subspecies are generally found in areas isolated by various geographical features, such as mountains, deserts or other features that confine a particular species.

The term "subspecies" should only be used for those honey bees showing certain definable characteristics. Sometimes the word "race" is used in place of "subspecies." However, when referring to most of the honey bees kept by beekeepers today, the word "stock" is a more appropriate term since these bees are a mixture of different subspecies and have been for a long time. In the search for improvement of certain characteristics, such as honey production, gentleness and disease resistance, honey bee species were transported around the world in the late 1800s and early 1900s. Subsequent breeding of these bees produced the mixture of genes of today.

SUE BEE HONEY HISTORY – The Sioux Honey Association Cooperative is the world's largest honey marketing organization. The Association was founded in 1921 in Sioux City, Iowa, by five local beekeepers, Edward G. Brown, Clarence Kautz, M.G. Engle, Charles S. Engle, and Noah Williamson, with $500 in capital including some 3000 pounds (1400 kg) of honey. Their original plant was in a space 25′ x 40′ on the west side of the basement of the Morningside Masonic Temple.

By 1977, they had 1009 active members located in 36 states producing over 50 million pounds (22,727,272 kg) of honey, which was marketed from six plants, located at Anaheim, California; Wendell, Idaho; Temple, Texas; Umatilla, Florida; Waycross, Georgia; with the home office and plant at Sioux City, Iowa.

Membership grew over the years to a peak of 1,200 but currently includes only about 315 members as operations have grown, consolidated, and modernized like other agricultural business. Most of the membership hails from the western two-thirds of the United States plus Florida and Georgia. Each member is responsible for supplying the organization with honey shipped to their processing plants in Sioux City, Iowa; Waycross, Georgia, and Anaheim, California. Combined, these plants can process 400,000 lbs/day (181,818 kg). Collectively, the membership produces around 40 million pounds of honey annually. Production consists of clover, orange, sage, and mixed floral.

In the early days, honey was marketed under the "Sioux Bee" label, but the name was changed in 1964 to "Sue Bee" to reflect the correct pronunciation more clearly. Over time, other lines of honey were added, including Clover Maid, Aunt Sue, Natural Pure and North American brands.

SUGARS IN HONEY – (see NECTAR, CONVERSION TO HONEY)

SUGAR FOR FEEDING BEES – (see FEEDING BEES)

SUPER – A super is a box that holds frames or comb honey sections for honey storage. As the name suggests, a super is placed above other boxes that make up a beehive. However, when the same size box is used to hold brood nest frames it is referred to as a brood chamber. (see EQUIPMENT FOR BEEKEEPING).

SUPERING COLONIES – It cannot be emphasized too strongly that it is important to place honey storage supers on colonies well before the honey flow. One obvious reason is that the date a honey flow may start varies from year to year. The starting dates of honey flows in the tropics and subtropics can be more variable than are those in cooler climates.

Also important is the fact that empty storage space stimulates foraging. If a colony has no storage space, and scout

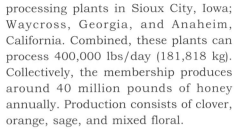

bees find food, they will not dance and recruit bees to forage if no space to store the incoming nectar can be found. The factors that stimulate or effect hoarding by honey bees have been studied by several researchers (see Free, J. B. and I. H. Williams. *Hoarding by honeybees.* Animal Behaviour 20: 327-334. 1972 and Rinderer, T. E. and J. R. Baxter. *Amount of empty comb, comb color, and honey production.* American Bee Journal 120: 641-642. 1980).

Bees prefer to store honey in old comb over new comb though there is an ongoing debate among beekeepers concerning what type of comb to use for honey storage (see HONEY, Effect of old and new comb on honey color). When given a choice bees prefer to deposit new honey in worker comb rather than drone comb. However, in some dry areas, such as parts of Arizona where the honey may be low in moisture because of the low humidity, beekeepers may use drone comb for honey storage supers because it is easier to extract the low-moisture honey from the larger drone cells. It is interesting that bees do not care to deposit pollen in drone cells.

When beekeepers are producing extracted (liquid) honey they usually top super, that is, the supers with empty combs are placed on top of those already in place. When one is producing comb honey the situation is quite different and the beekeepers bottom super, that is the new supers are placed under the storage supers already on the colony but still on top of the brood nest (see COMB HONEY, HOW TO MAKE).

Most beekeepers prefer to use fewer combs in honey storage supers than they do in the brood nest. When one is drawing foundation, that is making new combs, ten frames are always used in a 10-frame super. When producing extracted honey in drawn comb, eight or nine combs can be placed in honey storage supers. Frames can be spaced by hand, which is a slow process, or with frame spacers.

SWARMING IN HONEY BEES – Swarming occurs when a colony of honey bees divides and 30 to 70 percent of the bees, along with their queen, leave the hive to start a new colony. Swarming is a normal process that is necessary for the species to survive. Feral colonies in nests the bees have selected themselves, which are usually 40 to 50 liters in volume, about the same volume as a single, deep 10-frame Langstroth hive body, probably swarm once a year. Successful beekeepers prevent their colonies from swarming because bees that swarm may be considered lost in terms of honey production and pollination.

Although a swarm may emerge from a colony in a matter of minutes, the steps leading up to swarming take many weeks. **Swarm prevention** is the term given to those actions a beekeeper takes to prevent colonies from building queen cells. The production of such cells is a sign that swarming will take place. **Swarm control** is what a beekeeper will do once a colony has well-developed, but not capped, queen cells and is only a few days away from swarming. Swarm prevention is a fairly simple process that is much preferred over swarm control, which is a slow, time-consuming process that is not always successful. Under normal circumstances a swarm will not emerge from the parent colony until the first queen cells are capped.

Swarming is partly caused by congestion of the brood nest. Note that this does not mean congestion of all of the boxes or area a colony might occupy, or partially occupy, but merely the brood-rearing area itself. In a nest with a restricted amount of space this may be the entire nest, but in managed hives, where multiple boxes may be used, many factors may prevent the expansion of the brood nest.

It must also be remembered that bees naturally expand their nests upward. For some reason, bees have difficulty expanding their nests sideways and they never expand them downward, although they will move downward in the autumn. A brood nest may, for example, fill the top hive body in a stack of three or four. Although the bottom boxes may be empty, the nest itself is still congested since there is no room for upward expansion.

A congested nest will cause swarming for various reasons, for example the distribution of queen pheromones, which enable bees to recognize their queen's presence, is somehow disrupted. A sudden emergence of a large number of worker bees may compound the problem, as may poor ventilation.

Bees in colonies destroy all old queen cells and queen cups in the late autumn. Queen cups are usually not present in colonies during the winter and very early spring. Most queen cups are constructed using old beeswax, but they can have some white spots made by the addition of new wax scales. When these white specks are especially numerous, it is an indication that the colony is in the early stages of congestion. Beekeepers refer to this as the whitening of queen cups, and it is one sign that they should take steps to relieve congestion.

The number of queen cups in a colony may also be an indication of congestion, but no data indicates what number of cups equals what level of congestion. Some races and hybrids of bees naturally build many more queen cups and cells than do other races. As many as 15 to 20 cups can be found in a colony.

About eight to 10 days before a colony casts a swarm, and congestion is obviously advanced, several events occur at the same time. First, and clearly visible, the queen begins to lay eggs in the queen cups. Second, the queen begins to lose weight so that by the day the swarm emerges she has lost about one-third of her weight. Much of her weight loss is due to a reduction of egg production with a resulting decrease in ovary size; however, the blood

(Tew photo)

Queen cups, and thus queen cells, are often, but certainly not always located on the bottom bars of frames. The number in any given colony depends on the race of bees in the colony and the available resources in the environment.

When a swarm emerges from a colony the bees will initially fill the sky, with flight appearing random and chaotic. The bees soon group and organize and may suddenly seem completely ordered. They often, almost always, fly a short distance and become organized as a unit while scouts continue to explore already-discovered housing, or search for other locations if the initial finds do not excite enough other scouts. (Morse photo)

(hemolymph) appears to thicken, too. Third, some worker bees begin to engorge. Engorged workers do no work but slowly their wax glands develop and wax is secreted. Engorgement and motionlessness are necessary for wax production. The engorgement of house bees also reduces the number of bees that are available to receive and process incoming nectar. Scouts that can find few or no workers to take the nectar they have collected slow or cease foraging, and sometimes react by attempting to find a new home. At this time scout bees may be seen flying up and down tree trunks, especially along the edge of a forest or hedgerow, examining knotholes and seeking a hollow tree that might provide a new nest.

When the swarm emerges from the nest, about 30 to 70 percent of the worker bees, excluding the queen or any drones that accompany the swarm, will be engorged. The food they are carrying is needed to sustain the flight of the bees in the swarm to their new home, to continue to stimulate wax production, and to serve as a reserve if the swarm encounters inclement weather while enroute to or during its first few days in its new home. New comb must be built, and the queen must start egg laying as soon after arrival as possible, since a minimum of 20 days will elapse before young bees will be produced. It is even longer before these new bees can enter into the important activities undertaken by house bees, especially the feeding of young larvae. Obviously, physiological changes occur in both the queen and the worker bees when swarming is about to occur.

Typical first-landings

From the point of view of practical beekeeping it is important to prevent swarming. When a queen loses weight and slows egg laying, fewer bees are produced. In addition, engorged bees are clearly not working. When scouts turn their attention to seeking a home instead of food, the beekeeper loses too. Recognizing the signs of congestion requires hands-on experience and, probably above all, separates the best beekeepers from the rest.

Absconding swarms – Honey bees may totally abandon a nest for a variety of reasons. When all of the bees and their queen leave, even sometimes abandoning their brood, the process is called absconding. Honey bees of European ancestry are less inclined to abscond than are honey bees of direct African ancestry. *Apis cerana,* the Asian species of honey bees will abscond easily too, often with what appears to be little provocation.

Bees may abscond because of a lack of food, bad odors, wetting of the top of the nest in such a way that the bees cannot protect it, a heavy infestation of wax moths, small hive beetles or other pests, repeated attacks by bears and other predators, and sometimes diseases. If the above circumstances become too severe, the colony may not have the strength to abscond, and dies.

Afterswarms – When a colony swarms (divides), the old queen and 30 to 70 percent of the bees leave the hive and

A not so typical first landing.

establish a new colony. The number of bees that leave with the first swarm (called the primary swarm) varies greatly. If a swarm is captured and put back into the hive from which it emerged, the bees will swarm again the next day, but the number of bees that leave the next day will differ, and different workers may leave the second time.

After a colony has swarmed it may cast a second (secondary) or even a third or fourth swarm. These are called afterswarms and, of course, usually have many fewer worker bees. The afterswarms frequently have virgin queens.

Clipping queen's wings – Not too many years ago clipping one or both of a queen's wings was recommended as a swarm prevention or swarm control method. Queens with clipped wings cannot fly, and if a swarm attempts to emerge it will return to the parent colony when the bees discover that their queen cannot join them. The swarm will, however, depart with the first virgin queen that emerges. Thus, some time delay is gained. However, more important, it should be clear from the above discussion that clipping a queen's wings does nothing to halt the physiological changes such as queen weight loss and worker engorgement that take place during swarming. While clipping a queen's wings may stop swarming for a short while, the colony is still not producing honey and the foragers are not active as pollinators.

Cutting queen cells – Removing queen cells can keep a colony from swarming, providing it is not done within hours of when the swarm would normally emerge. Although removing queen cells will keep the swarm from leaving the parent colony, it does nothing to alter the basic problem. In addition, locating all of the queen cells in a congested colony is not easy. If even one cell is missed, the swarm will emerge.

Demareeing – In 1884 George Demaree described a drastic, but almost always effective, method of swarm control that still carries his name.

First, the queen is located and placed into the bottom brood chamber with one frame containing one-quarter frame or less of brood. It is important to leave a small amount of brood with the queen. The remaining combs in the bottom brood chamber are empty and thus the queen has plenty of room in which to lay eggs. All queen cells are found and destroyed. A queen excluder is placed over the bottom brood chamber and one or two boxes of empty comb are placed above it. The remaining brood is then placed in a hive body on top of these empty boxes. This separation of the queen from the brood is sufficient to stop all efforts at swarming, but none of the colony's strength is lost. The separation will, however, cause the worker bees tending the brood on the top of the colony to believe that their queen is lost and they will usually begin to rear a new queen. The brood area must be checked carefully after no longer than a week to destroy any queen cells that are present. The brood area will need to be checked again for queen cells. Fulton, as in 2003, advocates the use of the Demaree method for swarm prevention (Fulton, H. "Swarm prevention: using the Demaree method." *American Bee Journal* 143: 358-360. 2003). The article is profusely diagrammed to make it easier to follow.

Late swarming – In some years, and some areas, swarms will emerge during the late summer and early autumn. Swarms emerging after about mid-June usually fail to survive the following winter because they have so little time to build a new nest and collect the food needed for winter. Obviously, no autumn swarms will survive the winter, especially in the north. Spring swarming is probably more closely tied to temperature and increasing day length. However, an enigma that has not been studied is the swarming that may occur in southern Florida when the introduced melaleuca (cajeput tree) flowers in late October. This plant is an excellent producer of nectar that apparently stimulates brood production and therefore congestion at that time of year.

Honey-bound condition – Colonies are honey-bound when a band of honey above the brood deters the queen from moving upward or when too many frames are full of honey stored the previous autumn. The band of honey need only be half an inch wide to prevent the queen from moving upwards into an empty comb. Colonies that are honey-bound usually have congested brood nests and swarm readily. The best way to break a honey-bound condition is to remove one or more frames full of honey from above the brood nest area and replace them with empty comb.

Ringing bells and beating on pans – A popular old wives' tale says that beating on a pan, called tanging, or ringing a bell, will cause an airborne swarm to settle. This belief seems true simply because a swarm does not remain airborne forever, but certainly beating on a pan and bell ringing does not cause a swarm to settle. In fact, a swarm just emerging from a hive will settle within 50 to about 200 feet (15 to about 60 meters) of the parent colony with or without bells or pans. This has led some beekeepers to believe that the beating or ringing has an effect on the bees. The idea apparently originated in certain archaic laws that permitted beekeepers to trespass on another's land for the purpose of pursuing and capturing their own swarm. The beekeeper was to give warning of his presence by making a noise. Pan beating is illustrated in some of the beekeeping prints from a few hundred years ago.

"Tanging" a metal object, once believed to settle swarms.

Swarm control – When larvae are found in queen cups, one of three steps must be taken to stop swarming. One may remove the queen, remove the brood, or separate the queen from the brood; the latter method is called "Demareeing." Bees recognize the absence of their queen within a about 15 minutes; both her removal and removal of the brood dramatically affect their behavior. A fourth possibility is to divide the colony into several parts to make new colonies. This is, in effect, artificial swarming, and when queen cells are capped and swarming is imminent, it may be the most practical recourse.

Congestion in a brood nest may be relieved in a variety of ways. The easiest and most common method is to add a hive body of empty combs on top of the hive. A second method that is also effective is to reverse the brood chambers, assuming that the nest itself occupies a part of two boxes. The brood nest is divided, provided the weather is warm, thus providing more space for the queen to lay eggs. If the nest is in the top brood chamber only, reversing will allow room for upward expansion. Reversing must be done when the weather is warm enough to avoid chilling the brood.

If the colony occupies only a single brood chamber, then adding an empty hive body of drawn combs on top is an invitation for the queen to move upward. Adding a box with new frames with foundation does little to relieve congestion, but it is better than doing nothing. A queen cannot lay eggs in comb until the cells are at least partially drawn.

Placing one or two frames of drawn comb in the center of a box of foundation will attract the bees to work in the new brood chamber more rapidly. Empty comb is then substituted in the lower brood nest.

Swarm prevention – Up until about the time that the eggs in the cups hatch, swarming is more or less preventable. Before that time, relieving congestion will usually stop swarming and return the colony to normal. Once larvae have appeared in queen cells, especially larvae two and three days old, it is time to take more drastic swarm control measures.

The effect of queen age – Two-year-old queens are more than twice as likely as younger queens to head colonies that will swarm. Older queens produce smaller amounts of the pheromones by which they are recognized. Annual requeening will decrease swarming in addition to providing the colony with a more productive queen. Perhaps one of the greatest advantages for beekeepers

When queen cells are capped and swarming is imminent, artificial swarming may be the most practical method of swarm control.

Marked queens offer the advantage of being easier to find in a large colony, but also reliably advertise their age. Two-year-old queens are more than twice as likely as younger queens to head colonies that will swarm.

who migrate south with their bees is their ability to requeen their colonies before returning north. Colonies with young queens will produce more eggs and are less likely to swarm.

The old queen accompanies the primary swarm – The old queen usually accompanies the first swarm because she is ready to take flight before the new queens emerge from their cells. Data show that only about 20 percent of swarms survive the first winter, whereas about 80 percent of parent colonies winter successfully, living to produce a swarm the following year. From the point of view of natural selection and evolution, it therefore becomes practical that the greatest investment, the young queen, should stay with the unit that is most likely to survive, that is, the parent colony. However the old queen, the one accompanying the swarm will be superseded within four to six weeks. The colony will, therefore, go into the winter with a new queen.

When swarming occurs – Swarm season is different in different parts of the country, obviously climate-related. In New York it was found that the first swarms would be cast about May 15 and the last spring swarms about July 15 with a peak about June 15. In Maryland the spring swarming season was about one month earlier. In the area of Davis, California, 88 percent of swarming took place in April and May. In Florida, swarming may occur from late February through April, and perhaps longer.

In the northern states that are near to or border Canada, usually only four months are frost-free. In this area, because of the short season, the time when swarming occurs is much better defined. This makes it easier for the beekeeper to know when to take swarm prevention methods. In the southern states the swarming season is much less clearly defined and is subject to wide variations of weather. One aspect of successful beekeeping and swarm prevention is a thorough knowledge of the climate and how it will affect swarming.

SWARM BEHAVIOR AND MOVEMENT – When a swarm emerges from a colony it first settles within a short distance of the parent colony. If the queen is not present, for whatever reason, the bees will return to the hive but will try repeatedly to leave. When the swarm is settled in the first location scout bees can be seen dancing on the swarm surface, and, if the new nest site is known, scout bees will be seen visiting it.

Swarms have been seen to settle part way between their point of origin and their new home. The swarm may appear to rest for a few hours or perhaps overnight before moving on. As soon as the queen drops out of the flying mass her absence is detected by the bees and the movement of the swarm stops.

A swarm can and does select a new home close to the parent colony. The average, or even maximum distance that a swarm of European honey bees might move, remains unknown. Swarms fly at a speed of about seven or eight miles per hour (11 to 13 km per hour). However, when the bees first become airborne they move much more slowly. A few hundred feet (40 to 100 meters) before they reach their destination, which is known only to the scouts guiding the swarm, the scouts slow their flight and are followed by the rest of the bees in the swarm. When the swarm reaches its destination, the scouts move to the entrance of the new home, expose their scent gland, fan their wings, and attract the rest of the bees and the queen to the new home. In one test it was observed that only 30 minutes elapsed between the time a swarm became airborne and was completely settled in a new home 900 meters (yards) away (see PHEROMONES).

SWARM COLLECTING AND HIVING – (see BEGINNING WITH BEES)

SWARMING AND SUPERSEDURE – Supersedure is the process whereby bees replace their queen without swarming. In the case of supersedure, fewer queen cells are usually constructed and they tend to be located in or near

Supersedure cells are usually found near the middle of the brood nest, and usually on the face of a comb, rather than on the bottom bars of a frame. Fewer cells are usually produced for a supersedure than for swarming.

the middle of the brood nest. Most races of bees build a greater number of queen cells when swarming will occur; these cells tend to be on the bottom of the nest. Supersedure may take place at any time of the year, whereas swarming is more rigidly controlled by the season, with day length believed to be an important consideration.

The underlying causes of swarming and supersedure are probably the same. In both cases, the quantity of queen substance, or pheromones, produced by the queen is apparently less, or perceived to be less. Supersedure occurs when less pheromone is produced, usually because the queen is old or has been injured. When the distribution of pheromones breaks down because of congestion, with the inability of the workers to distribute the pheromone, then swarming occurs. Queens that are somehow injured, i.e., those that do not have normal body parts, especially legs or antennae, may also be superseded.

Interestingly, in the case of supersedure, the old queen may not be killed but may continue to lay eggs, although at a much reduced rate. No antagonism exists between the old and new queen. The old queen will continue to solicit food and is fed by the worker bees, though she may receive less food than previously. Normally it appears that when two queens are present the older one usually dies after a few months. Mother-daughter queens can be found in about 20% of colonies.

TASTE IN THE HONEY BEE – (see SENSES OF THE HONEY BEE)

TAYLOR, RICHARD (1919-2003) – Richard Taylor and his twin brother were born in 1919, in Charlotte, Michigan. Richard was fascinated with nature as a child, and he soon had a hive of bees, then two, then many. He worked with local beekeepers offering muscle and energy, and was paid in wisdom and experience.

Brief stints at college in Michigan were unproductive and he served in the Navy for four years, where he found the writings of the great philosophers.

He returned to college, Oberlin in Ohio, then received his Ph.D in Rhode Island in 1951.

His fame in the academic world enabled him to serve on the faculties of Brown University, Columbia University in New York and at the University of Rochester, in New York. He was also visiting professor at Swarthmore College, The Ohio State University in Columbus, Wells College, Hamilton College, Hobart and Smith, Princeton, Cornell, Hartwick and Union College.

During that time he published several books on his subject, including *Metaphysics; Action and Purpose; Good and Evil; Freedom, Anarchy and Law; With Heart and Mind; Having Love Affairs; Ethics, Fair and Reason; Restoring Pride;* and completed shortly before his death, *Understanding Marriage.*

As soon as Richard was able, after finishing his studies, he returned to beekeeping, having bees in the city on rooftops and balconies. But once settled in his ancient farmhouse near Ithaca, New York, he expanded his operation. For a time he was a successful sideline

Richard Taylor

beekeeper, selling honey from his house, by mail and to other beekeepers.

He continued as a liquid honey producer until his conversion to producing round-comb honey in the early 1970s. He never looked back. His articles in *Gleanings In Bee Culture* advocating this type of beekeeping undoubtedly were instrumental in the popularity this product has enjoyed ever since.

In 1966 Richard published his first article in *Gleanings* and in 1970 began his column "Bee Talk," which ran until 2002. His last piece described the beekeeping area around his home in Ithaca, New York, site of the 2002 EAS Conference that he would attend. At that conference he was awarded the first, and only, EAS 'Joys Of Beekeeping Award' for his 30 years of contributions to the beekeeping industry.

He published several well-known beekeeping books. *The How-To-Do-It of Beekeeping* was published in 1974 and was updated for four more editions. *How To Raise Beautiful Comb Honey* came out in 1977, with three updates, as well as *Beekeeping For Gardeners*, and *Beeswax Molding and Candle Making*. In 1974 Richard brought together the whole of his lives as philosopher, parent, naturalist and beekeeping in *The Joys Of Beekeeping*. No work, before or since, rivals the combination of simple explanation of the craft, elegant style, and the profound sense of satisfaction beekeeping plays in the lives of those who pursue it.

TEMPERATURE CONTROL AND THERMOREGULATION – Insects are said to be cold-blooded, that is they assume the temperature of their surroundings. This is certainly the case when foraging honey bees are caught away from the nest and stay out all night because it is too cool for them to fly back. You may sometimes see a motionless, cold bee covered with dew on a flower or leaf in the early morning. Such bees are easy prey for a number of animals. You may also see a bee that has been warmed by the early sun return to a hive before normal flight begins for the day.

Individual bees shiver and thus generate heat, which they do just before taking flight. During flight, the temperature of the muscles in the thorax may be as high as 40°C (104°F) and probably higher. High temperatures are difficult to measure in individual bees. They may also fly when the ambient temperature is as high as 46°C (114.8°F) and cool themselves in the process. A book for those interested in this area is Heinrich, B. *The Hot Blooded Insects*, Harvard University Press, Cambridge. MA 601 pages. 1993. However, research in this area with several insects, including honey bees is ongoing and new papers are appearing each year.

When it becomes cool, honey bees cluster and exercise their muscles to generate heat to keep the cluster interior and the brood warm. The cluster around a brood nest takes the form of a ball with a layer several bees thick. In addition to bees packing themselves closely together between the combs, the bees in the outer shell fill adjacent cells in the comb. The hair on their bodies helps to contain the warm air within the cluster. When the temperature becomes high, above brood rearing temperature, the bees will gather water, fan their wings to ventilate and cool the hive. Colonies of bees have been able to survive on deserts where the temperature may be well above 38°C (100.4°F) provided they are able to collect water.

Honey bees without brood form a

Honey bees fanning at a colony entrance, pulling warm air out of the colony, so cooler air will enter.

cluster at about 14°C (57°F) though they begin to huddle together in small groups as the temperature drops and approaches this level. As the temperature falls still lower the cluster becomes more compact. The temperature in the brood nest is 34 to 35°C (93 to 95°F). There is a small amount of data to indicate that when bees first start to rear brood in the early spring that temperature control may not be so precise but if it falls too low the brood will die.

Bees keep only the brood rearing area warm in cold weather; they do not warm the whole of the inside of the hive though obviously during cold weather a well-insulated hive interior is warmer than the outside temperature. In the northern prairie provinces of Canada, where some beekeepers overwinter their colonies, it has been noted that in a group of four colonies packed together that the individual clusters will form on the interior walls adjacent to each other thus allowing the individual clusters to gain some heat from one another. Efforts to heat colonies with electrical hive heaters have not been too successful thus heaters are not recommended. In cold climates some form of winter protection is advisable and will assist the bees and cause them to consume less food (see WINTERING).

TERRAMYCIN® – (see ANTIBIOTICS AND OTHER CHEMICALS FOR BEE DISEASE CONTROL, Oxytetracycline)

THAI SACBROOD – (see DISEASES OF THE HONSY BEE, Viruses, Sacbrood)

THELOTOKY – (see PARTHENOGENESIS)

THIXOTROPHY – (see HONEY, Chemical and physical properties of honey, thixotrophy)

THURBER, LOUISE W. (1917-1999) – Louise Thurber, wife of Roy Thurber died February 24, 1999.

She and husband Roy were active in the Puget Sound Beekeepers Association for many years. In 1968 her husband started writing the monthly newsletter and for 16 years Louise edited, typed, reproduced and mailed the newsletter to over 250 members. Many of those articles were published in the *American Bee Journal* and *Bee Culture* magazines.

Her silent interest in beekeeping led her to donate a considerable portion of their estate to Washington State University. This established the Thurber Chair and a full-time Professorship in the field of Entomology, specifically directed at honey bee research.

After her husband's death, Louise compiled and published a book, *Bee Chats, Tips and Gadgets*, containing all of Roy's numerous research and published bee articles.

THURBER, P.F. (ROY) (1916-1984) – Roy Thurber was an internationally known hobbyist beekeeper and writer from Kirkland, Washington. Roy wrote for the *American Bee Journal, Gleanings in Bee Culture* and *Speedy Bee*. He wrote a column called "Sadder But Wiser" for *ABJ* for many years under the pen name B. Luver.

He confined the bulk of his articles to useful beekeeping gadgets, and was instrumental in bringing about beekeeper concern about pesticide bee kills, especially those caused by microencapsulated methyl parathion, Penncap-M®. His many articles on bee disease prevention and treatment were also well known to readers.

Roy was recognized in 1981 by the Western Apicultural Society for his "Outstanding Service to Beekeeping."

He developed the world's first mobile fumigator in 1970 after successfully fumigating in a 55-gallon (208 *l*) drum. Ethylene oxide fumigation of diseased bee equipment became legal in Washington, as well as in several other states, and was used in a number of Canadian provinces.

After his death, his wife Louise, who started him in beekeeping, endowed a chair in his name at Washington State University, Department of Entomology in Apicultural Research and Extension in Pullman. Currently, Dr. Steve Sheppard holds that position.

TOADS – Several species of toads are serious pests of honey bees in the warmer parts of the world including the U.S. The most serious of these is *Bufo marinus*, a native of Central America. This toad has been purposefully transported to Hawaii and Australia for the control of noxious insects. In both places it has become a serious pest, especially of honey bees.

Several people have dissected toads that have been feeding on honey bees and have observed that a toad will consume a large number of bees in a single visit to a hive. The fact that they have been stung in their mouths and stomachs has not slowed toad feeding.

There is no good control for toads other than to fence apiaries or to elevate

Close-up of Bufo marinus.

Bufo marinus (University of Florida photo)

colonies, both of which techniques are commonplace in such areas as Bermuda and the Dominican Republic where toad feeding has been reported to have been unusually severe. Frogs, like toads, live largely on insects. However, it has been extremely rare that a frog has ever been reported feeding on honey bees.

TODD, FRANK EDWARD (1895–1969) – Todd was a native of northeast Ohio and earned a B.S. degree from Ohio State University. Early in his career he worked as an entomologist in several capacities in Arizona, Argentina, Spain, and California. He joined the USDA Bee Research group in Davis, California, in 1931. In 1942 he moved to Beltsville, Maryland, and served under James Hambleton. From there he was transferred to the USDA lab in Logan, Utah and Tucson, Arizona. In 1961 he returned to Beltsville where he became head of the Apiculture Research Branch of the USDA until his retirement in 1965. Todd was especially concerned with nutrition of the honey bee, the effects of pesticides on honey bees, and pollination, especially alfalfa pollination and seed set. The Todd dead bee trap, a modification of the Gary trap for collecting dead bees at a colony entrance, was a major contribution by him that enabled researchers to gather more accurate data on in-hive losses of honey bees. Todd authored (together with John Kenneth Galbraith, the well-known economist, and Professor Edwin C. Voorhies) two well-known bulletins published in 1933, one on the economics of the honey market and the other on the beekeeping industry.

TONSLEY, CECIL (1915-2003) – Cecil Tonsley was on eof the elder statesmen of British beekeeping.

Well-known internationally, Cecil was Vice President of Apimondia (1985-87) and President of the British Beekeepers' Association (1983-84) after serving as its General Secretary (1954-60) and on its National Executive thereafter. He joined William and Joseph Herrod-Hempsall on the staff of the *British Bee Journal* in 1951, taking over from them as Editor in 1953 until the Journal ceased publication in 1998.

TOWNSEND, GORDON FREDERICK (1915-1988) – Townsend, professor emeritus of the University of Guelph and an internationally known expert on beekeeping, died December 14, 1988.

He was instrumental in adapting sulfa to treat AFB during the Second Word War and helped develop the Ontario Agricultural College honey pasteurizer and honey strainer (the OAC strainer).

Professor Townsend joined the Ontario Ag College in 1938 and served as head of the apiculture department from 1942 to 1971. He also served as Ontario's

Frank Todd

Gordon Townsend

apiarist and was responsible for administering the Bees Act in Ontario.

He retired in 1980 but continued an active involvement with international development projects, and writing the history of Ontario Beekeeping.

TOP BAR HIVE – (see HIVES, TYPES OF)

TRACHEAL MITE – (see MITES PARASITIC, *Acarapis woodi*)

TRANSFERRING BEES FROM FIXED COMB HIVES – (see REMOVING FERAL COLONIES AND BOX HIVES)

TRANSITION CELLS – (see COMB, NATURAL)

TRAVEL STAIN – When foraging bees walk over newly capped honey or any surface, bits and pieces of pollen and propolis drop from their bodies and contaminate and color the surface; this is called travel stain. It is especially evident when there is comb with new, white cappings near an upper entrance. This staining is of little or no importance in liquid honey production since the cappings are cut from the comb before extracting. When cappings are rendered (melted), the travel stain materials give

Travel stain is evident on the outside of this colony around the upper entrance. The same can and will happen on internal furniture and wax cappings. When wooden section boxes were popular, beekeepers would cover exposed wood with tape or paraffin to keep them clean.

the wax some of its distinctive odor and color.

Travel stain can be disastrous for comb honey producers. The stain detracts from the otherwise white color of the cappings and makes it less attractive to consumers. In honey shows, sections of comb honey that are travel-stained lose points. When comb honey is produced, the colony is given no upper entrance. One reason is that the bees are forced to walk through the main colony entrance and it is much less likely that new comb cappings will be stained.

TWO-QUEEN SYSTEMS – It is generally agreed by beekeepers and supported by research that a colony containing 50,000 to 60,000 worker bees is more efficient in honey production than is a colony with a smaller population. However, data to support this position are scanty and/or not strongly reinforced with observations on large numbers of colonies carefully analyzed statistically. It has been suggested by some that building colonies with even greater populations would make the larger colony even more efficient. This could be done by uniting two larger colonies. This is the basis for founding the two-queen system of management, though data to support the thought that two-queen systems are worth the effort are limited.

A number of researchers have studied and recommended various forms of two-queen management. Despite the many articles and bulletins that have appeared only a small number of beekeepers have adopted a two-queen system of colony management. In talking to beekeepers it appears that the extra time involved in making and managing two-queen colonies is seldom worth the extra effort. Managing two-queen colonies requires a great deal of work, including lifting heavy supers. The standard reference for those who wish to pursue this subject is: Moeller, F.E., "Two queen system of honey bee colony management." *Production report 161. U. S. Department of Agriculture.* 11 pages. 1976.

Moeller writes that an effective two-queen system depends on having two queens living harmoniously in the same colony and producing eggs for about two months before the honey flow. It should be understood that this means that two-queen systems would be more effective in northern climates where the date the honey flow starts can be pinpointed more precisely than it can in warmer climates. Much of the research done on two-queen systems of management was undertaken in Massachusetts, Wyoming and Wisconsin.

One way to make a two-queen colony: the younger brood, about half of the bees and the old queen are placed in the lower brood chamber. An empty box of combs is placed above this brood chamber to give space for the brood nest to expand upwards. The older brood and about half of the bees are placed in a box on top of an inner cover which is placed on top of those lower brood chambers. The inner cover hole is screened with two pieces of screening, one above and one below the hole. This allows some heat to move upward from the bottom colony that is usually the stronger of the two units, thus saving some energy for the upper unit. An exchange of odor between the two colonies may be helpful, though the role of odor in this instance is not clear. A new queen is introduced, using any one of a variety of queen introduction techniques, into the third (top) brood chamber where the remaining half of the brood has been placed (see REQUEENING and PACKAGE BEES, INSTALLATION AND IMMEDIATE CARE). The upper unit must have its own entrance. A fourth box is usually not added until the new queen has been laying eggs for several days and the upper unit needs more room for expansion.

About two weeks after the second queen is introduced the inner cover separating the two brood areas is removed and is immediately replaced with a queen excluder. Moeller states that the two queens need to be separated by one queen excluder only, but some suggest that the two queens should be separated by two queen excluders with an empty box of combs between the two. Authors differ as to when the queen excluder(s) should be removed and the queens allowed to fight to produce a single queen colony. (It is possible that the bees may be responsible for killing one of the two queens; this is an area in which more research is needed to make the picture clear.) Presumably there is nothing to gain in having two queens in the colony after about a month before the honey flow ends; any eggs produced after this time will not result in bees that will forage or contribute anything meaningful to honey production (that is, the crop to be harvested). During the two months the two-queen hive is in existence the brood chambers in each are reversed as needed to prevent swarming. More honey supers are added on top of the hive as needed, keeping in mind that empty storage space encourages foraging.

UNITING COLONIES AND NUCLEI – It is often desirable to unite colonies of bees that might not be strong enough to survive a winter or other adverse conditions. A time-proven method of uniting two or more colonies, used universally by beekeepers at any time of year, is to place one brood chamber on top of another with a single sheet of newspaper between the units. By the time the newspaper has been chewed away by the bees there has been sufficient mingling of colony odors that the bees accept one another and there is little fighting. Three or four slits, each six to 10 inches long, can be cut in the paper to provide adequate ventilation and to help the bees start to chew away the paper. The upper colony may or may not be given an entrance. The stronger colony is placed on the bottom, the weaker one on top. When uniting colonies it is best to find and remove the queen from the weaker hive.

Perhaps one of the chief problems in uniting colonies is that some brood may be chilled if there are not enough bees to cover it. For this reason it is important

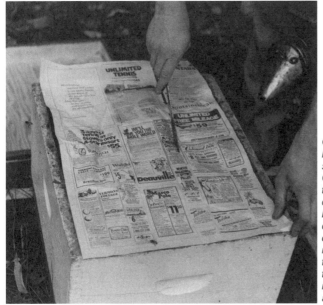

Once the two colonies to be joined have been selected, keep the stronger of the two on the bottom. Cover with a single sheet thickness of newspaper. Cut two or three short slits in the paper between frames. Place the brood chamber from the weaker colony on top of the newspaper and replace the covers. Depending on the strength of the two colonies the bees will have removed the newspaper and be united in about a week.

Uniting Colonies And Nuclei

Full-sized colonies or nucs can be united using the newspaper technique.

to combine colonies when the bees have time to rearrange their broodnest area and honey stores.

UNRIPE HONEY – The nectar that bees gather from flowers contains predominantly the sugar sucrose and water. To make this into honey the bees remove water and add enzymes that change the sugars in the nectar. The process of ripening the honey takes 24 to 48 hours. During this time the sugar-water solution is neither nectar nor honey and is given the name unripe honey. For details of what takes place during the process of ripening the nectar see NECTAR, CONVERSION TO HONEY.

UNSEALED HONEY – After the bees have ripened honey, they normally cover the cells with wax cappings. The cells that contain honey, but are not yet capped, are referred to as unsealed. As a general rule we think of unsealed

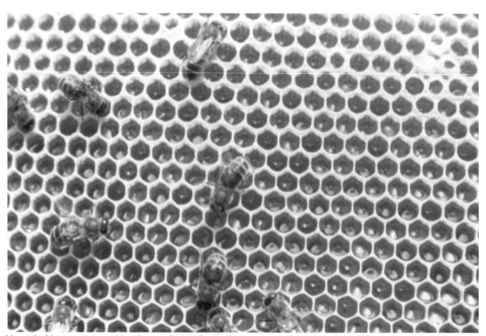

Unsealed honey may not be ripe, or it may have the proper moisture content. Warm, humid weather may play a significant role in sealing honey.

Sealed honey.

honey as not being ripe; however, this is not always true and unsealed cells can contain honey with a proper moisture content. Also, cells of honey covered with cappings can still take on or lose moisture since the cappings are quite porous; water vapor can and does move through the cappings. The role of the cappings is not clear but presumably they prevent the contamination of the honey by bees walking over the surface of the comb.

As a general guideline honey should not be removed from a colony and extracted until at least two-thirds or three-quarters of the honey cells are capped. Bees are not inclined to cap honey that is too high in moisture. If honey with high moisture content is extracted there is great danger of fermentation.

URBAN AND SUBURBAN BEEKEEPING – Many beekeepers keep their colonies in large cities, such as New York, Chicago and San Francisco, with good success. The number of colonies that may be kept in an urban area depends on the number of honey plants available just as it does in the country. Lawns filled with dandelions and clovers are good foraging areas. Streets in some cities may be lined with flowering trees such as maples, basswood or locust. Urban and suburban residents often grow plants that flower from early spring to frost and water is abundant nearly everywhere.

City beekeeping has a few requirements that are less important in the rural areas. The apiaries should be fenced or surrounded by a hedgerow. The bees will fly higher in the air as they leave and arrive at their hives in flight lanes well above the heads of people. Water should also be provided in the apiary at all times so that the bees will not be a nuisance at neighborhood swimming pools and bird baths. And that water source should *ALWAYS* be

Keep out the curious, or the foolhardy, with a good fence.

available during flight season. A dry source means bees will find a new source.

The beehives should be opened and examined only under favorable weather conditions and at the right time of day so that guard bees are not unduly aroused and there are fewer bees at home, and those that are, are busy taking care of incoming nectar. Keeping bees behind locked gates will make the apiary less accessible to pranksters and those who might be tempted to vandalize the colonies. A locked gate goes a long way toward convincing authorities you have safety foremost in mind.

Defensive colonies should be requeened with gentler stock and swarm control should be aggressively practiced so that swarms do not become a nuisance in the vicinity. Colonies should be requeened annually, without fail. Honey should be removed with great care to avoid disturbing the bees; bee escapes or fume boards are preferred since they do not seriously arouse bees.

A few municipalities have banned beekeeping within their borders (see ORDINANCES BANNING BEES). As long as forage, that is flowering plants, is found within city boundaries, bees will be present to take advantage of these food sources. It should be pointed out to people considering the enactment of such rules that well-managed colonies are preferred over feral colonies in trees and buildings, since the latter will swarm excessively and have been shown to be different in behavior and other traits. Furthermore well-managed colonies can compete favorably against the feral colonies for the limited forage.

The guidelines for finding an apiary site in a city are no different from those in rural areas (see APIARY LOCATIONS). European-type bee houses are quite satisfactory for city beekeeping and may even have the advantage of protecting the bees against vandalism. Keeping bees on rooftops, even those many stories high, poses no special problems other than those created by winds that may chill colonies in the winter or prevent or slow flight in the summer.

It is a good idea to belong to local and state beekeeping organizations so

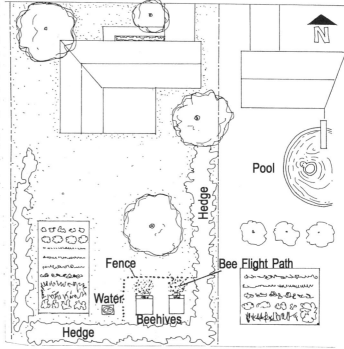

Placement of colonies in an urban area. The hedge on the sides and back both screens the colonies and keeps them out of sight. The fence, in front and on one side is made of material to reduce the visibility of the colonies and also not easily scaled. The fence and the hedge keep out the curious, vandals and keep the bees flying over the peoples' heads. A permanent water source is near the colonies. NEVER let it go dry. Water nearby will keep bees away from the neighbor's pool.

you can be apprised of local ordinances. Without doubt, it is a good idea to have liability insurance. For further reading see *Honey Bee Law* by Sylvia Ezenwa, published by the A.I. Root Company, 2005. It has excellent and current information on good neighbor beekeeping, municipal ordinances, beekeeper liability and apiary location precautions.

VARROA DESTRUCTOR – (see MITES, PARASITIC)

VARROA SENSITIVE HYGIENE – The mite-resistance trait called suppression of mite reproduction (SMR) can be explained by a form of hygienic behavior called *Varroa* sensitive hygiene (VSH). With VSH, adult honey bees remove worker-bee pupae from brood cells infested with *Varroa destructor*. The nearly continuous distribution of colony phenotypes from low to high suggests that most or all of the genes for VSH are additive, differing from the recessive genes that control hygiene for resistance to American foulbrood. When measuring both the removal of infested cells and the frequency of non-reproducing mites in all colonies, an increase in the rate of removal of infested cells was strongly related to a decrease in all categories of reproductive mites, even mites that produced eggs too late to mature. However, removal rates were not related to the number of mites that produced no progeny. This selective removal of egg-laying mites creates an increase in the proportion of mites that lay no eggs. Therefore, the simplest way to measure VSH is to measure the frequency of mites that lay no eggs. For example, a population of mites typically has those that enter cells but do not lay eggs. The average frequency of these nonreproducing mites is about 12%. When examining worker cells that are >7 days postcapping, a colony that has 12% of the mites with no eggs has had little or no removal of infested cells and probably has none of the genes that express VSH. With 45, 70 or 100% no eggs the colony has about 50, 75 or 100%, respectively, of the genes that express VSH. (Harbo & Harris)

VENOM, COMMERCIAL PRODUCTION OF HONEY BEE – It has been known since the early 1950s that if a worker bee is given an electric shock she will extend her sting and a drop of venom will form at the tip. A.W. Benton, while at Cornell University, was the first to develop and publish information on a practical method of collecting quantities of venom from worker honey bees (Benton, A.W., R.A. Morse and J. Stewart. "Venom collection from honey bees." *Science* 142:228-230. 1963). Using Benton's method one may collect about one gram of venom, the estimated amount produced by 10,000 bees, in about two hours. The system is so effective that one person, working for a short time each year, can supply the U.S. demand for honey bee venom.

A venom collector. Note the extreme protective gear worn by the beekeeper. Apiaries where venom is collected should be very isolated – from people, farm animals and roads. (Morse photo)

The device used to collect venom has a wooden frame over which fine steel or copper wires are stretched. The wires are alternately charged and grounded. When a worker bee touches two adjacent wires, the circuit is completed and the bee receives a slight electric shock. A car battery is the usual power source. The bee responds to the electric shock by bending her abdomen downward, perpendicular to the rest of her body, and driving the sting downward between the wires. Benton placed a piece of tightly stretched nylon parchment taffeta under the wires so that the sting passed through holes in the woven nylon. The nylon is slippery and the barbs on the sting do not catch on it. The current is turned off and on every few seconds. When the current is turned off the bee withdraws her sting and the droplet of venom is left on the underside of the taffeta. As thousands of stings penetrate the taffeta it becomes wet with venom. The venom is allowed to air dry and is then scraped off of the nylon and frozen. Later in his studies, Benton found that a very thin piece of plastic served better than taffeta and enabled the collection of a cleaner product. Again, the thin plastic is slippery and will not hold the sting. Very rarely will a bee be electrocuted. The final product is clear and crystalline in form.

Charles Mraz of Middlebury, Vermont, made modifications of the original collector. The bee's stinger passed through a thin rubber sheet and the venom was deposited on a glass plate. This manner of collection produced very clean venom.

Great caution must be exercised when collecting venom. One does not use smoke or try to calm the bees for two reasons: first, smoke particles might contaminate the venom, and second, one wants as many bees as possible to sting. However, using an electric shocking device in an apiary causes the bees to become very angry. They will range out from the apiary and sting any person or animal within several hundred yards (meters). Venom should be collected in remote locations only and even then extra care should be taken to make certain no one is in the vicinity. The person doing the collecting should be very carefully dressed. Two pairs of pants, two shirts, heavy boots, gloves, and a strong wire veil are advised. Dried venom is irritating to the mucous membranes and prolonged exposure can cause an allergic reaction, so one should wear a suitable respirator while collecting and preparing venom.

This device is satisfactory for collecting venom from honey bees in standard hives, but not other stinging insects. Still, in very populous colonies only a small percentage, perhaps five to 10 percent, of the bees present will sting.

VENOMS OF STINGING INSECTS – The primary function of insect venom, as delivered by a sting, appears to be to irritate and drive a potential enemy away from a nest. While they are foraging, most stinging insects flee an enemy unless they feel threatened or are abused. However, when an animal they perceive to be an enemy is in the vicinity of their nest they can become defensive.

Some stinging wasps use their venom to paralyze prey, usually other insects, and sometimes spiders, that are then carried back to the nest and deposited in cells where they become food for developing wasp larvae. Stinging prey in this manner may be thought of as a method of food preservation. The prey is alive, but cannot move, and stays fresh for many days, perhaps even weeks.

The chemical make-up of insect venoms varies from one species to the next. There have been several studies of insect venoms. One reference is: Piek, T., *Venoms of the Hymenoptera, Biochemical, Pharmacological and Behavioral Aspects*. Academic Press, New York. 570 pages. 1986. (see STINGS, HOW TO AVOID and REACTIONS TO BEE AND WASP STINGS)

VENTILATION – Several circumstances cause honey bees to move large volumes of air through their hive. When bees are ventilating strongly one will see several wing-fanning bees on the landing board at the colony entrance. At this time a small piece of tissue paper, perhaps an inch long and a quarter of an inch wide (about 2 cm long by 0.5 cm wide), may be held at the entrance with a forceps or on a toothpick and it will be seen that the paper will flutter inwardly on one side and outwardly on the other side of the entrance.

Honey bees ventilate furiously under several circumstances. These include the times when they are removing the moisture from nectar or unripe honey, when they are cooling a hive by evaporating droplets of water that have been scattered about the hive's interior, when they are removing a foul odor or smoke, when the moisture level is too high in the winter, and when the carbon dioxide level in the hive has become

Common stinging wasps include the yellowjacket, left; and the "bald-faced-hornet," right.

unacceptably high. When a need for ventilation arises wing fanning appears to be started by one bee who is then joined by others as needed. A question that remains unresolved is whether this effort is directed by one or more bees or whether it comes about because individual bees recognize a need.

VENTILATING SCREENS – (see MIGRATORY BEEKEEPING)

VIRGIN QUEEN – A queen honey bee that is not mated is called a virgin queen. Mating in honey bees normally takes place when the queen is three to five days old (see MATING OF THE HONEY BEE). As virgin queens age they appear to have greater difficulty in mating and after three to four weeks the mating process either cannot or does not take place. Bees can rear queens in the winter if they lose their queen but if the weather is not good for mating (flying) and drones are not available, these queens start to lay unfertilized eggs and produce only males. Once virgin queens start to lay eggs they make no attempt to mate.

From time to time a small number of queen breeders have attempted to sell virgin queens and because these queens are offered at a lower price, there have been some buyers. Normally, queen producers sell only queens that have mated and started to lay eggs. When one buys a virgin queen it is expected that it will be introduced into a colony in the normal manner and then mate. Experience with buying virgin queens has not been good. Many of these queens fail to mate naturally and are soon superseded or become drone layers. Mated queens can be assessed for their laying ability.

VISION IN THE HONEY BEE – (see SENSES OF THE HONEY BEE)

WADLOW, RALPH – Ralph Wadlow died October 4, 1992, in Fort Myers, Florida. His passing was a true milestone in Florida apiculture. Mr. Wadlow was one of the pioneers who brought beekeeping below the frost line in Florida. He was instrumental in convincing vegetable growers that pollination by honey bees increased yields. Mr. Wadlow cooperated on bee research projects at Cornell University, the University of Bogota in Colombia and the University of Florida.

He was a charter member of the Florida State Beekeepers Association, and a prime mover in that organization for decades. He attended almost every meeting and held many responsible positions. He held all the offices in the Southern States Beekeepers Federation, regularly attended Eastern Apicultural Society gatherings and was often one of Florida's delegates to the American Beekeeping Federation. He was also involved in international apicultural activities, consulting in South America (Bolivia) and attending several Apimondia conventions.

WASHBOARD BEHAVIOR – One will often see a hundred or more bees rocking back and forth at a colony entrance. Their heads are bent downward. Their hind four legs grip the surface firmly and the front legs scrape the entrance surface. It can be seen that the entrances to colonies are polished by bees; polishing is especially clear on trees. No one has been able to document why bees behave in this manner but it is assumed that this smoothing of the entrance surface serves to eliminate cracks and crevasses where noxious microbes might live, just as polishing the inside of the nest with propolis gives protection there.

A typical washboard activity and position.

WASPS – (see HORNETS, YELLOWJACKETS AND WASPS)

WATER COLLECTORS – (see WATER FOR HONEY BEES)

WATER FOR HONEY BEES – Bees use water to dilute honey for food, to cool the hive and to dissolve crystallized honey. In most areas water is not stored in the hive but is collected as needed; in hot and dry areas bees may collect and store small quantities of water in cells in the hive. A small number of bees may serve as "tank" bees and store water in their bodies for short periods of time. In the winter there may be an accumulation of water in the hive because of metabolic water (see WINTERING, Metabolic water) that results from bees consuming honey; if there is too much of this water it may pose problems for bees because the inside of the hive becomes damp.

Honey that is fed to larvae is diluted; the honey adult bees eat may be diluted, too. To cool a hive, bees will gather water and deposit it in droplets around the hive. A number of bees then act together, fan their wings and drive a large volume of air through the hive. As the exposed water is evaporated the hive is cooled. In desert areas where the temperature may rise above 100°F (38°C) a colony of honey bees may collect and evaporate over a gallon of water a day in order to cool the hive. It is likely that some water is used to maintain a more or less constant humidity in a hive but this has been little studied.

Most honeys crystallize in the winter though it is the glucose, not the fructose fraction that crystallizes. Water is collected for the purpose of dissolving these hard crystals so that the resulting honey can be consumed. In the spring it is not uncommon to see white glucose crystals in uncapped honey storage cells. The bees have removed the liquid portion during the winter and the white glucose crystals are left. One of the problems with feeding honey bees dry sugar for winter or spring food is that the bees may not have access to the water necessary to dissolve the crystals.

Behavior of water collectors – Some bees may devote their entire field life to water collection. Such a bee was observed by Robinson, G.E., B.A. Underwood and C. Henderson. "A highly specialized water-collecting honey bee." *Apidologie* 15: 355-358. 1984. They followed a water collector for 14 days after which time she was lost and presumably died. During a one-hour period this bee made an average of 7.1 trips collecting water from a creek about 0.5 km (0.3 mi) from her hive. On the

This research apiary in the desert southwest of Arizona has a large water tank on site to provide water at all times. Feral desert colonies can generally only be found near natural or man made water sources.

In temperate areas, open bodies of water – ponds, lakes or rivers – offer plenty of water when bees need it during the warmer months. The closer bees are to a natural source, that does not go dry, the less they must work to find, and gather water.

average she spent 1.0 minute flying to the creek, 1.1 minutes sucking up water, 1.2 minutes returning to the hive (it was an uphill flight) and 3.9 minutes in the hive. Other researchers have noted that water-collecting bees do not place the water in cells around the hive themselves but that they give it to house bees. Water collectors will dance as do other bees to indicate a source of water. A water collector will eat some honey before leaving her hive.

Water collectors are fewer when nectar is available. Bees may be able to obtain some of the water they need from nectar. It has been noted that bees will often collect water from around compost piles and puddles in barn lots. When bees were offered warmed water during cool weather it was much preferred over cooler water.

Providing water for bees – In temperate climates it may not be necessary to provide water for bees. However, in desert areas a water shortage can lead to the loss of both bees and brood. It must be remembered that once bees have learned where water can be obtained, that source must be constant. This becomes a problem for beekeepers who keep bees in irrigated areas where the water source may be available only on a schedule. It is imperative that beekeepers in urban areas provide water for their bees to prevent them from being a nuisance at bird baths and swimming pools.

A common way to water bees in remote areas is to use 55 gallon drums (208 *l*) with floats so that the bees will not

A Boardman-type feeder can be used to give plain water to a colony during warm weather. Some beekeepers add a drop or two of essential oils, or a commercial product (a mix of essential oils) to waterers so bees recognize them, and to make them easy to find.

In drier climates irrigation water may be all that is available, and then for only the growing season.

drown. Where running water is available from a faucet, beekeepers will sometimes allow a faucet to drip onto a board or into a receptacle with a float or stone from which the bees may drink. Pet or chicken waterers are especially useful in urban areas where beekeepers can attend to their hives on a daily basis. Boardman-type feeders work well during warm weather as the water is close to the bees and the level of remaining water can be observed continuously.

WATERY CAPPINGS – (see HONEY, Judging of)

WATSON, LLOYD RAYMOND (1876-1948) – Watson was the first to successfully instrumentally inseminate a queen honey bee. This insemination was the subject of his Ph.D. thesis at Cornell University in 1927. During the early part of his career he was an extension specialist in apiculture in Connecticut and Texas. From 1919 to 1921 he was a researcher with the USDA in Washington. Watson earned his B. S. degree from Alfred University in Alfred, New York, in 1905. He returned there as Professor of Chemistry and Director of Research in 1927 and remained there for the rest of his life.

Prior to Watson's studies, instrumental insemination of queens had been attempted by several researchers, some of whom claimed success. However, no one demonstrated a method, or developed equipment, that allowed repeatable successful inseminations until Watson developed his apparatus. Watson's laboratory and apiary in Alfred attracted researchers from around the world including such a well-known person as Walter C. Rothenbuhler, who wrote about Watson's latest syringe and syringe tip

Dr. Lloyd Watson

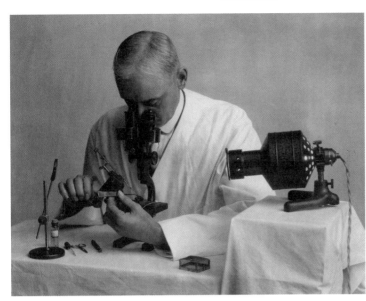

Dr. Lloyd Watson and some of the equipment he developed.

design after Watson's death (Rothenbuhler, W.C. and O.W. Park. "Posthumous contributions of Dr. Lloyd R. Watson to instrumental insemination of the honeybee." *American Bee Journal* 88: 248-249. 1948.)

Watson was praised in a series of eulogies in the April 1948 issue of *The American Bee Journal*. It was stated that he was painstaking and thorough, a friend of all and willing to share his knowledge with others. Many researchers have improved on Watson's original equipment and today instrumental insemination is an important tool in research and the development of new bees (see INSTRUMENTAL INSEMINATION OF QUEEN BEES).

WAX EXTRACTORS AND PRESSES – (see BEESWAX, Hot water pressing of cappings and slumgum)

WAX MOTHS – Among the moths whose larvae may eat honey, pollen and comb, two species are especially destructive: the greater wax moth, *Galleria mellonella*, and the lesser wax moth, *Achroia grisella*. Larvae of both of these moths may be found in active colonies, though bees are usually able to remove the larvae as they are found. Wax moth and other moth larvae are more commonly found in weak colonies and in stored combs. If one protects against the two moths then any other moth larvae that might be present will be controlled too.

Greater wax moth (*Galleria mellonella*) – This moth is present everywhere honey bees are found. It is a native of Asia where small numbers of larvae and pupae are found in colonies of all honey bee species, including the larger and presumably the most

Wax moth larva among the mess it creates.

Webbing.

Cocoons and last stage larva.

ferocious *Apis dorsata*. However, as travel and commerce between East and West intensified, the moths were transported to all continents. In the southern United States, and warm areas of the world, wax moths are a problem for beekeepers all year. Weak colonies are soon invaded and the combs destroyed. Stored comb, especially brood comb, is especially vulnerable to attack.

No life stage of the greater wax moth can survive freezing temperatures. However, moths may survive in any stage in a heated building where combs are stored. Also, wax moth larvae are sold as fish bait and may live through the winter in buildings where they are grown for that purpose. Wax moths are sometimes used as experimental insects and may be kept in some laboratories. The adult moths are good fliers and could escape from places where they are grown.

Lesser wax moth (*Achroia grisella*) –The lesser wax moth is widely distributed in the beekeeping world. However, it is much less common than the greater wax moth. Lesser wax moths are smaller, weighing only 15 to 20 percent that of the greater wax moths. When both moths infest a nest the greater wax moth larvae usually eat the larvae and pupae of the lesser wax moths and the latter do not survive. Lesser wax moths are found in live colonies much more often in tropical and subtropical areas than in the north. The larvae cause bald brood, a condition in which the cells are uncapped and the heads of larvae, usually in the later stages of development, are exposed. The presence of lesser wax moth larvae may be confirmed when bald brood is found by seeing fecal pellets distributed over the

Larva will chew out indentations on top bars or hive body walls when they spin their cocoons. Severe infestation can and will destroy combs and damage frames and other woodenware.

surface of the honey bee larval bodies. These fecal pellets are apparently deposited as the wax moth larvae move about and feed in the cells with the developing bees.

Wax moths, especially the greater wax moth, are often thought to be useful in the control of certain bee diseases. Any colony that dies, especially from American foulbrood, will soon have its comb reduced to a mass of webbing and moth feces within weeks. In this way it is thought, any American foulbrood spores that are present are destroyed or contained in the debris in such a way they are no longer a problem.

Unlike most animals, which cannot digest waxes, the wax moths can obtain some nutrients from beeswax. Wax moths do not attack and cannot live on pure cakes of beeswax or comb foundation. Like all animals they must have a complete diet to survive. They obtain the other nutrients they need from honey, pollen and any residues in the comb.

Other moths that have been found in stored combs, and sometimes in live colonies, include the dried-fruit moth, the Indian meal moth, the bumble bee wax moth, the Mediterranean flour moth and the death's-head sphinx moth (see DEATH'S-HEAD HAWK MOTH). No doubt others may be found too.

WEAVER FAMILY – The Weavers started keeping bees in 1888 when Florence Somerford (1867-1969) married Zachariah Weaver. Florence's brother, Walter, gave the Weavers 10 colonies as a wedding present. Florence had helped her brother keep bees for several years before she was married. Zach Weaver soon caught bee-fever and had expanded to 500 colonies by 1900. The Weavers acquired a foot-powered saw to cut out hive body, frames, and other wooden supplies. They also bought a foundation mill and built a homemade extractor.

In the early days basswood trees were the main nectar source. These trees were gradually replaced by cotton and other cultivated crops.

Zach and Florence Weaver raised six sons and three daughters. Three of the sons became beekeepers as have several of the grandsons and great-grandsons. One son, Carroll, was operating 600 colonies at Alto, Texas, when he passed away in 1965. Two other sons, Roy, Sr. and Howard, expanded the original Weaver Apiaries, then eventually formed two large separate queen and package bee operations.

Roy Weaver, Sr. assumed management of the apiaries in 1915 and in the early 1920s began migrating to the cotton fields in the blacklands of Central Texas. Then the Weavers began to suffer severe losses to their bees from the application of arsenical insecticides to cotton.

Roy Weaver, Sr.

Roy Weaver, Jr.

In 1925 Horace Graham of Cameron urged Weavers to rear queens and offered to buy a minimum of 1000 yearly to put in the package bees he was selling. Roy and Howard Weaver became partners in the queen raising business and raised both Italian and Caucasians.

Roy, Sr. and Howard operated as partners from 1927 until 1946 when they formed two separate companies with their respective sons as partners.

Weaver Apiaries was reorganized in 1946 by Roy Weaver, Sr. (1892-1978) with two of his sons, Roy, Jr. and Binford. Another son, Dr. Nevin Weaver, is a professor at the University of Massachusetts.

In 1977 the company was operated by Roy, Jr., Binford, and Roy, Sr.'s grandson, Richard.

In 1975 the Weavers operated 19,500 nuclei for queen production in Texas and produced 55,000 queens.

A honey-producing unit was operated in North Dakota and Roy, Jr. and Binford also own Texas Honey Farms at Jourdantown, Texas.

In 1976 Weaver Apiaries and Powers Apiaries jointly opened a new branch in Kona, Hawaii, called Kona Queen Company to raise queens.

The Weavers are active in state and national beekeeping associations. Roy, Sr. was president of both the Texas Beekeepers Association and Southern States Beekeepers Association. He was voted Texas "Beekeeper of the Year" in 1976.

Both Roy "Stanley" Weaver, Jr. and Binford Weaver have served as president of the Texas Beekeepers Association and the American Bee Breeders Association. Roy, Jr. was president of the American Beekeeping Federation from 1966-1967, and Binford was elected President of the Federation in 1981 and 1982.

This operation eventually split again into R Weaver, now run by Roy, Jr.'s son Richard, and B Weaver, run by Binford and his son, Danny.

Binford Weaver

WEIGHT OF HONEY BEES – Honey bees are sold by the pound in packages (see PACKAGE BEES, INSTALLATION AND IMMEDIATE CARE). When a beekeeper buys bees the question arises of how many bees are being purchased. The weight of individual honey bees is also of interest when we discuss pollination and observe that a bee must be of a minimum size and weight to operate (trip) the pollinating mechanism in certain of the larger flowers. Alfalfa, for example has a peculiar pollinating mechanism that requires a honey bee of a minimum size (see HONEY PLANTS, Alfalfa.)

Since honey bees may carry nectar, honey and water internally, their weight varies greatly. When the bees are carrying no food it is generally agreed that there are about 4000 bees per pound (8800 bees per kilogram). Normally fed bees, however, weigh in at about 3000 bees per pound (6600 bees per kilogram.) About 80 percent of the bees in a swarm will be fully engorged and thus, under these circumstances, the number of bees per pound will be fewer. The time of year when the bees are reared may also affect their weight; fall bees have a much larger fat body than do spring bees. The quality of the food, especially the pollen, may also affect an individual's weight because of nutritional differences. Drones weigh much more than workers, about 1800 drones per pound (4000 per kilogram).

The weight of queens varies much more than that of workers and drones. The size, weight and egg-laying ability of a queen has been shown to be closely correlated with the quality and quantity of food she receives during larval life. The feeding and rate of larval growth is reported in Nelson, J.A., A.P. Sturtevant, and B. Lineburg. "Growth and feeding of honeybee larvae." *U.S. Department of Agriculture Bulletin* 1222. 37 pages. 1924.

African and Africanized honey bees are about 10 percent smaller than European honey bees. Their weight is, accordingly, less.

WELLS, HORACE "LINC" (1913-1982) – On September 25, 1981 at the age of 69 "Linc" Wells of Riverhead, New York,.

The son of a Methodist Minister, Linc was a graduate of the NYS College of Agriculture at Cornell University and worked for 29 years as an agricultural agent and administrator with Cooperative Extension of Suffolk County on Long Island.

Linc was instrumental in the organization of the Suffolk Beekeepers' Association in 1949. This group later became known as the Long Island Beekeepers' Club. He guided the Beekeepers for 32 years.

WENNER, DARRELL C. (1942-1991) – Darrell Wenner, 49, of Glenn, California died January 24, 1991, at his home.

Darrell became active in his father's bee business in 1960. Clarence Wenner established his business in 1931 and was known worldwide as the "King of Queen Bees," a tradition Darrell continued. Wenner Honey Farms, Inc., is a business they both worked hard to be proud of.

Darrell was deeply involved in the activities of the beekeeping industry. On the state level, he was immediate past president of the California State Beekeepers Association and had served the association in various capacities. Likewise he was active in the California Bee Breeders Association and has represented it on several national panels. In 1984, the CSBA named him Beekeeper of the Year.

On the national level, Darrell was organizing chairman of the Tri-County Committee on Africanized Bees and Parasitic Mites. He served on the USDA-APHIS advisory committee for the U.S./Mexico cooperative program, the APHIS Technical Advisory Committee of Africanized Bees and Parasitic Mites, and the *Varroa* Mite Negotiated

Darrell Wenner

Rulemaking Committee. He made trips to Venezuela, Central America, and Mexico to investigate the Africanized bees. He represented California on the initial National Honey Board Nominations Committee.

A director to the American Beekeeping Federation for many years, he had served two terms on the ABF Executive Committee. He had also served as chairman of the ABF Research and Technical Committee and on the Convention Committee.

WET COMBS – (see COMB STORAGE)

WHITE, JONATHAN (1916-2001) – Jonathan Winborn White Jr., Ph.D, 84 of State College, Pennsylvania died September 2, 2001 at Brookline Village, State College. He was born in State College, September 29, 1916. He received a B.S. in agricultural chemistry from the Penn State University in 1937 and his M.S. and Ph.D degrees from Purdue University in 1942. In 1943 he married Rosalind Christman, who died in 1998.

He worked for the U.S. Bureau of Censorship during WWII. After that he joined the U.S. Department of Agriculture, Eastern Regional Research Center in Wyndmoor and worked there until his retirement in 1978. Dr. White spent most of his research career working on the chemistry of honey. He developed numerous unique methodologies for honey research. In 1986, Dr. White was honored with the Harvey W. Wiley Award, which is awarded annually by the Association of Official Analytical Chemists to recognize outstanding contributions to the development and validation of methods of analysis for foods and other related areas.

Dr. White was widely recognized as the world's most foremost authority on the analysis and composition of honey. His huge body of work on honey includes the discovery of four new sugars, new

Jonathan White

methods of separation and identification, finding gluconic acid to be the principal honey acid, the development and improvement of methods for examining honey adulteration, the characterization of honey's antibiotic principle, the demonstration of the nature of various honey enzymes and technical work on new processes and products.

Dr. White received many honors from the U.S. Department of Agriculture including three consecutive Outstanding Performance Awards, the Superior Accomplishment Award and the Superior Service Award. He has been honored by the International Bee Research Association, the American Beekeeping Federation, the Honey Industry Council, and in 1980 received the James I. Hambleton Award for Outstanding Research from the Eastern Apicultural Society.

Following retirement in 1978, Dr. White moved to Navasota, Texas, where he continued to conduct and publish research and collaborate with colleagues worldwide. He published well over 120 refereed journal articles including 19 in the Journal of the Association of Official Analytical Chemists that describe analytical methods for honey and beeswax. He has contributed chapters to more than 10 books and was active in his profession until his death. He was a member of the American Chemical Society, the American Association for Advancement of Science and the Institute of Food Technologists.

WILBANKS, WARREN GUY (1921-2006) – Warren Wilbanks, 85, died September 16, 2006. He was born January 28, 1921 in Banks County, Georgia. Warren attended Banks County schools and North Georgia College. He was a veteran of WWII serving in the Philippines and

Warren Wilbanks

Japan. Warren built a beekeeping business with his father and his sons to become one of the nations leading suppliers and a worldwide shipper of package bees and queens.

Warren Wilbanks started the present company in 1948 in Claxton. He became familiar with package bee and queen producers while working with the Georgia Department of Entomology.

However, his father, Guy T. Wilbanks (1895), and maternal grandfather, Gresham Duckett, both had kept bees in North Georgia.

Guy got started in beekeeping when his father-in-law gave him some bees as a wedding present. He was active with Warren in the bee business until he was 78.

Warren was past-president of Wilbanks Apiaries, Inc., past-president of the American Bee Breeders Association, past-president of the Georgia State Beekeepers Association, and past-president of the Southern States Beekeeping Federation.

His son, Reg and Warren's wife, Alva Wilbanks of Bellville continue the business.

WILLSON, R.B. (1894-1981) – A leading figure in the honey business for more than half a century, R.B. Willson, died at his home in Yonkers, NY, August 6,

1981, after a long illness. He was 87.

As a young man Mr. Willson specialized in biological sciences and entomology and finally beekeeping at Cornell University. After graduation and service in the Army during World War I, he taught at Mississippi State College, then returned to Cornell where he was on the faculty in the biology department as a specialist in apiculture.

Mr. Willson entered the honey business in New York in 1926, becoming vice president of the John G. Paton Company before he established his own business, R.B. Willson, Inc., in the spring of 1946.

He served as a director of the American Honey Institute, chairman of the Honey Industry Council of America, and president of the National Honey Packers and Dealers Association. He was an active participant in the world Apimondia Congresses, notably in Bucharest, Budapest, Prague, Rome, Vienna and Moscow. He presented a paper at the Moscow Congress in 1971 and was one of the first representatives of the American honey industry to visit China in 1973.

Mr. Willson often represented the honey industry in Washington on tariff matters and the establishing of quality standards. He remained active in R.B. Willson, Inc. until his 80th year and continued as Chairman of the Board.

WIND, EFFECTS OF – It is important to protect an apiary from wind for several reasons, chief among these being that a worker bee's life depends upon how much work she does. Work, especially flying, wears out flight muscles and brings about an early death. While we can do nothing to protect a bee while she is foraging, the task of entering and leaving the hive can be made easier. High winds also affect mating flights. Strong winds force the drones and queens to fly lower than normal.

During the winter it is important that honey bees take flights as often as possible both to void fecal matter and also to clean the hive of debris. If the environment in the immediate vicinity of the hive is protected these flights will be encouraged. High winds can cause snow to drift and block colony entrances or even to blow snow into colonies. Bees in hives in protected apiaries use less energy to keep the winter cluster warm.

Winds of about 12 miles per hour (19 km per hour) or more will stop bee flight. Bees will take flight on days when the weather is marginal but they are less efficient in gathering food and less effective for pollination. Some fruit growers, for a variety of reasons, including more effective pollination, are using windbreaks to protect their fields.

WINDBREAKS – The ideal apiary location faces and slopes south or southeast, has maximum exposure to the sun and is also protected from the wind. A natural windbreak, such as a hill or trees, gives the greatest protection but such windbreaks are not available

R.B. Willson

everywhere. Wooden fences may protect bees but they are costly both to build and maintain but in some locations a beekeeper may have no other choice.

An inexpensive and effective temporary windbreak can be made of bales of straw stacked up on the windward side of a group of colonies. Over time the bales become wet, heavy and solid.

Another fence can be made of landscape burlap and fence posts. The burlap can be fastened to the post with ratchet ties, wire or string. The fence can be taken down, rolled and reused for several years.

Plastic or wooden snow fence is adequate, also, as are simply piles of evergreen boughs or holiday trees.

Keeping bees in two or more locations allows one to test the qualities of an apiary site for a variety of factors including wintering. Some beekeepers have observed that certain locations are

A windbreak of evergreens provides year-round protection. Placed well, it can provide shade at the right time of day, at the right time of year, as well as snow and wind protection.

Even a bank of deciduous trees and shrubs can shield these colonies from the prevailing winter winds, to afford some winter protection.

This fence acts as a windbreak.

much better for wintering than others and will move their colonies to those for winter.

WING-FANNING – (see VENTILATION)

WING VENATION – (see ANATOMY AND MORPHOLOGY OF THE HONEY BEE)

WINTER BEES – Honey bees respond to changes in the season as do many other animals. In the autumn the enlargement of the fat bodies in worker bees is especially noticeable. We presume that this helps to provide energy for the winter cluster during the cold weather. Spring bees have almost no visible fat body. Since honey bees rear the least amount of brood in the autumn months the head glands of autumn worker bees are little used and they remain in good condition. It has been stated that fall bees live longer because of their enlarged fat body; however, they do less flying and other work.

WINTER BROOD REARING – For many years it was thought that winter brood rearing by honey bees was abnormal. If it occurred, it was said that the bees were suffering from a disease or unusual disturbance that raised the hive temperature and caused the queen to lay eggs. Once egg-laying had started the colony was forced to continue brood rearing and this, it was thought, might even cause the colony to perish. Poor food, which might cause an accumulation of fecal matter was also said to cause discomfort that in turn lead to winter brood rearing. In large part it would appear that many people thought that winter brood rearing was poor economy on the part of nature and it was therefore unreasonable that it should occur; this last appears to be a bit of armchair apiculture that we find has permeated the beekeeping literature frequently. In asking questions about bee biology it is important to devise experiments that are answered by bees, not by armchair logic.

Two studies in northern climates, with similar data, show that the least amount of brood rearing in colonies occurs in the fall months of October and November and that normal brood rearing starts in December and increases greatly during the cold months of January through March. The first of these studies was done in Scotland (Jeffree, E.P. "Winter brood rearing and pollen in honeybee colonies." *Insectes Sociaux* 3: 417-422. 1956.) and the second in Connecticut (Avitabile, A. "Brood rearing in honeybee colonies from late autumn to early spring." *Journal of Apicultural Research* 17; 69-73. 1978.)

Avitabile found that in Connecticut the average number of bees in a colony declined from about 21,000 in November to 12,000 in March at which time the population began to increase. The amount of brood rearing increased rapidly after the winter solstice. Colonies of bees with young queens had twice as much brood as did those with older queens. The number of eggs the average queen deposited in November and December was about 25 per day. This rose to 110 per day in January and 161 per day in February. These observations were made over a period of three years. Avitabile felt that brood rearing is controlled, at least in part, by day length.

WINTER CLUSTER – Honey bees survive the winter by forming a hollow, ball-shaped cluster over the combs but just under the stored honey. Bees cannot form a continuous cluster over combs filled with honey. A part of the solid ball of bees includes those that crawl into and remain in empty cells in combs on the outside of the cluster. The cluster surface will be several bees thick. When there is no brood in the autumn the cluster is well-formed when the temperature reaches 57°F (14°C). When brood is present the cluster forms any time the temperature falls below that required to keep the brood warm. Within the hollow cluster some bees exercise their flight muscles to generate heat that keeps the interior of the cluster warm. As the outdoor temperature drops the cluster shrinks and becomes more compact.

All insects, including honey bees, are cold-blooded and assume the temperature of the environment around them except that honey bees have the ability to raise both their own body temperature individually and as a group. Bees on the outside of the winter cluster become cold, in fact, they appear to be so cold that they cannot move. However cold they become, bees in a winter cluster are able to protrude their sting from a disturbance so that the cluster surface appears like the back of a porcupine and any animal that touches

A 'winter killed' colony. Bees seldom die from the cold, but often starve, running completely out of food or being unable to reach available food nearby. As the cluster moves up to food during the winter, when it reaches the top it is usually dangerously close to running out of food. The colony will die if emergency rations are not provided by the beekeeper.

it will be stung (see COLONY DEFENSE AT LOW TEMPERATURES).

The mechanics of the winter cluster have not been carefully studied but it is assumed that when bees on the outside become so cold that they cannot move they are pushed to the center of the cluster by warm bees from the interior who then take their place. No one has ever made precise observations on how long individuals may remain on the exterior of the cluster. There are no records of individual bees that are part of an active cluster dying because of cold exposure. There are northern limits where colonies of bees can survive but survival is limited by the amount of protection the colony has and its food reserves. Interestingly, it has been observed that in the Peace River district of the Canadian province of Alberta, close to the northern limit where colonies can survive, when four colonies are wintered together in a square group all four clusters will form in the corners near the center of the four colonies where each may gain some heat from the other.

Some measurements have been made of the amount of food in the honey stomach of bees in a winter cluster. Since the cluster forms below the honey, bees cannot engorge continually. Bees in a cluster would starve if it were held to a low temperature continuously and that cluster not allowed to break, move, and feed. A part of successful wintering is allowing the cluster to break periodically so that bees in the cluster may engorge on honey that will last them for a long period of time, a matter of days or weeks. In the autumn, worker bees have a fairly large fat body that can serve as a food reserve in emergencies. A second part of successful wintering is allowing the bees to take flights to void fecal matter. A small number of warm periods during the months of January and February when bees may both feed and fly are therefore critical for winter survival.

Colonies should be disturbed as little as possible in the winter so that clustering and winter feeding proceeds naturally. While it is possible to screen and move a colony in winter the disturbance will cause the cluster to break, chilling any brood and even the bees themselves. If it is necessary to move colonies during cold weather it should be done as close to spring as possible.

Moving bees in the winter in a northern climate will stimulate a colony to rear brood. On the other hand, migratory beekeepers move colonies south at any time of the autumn, winter or spring without causing any harm; however, under these circumstances the bees are usually being moved to a warmer climate where they would be stimulated to rear more brood anyway.

WINTER DEFENSE – (see COLONY DEFENSE AT LOW TEMPERATURES)

WINTERING – In the north, colonies of honey bees can benefit from some human assistance to survive the winter. While any one disease or mite may not destroy a colony, winter may complicate the problems caused by mites, *Acarapis woodi*, *Varroa destructor,* and diseases of viral origin, nosema, and American and European foulbrood. When a colony is plagued by two, three or more of these diseases, winter survival becomes more difficult. The process of inspecting, weighing, feeding and protecting colonies against the extremes of cold weather in the north is called, in loose terms, wintering.

It should be emphasized that cold

weather, by itself, is not the problem, at least for most of the northern climates that are inhabited by humans. In winter, bees form a round cluster that spreads across several combs and is just under the stored food. Not only do bees fill the spaces between the combs, but also many bees crawl into cells so that the outer shell of the cluster is compact and complete. Bees within the cluster generate heat by flexing their muscles. Cold, immobile bees on the outside of the cluster are periodically pushed to the center where they are warmed; at the same time warm bees move to the outside. This apparent rotation of bees prevents any from being lost because of cold itself.

Warm periods in winter are important to colonies for two reasons. First, the bees must break their cluster periodically and gorge on honey. It is honey and their fat bodies that provide the energy for warming the cluster interior. Warm periods in winter are also necessary so that bees may fly briefly and void feces. Many older bees will be found dead on the snow after such winter flights, but this is not considered serious since their advanced age, and sometimes diseased condition, indicates they had little to contribute to the economy of the colony. In the north, and into territory that has fewer warm spells in winter, the quality of the food becomes more important. The reason feeding of sugar (sucrose) for winter food is recommended is that sugar contains no indigestible ingredients. Darker honeys, with greater amounts of plant pigments, dextrins and other ingredients that give distinctive honeys their great flavor, do not make especially good winter food for honey bees. The farther north and the longer and harsher the winter period, the more important the quality of the winter food becomes.

Older winter bees are sometimes found dead in the snow after a flight.

Inspecting and weighing colonies for winter – There is no point in attempting to winter a colony that does not have a healthy queen, brood, sufficient bees and food for winter. A final inspection of colonies in the north should be made in September or October, at a time when at least some brood is still present. This time will be correspondingly later as one moves south. In the north bees rear the least amount of brood in October and November and often it is difficult to make a thorough fall inspection because there is no brood. However, even when brood is not present one may check for brood dead from a disease, especially American foulbrood. The queen need not be found in the fall inspection, or any inspection for that matter. Queens are assessed by the quality of their brood patterns (see BROOD PATTERNS).

If no brood is present the worker bees should be examined to determine if a sufficient number of young bees capable

One way to weigh a colony. You need a tare weight for equipment without bees, honey and pollen inside.

An easy way to weigh a colony. Obtain a tare weight of empty equipment first. Lift the front of the colony so it pivots on the back edge, and the back of the colony so it pivots on the front edge, add the two weights for the total. (Morse photo)

of surviving the winter are present. Young bees are characterized by having a full complement of body hair and wings that are not frayed. It is difficult to estimate the number of bees present in a colony except through experience. A colony should have a minimum of 15 to 20,000 bees, four to five pounds (two to three kg), in the fall to survive the winter.

One way to determine if a colony has sufficient food for winter is to weigh it. Many colonies have been lost because they were short only a small quantity of honey necessary to survive the winter. A variety of scales have been designed for weighing colonies. One method for weighing a two-story colony, which is the usual size of a wintering colony, is for two men to use a round scale mounted on a shoulder yoke. Scales are also available from equipment suppliers and large commercial scales can be found second-hand.

The most reliable way to determine winter stores is careful inspection of the hive, looking for the number of frames filled with honey. Pollen is difficult to assess because the bees will top off pollen cells with honey and cap them. Therefore cells with pollen will resemble honey stores.

Metabolic water – When insects eat sugars and fats for producing energy, water is produced as the food is disgested. This is called metabolic water and is a by-product of the digestive process. In technical terms metabolic water is that produced by the oxidation of organic matter (sugars and fats).

For insects that live in dry climates, or where their food is dry, such as stored grain, this water can be life saving. In the case of honey bees, in the north in the winter, too much water can be disastrous. A colony may be killed if this water condenses in quantity in the hive and wets the hive interior, honey, and even the bees themselves.

In the autumn, honey bees have a large fat body, not true of spring bees. The function of the fat body is to provide some of the energy needed to survive the winter. As bees use or digest this fat in winter 1.14 grams of water are produced for each gram of fat metabolized. The production of water from the consumption of sugar is less, being 0.55 grams for each gram consumed.

The use of one gram of fat releases 9500 calories while an equal amount of sugar releases 4200 calories. In January and February, when bees are rearing brood in the north, and holding a brood rearing temperature of 92 to 95°F (33 to 35°C), a great deal of food is consumed and metabolic water produced. This is the time when the ventilation of the colony is especially important.

Packing for winter – Experiments conducted in many of the northern states and Canadian provinces show

The result of metabolic water production and inadequate ventilation. Warm, moist air will rise and condense on the colder hive parts and freeze. Later, when outside temperatures rise, this ice will melt and drip down on the bees. Cold, wet bees will die.

Two colonies packed together and wrapped in roofing paper, then insulated with straw. This method has been used since shortly after the turn of the 19th century.

Three steps to wrap a single colony. Note the upper entrance, and the loose fold at the top so moisture can escape.

that packing or wrapping colonies for winter can be desirable. In the past, heavy wooden packing cases were used for wintering with considerable success. The cost of constructing these cases and the labor involved in packing and unpacking them is prohibitive. Furthermore, it has been found that colonies heavily packed in wooden cases do not warm up quickly or often enough to allow the necessary midwinter flights.

Equipment suppliers sell various types of plastic, cardboard and other wrapping materials so one can find something suitable. Light-weight tar paper (used by builders) is easy to handle. Heavy tar paper that forms a vapor barrier or seal should not be used because moisture will condense on the inside of the wrap and keep the hive wet.

The chief advantage of all these wrappings is the dark color. Heat is absorbed from the sun that warms the colonies during the winter months. Warmth helps to keep the colonies dry, permits the clusters to move upwards

Wrapping four colonies together has been practiced for years. Here, a layer of insulation is wrapped around the colonies first. Then, roofing paper is wrapped around that, and a large sheet of plywood covers them all. Each colony is provided with an upper entrance.

Cardboard, plastic-coated cardboard, and corrugated plastic sleeves are available that fit over one, two or four colonies. The covers are durable, lightweight, easy to store and effective.

In cold climates some beekeepers do not wrap colonies. Richard Taylor, who lived near Ithaca, New York, wintered his colonies like this. The colony, in late summer, had most of the bees in the medium-size hive body on top of a deep hive body. These boxes were reversed. The bees stored autumn honey above their nest, in the deep. The colonies were tilted forward to allow water to drain. A small piece of roofing felt was stapled over part of the entrance. This technique was successful for this location.

to new stores, and provides more opportunities for cleansing flights.

Various methods of packing colonies for winter have been studied. Two colonies can be wrapped together on a hive stand or four colonies on a pallet. Upper entrances should be provided on wrapped hives so that bees can exit to void feces on warmer days. Ventilation is also necessary to remove excess moisture.

Another type of plastic wrap is flexible with its own insulating properties. Easy to apply and inexpensive, this wrap serves as a windbreak and insulator.

Hive stands – Hives should be not be placed on the ground but on hive stands at a suitable height for the beekeeper. With the use of screened bottom boards, the screen needs to be two or more inches (five cm) above the ground or any other bottom surface. Some beekeepers are leaving screened bottom boards on their hive the year around. Various types of stands are used, for example, cement blocks or pallets, depending on the choice of the beekeeper.

Wintering in Canada – For many years in the past most beekeepers in the

Three wrapping techniques used by Canadian beekeepers (Dick photos)

1. Colonies individually wrapped together on a single pallet.

2. Four colonies wrapped together on a single pallet.

3. Pallets ganged together and colonies wrapped in bulk. Bottom entrances are provided for these colonies.

northern parts of the Canadian prairie provinces, especially the Peace River district, have purchased package bees that are installed in hives in mid-April through mid-May. The honey has been harvested and the bees killed in August and September. When tracheal and varroa mites were found in the U. S. and the Africanized bees were viewed as a threat by the Canadians, more thought was given to overwintering colonies. Two methods of wintering evolved, one method involves packing colonies in groups of four and substituting sugar syrup for honey for winter food. The second method of wintering was the use of specially constructed buildings for indoor wintering.

A study of the economics of overwintering colonies indicates it is more profitable to overwinter colonies than it is to buy packages. However, it cannot be emphasized too strongly that a high level of management skill is needed to do so successfully (MacDonald, D. and G. Monner, "An economic comparison of wintering and package bees in the Peace River region." *Report no. 821-14. Alberta Agriculture*).

WINTERING, INDOORS – Early indoor wintering occurred in what were basically large root cellars. The facilities were varied but had the following features: total darkness, steady temperature and adequate air exchange. Most cellars were partly or wholly below ground. A hillside of sandy soil was considered best because of the good water drainage. Temperatures were maintained at 4–10°C (39–50°F) and ventilation was usually provided via a convection (passive) chimney. Single-chambered colonies were most often

An early bee cellar arrangement. The cellar was dark, and ventilated with a passive convection chimney.

selected for wintering and the necessary food stores were estimated to be 9 kg (20 lbs.). The colonies were placed on stands at least 15 cm (six in.) off the floor with a 15 cm (six in.) space between colonies. The colonies were arranged in rows back to back and three to five colonies high if singles, or two high if doubles. Top ventilation was usually provided and some beekeepers replaced the summer covers with porous material such as sacking, felt or straw.

Wintering honey bee colonies in a modified environment has been practised in cold regions in North America since the early 1920s. During the last thirty years advancements have made indoor wintering more popular and more successful. There have been changes in management practices, new economic conditions, and most importantly, new technologies which enabled beekeepers to set up control systems for fans, heaters and alarms.

Wintering colonies indoors in above-ground buildings became possible with the advent of readily available temperature and ventilation control equipment, electronic control systems and inexpensive insulating materials. High capacity air conditioners were first used in the early 1960s, and by the late 1960s were being used by some beekeepers in the mid-western U.S. The use of high performance air conditioners which could control the temperature within a degree or two was fairly short-lived because the units were noisy, costly to maintain at a precise temperature, and produced tremendous air flow rates which often dried out the bees and reduced relative humidity to less than 25% in the building. In addition the units were often not large enough to control the temperature when colonies were first moved into the building or when spring temperatures rose during the day. Air conditioning was replaced by the use of exhaust fans in conjunction with multiple speed fans in plastic duct distribution systems, or with ceiling fans, all of which mixed the air to remove excess heat and prevent CO_2 build-up.

By the mid 1970s many beekeepers began indoor wintering in western Canada, Quebec and the northern and mid-western U.S. Interest in indoor wintering developed due to increased costs of package bees and the growing interest by beekeepers in self-sufficiency. In addition, some producers were interested in wintering smaller units and nuclei, which was possible in a modified environment provided by indoor wintering.

Colony basics – The most important aspect of successful indoor wintering is

Feeding sugar syrup in the fall. (Nelson photo)

the health and stores of the colony to be wintered. Many aspects of the building such as ventilation and temperature are important. However, without a strong, queenright, and disease- and pest-free colony, success will be variable and usually poor. The physical aspects of indoor wintering will not sustain a poor colony during the winter. Colony selection, requeening (if necessary) and feeding should start in August. The use of sugar syrup to boost winter stores seems to give the best results. Some honeys, particularly from canola, (which tends to granulate quickly) may cause stress and dysentery.

Many colonies wintered indoors are started as small colonies (from splits) with a queen cell in mid- to late June. The queen mates during good weather when there are excess drones, and colonies become strong by mid-August with a young queen. In most years, a queen excluder and super are placed on each unit and a super or more of honey can be obtained. Allowing colonies to store excess honey is important so that egg-laying by the queen is not restricted in the late summer, as these bees are very important to colony strength and wintering success. Five- and six-frame nuclei can also be wintered, but special care is required to have a good queen and young bees. Often the biggest problem with nuclei is the inability of the bees to vent excess metabolic moisture causing the cluster to become damp and unable to maintain temperature.

Single-chambered colonies should have an average gross weight (including the hive equipment) of 43 kg (95 lb), which will provide approximately 25 kg (55 lb) of stores. The average consumption is between 9 and 14 kg (20 and 30 lb) over the winter period. Double-chambered colonies should have an average gross weight of 61 kg (135 lb), which will provide approximately 32 kg (70 lb) of stores. The average consumption of stores is between 16 and 23 kg (35 and 50 lb) over the winter period.

Facilities and ventilation – Wintering facilities vary greatly. Facilities used for indoor wintering include insulated garages, honey hot rooms, poultry barns, as well as specially constructed winter/storage facilities. These facilities can winter as few as 20 colonies or as many as 5000. A general rule for the size of facility is to provide 15 ft^3 (0.4 m^3) for each single chambered colony and 25 ft^3 (0.7 m^3) for each double-chambered colony. These estimates assume that the colonies will be stacked about four high as singles and two high as doubles and allow space for aisles.

Indoor wintering facility in Canada. (Nelson photo)

Telephone and alarm for monitoring atmosphere of indoor wintering facility. (Nelson photo)

Bank of three exhaust fans set for different temperatures. (Nelson photo)

Wintering colonies indoors has been reduced to a few simple considerations: temperature control, air circulation and ventilation and light exclusion (use of light traps on all ventilation and fan ports). Honey bee colonies give off heat and metabolic by-products (water and CO_2). The ventilation system must remove these by-products by exhausting heat and CO_2 and bringing in fresh outdoor air. The temperature and humidity of the air are controlled for the survival of the bees.

The rate of air exchange varies with the time of year and the outside temperature. In the fall and spring as bees produce more heat, more outside air is required to maintain an even temperature and to keep the bees calm. During the winter when outside temperatures are cold, less air exchange is required to control the temperature. However, the moisture and CO_2 still must be exhausted, so a low level of continuous ventilation is required. The use of multiple-speed fans located in different parts of the building set to either control temperature or to run for a given period for air exchange can produce the desired result. A general recommendation for air exchange requirements during the winter is based

Two-story hives on pallets in indoor wintering facility. (Nelson photo)

One-story hives in indoor wintering facility. (Nelson photo)

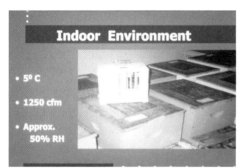

Hygro-thermograph to record humidity and temperature. (Nelson photo)

Ventilation tubes for bringing in fresh air. (Nelson photo)

on 0.1 cfm (cubic feet per minute) per lb of bees and 1.5 cfm per lb of bees during the fall and spring. An alternate method is based on the number of brood chambers in the building. Using this approach, the recommended rate is 0.5 cfm per brood chamber during the winter and 9.0 cfm per brood chamber during the autumn and spring.

Various temperatures have been evaluated for indoor wintering. In general, temperatures of 4 to 8°C (39–46°F) have been found to be optimum during the winter period. In early fall and in the spring it may not be possible to maintain these temperatures with air exchange alone. Some beekeepers have used heat pumps, pits full of stones or underground pipes (3–5 ft [1–2 m] below ground level) to provide cool air when temperatures increase above about 12°C (54°F). Most facilities are equipped to provide for up to 600–1000 watts of heat per 100 colonies using baseboard heaters or a source of added heat into the air mixing chamber for the building.

Relative humidity is usually not controlled, but some studies indicate that 50–60% is suitable. Because of the high volume of air exchanged during periods of extreme cold, humidity will drop much lower.

Comparisons of bee losses were made in buildings where the temperatures were maintained at 5°C (41°F) and 9°C (48°F). A steady increase in the weight of dead bees was observed from the first to the last month. Over the entire winter period (150 days), there were 1.3 lb (0.6 kg) and 2.2 lb (1.0 kg) of dead bees per colony at 5°C and 9°C, respectively. Several studies have shown that more bees are lost (30 to 50%) during the winter at warmer temperatures. Thus, 3–5°C (37–41°F) is recommended to minimize bee loss. The lower temperature also provides greater margins of safety in the event of warming trends or mechanical failures.

Moving colonies outside – Colonies are usually moved outside when weather conditions have improved and daytime highs are routinely 10-12°C (50-54°F). The bees usually become restless when the temperatures are higher in the building than outside. Depending on ground conditions (i.e. snow or mud) some producers move 1/3 to 1/2 of the colonies outside which allows the air exchange system to reduce temperatures because of the increased air volume in the building per remaining colony. This is strictly a judgment call. Eventually one learns what is workable in their particular region. In most of the northern regions of the U.S. and Canada this moving date is about April 5–10th (plus or minus a few days). The most important point to remember is not to move the colonies out while there is still snow on the ground. The snow may result in greater than normal bee losses as the bees cannot orient themselves.

Summary – Wintering of single- and double-chambered colonies can be very successful when the basic colony requirements are met; populous, queenright, and disease- and pest-free colonies with ample stores. The wintering facility needs to be completely dark and with the means to provide sufficient air exchange to cool the building to about 4°C (39°F) and to keep the air fresh. Colonies can be moved outside when weather conditions warrant but never when there is snow on the ground in the immediate vicinity of the hives. For additional reading see (McCutcheon, D. "Indoor Wintering of Hives." *Bee World* 65(1):19-37. 1984. *Beekeeping in Western Canada.* Edited by John Gruszka. Alberta Agriculture, 7000 –113 St., Edmonton, AB Canada T6H 5T6 . 172 pages. 1998. Nelson, D.L. Population Dynamics of Indoor Wintered Colonies. Proceedings, 36th International Apiculture Congress. Sept. 12–17/99, Vancouver, Canada. 309 pages, 1999. Nelson, D.L. Indoor wintering: Outline of basic requirements. NRG Pub. No. 821. Beaverlodge Research Station, Beaverlodge, Alberta. (2 pp.). 1982. Nelson, D.L. and G.D. Henn. Indoor Wintering: Research Highlights. NRG No. 77-10 Beaverlodge Research Station, Beaverlodge, Alberta. (9 pages. 1977.) - DLN

WINTER STORES, CRYSTALLIZATION OF – A sometimes serious problem in the northern states is that honey to be used by the bees for winter food may crystallize in the comb in the late autumn and winter. This is especially true of goldenrod, aster and a few other honeys. How rapidly a honey crystallizes depends in part on the amount of glucose present in the honey since the sugar glucose, not fructose, forms hard, white crystals. Thus, it is really only about half of the honey that forms these crystals and is not available to the bees. The winter cluster of bees forms under the honey stores, moves in an upward direction and consumes food. Crystallization is an important consideration in successful wintering. Beekeepers in the northern parts of the U. S. should be concerned about the quality of the food they leave for their bees for winter.

Bees can dissolve sugar crystals if they have water but in the winter this is usually not available to them. Not infrequently, in the spring and sometimes in colonies that have died, one will find open cells where the sugar crystals are exposed and the bees have removed and consumed the liquid part. Bees will often remove these crystals in the spring and dump then outside the hive.

In areas where the quality of winter food is likely to be poor, and the period of time during which the bees are confined is long, it may be advisable to feed sugar syrup in the late fall. Medication to help protect against nosema may be added to syrup. The bees usually store the syrup just above the place where the cluster will form and will be the first winter food the bees will consume. Honey consumption during the months of October, November and early December is usually not too great as this is the time of the year that bees have the least amount of brood. As the bees begin to rear brood in quantity and need much more food they will eat their way into the poorer quality food but this will happen in the spring when they will have access to water and can cope with poorer food. Feeding sugar syrup in late autumn may help with honeys that crystallize early and also with darker honeys containing excessive material that may cause dysentery.

WIRTH, ED. D. (Propolis Pete) – "Propolis Pete" whose interesting column appeared for many years in *Gleanings*, died in 1970, at Southold, Long Island, New York.

Ed. D. Wirth (his real name) possessed the rare ability which enabled him to write a monthly column for many years, yet each month had a new approach. His writings never once were stale or repetitious.

He was born in Brooklyn and started keeping bees on the roof of his home about 1930. After the bees swarmed, creating quite a show on a busy street, he moved them to his summer place. Here he added many colonies and faithfully studied the activity within the hives. His studies enabled him to write his authoritative articles for *Gleanings* which were sprinkled with humor, making them delightful.

He won quite a number of blue ribbons for his show honey which was sought by customers for miles around. He gave out advice and sold bee supplies.

WOOD PRESERVATIVES FOR BEEKEEPING EQUIPMENT –
Traditionally, most beekeepers have used a variety of methods to preserve their woodenware. This has included paint and hot wax and, in warm climates where the climate and insects (especially termites) are especially hard on hives, wood preservatives. The primary rule to follow to maximize woodenware life is to place hives on stands above ground to avoid moisture with subsequent decay, and termites. As more information becomes available concerning wood preservatives, it is clear that a beekeeper's choices are limited.

Painting beehives – Generally, two coats of a good quality latex paint are used on the outside only of woodenware used in beehive construction. The inside is not painted to allow the wood to absorb excess moisture produced by the bees. Special attention in painting hives must be paid to the joints and exposed end grain.

Caution – Most wood preservatives are classified as pesticides and may be injurious to humans or animals, plants, fish or other wildlife. All must be used according to the label that is the law. In the past some wood preservatives have contained traditional insecticides and labels should be read carefully and instructions followed.

WOODMAN, BAXTER (1908-2000) – Baxter Woodman was born July 15, 1908 in Grand Rapids, Michigan, and died at his home in Green Valley, Arizona, February 17, 2000.

Baxter's grandfather was a fruit grower who developed a sideline making bee equipment for other growers who had become aware that having colonies of bees in the orchard during bloom increased fruit yield. Baxter's father, A.G. Woodman, went full-time manufacturing and successfully built the business in Grand Rapids, Michigan. Over time they concentrated on metalware – tanks, extractors, uncapping equipment, smokers, etc. At one time they made a high percentage

Baxter Woodman

of the smokers used in the United States. The company usually operated two or three apiaries for testing and streamlining their equipment.

Baxter was brought up in this business and he gradually took control as A.G. retired. He regularly attended beekeepers' meetings and was well known and liked for his sense of humor and pleasant personality. He also contributed generously to door prizes.

About 1971 Baxter decided that it was time to retire, and Dadant bought the business. (Bert Martin)

WOODPECKERS – A small number of beekeepers have reported that woodpeckers have attacked their hives and made holes in the wood probably in an attempt to eat the bees within. Stacks of hive bodies and supers that have been stored outdoors, and colonies containing no bees, have been attacked by woodpeckers. Apparently, under these circumstances, they are after insects other than honey bees.

There is no good protection against woodpeckers. It is possible to cover holes they have made with wire mesh.

WOODROW, ALAN W. (1902-1987) – Dr. Alan W. Woodrow, a retired USDA apiculturist died February 20, 1987, in Tucson, Arizona. Born and raised in Xenia, Ohio, he received his B.S. degree from Ohio State University in 1927 and his Ph.D. in apiculture from Cornell University in 1935.

After graduation, he joined the USDA at its Intermountain States Bee Lab, Laramie, Wyoming, where he studied AFB and its transmission and effects. In 1936 he was transferred to the Pacific States Bee Culture Lab in Davis, California, where he began his research on pollination and bee behavior. In 1942, he was appointed apiculturist in charge at the U.S. Legume Seed Research Laboratory at Columbus, Ohio, where he was a member of a multidisciplinary team studying red clover seed production. In 1953 he joined the staff of the Bee Lab (now the Carl Hayden Bee Research Center) at Tucson, Arizona. Here he worked on pollination, foraging behavior, physiology and participated in the development of propionic anydride and butyric acid as substitutes for carbolic acid for repelling bees. He also carried on research on toxicity of pesticides, wax metabolism and other similar research problems. He also

Alan Woodrow

taught a course on the honey bee at the University of Arizona for a number of years. He retired in 1967.

WORKER (see ANATOMY AND MORPHOLOGY OF THE HONEY BEE; BEESWAX, Beeswax secretion by honey bees; BROOD; CASTES; COMMUNICATION; FANNING; FORAGING DISTANCE; FORAGERS; GLANDS ON THE HONEY BEE; GUARD BEES; LARVAE; LAYING WORKERS; NEPOTISM SENSES OF HONEY BEE; PHEROMONES) – Most of the bees in the hive are workers. These are also the bees we see on flowers, at water holes and gathering propolis. Except for egg laying, workers do all of the work in the hive. A worker's life is a series of chores, and, in effect, a worker graduates from one chore to the next as she ages (see LIFE STAGES OF THE HONEY BEE).

Worker honey bees are females but they are not fully developed. Workers are about half of the size and weight of queens, the only true females in a honey bee colony. Full growth in worker honey bees is inhibited by the food they are fed and the size of the cell in which they develop though the latter is much less important than the amount and quality of the food received. A worker honey bee lives only five or six weeks during the active season when she is required to fly since flying wears out body cells (see LENGTH OF LIFE OF THE HONEY BEE). During the winter season, when there is much less work to do, a worker may live for several months.

A worker bee's body, both internally and externally, is structurally very different from that of a queen or drone. Some of the major differences are the glands, especially in the head, that produce chemicals used to change nectar into honey and also to produce brood food. The crop is modified to carry nectar and water. The hind legs e designed to carry pollen and propolis. Honey bees are cold-blooded but their bodies are built so that they can generate heat by flexing their thoracic muscles. The worker bee's body is covered with branched hair (plumose) in which pollen is easily trapped and carried from one flower to the next. The bee's body and habits are designed for colony life; no bee, worker, drone or queen can live alone or even in a very small group. Under most circumstances a family of several thousand worker bees and a queen are needed for survival

WORKER CELLS – (see COMB, NATURAL)

WORKING A COLONY – Bees can indeed manage their own lives, but with careful, thoughtful management honey bees can be extremely productive pollinators and honey producers. Therefore, from time to time, beekeepers must examine colonies to determine the needs of the colonies.

Since bees will perform better with a minimum of disturbance, the very first step in working a colony is to have a plan in mind before even opening the colony. The main items that need to be observed are queen performance; food and its availability; and presence of disease, mites and small hive beetle. Seasonal and climatic conditions also enter into colony inspections.

Pre-inspection plans include being appropriately dressed and having smoker, smoker fuel and hive tool ready for use. The smoker should be well-stuffed and lit. Some type of record keeping is desirable. This can range from pen and paper to a small tape recorder or to a palm-type hand-held device to

Working A Colony

Work the colony from the side, avoiding the front.

After you remove frame "two" there is available space to move frames.

give data that can be entered into a computer. It is possible to bring a laptop computer into the apiary but caution must be taken to keep it clean of dust, honey, wax, propolis and other debris.

Hive placement will determine how easily you can inspect a colony. A hive that can be approached from the side is preferable since you are not interfering with bee flight at the front or causing disturbance from reaching over frames when entering at the back. Having a space for setting supers and hive bodies aside is essential.

If time permits, some minutes spent in observation of bees at the entrance of the hive will give some clues to any problems that may be found inside. A few puffs of smoke at the entrance and under the outer cover will initially confuse the bees, then cause them to eat some honey and remain calm. If possible, wait a minute before removing covers and entering the hive.

Oversmoking a colony will cause the bees to run aimlessly and become disorganized. Such a disruption can lead to stings. Good beekeepers have learned that quiet but deliberate movements do not disturb the bees.

The outer cover can be set on the ground or hive stand to serve as a platform for hive bodies and supers. If it is the only cover, inspect the underside for the queen. If a telescoping cover and inner cover are used, inspect the underside of the inner cover for the queen before setting this part aside.

If you are planning a thorough inspection of the colony, set the upper brood chambers aside and begin inspection with the bottom one. You will cause less disturbance of the colony.

As you stand facing the side of the hive, mentally number the frames starting with "one" closest to you. The frame you want to loosen with the hive tool and remove first is frame "two."

Remove frames by loosening them first then move them into the empty space made available and lift straight up, so as not to roll bees on either frame.

When bees are "looking at you," gently, apply smoke to head them back down and away from where you are working.

While holding this frame over the hive inspect both sides of the comb and its bees for the queen. If she is not present and open brood is not present, this frame can be set aside to allow space to be made between frames. By leaving space your chances of killing bees and the queen are greatly diminished. Furthermore the space allows you to inspect frames without jarring the hive. Bees become defensive when vibrations and jarring become excessive.

If you are planning to determine queen performance you will want to separate frames until you reach the brood area where you can observe larvae, pupae and perhaps eggs. It is not necessary to see eggs and indeed they are difficult to see. Neither is it necessary to see the queen. If young larvae are visible in the cells you know a queen was present at least a few days before.

The color and condition of the larvae will be an indication of colony health. Pearly white larvae indicate good health. Brood cappings should appear relatively smooth with color varying with age from tan to a deeper brown. Only a few missed cells should be visible. The cells in the brood area should not show a spotty pattern of empty cells. Scattered brood may indicate a problem with either the queen or with disease.

During the main brood-rearing season the bees will raise drones. If about 10% of comb cells are drone size the colony is functioning well. However, if large areas of damaged comb have been built back with drone cells the frame should be marked for replacing comb with foundation when appropriate.

A few capped drone cells can be opened, the pupae pulled out and inspected for the telltale reddish varroa mites. Tracheal mite (*Acarapis woodi*) presence cannot be detected visually without a microscope. The presence of wax moth larvae would be a signal that the colony is weak. The presence of small hive beetle larvae indicates that control measures must be taken immediately since disturbance of a colony causes an explosion in beetle numbers.

Inspection of the brood area will give you information on disease (see DISEASES OF THE HONEY BEE). The expertise of an apiary inspector (if your state has such a service) or that of an experienced beekeeper can help diagnose diseases such as American foulbrood, European foulbrood and chalkbrood.

Look for the honey and pollen stores above the brood area and that sufficient pollen and honey, as well as empty comb for laying, are available.

Throughout the inspection, lightly smoke any bees that have gathered between the frames and "are looking at you." When inspection is finished, lightly smoke the tops of the frames in order to replace those you have removed. Smoke the edge of the boxes before replacing them. Shake the bees off the inner and outer covers to avoid crushing bees. Keep a record of the inspection. - AWH

WORLD TRADE IN HONEY – (see BEEKEEPING IN VARIOUS PARTS OF THE WORLD)

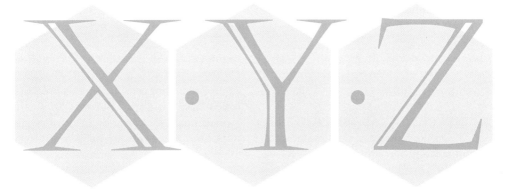

YEASTS IN HONEY – (see HONEY, and HONEY PRODUCTS; Fermentation of honey)

YELLOWJACKETS – (see HORNETS, YELLOWJACKETS AND WASPS)

YORK, HARVEY F. – Harvey F. York, Jr., 77, of Jesup, Georgia died August 12, 2002.

He was the owner and operator of York Bee Co., established by his father in 1924, an internationally known shipper of package bees and queen bees. He was a member and past president of the American Bee Breeders Association, and a member and former officer of the Southern States Beekeepers Federation, and a member of the American Beekeeping Federation and the Georgia Beekeepers Association. He was a contributing author to the 1975 edition of *The Hive and The Honey Bee*.

Mr. York was a close cooperator with Dr. G.H. (Bud) Cale in the commercial development and sales of the Dadant Starline and Midnite hybrid queen lines in the 1960s. Subsequently, he was a principle in Genetic Systems, which took over the hybrid lines from Dadant & Sons in 1976 and attempted to raise and sell instrumentally inseminated queens on a commercial basis.

During this time this large queen and package bee producer had between 10,000 and 12,000 colonies and about 12,000 queen mating nuclei, in three different areas of Georgia. He kept his three strains of bees separated; Italians in Jesup, Dadant's Starline Hybrids at Baxley, and Midnite Hybrids at Vidalia.

Eight employees were at each location. All colonies were kept on benches to make them easier to work.

Harvey York

The company manufactured all its own wooden goods except frames. An unusual feature was their continuing usage of pure beeswax in the manufacture of queen cell cups for forms making 100 cups.

ZBIKOWSKI, WLADYSLAW (1896-1977) – Zbikowski was the inventor of the plastic circular comb honey section. He was born in Beaver Falls, Pennsylvania, but educated in Russia and Poland. He began keeping bees in 1953, shortly after retiring from his practice of medicine. The following year he perfected the plastic rings and frames that are so widely used today with virtually no change from his original design. Experimentation with round sections made of wood and glass had been reported in *Gleanings in Bee Culture* in 1888 and 1889, but Dr. Zbikowski apparently conceived of his invention independently of any forerunners. He was one of the first to note that bees adapt more readily to plastic than to wood in building comb. However he apparently did not note the even greater advantage of his invention over the traditional rectangular wooden sections, which was that they conform to the bees' manner of comb building and thus, without corners to fill, result in a superior product. - RT

Biographies Of Noted Bee-Keepers

Believing that many of the ABC scholars would be interested in reading the biographical sketches of some of the prominent bee-men – men who have distinguished themselves in the line of apiculture – it is with no little pleasure that we now introduce them to you as far as it is possible to do so on paper. Dr. C.C. Miller, who, by reason of his natural fitness for the task, and who for long years has been more or less acquainted with the writings and doings of these men, has been detailed to write most of the sketches. The others are condensed from longer sketches that appeared in *Gleanings in Bee Culture*, written primarily by Ernest Root, unless otherwise noted.

Lorenzo Lorraine Langstroth

Lorenzo Lorraine Langstroth was born in Philadelphia, Pennslyvania, December 25, 1810. He graduated at Yale College in 1831, in which college he was tutor of mathematics from 1834 to 1836. After his graduation he pursued a theological course of study, and in May, 1836, became pastor of the Second Congregational Church in Andover, Massachusetts, which position ill health compelled him to resign in 1838. He was principal of the Abbott Female Academy, in Andover, in 1838-9, and in 1839 removed to Greenfield, Massachusetts, where he was principal of the High School for Young Ladies, from 1839 to 1844. In 1844 he became pastor of the Second Congregational Church in Greenfield; and after four years of labor here, ill health compelled his resignation. In 1848 he removed to Philadelphia, where he was principal of a school for young ladies from 1848 to 1852. In 1852 he returned to Greenfield; removed to Oxford, Ohio in 1858, and to Dayton, Ohio in 1887.

At an early age the boy Lorenzo showed a fondness for the study of insect-life; but "idle habits" in that direction were not encouraged by his matter-of-fact parents. In 1838 began his real interest in the honey-bee, when he purchased two stocks. No such help existed then as now, the first bee-journal in America being issued more than 20 years later, and Mr. Langstroth at that time had never heard of a book on bee culture; but before the second year of his bee-keeping he did meet with one, the author of which doubted the existence of a queen! But the study of bees fascinated him, and gave him the needed outdoor recreation while engaged in literary pursuits, and in the course of time he became possessed with the idea that it might be possible to so construct a hive that its contents in every part might be *easily* examined. He tried what had been invented in this direction, bars, slats, and the "leaf hive" of Huber. None of these, however, were satisfactory, and at length he conceived the idea of surrounding each comb with a frame of wood entirely detached from the walls of the hive, leaving at all parts, except the points of support, space enough between the frame and the hive for the passage of the bees. In 1852 the invention of the movable-comb hive was completed and the hive was patented October 5 of that year.

It is well known that, among the very many hives in use, no other make is more popular than the Langstroth, but it may not be so well known that, in a very important sense, every hive in use among intelligent bee-keepers is a Langstroth; that is, it contains the most important feature of the Langstroth – **the movable comb**. Those who have entered the field of apiculture within a few years may faintly imagine but can hardly realize what bee-keeping would be today, if, throughout the world, in every bee-hive, the combs should suddenly become immovably fixed, never again to be taken out of the hive, only as they were broken or cut out. Yet exactly that condition of affairs existed through all the centuries of bee-keeping up to the time when, to take out every comb and return again to the hive without injury to the colony, was made possible by the inventive genius of Mr. Langstroth.

As a writer, Mr. Langstroth took a high place. "Langstroth on the Hive and Honey-bee," published in May, 1853, is considered a classic; and any contribution from the pen of its author to the columns of the bee-journals was read with eagerness. Instead of amassing the fortune one would think he so richly deserved, Mr. Langstroth died not worth a dollar. He sowed, others reaped. At the date of his invention he had about 20 colonies of bees, and never exceeded 125.

In August, 1836, Mr. Langstroth was married to Miss Anna M. Tucker, who died in January, 1873. He had three children. The oldest, a son, died of consumption, contracted in the army. Two daughters still survive.

After his twentieth year, Mr. Langstroth suffered from severe attacks of "head trouble" of a strange and distressing character. During these attacks, which lasted from six months to more than year (in one case two years), he was unable to write or even converse, and he viewed with aversion any reference to those subjects which particularly delighted him at other times. Mr. Langstroth was a man of fine presence, simple and unostentatious in manner, cheerful, courteous, and a charming conversationalist.

The father of American bee-keeping has left the scenes of his labor. His death was entirely in keeping with his holy life. While administering the Lord's supper on Sunday morning, October 6, 1895, in his place of worship, in Dayton, Ohio, he died in his chair, without any previous warning. His last words were concerning the goodness of God, and were a fitting termination to one of the most exemplary and useful lives this world ever produced.

Although many years have passed since the death of father Langstroth, his impressive personality still lingers among us, inciting us, by the recollection of his struggles, to the attainment of a higher life.

Moses Quinby

Moses Quinby was born April 16, 1810, in Westchester County, New York. While a boy he went to Greene County and in 1853 from thence to St. Johnsville, Montgomery County, New York, where he remained till the time of his death, May 27, 1875.

Mr. Quinby was reared among Quakers, and from his earliest years was ever the same cordial, straightforward, and earnest person. He had no special advantages in the way of obtaining an education, but he was an original thinker, and of that investigating turn of mind which is always sure to educate itself, even without books or schools.

When about 20 years old he secured for the first time, as his own individual possession, sufficient capital to invest in a stock of bees, and no doubt felt enthusiastic in looking forward hopefully to a good run of "luck" in the way of swarms, so that he could soon "take up" some by the aid of the brimstone pit. But "killing the goose that laid the golden egg" did not commend itself to his better judgment, and he was not slow to adopt the better way of placing boxes on the top of the hive, with holes for the ascent of the bees, and these boxes he improved by substituting glass for wood in the sides, thus making a long stride in the matter of the appearance of the marketable product. With little outside help, but with plenty of unexplored territory, his investigating mind had plenty of scope for operation, and he made a diligent study of bees and their habits. All the books he could obtain were earnestly studied, and every thing taught therein carefully tested. The many crudities and inaccuracies contained in them were sifted out as chaff, and, after 17 years practical experience in handling and studying the bees themselves as well as the books, he was not merely a bee-keeper but a bee-master; and with that philanthropic character which made him always willing to impart to others, he decided to give them, at the expense of a few hours' reading, what had cost him years to obtain, and in 1853 the first edition of "*Mysteries of Bee-Keeping Explained*" made its appearance.

Thoroughly practical in character and vigorous in style, it at once won its way to popularity. From the year 1853, excepting the interest he took in his fruits and his trout-pond, his attention

Moses Quinby

was wholly given to bees, and he was owner or half-owner of from 600 to 1200 colonies, raising large crops of honey. On the advent of the movable frame and Italian bees, they were at once adopted by him, and in 1862 he reduced the number of his colonies, and turned his attention more particularly to rearing and selling Italian bees and queens. In 1865 he published a revised edition of his book, giving therein the added experience of 12 years. He wrote much for agricultural and other papers, his writings being always of the same sensible and practical character. The Northeastern Bee-Keepers' Association, a body whose deliberations have always been of importance, owed its origin to Mr. Quinby, who was for years its honored president – perhaps it is better to say its *honoring* president, for it was no little honor, even to so important a society, to have such a man as president. In 1871 Mr. Quinby was president of the North American Bee-Keeper's Association.

It is not at all impossible that the fact that so many intelligent bee-keepers are found in New York, is largely due to there being such a man as Mr. Quinby in their midst. The high reverence in which he was always held by the bee-keepers, particularly those who knew him best, says much, not only for the bee-master, but for the man.

Professor A.J. Cook

Albert J. Cook was born August 30, 1842, at Owosso, Michigan. Those who are intimately acquainted with the man will not be surprised to learn that his parents were thoroughly upright Christians. The daily reading of the Bible with comments by the father, re-enforced by the constant example of a chaste, honest, and industrious daily life, left its impress for life on the character of the son.

At the age of 15 he entered Michigan Agricultural College, where he graduated at 20, having been obliged during his course to suffer the sharp disappointment of suspending study a whole year on account of sickness, his health always having been rather delicate during his earlier years. Upon his graduation he went, on account of poor health, to California, where for three years he labored very successfully as a teacher. He then studied a portion of two years at Harvard University and Harvard Medical College with Agassiz, Hazen, and Dr. O.W. Holmes as teachers. In 1866 he was appointed instructor at Michigan Agricultural College, and in 1868 Professor of Entomology and Zoology in the same college

He has done and is doing a work unique in character, for he instructs the students, not only about insects in general, but about bees in particular. Every student that graduates goes all over the theory of bees, studies the bee structurally from tip of tongue to tip of sting, and goes through with all the manipulations of the apiary – that is, if there is any honey to manipulate; handles the bees, clips queens, prepares and puts on sections, extracts, etc. Probably in no other institution in the country, if in the world, is this done.

Prof. Cook is an active and influential member of the North American Bee-Keepers' Association, of which he has been president; was one of the originators of the Michigan State Bee-Keepers' Association, of which he was president for a number of years, and helped start the State Horticultural Society, being a member of its board for some years. He is widely known as a writer. His "Manual of the Apiary" has reached a sale of 15,000 copies, and "Injurious Insects of Michigan" 3,000 copies. He is also the author of "Maple Sugar and the Sugar-Bush," of which 5,000 copies have been published. He has written much for bee-journals, as also for the general press. He is a clear, practical writer, with a happy style.

In the battle waged against insect foes, he has rendered valuable service. Remedies which he first advised are now common, and he was probably the first to demonstrate the efficacy and safety of Paris Green for codling moth.

Lyman C. Root

Lyman C. Root was born in St. Lawrence County, New York, December 19, 1840. The better part of his education was obtained in "brush college" but before entering this he had two terms in the academy, two in St. Lawrence University, and a course in Eastman's Business College, where he graduated in 1865. The eight years following he was with Mr. Quinby, for the last five years his partner. It was his high privilege to be associated with him during what may be called the transition period of modern bee-keeping; during the time of the most rapid changes from box to frame hives; the time of the dissemination of the Italian bee, the introduction of the honey-extractor, the invention of the Quinby bee-smoker, the adoption of the one-comb section, and the perfecting of the new Quinby frame and hive. The various experiments that ended in the adoption of comb foundation were then in progress, and Mr. Quinby could have had no young man with him more enthusiastic and more helpful than the energetic L.C. Root, who released him from business cares, and gave him the needed leisure for study and invention. These were golden days for Mr. Quinby, well improved; and for Mr. Root nothing less, as he recalls the results obtained. Their supply-business rapidly grew to large proportions, and it was common for them to buy from three to five hundred colonies in box hives in the spring, transfer them to the new hive, and sell them to their customers in the different States. This necessitated a very large amount of exhausting work; but at this time Mr. Root knew nothing of sparing himself, and often did in one day what the average man would have taken two days for accomplishing.

In 1873 it was discovered that a rest was needed, and in the fall of that year he retired from the partnership and removed to Mohawk. But it seems impossible for a man of his temperament to rest, and we shortly find him extending his bee-business, going out in the early morning with his assistants to a bee-yard half a dozen miles away, and returning late at night with from two to three or more thousand pounds of extracted honey – the same process to be repeated the next day.

After the death of Mr. Quinby, Mr. Root took his supply-business. To all of this must be added his literary work as regular contributor to the *American Agricultural* and the *Country Gentleman*, with frequent articles to all the bee-journals of the country; his presidency of the North American Bee-Society, and of the Northeastern Association, with his long and laborious exertions in establishing the latter, and finally his re-writing Mr. Quinby's book – a task on which he expended a greater amount of careful, conscientious work, and which caused him greater anxiety, than though it had been entirely his own. For this last work Mr. Root was peculiarly fitted by his long residence with Mr. Quinby, and knowledge of his methods.

In keeping bees Mr. Root has preferred to raise extracted honey, and to keep about forty colonies in a yard. His crop was usually as much per yard as his neighbors' who kept twice the number in a place. The most of this success was due to skillful manipulations, improved honey-gatherers, and wise selection of locations; but after subtracting all these there probably remains something to be credited to moderate-sized yards. One fall he put into the cellar at the Hildreth yard forty stocks, took the same out in the spring without the loss of a single colony, and produced from them 9,727 lbs. of extracted honey, 4,103 lbs. of which was gathered in just seven days. Is better evidence needed that the author of the "New Bee-Keeping" is a practical bee-keeper?

P. H. Elwood, *Gleanings*, June, 1888

A.E. Manum

Augustin E. Manum was born in Waitsfield, Vermont, March 18, 1839. When the war broke out he enlisted in Co. G, 14th Vermont regiment, as a nine-months' man. He served at the battle of Gettysburg, where his comrades in line one either side were killed; his own gun was shattered, and he was hit four times.

In March, 1870, a friend desired to lend him "Quinby's Mysteries of Bee-keeping." Reading the book, his enthusiasm upon the subject was kindled, and he immediately purchased four colonies of bees and began the study of apiculture. Having a natural aptitude for the business, and a love for the bees, he was successful from the first. His apiary so rapidly increased, that, at the end of four years, when he had 165 colonies, he sold out his harness-business and began the pursuit as a specialist.

Since 1884 Mr. Manum has devoted all his energies to the production of comb honey, increasing his plant until his bees now number over 700 colonies in eight apiaries. He always winters his bees out of doors, packed in the "Bristol" chaff hive. For the eight years previous to 1887, his average loss in wintering for the entire time was only 3½ per cent. He uses exclusively a frame about 12¾ x 10 inches, outside measure, which he considers the best for practical purposes in his apiaries. His hive, the "Bristol," is almost entirely his own invention, being specially adapted to the perfect working of the system upon which his bees are managed. In 1885 his production was 44,000 pounds of comb honey, an average of 93¼ pounds per colony, all made in twelve days from basswood.

Because of the failure of the honey sources the past season, about 14,000 pounds of sugar syrup was fed the bees to prepare them for winter. He still has much faith in the pursuit, although the past three successive poor honey years have told heavily upon his enthusiasm.

J.H. Larrabee, *Gleanings,* page 301, Vol. XVII., 1889

Edwin France

Edwin France, of Platteville, Wisconsin, is noted as a producer of extracted honey on a large scale. He was born in Herkimer Co., New York, February 4, 1824. His father was a furnace-man, molding and melting iron; and, having a large family to support, had difficulty in making both ends meet. At the age of eight, young Edwin was sent to live with his mother's brother, returning home at 16. He then served an apprenticeship of four years at the furnace, when his father bought forty acres of timber, which they cleared up as a farm, working at the furnace winters. At the age of 24 his father died, leaving him the main stay of the family. He gave up the furnace, and worked part of the time making salt-barrels summers, and cutting sawlogs winters. About this time he got, and kept on this little place in the woods, a few hives of bees.

At the age of 32 he took the "Western fever," and settled on a 200-acre prairie farm in Humboldt Co., Iowa, marrying and taking with him a wife, leaving his mother in care of her older brother, a single man, amply able to care for her. Here again he kept a few bees. He lived here six years, farming summers and trapping winters, when the breaking-out of the war brought prices of farm products down to a ruinous point, and he went on a visit to Platteville, Wisconsin, intending to return when times brightened. Desiring some employment, he answered an advertisement, "Agents wanted, to sell patent bee-hives," and was soon the owner of the patent for his county. He made the hives himself; and as at that time nearly every farmer kept bees, the business paid well, and he soon bought two more counties. In his trades he got some bees, his starting-point as a bee-keeper. These he increased until 1871, when he went into winter quarters with 123 colonies, bringing out 25 in the spring, and 14 in the spring following. Enlarging his hives, and studying the wants of the bees, led to better success, reaching 500 colonies in the spring of 1888, kept in six apiaries. In 1886, from 395 colonies he took 42,489 lbs. of honey, increasing to 597. In 1885 his 323 colonies averaged 113 lbs. each. He owns eleven acres in the city limits of Platteville, devoted to garden truck and berries.

Mr. France and his son do all the work, except during a few weeks in the busy season, when he hires eight assistants from 12 to 18 years old. The whole then go to one of the different apiaries each day, making a sort of picnic, and returning at night. Mr. F. has not written much for the press; but what he has written bears the marks of ripe experience.

Gilbert M. Doolittle

Gilbert M. Doolitle was born April 14, 1846, in Onondaga Co., New York, not far from the home of his later years at Borodino, New York. During his childhood he often did duty by watching swarms from 10 to 3 o'clock, and at the age of eight was given a second swarm for the hiving. A thief, however, emptied the hive of its contents; and as foul brood prevailed in that region during several of the succeeding years it was not till the spring of 1869 he laid the foundation of his present apiary by purchasing two colonies of bees. Like many others he commenced with great enthusiasm, diligently studying all the books and papers obtainable, but, unlike many others, he has never allowed his enthusiasm to die out, and is to-day a diligent student of the ways of the busy bee. It is rare to find any one so familiar with what has been done and written relative to bee-keeping. As a business, Mr. D. has made bee-keeping a success, although he has never kept a large number of colonies, principally if not wholly because he prefers to keep no more than he can manage without outside help. In 1886 he wrote in the *American Bee Journal*, "From less than 50 colonies of bees (spring count) I have cleared over $1,000 each year for the past 13 years, taken as an average. I have not hired 13 days' labor in that time in the apiary, nor had any apprentices or students to do the work for me, although I have had many applications from those who wished to spend a season with me. Besides my labor with the bees, I take care of my garden and a small farm (29 acres); have charge of my father's estate, run my own shop a n d steam-engine, sawing sections, hives, honey-crates, etc., for myself and my neighbors; write for seven different papers, and answer a host of correspondence."

Mr. D. works for comb honey, and also makes quite a business of rearing queens for sale. Although a prolific writer, his fund of information never seems exhausted, and he is uniformly practical and interesting. His writings give evidence of the close and careful thinker.

G.M. Doolittle

Charles Dadant & Son

Charles Dadant was born in a village of the old province of Champagne (now department of Haute Marne), France, May 22, 1817. When a young man he was a traveling agent for a dry-goods firm, and afterward became a wholesale dry-good merchant himself, subsequently leaving this business to associate himself with his father-in-law in the management of a tannery. In 1863 he came to the United States, intending to make a business of grape-growing, with which business he had been familiar from childhood, as it was the leading business of his native place. He did not know a word of English at this time; but by the aid of a dictionary he became acquainted with it, so that, four years later, he could write articles for the papers, but he never learned to pronounce English correctly.

In 1864, a love for bees, which had shown itself in childhood, asserted itself anew and he obtained two hives of bees from a friend. After trying movable-frame hives side by side with the old European "eke" horizontally divided hives, the latter were cast aside, and in 1868 he tried to get the French apiarists to try the Langstroth system, but was rebuked by M. Hamet, the editor of a French bee-journal, who has never ceased trying to fight against the invading progress of movable fames, although other bee-magazines have started in France which have done the work he might so well have done. About this time Mr. D. tried to import bees from Italy. In 1873 he went in person to Italy, but was not entirely successful till 1874, when he succeeded in importing 253 queens. These importations were kept up for years. Earlier, in 1871 he started an out-apiary, and steadily increased the number of his colonies from year to year. In 1874 he took into partnership his son, Camille P. Dadant, then 23 years old, who had been raised in the business. Since 1876 they have kept five apiaries, of 60 to 120 colonies each. They have built up a large trade in extracted honey – the product of their bees in 1884 having been 36,000 lbs. Messrs. Dadant & Son are among the largest, if not the largest, manufacturers of comb foundation in the world. Commencing with 500 lbs. in 1878, they reached in 1884 the enormous amount of 59,000 lbs. Both father and son have written no little for the American press. Mr. C. Dadant is better known as a writer for European publications, and has been one of the main expounders of American methods in Europe; and the Langstroth-Quinby-Dadant hive, introduced by him into the Old World, is largely used under the name of the Dadant hive. He published a *Petit Cours d'Apiculture Pratique* in 1874, in France. To him was committed the task of preparing a revised edition of Langstroth's book, and this he has also translated for publication in the French language.

D.A. Jones

Most prominent among the bee-keepers of Canada is Mr. D.A. Jones, of Beeton, Ontario. If for no other reason, his name deserves a place in the history of bee-keeping as the man who undertook to scour foreign lands and the isles of the seas for new races of bees. Few would have undertaken such a daring enterprise as that of Mr. Jones, when, in 1879, he set out in person, at great expense, and amid dangers and exposures, visited Cyprus and Palestine in search of the races of bees which he not only sought but found. As a fitting adjunct to this undertaking he established, on separate islands in the Georgian Bay, apiaries where the different races might be kept in purity, or crossed at will. Such things as these, of which the public enjoys the benefit, are usually undertaken by government; but Mr. Jones drew on his private purse, and estimates that he was poorer by several thousands dollars for the operation.

October 9, 1836, D.A. Jones was

D.A. Jones

born near Toronto, Canada. Until of age he worked on the farm with his father. He then engaged in different occupations, bringing up in Illinois about 1860, where he worked a few months with a stockman. In the fall of the same year he attended a large exhibition at Chicago, where he was intensely interested in seeing a man exhibiting the Langstroth hive, manipulating the combs covered with bees, and explaining the advantages of movable combs. Mr. Jones took measurements of the parts of the hive, a fresh interest being awakened, for his father had been a bee-keeper, and among his earliest recollections was that of being carried by his father to the hives to watch the bees. At the age of five he was fairly versed in what was then generally known as to the habits of bees; and before the age of fifteen he hunted and captured bees, without the aid of his father.

Mr. Jones married and settled in Beeton, where he engaged in merchandising, afterward becoming so much interested in real-estate affairs and improvement of his village that he sold out his store, and thus had leisure to gratify his taste for bees, and commenced with two colonies in Langstroth hives. Afterward he established a much larger store, became profitably interested in railroads and other matters, but still found time to give attention to bees, until his two colonies became several apiaries. He has built up a large trade in extracted honey, and has given great impetus to exhibitions of honey at fairs, especially in very small packages.

In 1878 he commenced in a small way to manufacture supplies, and about six years later built a large factory. In 1886 the business had grown to such proportions that a company was chartered, with the title, "The D.A. Jones Co., Limited," and a capital of $40,000.

The *Canadian Bee Journal*, the first dollar weekly in the world, is another child of Mr. Jones, in which he may justly take pride.

Thomas G. Newman

For 15 years the *American Bee Journal* has remained under the management of one man; and, aside from being edited, its general make-up and clean typographical appearance impress one strongly, that, somewhere connected with it, is a man who is well up in the art preservative of all arts. The secret of it is, that Thomas Gabriel Newman, its proprietor, is himself a thorough practical printer. Born near Bridgewater, in Southwestern England, September 26, 1833, he was left fatherless at ten years of age, with three older brothers and a sister, the mother being a penniless widow by reason of the father's endorsing for a large sum.

The boys were all put out to work to help support the family. Thomas G. chose the trade of printer and bookbinder, serving an apprenticeship of seven years, and learning thoroughly every inch of the business from top to bottom, in both branches.

Early in 1854 he came to Rochester, New York, where he had relatives; and before noon of the day of his arrival he secured a permanent situation in the job-room of the *American*. Within two months he took the position of assistant foreman on the *Rochester Democrat*, then the leading Republican paper of Wester, New York. Later on he spent seven years editing and publishing a religious paper, called the "Bible Expositor and Millennial Harbiner," in New York, and published a score or more of theological works, some written by himself. In 1864 he moved it to Illinois, sold out the business, and, for a "rest," took his family to England. Returning in 1869 he located at Cedar Rapids, Iowa, where he published and edited its first daily paper. In 1872 he sold this and removed to Chicago, where he embarked in the business of publishing *The Illustrated Journal*, a literary serial printed in the highest style of the art, and magnificently embellished. The panic of 1873 ruined this luxury, bringing upon him a loss of over $20,000. It was revived in 1889 under the name of the *Illustrated Home Journal*.

In 1879 he went to Europe, at his own expense, as American representative to the various bee-keepers' societies, and attended conventions in England, France, Italy, Austria, Germany, etc., and was awarded several gold medals for exhibitions of American apiarian implements. He has been elected an honorary member of 14 bee-keepers' associations, and is also life member of the North American Bee-Keepers' Society (of which he was twice elected president), and treasurer of the Northwestern Bee-Keeper's Association.

In 1885 he was elected the first manager of the National Bee-Keepers' Union, which, under his management, has successfully defended a number of bee-keepers in suits at law brought against them. His successive re-election each year gives evidence of the satisfactory manner in which he has performed the duties of his office.

Mrs. Lucinda Harrison

Among women, no bee-keeper is more widely or favorably known than Mrs. Lucinda Harrison. Born in Coshocton, Ohio, November 21, 1831, she came, in 1836, to Peoria Co., Illinois, her parents, Alpheus Richardson and wife, being pioneer settlers. Public schools in Peoria at that time were underdeveloped, and educational advantages few; but her parents gave her the best that could then be had in private schools. Her brother Sanford was a member of the first class that graduated from Knox College, Galesburg, Illinois, and she then spent a year at an academy taught by him at Granville, Illinois. She taught school from time to time till 1855, when she married Robert Dodds, a prosperous farmer of Woodford Co., Illinois, who died two years later, leaving her a widow at 25. In 1866 she married Lovell Harrison, one of the substantial citizens of Peoria, from that time making Peoria her home.

Mrs. Harrison thus describes her entrance into the ranks of bee-keepers:

"In 1871, while perusing the Reports of the Department of Agriculture, I came across a flowery essay on bee culture, from the graceful pen of Mrs. Ellen Tupper. I caught the bee-fever so badly that I could hardly survive until the spring, when I purchased two colonies of Italians of the late Adam Grimm. The bees were in eight-frame Langstroth hives, and we still continue to use hives exactly similar to those then purchased. I bought the bees without my husband's knowledge, knowing full well that he would forbid me if he knew it, and many were the curtain lectures I received for purchasing such troublesome stock. One reason for his hostility was that I kept continually pulling the hives to pieces to see what the bees were at, and kept them on the war-path. Our home is on three city lots, and at the time I commenced bee-keeping our trees and vines were just coming into bearing, and Mr. Harrison enjoyed very much being out among his pets, and occasionally had an escort of scolding bees. Meeting with opposition made me all the more determined to succeed. 'Nothing succeeds like success.' I never wavered in my fixed determination to know all there was to know about honey-bees; and I was too inquisitive, prying into their domestic affairs, which made them so very irritable."

Her perseverance was rewarded. In time Mr. H. ceased opposition, became himself interested in the bees, and helped take care of them, saying he believed that bee-keeping would add ten years to their life. For a number of years her apiary has contained about 100 colonies, she being prevented from doing as much with the bees as she otherwise would, by ill health and family cares; for, although childless herself, she has been a mother to several orphan children.

Mrs. H. is best known as a writer, her many contributions to the press being marked by vigor and originality, with a blunt candor that assures one of her sincerity. She has been bee-editor of the *Prairie Farmer* since 1876, and has written for Colman's *Rural World*, and occasionally for other papers. She has held important offices in the North American Bee-Keeper's Association., and also in other societies. She credits bee-keeping with making life more enjoyable, opening up a new world, and making her more observant of plants and flowers.

Dr. C.C. Miller

One among the very few who make bee-keeping their sole business is Dr. C.C. Miller, of Marengo, Illinois. He was born June 10, 1831, at Ligonier, Pennsylvania. With a spirit of independence, and a good deal of self-denial sometimes bordering upon hardship, young Miller worked his way through school, graduating at Union College, Schenectady, New York, at the age of 22. Unlike many boys who go through college self-supported, running into debt at the end of their course, our young friend graduated with a surplus of some seventy odd dollars, over and above his current expenses at school; but, as we shall presently see, it was at the expense of an otherwise strong constitution. He did not know then, as he does now, the importance of observing the laws of health. Instead of taking rest he immediately took a course in medicine, graduating from the University of Michigan at the age of 25. After settling down to practice, poor health, he says, coupled with a nervous anxiety as to his fitness for the position, drove him from the field in a year. He then clerked, traveled, and taught. He had a natural talent for music, which by hard study he so developed that he is now one of the finest musicians in the country. If you will refer to the preface to Root's Curriculum for the Piano (a work, by the way, which is possessed or known in almost every household where music is appreciated), you will see that this same Dr. Miller rendered 'Much and important aid" to the author in his work. In this he wrote much of the fingering; and before the Curriculum was given to the printers for the last time, Mr. Root submitted the revised proofs to the doctor for final correction.

His musical compositions are simple

and delightful and you would be surprised to learn that one or two of the songs which are somewhat known were composed by Dr. Miller. Speaking of two songs composed by friend M. especially to be sung at a bee-keepers' convention. Dr. Geo. F. Root, than whom no one now living is better able to judge, said, "They are characteristic and good." Dr. Miller also spent about a year as music agent, helping to get up the first Cincinnati Musical Festival in 1873, under Theodore Thomas. Dr. M. is a fine singer, and delights all who hear him. Upon hearing and knowing of his almost exceptional talents for music, we are unavoidably led to wonder why he should now devote his attention solely to bee keeping; and this wonder is increased when we learn that he has had salaries offered by music publishing houses which would dazzle the eyes of most of us. But he says he prefers God's pure air, good health, and a good appetite, accompanied with a smaller income among the bees, to a larger salary indoors with attendant poor health.

As has been the case with a good many others, the doctor's first acquaintance with bees was through his wife, who in 1861, secured a runaway swarm in a sugar-barrel. A natural hobbyist, he at once became interested in bees. As he studied and worked with them he gradually grew into a bee-keeper, against the advice and wishes of his friends. In 1878 he made bee-keeping his sole business. He now keeps from 200 to 400 colonies, in four out apiaries. All the colonies are run for comb honey, and his annual products run up into the tons. He is intensely practical, and an enthusiast on all that pertains to his chosen pursuit. Though somewhat conservative as to the practicability of "new things," he is ever ready to cast aside the old and adopt the new, providing it has real merit. Although he claims no originality, either of ideas or of invention, he has nevertheless given to the bee-keeping world not a few useful hints, and has likewise improved devices or inventions otherwise impracticable.

As a writer he is conversational, terse, and right to the point. Not unfrequently his style betrays here and there glimmerings of fun, which he seems, in consequence of his jolly good nature, unable to suppress. His book, "Year Among the Bees" (see Book Notices), his large correspondence for the bee-journals, and his biographical sketches preceding this, as also his writings elsewhere in this work, are all characteristic of his style.

Ernest Root

Dr. John Dzierzon

Probably few readers of English have come across this name for the first time without stopping to look at it in order to ascertain what to call it. The Germans have had the same difficulty, and got around it by calling it Tseer-tsone; and as this pronunciation is pretty well established, perhaps it would be well to stick to it. There is little doubt, however, that it should be looked at from a Polish standpoint, and called Jeer-zone. As a considerable part of this book is taken up in the explanation of the genesis of the bee, and as this necessarily involves the theory which has made this man famous for all time, more as a naturalist than as a bee-keeper, I will not stop to dwell on what is now called the Dzierzon theory – a theory so well established that it is no theory at all, more than is the rotundity of the earth.

This eminent man was born in

Lokowitz, Upper Silesia, January 16, 1811, just three weeks after the birth of Mr. Langstroth. Thus we see these two great lives starting out like two rivers at the same time, and running nearly parallel with each other for 85 years, each a worthy example for all time to come. Dr. Dzierzon, like Mr. Langstroth, plainly showed from his earliest youth a great love for the works of nature; and the most interesting of all things to him was the observation of the habits of bees, his father keeping a few in log skeps. He early manifested a deeply religious turn of mind, which his father cultivated with care, sending him to the public school at Pitschen.

In order to have more time and means to pursue this branch of natural history, Dr. Dzierzon chose the clerical profession, and here again is a remarkable similarity to the career of Langstroth.

The hives in vogue when Dr. Dzierzon was a boy were simply four-sided boxes, and with them he began apiculture in earnest in 1835, just as he began his pastorate in Karismarkt. He was not slow to discover the gross defects of such hives, and the first thing he did was to devise a straw cover which would not allow so much moisture to be precipitated as was the case with hives covered with boards alone. To quote a German writer, "In order that this straw cap might be lifted off without injury to the combs he put on as many inch-wide bars, spaced a finger-breadth apart, as were required to cover the hive. This being done, and the bees, having built regularly to these bars, he fastened to each bar a piece of comb saved from old hives. This was the first step toward the invention of movable combs, for thereby was the master enabled to remove from the hive each individual comb."

The idea of the mobility of frames being established, Dr. Dzierzon gave himself no rest in his desire to unlock the inner mysteries of the hive; and while working with the bees he cast a glance at them whenever he could. "By this research many other mysteries were cleared up – pre-eminent among which was one that revolutionized the teachings in natural history in certain classes of zoology, namely Parthenogenisis." In establishing this theory, the Italian bee was the chief factor. Its color itself forming a proof of the theory. Like most truths, the contrary of which has been held, this theory met with great opposition; but it was finally settled for all time by an appeal to the dissecting-knife and microscope. Dr. Dzierzon's triumph was so complete that he was soon decorated with medals of honor from different potentates and associations; but his natural modesty prompts him rather to conceal them than to display them.

Dr. John Dzierzon

Francis Huber

(In view of the many animated discussions that have been held in regard to the benefits arising from Huber's investigations, we deem it no more than fair to state that his efforts, as the writer of the article suggests, were directed mainly toward the habits of the bee rather than toward any particular method of securing large amounts of honey; but his labors, nevertheless, will always be held in very high esteem by the world at large.

The sketch below was written in the German language by Mr. T. Kellen, of Luxemburg, and first appeared in Gravenhorst's Illustrated Bee Journal. *– Ernest Root.)*

Francis Huber, by his investigations and researches in apiculture, did more to promote that science than all his predecessors who had employed themselves in the study of this interesting insect. It was his discoveries alone that marked that golden age in the history of apiculture which is destined to remain for all ages. Huber's observations are not only of the greatest importance of themselves, but wonderful for the manner in which they were all made; for Huber was blind.

This distinguished man was born in Geneva, July 2, 1750. He was the son of a prosperous and respectable family, which as early as the 17th century were celebrated for their knowledge of the arts and sciences. His father, John Huber (born in 1722, died in 1790), was well known on account of his attachment to the celebrated French philosopher Voltaire.

From his earliest youth Huber showed a passionate predilection for natural history, and he applied himself to study with such zeal as to endanger his health, so that at the age of fifteen the reflection of glary snow destroyed his sight. If ever a man utterly deplored the loss of eyesight, that man was Huber. But his misfortune did not hinder him from applying himself to the study of those insects for which he had an especial liking; namely, the bees. It was this little insect that turned the darkness

Francis Huber

of the investigator into day; for Huber was the first to see clearly into that domain which to the best eyes had previously remained in darkness.

Huber did not lose his vigor of mind, for he went forward in the study of bees; but he could do this only by the help of his wife, Marie Aimie Lullin; his niece, Miss Jurine, and, above all, his servant Burnens. He himself manifested the most untiring perseverance and the greatest ingenuity, so that, by Burnens' sagacity, all of Huber's experiments with bees were practically demonstrated. Miss Jurine, who loved natural history above all else, supplemented Huber's work all she could, fearing not to take up the dissecting-knife and microscope in his aid. She was the first after Swammerdam to demonstrate that worker-bees are females. She it was, too, who, with Huber, established the principles on which the sages of our century grounded the doctrine of parthenogenesis. Besides that, Miss Jurine was Huber's secretary, full of willingness and self-devotion. Every day she noted down the results of the new investigations, and she also wrote the letters which Huber dictated to Charles Bonnet and his friends, and imparted to him the results of his labors, and directed their attention to numerous questions relating to bees.

Huber's interest in bees was greatly enhanced by the researches and writings of Swammerdam, Reaumur, Schirach, and probably also of the celebrated Swiss bee-keeper Duchet de Remauffens, and the Messrs. Gelieu. As a conclusion to the investigations of these men, it was possible for him, in spite of his unfortunate surrounding, to add greatly to the realm of apiculture; hence we may not forget that he everywhere encouraged and helped others by the nobility of his life.

In his later days he lived retired, but in peace, at Lausanne, where he died, December 22, 1831, aged 81 years.

Huber's discoveries are known to scholars through his Letters to Charles Bonnet; and they made his name so celebrated in all Europe, and even in America, that for many years he was recognized as the greatest apicultural genius; and even yet Mr. Hamet calls him the greatest of the lovers of bees (*le plus grand des apiphiles*). It was in 1796 that his first epoch-making work was brought to light, bearing the title. *Nouvelles Observations sur les Abeilles* (New Observations on Bees). His son, Peter Huber, in 1814, published the work in two editions, and added thereto an appendix in regard to the origin of wax.

Huber's work is, not only on account of its contents, but for the peculiar circumstances under which it was first brought to light, entirely without parallel in scientific literature. The recognition it received was universal, so that, after the first appearance of the work, Huber was received into the French Academy of Sciences and other scientific bodies.

The *New Observations* was translated early into every European tongue. The Saxon commissariat Reim, in Dresden, translated it into German in 1798, and Pastor Kleine, of Luethorst, translated it again in 1856, and published another edition in 1869, with notes.

Huber, by his observations on the secrets of bee-life, made clear what the most sagacious and learned observers from the time of Aristotle and Aristomachus down to Swammerdam and Reaumur had sought for in vain; and it is to be the more regretted that some German bee-keepers of great influence, such as, for instance, Spitzner and Matuschka, gave him no recognition.

He gave interesting explanations in regard to the habits of bees, their respiration, the origin of wax, the construction of comb, etc. He confirmed Schirach's proposition, that, by a change in the mode of treatment and food of larval bees, queens could be reared from worker eggs, and showed, likewise, the influence which the cell exerts on the insect. He showed further, that not only the queen but a certain species of worker-bee could lay fertile eggs, and showed, likewise, the function of drones. In opposition to Braw, Hattorf, Contardi, Reaumur, and others, who held very peculiar opinions in regard to the fertilization of queens, Huber showed that the fertilization takes place outside of the hive, at the same time that drones are flying, and that the union is effected in the air, and that the queen, on her return from the flight, has adhering to her body the evidences of fertilization and that egg-laying takes place about 46 hours afterward. These and numerous other experiments he often proved in his works with the utmost exactness; and especially did he lay down the most important and interesting information in regard to feeding bees, their method of building, the leaf-hive, foul brood, etc., in his letters to an eminent apiculturist in Switzerland, Mr. C.F.P. Dubied. These eighteen very long letters of Huber, the first of which was dated October 12, 1800, and the last August 12, 1814, were written partly by

Huber himself, partly by his wife or daughter, to whom he dictated. So far as I know, this correspondence has never been translated into German.

Julius Hoffman

Julius Hoffman, the inventor of the frame bearing his name, and which has a wide sale throughout the United States, was born in Grottkau, province of Silesia, Prussia, October 25, 1838, only a few miles from where Dr. Dzierzon spent some of the best portion of his life among the bees. Indeed when a young man he visited the doctor and from this great bee-master he learned much. No wonder with such a Gamaliel for an instructor the name Hoffman has come to be known in almost every bee-man's home.

In 1862 Mr. Hoffman left his native country and settled in London, England. Four years later he came to America, settling in Brooklyn, New York, where he accepted employment in the organ and piano business, but bees he had to have, for his heart and soul were in the bee business. As there was hardly room in such a crowded city to keep many bees, he moved to Fort Plain, New York, where he settled in the spring of 1873. In a few years he had increased his stock of bees up to 400 colonies. At the time of the writer's visit in 1890 at his home he had something over 700, all of them on Hoffman frames, and in his quiet way had used them for ten or twelve years. He told me that, with these frames, he could handle two or three times as many colonies, with the same labor, as he could on the old-style unspaced Langstroth frames, and by way of proof he showed how he could handle such frames in lots of twos, threes, and fours, picking them all up at once; how he could slide them from one side of the hive to the other; how easily he could handle his colonies in halves or quarters instead of one frame at a time.

At this time (1890) nothing but the unspaced Langstroth frame was used by modern bee-keepers except, perhaps, the closed-end Quinby in certain portions of New York; and the idea of a self-spacing frame seemed to be utterly impracticable in the mind of the average bee-keeper; but so thoroughly impressed was I with the importance and the value of the Hoffman self spacing frame as a labor-saver that I came home and wrote it up for our journal, *Gleanings in Bee Culture*. We soon adopted it in our yards; and my own personal experience with it showed it could be handled rapidly and easily; and that for many operations in the apiary much valuable time could be saved with it over the handling of the old-style unspaced frames. The bee-keeping world, not sharing in my enthusiasm, took hold of the Hoffman slowly at first; but now, 1902, twelve years later, it has come to be the leader in nearly all the hives put out by the principal manufacturers of the United States. These frames are discussed at length in this edition of the ABC. It should be stated, perhaps, that while the Hoffman frame of today differs in detail from the original frame the main self-spacing feature has been retained.

H.R. Boardman

H.R. Boardman was born April 2, 1834, in Swanzey, New Hampshire, and at about one year of age he was taken to what was then the wilderness West, and during nearly all his life his present place of residence, East Townsend, Ohio, has been his home. The district school was his only college, unless we take into account the opportunities for development afforded by an acquaintance with the wild woods, abounding in deer, turkeys, and other wild game. Mr. Boardman says, "The wild woods have ever possessed a charm for me. The pages of Nature's great open book have furnished me much with which to make life pleasant; and it is this aesthetic taste, no doubt, that has led me to my present occupation of bee-keeping." Mr. B. has a cabinet of mounted specimens of birds, prepared by his own hands, in which he takes a pride next to that which he takes in his apiaries.

Mr. Boardman's training as a bee-keeper commenced at a very early age. His father was a bee-keeper of the old school and a very successful one. By means of box hives and the brimstone-pit he secured honey for the family table, and also some to sell, nearly every season. Later on, boxes were put on top, the boxes sealed around with lime mortar or moist clay, to exclude the light entirely, in order to induce the bees to commence work in them. One year his father bought 25 colonies of bees early in the season, away from home; and as there was no one to watch them at swarming time, he tiered them up by putting an empty hive over each colony, there being a hole through which the bees could pass into the hive above. In the fall the bees were brimstoned, and the honey hauled home, nearly a ton! Considerable *wild* honey was also obtained from the trees. The abundance of these wild bees before tame bees were abundant, suggested, Mr. B. thinks, that they were native.

Mr. B. is a careful observer, doing his own thinking, and adhering to plans which he has found successful. He produces comb honey, and keeps 400 or 500 colonies in four apiaries. He is remarkably successful in wintering. He aims to secure a moderate yield with moderate increase, and has thus carried on a profitable and increasing business.

W.Z. Hutchinson

W.Z. Hutchinson is one of the many, who, although born in the East, have spent in the West all of life that can be remembered. Born in Orleans Co., New York, February 17, 1851, he was taken, four years later, with his father's family, to the dense forests of Genesee Co., Michigan, where his father literally hewed out a farm. W.Z. had the full benefit of pioneer backwoods life; and although hunting, trapping, etc., had a full share of his time, his natural bent was toward machinery. This passion for machinery was, as he advanced in his "teens," put to practical use by building a turning-lathe, and beginning the manufacture of spinning-wheels and reels. These he continued to make for several years, peddling them out in the surrounding country. At eighteen he began teaching school winters. While thus "boarding around," a copy of King's "Text-Book" fell in his way. It was to him a revelation. He learned that the book's owner had about 50 colonies of bees down cellar, which he was not long in asking to see, and for the first time he looked upon a movable-comb hive – the American. The next season, in swarming time, he visited this friend, and the charms of bee-keeping appeared greater than those of any other business. Although not really owning a bee till the lapse of many months, he became then and there in spirit a bee-keeper, reading all he could find on the subject, and visiting bee-keepers. The introduction of woolen-factories compelled him to abandon the spinning-wheel trade and one afternoon in June, while peddling out his last lot, he made a sale to a farmer about 16 miles from home; and although it was only about four o'clock, he begged to be allowed to stay all night, urged thereto by the sight of a long row of brightly painted hives. This bee-keeper had an only daughter, and the reader can weave his own romance, upon being told that the father, Mr. Clark Simpson, became the father-in-law of Mr. Hutchinson.

In 1877 he began bee-keeping with four colonies, and an excellent theoretical knowledge of the business. Mr. H. has never kept a very large number of colonies, but has made a comfortable living by the sale of comb honey. In 1877 he moved from Rogersville to Flint, MI, where he established the *Bee-Keeper's Review*, which fills a place not previously occupied, and is edited with the ability that might be expected from one who has been so favorably known through his many articles published in the bee-journals.

George W. York

George W. York is better known as the editor of a bee-journal than as a bee-keeper. To edit and publish each week a journal in so able a manner as that in which Mr. York edits and publishes the *American Bee Journal* leaves time for bee-keeping on only a very limited scale.

George Washington York was born February 21, 1862, at Mount Union, Stark Co., Ohio, where his father John B., was completing his studies at Mount Union College.

In 1882 he was graduated from the commercial department of Mount Union College, and continued there for a time as instructor in penmanship, mathematics, and book-keeping. A subsequent engagement at the same school he had first taught led to acquaintance with T.G. Newman, editor and publisher of the *American Bee Journal*, and on April 1, 1884, Mr. York went to Chicago to work in any part of Mr. Newman's business or in that of his son (a supply-dealer) in which they might desire his services. That ranged from sweeping out the office to reading proof, including setting type, washing the windows, acting as shipping clerk, etc. It was precisely the training to fit him for the position he has so well filled these later years. His remarkable memory soon made him as good as a cyclopedia to his employer, who could depend upon him for names, addresses, or to find any item that had appeared in the journal. In an editorial in 1892, Mr. Newman said, "Step by step he advanced to positions of responsibility and confidence, until, during our late and long-continued indisposition, he has had the entire editorial management of this journal."

At this date, 1892, Mr. York bought out the journal, almost his sole capital being his experience, having enough to pay for a third and going in debt for the rest. Six years saw him clear of debt, and seven with a subscription list 40 per cent larger than when he took it.

A very pleasing manner, united with real executive ability, makes his office work move without friction, a strong bond uniting together his office force in unusual loyalty to the employer. His constant study is for some fresh improvement for his beloved journal. The clock-work regularity of its weekly appearance is something remarkable.

Since 1878 an active worker in the M.E. Church, he has been prominent in Sunday-school and League work, and his wife and he, both good singers, have rendered efficient service with their voices. He is an officer in the church at

Ravenswood (a suburb of Chicago, where he has a delightful home), and since 1896 superintendent of its Sunday-school of 600 members. For two years in succession he was honored with the presidency of the North American Bee-keeper's Association, which office he has filled with the same characteristic faithfulness and energy that have marked his career as editor and publisher.

O.O. Poppleton

O.O. Poppleton was born near Green Springs, Seneca Co., Ohio, June 8th, 1843. When four years old his parents removed to Napoleon, Henry Co., Ohio, where, two years later, his father died, leaving his mother a widow with two sons, in straitened circumstances. Two years later his mother married Mr. Joseph George, of Clyde, Ohio, and settled in Sandusky Co., Ohio, After living there a few years the great inducements of the West influenced his stepfather to move to Northern Iowa, where they settled in Chickasaw Co., when Mr. Poppleton was 12 years of age. This was his home until 1887, when he removed to Florida on account of his health.

In October, 1861, he enlisted as a private in the 7th Iowa infantry, and re-enlisted as a veteran in 1863. In February, 1864, he was promoted to a lieutenancy in the 111th U.S.C. Inf., and a few months later he was made regimental adjutant. It was while performing the duties of this office, and also at the same time those of post-adjutant at Murfreesboro, Tennessee, that overwork resulted in the eye-trouble that has so seriously affected his health ever since, and which compelled the refusal of an excellent offer of employment at the time of mustering out. He served his country faithfully for five years; and though he received no scar upon his body, yet the smell of smoke was strong upon his garments. He was in several hard-fought battles, and taken prisoner once, but was held only a few weeks, when he was released or exchanged.

On account of having such poor health he made no effort to do a large business, but confined himself to a simple apiary varying from 75 to 150 colonies, spring count, and to the almost exclusive production of extracted honey. For the last ten years that he lived in Iowa, his annual crop of honey averaged 110 lbs. per colony. His half-brother Mr. F.W. George, has had charge of his apiary since his removal to Florida.

He now practices migratory bee-keeping. The bees are loaded on a raft, and drawn from one pasturage to another by means of a gasoline-launch.

Some fourteen or fifteen years ago he discovered the value of chaff as a winter protection for bees, without knowing that any one else, notably Mr. J.H. Townley, of Michigan, had previously made the same discovery. He also invented the solar wax-extractor about the same time. He was vice-president for several years of the North American Bee-Keeper's Association.; president of the Iowa Bee-Keeper's Society., and honorary member of the Michigan State Bee-keeper's Society.

Mrs. M. George

Eugene Secor

Eugene Secor was born in Putnam Co., New York, in 1841, and it was his good fortune to be kept there on a farm until he attained his majority. In 1862 he went to Iowa, entering Cornell College at Mount Vernon. A brother who was county treasurer and recorder, as well as postmaster, enlisted to hold up his country's flag, and Eugene abandoned his college course to take charge of his brother's business, thus occupying two years. Had his health been more robust, he probably would have borne his brother company in the army.

Asked what his business is, aside from bee-keeping, Mr. Secor replies, "When the bees are not swarming, and no public duty calls me, I 'recreate' by running a real-estate and abstract office in the daytime, and writing for the papers at night."

In spite of his special interest in apiculture he has a leading hand in agricultural matters, having organized the agricultural society of his county (Winnebago), of which society he was president for two years, and in 1888 he was elected by the State legislature one of the board of trustees of the State Agricultural College, to serve a term of six years.

The State Horticultural Society showed its appreciation of his services by re-electing him as president thereof and giving him charge of one of its experiment stations. The State Bee-keepers' Society elected him president in 1891 and 1892.

In 1896 Mr. Eugene Secor was elected to the position of General Manager of the United States Bee-keepers' Union. When the National and the United States Bee-keepers' unions were amalgamated into what is now known as the National Bee-keepers' Association, Mr. Secor, by the act of amalgamation, was made General Manager of the new organization. Ever since, he has been unanimously re-elected to the position. The office has required tact, executive ability, and general business qualifications. In all of these points Mr. Secor has filled the bill.

One of the strong characteristics of his make-up is geniality. There is no more popular bee-keeper in the ranks of all beedom than "Genial Gene." His popularity, coupled with his other qualifications, has placed him and maintained him in the position he has held with such credit.

John H. Martin

John H. Martin, better known, perhaps, as "Rambler," was born in the town of Hartford, New York, December 30, 1839, and died in Cuba, January 13, 1903.

In 1868 he married Miss Libbie C. Edwards, who died in 1881, leaving no children. She was an estimable lady, and her death was a great loss to all.

For many years Mr. Martin followed agricultural pursuits on his father's farm; but owing to a rather frail constitution, and the death of both his parents, he gave up the farm entirely; and bee culture, which had formerly been a side issue was given all his time and attention.

As early as 1874 we find him with 55 colonies of bees, and a contributor to *Gleanings in Bee Culture*. Since that time his apicultural career has been plainly indexed by his contributions to that journal. While he resided in New York it was his method to keep from 200 to 300 colonies, running them for extracted honey, and doing all the work himself, except during the extracting season. One season his crop was 16,000 lbs. of honey, and his average for twelve or fifteen years was about 7,000 lbs. of extracted honey per year. After the advent of the Heddon hive he adopted it and its methods, and the chaff hives and outdoor wintering were discarded.

In person Mr. Martin was quite tall and slender. There was not an ounce of spare flesh about him. In manner he was very modest and quiet; yet continually, through his eyes and his words, one could see the humor of the man. He had great love for the quaint and humorous side of humanity, yet his humor never offended by its coarseness nor galled by its acidity. The series of articles written during the last few years under the *nom de plume* of "Rambler," in *Gleanings in Bee Culture*, have made him known to bee-keepers generally. His method of combining the entertaining and instructive in a manner to make it read by all was very characteristic.

Mr. Martin's first article under his now favorite *nom de plume* was published in *Gleanings* for June 1, 1888. His first rambles covered the territory of Eastern New York, but they gradually enlarged till they took in the bordering States. But John was a rover, and could not be held down, and the circle of his wanderings kept on enlarging until he reached the land of the setting sun. From 1891 to 1902 his rambles were confined particularly to California, Washington, and Oregon. In 1902 he moved to Cuba, whence he sent out an interesting series of articles. Assisted by pencil and camera he made his travels among bee-keepers particularly graphic. Everywhere he went he was sure to be welcomed, and sometimes was recognized on sight by bee-keepers, even though they had never seen him. His long lank appearance, his striped pants, his characteristic long-tailed coat, his ever present umbrella and camera, were exhibited in a series of articles in hundreds of different poses. These, and his quaint way of writing, made him one of the most popular writers in bee culture. Indeed, he might almost be styled the Mark Twain of beedom.

Francis Danzenbaker

Francis Danzenbaker was born January 8, 1837, near Bridgeton, New Jersey. His interest in bees began at an early age. Being of an inventive turn of mind, he set to work experimenting with various improved devices; but it was not until he had been a bee-keeper for more than thirty years that he came prominently before the bee-keeping world, and that was in the summer of 1888. At this time he called the attention of the A.I. Root Co. to the value of the dovetailed or, more properly speaking, lock corner, in hives – a feature that the company subsequently adopted, since which time it is used universally by all the manufacturers of bee-keepers' supplies. While it was conceded at that time that this joint would be satisfactory for packing boxes. It was feared it would hardly be suitable to stand the weather. But experience during all these years has not only shown that it does stand, but it makes the strongest possible joint that can be devised outside of the true dovetail, a corner which would be impractical by reason of the expense of making.

And so Mr. Danzenbaker became prominent, not because he was an extensive bee-keeper, or produced large crops of honey, but because of the fact that he introduced a number of valuable improvements in hives outside of the one already mentioned – the lock corner.

A believer in thinner combs, he at first advocated sections 4½ inches square and 1-5/8 inches thick to those 4¼ square and 1-7/8 thick; but after having visited Capt. J.E. Hetherington, he became convinced that a box taller than broad was not only more artistic, and more in keeping with objects around us, but economized space in the hive, so that more sections could be used per super. These he subsequently adopted for his hive.

Later on he introduced his shallow-brood-chamber hive, and afterward discarded this for what he now calls the Danzenbaker, making use of closed-end frames, plain sections, and slatted separators or fences.

Mr. Danzenbaker is a firm believer in a 4x5 section, and has proven to his own satisfaction, and that of his friends and followers, that it is a better seller than the regular 4¼, looks handsomer, and is less liable to break during shipment.

He has traveled extensively over the country, visited many bee-keepers of note, with the view of bringing his hive to still greater perfection if possible. He has attended many conventions, is prominent in the discussions, and is ever the persistent advocate of closed-end frames, shallower hives, and taller sections. He is a user of neither tea nor coffee, liquor nor tobacco. Although 65 years old, when the faculties of most begin to fail, our friend is full of vigor and enthusiasm.

W.L. Porter

Mr. Porter is what may be called a specialist bee-keeper in the strictest sense – that is to say, his sole means of livelihood is derived from bees. He has had as many as 700 colonies, but I believe he is now operating only 500, located in five or six different apiaries. He produces both comb and extracted – about an equal amount of each.

Mr. Porter has had a varied experience in beekeeping. Born in West Virginia in 1850, he migrated with his parents northward to Michigan in 1864. His parents assumed the life of pioneers clearing off the forests. Young Porter, with the rest of the boys, was detained from school to help the father, and the consequence was their early school advantages were limited.

In a short time afterward we find him at the Michigan Agricultural College, under Prof. Cook, as a student. His general aptitude for the bee business resulted in his being placed in charge of the apiary of the Agricultural College. He used well his opportunities, and finally became the possessor of some bees of his own. He suffered many reverses, but made the best of some assistance to him financially in helping him through college. Ill health and a lack of funds finally compelled him to give up his course before he had completed it.

He subsequently drifted to Wisconsin, and formed a partnership with Miss Allyn – a partnership which he says was "very happy and successful." He very soon engaged in bee-keeping again, meeting with his usual success. But again ill health caused him and his wife to move to the land of gold, sunshine, and alfalfa honey in 1881 and here he has cast his lot and his fortune; and if I may judge from general appearances he has secured a fair share of the sunshine, of the alfalfa honey, and the gold which it brings.

Amos Ives Root

Up till the edition of this work had reached the 75th thousand, there had been no biographical sketch in it of A.I. Root. Now that the authorship of the present work has passed largely out of his hands, it seems appropriate in the book which he named, and which he originally wrote, that at least a brief sketch should appear.

A.I. Root was born in a log house, in December, 1839, about two miles north of the present manufacturing plant of The A.I. Root Company. He was a frail child and his parents had little hope of raising him to manhood, although some of the neighbors said his devoted mother would *not let him die*. As he grew older his taste for gardening and mechanics became apparent. Among his early hobbies were windmills, clocks, poultry, electricity, and chemistry – anything and everything in the mechanical line that would interest a boy who intensely loved machinery. Later on we find him experimenting in electricity and chemistry; and at 18 he is out on a lecturing-tour with a fully equipped electrical apparatus of *his own construction*.

We next find Mr. Root learning the jeweler's trade, and it was not long before he decided to go into business for himself. He accordingly went to an old gentleman who loaned money, and asked him if he would let him have a certain amount of money for a limited time. This friend agreed to lend him the amount, but he urgently advised him to wait a little and earn the money by working for wages. This practical piece of advice, coming as it did at the very beginning of his career, was indeed a God-send, and, unlike most boys, he decided to accept it. Imbued with a love for his work, and having indomitable push, he soon *earned* enough to make a start in business, without borrowing a dollar. The business prospered till A.I. Root & Co. were the largest manufacturers of *real* coin-silver jewelry in the country. From $200 to $500 worth of coin was made weekly into rings and chains, and the firm employed something like 15 or 20 men and women.

It was about this time in 1865 that a swarm of bees passed over his shop; but as this incident is given so fully in the introduction I omit it here. Not long after he became an ABC scholar himself in bees, he began to write for the *American Bee Journal* under the *nom de plume* of "Novice." In these papers he recounted a few of his successes and many of his failures with bees. His frank confession of his mistakes, his style of writing, so simple, clear, and clean-cut, brought him into prominence at once. So many inquiries came in that he was finally induced to start a bee-journal, entitled *Gleanings in Bee Culture*. Of this, how his business grew to such a size that the manufacturing plant alone covered five acres, and employed from 100 to 200 men – all this and more is told in the introduction by the writer.

As an inventor Mr. Root has occupied quite a unique field. He was the first to introduce the one-pound-section honey-box, of which something like 50,000,000 are now made annually. He made the first practical all-metal honey-extractor. This he very modestly styled the "Novice," a machine of which thousands have been made and are still made. Among his other inventions may be named the Simplicity hive, the Novice honey-knife, several reversible frames, and the metal-cornered frame. The last named was the only invention he ever patented, and this he subsequently gave to the world long before the patent expired.

In the line of horticultural tools he invented a number of useful little devices which he freely gave to the public. But the two inventions which he considers of the most value is one for storing up heat, like storing electricity in a storage battery, and another for disposing of sewage in rural districts. The first named is a system of storing up the heat from exhaust steam in Mother Earth in such a way that greenhouses and dwelling-houses can be heated, even after the engine has stopped at night, and for several days after. The other invention relates to a method of disposing of the sewage from indoor water-closets so that "Mother Earth," as he calls it, will take it automatically and convert it into plant life without the least danger to health or life, and that, too, for a period of years without attention from any one.

Some of the secrets of his success in business may be briefly summed up by saying that it was always his constant aim to send goods by return train, and to answer letters by return mail, although, of course, as the business continued to grow this became less and less practicable. He believed most emphatically in mixing business and religion – in conducting business on Christian principles; or, to adopt a modern phrase, doing business "as Jesus would do it." As might be expected, such a policy drew an immense clientage, for people far and wide believed in him. But how few, comparatively, in this busy world, go beyond the practice that honesty is the best policy! While A.I. Root believed in this good rule he did not think it went far enough, and accordingly, tried to adopt and live the Golden Rule.

The severe strain of long hours of work, together with constantly failing health, compelled Mr. Root to throw some of the responsibilities of the increasing business on his sons and sons-in-law. This was between 1880 and 1890. At no definite time could it be said that there was a formal transfer of the management of the supply business and the management of the bee department of *Gleanings* to his children; but as time went on they gradually assumed the control, leaving him free to engage in gardening and other rural pursuits, and for the last ten years he has given almost no attention to bees, devoting nearly all his time to travel and to lighter rural industries. He has written much on horticultural and agricultural subjects; indeed, it is probable that he has done more writing on these subjects than he ever did on bees.

For the last twenty-five years he has been writing a series of lay sermons, touching particularly on the subject of mixing business and religion, work and wages, and, in general, the great problem of capital and labor. An employer of labor he had here a large field for observation, and well has he made use of it. Perhaps no series of articles he ever wrote has elicited a more sympathetic response from his friends all over this wide world than these same talks; and through these he has been the means of bringing many a one into the fold of Christ.

Mr. Root, ever since his conversion, in 1875, has been a most active working Christian. No matter what the condition of his health, he is a regular attendant at church and prayer-meeting. He takes great interest in all lines of missionary work, and especially in the subject of temperance. He annually gives considerable sums of money to support the cause of missions, and to the Ohio Anti-saloon League; and now that the heavier responsibilities of the business have been lifted from his shoulders he is giving more and more of his time and attention to sociological problems. – *Ernest Root.*

Before the foregoing was given to the printers, my son Ernest asked me to take time to read it over, and there is just one thing I wish to add. Since the "boys" have kindly relieved me from business, and permitted me to take wheelrides to visit successful gardeners, fruit-growers, and bee-keepers, I have enjoyed my vacation fully as much as I ever enjoyed any work or play in my boyhood days. It has been suggested by some that, as we grow older, we lose interest in things around us. It has not been so in my case. In fact, I have thanked God again and again for the liberty that I now enjoy (in this the 62nd year of my age), of being able to take up and follow out, without interruption from business, any wonderful line of industry that I may come across or hear about in this busy world of ours. Many thanks to the younger members of our firm – not my two boys alone, but my two sons-in-law as well. – A.I. Root

GLOSSARY

A

AAPA – American Association of Professional Apiculturalists.

ABDOMEN – The posterior or third segmented region of the body of the bee that encloses the heart, honey stomach, stomach proper, intestines, sting and reproductive organs.

ABF – American Beekeeping Federation.

ABSCONDING SWARM – A swarm consisting of the entire colony including the queen; a result of unfavorable conditions.

ABSORBENTS – Porous materials placed on top of the hive to absorb moisture.

ACARAPIS WOODI – The scientific name of the parasitic mite that infects the tracheae of bees; common name: tracheal mite.

ACAROLOGY – The study of mites and ticks.

ACHROIA GRISELLA – The lesser wax moth, a minor pest of honeycomb.

AGE POLYETHISM – The changing of labor activities by colony members as they age.

AFB – American foulbrood.

AFRICANIZED BEE – The term used for the honey bee, *Apis mellifera scutellata*, introduced into Brazil and subsequently spread into North America.

AFTERSWARMS – Swarms that leave a colony with a virgin queen after the primary swarm of the same season has already left.

AHPA – American Honey Producers Association.

AIA – Apiary Inspectors of America.

ALARM ODOR – A pheromone (n-heptanone) given off by guard bees to alert the colony of danger; a pheromone (isopentyl acetate) released by a worker bee at time of stinging.

ALBINO – A mutant (bee) that lacks normal pigmentation and appears white.

ALIGHTING BOARD – The projection before the entrance to a hive to make it easier for bees to land.

ALLERGIC REACTION – A systemic or general reaction to some compound, such as bee venom, characterized by itching all over (hives), breathing difficulty, sneezing, rapid loss of blood pressure, loss of consciousness or death.

AMERICAN FOULBROOD – A contagious bacterial disease of bees that affects the larval and prepupal stages and is caused by *Paenibacillus larvae* subsp. *larvae*.

AMINO ACIDS – The building blocks of proteins.

AMOEBA DISEASE – caused by the unicellular animal *Malpighamoeba mellificae*.

ANTENNA (plural **ANTENNAE**) – Paired, slender, jointed taste, touch, smell and sound receptors on the head of insects.

ANTHER – In seed plants, the part of the stamen that develops and contains pollen.

APHID – An insect, belonging to the order Homoptera, that secretes a sugary liquid, termed honeydew, that bees collect like nectar.

APIARIST – A beekeeper.

APIARY – A collection of colonies of bees; also the yard or place where bees are kept. A beeyard.

APICULTURE – The science and art of raising honey bees for man's economic benefit.

APIS – The genus to which honey bees belong.

APIS CERANA – The scientific name of a cavity-nesting eastern honey bee. Other Asian honey bees in this group are: *Apis nuluensis, Apis koschevnikovi* and *Apis nigrocincta*.

APIS DORSATA – The scientific name of a giant Asian honey bee.

APIS FLOREA The scientific name of a dwarf Asian honey bee.

APIS MELLIFERA – The scientific name of the western honey bee.

APIS MELLIFERA MELLIFERA – The scientific name of the German black honey bee.

APIS MELLIFERA SCUTELLATA – The scientific name of an African honey bee.

APIS MELLIFICA – Antiquated scientific name for honey bees, replaced by *Apis mellifera*.

APISTAN® - Commercial name of fluvalinate, packaged in plastic strips, used for treatment of *Varroa* mites.

APITHERAPY – The use of hive products for therapeutic purposes; includes bee venom, honey, pollen, propolis, wax and royal jelly.

ARS – Agricultural Research Service.

ARTIFICIAL CELL CUP – see CELL CUP.

ARTIFICIAL INSEMINATION – The act of depositing semen into the oviduct of a queen by the use of instruments; also called Instrumental Insemination.

ARTIFICIAL PASTURAGE – Plants purposely cultivated for their nectar.

Glossary

ARTIFICIAL POLLEN – see POLLEN SUBSTITUTE.

ARTIFICIAL SWARM – A small colony of bees made by dividing a full colony.

ASH – The residue remaining after incineration analysis of honey. The main mineral constituents of honey ash are potassium, sulfur, and iron; other minerals are present in small quantities. The average value of ash in U.S. honey is about 0.17%.

B

BABY NUCLEUS – A miniature hive containing 200 to 300 bees used for the mating of queens. It is smaller than a normal nucleus.

BAIT HIVE – An empty hive or a commercial bait hive used to attract and capture swarms.

BALLING A QUEEN – For a variety of reasons, several to many workers will form a ball around a queen and attempt to bite and sting her. If the queen survives she may be injured and may be replaced.

BANKING A QUEEN – To place a caged queen or queens into a queenless colony in order to keep them until needed.

BEE BLOWER – A gas- or electrically-driven blower used to blow bees from supers full of honey.

BEE BREAD – A common name given to pollen stored in the comb. See POLLEN.

BEE BRUSH – A device for brushing bees from their combs.

BEE CELLAR – An underground room used for storing colonies during cold winter months.

BEE CULTURE – The care of bees. Also, the name of the beekeeping magazine published by the A.I. Root Company in Medina, Ohio. Formerly titled *Gleanings In Bee Culture*.

BEE DANCE – The worker bees in a normal colony perform various dance-like movements for communication. The dances indicate the direction, distance, and richness of food, and the desirability of new home sites.

BEE ESCAPE – A device to remove bees from supers or buildings, constructed to allow bees to pass through in one direction but prevent their return.

BEE GLOVES – Gloves worn to protect the hands from stings and from becoming sticky with propolis.

BEE GLUE – see PROPOLIS.

BEE GO® – A chemical repellent to bees and used with a fume board to clear bees from honey supers.

BEE GUM – A colloquial term meaning a hive of bees, usually a hollow-log hive.

BEEHIVE – A box or other container for holding a colony of bees. See HIVE.

BEE HOUSE A building constructed to house hives of bees, allowing them to fly and forage.

BEELINE – The shortest distance between two points; as the bee flies.

BEELINING – Tracking worker bees from flower sources back to their hive.

BEE LOUSE – A commensal, found chiefly on queens, young bees and drones. Comparatively harmless, the bee louse (*Braula coeca*) is a wingless fly (Order Diptera, Family Braulidae). Only a single species is known. The larvae cause damage to comb honey by burrowing in the cappings.

BEE METAMORPHOSIS – Bees exhibit complete metamorphosis: egg, larva, pupa, adult.

BEE MOTH – see WAX MOTH.

BEE PARALYSIS – A term that covers several diseases of the honey bee, characterized by various symptoms including the inability to fly.

BEE PASTURE – Plants and trees from which bees gather pollen and honey.

BEE PLANTS – Honey plants. Common plants which yield nectar available to honey bees in quantity sufficient to produce surplus honey.

BEE SPACE – An open space, about one cm or 3/8 inch, in which bees build no comb or deposit a minimum of propolis. It forms a passage between combs in a feral colony or between parts in a man-made hive.

BEE SUIT – A pair of coveralls, usually white, made for beekeepers to protect them from stings and keep their clothes clean; some come equipped with zip-on veils.

BEE TENT – Tent of screen or netting large enough to contain a hive and the operator in which bees may be manipulated without being troubled by robber bees.

BEE TREE – A hollow tree occupied by a colony of wild bees.

BEE VEIL – A net covering for protecting the head from the attack of bees.

BEE VENOM – The liquid secreted by special glands attached to the sting of the bee.

BEEYARD – see APIARY.

BEESWAX – The wax secreted by honey bees from eight glands within the ventral abdominal segments and used in building their combs. It is composed of about 300 different compounds. Bees may consume about seven pounds (three kg) of honey to secrete one pound (500 g) of beeswax.

BEEWAY SUPER – The shallowest section super used with wooden section boxes to make comb honey; has a built-in beeway or bee space. Rarely used today.

BLACK BEE – Also known as German black bee; the first bee brought to America. *Apis mellifera mellifera.*

BLACK SCALE – Refers to the appearance of a dried larva or pupa that died of a foulbrood disease.

BOARDMAN FEEDER – A metal or plastic base holding a jar that fits into the entrance of a beehive; used for feeding syrup or water.

BOTTLING TANK – A plastic or stainless steel tank equipped with a honey gate to fill honey jars.

BOTTOM BOARD – The floor of a beehive.

BOX HIVE – A plain box used for housing a colony of bees. Illegal in the U.S. because it does not have moveable frames.

BRACE COMB – The terms "brace comb" and "burr comb" are often used interchangeably. More exactly, a brace comb is comb built between two adjacent parts fastening them together; burr comb is excess comb built without connecting parts.

BRAULA COECA – see BEE LOUSE.

BREATHING PORES – see SPIRACLES and TRACHEAE.

BRIMSTONING – The old-time operation of killing a colony of bees with sulfur fumes; used in colonies without moveable frames where killing was the only way to remove the honey.

BROOD – Young developing bees in the egg, larval, and pupal stages, not yet emerged from their cells.

BROOD CHAMBER – That part of the hive in which the brood is reared and stores are maintained for the survival of the colony.

BROOD COMB – One of the combs in the brood chamber. See BROOD.

BROOD DISEASES – Diseases that affect only the immature stages of bees, such as American or European foulbrood.

BROOD FOUNDATION – Wired, heavier wax sheets or plastic embossed sheets used in frames in the brood nest.

BROOD NEST – That part of the brood chamber occupied by the various stages of developing brood.

BROOD REARING – Raising bees from the egg to the adult.

BRUSHED SWARM – An small colony made by brushing or shaking part of the bees of a colony into an empty hive to prevent a natural swarm. Sometimes referred to as a "shook swarm."

BUMBLE BEE – A large, hairy social bee of the genus *Bombus*.

BUCKFAST HYBRID - A strain of bees developed by Brother Adam at Buckfast Abbey in England, bred for disease resistance, disinclination to swarm, hardiness and comb building.

BURR COMB – see BRACE COMB.

C

CAGE, SHIPPING – Also called a package. A screened box filled with two to five pounds (1 to 2 kg) of bees, with or without a queen, and supplied with a feeder can; used to start a new colony or to boost a weak one.

CANDIED HONEY – See GRANULATED HONEY.

CANDY PLUG – A fondant-type candy placed in one end of a queen cage to delay her release.

CAP – (noun) The wax covering which closes cells containing brood or honey; the capping, the seal. (verb) To cover a cell with a capping; to seal.

CAPPED BROOD – see SEALED BROOD.

CAPPED HONEY – Combs of ripened honey in which the bees have closed a full cell with a thin layer of wax. Also called sealed honey.

CAPPINGS – The thin wax covering over honey; once cut off of extracting frames they are referred to as cappings and are a source of premium beeswax.

CAPPINGS SCRATCHER – A fork-like device used to remove wax cappings covering honey so it can be extracted.

CARBOHYDRATE – A food (organic compound) composed of carbon, hydrogen, and oxygen.

CARNIOLAN BEES – A grayish-black race of very gentle bees from southern Austria and Yugoslavia.

Glossary

CAST – A second swarm having a virgin queen.

CASTE – Workers and queens, the two types of female bees. Drones are not a caste.

CAUCASIAN BEES – A gentle race of black or dark-colored bees introduced into America from the Caucasus Mountains.

CELL – One of the hexagonal compartments of a honeycomb. Worker cells are approximately five per linear inch (20 per 10 cm), drone cells about four per inch (16 per 10 cm).

CELL CUP – A queen cell when it is only about as deep as it is wide. Artificial cell cups are made for raising queens.

CELL PROTECTOR – A receptacle made of wire or plastic that protects the sides of a queen cell from the attacks of bees, but leaves the apex of the cell uncovered.

CELLAR WINTERING – To winter bees by putting them in a cellar or sheltering building during the winter months.

CHALKBROOD – A disease affecting bee larvae cause by the fungus *Ascosphaera apis;* larvae eventually turn into hard, chalky white "mummies."

CHILLED BROOD – Larva that have died from being too cold.

CHORION – The membrane or shell covering the egg.

CHUNK HONEY – A type of honey pack containing one or more pieces of comb honey covered with liquid honey in the same container.

CLARIFYING – The removal of foreign particles from liquid honey or wax by straining, filtering or settling.

CLEANSING FLIGHT – The flight of the bees from the hive after long confinement in order to void their feces in the air.

CLIPPED QUEEN – A queen with a portion of one or more wings removed to prevent her flight.

CLUSTER – The hanging together of a large group of bees, one upon another, See WINTER CLUSTER.

COCOON – A thin silk-like covering secreted by larval honey bees in their cells in preparation for pupation.

COLONY – A community of bees having a queen, some thousands of workers, and during part of the year a number of drones; the bees that live together in a social unit.

COLOR COMPARATOR – A device used for grading the color of honey; water white, extra white, white, extra light amber, light amber, amber and dark amber.

COMB – see HONEYCOMB.

COMB BASKET – That part of a honey extractor in which the combs are held. See EXTRACTOR.

COMB CARRIER – A receptacle in which one or more combs may be placed and covered, so as to be easily carried and protected from robbing bees.

COMB, DRAWN – Wax foundation with the cell walls drawn out by the bees, completing the comb.

COMB FOUNDATION – Thin sheets of beeswax or plastic embossed to form a base on which the bees will construct a complete comb of cells.

COMB FOUNDATION MACHINE – A machine for embossing smooth sheets of wax.

COMB HONEY – Honey in the comb, not extracted.

COMMERCIAL BEEKEEPER - One whose business is beekeeping, including package and queen production, honey production, pollination, wax or other product production or some combination of these.

CONICAL ESCAPE – A cone-shaped bee escape that permits bees a one-way exit; used in a special escape board to free honey supers of bees.

CORBICULA – see POLLEN BASKET.

CORN SYRUP – Mixture of dextrin, maltose, glucose and water in nearly equal parts, formed by hydrolysis of cornstarch. Not suitable for bee feed.

CREAMED HONEY – Honey that has undergone controlled granulation to produce a very finely crystallized honey which spreads easily at room temperature. Also called cremed honey.

CROSS – When races of bees are bred together the resulting progeny is called a cross, e.g. cross breed, or hybrid.

CROSS-POLLINATION – The transfer of pollen from an anther of one plant to the stigma of a plant of different genetic makeup.

CRYSTALLIZATION – see GRANULATED HONEY.

CUT-COMB HONEY – Comb honey cut into various sizes, the edges drained and the pieces wrapped or packed individually.

CUTICLE – The exoskeleton of an insect, the waxy outermost layer.

D

DANCING – see BEE DANCE.
DEARTH – A period of time when there is no available forage for bees, due to weather conditions (rain, drought) or time of year.
DECOY HIVE – See BAIT HIVE.
DEEP BOX – Hive box that measures 9-5/8 inches (24.4 cm) high.
DEMAREE – A method of swarm control that consists of separating the brood from the queen.
DEQUEEN – To take the queen from a colony of bees.
DEXTROSE – A term no longer in use. See GLUCOSE.
DIASTASE – An enzyme that helps to convert starch to sugar.
DIPLOID – Having a complete set of chromosomes, as in workers and queens, having been produced from fertilized eggs.
DISEASE RESISTANCE – The ability of an organism to avoid a particular disease.
DIVIDING – Separating a colony to produce two or more colonies. See ARTIFICIAL SWARM.
DIVISION BOARD – Any device designed to separate two parts of a hive, making two separate units. Also called dummy board or follower board.
DIVISION BOARD FEEDER – A wooden or plastic container that is hung in a hive, taking the place of one or more frames and contains syrup for feeding.
DIVISION SCREEN – The size of the inner cover, a screened board used to separate two parts of a colony. Also called a Snelgrove board.
DOUBLE STORY – Referring to a beehive comprised of two hive bodies.
DOVETAILED HIVE – An incorrect term used to describe hive bodies having a box joint.
DRAWN COMBS – Completed brood or honey comb.
DRIFTING BEES – Bees that do not return to their own hive but enter other hives in an apiary.
DRONE – Male bee.
DRONE BROOD – Brood from unfertilized eggs that mature into drones, reared in larger cells than worker bees.
DRONE COMB – Comb having cells measuring about four cells to the linear inch (16 per 10 cm). Drones are reared in drone comb; honey can be stored in it, but not pollen.
DRONE CONGREGATION AREA (DCA) – A specific area to which drones fly to wait for virgin queens. Generally the areas are found around identifiable landscape marks, such as tree lines and land contours; DCAs persist from one year to the next.
DRONE LAYER – A queen that can lay only unfertilized eggs that develop into drones.
DRONE TRAP – see QUEEN TRAP.
DRUMMING – Pounding rhythmically on the sides of a hive to make the bees ascend into another hive placed over it.
DUMMY – A thin board of the same size as a frame with a top bar. See DIVISION BOARD.
DWARF HONEY BEES – A group of Asian dwarf honey bees: *Apis florea* and *Apis andreniformis*.
DWINDLING – The rapid dying-off of bees.
DYCE PROCESS – A patented process used to produce creamed honey; controlled crystallization.
DYSENTERY – The excessive discharge of fecal matter by the bees. Many conditions may contribute to this condition including fermented honey, nosema, and honeydew.
DZIERZON THEORY – The theory that honey bees are parthenogenic. See PARTHENOGENESIS.

E

ECLOSION – Hatching from the egg.
EGG – The first phase in the bee life cycle, usually laid by the queen, is the cylindrical egg 1/16 inch (1.6 mm) long; it is enclosed with a flexible membrane or chorion.
EMBED – To force wire into wax comb foundation by heat, pressure or both for the purpose of strengthening the resulting comb.
EMBEDDER – The tool, either mechanical or electrical, used to force wire into wax foundation for the purpose of adding strength to the finished comb.
EMERGING BROOD – Young bees in the act of gnawing their way out of their brood cells.
ENTRANCE – Any opening in the hive permitting the passage of bees. Standard hives have a bottom board entrance and may have other smaller openings above.
ENTRANCE REDUCER – A notched wooden strip for regulating the size of the bottom board entrance.

ENZYMES – Catalysts produced by both plants and animals, including the honey bee, essential to the chemical reactions in metabolic processes. See DIASTASE and INVERTASE.

ESCAPE – A device that allows bees to pass through an exit but not return. Used for removing bees from filled supers. See BEE ESCAPE.

ESCAPE BOARD – A board having one or more bee escapes in it; used to remove bees from supers.

EUROPEAN FOULBROOD – An infectious larval disease of bees caused by the bacterium *Melissococcus pluton*.

EXCLUDER – see QUEEN EXCLUDER.

EXTENDER PATTIES – Vegetable shortening mixed with sugar and Terramyacin®, formed into a patty, used in the hive to suppress, but not cure, the symptoms of American and European foulbrood. No longer recommended as a medication.

EXTRACTED HONEY – Honey that has been removed from the comb by an extractor.

EXTRACTING – The act of removing honey from the comb by means of a centrifugal extractor.

EXTRACTOR – A machine for removing honey from the comb, consisting of a round can in which is mounted a revolving reel carrying a set of combs from which the cappings have been removed. The honey is thrown out by centrifugal force without destroying the combs.

EYELETS, METAL – A small metal piece fitting into the wire holes of a frame's end bar, used to keep the reinforcing wires from cutting into the wood.

F

FECES – Excreta (of bees).

FEEDERS – Containers for feeding bees sugar syrup.

FERAL BEES – Wild, non-managed nest of honey bees, as in a tree.

FERMENTATION – A chemical breakdown of honey caused by sugar-tolerant yeasts. Associated with honey having a high moisture content.

FERTILE – A fertile queen is one that has mated with 12 to 20 drones and has a supply of spermatozoa in her spermatheca.

FERTILIZE – A queen's eggs that are to produce workers or queens are fertilized on their outward passage by receiving one or more of the spermatozoa contained in the spermatheca of the queen. Drone eggs are unfertilized.

FESTOONING – The activity of young bees, engorged with honey, hanging on to each other and secreting beeswax.

FIELD BEES – When worker bees become about 16 days old, they begin the work of flying to collect nectar, pollen, water and propolis and are then called field bees.

FILTERED HONEY – Honey that has been flash heated, passed through a micropore filter under slight pressure and immediately cooled to prevent crystal growth.

FLASH HEATER – A device for heating honey very rapidly to prevent it from being damaged by sustained periods of high temperature.

FLIGHT PATH – The direction bees fly when leaving their colony.

FOLLOWER BOARD – See DIVISION BOARD.

FONDANT – A soft candy used for feeding bees in winter or for queen or shipping cages. Made from sucrose and high fructose corn syrup. See QUEEN CANDY.

FOOD CHAMBER – A hive body filled with honey for winter stores.

FORAGE – (noun) Natural food source of bees (nectar and pollen) from wild and cultivated flowers; (verb) to search for food.

FOREIGN MATTER – Bee bodies, particles of wax, or other objectionable debris found in honey.

FOULBROOD – See AMERICAN FOULBROOD AND EUROPEAN FOULBROOD.

FOUNDATION – see COMB FOUNDATION.

FOUNDATION FASTENER – Several types of devices for fastening foundation in brood frames or sections.

FOUNDATION, WAX – Thin sheets of beeswax, with or without wires, embossed with the base of worker or drone cells on which bees will construct a complete comb (called drawn comb).

FOUNDATION, WIRED – Wax foundation with embedded evenly-spaced vertical wires for added support; used in brood or extracting frames.

FOUNDATION, UNWIRED – Wax foundation without vertical wires.

FOUNDATION, THIN OR THIN SURPLUS – Very thin unwired wax foundation used for comb honey.

FRAME – Four pieces of wood joined at the ends to form a rectangular shape for holding honey comb. It consists of one top bar with shoulders, one bottom bar and two end bars. A series of frames are held a bee space apart in a vertical position in a hive invented by Langstroth in 1851.

FRUCTOSE – One of the two main sugars of honey. The other is glucose.

FUMAGILLIN – An antibiotic used in the elimination of *Nosema apis*, and *Nosema ceranae* which causes nosema disease.

FUME BOARD – A top cover that has an absorbent pad on the underside, used with a liquid repellant that causes the bees to move from the honey super.

FUMIGATE – To submit beekeeping equipment to the fumes of a toxic chemical for the purpose of destroying organisms or pests.

G

GALLERIA MELLONELLA – The scientific name of the greater wax moth.

GAMETE – The mature sexual reproductive cell; the egg or the sperm.

GIANT BEES, *APIS DORSATA* – Native of India. One of the largest honey bee species in the world. They build huge combs in the open air, often from five to six feet (two meters) in length and from three to four feet (one meter) in width, which they attach to overhanging ledges of rock or to large limbs of trees. During periods of dearth they migrate to more favorable locations. Not kept in hives. Another Asian giant bee is *Apis laboriosa,*

GLOVES – Leather, cloth, plastic or rubber gloves worm while inspecting bees.

GLUCOSE – One of the two main sugars in honey. The other is fructose.

GLUCONIC ACID – The principal acid in honey.

GOLDEN BEES – Italian bees in which the workers show four to six bright yellow bands on their upper abdomens.

GRAFTING – The process of transferring a newly-hatched worker larva from its brood cell into special queen cups used for queen rearing.

GRAFTING TOOL – A needle, probe hook or scoop used for transferring the larva in grafting.

GRANULATED HONEY – Honey that has crystallized either naturally or by the Dyce process.

GREASE PATTY – Vegetable shortening mixed with sugar, formed into patties, and placed above brood area for control of tracheal mites.

GREEN HONEY – See UNRIPE HONEY and RIPE HONEY.

GROOMING – The cleaning of the body surfaces of one's self or nestmates by licking with the tongue and stroking with the legs

GUARD BEES – Worker bees about three weeks old, that have their maximum amount of alarm pheromone and venom; they challenge all incoming bees and other intruders.

GUM – A hollow log beehive.

GYNANDROMORPH – An abnormal bee having structural characteristics of worker (female) and drone (male).

H

HALF-DEPTH SUPERS – A honey super, height 6-5/8 inches (16.8 cm). See also SHALLOW SUPER.

HAPLOID – Having a single set of chromosomes, as a drone, that is produced from an unfertilized egg.

HAPLODIPLOIDY – The genetic condition whereby female workers and the queen have diploid body cells but males develop from unfertilized eggs so body cells are haploid.

HEAD – The front portion of an insect containing mouthparts, antennae, glands and eyes.

HIGH FRUCTOSE CORN SYRUP (HFCS) – A sugar syrup, made from corn, that contains glucose and fructose, used for feeding honey bees.

HIVASTAN® – A medicine used for the treatment of *Varroa* mites.

HIVE – (noun) Home for bees furnished by man. The modern hive includes a bottom board, cover, and one or more brood chambers and honey supers stacked one above the other. Inside each box or hive body is a set of moveable frames of comb or foundation held in a vertical position a bee space apart. (verb) To put a swarm or package of bees in a hive.

HIVE BEE – An adult worker who performs duties in the hive (cleaning, nurse duties, guarding) before becoming a field bee.

Glossary

HIVE BODY – A box of wood or plastic that holds 10 frames and serves as a home for bees. The "deep" size is 9-5/8 inches (24.4 cm) high and is usually used for brood. Shorter hive bodies are often used for honey storage.

HIVE ODOR – A distinctive smell characteristic of individual colonies that results from food odors and pheromones of the hive.

HIVE SCALE – Used to weigh beehives to monitor amount of weight gained or lost.

HIVE STAND – A structure serving as a base for a beehive; it helps extend the life of the bottom board by keeping it off damp ground, and makes a convenient working height.

HIVE STAPLES – Large U-shaped metal nails, hammered into the wooden hive parts to secure the hive boxes before moving a colony.

HIVE TOOL – A tempered metal device with a scraping surface at one end and a flat blade at the other, used to open hives, pry frames apart, clean the hive among other uses.

HOBBYIST BEEKEEPER – One who keeps bees for pleasure without intent to profit.

HOFFMAN FRAMES – Self-spacing frames having end bars wide enough at the top to provide the proper spacing when the frames are placed in contact.

HOMEOSTASIS – Refers to the characteristic of the bees that enables them to respond to the ambient temperature and maintain a constant temperature in the cluster.

HOME YARD – An apiary closest to the beekeeper's home.

HONEY – A sweet, viscous material produced by bees from the nectar of flowers, composed largely of a mixture of the two sugars glucose and fructose dissolved in about 17% water. It also contains very small amounts of sucrose, minerals, vitamins, proteins and enzymes.

HONEY BEE – A social, honey-producing bee in the class Insecta, order Hymenoptera, superfamily *Apoidea* and family *Apidae*. In 1758 Linnaeus named the honey bee *Apis mellifera* (honey-bearer), and three years later, 1761, changed the name to *Apis mellifica* (honey-maker). The American Entomological Society has ruled the former will be the correct scientific name for the honey bee. Races, subspecies or varieties of the honey bee are also distinguished by the names of the geographical localities in which they occur and from which they have been exported, such as Italian, Carniolan, Syrian, Cyprian, Banat, Caucasian and Tunisian.

HONEY COLOR – Honey colors are classified between water white to dark amber, with a total of seven gradations. Formerly graded by a Pfund grader, now by a digital color grader using Pfund units.

HONEYCOMB – The mass of hexagonal cells of wax built by honey bees in which they rear their young and store honey and pollen. The cells are built back to back with a common wall. See DRONE COMB and WORKER COMB.

HONEYDEW – A sugary liquid excreted by aphids and scale insects, usually in the order Homoptera. Bees will collect it and ripen it into a type of honey.

HONEY EVAPORATOR – A machine for removing water from honey.

HONEY EXTRACTOR – See EXTRACTOR.

HONEY FLOW – A time when nectar is plentiful and bees produce and store surplus honey.

HONEY GATE – A faucet used for drawing honey from drums, buckets or extractors.

HONEY HOUSE – A building used for honey extraction, storage, and bottling.

HONEY KNIFE – See UNCAPPING KNIFE.

HONEY PLANTS – Plants whose flowers (or other parts) yield enough nectar to produce a surplus of honey; examples are asters, basswood, citrus, eucalyptus, goldenrod and tupelo.

HONEY PUMP – A pump for moving honey from a honey extractor or tank into another tank.

HONEY SAC or **STOMACH** – An enlargement of the posterior (back) end of a bee's esophagus but lying in the front part of the abdomen, capable of expanding when filled with nectar or water.

HONEY SUMP – A baffle or clarifying tank into which honey from the extractor, uncapping knife and uncapped combs runs by gravity. It removes pieces of broken comb and particles of wax from the honey.

HONEY SUPERS – Hive boxes placed on top of the brood chamber and used for storage of surplus honey.

HORNET – The only true hornet in the U.S. is the imported European hornet, *Vespa crabro*. It is carnivorous and will prey on many insects including honey bees.

HOUSE APIARY – See BEE HOUSE.
HYBRIDS – The offspring resulting from a cross between different subspecies of honey bees.
HYGROSCOPIC – Refers to the ability of honey to absorb moisture from the air.
HYMENOPTERA – The scientific classification Order to which all bees belong, including ants, wasps and certain parasitic insects.
HYPERSENSITIVE – A condition in which reactions to any environmental stimulus, such as honey bee venom, is life-threatening.
HYPOPHARYNGEAL GLANDS – Located in the head of worker bees, a pair of organs that produce royal jelly and brood food.

I

INCREASE – To start new colonies with the purpose of adding to the total number of colonies by dividing established colonies, installing package bees, or hiving natural swarms.
INFERTILE – Incapable of producing a fertilized egg, as a laying worker.
INJECTIONS, DESENSITIZING – A series of injections given to persons with allergies, such as to bee venom, so they become immune.
INNER COVER – A cover fitting on top of the uppermost hive box but underneath the outer, telescoping cover, with an oblong or round hole in the center.
INSECTICIDE – Any chemical that kills insects.
INSPECTORS, STATE – Persons usually employed by state agriculture departments to inspect and regulate colonies of bees for diseases and pests.
INSTAR – In insects any stage between molts (casting off the outgrown skin) during the course of development.
INSTRUMENTAL INSEMINATION – The act of depositing semen into the oviduct of a queen by the use of a man-made instrument. See ARTIFICIAL INSEMINATION.
INTRODUCING – See QUEEN INTRODUCTION.
INTRODUCING CAGE – A small wood and wire cage (Benton) or plastic or wire cage used to transport queens and/or introduce and release queens into the colony.

INVERT SUGAR – Once used for feeding honey bees; replaced by high fructose corn syrup (HFCS).
INVERTASE – An enzyme produced by bees that converts the sucrose in nectar to glucose and fructose.
IPM – Integrated Pest Management.
ISLE OF WIGHT DISEASE – An outdated name for tracheal mite infestation. See TRACHEAL MITE.
ISOMERASE – A bacterial enzyme used to convert glucose in corn syrup into fructose.
ISOPENTYL ACETATE – The alarm odor produced by a worker bee's sting glands. Also known as isoamyl acetate.
ITALIAN BEES – *Apis mellifera ligustica*. The most common race of bees for honey production. They were first successfully introduced into the U.S. about 1860. The first three (sometimes four or five) dorsal segments of the abdomen are banded with yellow. Excellent honey and brood producers.

J

JUMBO FRAME – A frame of standard length but 11 inches (24.7 cm) in depth.
JUMBO HIVE – A standard Langstroth hive but having a depth of 11-3/4 inches (26.4 cm). It uses the same covers, bottoms and supers as the normal Langstroth hive.
JUVENILE HORMONE – A hormone that controls the development of the bee.

L

LANGSTROTH, L.L. – A Philadelphia native and minister (1810-1895). He lived for a time in Ohio where he continued his studies and writings of bees, recognized the importance of bee space, resulting in the development of the moveable frame hive.
LANGSTROTH FRAME – Most common frame, measuring 19 inches (48.3 cm) long and 9-1/8 inches (23.2 cm) deep; also referred to as the standard frame.
LANGSTROTH HIVE – A hive having frames 19 inches (48.3 cm) long and 9-1/8 inch (23.2 cm) deep. Any movable-frame hive could be considered a Langstroth-type hive, since Langstroth made the moveable frame popular.
LARVA (plural **LARVAE**) – Second stage of bee metamorphosis. A developing bee in the "worm" stage; unsealed brood.

LAYING WORKER – A worker that can lay eggs that produce only drones. Laying workers appear in colonies that are hopelessly queenless.

LEG BASKETS – See POLLEN BASKETS.

LEGUME – Any species of the Leguminosae, or pulse family, is often called a legume. The fruit of this family, which are two-valved pods with the seeds borne on the ventral suture only, include clover, alfalfa, beans and peas and a host of annuals, perennials, shrubs and trees. These plants can fix atmospheric nitrogen into the soil by means of their root nodules.

LEVULOSE – Outdated name for fructose.

LIGURIAN BEE – Italian bee, named for the district in which the Italian bee originated.

LINING BEES – Watching the direction of the flight of bees from flowers to trace them to their home.

M

MANDIBLES – The jaws of an insect. In the honey bee and most insects the mandibles move laterally.

MARKED QUEEN – Queen with a paint dab or clipped wing to make it easier to find her in the colony, document her age and/or confirm genetic history.

MATED QUEEN – A queen that has been successfully inseminated, naturally by drones or artificially by humans.

MATERNAL – From the mother's side of the family.

MATING FLIGHT – The flight taken by a virgin queen during which she mates in the air with one or more drones. Normal queens mate with 12-20 drones during two or more mating fights.

MEAD – Honey wine.

MEDIUM BROOD FOUNDATION – Wax foundation that is heavy enough to be used in the brood chambers. See COMB FOUNDATION.

MELIPONA – A genus of stingless bees native to South and Central America. Some bite persistently but do not sting.

MELISSOCOCCUS PLUTON – The bacterium that causes European foulbrood.

MENTHOL CRYSTALS – Essential oil of mint, in crystalline form, used to control tracheal mites.

METAMORPHOSIS – The developing process of a honey bee in four stages: egg, larva, pupa, and adult. Honey bees have a compete metamorphosis. See BEE METAMORPHOSIS.

MIDGUT – The middle portion of the alimentary tract where the products of digestion are absorbed.

MIDNITE HYBRID – A hybrid of Caucasian and Carniolan bees. No longer available.

MIGRATORY BEEKEEPING – Moving colonies of bees from one locality to another to take advantage of the honey flow in another location, to pollinate crops or to locations where winter weather allows splitting colonies.

MIGRATORY COVER – An outer cover used without an inner cover, that does not telescope over the sides of the hive; used by commercial beekeepers who frequently move hives.

MITE – See *ACARAPIS WOODI* and *VARROA DESTRUCTOR*.

MOISTURE CONTENT – In honey the percentage of water should be no more than 18.6; a higher percentage will allow the honey to ferment. The remaining material, referred to as solids, is composed of sugars, ash and other components.

MORPHOLOGY – The study of the form and structure of an organism.

MOVEABLE FRAME – A frame of comb which can be easily removed from the hive because it is so constructed to maintain a proper bee space with all other surrounding surfaces to prevent bees from adding comb or propolis.

MOVING BOARD – A framed screen that fits on top as a hive cover; used to move bees in hot weather to provide sufficient ventilation to keep bees from suffocating and wax from melting.

N

NASONOV GLAND – Commonly called the scent gland. The odor attracts workers to food, a nest site, a new home. Also used in swarm movement. Found on the basal part of tergum seven of the worker bee.

NATURAL HONEY – Unfiltered and unheated honey

NATURAL SWARM – A swarm of bees issuing from a parent hive to form a new colony. The old queen leaves with the swarm a few days before a virgin queen emerges.

NECTAR – A sugary liquid secreted by special glands called nectaries located chiefly in flowers and also from extrafloral nectaries.

NECTARIES – Organs of a plant composed of specialized tissues that secrete nectar.

NEWSPAPER METHOD – A technique using a newspaper barrier to join together two different colonies.

NOSEMA APIS – The scientific name of one protozoa that causes a nosema disease.

NOSEMA CERANAE – The scientific name of one protozoa that causes a nosema disease.

NOSEMA DISEASE – A malady of adult bees caused by a micrsporidian parasite, *Nosema apis* or *Nosema ceranae* which infects the mid-gut of the honey bee.

NUCLEUS (plural **NUCLEI**) – Commonly referred to as a nuc. A small hive of bees, usually having two to five frames of comb. Nucs can be used for rearing or storing queens, or starting new colonies.

NURSE BEES – Young worker bees, three to ten days old, that feed the larvae and do other work inside the hive.

O

OBSERVATION HIVE – A hive largely of glass or clear plastic to permit observing the activities of the colony.

OCELLUS (plural **OCELLI**) – One of three simple eyes on the top of the honey bee's head, used primarily as light sensors.

OMMATIDIUM (plural **OMMATIDIA**) – One of the visual lenses that comprise the compound eye.

OSMOPHILIC YEASTS – Naturally occurring in honey, the yeasts are responsible for fermentation in honey that contains more than 18% water.

OSMOTIC PRESSURE – The minimum pressure that must be applied to a solution to prevent it from gaining water when it is separated from pure water by a permeable membrane; in honey, its ability to absorb water from the air or other microscopic organisms.

OUT-APIARY – An apiary kept at some distance from the home of the beekeeper.

OUTER COVER – The last cover that fits over a hive to protect it from rain; the two most common kinds are telescoping and migratory covers.

OUTYARD – Also called out-apiary. It is an apiary kept at some distance from the home or main apiary of a beekeeper.

OVARY – The egg-producing part of a plant or animal.

OVERSTOCKING – A condition reached when there are too many bees for a given locality.

OXYTETRACYCLINE – An antibiotic sold under the trade name Terramycin®; used to control American and European foulbrood diseases.

P

PACKAGE BEES or COMBLESS PACKAGE – From two to five pounds (one to two kilograms) of adult bees, with or without a queen and usually with a can of sugar syrup, contained in a ventilated shipping case.

PARAFOULBROOD – Generally accepted as European foulbrood, no longer used to identify a disease.

PARALYSIS – see BEE PARALYSIS

PARENT STOCK – The original colony that has cast a swarm.

PARTHENOGENESIS – Production of a new individual from a virgin female without intervention of a male; reproduction by means of unfertilized eggs. In bees the unfertilized eggs produce only males. An unmated queen, and sometimes a worker, may lay unfertilized eggs that will hatch, producing drones. Also termed haplodiploidy.

PATRILINE – The members of a social insect colony who share the same father.

PDB – Paradichlorobenzene; a white crystalline substance that changes into a heavy gas, used to fumigate combs to protect them from wax moth larvae and adults.

PESTICIDE – Any chemical applied to a crop to protect it from pests.

PFUND GRADER – A grader for honey color; now replaced by a digital color grader.

PHEROMONE – A substance secreted by insects which when sensed or ingested by other individuals of the same species causes them to respond by a definite behavior or developmental process. See QUEEN SUBSTANCE.

PIPING – A series of sounds made by a queen, louder than any sound made by a worker, consisting of a loud, high-pitched tone for two seconds, followed by a series of quarter-second toots. A laying queen is seldom heard to pipe; a virgin queen may pipe at intervals after emerging from her cell, and in response to her piping may be heard the "quacking" of one or several virgins in their cells. The "quacking" is in a lower key and in a more hurried manner than the piping.

PISTIL – The pistil is the female reproductive part of a plant consisting of the stigma, style and ovary.

PLAY FLIGHTS – Short flights taken in front of and nearby the hive to acquaint young bees with their immediate surroundings, Sometimes mistaken for robbing or preparation for swarming.

POISON SAC – Large oval sac containing venom and attached to the anterior (front) part of the sting; stores venom produced by the poison gland.

POLARISCOPE – An optical instrument used to determine the presence of crystals, debris and air bubbles in honey. A polariscope uses two pieces of polaroid film and a light source with the honey between the two pieces of film.

POLLEN – Dust-sized grains formed in the anthers of flowering plants; the male germ cells. The protein food essential to bees for raising brood.

POLLEN BASKET – A flattened depression surrounded by curved spines or hairs located on the outer surface of the bee's hind legs adapted for carrying pollen to the hive. The corbicula.

POLLEN CAKE – A cake of pollen substitute, pollen supplement or pollen, mixed with sugar syrup or honey.

POLLEN INSERT – A device inserted in the entrance of a hive into which hand-collected pollen is placed. As the bees leave the hive they have to pass through the insert where some of the pollen adheres to their bodies and is carried to the blossoms resulting in cross-pollination.

POLLEN PELLETS – The cakes of pollen packed in the pollen baskets of bees and transported back to the colony.

POLLEN SUBSTITUTE – Material such as brewer's yeast, powdered skim milk, or soybean flour, or a mixture of these used to stimulate brood rearing.

POLLEN SUPPLEMENT – A mixture of natural pollen and pollen substitute materials.

POLLEN TRAP – A device for collecting pollen by removing from incoming field bees.

POLLEN TUBE – A slender thread-like growth, containing two sperm cells, which penetrates the female tissue (stigma) of a flower, grows and eventually reaches the ovary; there the sperm cells unite with the ovule.

POLLINATION – The transfer of pollen from an anther to a stigma of a flower.

POLLINATOR – The agent that transmits the pollen; e.g. a honey bee.

POLLINIZER – The plant that furnishes pollen for pollination.

POLYMORPHISM – The coexistence of two or more functionally distinct types of colony members of the same sex.

PORTER BEE ESCAPES – Introduced in 1891, the escape is a device that allows the bees a one-way exit between two thin and pliable metal bars that yield to the bees' push; used to free honey supers of bees.

PRIME SWARM – The first swarm to issue from the parent colony, usually containing the old queen.

PROBOSCIS – The tongue or combined maxillae and labium of the bees. The mouth parts of the bee that form the sucking tube and tongue.

PROPOLIS – Sap or resinous material collected from the buds or wounds of plants by bees, used to strengthen wax comb, seal cracks, reduce entrances, and smooth rough spots in the hive.

PROTEIN – Naturally-occurring complex organic substances, such as pollen, composed of amino acids.

PUPA – The third stage of a developing bee, during which it is inactive, encased in a cocoon and sealed in its cell.

Q

QUEEN – A fully-developed female bee capable of reproduction and pheromone production. Larger than worker bees.

QUEENLESS – Having no queen.

QUEENRIGHT – Having a laying queen.

QUEEN CAGE – A small box of wire and wood or plastic in which queens are shipped and introduced to new colonies.

QUEEN CANDY – Candy made by kneading powdered sugar into sugar syrup until it forms a stiff dough; used as feed in queen cages. See QUEEN INTRODUCTION. See FONDANT.

QUEEN CELL – A cell in which a queen is reared, having an inside diameter of about 1/3 inch (8 mm), hanging downward and about an inch (2.5 cm) long.

QUEEN CUP – A round, cup-shaped structure that workers build on the face and bottom edge of comb that may become a future queen cell. The current queen must place an egg in the cup before the workers begin building the rest of the queen cell.

QUEEN EXCLUDERS – Any device having openings permitting the passage of worker bees but excluding the passage of drones and queen bees. Prevents the queen from entering honey supers.

QUEEN INTRODUCTION – Giving a new queen to a queenless or queenright colony of bees.

QUEEN SUBSTANCE – Pheromones produced by a queen that the attendant worker bees collect and pass to the rest of the colony. If the queen's supply of the secretion is not adequate or ceases entirely, the colony will be motivated to supersede its queen.

QUEEN TRAP – A grid attached to the entrance of a hive, allowing workers to leave but stopping any queen or drone that attempts to leave. Also called a drone trap.

R

RABBET – A recessed ridge in the upper inside edge of a hive to hold the top bars of the frames.

RACES OF BEES – Race is a commonly-used term for subspecies of honey bees.

RADIAL EXTRACTOR – A centrifugal force machine used to throw out honey but leave the combs intact; the frames are placed like spokes of a wheel to take advantage of the slope of the cells.

RAW HONEY – See NATURAL HONEY.

REFRACTOMETER – A precision instrument for determining the moisture content of honey.

RENDERING WAX – The process of melting combs and cappings to separate the wax from its impurities, done by means of hot water, a solar wax melter or other equipment.

REQUEENING – The act of introducing a queen to a colony of bees made queenless, either intentionally or by accident.

REVERSING – The act of exchanging places of different hive bodies of the same colony usually for the purpose of nest expansion. A brood chamber with the queen is placed below an empty brood chamber to give the queen extra laying space as she moves up.

RIPE HONEY – Nectar that has been changed (enzymes added and dehydrated) to a level the bees consider mature, water content of 18.6% or below, and will be capped. Capped honey is used synonymously with Ripe Honey

ROBBING – As applied to bees, taking honey by stealth or force from other colonies.

ROPY CHARACTERISTICS - A diagnostic test for American foulbrood in which the decayed larva forms an elastic rope when drawn out with a toothpick.

ROUND SECTIONS – Sections of comb honey in plastic round rings instead of square wooden boxes.

ROYAL JELLY – A milky white, thick liquid secreted from the hypopharyngeal glands of nurse bees, used to feed developing larvae and the queen.

S

SACBROOD – A virus disease of bee larvae transmitted by infected adult bees.

SCREENED BOTTOM BOARD – A framed screen used instead of a solid bottom board to improve air circulation. Also, allows *Varroa* mites and other pests and debris to fall through and exit the colony.

SCREENED VENTILATION BOARD – A framed screen used to cover the top of a hive being moved in hot weather.

SCOUT BEES – Worker bees searching for nectar, water, pollen or other needs, including a suitable location for a swarm to nest.

SEALED BROOD – Brood that has been capped or sealed in the brood cells by the bees with a somewhat porous capping made of wax, propolis, and other hive material.

SECTION – A small basswood frame that is placed on a hive to produce surplus comb honey; a section box. Also, the honey contained in a section box. Used rarely now.

SELF-POLLINATION – The transfer of pollen from the male to the female parts of the same plant.

SELF-SPACING FRAMES – Frames that will provide bee space between the combs.

SELF-STERILE – The inability of a flower to be fertilized by its own variety. It is only fertilized by pollen from another variety.

SEPTICEMIA – A blood disease of adult bees. Devastating to queen rearing operations using artificial insemination. Caused by *Pseudomonas aeruginosa*.

SETTLING TANK – A large capacity container used for extracted honey. Air bubbles and debris will float to the top clarifying the honey. Usually shallow with a large surface area.

SHAKING BEES – Removing bees from combs by jarring the frames or the hive box.

SHALLOW SUPER – A honey super that is 5-11/16 inches (14.4 cm) deep.

SHIPPING CAGE – A container made of wood and screen used for shipping bees. See PACKAGE BEES and QUEEN CAGE.

SIDE BARS – The wooden pieces on the ends of frames, also called end bars.

SIDELINE BEEKEEPER – One who keeps bees for monetary gain but has other means of income.

SKEP – A dome-shaped old-fashioned beehive without moveable frames, made of reeds or straw, wood or other material.

SLUMGUM – The refuse left after old combs have been rendered. Consists of brood cocoons, pollen, propolis and a small amount of wax.

SMALL HIVE BEETLE – A destructive beetle, *Aethina tumida*, originally from South Africa, that is a bee hive and honey house pest.

SMOKER – A device which burns organic fuel to generate smoke for the purpose of subduing bees during colony manipulation.

SOCIAL INSECTS – Insects that live in a family society, with parents and offspring sharing a common dwelling place and exhibiting some degree of mutual cooperation; e.g. honey bees, ants, termites. Sociality involves cooperative brood care, reproductive castes and generation overlap.

SOLAR WAX EXTRACTOR – A glass-covered box for rendering beeswax using the heat of the sun as an energy source.

SPERM CELLS – The male reproductive cells (gametes) which fertilize eggs; also called spermatozoa.

SPERMATHECA – A small sac attached to the oviduct of the queen in which are stored spermatozoa received from the drones with which she mated.

SPERMATOZOA – The male reproductive cells that fertilize the eggs.

SPIRACLES – The external openings of the tracheae, located on the sides of the thorax and abdomen, are called spiracles. They connect to the system of internal tubes called tracheae used to move air into and out of an insect.

SPLIT – To divide a colony for the purpose of increasing the total number of hives.

SPREADING BROOD – Putting a comb without brood between two combs of brood to induce the queen to lay in the former.

SPRING DWINDLING – A term that refers to the weakened condition of a colony in the spring.

STAMENS – The pollen producing organs of flowers.

STARLINE HYBRID – A four-way Italian bee hybrid known for vigor and honey production. No longer available.

STARTER – 1. A small piece of comb or foundation fastened in a frame or section to start the bees building comb correctly. 2. Finely crystallized honey used to seed liquid honey that then crystallizes into very fine crystals.

STEAM HONEY KNIFE – See UNCAPPING KNIFE.

STIGMA – That part of the pistil of a flower that receives the pollen for the fertilization of the ovules; the end of the pistil.

STING – The queen's and worker bee's weapon of defense. It is an ovipositor modified to form a piercing shaft through which venom is injected into the tissues of the victim.

STING SAC – See POISON SAC.

STRAINING SCREEN – A metal or plastic screen through which honey is strained; also serves as a base for other, finer screening material.

SUCROSE – One of the five important sugars. Refined white table sugar, either cane or beet, is pure sucrose. The main sugar in nectar. A 12 carbon sugar.

SUGAR – The term sugar generally refers to sucrose, which is the sole constituent of refined white sugar, cane or beet. The five important food sugars are classified as follows:

Name	Synonyms	Where Found
Sucrose	Saccharose	Cane or beet sugar or in maple sugar
Lactose	Milk sugar	Milk
Maltose	Malt sugar	Malt products and corn syrup
Glucose	Grape sugar	Honey, corn syrup, fruits
Fructose	Fruit sugar	Honey, fruits

SUPER – (noun) The hive body in which bees store surplus honey; so-called because it is placed above the brood chamber. (verb) To add supers in expectation of a honey flow; also called supering.

SUPERSEDURE – The natural replacement of an established queen by a daughter queen.

SURPLUS or **SURPLUS HONEY** – An excess amount of honey above that needed by the bees for their own use; may be harvested by the beekeeper.

SWARM – The aggregate of worker bees, drones and queen(s) that leave the mother colony to establish a new colony. Swarming is the natural method of propagation of the honey bee colony.

SWARM CELL – Queen cell usually found on the bottom of frames in preparation for swarming.

SWARMING SEASON – The period of the year when swarms usually issue, generally in the spring.

T

TARSUS – The fifth segment of a bee's leg.
TERRAMYCIN® - See OXYTETRACYCLINE.
TESTED QUEEN – A queen whose progeny show she has mated and has the qualities which would make her a good colony mother.
THIN SURPLUS FOUNDATION – Very thin wax foundation used for comb honey production.
THIXOTROPIC – A peculiarity of heather, grapefruit and some other honeys. The honey jells in the comb but upon being agitated it becomes fluid. Usually associated with a high protein content.

THORAX – The middle part of a bee, between the head and abdomen. Contains the muscles to which the wings and legs are attached.
TOP BAR – The top part of a frame.
TRACHEA (plural **TRACHEAE**) – The breathing tubes of an insect that open to the spiracles. See SPIRACLES.
TRACHEAL MITE – *Acarapis woodi*; a mite that inhabits the tracheae of adult honey bees.
TRANSITION CELLS – A comb cell with an irregular shape or size.
TRAVEL STAIN – The darkened appearance on the cappings of comb honey when left on the hive for some time. Caused by bees tracking propolis and pollen over the surface.
TROPHALLAXIS – The exchange of food between nestmates. At the same time queen pheromone is being exchanged giving information of the queen's presence.
TYLOSIN® – An antibiotic used to suppress but not cure American foulbrood.

U

UNCAPPED BROOD – Brood from one to five days old and not yet covered by the wax capping.
UNCAPPING KNIFE – An implement with a sharp blade heated by electricity or rarely steam to remove the cappings from combs before extracting honey.
UNCAPPING PLANE – A device with an electrically heated blade that fits a frame. When moved over the frame the blade removes the cappings from combs of honey before extracting.
UNCAPPING TANK – A container over which frames of honey are uncapped.
UNFERTILIZED – An egg which has not been united with a sperm.
UNITING – Combining two or more colonies to form one larger colony. Usually special precautions must be taken to minimize fighting between the two colonies as they are united.
UNRIPE HONEY – Honey that is not ripe, i.e. containing more than 18.6% water. See RIPE HONEY.
UNSEALED BROOD – Brood not yet sealed over by the bees. See UNCAPPED BROOD.

Glossary

V

VARROA DESTRUCTOR – An external mite parasitic on honey bees.

VARROA JACOBSONI – The previous name of *Varroa destructor*. A separate species found on Asian bees.

VEIL – A protective netting that covers the face and neck, allows ventilation, easy movement and good vision.

VENTRICULUS – The midgut or stomach, located in the abdomen between the honey stomach and the hindgut.

VIRGIN QUEEN – An unmated queen.

VISCOSITY – The property of liquid honey that causes it to flow slowly. As honey is cooled it becomes more viscous and its rate of flow decreases.

W

WARMING CABINET – An insulated box or room heated to liquefy honey.

WASP – Carnivorous social insects in the family Vespidae.

WAX – See BEESWAX.

WAX BLOOM – A powdery coating which forms on the surface of beeswax composed of the most volatile components of beeswax, especially when stored in cool temperatures. It disappears when the wax is warmed.

WAX EXTRACTOR – An appliance for rendering wax by heat, or by heat and pressure.

WAX GLANDS – The eight glands of a honey bee which secrete beeswax after the bees have gorged on honey. They are located in pairs on the last four visible ventral abdominal segments.

WAX MOTH – A moth whose larvae destroy honeycombs by boring through the wax in search of food and leaving trails of webbing. There are two – the greater and lesser, that attack honey bee combs.

WAX PRESS – A press in which the wax is squeezed out of combs using heat and water.

WAX SCALE – A drop of liquid beeswax as secreted by the wax glands. The liquid hardens into a scale upon contact with air; the scale can then be shaped into comb.

WAX TUBE FASTENER – A tube for applying a fine stream of melted wax along the edge of a sheet of foundation to cement it to the top bar of a brood frame or the top of a section. Rarely used since wood section honey comb is no longer used.

WILD BEES – Bees living in hollow trees or other abodes not prepared for them by man. Also called feral bees. Also bees other than honey bees.

WINDBREAKS – Either specially constructed fences or barriers composed of growing trees to reduce the force of the wind.

WIND-POLLINATED – Plants whose flowers manufacture light-weight pollen (and usually no nectar) that is released into the air and blown to fall by chance on a receptive stigma; examples include the grasses (corn, oats) and conifers (pines).

WINTER CLUSTER – The oval-shaped group of bees that forms in the hive when the temperature is below 57°F (14°C).

WINTER HARDINESS – The ability of some strains of honey bees to survive long winters by frugal use of stored honey.

WINTERING – The care of bees during the winter and in preparation for winter to insure their survival.

WIRE CONE ESCAPE – A one-way cone formed by window screen mesh used to direct bees from a house or tree into a temporary hive.

WIRED FOUNDATION – Comb foundation that has wires embedded vertically during manufacture for the purpose of preventing the finished comb from sagging in hot weather.

WIRED FRAMES – These are brood or honey frames having wires stretched across them, either vertically and/or horizontally, for the purpose of holding the wax foundation, and later the comb, solidly in position.

WIRING FRAMES – The act of stringing wires through holes in frames to hold wax foundation in place.

WORKER BEE – A female bee whose organs of reproduction are undeveloped; workers do all the work of the colony except laying eggs.

WORKER COMB – Comb having cells that measure about five to the inch (20 per 10 cm), in which workers may be reared and honey or pollen stored.

WORKER EGG – A fertilized egg laid by a queen that may produce either a worker or a queen.

Y

YELLOWJACKET – A carnivorous social wasp of the family Vespidae; often mistaken for honey bees.

INDEX

A
A COMPARISON OF SYMPTOMS OF VARIOUS BROOD DISEASES 196
A.M. LIGUSTICA 711
ABNORMAL BEES 1
ABRAMS, GEORGE 1
ABSCONDING SWARM 2
ACARAPIS WOODI 332, 540, 541
 Control 545
 Detection 542
 Dispersal 545
 Distribution 542
 Life cycle 23, 544
 Pathogenicity 542
 Taxonomic position 542
ACARINE DISEASE (*Acarapis woodi*) 2
ADDLED BROOD 2
ADEE HONEY FARMS 2
 Adee, Richard 2
 Adee, Vernon 2
AEBI, HARRY & ORMAND 2
AESCULUS CALIFORNICA 2, 439, 442
AFRICA 2
AFRICAN HONEY GUIDES 3, 139
 The Boran people 3
AFRICANIZED HONEY BEES (AHB) 4, 93, 481, 709
 absconding 5, 787
 afterswarms 5, 11, 789
 Colony invasion by migrating swarms 6
 Defensiveness 6, 52
 Dressing for defensive bees 7
 Effect of feral colonies 7
 Honey production 8
 Identification of 10
 In The Americas 11
 Killer bees 481
 Range expansion 10
 Requeening and finding queens 8
AFTERSWARMS 5, 11, 787
AGING IN HONEY BEES 11
ALARM ODOR 11, 13, 36, 283, 288, 632, 782
ALFALFA LEAFCUTTING BEE 11, 492
ALIMENTARY SYSTEM 11
ALKALI BEE 11
 Pollination 11
ALLERGIES TO STINGS 12, 779, 811
AMERICAN ASSOCIATION OF PROFESSIONAL APICULTURISTS (AAPA) 60
AMERICAN BEE JOURNAL 6, 99, 153, 191, 291, 863, 872
AMERICAN BEE RESEARCH CONFERENCE (ABRC) 61
AMERICAN BEEKEEPING FEDERATION (ABF) 60
AMERICAN HONEY PRODUCERS' ASSOCIATION (AHPA) 61
AMITRAZ 12
AMOEBA DISEASE 197, 200
ANATOMY 725
ANATOMY AND MORPHOLOGY OF THE HONEY BEE 12, 161, 163, 298
 A Scanning Electron Microscope Atlas of the Honey 13
 Alimentary canal 11, 23, 33, 34

Anatomy and Dissection of the Honey Bee by H.A. Dade 13
Antennal sense organs 19, 752
chorion 14, 15, 228
crop 23, 34
drone legs 30
Drone prepupa 16, 216
Drone Reproductive organs 26, 40, 216, 725
Drones – Head & Thorax 24, 729
Egg Development 14
Flight 20, 831
Form and Function in the Honey Bee 13
glands 11, 13
Larva shedding skin 16, 490, 530
major glands 23, 281
mandibles 24, 27, 37
nervous system 22
proventriculus 34
Queen – Head 37, 676
Queen Reproductive organs 38, 468, 682, 725, 772
scent gland 34, 245, 565, 635
sting 11, 36
tracheal structure 23, 540, 800
ventriculus 34
wax glands 35, 107, 108
Worker – Abdomen 33, 106, 201
Worker – Thorax 29, 646, 729
Worker - Head 27
worker legs 30
Worker development 13, 16, 585, 638, 847
ANDERSON, EDWIN J. 41
ANIMAL AWARENESS 41
ANNUAL CYCLE OF HONEY BEE COLONIES 42
ANTIBIOTICS & OTHER CHEMICALS FOR BEE DISEASE CONTROL 43, 197, 241
 Fumagillin 44, 205
 Government regulation and monitoring of pesticides 46
 Oxytetracycline 44, 197
 Tylan® 43, 45, 197, 200
ANTS - A PROBLEM FOR HONEY BEES 12, 49, 306
APIARY INSPECTORS OF AMERICA (AIA) 61
APIARY LOCATIONS 50, 215, 306, 589, 596
 With African honey bees 4, 52
APIMONDIA 53, 62, 100
APIMYIASIS 53
APIS, DISTRIBUTION AND SPECIES 53
 A. binghami 57
 A. breviligula 57
 A. cerana 58, 88
 A. florea 54
 A. koschevnikovi 58
 A. laboriosa 56
 A. mellifera 87, 709
 A. mellifera scutellata 4
 A. nigrocincta 58
 A. nuluensis 58
 Cavity-Nesting Honey Bees 57
 Giant Honey Bees 56
APISTAN® 59, 208, 540, 811
APITHERAPY 12, 59, 107
ARTIFICIAL SWARMING 59, 784

Index

ASCOSPHAERA 201
ASSOCIATIONS FOR BEEKEEPERS 59
ATKINS, E. LAURENCE 67, 622, 625
AUSTRALIA 67, 88
AVERAGE DEVELOPMENT TIMES FOR QUEENS, WORKERS AND DRONES 13

B

BACON, MILO R. 69
BAILEY, SEYMOUR E. 69
BAIT HIVES 69, 138, 784, 792
BATIK 121
BEARS 72, 73
BECK, DR. BODOG F. 75
BEE BEARDS 75, 275
BEE BOLES 78
BEE BREAD 78, 145, 258, 585, 648
BEE CELLAR 840
BEE, DEFINITION OF 78, 331, 656
BEE ESCAPES 80
BEE HOUSES 80
   drifting 80
   supering 80
   the weather 80
BEE LINING 104
   A bee-lining box 104
   Triangulation 105
BEE LOUSE 105, 143
BEE MILK 78, 105
BEE PARALYSIS 201, 210
BEE RESEARCH 105, 195, 241, 728
BEE SPACE 105, 150, 304
   America's Master of Bee Culture: The Life of L.L. Langstroth 106
   Langstroth, L.L. 105, 304, 855
   Langstroth on the Hive and the Honey Bee: A Beekeeper's Manual 106
   modern Langstroth 106
   natural nest 106, 566
BEE STINGS, FIRST AID 12, 106, 285
   Honey bee venom 59, 107
   poison sac 33, 106, 779
BEEKEEPER INDEMNITY PAYMENT PROGRAM 80
BEEKEEPING ASSOCIATION DIRECTORY 62, 99
BEEKEEPING IN ANTIQUITY 81, 188
   forest and skep beekeeping 83
   rock paintings 82
   The Egyptians 82, 93
   The Greeks and the Romans 83
BEEKEEPING IN VARIOUS PARTS OF THE WORLD 83
   Africa 2, 83
   Africanized bees 4, 96, 709
   Apis cerana 58, 88
   Apis mellifera 87, 709
   Argentina 85
   Asia, Southeast 85
   Australia 67, 88, 622
   Brazil 91
   Canada 91, 619
   China 92
   Egypt, traditional beekeeping in 82, 93
   Europe 78, 80, 93
   Honey hunting 87
   Mexico 95
   New Zealand 98
   South Africa 99
BEEKEEPING JOURNALS 99
   Alberta Bee News 91, 100
   American Bee Journal 6, 99, 191, 620, 621, 863, 872
   Australian Bee Journal 67, 88, 104
   Apiacta 53, 100
   Bee Craft 104
   Bee Culture 99, 620, 621
   Beekeepers Quarterly 99
   Bees For Development 83, 100
   Gleanings in Bee Culture 99, 194, 562
   Hivelights 91, 104
   Honeybee Science 100
   Insectes Sociaux 100
   Journal of Apicultural Research 65, 100, 188, 470
   L'Abeille De France 100
   Manitoba Beekeeper 91, 100
   Mellifera 100
   Scottish Beekeeper 104
   South African Bee Journal 2, 83, 104
   Speedy Bee 99
   Teknik Aricilik 100
   The Australasian Beekeeper 67, 88, 104
   The New Zealand Beekeeper 100
BEESWAX 107, 150, 742, 772, 820
   Beeswax foundation 109, 116, 264
   Beeswax, Production, Harvesting, Processing 108, 125
   Beeswax secretion by honey bees 35, 108
   best metals and effects on 110
   candles 1092 121, 126, 154
   Chemical extraction 111
   chemical solvents 111
   extruded 109
   Harvesting 110
   honey needed to produce a pound 108
   Infrared and other overhead cappings
   physical and chemical properties 115
   pounds of beeswax for 100 pounds of honey 110
   pressing of cappings and slumgum 112
   reducers 112
   Rendering beeswax cappings 110, 156, 559
   screws or hydraulic cylinders 112
   slumgum 112
   Solar wax melter 113, 772
   Stainless steel 110
   wax glands 35, 107, 283
   Wax-producing bees 35, 108
   melts at 109, 125
BEESWAX FOUNDATION 109, 116, 264
   Aluminum & plastic core foundation 118
   Cell diameter in manufactured foundation 120, 159, 511
   foundation mills 120
   Mehring, J. 118
   rolls 119
   sheeting 119
   Washburn, A. 118
   wire embedded 119
BEESWAX IN ART AND INDUSTRY 120
   Batik 121
   Beeswax figures 121, 131

Index

Candles 109, 121, 154
  Cosmetics 129
  Dipped candles 129
  Encaustic painting 130
  Foundation candles 128
  Grafting Wax 131
  Japan wax 121
  Lost-wax casting 131
  Melting beeswax 109, 125
  Metal molds 126
  Polyurethane Mold Candles 127
  Preparing wax 108, 125
  Sealing wax 133
  Wicking 126, 154
BEGINNING WITH BEES 133, 213, 233, 235, 291, 312, 321, 398, 694, 773, 847
  Buying established or secondhand colonies 134, 193,
  Capturing swarms 69, 138, 792
  Capturing swarms with bait hives 69, 138, 784, 792
  Nucleus colonies 136, 578
  Number of colonies to buy or manage 139
  Package bees 134, 599
BENTON, FRANK 139
  Benton Queen Cage 140, 696, 725
BIOGRAPHIES OF NOTED BEE-KEEPERS 855
  A.E. Manum 859
  Amos Ives Root 876
  Charles Dadant & Son 191, 861
  D.A. Jones 862
  Dr. C.C. Miller 175, 538, 865
  Dr. John Dzierzon 224, 866
  Edwin France 271, 860
  Eugene Secor 873
  Francis Danzenbaker 875
  Francis Huber 177, 867
  George W. York 6, 99, 191, 872
  Gilbert M. Doolittle 213, 861
  H.R. Boardman 248, 870
  John H. Martin 874
  Julius Hoffman 265, 317, 870
  Lorenzo L. Langstroth 105, 304, 489, 855
  Lyman C. Root 746, 858
  Moses Quinby 766, 856
  Mrs. Lucinda Harrison 864
  O.O. Poppleton 873
  Professor A.J. Cook 61, 857
  Thomas G. Newman 6, 99, 191, 863
  W.L. Porter 80, 715, 876
  W.Z. Hutchinson 871
BIRDS 3, 139, 846
BLOOM 109
BOCH, ROLF 141
BONNEY, RICHARD (DICK) 142
BOTULISM 142, 374, 463
BOX HIVES 142, 311, 725
BRAND MELTERS 107, 111, 116, 120, 143
BRANDING HIVES 143
BRAULA COECA (Bee Louse) 105, 143
BREEDING BEES 69, 145, 276, 380, 687, 694, 701
BROOD 145,633
  Chamber 145, 566
  Development 14, 16, 113, 216, 500, 530
  Diseases 43, 44, 197, 201, 203, 208, 230

Food 145, 156, 258, 585, 646, 648, 651
  Nest 146, 154, 167, 681
  Patterns 147, 676, 680
  Rearing 147, 161, 281, 288
BROTHER ADAM 148, 713
BURLESONS, INC. 149
BURR AND BRACE COMB 105, 107, 150

C

CALE, G.H., (BUD) JR. 99, 153, 192, 278
CALIFORNIA BUCKEYE 2, 439, 442
CANADIAN ASSOCIATION OF PROFESSIONAL APICULTURISTS (CAPA) 62, 91
CANADIAN HONEY COUNCIL (CHC) 63
CANDLES 109, 121, 126, 127, 128, 129, 154
CANDY, QUEEN CAGE 139, 140, 154, 696, 725
CAPPING OF CELLS 145, 154
CAPPINGS, TREATMENT AND RENDERING OF 110, 156
CARBOHYDRATES 156, 247, 300, 321, 585, 784
CASTE 156, 216, 260, 293, 498, 500, 527, 585, 676, 847
CASTE DETERMINATION 38, 158, 276, 680, 682
CAUCASICA 87, 709, 710
CELL SIZE 120, 159, 218, 511, 554
  Measuring cell size 160
  Size of natural cells 160
CELLAR WINTERING 160, 840
CHALKBROOD DISEASE 161, 201
  alfalfa leafcutting bees 202
  Ascosphaera 201
  Distribution 202
  first record of chalkbrood 202
  Life cycle 202
  Megachile rotundata Fabr. 201
  Treatment 202
CHECKMITE+® 12, 161, 187, 554
CHILLED BROOD 147, 161
CHITIN 12, 161, 241
CHRYSLER, CHESTER E. 161
CHUNK HONEY 161, 383, 448
CIRCULATORY SYSTEM 12, 163
CLIPPING QUEEN'S WINGS 691, 788
CLOSED POPULATION BREEDING 145, 276, 279, 486, 687
COCOONS 163, 490, 821
COGGSHALL, WILLIAM L. 164
COLONY 165
COLONY DEFENSE AT LOW TEMPERATURES 11, 165, 194
COLONY ENVIRONMENTAL CONTROL 34, 166, 245, 565, 635, 813
COLONY ODOR 166
COLONY ORGANIZATION 146, 167, 312, 317, 665, 800, 833
COMB 107, 168, 218, 383, 591, 800, 847
COMB FOUNDATION 168, 591
  Traditional wooden sections 173, 383
COMB HONEY 168, 173, 383, 800
  Assembling Section Comb Boxes 170
  Comb Honey Era 168
  Comb honey, how to make 169
  dehumidifiers 174
  Foundation & the preparation of sections & supers 171
  Preparing comb honey sections for market 174

899

Round sections 173
Sections for comb honey production 172
Shook swarming 175
Sizes of section boxes 169
split wooden sections 172
COMB HONEY PRODUCTION, REARING QUEENS FOR 175
Miller, Dr. C.C. 175, 538, 865
Propolis 176
watery 176
COMB, NATURAL 176
Huber 177, 460, 867
COMB STORAGE 108, 178, 273, 642, 827
animals that would destroy 178, 759, 821
controlled atmosphere 180
maximum air flow 179
Pollen mites 179
COMMUNICATION AMONG HONEY BEES 180, 184, 192, 271, 493, 596, 661, 847
Migration Dance 183
Odor recruitment to food and the genome analysis 184
The dance language for food 182, 271
Vibration signal 184
CONSERVATION, THE IMPORTANCE OF THE HONEY BEE IN 185
CONTROLLED MATING 185, 515, 687, 701
CONTROLLING NATURAL QUEEN MATING BY DRONE FLOODING 515
COOK, PROFESSOR A.J. 61, 857
COOKING WITH HONEY 187, 332, 453
CORRIGAN, RICHARD 187
COUMAPHOS 12, 187, 554
Apistan® 12, 187, 257, 554
Bayer Corporation 187, 554
CheckMite+® 187, 554
COWAN, THOMAS W. 188
CRANE, EVA 65, 188, 100, 470
CUT-COMB HONEY 188

D

DADANT FAMILY 191
American Bee Journal 191
Camille Pierre 191
Charles C. Dadant 192, 861
crimp-wired foundation 116, 191, 264
Dr. G.H. Cale, Jr. 99, 153, 192, 461
G.B. Lewis Company 192
Henry (H.C.) 191
Louis (L.C.) 191
Maurice (M.G.) 191
Midnite Hybrids 192, 278, 461
Nicholas J. Dadant 192
Starline 192, 278, 461
The Hive and The Honey Bee 191
Thomas G. Ross 192
Timothy C. Dadant 192
DADANT HIVE 192, 314
DADE, H.A. 13
Anatomy & Dissection of the Honeybee 13
DANCE COMMUNICATION BY HONEY BEES 180, 184, 192, 271
DEAD BEES, REMOVAL OF 192
undertaker 192
DEATH'S-HEAD HAWK MOTH 193
Acherontia atropos 193

DECONTAMINATION OF HIVE EQUIPMENT 193, 312
American foulbrood 193
bacteria 194
chalkbrood 193
European foulbrood 193
insects 194
mites 194
nosema 193
viruses 194
DEFENSIVE COLONIES, HOW TO TREAT 11, 165, 194
DEMAREE METHOD OF SWARM CONTROL 194, 494, 788
DEMUTH, GEORGE S. 99, 194, 733
Commercial Comb Honey Production 194
Gleanings in Bee Culture 195
DEYELL, JOHN MOSSOM 195, 733
Gleanings in Bee Culture 195
head apiarist & manager for Root apiaries 195
Talks to Beginners 195
DIPTERA PESTS OF HONEY BEES 53
DISAPPEARING DISEASE 195, 203
DISEASE DIAGNOSTIC SERVICE 195, 728
Adult diseases 197
Brood samples 195
DISEASE RESISTANCE 211, 554, 811
Park, O.W. 612
Rothenbuhler, Prof. Walter C. 211, 748
DISEASES OF THE HONEY BEE 43, 197, 201, 850
American foulbrood disease 43, 44, 197, 519
Amoeba Disease 200
Chalkbrood 161, 201
Disappearing disease 195, 203
Dysentery 203, 205, 223
European foulbrood 203
Melissococcus pluton 203
Nematodes 205
Nosema 197, 205, 223, 578
Nosema ceranae 206
Paenibacillus larvae subsp 198
Parafoulbrood 208
Parasitic mite syndrome, bee 208
Viruses 208, 797
DIVELBISS, CHARLES A. 211
DIVIDING COLONIES (SPLITS) 212, 229, 701
DIVISION OF LABOR 156, 213, 676, 847
DOOLITTLE, G.M. 213, 861
Gleanings in Bee Culture 213
Scientific Queen Rearing 213
DOUBLE GRAFTING 687, 692, 748
DRESSING FOR THE APIARY 213, 756
DRIFTING 50, 215, 306
common bee diseases 43, 197, 216
landmarks 215
packages 216, 599
DRONE 16, 26, 40, 216, 465, 514, 654
Apistan® 217
Breeder Colonies 217
Cells 218
Checkmite+® 217
Comb Foundation 159, 218
Congregation Areas 218, 637
Essential oils 218
Extraction supers 218

Index

Lower honey yields 218
Miticides 217
Organic acids 218
Population Control 218
Trap 218, 654
Varroa destructor 218
DRUMMING 219, 311, 725
package bees 220
rhythmic 219
substrate-borne vibrations 220, 752
DUMMY BOARDS 220, 258, 264
DWARF HONEY BEES 53
A. andreniformis 54
A. florea 54
DYCE, ELTON JAMES 220
DYCE PROCESS FOR MAKING CRYSTALLIZED HONEY 220, 326, 449
DYSENTERY 203, 223
confinement 223
nosema 223
Poor food 224
wintering 223
DZIERZON, H.C.J. 224, 866

E

EAS Honey Show Rules 63, 383, 394
EASTERN APICULTURAL SOCIETY (EAS) 63
ECKERT, JOHN E. 227
ECOLOGY OF THE HONEY BEE 227
EGG 14, 228
ELECTRIC FENCE 72, 73
EMERGENCY QUEEN CELLS 228, 676, 682, 698, 785, 788, 792
ENEMIES OF THE HONEY BEE 3, 4, 49, 105, 72, 139, 143, 193, 197, 228, 306, 457, 530, 575, 622, 676, 682, 698, 731, 758, 759, 785, 788, 792, 798, 811, 821, 846, 853
ENVIRONMENTAL PROTECTION AGENCY (EPA) 46, 625, 630
ENVIRONMENTAL QUALITY INDICATOR SPECIES – HONEY BEE 228
EQUALIZING COLONIES 70, 212, 229
brood 145, 230
dearth 230, 573
exchange the positions 50, 230
EQUIPMENT FOR BEEKEEPING 99, 213, 233, 264, 310, 654, 683, 765, 784
commercial beekeepers 234
interchangeable 233
Plastic 235, 642
standardize equipment 233
EQUIPMENT MAINTENANCE 235
screws 236
wood preservatives 236, 845
ESSENTIAL OIL BASED COMPOUNDS 237, 547
ApiGuard 237
Apilife Var 237
EVOLUTION OF THE HONEY BEE 238
Amber 238
fossil honey bee-like insects 238
EXAMINING COLONIES 238
Dressing for the Apiary 239
EXHIBITING HONEY 241, 383, 394
EXOSKELETON 12, 161, 241
chitin 12, 161, 241
EXTENDER PATTIES 43, 47, 241

EXTENSION APICULTURE 105, 195, 241, 728
EXTRACTOR HISTORY FROM PAST ABC EDITIONS 345, 360
Advantages of the Radial Extractor Over The Reversible extractor 355
Bohn's Extractor 353
Control Pivot Reversing Extractor 351
Cowan Reversible Extractor 351
Extracting Without Reversing 353
How Both Sides Of A Comb Are Extracted At Once 356
parallel/radial extractors 358
Power vs. Hand Machines 357
The Root Multiple Reversing Extractor 351
Extractor, How To Make circa 1880 347

F

FANNING BY HONEY BEES 34, 245, 565, 635
Nasonov 34, 245, 565, 635
FAT BODY 167, 245, 833
glycogen 245
FDA 46, 378, 444, 485
FEDERAL INSECTICIDE, FUNGICIDE, AND RODENTICIDE ACT 46, 622
FEEDING ANTIBIOTICS 43, 47, 187, 241, 797
Bulk feeding 47
Dusting 47
Extender patties 47, 241
Honey flows and bee disease control 48
pollen flow 48
Pre-mixes 48
FEEDING BEES 156, 247, 300, 580, 655, 773, 784, 819
Boardman feeders 248, 870
Bulk outdoor feeding of syrup 248
Different methods of feeding syrup 248, 539
Division board feeders 249
Feeding combs of honey 247
Feeding dry sugar 249
Feeding liquid honey 247
Feeding under the brood nest 250
Feeding with jugs/cans through the cover 251
Filling combs with syrup 251
Fondant 251
Jar and pail feeding 253
Miller-type feeders 254
Open feeding. 249
Plastic Bags 254
Pollen Supplements & Substitutes 651
FIRE BLIGHT 255, 656
Erwinia amylovora 255
FLIGHT DISTANCE 255, 260, 263, 497
FLIGHT OF THE HONEY BEE 255, 260, 263
Aristotle 257
Eckert, John E. 256
flight range 256
temperature 256
FLOREA NESTS 54
FLOWER FIDELITY 256
FLUVALINATE 12, 187, 257, 554
Apistan® 187, 257, 554
residues of 257
FOLLOWER BOARDS 220, 258
FOOD CHAMBER 258, 556
FOOD EXCHANGE BETWEEN ADULT WORKER BEES 259, 568

901

Index

pheromone movement 260
FOOD OF THE HONEY BEE 78, 156, 258, 527, 567, 585
  trophallaxis 259, 568
FOOTPRINT PHEROMONE 260, 634
FORAGERS 260, 818
  Castes 156, 216, 260, 676, 847
FORAGING BEHAVIOR 255, 260, 263, 497, 847
  Aristotle 261
  foragers 261
  honey bee cooperation 260
  Honeybee Ecology 260
  scouts 261
  Seeley, T.D. 260
  Water for honey bees 261
FORAGING DISTANCE 255, 263
  Distance flown 262
FORMIC ACID 12, 49, 263, 554
FOUNDATION 107, 116, 191, 264
FRAME SPACERS 270, 780
  Irwin A. Stoller 270
FRAMES 220, 233, 264, 319
  brad driver 268
  Depth of frames 265
  favored wood 265
  Frame Assembly Pointers 269
  Hoffman frames 265
  How to nail 265
  lack of ventilation 265
  Propolis 264
  spur embedder 267
  support pins 268
  wedge bar 268
  Wiring a frame 267
FRANCE, N.E. 271, 860
FRISCH, PROFESSOR KARL VON 192, 271, 615
  A Biologist Remembers 272
  Bees, their Vision, Chemical Senses and Language 272
  language of bees 272
  Nobel Prize 272
  round dance 272
  The Dance Language & Orientation of Bees 272
  wag-tail dance 272
FRUIT DAMAGE BY HONEY BEES & WASPS 272
  apples 272
  ripe grapes 272
FRUIT TREES 273, 656
FUMIGATION & STORAGE OF COMBS 178, 273
FURGALA, BASIL 60, 273

G

GAMBER, RALPH 275
GARY, NORMAN 75, 275, 513
  drone congregating 275
GENETIC ANOMALIES 279
  cordovan 279
  eye mutations 281
GENETICS OF THE HONEY BEE 276, 575, 619
  Closed population breeding 279
  Colony structure 276
  drone honey bees 277
  Gregor Mendel 276
  H. H. Laidlaw 277, 463
  Inbred-hybrid breeding programs 153, 192, 278

instrumental insemination 277
  L. R. Watson 277, 463
  Maternal mother-daughter mating 278
  Mating behavior 276
  O. Mackensen 277, 463
  queens 277
  Selection of breeding stocks 277
  Stock improvement 277
  super-sister 277
  Super-sister–super-sister-mating 278
  workers 277
GIBSON, GLENN 281
  American Beekeeping Federation 60
  Ameircan Honey Producers 61
GLANDS IN THE HONEY BEE AND THEIR SECRETIONS 23, 281, 475, 513, 632, 847
  10-hydroxydec-2-enoic acid 286, 634, 687
  2-heptanone 286, 632
  Endocrine glands 281
  Exocrine glands 282
  Glands of head 290
  Glands of the head and thorax 287
  Head and thoracic glands 142, 288
  hypopharyngeal plate 287
  isopentyl acetate 5, 11, 288, 632, 787
  Mandibular glands 285
  Nasonov gland 285, 565
  royal jelly 286, 692, 748
  Silk glands 288
  Sting glands 11, 283, 787
  The alarm odor and the sting scent gland 11, 13, 283, 285, 288, 361, 632, 782, 811
  The endocrine organs 282
  The sting apparatus 33, 284
  Wax glands 35, 107, 283
GLUCOSE OXIDASE 291, 470, 569
GOOD NEIGHBOR BEEKEEPING 133, 291, 589, 596, 672, 807
  Flight Patterns 292
  How Many? 293
  Placing Colonies 292
  Providing Water 292
  Working Bees 293
GOODMAN, LESLEY 13
GRADES OF EXTRACTED HONEY 383, 445
GRAFTING 291, 692, 748
GRAFTING TOOLS 291
GRANULAR AND MUSHY WAX AND EMULSIONS IN BEESWAX 107, 109
  Emulsions 109
  Granular wax 109
  water in wax 109
  wax in water 109
GREENHOUSE POLLINATION 291, 659
GROUND-NESTING BEE 11, 853
GROUT, ROY A. 291
  American Bee Journal 291
  Langstroth on the Hive and the Honeybee 291
  the size of the cell 291
GUARD BEES 27, 292, 632, 779, 847
  Caste 156, 293
  winter cluster 293, 831, 832
GUMS 294, 312
  gum wood trees 294
  Removing feral colonies 294

Index

GYNANDROMORPHIC BEES 1, 276, 295
   causes of 295

H

HAMBLETON, JAMES I. 297
   Eastern Apicultural Society 298
HAY HITCH KNOT 529, 534
HEARING IN HONEY BEES 298, 752
HEARTLAND APICULTURE SOCIETY (HAS) 65
HEMOLYMPH 12, 298
   clotting factor 298
HEWITT, PHILEMON J. 299
   Eastern Apicultural Society of North America 299
HIDALGO, TEXAS 4, 11
HIGH FRUCTOSE CORN SYRUP 247, 300, 784
   feeding 300
   imitation honey 300
   Isomerose 300
   using acid 300
HISTORY OF BEEKEEPING 300
   bee space 304, 489
   beeswax 107, 301
   Dark Ages 301
   Egyptians 93, 301
   forest beekeeping 93, 301
   Greeks and Romans 301
   Koran 301
   Propolis 301
   rock painting 301
   skep beekeeping 302, 756
   The Archeology of Beekeeping 188, 305
   the Christian Bible 301
   The Feminine Monarchie 302
   The Middle Ages 302
   The World History of Beekeeping and Honey Hunting 305
HIVE CARRIERS 305, 534, 561
HIVE PRODUCTS 107, 305, 321, 669, 748, 811
HIVE STANDS 49, 306, 758
   Bee boles 78
HIVE TOOLS 233, 310
HIVELIGHTS 63, 91
HIVES, DEVELOPMENT OF 311
   box hives 142, 219, 311, 725
HIVES, TYPES OF 80, 137, 294, 312
   Concrete 316
   Dadant hive 192, 314
   Deep hive 314
   Eight-frame hive 313
   Other than wooden hives 316
   Plastic supers, frames, feeders, inner covers, 316, 642
   polystyrene foam 316
   Ten-frame Langstroth hive 313
   Top bar hives or long hives 314, 481, 800
   Warm-way, cold-way construction 316
HOARDING BY HONEY BEES 317, 566
   Supering colonies 167, 317
HOFFMAN FRAMES 265, 319
HOFFMAN, JULIUS 317, 870
HONEY AND HONEY PRODUCTS 321
   A Pfund grader 324
   Acidity of honey 322
   age of the combs 323
   blue to purplish color 324

Caloric value 322
Chemical and physical properties of honey 322
citrus honeys 322
Color 323
Color grading 324
Crystallization 220, 326, 449
diastase 322
Fermentation of Honey 326
Hanna Color Analyzer® 325
Heating 326
Honey A Comprehensive Survey 321
Honey, Extracted pasteurization 327
hydrometers 328
Hygroscopicity 329, 596
increase the HMF reading 322
Lovibond® grader 325
Moisture content 328
Nectar 156, 247, 300, 321, 527, 564, 585, 784
Osmotic pressure of honey 325
pale green honey 324
proteins 323
refractometers 328
Specific gravity 330
Thixotrophy 330
Townsend, Gordon F. 323, 799
USDA grades 325, 326, 383
HONEY BADGERS 4, 228
HONEY BEE, DEFINITION OF 78, 331
HONEY BEE, HOW TO SPELL 332
HONEY BEE TRACHEAL MITE 332, 540, 541, 800
HONEY COOKERY 187, 332, 453
   Substituting Honey For Sugar 333
HONEY, EXTRACTED 333, 501, 562
   Blending honey 377
   Bulk storage of honey 373
   canning jars 377
   Cleaning jars 377
   Extracting room 338
   Filtering 370
   flash heater 374
   Harvesting 80, 338, 827, 876
   Honey gates 371
   Honey houses 334, 344
   Honey processing area 368
   Honey pump 372
   Honey tanks 373
   Hot rooms 335
   Labeling varietal honey 46, 378
   Liquefying honey 377
   low moisture honey 338
   milk tanks 376
   O.A.C. HONEY STRAINER 370
   Off-loading dock or room 334
   Other considerations in honey house construction 361
   Pasteurization of honey 376
   Plastic containers 379, 642, 644
   Plastic totes 374, 644
   Preparation of honey for market on a small commercial basis 142, 374, 729
   Radial extractors 345
   Selecting a cap 378
   Selecting a jar for packing 377
   Shelf life 381
   Straining 368
   Super storage area 361

Index

tangential extractors 345, 360
HONEY FORMS, PACKAGED 382
  Chunk Honey 161, 383, 448
  Comb Honey 168, 173, 383, 748
  Cremed Honey 220, 226, 449
  Cut-Comb Honey 188, 383
  Honey products 448
HONEY JUDGING 241, 325, 383, 445
  Color grading 324
  EAS Honey Show Rules 394
  refractometer 383, 385
  Score Cards 387
HONEY PLANTS 398, 504, 571
  Alfalfa – *Medicago sativa* 398, 493
  American Holly, *Ilex opaca*. 430
  Annise Hyssop, *Agastache foeniculum* 430
  Apples, *Malus* sp. 430, 648
  Aster – *Aster* spp. 400, 649
  Basswood - *Tilia* spp. 400
  Bee Bee Tree, *Evodia danelli*. 430
  Birdsfoot Trefoil, *Lotus* sp. 430
  Black locust *Robinia pseudoacacia* 405, 648
  Blackberries – *Rubus* spp. 403
  Brazilian pepper *Schinus terebinthifolius* 405
  Buckwheat – *Fagopyrum esculentum* 406
  Canola – *Rapus spp.* 407
  Century Plant – *Agave sp.* 430
  Citrus – *Citrus spp.* 407
  Clover – *Melilotus* spp. 407, 648
  Clover – *Trifolium* spp. 407, 648
  Cotton – *Gossypium* spp. 410
  Dandelion, *Taraxicum officinale* 431, 649
  Eucalyptus – *Eucalyptus spp.* 412
  Figwort, *Scrophularia matilandica* 431
  Fireweed – *Epilobium angustifolium* 413
  Gallberry – *Ilex glabra* 413
  Goldenrod – *Solidago* spp. 414
  Heather – *Calluna vulgaris* 415
  Knapweed, Star Thistle, *Centaurea* spp. 432
  Lowbush Blueberry, *Vaccinium pennsylvanicum* 433
  Mangrove, black – *Avicennia nitida* 416
  Maples, *Acer* sp. 433, 648
  Mesquite – *Prosopis* spp. 416
  Milkweed, *Asclepias* sp. 433, 538
  Peaches, Plums, *Prunus* sp. 433
  Poisonous Plants 2, 439, 442, 573, 646
  Purple loosestrife – *Lythrum salicaria* 418
  Rape – *Brassica spp.* 420
  Raspberry – *Rubus* spp. 421
  Sage – *Salvia* spp. 421
  Saltcedar – *Tamarix* spp. 422
  Saw palmetto – *Serenoa repens*. 422
  Sourwood – *Oxydendrum arboreum* 424
  Soybean – *Glycine max* 424
  Spanish needle – *Bidens* spp. 424
  Sumac – *Rhus* spp. 425, 648
  Sunflower – *Helianthus* spp. 425
  Tallow tree – *Sapium sebiferum* 427
  Thyme – *Thymus* spp. 428
  Tulip poplar – *Liriodendron tulipifera* 428
  Tupelo – *Nyssa* spp. 428
  Vetch *Vicia* spp. 429
  Willows, *Salix* spp. 438, 649
  Yellow Rocket, *Cruciferae* family 438, 649

HONEY PRODUCTS 448
  Chunk Honey 161, 383, 448
  Comb Honey 168, 173, 383
  Cremed honey (DYCE PROCESS) 220, 326, 449
  Cut-Comb Honey 188, 383
  Flavored honey 449
  Honey beer 448
  Honey butter 448
  Honey candy 448
  Honey ice cream 449
  Mead 450, 482, 483
  Soft drinks 451
  Vinegar 451
  Yogurt 453
HONEY PURE FOOD & DRUG LAWS 46, 444, 495
HONEY RECIPES 187, 332, 453
  Baked Acorn Squash Halves 455
  Bar-Le-Duc Preserves 455
  BBQ Beer 454
  Cherry Pie 454
  Cranberry Fruit Salad 455
  Easy Chicken 455
  Grape Drink 454
  Hot Fruit Punch 455
  Skillet Baked Beans 454
  Tangy Cole Slaw 454
HONEY VINEGAR 452
HONEYDEW 456
HORNETS, YELLOWJACKETS & OTHER WASPS 457, 817, 853
HOUSE BEES 13, 16, 458, 476, 630
HRUSCHKA, MAJOR F. 458
HUBBARD APIARIES, INC. 458
HUBER, FRANCOIS 177, 460, 867
HUMIDITY CONTROL IN NEST 460, 566
HYBRID BEES 99, 153, 192, 461, 713

I

IMITATION HONEY 300
IMPORTING BEES 463, 495
INDICATOR INDICATOR 3, 139
INFANT BOTULISM 142, 463
INSTRUMENTAL INSEMINATION 277, 463
  Drones 216, 465
  drone's endophallus 466
  Harbo device 469
  Harbo large capacity syringe 466
  Harry Laidlaw 463, 486
  instruments 465
  Ken Tucker 464
  Lloyd Watson 463, 820
  Mackenson instrument 465
  Otto Mackensen 464, 509
  The valvefold 38, 468
  Walter Rothenbuhler 464, 748
INTEGRATED PEST MANAGEMENT (IPM) 46, 470, 622, 628
INTERNATIONAL BEE RESEARCH ASSOCIATION 65, 100, 188, 470, 728
INVERT SUGAR 470, 567, 784
  fructose 470
  glucose 291, 470
  Nectar, conversion to honey 470, 568
INVERTASE 471

Index

J

JARVIS, DR. D.C. 473
   Folk Medicine: A Vermont Doctor's Guide to Good Health 473
JAYCOX, ELBERT RALPH, "JAKE" 41, 60, 99, 474
   Beekeeping in the Midwest 474
   Beekeeping Tips and Topics 474
JOHANSSON, TOGE S.K. 475
JUVENILE HORMONE 281, 290, 475
   Caste determination 158, 475
   division of labor 13, 16, 458, 476, 630

K

KELLEY, WALTER T. 479
KELTY, RUSSELL 480
KENYA TOP BAR HIVE 314, 481, 800
KERR, WARWICK E. 4
KILLER BEES 4, 52, 96, 481, 709
   Africanized HONEY bees 4, 52, 96, 481, 709
KILLION, CARL E., SR. 175, 481
   Honey in the Comb 481
   The Covered Bridge 481
KILLION, ELIZABETH 482
KIME, ROBERT W. 450, 482

L

LABELS FOR HONEY JARS 46, 485
   Back or second label 486
LAIDLAW JR., HARRY HYDE 279, 463, 486
   closed population breeding program 488
   instrumentally inseminating queens 487
   J.E. Eckert 488
   M.E. Kühnert 488
   Queen Rearing 488, 687
   W.J. Nolan 487
LAND BASED HONEY PRODUCTION 488
LANGSTROTH HIVES 490
   Hives, types of 490
LANGSTROTH, LORENZO LORRAINE 105, 304, 489, 855
   A. I. Root Company 489
   America's Master of Bee Culture: The Life of L.L. Langstroth 106
   Dadant and Sons 489
   discovery of bee space 489
   Langstroth on the Hive and the Honey Bee: A Beekeeper 489
LARVA (s), LARVAE (pl) 145, 163, 490, 847
   Life stages of the honey bee 16, 490
   Varroa 490, 540, 811
LAWS RELATING TO BEEKEEPING 491
   Legislation 491
LAYING WORKERS 491, 847
   Diagnosing laying worker colonies 491
   Parthenogenesis 491, 619
LEAFCUTTING BEES 11, 492
   Alfalfa leaf cutter bee management in western Canada 493
   alfalfa leafcutting bee, Megachile rotundata 492
   blister beetles 493
   chalcidoids 493
   cuckoo bees 493
   dermestids 493
   dried fruit moth 493
   synchronize bee emergence 492
LEARNING IN THE HONEY BEE 180, 493

   find the location of their hive 494
   swarm control 494, 792
   tripping mechanism of alfalfa flower 398, 493
LEGISLATION IN THE UNITED STATES RELATING TO HONEY 444, 495, 496, 596
   Composition of American Honeys 495, 568
LEGISLATION IN THE UNITED STATES RELATING TO HONEY BEE 444, 495, 496, 596
LENGTH OF LIFE OF THE HONEY BEE 497
   Castes 156, 498
   Drone honey bees 24, 216, 498
   Field work (flying) 255, 260, 263, 497
   Queen honey bees 498, 676
   Winter bees 497, 831
LEVIN, MARSHALL D. 498
   National Research Program Leader 498
   research center at Tucson 499
   USDA Legume Seed Research Lab. 498
LIFE CYCLE OF THE HONEY BEE 145, 500, 530, 561, 610
   Development time 156, 500
LINDNER, JOHN V. 501
LIQUID HONEY 333, 370, 373, 376, 377, 378, 381, 383, 501
LITTLEFIELD, HOOD 502
LOVELL, HARVEY BULFINCH 502
   John Harvey Lovell 503
LOVELL, JOHN HARVEY 398, 504
   ABC and XYZ of Bee Culture 505
   The Flower and the Bee, Plant Life and Pollination 505
   The Honey Plants of North America 505
   The Maine Naturalist 506

M

MACKENSEN, OTTO 463, 509
   carbon dioxide to immobilize a queen 509
   diploid males 509
   instrumental insemination 509
MACLEOD, HUGH JOHN 510
MAGNETIC AND ELECTRICAL FIELDS, EFFECTS ON BEES 510
   Circadian rhythm 510
   Comb building 120, 159, 511
   Horizontal dancing 510
   Malpighamoeba mellificae 197, 201
   malpighian tubules 33, 201
MASON BEES 511
   Osmia cornifrons 512
   Osmia cornuta 511
   Osmia lignaria 511
   Osmia rufa 511
MATING NUCLEUS 518, 578, 694, 699,
MATING OF THE HONEY BEE 513, 676, 687, 814
   Distances queens & drones fly for mating 517
   Dr. N.E. Gary 75, 275, 513
   Drone Congregation Areas 216, 514
   Mating frequency 517
   Mating in confinement 518
   queen's mandibular gland 281, 513
MAXANT, WILLIAM T. 358, 518
MAY, GEORGE W. 519
McEVOY, WILLIAM 519
   shaking treatment for American foulbrood 197, 519
McGINNIS, DAVID K. 520

Index

Tropical Blossom Honey Co., Inc. 520
McGREGOR, SAMUEL E. 520, 656
   Carl Hayden Bee Research Center, Tucson, Arizona 520
   Insect Pollination of Cultivated Crop Plants 521
MECHANICAL HANDLING OF COLONIES 522, 530, 535, 611
   banding 528
   Hay Hitch Knot 529
   hive lifters 527
   mechanical booms 524
Mehring, J. 118
   Melissococcus pluton 203
MELLIFERA 711
METAMORPHOSIS 145, 500, 530
   Life stages of the honey bee 16, 490, 530
MEYER, A.H. & SONS 530
MICE 228, 530
   stored supers of comb 531
MICROENCAPSULATION OF PESTICIDES 533, 798
   Penncap-M® 533
   Pesticides, effects on HONEY bees 534
MIGRATORY BEEKEEPING 522, 529, 534, 561, 661, 714
   carbon dioxide 537
   closed van 537
   nylon net 535
   pallet clips 535
   Pallets 534, 611
   Refrigerated trucks for moving bees 536
   Staples or straps 535
   the early 1930s 534
   truck-mounted booms 522, 535
   Ventilation 537
MILKWEED POLLINIA 433, 538
MILLER, CHARLES C. 175, 538, 596, 865
   E.F. Phillips 539, 638
   E.R. Root 539, 740
   Fifty Years Among the Bees 538
   Stray Straws 99, 539, 733
   sugar syrup feeder 248, 539
   The Miller method of queen rearing 687, 703
MILLER, NEPHI (1873-1940), AND MILLER HONEY COMPANY 539
MITES ASSOCIATED WITH HONEY BEES 59, 208, 332, 490, 540, 811
   Other parasitic mites 556, 800
   Phoretic mites 541
   Varroa destructor 547, 811
MOELLER, FLOYD E. 557, 801
MOFFETT, JOSEPH O. 558
   Some Beekeepers and Associates 558
MOLD ON COMBS 110, 559, 772
   Penicillium waksmanii 559
MORSE, ROGER ALFRED 559
   ABC & XYZ xix
   Gleanings in Bee Culture 560
   Research Review 560
MOULTING 500, 561
MOVING BEES LONG DISTANCES 305, 534, 561
MOVING BEES SHORT DISTANCES 133, 305, 561
MRAZ, CHARLES 562, 811
   American Apitherapy Society 562
   bee venom therapy 562, 779

carbolic acid 333, 562
fume board 562
Gleanings In Bee Culture 99, 562
Health and the Honey Bee 563
MUTH, FRED W. 563
MUTH HONEY JAR 563

N

NACHBAUR, ANTON J. 565
NASONOV PHEROMONE 34, 245, 285, 565, 630, 635
NATIONAL HONEY BOARD 565
NATURAL NEST OF THE HONEY BEE 145, 258, 259, 460, 566, 796
   Nest architecture and structure 567
NECTAR, COMPOSITION OF 156, 321, 470, 527, 567, 585
   sweetness scale 567
   water content 383, 568, 714
NECTAR, CONVERSION TO HONEY 47, 259, 317, 470, 495, 568
   Glucose oxidase 291, 569
   Invertase 471, 569
NECTAR GUIDES 398, 571
NECTAR, POISONOUS 439, 573
NECTAR SECRETION, CONTROL OF 230, 398, 573
   moisture 574
   nectary tissue 573
   phloem of the plant 574
NECTARIES 568, 572
NEMATODES 228, 575
NEPOTISM 276, 279, 575, 687, 847
   Charles Darwin 575
   half-sisters 575, 847
   kin selection 575
   super-sisters 575, 847
NEST HYGIENE 566, 576
   Pollen 577, 646
   Propolis 577, 669
   Uncapped honey 328, 577
NITROUS OXIDE 578, 765, 771
NOMIA MELANDERI 11
   Alkali bee 11
NOSEMA DISEASE 34, 43, 197, 203, 205, 223, 578
NUCLEUS COLONY 136, 518, 578
   Feeding 247, 580
   how to make 581
   Joining two nucs 580, 805
   skunks 580, 758
   The Mississippi Split 582, 701
NURSE BEES 16, 585
   Castes 156, 585
   Royal jelly 585, 748
NUTRACEUTICALS 585
NUTRITION OF THE HONEY BEE 78, 156, 258, 585, 646
NYE, WILLIAM (BILL) PRESTON 585

O

OBSERVATION HIVES 589
   Comb construction 168, 591
   Location 50, 291, 589
   Make 590
   traveling hive 595

Index

OERTEL, EVERETT 595
OLFACTORY SENSE 596, 752
ORDINANCES BANNING BEES 291, 495, 596
 banning beekeeping 596
 Honey Bee Law 596
LEGISLATION IN U.S. RELATING TO HONEY BEES 596
ORIENTATION BY HONEY BEES 180, 596, 645
OSMOTIC PRESSURE OF HONEY 329, 564, 573, 596
 Nectar 596
OUT-APIARIES 50, 538, 596
OUTYARDS 50, 538, 596
 Apiary Locations 50, 538, 597
OVERCROWDING OVERSTOCKING 50, 597

P

PACKAGE BEES, INSTALLATION & IMMEDIATE CARE 134, 216, 599, 801, 825
 Balling 609, 686
 Beekeeping equipment for installing package bees 233, 604
 Care of package bees before installation 604
 funnels 233, 606
 Growth of a package bee colony 500, 610
 Kinds of packages available 599
 Package colony management 606
 Preparation of package bees 600
 Shipping packages 604
 Time of day to install packages 605
 Weight of the honey bees 599
PADDOCK, FLOYD B. 611
 Apiary Inspectors of America 611
 Iowa Beekeepers' Association 611, 748
PAENIBACILLUS LARVAE SUBSP 43, 198, 519
PAGE, ROBERT 279, 575, 687
PALLETS FOR MOVING BEES 522, 534, 611
PARADICHLOROBENZENE 179, 821
PARAFOULBROOD 197, 208
PARASITIC MITE SYNDROME, BEE 59, 187, 208, 547, 811
 Varroa destructor 208, 811
PARK, HOMER E. 612
PARK, OSCAR WALLACE 211, 612
 American foulbrood 197, 615
 bee dances 180, 615
 C.E. Bartholomew 613
 Ellen Tupper 613
 Frank C. Pellett 613, 620
 natural resistance 187, 615
 Rothenbuhler, Walter C. 748
 von Frisch 271, 615
PARKHILL, JOE M. 618
 Nutrition of Pollen, Inc. 619, 646
PARTHENOGENESIS 276, 491, 619, 797
 Arrhenotoky and Thelytoky Parthenogenesis 619
PATENTS IN BEEKEEPING 619
PEER, DON 91, 619, 839
PELLETT, FRANK CHAPMAN 99, 620
 A Living with Bees 620
 American Honey Plants 620
 Beginner's Bee Book 620
 Birds of the Wild 620
 Flowers of the Wild 620
 History of American Beekeeping 620

 How to Attract Birds 620
 Practical Queen Rearing 620
 Practical Tomato Culture 620
 Productive Beekeeping 620
 Success with Wild Flowers 620
 The Romance of the Hive 620
PELLETT, MELVIN A. 99, 621
 American Honey Plants 621
PENDER, W.S. 622
 Australasian Beekeeper 88, 622
PESTICIDES, EFFECTS ON HONEY BEES 46, 470, 622
 arsenicals 622
 Beekeeper indemnity program 628
 Colony recovery 629
 Combining insecticide with repellent 624
 Contamination of hives w/pesticides 627
 DDT 624
 E.L. Atkins 67, 625
 Environmental Protection Agency 46, 625
 Integrated Pest Management 470, 628
 Night spraying with aircraft 626
 Protecting colonies from pesticides 629
 Pyrethrums 622
 Reporting and preventing losses 46, 630
 Rotenone 622
 Shaw, F.R. 227, 622
 Sulfur compounds 622
PHEROMONE LURE 71, 792
PHEROMONES 458, 630
 (E)-9-oxo-2-decenoic acid (9-ODA) 634
 2-heptanone 286, 632
 9-ODA 634, 637
 Alarm pheromone 11, 13, 36, 283, 285, 288, 292, 632, 787, 811
 Brood and comb pheromones 145, 633
 Dr. C. Zmarlicki 637
 Dr. N.E. Gary 275, 637
 Drone Congregation Areas (DCA) 216, 637
 Footprint pheromones 260, 634
 isopentyl acetate 11, 13, 36, 288, 632
 Laying workers 637, 638
 mandibular glands 281, 632
 Mating of the honey bee 216, 637, 676
 Nasonov 34, 245, 565, 635
 primer pheromones 631
 Queen substance 634, 687
 queen's mandibular glands 634, 676
 release pheromones 631
 Scent gland pheromone 34, 245, 565, 635
 Sex attractant 286, 637, 687
 Worker ovary development 16, 638
PHILLIPS, EVERETT FRANK 539, 638
 Anatomy of the Honey Bee 638
 Cornell University 639
 E.R. Root 638, 732
 George S. Demuth 99, 194, 639
 Gleanings in Bee Culture 99, 194, 638, 733
 R.E. Snodgrass 13, 638, 771
 The library of beekeeping at Cornell 641
 USDA 638
 wounded soldiers, training 641
PIPING BY QUEENS 641, 687
PLASTIC BEEKEEPING EQUIPMENT 178, 233, 316, 379, 642, 680
 Plastic Containers 374, 379, 644

907

Index

PLAY FLIGHTS 596, 645
POISONOUS HONEY 646
  Honey plants 439, 646
POISONOUS PLANTS 439
  California Buckeye (*Aesculus californica*) 2, 439, 442
  purple brood 440
  Rhododendron 440
  Summer Titi - *Cyrilla racemiflora* 440
  The Tutu Mystery 443
  Yellow Jessamine – *Gelsemium sempervirens* 442
POLLEN 145, 577, 646
POLLEN GATHERING BY HONEY BEES 585, 646
  Depositing the loads in the hive 78, 648
  How bees collect and pack pollen 29, 646
POLLEN GRAINS 648, 656
  Apple 430, 648
  Aster 400, 649
  Black Locust 405, 648
  Dandelion 431, 649
  Maple 433, 648
  Mustard 438, 649
  Red Clover 407, 648
  Sumac 425, 648
  White Sweet Clover 407, 648
  Willow 438, 649
POLLEN MITES 228, 651
POLLEN SUBSTITUTES 145, 258, 651
POLLEN SUPPLEMENTS 145, 258, 651
POLLEN TRAPS 233, 654
  brood 654
  drones 654
  entrance 654
  fungus 656
  hardware cloth 654
  precisely punched holes 654
POLLINATION 520, 656
  Agricultural pollination 255, 273, 656
  Almond pollination 658
  Pollen Grains 648
  Greenhouse pollination 291, 659
  Hand pollination 660
  hand-collected 661
  Migratory beekeeping 534, 661
  Orientation of bees moved 180, 661
  Plant attractiveness 662
  pollen dispenser 661
  Pollination contracts 658
  Pollination fees 658
  pollinizers 661
POLLINATION IN HOME GARDENS 663
POLLINATION FOR WILDLIFE 663
POLLINATION SHORTCUTS 664
  sugar syrup 247, 665
POLLINIA 435, 538
POPULATION DYNAMICS OF HONEY BEE COLONIES 146, 167, 312, 665, 833
POWERS APIARIES, INC. 666
PRICE SUPPORT 667
  Food Security Act 667
PRITCHARD, MELVIN T. 668
  A.I. Root Company 669
PROPOLIS 577, 669
  entomb large animals 670
  flavones 669
  human medicine 671
  Races of honey bees 669, 709
  road tar, drying or soft paint, caulk 671
  varnish 671
  waterproofs 669
PROPOLIS PETE 99, 845
PUBLIC RELATIONS FOR BEEKEEPERS 291, 672
PURPLE BROOD 440

Q

QUEEN 37, 156, 498, 513, 676, 772
  Assessing quality of 147, 676, 685
  Cells 228, 676
  Drone-laying queens 145, 680
  Effect of queen age 680
  Egg laying rate of a queen 146, 681
  Emergency queen cells 228, 682, 792
  Excluders 233, 683
  Length of life of a queen 676, 685, 694
  Locating queens 676, 685, 847
  looking for a queen 676, 847
  Plastic queen cell cups 642, 680, 687
  Queen cups 642, 678, 682, 785, 792
  Queen introduction techniques 609, 686, 689
  Queen piping 641, 686
  Queen substance 286, 634, 687
  Reproductive organs 38, 682, 772
  ventilation 684, 792, 813
  Virgin queen 513, 641, 687, 814
  Wax cell cups 679, 687
QUEEN REARING: GRAFTING & COMMERCIAL QUEEN PRODUCTION 145, 246, 687, 703
  Balling 686, 689
  Bank colonies 687
  Benton cages 140, 696
  Breeder colonies 145, 690
  Cell incubators 145, 690
  Clipping queen's wings 691, 788
  color-coded marks 685, 694
  Commercial queen rearing areas 691
  Crowding to force cell production 701
  Double grafting 286, 692, 748
  Equipment for queen rearing 692
  Finishing colonies 694, 698
  Grafting tools 687, 694
  H.H. Laidlaw, Jr. 486, 687
  International Color Code 695
  Marking queens 133, 676, 691, 694, 791
  Queen cages 140, 696
  queen mating nucs 145, 518, 694
  Queen Rearing and Bee Breeding 486, 687
  R.E. Page 279, 575, 687
  small battery box 697
  Splitting colonies 701
  Starter colonies 694, 698
  Stocking mating nucs 518, 699
  The Alley method 703
  three-hole queen cage 251, 696, 697
  Using swarm and supersedure cells for increase 687, 704
  Wet grafting 692, 700
QUEEN REARING: RAISING A FEW QUEENS 145, 701
QUEENLESS COLONY 705

Index

R

RACES OF HONEY BEES 669, 709, 749
  African and Africanized honey bees 4, 52, 96, 481, 709
  Buckfast 148, 461, 713
  Cape bees 710
  Carniolan honey bees 710
  Caucasian honey bees 87, 709, 710
  Cordovan 712
  German or black bees 711
  Italian honey bees 711
  Russian 713
REACTIONS TO BEE AND WASP STINGS 713, 813, 817, 83
REFRACTOMETERS 328, 568, 714
REFRIGERATED TRUCKS FOR MOVING BEES 534, 714
REMOVING BEES FROM BOXES AND FIXED COMB HIVES 219, 312, 725
REMOVING BEES FROM SUPERS 714
  Bee escapes 715
  Bee Go® 718
  Bee Quick® 719
  Benzaldehyde 717
  Blowers 716
  Butyric anhydride 718
  Porter Bee escape 715, 876
  Repellents 562, 717
  screened escape board 716
  Smoking, shaking and brushing 719
  triangle escape board 716
REMOVING FERAL COLONIES 720
  Removing bees from buildings 720
  Removing bees from trees 724
REPRODUCTIVE SYSTEM 38, 40, 725
REQUEENING 140, 696, 725, 801
REQUEENING WITHOUT DEQUEENING 727, 801
RESEARCH IN APICULTURE 105, 195, 241, 470, 728
  Resistance management 47, 187, 208, 615, 625
RESPIRATORY SYSTEM 24, 29, 729
REVERSING FOR SWARM PREVENTION 494, 729, 784
RINDERER, T. 8, 713, 728
ROADSIDE SALES 374, 381, 448, 729
  Examples of Roadside Stands 729
  good signage 730
ROBBING BY BEES 731
  Progressive robbing 731
ROOT FAMILY HISTORY 732
  Alan Root 742
  Brad Root 745
  Ernest R. Root 539, 740
  Gleanings In Bee Culture 99, 194, 733
  Honey packing 741
  Huber Root 742
  John Root 745
  Rev. Langstroth 735
  San Antonio, Texas 744
  Sheeter machine 739
  smooth roll machines 739
  Stuart Root 745
  The original A.I. Root Jewelry factory 734
  windmill 737
  wire embedded foundation 739

Wright Brothers 736
ROOT, LYMAN C. 746, 858
ROSS, THOMAS AND ROSS ROUNDS 746
  Cabana Sections 746
  Comb Honey, How To Make 746
  Dr. Zbikowski 746
  Lloyd Spear 747
  Mr. Chambers 747
  Rambler's 747
  Tom Ross 747
  Taylor, Richard 795
ROTHENBUHLER, WALTER C(HRISTOPHER) 211, 464, 612, 748
ROUND SECTIONS 168, 173, 186, 383, 748
ROYAL JELLY 585, 692, 748
RUTTNER, FRIEDRICH 709, 749

S

SACBROOD DISEASE 208
SCALE HIVES 751
SCENT GLAND 34, 245, 565, 635
SECHRIST, EDWARD LLOYD 751
SENSES OF THE HONEY BEE 12, 19, 220, 298, 596, 752, 795
  Carbon dioxide detection 755
  Color vision 754
  Johnston's organ 753
  Odor perception 755
  Sound perception 752
  Taste 752
SEX DETERMINATION IN THE HONEY BEE 156, 755
SHAPAREW, VLADIMER 755, 813
SHERRIFF, B.J. 213, 756
SIGHT IN THE HONEY BEE 752, 756
SKEPS 302, 756
  Skep Construction 757
SKUNKS 306, 580, 758
  skunk predation 758
SMALL HIVE BEETLES 178, 759
  Aethina tumida 760
  Control 762
  Distribution 761
  Life Cycle 762
  propolis prison 763
  Taxonomic position 760
SMITH, JAY 687, 764
SMOKE, EFFECTS OF 764
SMOKERS 233, 578, 765
  Bingham Smoker 766
  Bingham Smoker, large size 768
  Original Cold Blast 766
  Original Quinby smoker 766, 856
SMOKING HONEY BEE COLONIES 768
  Fuel for smokers 771
  Nitrous oxide and laughing gas 578, 771
  Tobacco smoke 771
SMR (SUPPRESSED MITE REPRODUCTION) 771, 811
SNELGROVE, L.E. 771, 791
SNODGRASS, ROBERT E. 13, 638, 771
SOLAR WAX MELTER 113, 772, 821
SOUNDS MADE BY HONEY BEES 752, 772
SPERM STORAGE 38, 772
SPLITTING COLONIES 212, 229, 582, 701

Index

SPRING NECTAR FLOW, MANAGEMENT FOR 398, 773
STANDIFER, LONNIE NATHANIEL 773
STARTING IN BEEKEEPING 133, 773
STARVATION 247, 773, 833
STEPHEN, W.A. 774
STINGLESS BEES 774
   Meliipona spp. 775
   Trigona spp. 777
STINGS 12, 106, 285, 562, 779
STINGS, FIRST AID FOR 12, 779
STINGS, HOW TO AVOID 12, 779
STOLLER HONEY 270, 780
STONEBROOD 208
STRACHAN, DONALD J. AND STRACHAN APIARIES 781
STRAUB, WALTER AND W.F. STRAUB AND CO. 781
SUBMISSIVE BEHAVIOR 782
SUBSPECIES 783
SUE BEE HONEY HISTORY 783
SUGAR FOR FEEDING BEES 156, 258, 300, 527, 567, 568, 585, 784
SUGARS IN HONEY 470, 784
SUPER 233, 784
SUPERING COLONIES 784
SWARM BEHAVIOR AND MOVEMENT 791
SWARM COLLECTING AND HIVING 69, 138, 792
SWARM CONTROL 194, 494, 788
SWARMING IN HONEY BEES 59, 71, 494, 684, 784, 792
   Absconding swarms 5, 787
   Afterswarms 5, 11, 787
   Clipping queen's wings 691, 788
   Cutting queen cells 228, 788
   Demareeing 194, 788
   Honey-bound condition 789
   Late swarming 789
   Marked queens 694, 791
   Old queen accompanies primary swarm 791
   Queen cups 228, 678, 785
   Supersedure 228, 682, 792
   Swarm control 729, 784, 790
   Swarm prevention 729, 784, 790
   Tanging 789
   The effect of queen age 790

T

TASTE IN THE HONEY BEE 746, 752, 795
TAYLOR, RICHARD 795
   Gleanings In Bee Culture 796
   Joys Of Beekeeping Award 796
TEMPERATURE CONTROL AND THERMOREGULATION 566, 796
   cluster 796, 832
   Individual bees 796
TEN-FRAME LANGSTROTH HIVE 313
TERRAMYCIN® 200, 797
   Antibiotics 47, 797
THAI SACBROOD 208, 797
THE "AFRICAN" HONEY BEE 11
THELOTOKY 619, 797
THELYTOKY PARTHENOGENESIS 797
THURBER, LOUISE W. 797
THURBER, P.F. (ROY) 798
   Penncap-M® 533, 798

Western Apicultural Society 66, 798
TOADS 228, 798
TODD, FRANK (Edward) 799
TONSLEY, CECIL 799
TOP BAR HIVE 481, 800
TOWNSEND, GORDON FREDERICK 799
TRACHEAL MITE 23, 332, 541, 556, 800
TRANSITION CELLS 168, 800
TRAVEL STAIN 168, 669, 800
TWO-QUEEN SYSTEMS 167, 800
   Moeller, F.E. 557, 801
   Package Bees 599, 801
   Requeening 725, 801
TYLAN 43, 45, 197, 200, 203
TYLOSIN 43, 45, 197, 200, 203

U

ULTRAFILTRATION 450, 483
UNCAPPERS 80, 338, 876
   uncapping knives 339
UNITING COLONIES AND NUCLEI 580, 805
   newspaper 805
UNIVERSAL PRODUCT CODE 485
UNRIPE HONEY 328, 806
   Nectar, conversion to honey 806
UNSEALED HONEY 806
URBAN AND SUBURBAN BEEKEEPING 291, 807
   Apiary locations 808
   Honey Bee Law 809
   Ordinances banning bees 808
   Placement of colonies 809

V

VARROA DESTRUCTOR 59, 208, 490, 547, 811
   Control 12, 159, 161, 187, 211, 263, 554
   Detection 549
   Dispersal 554
   Distribution 547
   Life cycle 550
   Pathogenicity 548
   Sticky Board Method For Detecting *Varroa* Mites 550
   V. jacobsoni 547
   Varroa disease 208
VARROA SENSITIVE HYGIENE 211, 811
   mite-resistance 811
VENOM, COMMERCIAL PRODUCTION OF HONEY BEE 12, 562, 779, 811
   Benton's method 811
VENOMS OF STINGING INSECTS 12, 107, 779, 813
VENTILATING SCREENS 814
VENTILATION 34, 166, 245, 565, 635, 684, 755, 813
VIRGIN QUEEN 687, 814
   Mating of the honey bee 513, 814
VIRUS DISEASES 208
   acute bee paralysis 201, 210
   Bee paralysis 201, 210
   Black queen cell virus 210
   chronic bee paralysis 210
   Deformed wing virus 210
   Filamentous virus 211
   hairless black syndrome 210
   Sacbrood 209
   slow paralysis 210

Index

VISION IN THE HONEY BEE 814
   Senses of the honey bee 814

W

WADLOW, RALPH 817
WASHBOARD BEHAVIOR 817
WASHBURN, A. 118
WASPS 457, 817
WATER COLLECTORS 817
WATER FOR HONEY BEES 291, 807, 818
   Behavior of water collectors 260, 818
   Boardman-type feeder 247, 819
   Providing water for bees 291, 819
WATERY CAPPINGS 107, 820
WATSON, LLOYD RAYMOND 463, 820
   instrumental insemination 820
   Walter C. Rothenbuhler 820
WAX EXTRACTORS AND PRESSES 772, 821
WAX MOTHS 163, 178, 179, 821
   Greater wax moth 821
   Lesser wax moth 823
WEAVER FAMILY 824
   Kona Queen Company 825
   Roy Weaver, Sr. 824
   Weaver Apiaries 825
WEIGHT OF HONEY BEES 826
   African and Africanized honey bees 4, 826
   Package bees 599, 825
WELLS, HORACE "LINC" 825
WENNER, DARRELL C. 826
WESTERN APICULTURAL SOCIETY OF NORTH AMERICA (WAS) 66, 798
WET COMBS 338, 827
   Comb Storage 178, 334, 827
WHITE, JONATHAN 827
WILBANKS, WARREN GUY 828
WILLSON, R.B. 828
WIND, EFFECTS OF 829, 833
WINDBREAKS 829
WING VENATION 20, 831
WING-FANNING 831
   Anatomy 831
   Ventilation 813, 831
WINTER BEES 293, 497, 831
WINTER BROOD REARING 831
WINTER CLUSTER 293, 796, 832
   colony defense at low temperatures 833
WINTER DEFENSE 833
WINTER STORES, CRYSTALLIZATION OF 844
WINTERING 167, 245, 833
   Brood patterns 835
   Hive stands 49, 306, 758, 838

   Inspecting & weighing colonies for winter 835
   Metabolic water 836
   northern climates 834
   Packing for winter 836
   Warm periods 834
   Wintering in Canada 838
WINTERING, INDOORS 619, 839
   air exchange 842
   alarm for monitoring 842
   bee cellar 160, 840
   Facilities and ventilation 841
   Relative humidity 843
   Ventilation tubes 843
WIRTH, ED. D. 845
   Propolis Pete 99, 845
WOOD PRESERVATIVES FOR BEEKEEPING EQUIPMENT 845
   Painting beehives 236, 845
WOODMAN, BAXTER 845
WOODPECKERS 139, 846
WOODROW, ALAN W. 846
WORKER 156, 575, 847
   Anatomy 12, 27, 284, 292, 847
   Communication 180, 847
   Fanning 847
   Foragers 260, 847
   Foraging Distance 260, 847
   Glands 281, 847
   Guard Bees 292, 847
   Larvae 16, 490, 847
   Laying Workers 491, 847
   Life stages of the honey bee 847
   Pheromones 630, 847
WORKER CELLS 168, 847
   Comb 168, 847
WORKING A COLONY 133, 676, 685, 847
   Diseases Of The Honey Bee 197, 850
   Pre-inspection 847
   queen performance 676, 849
WORLD TRADE IN HONEY 850
WRIGHT BROTHERS 736

Y

YEASTS IN HONEY 853
   Fermentation of honey 853
YELLOWJACKETS 11, 457, 853
YORK, HARVEY F. 853
   Starline and Midnite hybrid 853

Z

ZBIKOWSKI, WLADYSLAW 854
   circular comb honey section 854